Paleolimnology

PALEOLIMNOLOGY
The History and Evolution of Lake Systems

Andrew S. Cohen

OXFORD
UNIVERSITY PRESS
2003

OXFORD
UNIVERSITY PRESS

Oxford   New York
Auckland   Bangkok   Buenos Aires   Cape Town   Chennai
Dar es Salaam   Delhi   Hong Kong   Istanbul   Karachi   Kolkata
Kuala Lumpur   Madrid   Melbourne   Mexico City   Mumbai   Nairobi
São Paulo   Shanghai   Taipei   Tokyo   Toronto

Published by Oxford University Press, Inc.
198 Madison Avenue, New York, New York, 10016

www.oup.com

Oxford is a registered trademark of Oxford University Press

Library of Congress Cataloging-in-Publication Data
Cohen, Andrew S., 1954–
    Paleolimnology: the history and evolution of lake systems / Andrew S. Cohen.
       p. cm.
Includes bibliographical references and index.
    ISBN 0-19-513353-6
    1. Paleolimnology. I. Title.
    QE39.5.P3 C63 2003
    551.48–dc21   2002002722

9 8 7 6 5 4 3 2 1

Printed in the United States of America
on acid-free paper

This book is dedicated to

Chester Cohen, who taught me how to love science,

and to

Kerry Kelts, who helped the paleolimnology community to love mud.

Preface

As I began writing this book in late 1998, I was reflecting on the fact that the year past had been the warmest year of the warmest decade on record, consistent with the pattern of global warming that we have been experiencing for at least the past few decades. The nature and rate of global climate change will be debated by scientists in years to come. Ideally, science also gives politicians, planners, and economists a logical basis for debating the extent to which modern conditions deviate from the "norms" of environmental variability that our planet has experienced in the recent past. One important source of information for this evolving debate is paleolimnology, the study of lake history from preserved geological records.

Paleolimnology is a field whose intellectual roots can be traced to diverse disciplines with varying goals. Different scientists have viewed these goals through the filters of their own areas of interest, expertise, and times. Biologists have defined paleolimnology as "the interpretation of past limnology from changes that occurred in the ecosystem of the lake, and their probable causes," with much emphasis placed on the resolution of explicitly biological questions, for example about the eutrophication of lakes (Frey, 1988a). In contrast, geologists and pre-Quaternary paleoecologists have defined paleolimnology in more geological terms, emphasizing sedimentological or geomorphic information useful for understanding the physical history of lake basins or biological evolution in lakes (Reeves, 1968; Gray, 1988a). Many developments in these areas mirrored the burgeoning interest among geologists in the 1970s and 1980s in the potential of lake records for providing information on climate change, and for expanding petroleum exploration in ancient lake beds at that time. In the 1980s and 1990s we saw a proliferation of *green* applications of paleolimnology, in recognition of the potential of the field to address acute and rapidly changing problems associated with human activity (O'Sullivan, 1991). In this book I will try to encompass all of these viewpoints about the subject, by synthesizing the kinds of information that can be gleaned from lake deposits and explaining how that information can be put to work to address questions concerning lake and earth history.

Many subjects can be investigated using paleolimnology. Processes operating on a variety of temporal and spatial scales, such as watershed disturbance, global climate change, geological history of continental basins, and biological evolution, can all take advantage of what is perhaps the unifying theme of the study of lake deposits: their usefulness as records of earth history at both high resolution and over long intervals of time. The conveners of a recent Geological Society of America annual meeting symposium eloquently referred to this as the "Power of Paleolimnology."

Paleolimnology has been a growth industry in science in recent years, resulting in numerous scientific congresses, symposia, workshops, review papers, and special edited volumes. The development of several International Geological Correlation Program projects devoted to lakes and lake history, the spectacular success of the *Journal of Paleolimnology* over the past decade, and the development of an International Association of Limnogeology are all testimonies to the intellectual vibrancy of the field. Curiously, it has been many years since the "Power of Paleolimnology" was last generally reviewed in book form, and this served as my primary motivation for writing this book. Frey (1974), Binford et al. (1983), and Smol (1990) wrote major reviews of the field, emphasizing the biotic aspects of lake history. Similarly, Talbot and Allen (1996) compiled a general treatment of the sedimentological aspects of the field. Several books and edited journal issues in the 1970s, 1980s, and 1990s, notably those edited by Lerman (1978), Håkanson and Jansson (1983),

Haworth and Lund (1984), Gray (1988a), Battarbee et al. (1990), and Lerman et al. (1995), examined specific components of the field, but were not general syntheses. Other books provide "how to" compendia for the study of lake deposits, particularly for the Quaternary science community (Berglund, 1986; Miskovsky, 1987; Warner, 1990a; the notable, new multivolume series edited by Last and Smol, 2001). But we have to go back to Reeves (1968) to find a general text in the field.

General syntheses in paleolimnology have been deterred by the disparate intellectual origins from which the field originated. The investigation of lake deposits and their environmental interpretation can be traced to the early to mid nineteenth century (Sedgwick and Murchison, 1829; Lyell, 1830; Whittlesey, 1838, 1850; Agassiz, 1850). However, it wasn't until the late nineteenth century that the linkages between Pleistocene lacustrine sediments and climatic and hydrological history became clear. This culminated with classic studies on the Quaternary, glacial lakes of Europe (Ramsay, 1862), the North American Great Lakes region (Chapman, 1861; J.W. Spencer, 1890; Tyrrell, 1892; Leverett, 1897, 1902), and Western North America (Russell, 1885; Gilbert, 1890). In the early twentieth century considerable interchange existed between researchers interested in the biological, paleoclimatological, and geological aspects of lake deposits (Livingstone, 1991). During the second half of the twentieth century, however, very different communities of researchers carried out these investigations. Much of this division has followed the boundaries and interests of the professional societies and congresses that have promoted the study of lake deposits (R.B. Davis, 1989). Sedimentologists have investigated modern lake depositional processes and the pre-Quaternary history of lake deposits to reconstruct paleoclimates and to develop predictive facies models for hydrocarbon and minerals exploration. The roots of this strand can be traced to studies like the classic investigations of the Green River Formation paleolakes by W.H. Bradley (1929), early sedimentological studies of modern lakes (Forel, 1885; Nipkow, 1920), and the interpretation of lake deposits (e.g., De Geer, 1912). A paleobiological strand of paleolimnology concerned with the ecological and evolutionary history of lakes has been promoted quite independently by paleontologists with earth science backgrounds, and by some biologists. A Quaternary geology strand has focused on Pleistocene and Holocene paleoclimate and watershed paleoenvironmental reconstruction using the records provided by lake deposits. The

Quaternary ecology of lakes and their watersheds developed as an extension of early paleoclimate research, notably the studies of Swedish varve chronologies (De Geer, 1940). Studies of lake ecological history were pioneered in the late 1920s to 1940s by Hutchinson and Deevey in New England (Deevey, 1942), by Lundqvist (1927), Gams (1927), and Groshopf (1936) in continental Europe, and by Mortimer, Pennington, and others in England (Pennington, 1943). More recently, investigation of "anthropogenic impacts" with a focus primarily on the postindustrial revolution period has been promoted by a mix of scientists, from environmental studies, watershed management, biology, geology, and engineering backgrounds. Interaction between these various strands of paleolimnology has often been limited, and a major motivation for me in writing this book has been to illustrate the linkages between these various approaches to lake history.

This book is intended to serve as an introduction to the fields of paleolimnology and limnogeology for graduate students and advanced undergraduates. Chapters 1 and 2 explain general principles of paleolimnology and how lake systems form and evolve to house the sedimentary archives we study. Chapters 3–5 are brief reviews of aspects of the physics, chemistry, and biology of lakes that are important to understanding paleolimnological records. These chapters are particularly intended for students with little or no prior background in limnology. Chapters 6–11 discuss the types of geochronological, sedimentological, geochemical, and biological archives available for paleolimnologists to study. And finally, chapters 12–15 approach paleolimnology from the point of view of important scientific questions that the discipline can address, from the shortest time scales (anthropogenic problems) to the longest (evolution of lake ecosystems), and with a brief consideration of future directions in the field.

Acknowledgments

I would like to thank the many people who have helped me with this book. I thank Glen Berger, Kevin Bohacs, Owen Davis, David Dettman, Paul Fitzgerald, Darrell Kaufman, Kiram Lezzar, John Lundberg, Peter McIntyre, Ellinor Michel, Hank Mullins, Rob Negrini, Jonathan Overpeck, Pierre Denis Plisnier, Peter Reinthal, Chris Scholz, and John Tibby, all of whom reviewed specific sections of the book and made numerous helpful suggestions. The students of my 2001 and 2002 paleolimnology seminars—Simone Alin, Jake Bailey, Debbie Balch, David Goodwin, Kathy Hurlburt, Ned Kruger, Naomi Levin, Christine Lewis, Catherine O'Reilly, Rebecca Prescott, and Jeff Pigati—all read and provided invaluable comments on earlier drafts of the book. Pere Anadon, John Anderson, Victor Baker, Glen Berger, Larry Benson, Mark Brenner, Bernadette Coleno, David Dettman, Pete DeCelles, Bill Dickinson, Bill DiMichele, Dan Engstrom, Dana Geary, Andy Gleadow, Simon Haberle, Nick Harris, Richard Hay, Roy Johnson, Tom Johnson, Peter Kershaw, Peter Kresan, Conrad Labandeira, Peter Leavitt, David Loope, Andy Lotter, Hank Mullins, Jack Oviatt, Tom Phillips, Dave Richards, Tami Rittenour, Gar Rothwell, Chris Scholz, Mike Soreghan, Joris Steenbrink, Mike Talbot, Jim Teller, Jean Jacques Tiercelin, John Tibby, Dirk Verschuren, Robert Wetzel, and Bruce Wilkinson all graciously provided information and/or illustration materials for the book. I thank Nick Mathieu, Susie Gillatt, Shannon Mattinson, and Ken Chandler for help with the many illustrations. Special thanks go to my friends in the Department of Geography and Environmental Sciences at Monash University who hosted me while much of this book was being written, especially Peter Kershaw, John Tibby, and Simon Haberle. I thank the staff of Oxford University Press, particularly Joyce Berry, Cliff Mills, and John Rauschenberg, for their ongoing support of this project. I also thank the US National Science Foundation and the UN Global Environmental Facility's Lake Tanganyika Biodiversity Project for their generous financial support of my research over the last 20 years, which in large part has given me the perspective to write this book. And finally, I thank my family: my parents, Chester and Evelyn Cohen, for everything, Alex and Zach Cohen for keeping me smiling, and Debbie Gevirtzman for her extraordinary help, editing prowess, and moral support at every step of this long journey.

Credits

Figure 1.2: a, Courtesy of Nick Harris. Copyright © 1998. e, Courtesy of Tom Johnson. Copyright © 1999. h, Courtesy of André Lotter. Copyright © 1997.

Figure 1.3: Profile courtesy of AMOCO and Roy Johnson.

Figure 1.7: From Leavitt, P.R. et al. An annual fossil record of production, planktivory, and piscivory during whole-lake manipulations. *Jour. Paleolim.* 11 (1994):139 (fig. 5). With permission from Kluwer Academic Publishers.

Figure 1.8: From Overpeck, J. et al. Arctic environmental change of the last four centuries. *Science* 278 (1997):1251–1256. With permission. Copyright © 1997, American Association for the Advancement of Science.

Figure 1.9: From Currey, D.R. and Oviatt, C.G. Durations, average rates and probable cause of Lake Bonneville expansions during the last deep-lake cycle, 32,000 to 10,000 years ago. In Kay, P.A. and Diaz, H.F. (eds.), *Problems of and Prospects for Predicting Great Salt Lake Levels.* With permission from the Center for Public Affairs and Administration, Utah University. Copyright © 1985.

Figure 1.11: From Olsen, P.E. A 40 million year lake record of early Mesozoic orbital climatic forcing. *Science* 234 (1986):842–848. With permission. Copyright © 1997, American Association for the Advancement of Science.

Figure 2.2: Modified from Brodzikowski, K. and van Loon, A.J. *Glacigenic Sediments.* With permission from Elsevier Science, Amsterdam, p. 111. Copyright © 1991.

Figure 2.3: From Teller, J.T. Proglacial lakes and the southern margin of the Laurentide Ice Sheet. In Ruddiman, W.F. and Wright, H.E. (eds.), *North America and Adjacent Oceans During the Last Deglaciation. The Geology of North America.* Modified with permission of the Geological Society of America, Boulder, CO.

Copyright © 1987, Geological Society of America.

Figure 2.4: Courtesy of Jim Teller. Copyright © 2000.

Figure 2.5: Courtesy of Hank Mullins. EROS Satellite Photo E-1234-15244-5-02.

Figure 2.6: From Mullins, H.T. et al. Seismic stratigraphy of the Finger Lakes: a continental record of Heinrich event H-1 and Laurentide ice sheet instability. In Mullins, H.T. and Eyles, N. (eds.), *Subsurface Geologic Investigations of New York Finger Lakes: Implications for Late Quaternary Deglaciation and Environmental Change.* Modified with permission of the Geological Society of America, Boulder, CO. Copyright © 1996, Geological Society of America.

Figure 2.8: Courtesy of Jean Jacques Tiercelin.

Figure 2.10: Modified from Horton, B.K. and DeCelles, P.G. The modern foreland basin system adjacent to the Central Andes. *Geology* 25 (1997):895–898. With permission of the Geological Society of America, Boulder, CO. Copyright © 1997, Geological Society of America.

Figure 2.12: Modified from Garfunkel, Z. The history and formation of the Dead Sea Basin. In Niemi, T.M. et al. (eds.), *The Dead Sea: The Lake and Its Setting.* With permission from Oxford University Press, Oxford, pp. 36–56, 1997.

Figure 2.13: Modified from Gardosh, M. et al. Hydrocarbon exploration in the southern Dead Sea area. In Niemi, T.M. et al. (eds.), *The Dead Sea: The Lake and Its Setting.* With permission from Oxford University Press, Oxford, pp. 57–72, 1997.

Figure 2.14: Courtesy of Victor Baker. Copyright © 2000.

Figure 2.16: Courtesy of Peter Kresan. Copyright © 2000.

petroleum source rock or metal ore deposition, or both? In Fleet, A.J. et al. (eds.), *Lacustrine Petroleum Source Rocks*. Geol. Soc. Lond. Spec. Publ. 40 (1988):45–58. With permission of the Geological Society, London.

Figure 8.14: Courtesy of Dave Richards. Copyright © 2001.

Figure 8.15: From Eugster, H.P. and Hardie, L.A. Saline lakes. In Lerman, A. (ed.), *Lakes: Chemistry, Geology, Physics*. (After Jones, 1965.) With permission from Springer-Verlag Publishers, New York, chap. 8 (fig. 25). Copyright © 1978, Springer-Verlag Gmbh and Co. KG.

Figure 8.16: From Renaut, R.W. and Tiercelin, J.J. Lake Bogoria, Kenya Rift Valley—a sedimentological overview. In Renaut, R.W. and Last, W.M. (eds.), *Sedimentology and Geochemistry of Modern and Ancient Saline Lakes*. Soc. Econ. Paleo. Mineral. Spec. Publ. 50 (1994):101–123. With permission from SEPM.

Figure 8.17: From Eugster, H.P. and Hardie, L.A. Saline lakes. In Lerman, A. (ed.), *Lakes: Chemistry, Geology, Physics*. With permission from Springer-Verlag Publishers, New York, chap. 8 (fig. 40). Copyright © 1978, Springer-Verlag Gmbh and Co. KG.

Figure 8.18: From Li et al. *Palaeogeog., Palaeoclim., Palaeoecol*. 123 (1996):179–203. With permission from Elsevier Science. Copyright © 1996.

Figure 8.19: From Salvany, J.M. and Orti, F. Miocene glauberite deposits of Alcanadre, Ebro Basin, Spain: sedimentary and diagenetic processes. In Renaut, R.W. and Last, W.M. (eds.), *Sedimentology and Geochemistry of Modern and Ancient Saline Lakes*. SEPM Spec. Publ. 50 (1994):203–215. With permission from SEPM.

Figure 8.20: From Zolitschka, B. and Negendanck, J.F.W. *Quat. Sci. Rev*. 15 (1996):101–112. With permission from Elsevier Science. Copyright © 1996.

Figure 8.21: Modified from Nelson, C.H. et al. The volcanic, sedimentologic, and paleolimnologic history of the Crater Lake caldera floor, Oregon: evidence for small caldera evolution. *Geol. Soc. Amer. Bull*. 106 (1994):684–704. With permission of the Geological Society of America, Boulder, CO. Copyright © 1994, Geological Society of America.

Figure 8.22: From Nelson, C.H. et al. The volcanic, sedimentologic, and paleolimnologic history of the Crater Lake caldera floor, Oregon: evidence

for small caldera evolution. *Geol. Soc. Amer. Bull*. 106 (1994):684–704. With permission of the Geological Society of America, Boulder, CO. Copyright © 1994, Geological Society of America.

Figure 8.23: From Larsen, D. and Crossey, L.J. Depositional environments and paleolimnology of an ancient caldera lake: Oligocene Creede Formation, Colorado. *Geol. Soc. Amer. Bull*. 108 (1996):526–544. With permission of the Geological Society of America, Boulder, CO. Copyright © 1996, Geological Society of America.

Figure 9.1: From Colman, S.M. et al. Continental climate response to orbital forcing from biogenic silica records in Lake Baikal. *Nature* 378 (1995):769–771. Copyright © 1995, Macmillan Magazines Ltd.

Figure 9.3: From Brezonik, P.L. and Engstrom, D.R. Modern and historical accumulation rates of phosphorus in Lake Okechobee, Florida. *Jour. Paleolim*. 20 (1998):31–46 (fig. 11). With permission from Kluwer Academic Publishers.

Figures 9.4 and 9.5: From Olsson, S. et al. Sediment–chemistry response to land use change and pollutant loading in a hypereutrophic lake, southern Sweden. *Jour. Paleolim*. 17 (1997):275–294 (figs. 7 and 12). With permission from Kluwer Academic Publishers.

Figure 9.6: From Norton, S.A. et al. Stratigraphy of total metals in PIRLA sediment cores. *Jour. Paleolim*. 7 (1992):191–204 (fig. 2). With permission from Kluwer Academic Publishers.

Figure 9.7: From Bischoff, J.L. et al. Responses of sediment geochemistry to climate changes in Owens Lake sediment: an 800-k.y. record of saline/fresh cycles in core OL-2. In Smith, G.I. and Bischoff, J.L. (eds.), *An 800,000 Year Paleoclimate Record from Core OL-2, Owens Lake, Southern California*. Geol. Soc. Amer. Spec. Pap. 317:37–47. With permission of the Geological Society of America, Boulder, CO. Copyright © 1997, Geological Society of America.

Figure 9.8: From Katz, B.J. Clastic and carbonate lacustrine systems: an organic geochemical comparison (Green River Formation and East African lake sediments). In Fleet, A.J., Kelts, K., and Talbort, M.R. (eds.), *Lacustrine Petroleum Source Rocks*. Geological Society of London Special Publication 40. With permission from Blackwell Science Ltd., pp. 81–90. Copyright © 1988.

Figure 9.9: From Meyers, P.A. Organic geochemical proxies of paleoceanographic, paleolimnologic and paleoclimatic processes. *Org. Geochem.* 27 (1997):213–250. With permission.

Figure 9.10: From Bourbonniere, R.A. and Meyers, P.A. Anthropogenic influences on hydrocarbon contents of sediments deposited in eastern Lake Ontario since 1800. *Envir. Geol.* 28 (1996):22–28. Copyright © 1996, Springer-Verlag Gmbh and Co. KG.

Figure 9.11: From Leavitt, P. et al. *Can. Jour. Fish. Aquatic Sci.* 51 (1994):2411–2423. With permission from the National Research Council of Canada.

Figure 9.12: From Curtis, J.H. and Hodell, D.A. An isotropic and trace element study of ostracodes from Lake Miragoane, Haiti: a 10,500 year record of paleosalinity and paleotemperature changes in the Caribbean. In Swart, P. et al. (eds.), *Climate Change in Continental Isotopic Records.* AGU Monograph 78, p. 127. Copyright © 1993, American Geophysical Union.

Figures 9.13 and 9.14: From Dettman, D. and Lohmann, K.C. Seasonal change in Paleogene surface water $\delta^{18}C$: fresh-water bivalves of western North America. In Swart, P.K. et al. (eds.), *Climate Change in Continental Isotopic Records.* AGU Monograph 78, pp. 154, 159. Copyright © 1993, American Geophysical Union.

Figure 9.15: From Li, H.C. et al. Stable isotope studies in Mono Lake (California). 1. $\delta^{18}O$ in lake sediments as proxy for climate change during the last 150 years. *Limnol. Oceanogr.* 42 (1997):230–238. With the permission of the American Society of Limnology and Oceanography.

Figure 9.16: From Ricketts, R.D. and Anderson, R.F. A direct comparison between the historical record of lake level and the $\delta^{18}O$ signal in carbonate sediments from Lake Turkana, Kenya. *Limnol. Oceanogr.* 43 (1998):811–822. With the permission of the American Society of Limnology and Oceanography.

Figure 9.17: From Talbot, M.R. and Johannessen, T. A high-resolution palaeoclimatic record for the last 27,500 years in tropical West Africa from carbon and nitrogen isotopic composition of lacustrine organic matter. *Earth Planet. Sci. Lett.* 110 (1992):23–37. With the permission of North Holland Publishing Co.

Figure 10.4: From Wolfe, A.P. Late Wisconsin and Holocene diatom stratigraphy from Amarok Lake, Baffin Island, N.W.T. *Jour. Paleolim.* 10 (1994):129–139 (fig. 4). With permission from Kluwer Academic Publishers.

Figure 10.6: From Anderson, N.J. Using the past to predict the future: lake sediments and the modeling of limnological disturbance. *Ecol. Modeling* 78 (1995):149–172. With permission from Elsevier Science. Copyright © 1995.

Figure 10.7: From Wilson, S.E. et al. A Holocene paleosalinity diatom record from southwestern Saskatchewan, Canada: Harris Lake revisited. *Jour. Paleolim.* 17 (1997):23–31 (fig. 3). With permission from Kluwer Academic Publishers.

Figure 11.1: From Bertrand-Sarfati, J. et al. Microstructures in Tertiary nonmarine stromatolites (France). Comparison with Proterozoic. In Bertrand-Sarfati, J. and Monty, C. (eds.), *Phanerozoic Stromatolites II.* With permission from Kluwer Academic Publishers, Dordrecht, pp. 155–191, 1994.

Figure 11.2: Courtesy of John Tibby. Copyright © 2000.

Figure 11.3: Modified from Barker, P. et al. Experimental dissolution of diatom silica in concentrated salt solutions and implications for paleoenvironmental reconstruction. *Limnol. Oceanogr.* 39 (1994):99–110. With the permission of the American Society of Limnology and Oceanography.

Figure 11.4: From Charles, D.F. et al. Paleoecological investigation of recent lake acidification in the Adirondack Mountains. *Jour. Paleolim.* 3 (1990):195–241. With permission from Kluwer Academic Publishers.

Figure 11.5: Modified from Renberg, I. A 12,600 year perspective on the acidification of Lilla Öresjön, southwest Sweden. *Phil. Trans. R. Soc. London, Ser. B* 327 (1990):357–361. With permission from The Royal Society.

Figure 11.6: From Laird, K.R. et al. A diatom-based reconstruction of drought intensity, duration, and frequency from Moon Lake, North Dakota: a sub-decadal record of the last 2300 years. *Jour. Paleolim.* 19 (1998):161–179. With permission from Kluwer Academic Publishers.

Figure 11.7: From Anderson, N.J. et al. A comparison of sedimentary and diatom-inferred phosphorus profiles: implications for defining pre-disturbance nutrient conditions. *Hydrobiologia* 253 (1993):357–366. With permission from Kluwer Academic Publishers.

Figure 11.8: From Anderson, N.J. and Rippey, B. Monitoring lake recovery from point-source eutrophication: the use of diatom-inferred epilimnetic total phosphorus and sediment chemistry. *Freshwater Biol.* 32 (1994):625–639. With permission from Blackwell Science Ltd. Copyright © 1994.

Figure 11.9: From Charles, D.F. et al. Variability in diatom and chrysophyte assemblages and inferred pH: paleolimnological studies of Big Moose Lake, New York. *Jour. Paleolim.* 5 (1991):267–284. With permission from Kluwer Academic Publishers.

Figure 11.10: From Huber, J.K. A post-glacial pollen and nonsiliceous algae record from Gegoka Lake, Lake County, Minnesota. *Jour. Paleolim.* 16 (1996):23–35. With permission from Kluwer Academic Publishers.

Figure 11.11: From Hannon, G. and Gaillard, M. J. The plant microfossil record of past lake level changes. *Jour. Paleolim.* 18 (1997):15–28. With permission from Kluwer Academic Publishers.

Figure 11.12: From Eisner, W.R. et al. Paleoecological studies of a Holocene record from the Kangerlussaq (Søndre Strømfjord) region of West Greenland. *Quat. Res.* 43 (1995):55–66. With permission from Academic Press. Copyright © 1995, The University of Washington.

Figure 11.13: From Birks, H.H. Aquatic macrophyte vegetation development in Kråkenes Lake, western Norway, during the late-glacial and early Holocene. *Jour. Paleolim.* 23 (2000):7–19. With permission from Kluwer Academic Publishers.

Figure 11.14: From Frey, D.G. Littoral and offshore communities of diatoms, cladocerans, and dipterous larvae, and their interpretation in paleolimnology. *Jour. Paleolim.* 1 (1988):179–191. With permission from Kluwer Academic Publishers.

Figure 11.15: From Bos, D.G. et al. Cladocera and Anostraca from the Interior Plateau of British Columbia, Canada, as paleolimnological indicators of salinity and lake level. *Hydrobiologia* 392 (1999):129–141 (fig. 6). With permission from Kluwer Academic Publishers.

Figure 11.16: From Smith, A.J. Lacustrine ostracodes as hydrochemical indicators in lakes of the north-central United States. *Jour. Paleolim.* 8 (1993):121–134. With permission from Kluwer Academic Publishers.

Figure 11.17: From Porter, S.C. et al. The ostracode record from Harris Lake, southwestern Saskatchewan: 9200 years of local environmental change. *Jour. Paleolim.* 21 (1999):35–44. With permission from Kluwer Academic Publishers.

Figure 11.18: Modified from Cohen, A.S. Linking spatial and temporal changes in the diversity structure of ancient lakes: example from the ostracod ecology and paleoecology of Lake Tanganyika. In Rossiter, A. and Kawanabe, H. (eds.), *Ancient Lakes: Biodiversity, Ecology and Evolution.* Advances in Ecological Research 31. With permission from Academic Press, New York, pp. 521–537, 2000.

Figures 11.19 and 11.20: From Brooks, S.J. and Birks, H.J.B. Chironomid-inferred late-glacial and early Holocene mean July air temperatures from Kråkenes Lake, western Norway. *Jour. Paleolim.* 23 (2000):77–89. With permission from Kluwer Academic Publishers.

Figure 11.21: From Verschuren, D. et al. Effects of depth, salinity, and substrate on the invertebrate community of a fluctuating tropical lake. *Ecology* 81 (2000):164–182. With permission of the Ecological Society of America.

Figure 11.22: From Miller, B.B. and Thompson, T.A. Molluscan faunal changes in the Cowles Bog area, Indiana Dunes National Lakeshore, following the low-water Lake Chippewa phase. In Schneider, A.F. and Fraser, G.S. (eds.), *Late Quaternary History of the Lake Michigan Basin.* With permission of the Geological Society of America, Boulder, CO. Copyright © 1990, Geological Society of America.

Figure 12.2: From Gaillard, M.J. et al. A late Holocene record of land use, soil erosion, lake trophy and lake level fluctuations at Bjärsejösjön (South Sweden). *Jour. Paleolim.* 6 (1991):51–81. With permission from Kluwer Academic Publishers.

Figure 12.3: From Lipiatou, E. et al. Recent ecosystem changes in Lake Victoria reflected in sedimentary natural and anthropogenic organic compounds. In Johnson, T.C. and Odada, E. (eds.), *Limnology, Climatology and Paleoclimatology of the East African Lakes.* With permission from Gordon and Breach, Amsterdam, 1996.

Figures 12.4 and 12.5: From Hall, R.I. et al. Effects of agriculture, urbanization and climate on water quality in the Northern Great Plains. *Limnol. Oceanogr.* 44 (1999): 739–756. With permission from the American Society of Limnology and Oceanography.

Figure 12.6: From Engstrom, D.R. and Swain, E.B. Recent declines in atmospheric mercury deposition in the Upper Midwest. *Envir. Sci. Tech.* 31 (1997):960–967. With permission. Copyright © 1997, American Chemical Society.

Figure 12.7: From Renberg, I. et al. Atmospheric lead pollution history during four millennia (2000 BC to 2000 AD) in Sweden. *Ambio* 29 (2000):150–156. With permission.

Contents

Paleolimnology

I

Lakes as Archives of Earth History

For several months each year I work in central Africa collecting sediment cores and fossils from a large rift lake, Lake Tanganyika. Periodically my nonscientist friends ask me why I do this. They usually mean both "why would someone collect mud from the bottom of a lake"? and perhaps as an even greater challenge to my sanity, "why would one travel halfway around the world to do this"? The answer to these questions (and the theme of this book) is deceptively simple. Paleolimnologists study lake deposits because they provide science with *archives* of earth and ecosystem history that are both highly resolved in time and of long duration. In the particular case of Lake Tanganyika, this combination, in principle, permits us to study events as closely spaced in time as annual events over the lake's 10-million-year history. Few other records of earth history beyond those found in lake muds provide this combination of duration and resolution.

The range of questions that can be examined with these archives is enormous. Paleolimnologists provide constraints on the timing of past climate change, determine rates of evolutionary change in species, and investigate the timing of pollutant introduction into watersheds. One might reasonably ask what good could come from trying to synthesize these disparate questions. I believe that the unifying factor behind all of these fields of study lies in the character of lake sediment archives. Lakes are attractive targets for study by such different fields of investigation because of the special nature of their depositional environment. Considering the contents of lake archives and their characteristics is the best place to start thinking about what makes paleolimnology a distinctive discipline.

What Are Lake Archives?

An archive is a singularly appropriate term to describe the foundations of paleolimnological research. The term archive can refer both to a historical record, its content, and to the place where such records are housed, its container. Both content and container are critical elements in deciphering historical records because the content must be information-rich and the container must be durable in order for meaningful records to be available for study and interpretation. All lakes are subject to a variety of extrinsic and intrinsic forcing variables that regulate the subsequent history of the lake, such as climate, watershed bedrock composition, tectonic and volcanic activity, vegetation, aquatic biota, and human activities (figure 1.1). Keep in mind that these variables are highly interactive, for example watershed vegetation can be greatly affected by human activity. Typically it is the history of change in these variables that we wish to reconstruct using paleolimnology.

Forcing-variable records are housed in container archives. It is common to think of these "containers" solely in terms of the sedimentary pile at the bottom of a lake, but strictly speaking this is untrue. Paleolimnological records are contained in three distinct types of archives. The first is the lake water itself. Water and its contents have finite *residence times* in a lake, the time required for the average molecule of water or a solute to cycle through the system. In lakes with long residence times (measured in hundreds to thousands of years), the water itself may provide important paleolimnological clues to lake history. A second, more durable form of container archive is the geomorphology of a lake basin, particularly its shoreline and shape characteristics. Such features can persist for many thousands of years, long after the lake that they were associated with is gone. Undoubtedly, however, the most important container archives for paleolimnology are accumulated sediments. Sediments provide the most durable containers since they may persist long after the lake itself or its geomorphology are obliterated.

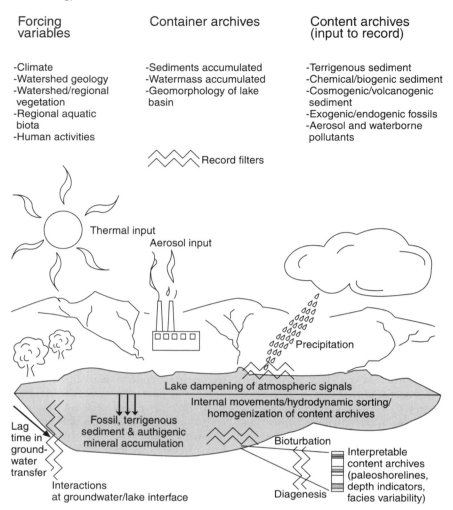

Figure 1.1. A simple model of important factors controlling the formation of sedimentary archives in lakes. *Forcing variables* are external factors that regulate the formation, persistence, and characteristics of the lake in question. *Container archives* are the recording media that house the indicators of external environmental change. *Content archives* are the physical records themselves (fossils, sediments, geochemistry) from which a paleolimnological history must be reconstructed. Zigzag lines highlight major points of record filtering.

Content archives in paleolimnology are the physical sedimentary inputs to the lake's record. These include terrigenous, chemical, and biogenic sediments, cosmogenic and volcanogenic particles, fossils that originated outside of the lake, like pollen, or in the lake, like fish, and aerosol and waterborne pollutants. These content archives have varying value for answering questions about past environmental conditions. Content archives are often misleadingly referred to as *proxies*, in the sense that they are assumed to be substitutes for a direct measurement of the original forcing variable controlling them. Despite the common usage of this term, it is far more accurate to refer to these sensitive archives as *indicators* of environmental conditions, since all lake archives are subject to modification by processes apart from the ones we are interested in recording.

Archives are not complete or completely accurate depictions of history. They have been filtered in ways that are both beneficial and detrimental to reconstructing history. Some filters obliterate the meaning of an archive, making its interpretation dubious. Other filters change information in such a way as to dampen the record of short-term modulations in the original mechanism driving the creation of the archive. These types of filters can actually be useful for the paleolimnologist, since they may remove uninteresting "noise" in the pattern being interpreted. For example, bioturbation of sediments to shallow depths may homogenize a very erratic record of paleocommunity change that otherwise might be very difficult to interpret. If the object of the study is to look at variation on a decadal time scale, then this type of filter might actually be desirable. And yet other filters serve to create lag times between the occurrence of the event creating the record and the deposition of the recording medium. The movement of atmospheric inputs, such as pollutants, into a lake by way of groundwater is an example of this type of filter, since the original signal, precipitation and its solute load, is significantly delayed in its arrival at the lake.

Paleolimnologists have an extensive toolkit to sample the container archives of lakes. These tools differ in their potential for resolving different types of paleolimnological problems, making it worthwhile to briefly examine their respective strengths and weaknesses.

Sedimentological Archives from Cores

Sediment coring is the most common method of obtaining paleolimnological data. Cores are effectively one-dimensional columns of sediment; even the collection of several cores only allows a limited perspective on the spatial variability of sedimentary archives in a lake (figure 1.2a). The diameter of a core barrel also places a limit on the quantity of sample available at any stratigraphical level. In existing lakes, cores are the only practical means of obtaining physical samples. However, core sampling also plays an important role in the interpretation of paleolakes, because the location for obtaining samples can be chosen by the scientist.

Because of the tradeoffs of cost and effectiveness under varying conditions, no single coring tool is appropriate for all circumstances (Aaby and Digerfeldt, 1986; Smol and Glew, 1992). As a result, different coring tools have been developed to serve different needs. *Chamber samplers* are primarily used to sample peats. The sampler is driven into the sediment while it is closed, and then opened using a rotating side aperture to allow the sample to enter the chamber. Hand-operated or motorized *augers* employ rotation of a hollow screw into the sediment, collecting a sediment column inside the driving screw (figure 1.2b). *Gravity, box,* and *multi-*

ticore corers use the corer's weight to force themselves into the sediment (figures 1.2c,d). They are lowered to the lake floor under controlled conditions to ensure vertical entry into the sediment and to prevent the corer from overpenetrating, becoming completely buried, and thereby not sampling the surficial sediments. A closing mechanism at the top of the core eliminates downward hydrostatic pressure after core collection, causing the core to be held within the barrel as it is retrieved. A jaw-like device, called a core catcher, is located just inside the cutting end of the corer, preventing the sample from falling out of the bottom during retrieval. Gravity cores are limited to a few meters in length. A *box corer* is a specialized type of gravity corer that takes a larger box-like sample, rather than a cylinder, allowing some spatial variation to be seen and larger samples to be collected. A multicorer collects several adjacent short cores (< 1 m) from a single device, allowing large sample size, precise collection of the sediment–water interface, and the observation of some spatial variation.

Piston corers are designed to overcome hydrostatic pressure problems that affect gravity corers (figure 1.2e). A piston corer uses the suction created by a sealed piston, located within the core barrel and immediately above the sediment, and driven either mechanically or hydraulically to eliminate internal hydrostatic pressure over the sediment sample. As the core barrel penetrates the sediment, the stationary piston allows sediment to enter the barrel without the sediment-deforming effects of internal hydrostatic pressure. In this way a longer, undisturbed sample can be obtained. A single drive of a piston corer is typically anywhere from 1 to 15 m, although longer cores can be collected in this way through the use of multiple drives of the coring device, for example by progressively sampling the sediment, withdrawing the core sample, and extending the corer further.

Piston and gravity corers will not work in lithified or coarse-grained sediments. *Vibracoring* can be used for the penetration of coarse-grained, unconsolidated sediments (figure 1.2f). A vibracorer has a large, vibrating motor head attached to the top of the core barrel. When the motor is turned on, the vibrations drive the barrel into the sediments. In lithified sediments, *rotary drilling* techniques must be used to obtain sediment cores (figure 1.2g).

Freeze sampling is used for a similar purpose to multicoring, to collect the sediments immediately below the sediment–water interface without disturbance, particularly in sediments with high water content (figure 1.2h). Freeze samplers operate by

Figure 1.2. Photos from the paleolimnologist's toolkit. (a) A typical lithified sediment core from Cretaceous lake beds, in a petroleum exploration well from the eastern Atlantic (offshore West Africa), showing the highly resolved textural and compositional variability that makes core analysis the primary tool of paleolimnology. Scale bar = 10 cm. (b) Split spoon auger coring at Summer Lake, Oregon. The auger is driven by a truck-mounted rig. (c) Gravity coring at Lake Tanganyika (Burundi). (d) Multicoring at Lake Tanganyika. Four short cores are obtained from the barrels in the center of the device, which is lowered to the lake floor.

(e)

(f)

(g)

(h)

Figure 1.2. (*continued*) (e) Kullenberg piston coring at Lake Turkana. (f) Vibracorer, about to be deployed at Lake Tanganyika. The motorized vibrating head is the black cylinder at the top of the core barrel. (g) GLAD800 drilling rig at the Great Salt Lake, Utah. This device collects continuous cores in 3-m segments using a combination of hydraulic piston coring and rotary drilling. (h) A freeze core, obtained to collect an accurate representation of surficial, unconsolidated sediments.

(i)

(j)

(k)

Figure 1.2. (*continued*) (i) An outcrop exposure of the Cretaceous Candeis Formation, Reconcavo Basin, Brazil, showing vertical and lateral variability in deep-water turbidite deposits. (j) Lake terraces surrounding the Great Salt Lake, representing the elevations of prior shorelines. (k) Deployment of multichannel hydrophone streamer for reflection seismic profiling at Lake Tanganyika.

allowing sediment to adhere by freezing to the outside surface of a hollow chamber that has been previously filled with a dry ice/alcohol mixture.

Sedimentological Archives from Outcrops

Outcrop analysis is another common means of obtaining paleolimnological records. Outcrops

allow much larger samples to be obtained than cores, and allow at least two-dimensional, and frequently three-dimensional, reconstructions of variation in sedimentary deposits (figure 1.2i). Outcrop analysis is limited to either paleolakes or deposits exposed around existing lakes. Limitations are also imposed by the pattern of exposure, from either natural outcrops or road cuts, which may not suit the needs of the investigator. Also, lake beds that have been exposed to air or near-surface conditions will have undergone additional diagenetic and pedogenetic (soil-forming) processes that may alter important aspects of their content archives.

Geomorphological Archives from Lake Paleoshorelines

Terraces commonly form at or near a lake's surface, through both erosional and depositional processes. They are most prominently developed on the exposed coasts of large lakes, where wave attack on a shoreline is strongest (Adams and Wesnousky, 1998). When lake levels fall, such terraces can be exposed, allowing their elevations to be measured (figure 1.2j). To a lesser extent such paleoshorelines may also be observed below the current lake level, through analysis of bathymetric profiles or side-scan sonar images. The unique feature of these terraces is that they provide information on what, at their time of formation, were originally horizontal surfaces, often preserved over a considerable lateral extent around the paleolake basin. When such shorelines are undeformed by subsequent tectonic activity, their elevations provide a means of estimating hydrologic parameters for the paleolake, and can also be useful in estimating ages of otherwise undated or undatable lake levels (Sack, 1995; Avouac et al., 1996). In areas where paleoshorelines have been deformed by secondary tectonic activity, they can also provide information about the viscous properties of the earth's mantle and isostatic rebound histories following deglaciation (Clark et al., 1994).

Geophysical Archives from Seismic Records

Seismic reflection profiles provide another important type of paleolimnological archive. Although their application is limited by expense, the information provided by seismic stratigraphy directly complements that of coring and outcrop studies, and provides three-dimensional data on the geometry of sedimentary deposits that cannot be obtained in any other practical way (Anstey, 1982; Sheriff and Geldart, 1995). Seismic reflection profiling relies on the fact that different types of sediments have different acoustical properties. Acoustical impedance can be defined for any material as the product of its density times the velocity at which it transmits sound waves. At an interface between two sediment types a seismic reflection is created whose strength is proportional to the difference in acoustical impe-

dance between the two materials. Reflections are shown graphically as a "wiggle-trace," a curve depicting the various reflections occurring vertically at a single location. Shading conventions are used to display strong acoustical impedance differences caused by the juxtaposition of materials with differing densities. They are depicted proportionally to the time of their arrival; thus the vertical axis on a seismic stratigraphical profile is one of travel time, generally displayed as the round trip or *two-way* travel time, rather than depth. Depth can only be inferred from this through some a priori knowledge of the actual velocity of sound waves at different depths.

Vibrations from a source, such as compressed air, ground-shaking equipment, or explosives, are differentially reflected off sedimentary layers in the subsurface (figure 1.2k). These reflections are collected using geophones or hydrophones, amplified, stacked (incorporating data from multiple geophones), and processed to eliminate various artifacts. The resulting data are displayed as a series of vertical traces aligned along the transect from which the data were obtained. Because similarly shaded reflections from adjacent vertical profiles are aligned, they run together to produce the appearance of horizontal or dipping strata, which are then available for interpretation on the composite seismic profile (figure 1.3).

Seismic profiles provide a two-dimensional view of the underlying geometry of sedimentary layers, their thicknesses, assuming travel time can be reasonably converted to true depth, and convergence or divergence patterns. Differences in the acoustical properties between different types of sediments allow researchers to interpret probable lithologies from seismic data. When two-dimensional transects are combined in a grid, it is also possible to obtain a three-dimensional view of the underlying mass of sediments. From these, seismic stratigraphers can interpret the relative history of depositional and tectonic events, such as the magnitude of past lake level fluctuations. If these can be coupled with core data, then an historical record of paleolimnological change can be established.

Other Geophysical Archives

A variety of other geophysical imaging tools are available to determine the size and shape of sedimentary deposits, either on a lake's floor, or in the subsurface. Side-scanning sonar can be used to obtain acoustical images of depositional features on a lake's floor (T.C. Johnson and Ng'ang'a,

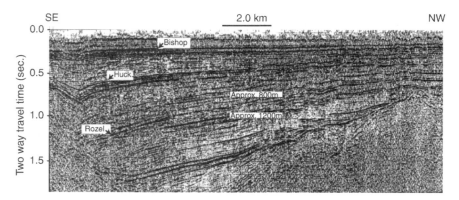

Figure 1.3. Reflection seismic profile of a cross-section in the north basin of the Great Salt Lake. Reflection seismic imaging provides a powerful tool for interpreting basinwide patterns of sediment accumulation in lakes, detecting climatic and tectonic controls on deposition, and, in an exploratory sense, determining the sites where cores should be obtained for detailed study. The vertical scale is in two-way travel time (sec) for seismic reflections derived from a surface source and reflected off a subsurface horizon (approximate depths in meters are also shown for specific horizons). Dated horizons based on volcanic ashes obtained from wells that penetrate this section indicate the great age of this basin. Bishop = Bishop Ash (0.759 ma), Huck. = Huckleberry Ash (2.06 ma), Rozell = Rozell Basalt (3.1 ma).

1990). Ground-penetrating radar can also be used to generate cross-sectional profiles that look very similar to those created by seismic sources (Sten et al., 1996; D.G. Smith and Jol, 1997).

Down-hole geophysical logging tools are commonly used in the petroleum exploration industry for determining sediment properties such as electrical conductivity, gamma-ray production, or temperature. Similar logs can be obtained by passing previously collected cores through an array of sensors. Different sediment types produce differences in various geophysical signals, and although the results of logging such signals rarely pinpoint a lithology exactly, they do constrain the range of possible lithologies present at any given stratigraphical level being recorded. Such tools have tremendous potential for providing auxiliary information about lake bed history and lithology when collected in conjunction with coring or drilling.

Modeling Approaches to Paleolimnology

Although descriptive and mathematical models of paleolakes are not archives of historical information, their study is proving increasingly useful in making sense of such archives. Simulations of lake processes are particularly useful in guiding researchers, by suggesting likely archivable responses to some forcing phenomenon, some of which may not have been previously investigated

in the lake basin being modeled. The most important applications of this approach to date come from paleohydrological models that relate particular assumptions about inflow and discharge to predicted lake levels (Tinkler and Pengelly, 1995; Bengtsson and Malm, 1997). Lake level data can be compared with such models to determine if assumed inputs and outputs are consistent with observation, and if not, to adjust the models to better fit available data sets. Modeling of sediment accumulation in lakes also allows patterns of deposition to be simulated under relatively realistic conditions and compared with observations (Schlische, 1991).

Time and Event Resolution

Lakes, by definition, are enclosed bodies of standing water surrounded by land. Most lakes are surrounded by watersheds that are large relative to the lake's area. This makes lakes quite different from the world's oceans (or even large marginal seas for that matter), where the ratio of land supplying sediment, or dissolved salts that can become sediment, to water is small. The implication of this difference is that lakes are environments of relatively rapid sediment accumulation in comparison with the oceans or many other terrestrial settings. Accumulation rates in lakes are typically one or two orders of magnitude faster than in comparable settings in the oceans, over comparable durations of observation. Even the largest and deepest lakes like Lake Baikal accumulate sediment on their abyssal plains at rates of 0.3 to 6 mm/yr (0.007 to 0.031 g cm^{-2} yr^{-1}) (Appleby et al., 1998), much faster than comparable basinal settings in the deep sea. Furthermore, terrestrial or coastal settings that display short-term accumulation rates that are comparable to lakes rarely display the continuity of deposition that is a hallmark feature of many lakes.

The ability to discriminate historical events in a sediment archive is not strictly a result of sediment accumulation rate. It is also a consequence of secondary modification, through transport or bioturbation, that results in *time-averaging* of a deposit, placing a lower limit on the duration of what can, in principle, be resolved. In this respect lakes are favorable settings for obtaining highly resolved records, since depths of bioturbation in lakes, in the millimeter to few centimeter range, tend to be considerably shallower than in terrestrial or marine settings, and sedimentation rates in lakes are also on average fairly rapid. This means that decadal or even annual events can be discriminated, an impossibility in many other depositional environments. In

fact, since the pioneering work of De Geer (1912) on annually laminated lake sediments, geologists have recognized the extraordinary potential for obtaining high-resolution historical records from lake beds.

Another factor that differentiates lake sedimentation from most other depositional settings is the greater degree of depositional continuity evident in lakes. For many years it has been recognized that the stratigraphical record is punctuated by abundant hiatuses at all time scales (e.g., Barrell, 1917; Ager, 1973). Sadler (1981) demonstrated that for most depositional environments the measured "rate" of sediment accumulation in a deposit is inversely correlated with the duration over which the measurement of the rate is made (figure 1.4). Over short time intervals deposition appears relatively rapid and over geologically long intervals that rate appears to slow. This relationship results from the fact that depositional "processes" are, in reality, mostly discrete events. As our yardstick of time and rate moves from direct observation, such as what collects in a sediment trap, to radiometric measurements on two widely separated stratigraphical horizons, our measured interval incorporates a greater and greater percentage of intervals of nondepos-

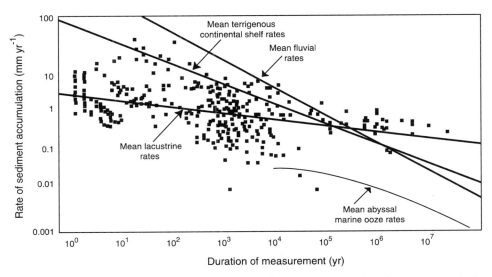

Figure 1.4. Regression curves showing the relationship between measured sediment accumulation rates and duration over which those rate estimates are made. Sedimentation "rates" are an inverse function of the time over which they are recorded. For most sedimentary environments rates decline quickly as the duration of study increases, as a result of the erratic nature of deposition and erosion. Lakes, shown by regression curves and data points (squares), provide a more continuous record of sedimentation, reflected by a lower slope, and overall higher rates of accumulation. Both factors provide great advantages in the use of paleolimnological records to reconstruct past events at high resolution. Adapted from Sadler (1981) with the addition of new lake data.

ition or active erosion. Measured rates of accumulation "slow" as a result. Strictly speaking, a depositional rate only has meaning within the context of the duration over which it was measured. Certain depositional environments are more prone to this rate measurement effect, largely as a result of their location as zones of temporary storage and secondary removal of sediments, for example in river valleys, or over longer time intervals, because of their tectonic setting on the earth's surface. This expresses itself in figure 1.4 as a steep inverse slope on the rate/duration curve. In contrast, lake sediment accumulation rate data have a relatively small dependency on measurement duration. Much of the shallow slope to the lake regression can be attributed to compaction, since most short-duration rates are measured from very recent, soupy sediments. Lake deposits display a lower degree of rate dependency on measurement duration in comparison to most other depositional settings. This results from the position of most lakes as terminal sumps for sediment deposition. Even open basin lakes, those with an outlet, discharge relatively little of their sediment input through their outlets. A high degree of depositional continuity at varying time scales is a hallmark feature of most lakes. This factor, coupled with high average rates of deposition, contributes to the usefulness of lake deposits as historical records.

Lakes, being relatively small bodies of water in comparison with the ocean, can respond relatively quickly to an external forcing variable. Changes in air temperature or regional land use can be reflected within months to decades in the inputs to the lake's sedimentary record. The rapidity of this response varies both within a lake, for example from nearshore to offshore environments, and between lakes. Typically small lakes will respond more quickly and more definitively to a perturbation than large lakes. We can put this difference to work through regional comparisons of different sized lakes and their response to environmental change. This difference provides us with a good departure point for considering one of the fundamental limitations to interpreting paleolimnological records, the *hydroclimate filter*.

The Hydroclimate Filter

The goal of paleolimnological research is typically paleoenvironmental reconstruction. What is unstated in this point is precisely what environment we are trying to reconstruct. The vast majority of paleolimnological studies published in recent years

have been directed at reconstructing external environmental variables such as climate, as opposed to being interested in the lake's history for its own sake. However, the lake itself interposes an important filter between our reconstruction of past external environments and the sedimentary record we use to derive this reconstruction. This hydroclimate filter incorporates all of the physical, chemical, and biological processes in lakes that transform input forcing variables into something other than a simple output recording.

A good example of the concept of the hydroclimate filter can be illustrated by examining seasonal cycles in temperature at Mono Lake, California (figure 1.5). Annual air temperature variability near the lake is on the order of $30°C$. However, because of the high specific heat of water, thermal variability in the lake is greatly reduced from that of the air. In the surface waters of Mono Lake the annual temperature variation is about $16°C$. Deeper in the lake (15 m) the variation is further reduced, to about $9°C$. Also note that as a result of continued heat exchange from the surface to deeper waters of the lake, the deeper water is continuing to warm during the late summer even as the air and surface water begin to cool.

Now consider three hypothetical indicators of temperature that are preserved in lake sediments in a lake with the thermal structure described above. The indicators are, respectively, a pollen species blown into the lake from the vegetational community surrounding the lake, a geochemical signal produced in the lake's surface waters in response to surface water temperature, and fossils from a benthic organism, living on the lake floor at 15 m. Assume that all three indicators are true proxies, that is, we can write an equation that precisely and quantitatively describes their abundance or occurrence patterns with respect to temperature. Furthermore, let's assume that all are preserved in an extremely highly resolved (monthly resolution) core record taken in 15-m water depth. What would we observe for an annual cycle in temperature? The benthic organism will record a much smaller variation in temperature than the surficial geochemical signal, and that signal in turn would be more attenuated than the pollen signal. The lake signals are responding to hydroclimate variation rather than to direct atmospheric variation. In some cases variation in lake temperature may closely follow that of air temperature (Livingstone and Lotter, 1998), but we rarely know this to be a fact a priori. We might presume from this observation that the externally produced pollen blowing into the lake is giving us a "better" signal of true

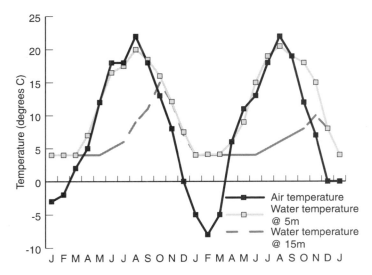

Figure 1.5. An example of a *hydroclimate filter*. Seasonal air temperature variability at Mono Lake, California, is reflected in water temperature variability at 5-m water depth but with more moderate seasonal shifts. At 15-m water depth this seasonal variation is even further dampened and the timing of maximum heating is delayed. Each of these sets of seasonal temperature change could be reflected in a paleolimnological signal preserved in equivalent-aged sediments, depending on where and how the content archive was derived. Adapted from data in Romero and Melack (1996).

climatic variation. However, there are several reasons why this may not be true. First, we have to better define what forcing variable we are really interested in. Is it regional or local climatic conditions that we wish to record? If it is the former, then artifacts may be introduced into our pollen record. For example, large lakes moderate their local climates. The vegetation near the lake that is providing the pollen may be unrepresentative of the regional vegetation. Furthermore, in many situations we may actually want more attenuated indicators of environmental change. If we are interested in longer-term climate change, the record of an overly sensitive indicator may provide too much information on short-term variability at the expense of clearly illustrating longer-term change. Suppose you were interpreting long-term temperature change from a tropical lake, using an indicator that was so sensitive and highly resolved that it actually recorded diurnal temperature variation. Long-term trends (even seasonal ones) would certainly be swamped in the "noise" of the night-to-day differences. The solution is to choose indicators whose sensitivity is appropriately scaled to the scientific question being addressed.

Hydroclimate can be thought of in a broader sense as the full panoply of environmental variables present in the lake environment. For example, ambient light and solute availability are aspects of the hydroclimate in a lake. Both of these may reflect an external forcing variable, such as local solar irradiance, or watershed solute input, but the lake itself imposes strong filters through its own internal processes, modifying the original forcing variable and thus the output of even the most sensitive of paleolimnological indicators.

Examples of Lake Archives Solving Earth History Problems

At this point it is useful to look briefly at a few case studies that illustrate the themes I have stressed up to this point. The studies focus on different time scales, and utilize different methodologies. What they all do in common is to take advantage of the qualities of the lacustrine stratigraphical record to address important questions about earth and biotic history.

Paleolimnology as a Tool for Lake Management and the Analysis of Ecological Experiments

The extremely high-resolution records that can be obtained from lake deposits make them ideal for application to problems in ecology. The combination of temporal and spatial scales available for study in lake deposit records allows them to be dovetailed with ecological experiments and ecosystem monitoring (figure 1.6). In comparison with other depositional systems, lakes are ideally suited for addressing questions about ecosystem dynamics on scales comparable with those of ecology.

Ecologists are frequently confronted with problems for which experimental or monitoring data cannot be readily collected. The "experiment" of interest might be some form of human impact on an ecosystem that has already occurred and cannot be reversed, or the process may occur over a longer time than monitoring will permit. A retrospective, paleolimnological approach to problem solving is well suited to these cases. Leavitt et al. (1994a) used a paleolimnological approach to examine the response of food webs and the distribution of lake organism sizes in Tuesday Lake (Wisconsin) to several human impacts. *Trophic cascade* ecological models predict that changes in the abundance of top predators such as large fish can have consequences that "cascade" down a food web. For example, an increase in the population of a top predator might be expected to result in a reduction in numbers of its prey, and that

species' food items, now released from predation pressure, might be expected to increase as a result of the cascading effect. Similarly, large zooplankton species, like some cladoceran crustaceans that may be favored targets of predation by certain zooplanktivores, might be expected to increase in number if the species that preys upon them are selectively eliminated because of a species introduction or other human impact. Tuesday Lake was subjected to road building within its watershed and various experimental treatments (aeration of its deep-water mass, additions of lime, and the introduction of largemouth bass and rainbow trout) over about a 30-year period. Through a detailed analysis of fossil cladocera and fossil algal pigments, Leavitt and his colleagues were able to demonstrate responses of the lake's food web to these impacts (figure 1.7). Temporal resolution in the laminated sediments of Tuesday Lake was sufficient to allow these investigators to sample fossil crustaceans and pigments at a seasonal level over the time period in question. This in turn allowed them to address questions of community structure in the lake that had not been investigated at the time the whole-lake experiments were being conducted.

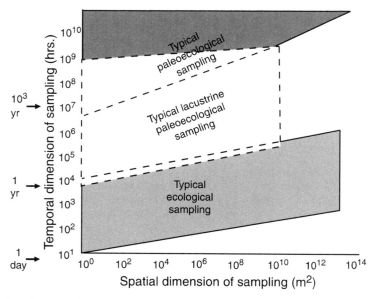

Figure 1.6. Sampling domains for ecology and paleoecology. The duration of any type of sampling generally increases as the area being sampled increases. Most paleoecological samples *time-average* content archives that formed over a much longer time interval than is typical of ecological sampling schemes. Paleoecological samples from lake beds, however, frequently overlap the time-averaging ranges of ecology, making them appropriate for direct comparisons.

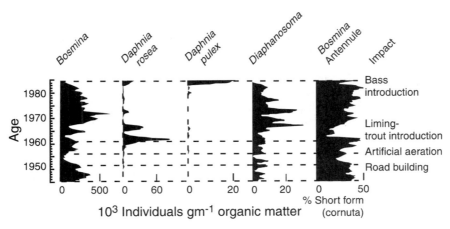

Figure 1.7. Annual changes in cladoceran crustaceans recorded in a core taken in Tuesday Lake, Wisconsin. Species and community responses to a variety of intentional and unintentional "experiments" altering the lake's ecosystem, including fish introductions, addition of lime to manipulate pH, artificial aeration of the water column, and road building can be inferred from their paleolimnological records. "Top-down" ecological effects of the introduction of predators are demonstrated by these data. Increases in the abundance of trout and bass reduced the abundance of their prey (smaller fish), which in turn allowed large cladocera (*Daphnia*) and benthic dipteran predators (*Chaoborus*) to increase in abundance. Small grazing cladocera (*Bosmina* and *Diaphanosoma*) were reduced in abundance by these predatory insects. This study illustrates the *power of paleolimnology* to provide sufficiently time-resolved data sets to answer questions of importance to lake and watershed management. From Leavitt et al. (1994a).

Lake Deposits as Sensitive Barometers of Regional and Global Change

One of the most exciting applications of paleolimnology today involves the exploration of rapid change in atmospheric composition and climate driven by human activities. In this area of study, high temporal resolution, rapid response of indicators to forcing variables, and indicator sensitivity are all very important. Individual lakes provide records with these characteristics, but one of the most valuable aspects of the lake archive comes from the use of records from multiple lakes within a region to examine an impact. Lakes in effect serve as unintentional, repeated *experiments* that evaluate human impacts on a wide variety of chemical and biological systems.

Both modern and historical data show that lake systems are undergoing profound changes because of human impacts from local to global scales. On the local scale, paleolimnologists have documented the timing of human activities such as deforestation, and have shown how these activities, along with climate change, alter lake systems (Hodell et al., 1995; Curtis et al., 1998). On more regional scales vast increases in atmospheric acid deposition from fossil fuel burning in the northeastern United States and northern Europe, primarily during the twenti-

eth century, have led to marked declines in lake pH. Increases in the use of fossil fuels on a global scale have been widely implicated in the sharp increase in the concentration of atmospheric CO_2 during the twentieth century. The vast majority of climate scientists believe this rise is largely responsible for the extraordinary warming of our atmosphere that has occurred during this same time period. Lakes can and have responded to this changing climate in various ways, such as warmer water temperatures and longer periods of ice-free conditions at mid- and high latitudes (Hanson et al., 1992; Assel and Robertson, 1995; D.W. Schindler et al., 1996), or warmer water surface temperatures in the tropics, leading to more stable water column stratification (Plisnier et al., 1999). All of these lake changes have consequences that can and have been recorded in lake sediment archives.

The Arctic is one of the most sensitive regions of the earth to atmospheric perturbations affecting climate, so it is not surprising that attention has been paid to its lakes for signals of global change. Overpeck et al. (1997) have summarized evidence from a wide variety of sources (lake deposits, tree rings, ice cores, etc.). These data indicate significant and perhaps unprecedented warming during the late nineteenth and twentieth century over much of the polar region in comparison with the last

5000 years (figure 1.8). Fossil diatoms from cores taken in neighboring Arctic Canadian lakes show that the responses to warming conditions vary between lakes with differing limnological and climatological characteristics. Recent responses to warming include increases in shallow-water species, increased biomass, and for the highest lakes, recency of colonization as permanent ice cover ended.

Lakes as Records of "Deep" Earth History

Paleolimnological archives provide insight into earth and biotic history that extends back to the Precambrian. In this book, I do not make a distinction between the study of Quaternary and pre-Quaternary lakes, since many of the barriers between these fields are artifacts of scientific cul-

ture. However, there are several real differences in methodologies between the study of older and younger lake deposits that we need to consider. Study of the paleolimnology of the Late Quaternary (what one of the great practitioners of paleolimnology, Edward Deevey, once referred to, somewhat tongue-in-cheek, as "neo-paleolimnology") has historically had a strongly biological flavor. In contrast, the study of pre-Late Quaternary age, Deevey's "paleo-paleolimnology," has been strongly dominated by geological interests.

Deevey's use of the terms above was an allusion to the fact that a methodological break exists between how one goes about investigating the history of a lake that is still a lake versus one that is today only represented by lake beds and outcrops. In the former case, most investigations are based upon the analysis of geophysical data and cores, analogous to marine geology. In contrast, outcrops

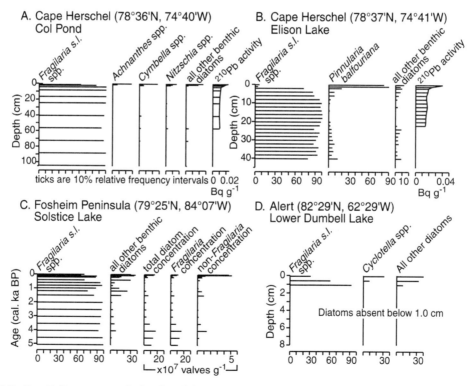

Figure 1.8. Fossil diatom records for four lakes in the high Arctic (Ellesmere Island, Canada) illustrating the use of paleolimnology to demonstrate an increase in temperatures across the polar region over the past 150 years. Col Pond, Ellison Lake, and Solstice Lake all show increased abundances of littoral and periphytic taxa and Solstice Lake shows an increase in algal biomass (numbers of sedimented valves per gram of sediment). The most northerly lake, Lower Dumbell Lake, shows a transition from permanent ice cover to one in which diatoms can reproduce (indicating periodic ice-free conditions) over the same interval. From Overpeck et al. (1997).

are the bread and butter of paleolimnologists interested in older lake deposits. Clearly this distinction is not watertight (pun intended). Outcrops representative of higher lake levels or fault blocks of raised lake beds surround many extant lakes. Similarly the value of continuous coring through the deposits of extinct lakes is clear from numerous studies. But paleolimnologists need to keep in mind the spatial differences in archives primarily derived from cores versus outcrops. Also, the indicators of paleoenvironmental change and lake history are undoubtedly better preserved on average in younger lake sediments. Again this distinction does not cut cleanly between the deposits of extant lakes and paleolake deposits, but is gradational. Fossils of organisms that are weakly skeletonized are much better represented in recent deposits, and geochemical archives can become obscured through progressive diagenesis.

In Late Pleistocene and Holocene lake deposits fossil assemblages comprise organisms that, for the most part, are still extant. Many of these species are extremely well characterized in terms of their ecological and limnological requirements. Often these environmental relationships are sufficiently well known to allow us to relate a particular fossil assemblage to a particular quantitative environmental variable (for example mean annual temperature or pH). Our ability to infer former ecological requirements of fossils weakens with time, and in an assemblage dominated by extinct taxa we are forced to rely on much weaker inferences from the requirements of related species. Hence pre-Quaternary paleolimnological reconstructions that are based on fossil assemblages tend to be more qualitative in nature than reconstructions based on younger assemblages.

These three distinctions between "neo-paleolimnology" and "paleo-paleolimnology" notwithstanding, I believe that the similarities between the fields warrant common discussion. Two examples can illustrate this potential for synthesis. The Great Salt Lake (Utah) and its precursor, Lake Bonneville, have been the object of paleolimnological research since the late nineteenth century (Gilbert, 1890). Extensive investigations of Late Pleistocene and Holocene sediments from both core and outcrop form the basis of much of our understanding of climate and hydrologic history in the Lake Bonneville basin, and, more generally, in the Great Basin of the United States (R.J. Spencer et al., 1984; Oviatt, 1987; Currey, 1990; Oviatt et al., 1999). These studies have documented in considerable detail lake level changes and associated sedimentological and geochemical responses (e.g., delta and soil surface formation and isotopic fluctuations), particularly over the past ~30,000 years. They also show the important linkages that can be drawn between the history of this lake and regional climate during the last glacial maximum and the period of deglaciation (figure 1.9).

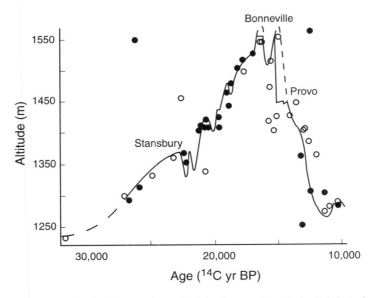

Figure 1.9. Late Pleistocene lake-level curve for Paleolake Bonneville, Utah. Solid circles represent radiocarbon dates on organic materials and open circles represent radiocarbon dates on carbonates. From Currey and Oviatt (1985).

Scientists have also recognized that a rich archive of lake history predating the Late Pleistocene exists in the Bonneville Basin (Eardley et al., 1973; Oviatt and Currey, 1987; Oviatt, 1988; O.K. Davis, 1998; Kowalewska and Cohen, 1998; Mohapatra and Johnson, 1998) (figure 1.3). These records demonstrate that multiple, lake-level cycles of the type recognized for Lake Bonneville in the Late Pleistocene to Holocene are characteristic of much of the lake's Middle–Late Pleistocene history. These 100,000-year cycles correspond in time with changes in the oxygen isotope record of deep-sea sediments, assumed to be a record of global ice volume. This suggests that Lake Bonneville fluctuated in large part because of climate changes that were global in nature.

Interestingly, the paleolimnological record for the Bonneville Basin indicates that this cyclicity only began about 750,000 years ago. Prior to that time paleolimnological records indicate that the

Bonneville Basin was occupied by a large marsh system, with only sporadic indications of open-water lake conditions (figure 1.10). The change in state of the lake at 750,000 years is also consistent with changes in the marine isotope record, again suggesting a global climate mechanism (Raymo et al., 1986).

Global climate change and its influence on lake and biotic history are also illustrated by the much older Triassic and Jurassic lake deposits of eastern North America. These deposits formed in a series of rift valley lakes that developed during the break up of the Pangaean supercontinent over a period of about 40 million years (Olsen, 1986; Olsen et al., 1996). However, individual lakes formed and dried over much shorter time intervals (Van Houten, 1962; Olsen, 1986) (figure 1.11). Statistical analyses of the thicknesses of individual phases of high and low lake-stand deposits suggest that individual lake-level fluctuations occurred at very reg-

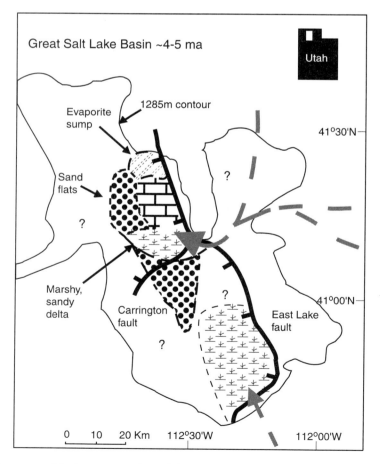

Figure 1.10. Paleogeographical reconstruction of the Great Salt Lake Basin about 4–5 million years ago.

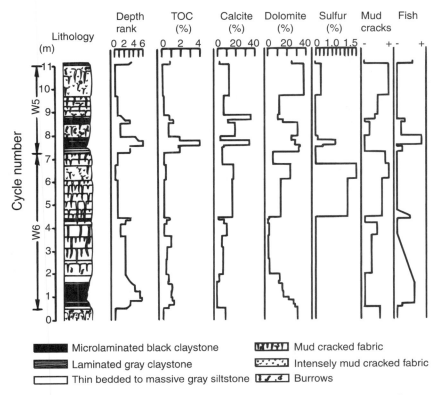

Figure 1.11. Stratigraphy of an outcrop from the Lockatong Fm., Tradesville, PA, illustrating typical cyclical features of Late Triassic–Early Jurassic (Newark Supergroup) lake deposits. Olsen (1986) has referred to these as Van Houten cycles, in honor of Franklin Van Houten, who first recognized the cyclical nature of these deposits and their climatic and possible orbital significance. Black claystone units with associated high total organic carbon content (TOC), absence of mudcracks, and abundant articulated fish fossils are indicative of deep lake phases, whereas thin-bedded to massive siltstones, low TOC and sulfur content, abundant mudcracks, and absence of fish fossils are indicative of low lake stands and more arid intervals. From Olsen (1986).

ular intervals. One probable mechanism for explaining such cyclicity is that changes in solar insolation, driven by cyclical changes in the earth's orbital parameters (the Milankovitch Cycles), have predictable effects on lake filling and drying. This type of cyclicity has long been recognized from Quaternary lake sediments, but that such cycles can be inferred from lake deposits 200 million years old is a testimony to the extraordinary resolution of events available for study to the pre-Quaternary paleolimnologist as well.

Summary

Our conclusions from this chapter can be summarized as follows:

1. The *high-resolution archives* of lake deposits provide a unifying theme for the various approaches to paleolimnology.
2. Paleolimnological archives are subject to various information *filters*, which operate both within and outside of the lake.
3. Sources of paleolimnological archives include sediment cores, outcrops, and geomorphic and geophysical studies. The interpretation of these data can be enhanced by the use of modeling approaches.
4. The historical resolution that can be obtained from lake archives is constrained by such factors as rates and continuity of deposition, *time-averaging* processes affecting deposits, and the *hydroclimate* filters that come between any lake archive and

the forcing environmental variable of interest.

5. A wide array of earth history questions can be studied using paleolimnological approaches. On short time scales paleolimnological data interfaces with or extends the ecological monitoring of lakes for management purposes and the determination of timing and rates of human impacts. On longer time scales paleolimnology can be used to provide extremely detailed records of the climatic, tectonic, and biotic evolution of the earth's surface.

2

The Geological Evolution of Lake Basins

Two things are required in order for a lake to exist on the earth's surface: a topographically closed hole in the ground and water. The subject of how topographical depressions form on the earth's continental crust has frequently been cast as one of lake *origins*, emphasizing the hole's initial formation. However, it is important to realize that the hole itself has a history, which is partly independent of the lake that fills it, and that this history interacts with that of the water body. This chapter will emphasize this dynamic interplay that occurs throughout a lake's history between a lake and its underlying substrate. In this sense, *lake basin evolution* is a more useful concept than the more static one of lake origin. The basin evolution process is manifest in everything from the three-dimensional geometry of the lake deposits that underlie the lake, to the rates of sediment accumulation, and the probable history and life span of the lake. Furthermore, different types of lakes are better or worse suited to answer specific paleolimnological questions. Some evolutionary mechanisms predispose lakes to persist for millions of years. Records in these lakes are ideally suited to answer questions that require long temporal records. Other questions require high-resolution records of short duration, which may be better represented in lakes formed by different mechanisms. And still other mechanisms result in the formation of numerous lakes with similar characteristics within a region, ideally suited for comparative studies. Understanding lake basin evolution is therefore an essential element in the design of a paleolimnological study, because the quality of paleolimnological records is directly linked to the mechanisms of basin evolution.

The formation of lakes has intrigued earth scientists for more than 100 years (W.M. Davis, 1882, 1887; Penck, 1882, 1894; Russell, 1895; Supan, 1896). Hutchinson (1957) elaborated on these earlier works, recognizing 11 major categories of lake origins and 76 subcategories. Numerous advances in understanding basin evolution have been made since Hutchinson's work, especially from improved radiometric dating techniques, seismic stratigraphy, and lake drilling over the past 50 years. The classification scheme used here (table 2.1) incorporates major elements of Hutchinson's scheme, updating that scheme where appropriate. No lake classification system can be entirely hierarchical because lake basins often evolve through multiple processes. For example, Lake Kivu in central Africa occupies a Neogene rift basin so it could be classified as a tectonically formed lake. However, during the Late Pleistocene, volcanic eruptions dammed its prior outlet, giving the lake a new and different *morphometry* (basin shape) and outlet level.

Lake basin evolution is interpreted from a combination of geomorphic and sedimentological clues. The shape and bedrock lithology of lake outlets (spillways) is used in extant lakes, or in Pleistocene lakes where geomorphic evidence is still fresh. Depositional features such as moraines or basalt flows that dam pre-existing drainage may unambiguously demonstrate a mode of origin. Seismic reflection data may provide clues from the shape of bedrock surfaces that underlie a lake's deposits. Where the constructional feature damming the outlet is higher than the ultimate spillway, the mode of origin may be more uncertain, because the relative timing of the purported dam's construction and spillway incision may not be known. The most difficult origins to interpret are those associated with lake deposits where the geomorphology of the basin is no longer evident. In pre-Pleistocene lake deposits tectonic controls may be evident from the association of lake deposits with a structural basin. But a secondary feature, for example an ice dam, that has left no permanent mark may have controlled the actual spillway elevation of a paleolake.

Most lakes fall into one of three of Hutchinson's categories, those formed by glacial, tectonic, or flu-

Table 2.1. Paleolimnologically significant lake-forming processes and properties of their sedimentary archives

Basin Closure Mechanism	Temporal Duration[a]	Potential Time Resolution[b]	Depositional Continuity[c]	Spatial Scale[d]	Comments
1. Glacial and Periglacial					Archive subject to scour and elimination during glacial readvances and isostatic rebound
Glacial lake basins–ice contact					
Englacial	S?	L	L	V (scales with ice-shed area)	Stratigraphical architecture poorly known
Subglacial	Potentially L under ice sheets	?	?	V (scales with ice-shed area)	Stratigraphical architecture poorly known
Supraglacial	S	H	L	S	Subject to rapid catastrophic draining
Proglacial	S–I	H	V	V (e.g., alpine glacier vs. ice-sheet dams), S to VL	Subject to rapid catastrophic draining. Maybe very large
Glacial rock basins					
Cirque lakes (tarns)	S	H	L	S	
Valley rock basin lakes	S	H	L	S	
Glint lakes	S	H	L	S	
Fjord lakes	S	H	L	S–M	
Piedmont (shield) lakes	S–I	H	Variable (H in deep lakes)	Potentially L to VL	
Glacial deposit-dammed basins					
Moraine-dammed lakes	S	H	L	S–M	
Kettle lakes (ice collapse)	S	H	L	S	
Dammed subglacial ice tunnels	S	L?	L	S	
Periglacial basins (thermokarst)	S	I–H	L	S	
2. Tectonic					
Basins associated with plate divergence					
Rift lakes	L–VL	I–H	V (VH in deep lakes)	Usually L (scales with lake area)	Often extremely deep and buffered from climatic drying

Table 2.1. (*continued*)

Basin Closure Mechanism	Temporal Duration[a]	Potential Time Resolution[b]	Depositional Continuity[c]	Spatial Scale[d]	Comments
Basins associated with plate convergence					
Foreland basin lakes	L–VL	I–H	V (but generally lower than rifts)	L	Frequently shallow lakes and/or marsh systems. Evolve to fluvial systems under high P/E conditions
Back-arc basin lakes	L–VL	H	H	VL	Evolve from prior marine conditions as marine exclosures. The most deeply subsiding lake basins because of formation on oceanic crust
Transtensional and transpressional basins	L	I–H	V	M to L	Often very deep. Older deposits migrate away from active depocenter
Cratonic basins	VL	L	L	L to VL	Extremely long-lived systems but sporadically filled by lakes
3. Fluvial					
Erosional basins					
Plunge pools	S	I–H	L–I	S	
Depositional/ subsidence-related basins					
Floodplain lakes	S–I	L	L	Extremely V (scales with river size)	
Abandoned channel lakes	VS–S	L	L	S	
Competitive fluvial aggradation lakes[e]	S–I	L–I	L–I	V	Volcanoclastic systems provide more complete and longer records
Delta plain lakes	S–I	L	L	M–L	Subject to extensive bioturbation and sea-level rise inundation
4. Coastal					
Back-barrier lakes	S	L	L	V	Modern systems date from mid-Holocene sea-level stabilization

continued

Table 2.1. (*continued*)

Basin Closure Mechanism	Temporal Duration[a]	Potential Time Resolution[b]	Depositional Continuity[c]	Spatial Scale[d]	Comments
5. Volcanic					
Crater lakes	I–L	VH	H[f]	S–M	May drain catastrophically
Lava-dammed lakes	S–I	L–I	L–I	V	May drain catastrophically during breaching
6. Wind (eolian)					
Deflation basin lakes	S–I	L	L	S	Sedimentation typically discontinuous as a result of ongoing deflation
Dune field-dammed lakes	S	L	L	S	
7. Solution					
Sinkholes/karst	Extremely V (S–L)	H	H	S–M	More persistent in tectonokarstic terrains
8. Landslide damming	VS	H?	?	V	Prone to catastrophic draining
9. Meteorite impact	L	VH when sheltered	H	S–M	Similar characteristics to volcanic crater lakes
10. Artificial reservoirs	VS?	H (high sedimentation rates)	H when undredged	Highly V	Duration unknown; no L reservoirs predate twentieth century

[a]Temporal duration defined as very short (VS) means lakes which typically persist for less than 10^3 years, short (S) means lakes which typically persist for less than 10^4 years, intermediate (I) 10^4 to 10^5 years, long (L) 10^5 to 10^7 years, and very long (VL) more than 10^7 years.

[b]Potential time resolution. Low (L) resolution lake systems are those whose records are generally resolvable to decades or centuries, whereas high (H) resolution systems are those that can be resolved to the annual scale or better.

[c]Depositional continuity and stratigraphical completeness of record, largely a function of water depth and probability of desiccation.

[d]Spatial scale and areal integration of archive. Small (S) watersheds are defined as being $< 10^2$ km^2, medium-sized (M) watersheds are 10^2–10^4 km^2 in area, large (L) watersheds are 10^4–10^5 km^2, and very large (VL) watersheds are $> 10^5$ km^2. Some lake classes vary considerably in this regard (V).

[e]Includes volcanoclastic aggradation.

[f]Frequently sheltered and deep.

vial processes (table 2.2). Of these, glacial and tectonic lakes occupy the greatest area, with tectonic lakes, which are often quite deep, most important in terms of volume (Herdendorf, 1990; Meybeck, 1995). Other lake types account for a much smaller fraction of the number, area, or volume of the world's present-day lakes.

The global abundance of modern lakes is clearly a product of extensive glaciation during the Pleistocene. Prior to the Quaternary, an accounting of the percentage of lakes that would fall into each category would certainly have been very different from today, although it is very difficult to estimate what such proportions might have

Table 2.2. Aggregate areas and volumes of the major lake classes at the global scale[a]

Lake Type	Aggregate Area (km²)	Aggregate Volume (km³)
1. Glacial and periglacial lakes	1,247,000	38,400
2. Tectonic lakes[b]	524,000 (893,000)	56,600 (134,900)
3. Fluvial lakes	218,000	580
4. Coastal lakes	40,000	130
5. Volcanic lakes	3,150[c]	580[c]
6. Other lake types	88,000	300

[a]After Meybeck (1995) and Herdendorf (1990).
[b]Values including the Caspian Sea are shown in parentheses.
[c]Crater lakes only.

been. Because of their mechanisms of formation and low-lying sites of accumulation on the earth's crust, there is a strong preservational bias favoring lake beds formed in tectonic basins, and a simple count of pre-Quaternary lake deposits would be overwhelmingly dominated by such basins (T.C. Johnson, 1984).

Modern lakes are distributed in latitudinal belts that reflect both the historical outcome of Pleistocene glaciation and modern climate effects (Talbot and Kelts, 1989) (figure 2.1). The vast majority of lakes are concentrated at mid-latitudes, between ~40–60°N and 40–50°S. Relatively low air pressure systems which generate abundant rainfall and snow, combined with relatively low evapotranspiration rates, give these regions a high ratio of precipitation to evaporation (P/E), critical for the production of abundant lakes. These latitudes also retain many relict glacial landform "holes," produced by moraines, glacial scour, and other

Pleistocene glacial features, that predispose them to be lake-rich.

Modern lakes are also abundant near the equator. The equatorial belt is a region of overall low pressure, with a positive precipitation balance (P/E > 1). These areas were not affected directly by Pleistocene glaciation, so the production of topographical closure necessary for lake formation requires other mechanisms. At these latitudes tectonic and fluvial processes dominate.

The high-pressure subtropics, from about 15–35°N and S, receive very little precipitation, and at the same time experience high rates of evapotranspiration (P/E ≪ 1). Consequently, this region has few lakes, and what lakes do form are dominantly *closed basins*, lakes with no surficial outlet. These lakes most commonly have tectonic origins, because they were never directly affected by extensive glacial activity and river systems are limited in these dry latitudes. Notwithstanding their limited numbers,

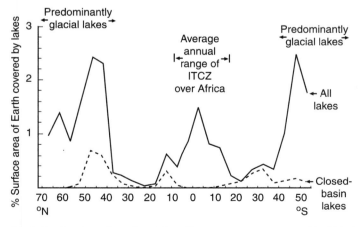

Figure 2.1. Latitudinal distribution of modern lakes. Percentages are grouped in 5° latitudinal strips. ITCZ is the Intertropical Convergence Zone, the equatorial belt of seasonally strong low pressure. Modified from Talbot and Kelts (1989), after Schulling (1977).

these lakes can be very important for paleolimnology, because of their sensitivity to slight changes in climate (Street-Perrott and Roberts, 1983; Gasse, 1990).

Beyond these first-order climatic controls there are a host of secondary, regional factors (for example, regional atmospheric circulation) that determine the likelihood of lake formation. These factors can vary both in space and time. Orographical effects, involving the creation of both regional rain traps and shadows by mountain belts formed during major orogenic events, profoundly influence the evolution of lake basins (Chenggao and Renaut, 1994).

Under conditions of different seasonal solar insolation the specific areas where lakes have an adequate P/E balance to form abundantly will vary. Regions of the Sahara Desert that are hyperarid today show evidence of abundant, large lakes during the Late Pleistocene and Early Holocene (Street-Perrott and Roberts, 1983; Gasse et al., 1990; Fontes and Gasse, 1991). Many pre-Quaternary lake deposits indicative of very large lakes occur today at latitudes or in locations far from where we would expect large lakes to form, indicative of the prior positions of mountain belts and continents. Computer models can be used to investigate both the likelihood of lake formation under past continental configurations and climate regimes and the likely interactions of those paleolakes with their local climate regimes and watersheds (Barron, 1990; Sloan, 1994; Sloan and Morrill, 1998).

Glacial Lakes

Lakes of glacial origin make up about 48% of the area and 22% of the volume of all lakes on earth (Meybeck, 1995). Almost half of the world's large lakes ($> 500 \, km^2$) are glacial, and these water bodies account for nearly a third of the area of large lakes (Herdendorf, 1990). Most glacially formed lakes lie north of 40°N, corresponding to the southern extent of the Pleistocene Laurentide (North American) ice sheet. Comparatively few glacial lakes are found at high southern latitudes because of a combination of limited land area and the extremely severe climatic conditions of the polar plateau of Antarctica that precludes the formation of surficial lakes.

Pre-Quaternary glacial lake deposits are comparatively rare and largely restricted to deposits that accumulated within tectonic basins during "ice-house" worlds, earlier periods of extensive glaciation (J.M.G. Miller, 1994). Lakes formed by alpine glaciers, as opposed to continental-scale ice sheets, have probably formed at many times in earth history, but their deposits have very low preservation potential, both as a result of their altitude and small size, making them very susceptible to erosion. Even during the Quaternary, multiple glacial advances and retreats have had the effect of erasing much of the record of earlier glacial lakes. For example, large lakes repeatedly occupied the glacially scoured Laurentian Great Lakes basins of North America, but their stratigraphical record has largely been obliterated. The lake deposits in this region overwhelmingly record the most recent period of deglaciation, with only minor records of older lakes (e.g., Dreimanis, 1992; Eyles and Williams, 1992).

Glacial lakes form in several ways, in direct contact with ice (or adjacent to ice masses), within basins eroded during prior glaciation, and behind or within depositional features associated with prior glaciation (Ashley, 1995). A distinction can also be made between lakes whose origins derive from glacial activity but which do not currently receive glacial input (for example the Laurentian Great Lakes) versus those which are under the current influence of glacial activity (N.D. Smith and Ashley, 1985). Lakes frequently change between these two states over their lifetimes. Additionally, lakes may form through a related category of *periglacial* processes, associated with permafrost ice and ice melting (thermokarst), although not necessarily with glacial activity.

Ice-contact lakes include lakes formed on the ice surface (*supraglacial*), within (*englacial*) or underneath ice caps (*subglacial*, Oswald and Robin, 1973; Dowdeswell and Siegert, 1999), or adjacent to ice sheets and glaciers (*proglacial*). The evolution of these types of lakes is highly dependent on glaciotectonic activity, subglacial geothermal gradients, and climate. Supraglacial lakes include melt-out depressions, thermokarstic depressions, wide areas in meltwater channels, crevasses, and areas where ice is blocked by bedrock obstructions or tills (Brodzikowski and van Loon, 1991). Following glacial ablation, deposits from such lakes may persist intact, although englacial deposits are frequently highly deformed by ice flow and shear.

Of the categories above it is the proglacial lakes that provide the most extensive paleolimnological record. Some complexity exists in the terminology of lakes formed in front of glaciers and ice sheets. Ashley et al. (1985) refer to lakes formed in direct contact with glacial ice as *ice-contact* lakes, and lakes formed away from direct ice contact as *non-contact* or *distal proglacial* lakes. Brodzikowski and van Loon (1991) differentiate a *terminoglacial* lake environment from a *proglacial lake* environment, the distinction lying in whether the lake is in direct contact with contemporaneous ice or not. In this sense the term *proglacial* is strictly reserved for distal noncontact settings (figure 2.2).

Lakes may be ice-dammed over a wide range of size scales. Small valley glaciers may impound lakes derived from drainages in adjacent tributaries, or block coastal outlets of other river systems. These lakes are typically ephemeral. Ice and ice-deformed sediments can rapidly override an existing lake. Also, ice dams can be breached by overflow or melting, and they often drain catastrophically, producing dramatic floods. The history of all ice-dammed lakes, large or small, is marked by frequent and sudden changes in water level, as outlets are breached, glaciers surge, subglacial meltwater expelled, or ice fronts retreat.

Proglacial lakes formed in contact with continental ice sheets are relatively rare today, and most modern ones are small (Gustavson, 1975; Donnelly and Harris, 1989; Bennett et al., 1998; Syverson, 1998). However, during the Late Pleistocene immense lakes of this type formed along the southern margins of the major Northern Hemisphere ice sheets (Teller, 1987; Fisher and Smith, 1994; Björck, 1995; Clark et al., 2001; Teller et al., 2001). In North America, paleo-Lake Agassiz is one of the best studied of the numerous proglacial lakes that formed a complex and often interconnected network along the southern Laurentide ice margin (figure 2.3). This lake was impounded in what is today the upper midwestern and Great Plains region of Canada and the United States. At its maximum extent, about 9200 cal yr BP, this lake was the largest Pleistocene lake in North America, covering an area of about 841,000 km^2, and was over 100 m deep. Individual phases of Lake Agassiz were ephemeral, persisting for only hundreds of years during the ice sheet's retreat. Ice dams migrated and wasted away through the various stages of the lake's evolution, generating rapid, and sometimes catastrophic, breaching and drawdown of the lake, with discharges up to 163,000 km^3, and continuous isostatic readjustments (the crustal uplift associated with ice sheet melting) (Teller and Thorliefson, 1983; Birchfield and Grumbine, 1985) (figure 2.4). Changes in the ice dam position were also associated with major changes in the position of the lake's outlet, between the Mississippi drainage and via the Lake Superior basin and proto-Laurentian Great Lakes to either the St. Lawrence or Hudson River systems. These pro-

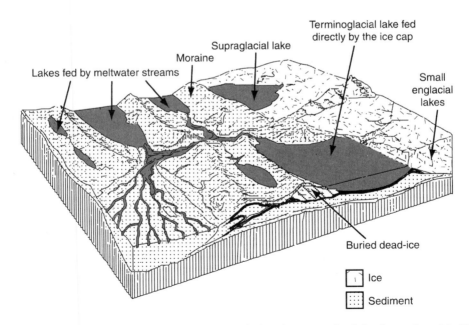

Figure 2.2. Schematic representation of a variety of glacial settings for lake formation. Modified from Brodzikowski and van Loon (1991).

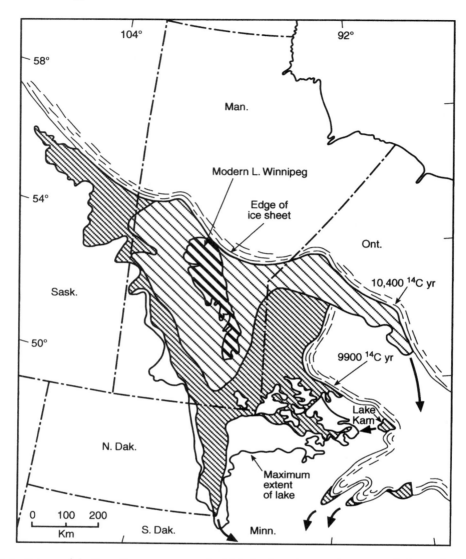

Figure 2.3. Extent of the Late Quaternary, proglacial Lake Agassiz at various stages in its evolution. From Teller (1987).

cesses, accompanied by periods of glacial surge, are manifest in dramatic lateral and vertical facies changes in the lake's deposits.

The erosion of land surfaces by ice and subglacial meltwater produces *glacial rock basins*, with topographical depressions formed as byproducts of this scour. A lake occupies this depression at a later time, either proglacially and/or after the complete retreat of glacial ice. Several mechanisms of glacial lake basin scour exist. *Ice scour lakes* form through the direct removal of both bedrock and loose material, particularly along pre-existing fractures and joint planes. *Cirque lakes (tarns)* form in arcuate amphitheaters at the heads of glacial valleys; their distribution and shape is not necessarily controlled by pre-existing topography or fracturing. Continuous freezing and thawing causes rock fragments to spall off, and this material is pushed or carried over the floor of the basin, often resulting in a secondary moraine dam controlling the actual spillway elevation of the future lake. Where cirques are dammed by bedrock they are commonly more deeply eroded at the outlet (down-valley) lip. *Valley rock basins* form as larger valley-filling glaciers move below the permanent snow line, excavating bedrock in the process. Where tributary glaciers

Figure 2.4. Ouimet Canyon, Ontario (Canada). Outlet channel for catastrophic outburst from Lake Agassiz into the Lake Superior Basin.

meet they accelerate, which causes local increases in erosion and deepening of the valley floor (MacGregor et al., 2000). This process, or encounters between ice and more resistant rock units, can result in the formation of hanging valleys with bedrock dams, which after glacial retreat will form a topographical barrier for individual lake basins. The lake chains are sometimes referred to as *paternoster lakes*, because of their resemblance to rosary beads! When ice is channeled through pre-existing valley topography it commonly impounds behind constrictions. *Glint lakes* are the remnant basins formed when these constrictions accelerate the rate of scour in the upstream valley. Where valley glaciers have excavated deep rock basins at low elevations they give rise to *fjord lakes*. Lakes of this type were often excavated below present-day sea level by maritime glaciers during the Pleistocene. Initially many of these valleys were drowned by Early Holocene sea-level rise. However, as isostatic rebound of previously ice-depressed crust proceeds, these marine embayments become progressively isolated from marine influences and eventually become freshwater lakes (Hobbie, 1984). At the largest scale of glacial rock basin formation, *piedmont lakes* and *shield lakes* have been formed by the extensive excavation of topographical basins by ice sheets, either adjacent to or distant from mountains.

The various categories of glacial rock basins all intergrade. A rock scour lake may evolve through the interaction of several mechanisms, and may also incorporate depositional processes in the formation of the lake's spillway elevation. Lakes that are dammed by both the lips of glacially scoured rock basins and moraines are particularly common, and many glacial rock basins are later inundated by much larger proglacial lakes. The interaction between pre-existing structure or ongoing tectonism and glaciation provides another intriguing example of these multiple mechanisms. *Overdeepened* lakes are the result of repeated episodes of glacial scour in older fluvial valleys, which are particularly prone to erosion, often as a result of prior tectonic history (Hsü et al., 1984). Pre-existing structures can determine both the overall morphometry of glacial scour and also the orientation of sedimentation systems in glacial lakes (Schneider et al., 1986; Schlüchter, 1987).

Investigation of the evolution of deeply scoured glacial lakes is frequently hampered by the erosion of subsequent ice advances. For example, the Finger

Lakes of New York are a series of elongated lake basins whose bedrock (subsediment) topographies are well known from extensive seismic reflection profiling (Mullins and Eyles, 1996; Mullins et al., 1996) (figures 2.5, 2.6). The oldest sediments mantling this erosional surface are Late Wisconsin (~14,400 ^{14}C yr BP) in age, coincident with the irregular lobate advance of the Laurentide ice sheet. Therefore no direct evidence of the lake's earlier history is preserved. However, the fact that these basins were scoured to as much as 306 m below sea level argues for significant glacial excavation during earlier episodes of ice advance.

Evidence of multiple phases of excavation can also be seen in the North American Great Lakes. These lakes form the largest contiguous series of glacial rock basins in the world (Karrow and Calkin, 1985). Unfortunately, very little evidence persists of their early excavation history because

of the erosion of older surfaces; in Lake Superior and most of Lake Huron, all sedimentary fill of the extant lakes appears to coincide with or postdate the last ice retreat (Wold et al., 1982; Dobson et al., 1995). It is clear, however, from the morphology of underlying bedrock surfaces that the latest Pleistocene lake deposits represent only the most recent phase of the Great Lakes' history (Wold et al., 1982). Lakes Superior, Michigan, and Huron all occupy glacial troughs that probably follow pre-existing river drainages (LaBerge, 1994). Stratigraphical evidence that the Great Lakes were already excavated prior to the Wisconsinan ice advance comes from scattered outcrops of older large lake deposits along the margins of Lakes Ontario and Erie (Dreimanis, 1992; Eyles and Williams, 1992).

During the Late Quaternary, spillway elevations in the Great Lakes have been controlled by a com-

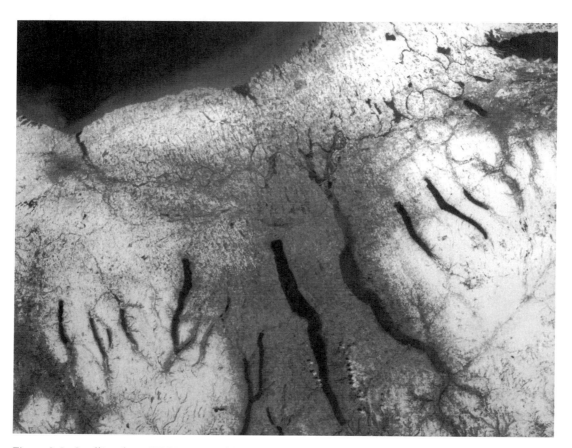

Figure 2.5. Satellite photo USGS EROS of the glacially scoured (overdeepened) Finger Lakes, north central New York State. Photo area is approximately 160 km (E–W) by 120 km (N–S).

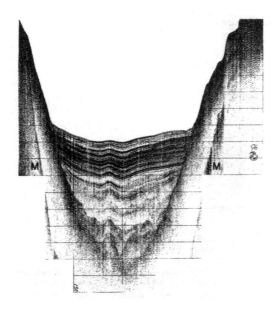

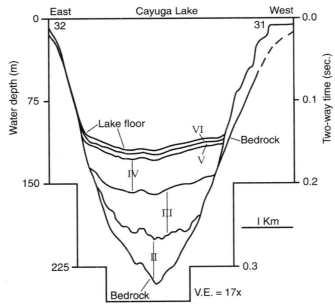

Figure 2.6. Transverse seismic reflection profile and interpretation of stratigraphical sequences in a cross-section of Cayuga Lake, one of the New York Finger Lakes. Interpretations of the numbered units are as follows: (II) tills or subglacial deposits into a proglacial lake; (III) thin, discontinuous unit of unstated origin; (IV) thick unit, tapering toward the south, representing proglacial lake deposits derived from subglacial meltwater channels (Petruccione et al., 1996); (V) thick unit tapering toward the north, representing a drainage reversal after ice retreat from the north end of the lake's valley; (VI) recent sediment cover. The entire series of numbered units are interpreted to be of Late Wisconsin to Holocene age (14,400 BP–Recent). From Mullins et al. (1996).

bination of water inflow rates, a diminishing ice sheet, and isostatic rebound (Lewis and Anderson, 1989). Periodic increased fluxes of water from upstream lakes were critical in establishing the outlet elevations of the lakes during and after the period in which these lakes were proglacial. Following glacial retreat the northerly portions of the lake basins uplifted at a faster rate than the southern ends, causing periodic abandonment of older established outflows.

The final major category of glacial lakes is those formed behind *glacial deposit dams*, including moraines, sediment around kettle holes, and glacial outwash in front of subglacial ice tunnels. Moraines are unconsolidated piles of rubble that can effectively block drainage basins. These dams initially impound glacial meltwater, but at a much later time they may serve to topographically enclose a nonglacial lake as well. The effectiveness of a moraine as a dam is enhanced when it sits on a large or deep, previously excavated valley. Kettle holes develop when stagnant ice blocks accumulate within or below moraines or outwash plains. After the ice melts a collapse structure forms which can infill with water. Lakes of this kind are extremely abundant on the margins of glaciated terrains of Europe and North America (Huddart, 1983; Ashley et al., 1985; Pedersen and Noe-Nygaard, 1995). Subglacial ice tunnels form as water flows along the base of a glacier or ice sheet. Where such tunnels discharge onto outwash plains or moraines they may become impounded, generating small lakes.

Many glacial deposit dam lakes are superimposed on basins formed by prior glacial rock basin excavation, with the sediments serving to raise the elevation of the spillway. Lakes formed in this way have subsequent infill histories that are strongly linked to the distribution of more and less easily erodible materials (Bernardi et al., 1984; Behbehani et al., 1986).

Periglacial lakes occur in areas of permafrost, and may develop following subsurface ice melting, or as a result of isostatic rebound of irregular land surfaces following deglaciation (King, 1991). For example, extensive areas of the North American Arctic are covered by lakes imbedded in permafrost. Most are small water bodies associated with surface cracking and ice wedging in permafrost. In many areas these smaller lakes coalesce into larger oriented lakes. The origin of these lakes, and specifically their parallel orientation, has been controversial (Carter et al., 1987). Although permafrost melting is clearly a prerequisite for their collapse, their orientation can also

result from underlying structural control and/or wind erosion.

Following deglaciation, innumerable small lakes in North America and Eurasia became filled by terrigenous sediments and vegetation, resulting in the formation of a variety of wetland environments and peat bogs. Bogs in particular form a special depositional and evolutionary class of "lake" because they can accumulate peat in the absence of a topographical depression.

Tectonic Lakes

Tectonic processes are responsible for the evolution of a vast number of lakes, including the largest, deepest, and oldest lakes on the planet. Although they account for a smaller proportion of total lake area (40%) than glacial lakes, their greater average depth makes them by far the largest class of lakes by volume (75%). Because they occupy structural as well as topographical basins, tectonically formed lakes are extremely well represented in the stratigraphical record in comparison with other types of lake deposits. This disparity is accentuated with age, as the vast thickness of deposits accumulated in tectonically formed lakes becomes a factor.

In this book I subdivide tectonic lake basins into five major categories, based on the classification schemes of Dickinson (1974) and Miall (2000). *Lakes associated with lithospheric divergence* form as a result of thinning of the earth's crust and thermotectonic subsidence during the early stages of rifting and subsequent separation of lithospheric plates. This process culminates either in sea floor spreading, or in rifts that fail to open completely. *Lakes associated with crustal convergence*, often though not always located at convergent continental margins, form as a consequence of various interactions between a subducting oceanic plate and its overriding continent. A variety of basins can form from the flexural downwarping of subducting slabs. This occurs as a result of cessation of convergence following the formation of major transform fault systems, from superimposed thrust loads, or from flow processes in the underlying aesthenosphere, such as coupling of the overlying continental plate with the subducting plate (DeCelles and Giles, 1996). *Transform fault basins* form along major transform fault systems where plates are slipping past one another, such as the movement of the Pacific Plate relative to the North American Plate along the San Andreas Fault System in California. However, purely strike–slip motion along such plate boundaries is rare, and most transform fault basins form along fault bends associated with some component of compression or extension. Basins formed during *continental collision* include a complex combination of the three categories mentioned above. *Cratonic basins* form in the interiors of continental plates for reasons that are not clearly related to plate boundary interactions. Their origins are more likely related to continental-scale subsidence caused by plate motion across irregularities in the earth's geoid (underlying shape) or mantle upwelling.

Tectonic lakes are fundamentally features of the continents, and frequently form far from plate boundaries. Therefore "pigeon-holing" of lakes into specific plate-tectonic categories has the danger of obscuring complex, and often interesting, modes of tectonic lake formation that are difficult to ascribe to specific plate interactions. Most large, tectonically formed lakes have such hybrid origins, and their placement in one category or another is somewhat arbitrary. Simply put, categorizing tectonic basins by plate-tectonic or morphotectonic setting has the effect of combining basins whose underlying mechanisms of formation are vastly different (Dickinson, 1993). For example, the broad classes of basins that form at convergent margins or in intracratonic settings include a wide range of morphotectonic subcategories. Fundamentally, the subsidence involved in almost all tectonic basins that enclose lakes can be thought of as resulting from either the thinning or loading of the lithosphere. In this spirit, the categorization of tectonic lake basins used here is kept to a minimum, and where possible, is placed within a process-oriented context.

Lakes associated with lithospheric divergence encompass a range of basin types that form during the early phases of continental rifting. They occur both deep within continental interiors (e.g., Lake Baikal) and along the continental extensions of mid-oceanic ridge systems (e.g., the East African rift valley lakes). Most rift lake basins are characterized by slow rates and small total amounts of extension, coupled with extremely deep subsidence, with up to 7 km of sediment fill (Friedman and Burbank, 1995; Leeder, 1995; Wescott et al., 1996). Such basins can persist for up to tens of millions of years. They account for an extremely important, and perhaps the largest component, of pre-Pleistocene lake deposits, including such important examples as the Triassic–Jurassic rift lakes of eastern North America, and major paleolakes formed during the Early Cretaceous opening of the South Atlantic Ocean.

Extensive seismic reflection profiling of the East African rift lakes during the 1980s revealed the structure of rift lake basins, and has served as an important comparative model for understanding continental rift history (Rosendahl, 1987). Rift lakes occupy the deepest depressions within high-relief basins bounded by steeply dipping normal faults. These faults often undergo some strike–slip motion as well, leading to greater structural complexity. The position of border faults determines the location of the maximum subsidence within basin segments, the areas of thickest sediment accumulation, and the deepest bathymetry of individual lake basins. Although border faults are largely extensional faults during rifting, they are commonly reactivated older faults, in some cases remnants of earlier rifting phases, but in some cases of compressional origin as well.

In most rift lakes, the primary structural unit is an asymmetrical *half-graben*, a basin in which a major bounding fault only occurs on one side of the lake, the escarpment margin, although numerous smaller faults may exist elsewhere around the basin (figures 2.7, 2.8). The opposing or *flexural* margin of the basin, away from the major boundary fault, shoals more gradually, resulting in shallower water depths and lesser amounts of accumulated sediment in these areas. Half-graben basins within rifts have dimensions on the order of tens to hundreds of kilometers long, with considerably narrower widths. A rift system encompasses a string of such half-graben basins, only some of which are occupied by lakes at any given time. Larger rift lakes may straddle several of these half-graben basins, producing the characteristic narrow but elongate shape associated with both modern and ancient rift lakes. Adjacent half-graben basins often display an alternating orientation of their major border faults. For example, in a north–south oriented rift lake, the border fault may be located on the west side of the lake in one segment and on the east side in the next. Where two half-grabens adjoin, displacement along faults is taken up by structures known as *accommodation zones*. These features are commonly, though not always, topographical highs, and may therefore result in the segmentation of a rift lake into distinct bathymetric basins, or if high enough relative to lake level, entirely separate lakes (Lezzar et al., 1996).

Rift lakes commonly evolve through predictable stages, with consequences for the bathymetry of their individual half-graben basins (Lambiase, 1990; Schlische and Olsen, 1990; Scholz et al., 1998). During early rifting, basins are small, sediment supply is adequate to keep pace with subsi-

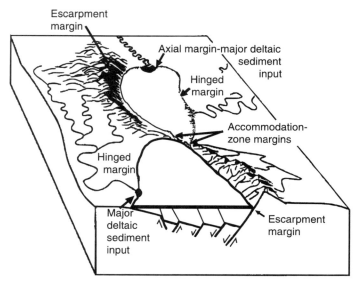

Figure 2.7. Plan view model and cross-section of a rift lake, showing major structural controls on sediment input. After Soreghan and Cohen (1993).

dence, and the basin remains filled with sediment. At this stage rivers, shallow lakes, and swamps typically occupy rift valleys, such as the Okavango Delta system in the modern African rift (Tiercelin and Mondeguer, 1991). As the basin continues to deepen and the rift widens, sediment input often does not keep pace with subsidence and the lake will deepen dramatically. Lake Malawi is representative of this stage of development. During the late stages of rift lake development, subsidence generally slows and sedimentation will once again cause the half-graben basin to infill and become shallower, often through the influence of major axial river deltas, as in the northern basin of Lake Tanganyika. The entire process of rift lake evolution can last millions to tens of millions of years, and may go through multiple episodes of activity as a result of reactivation of major boundary faults.

In ancient rift basins it is possible to observe situations where rifting associated with continental breakup generated basins extending at high angles into the interiors of continents that failed to separate. Such failed rifts or *aulacogens* are frequently filled by lake systems. Good examples can be observed in the Early Cretaceous lakes of failed rifts along the Brazilian continental margin (Magnavita and Da Silva, 1995).

Lakes associated with convergent continental margins develop from interactions between subducting oceanic and continental plates. Many types of basins form in this type of convergence,

but only a limited subset are important in the formation of lakes. Where oceanic crust is subducted under continents, *magmatic arcs* develop. These arcs are mountain chains built upon the continent, close to its margin, from volcanoes and their underlying solidified magma sources. On the landward side of these arcs, lakes may develop in several contexts.

First, during convergence compressional forces often generate major zones of thrust faulting and folding in the crust, directed toward the continental interior. These *thrust–fold belts* produce loads on the crust, generating a complex pattern of flexure and subsidence at their frontal edges, called *foreland basins* (DeCelles and Giles, 1996) (figure 2.9). Often the subsiding depressions in front of the thrust sheet are themselves deformed into distinct foredeep and back-bulge areas, which can form the loci of separate, closed depressions and lake or swamp systems. The entire system migrates as the thrusting propagates toward the foreland, causing older deposits to become bound up in the advancing thrust sheet. Basins also develop on top of the complex topography of the overriding thrust belt itself, referred to as *piggyback basins*.

Basins also form inboard from magmatic arcs as a result of changes in the dip angle of the subducting slab. Shallow subduction is thought to cause deformation in the crust of the overlying plate, resulting in the production of basement rock uplifts (Dickinson and Snyder, 1978; Jordan and

Figure 2.8. Digital elevation model of Lake Bogoria, Eastern Branch of the African Rift Valley (Kenya). Coverage of main image is 30 km (N–S) by 13 km (E–W). Note escarpment margin on the east and flexural margin on the west sides of the lake. Oblique inset views (A looking south and B looking north) from perspective points shown on the main image.

Allmendinger, 1986; Bird, 1998). These uplifts also cause loading on the surrounding terrain, generating local foreland basins in the process that may enclose lakes.

The large volumes of sediment generated in thrust–fold belts often preclude the formation of topographically deep basins. Whereas subsidence may be considerable in these systems, on the order of several kilometers or more, sedimentation by large alluvial fans and rivers is usually able to "keep up" with subsidence, preventing deep topographical enclosures from forming. For this reason most foreland basin lakes are shallow (tens of meters of water depth) compared with rift lakes,

and many are for much of their history very shallow swamp systems rather than open water lakes (Castle, 1990).

The foredeep portions of foreland basins often undergo accelerating subsidence accompanying the flexure of adjacent crust in contrast with rift basins, where subsidence typically slows over time. This may be reflected in bathymetric deepening of the lake during the interval of thrust loading if sedimentation is incapable of "keeping up" with subsidence. For this reason foreland basin lakes are best developed in relatively arid climates where sediment is not delivered efficiently to central regions of basins and where topographical closure can develop. In more humid climates, foreland basins are most frequently characterized by the development of through-flowing river systems with only marsh or wetland formation rather than open water lakes. This pattern of lake formation only under relatively arid conditions is in contrast with rift system lakes, where sediments derived from marginal uplifts are largely diverted away from the basin and humid conditions are commonly associated with deep lakes rather than swamp systems or through-flowing rivers. The deepest and most persistent foreland basin lakes form at distances of tens of kilometers in front of the thrust-loaded regions, near the surficial exposure of thrust belts or basement uplifts, where topographical closure can be generated most effectively. Closer to thrust fronts sediments from rivers or alluvial fans tend to accumulate and little or no topographical closure is generated. This differentiates foreland basin lakes from rift lakes; in the latter border faults are commonly adjacent to the shoreline, forming fault coastlines.

Along the margins of lakes bounded by basement blocks, deeper conditions may develop in response to rapid subsidence coupled with locally inadequate sediment supply (Liro and Pardus, 1990). In contrast, more peripheral areas of subsidence, for example the back-bulge region, tend to infill with sediment more readily, and as a result house shallower and more ephemeral lakes and swamps. The broad amplitude of crustal deformation and subsidence on top of or in front of thrust sheets, or adjacent to basement-cored uplifts, produces much wider basins than the narrow rift valleys. This gives foreland basin lakes and swamps a very distinctive, circular to ovate shape, which is quite different from elongate rift lakes.

Large lake and swamp systems, as well as numerous paleolakes, are known from a variety of foreland basin systems. One well-studied, modern foreland basin system occurs east of the modern central Andes (Horton and DeCelles, 1997) (figure

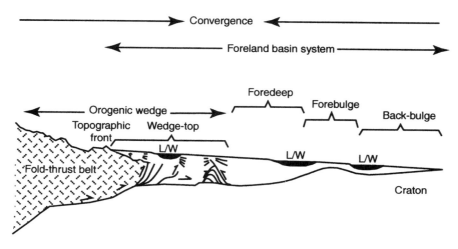

Figure 2.9. Schematic cross-sectional view of the relationship between a fold–thrust belt, foreland basin system, and positions of lake and wetland (L/W) development associated with wedge-top (piggyback basins), distal foredeep, and back-bulge topographical depressions. Arrows indicate sense of motion on faults. Vertical scale is greatly exaggerated, particularly of lake depth, and in the thin back-bulge basin. The distance from the fold–thrust belt to the forebulge depends on the structural rigidity of the rocks; a stiffer upper plate will result in a longer distance to the bulge and a gentler bulge dome, producing shallower accommodation space for lakes. Open lake formation in any of these settings requires limited sediment supply, typical of arid–semiarid conditions. Otherwise wetlands or fluvial sediment ramps will develop. Modified from DeCelles and Giles (1996).

2.10). Eastward-directed thrust faulting in southern Bolivia and northern Argentina has produced subsidence both on top of and to the east of the thrust belt, with piggyback basins located close to the eastern Subandean foothills, and broad ephemeral swamps in the Gran Chaco of Argentina and Bañados de Izozog of the Southern Bolivian Chaco Plain. East of the Bañados de Izozog lies another depression situated above a back-bulge basin. This region is currently occupied by the extremely shallow but broad Pantanal wetlands of southwestern Brazil. Late Tertiary and modern lakes (e.g., Salinas del Bebedero and Salinas Grandes) surrounded by basement-cored uplifts occur further south, especially in Argentina (Jordan and Allmendinger, 1986).

Probably the best studied paleolakes associated with basement-cored uplifts formed in the western United States during the Late Cretaceous to Early Tertiary, ~67–35 ma (Dickinson et al., 1988; Yuretich, 1989). The most extensive and long-lasting lake systems were those ponded directly adjacent to the frontal flank of the major Sevier thrust belt or surrounded by basement uplifts. These include the famous Early to Middle Eocene Uintah and Green River Basins of Wyoming, Utah, and Colorado, which formed extensive and at times deep lake systems that persisted for over 20

million years in some areas (Surdam and Stanley, 1979). During episodes of maximal subsidence and/or aridity the Green River lake remained "underfilled" by sediments (sensu Carroll and Bohacs, 1999), and regional precipitation was insufficient to cause the relatively deep basin to overflow, resulting in a closed basin, saline/alkaline lake. As subsidence diminished and sediment flux from more northerly basins increased, the Green River Basin "overfilled" and a hydrologically open lake with freshwater conditions developed.

More complex lake origins occur where thrust faulting bounds both sides of a basin, creating topographical closure and blocking external drainage. This happened in the evolution of the South American Altiplano, an internally drained basin in the middle of the Andean Plateau that houses numerous large lake basins (Titicaca, Poopo, and the Uyuni and Coipasa playas) (Servant and Fontes, 1978; Lamb et al., 1997).

Back-arc basins also form in association with lithospheric plate convergence, but as a response to very different processes than foreland basins, involving thinning and extreme subsidence of the lithosphere, and sometimes formation of oceanic crust. The overriding plate at a zone of plate convergence is often subject to extension in the interior area behind the magmatic arc, away from the sub-

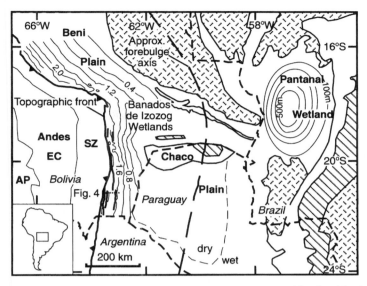

Figure 2.10. Map showing the relatively distal (sediment-poor) positions of foreland basin system wetlands of the Chaco Plain (Bañados de Izozog) foredeep and the backbulge Pantanal Wetlands basin, relative to the thrust structural front (hatched markings) of the Andean foreland. Sediment thickness contours are in kilometers near the Andean topographical front and meters in the Pantanal. Note the broadly elliptical shape of these basins in contrast to the elongate shape of rift lake basins. The much thicker deposits of the portions of the foredeep basin nearest the Andes (W. Beni Plain) form a continuous topographical ramp, not conducive to topographical closure and lake formation under humid climate conditions. True lakes in this setting are restricted to the more arid regions (lower sediment input) of the Andean foreland in Argentina, south of the map area. Modified from Horton and DeCelles (1997).

duction zone. This is thought to be caused by an increase in the angle of descent of the downgoing slab of subducting plate material, coupled with enhanced mantle upwelling. When the overriding plate is oceanic crust, back-arc spreading results in the formation of new crust similar to what is formed at mid-ocean ridges. Back-arc spreading on a continent causes extension, and both thinning and subsidence of the continental rocks. If the processes persist long enough, however, thin oceanic crust may eventually begin to form. Isostasy dictates that this thin oceanic crust will "float" much lower on the plastic portion of the earth's mantle, resulting in rapid subsidence to great depths, comparable to the abyssal plains of ocean basins.

Lakes form in association with back-arc basins in two ways, either during initial extension and subsidence of continental crust, or when oceanic back-arc basins become enclosed by continents after their formation. Unifying characteristics of such lakes are their large area and potential for rapid subsidence and extreme water depth (> 1 km), particularly if extension leads to the formation of oceanic crust within the basin. Another characteristic of back-arc basin enclosures is their frequent transitions

between lake and restricted marine conditions, resulting from their proximity to continental margins during their formation.

Large and extremely deep back-arc lake basins, like the Black and Caspian Seas, formed in response to multiple episodes of collision, back-arc spreading, and basinal enclosure during the convergence of Africa, Arabia, and India with Eurasia, the closing of the Tethyan Seaway, and the subsequent formation of the Mediterranean Sea (Zonenshain and Le Pichon, 1986; Mamedov and Babaev, 1995; Nikishin et al., 1997). The Black Sea, an extremely deep basin (> 2200 m), has a tenuous marine connection today, but for much of its Late Tertiary history it was a completely enclosed lake (Hsü, 1978) (figure 2.11). The Black Sea began forming in Jurassic–Early Cretaceous time along the southern Eurasian continental margin, north of a developing magmatic arc (Görür et al., 1993). Early Cretaceous block faulting and rifting gave way to rapid subsidence and the formation of back-arc oceanic crust in the basin by the Late Cretaceous (~75 ma). True oceanic crust occupies the central regions of the two prominent basins, and remarkable quantities of sediment have accumulated on

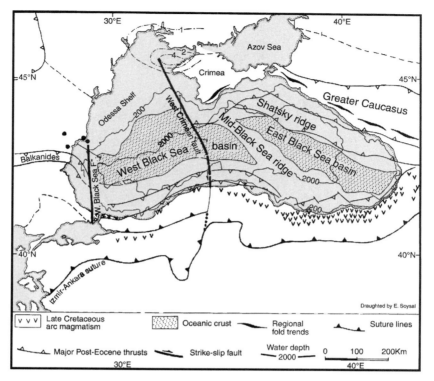

Figure 2.11. Tectonic map of the Black Sea region, showing major faults and extent of oceanic crust. Contours are water depths (m) in the modern Black Sea. Modified from Okay et al. (1994).

top of both of these depocenters (16–18 km in the western basin and 12–14 km in the eastern basin) (Chekunov et al., 1994; Kutas et al., 1998). Spreading of the western basin appears to have ended in the Early Eocene with the collision of the Istanbul zone microcontinental fragment with the Sakarya zone. During the Tertiary subsidence has been episodic, with maxima during the Eocene/Oligocene and again over the past 10 ma (Robinson et al., 1996; Nikishin et al., 1997).

Periods of restricted marine conditions appear to have alternated with open marine connections throughout the Cretaceous and Early Tertiary, but at no time in this early history was the basin lacustrine. Indications of persistent brackish conditions appear in the Middle Miocene, when the Black Sea was probably interconnected with a series of back-arc basins extending from eastern Europe to the Caspian Sea, the so-called Para-Tethys (Hsü, 1978). A series of large and extremely deep lakes evolved in the western basins of Para-Tethys from marine precursors. This process appears to have begun in the Late Miocene in the Pannonian Basin of east central Europe, which was undergoing back-arc-like extension at that time and had itself devel-

oped into the large Lake Pannon by about 12 ma (Royden, 1988; Magyar et al., 1999). By the Early Pliocene the Black Sea had become a completely isolated lake. However, tenuous marine interconnections developed periodically through the Black Sea's Late Tertiary history, including a catastrophic inundation during the Holocene that resulted in the modern marine interconnection.

Probable examples of lakes formed in back-arc basin settings are also known from ancient lake deposits. The Junggar Basin of northwest China formed initially in the Late Paleozoic in either a back-arc or ocean basin setting (Hsü, 1988; Carroll et al., 1990; Graham et al., 1993). Like the Caspian and Black Seas this large basin evolved from marine to lake conditions during the Triassic–Jurassic, with a complex subsequent history that produced many types of lakes.

Lakes associated with transform margins form in basins immediately adjacent to major strike–slip fault systems, and in basins that develop in response to the evolution of this type of plate boundary. Where offsets or bends occur along transform faults, *pull-apart basins*, zones of localized extension, may form. Also, motion of lithospheric plates past one another usually involves some component of oblique stress, particularly at "kinks" in the fault line, where plate interactions result in a combination of lateral transform motion, parallel with the major fault line, with either compression or extension (figure 2.12). Basins may form in either *transpressional* or *transtensional* contexts, and either setting can generate topographical closure and lake formation. Generally extension and compression occur in a near-parallel orientation to strike–slip motion, but examples of lake basins in which extension is occurring perpendicular to strike–slip motion (i.e., with normal faults oriented parallel to the strike–slip system) are also known (Applegate, 1995). Transtensional basins, also referred to as *pull-apart basins*, are frequently sites of volcanism, so lakes formed in these settings often have both volcanic and fault controls on their spillway elevation.

Transform-fault-related basins and their associated lakes have some unique evolutionary characteristics that result from the lateral motion of their surroundings. If the lake straddles the fault system it may migrate away from its original location, leaving behind a "trail" of progressively more deformed sediments recording its prior positions. Also, basin subsidence in transpressional or transtensional settings tends to be both very rapid and localized to the immediate area around the fault bend that generated the basin (Crowell and Link, 1982). These features combine to produce lakes that may be relatively deep compared to their area, and which often have steep basin margins located along bounding faults (e.g., the Neogene phase of development in Lake Issyk-Kul, Klerkx et al., 1999). However, the maximum size of basins associated with transform faults is small in comparison with the other tectonically formed lake basins considered to this point.

The largest modern transform fault lake is the Dead Sea, occupying the lowest topographical depression along the 1000-km-long Dead Sea transform fault system, the boundary between the Arabian and Sinai Plates, the latter a "subplate" of Africa (Ben-Avraham, 1997; Garfunkel, 1997) (figure 2.13). Over 100 km of offset has occurred along this fault, mostly during the Miocene, with less than 40 km of left lateral motion having occurred in the last 5 million years. The modern Dead Sea basin is about 15–17 km wide and 150 km long, although arid conditions restrict the modern lake to only a third of this length. A combination of normal and strike–slip faults bounds the highly asymmetric basin. Although the lake basin superficially resembles a normal fault bounded "rift," and is often referred to as such, it is actually quite distinct structurally, comprising a series of pull-apart basins and intervening topographical highs, representing zones of compression along the transform.

Up to 10 km of Neogene sediments fill this basin, although thickness and subsidence rate vary greatly over distances of only a few kilometers. During most of the Miocene the basin appears to have been occupied by through-flowing rivers with

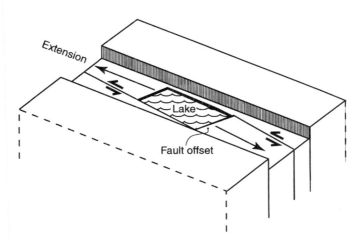

Figure 2.12. Model of basin formation along major transform fault systems. Pull-apart basins (a major subgroup of this category) are frequently infilled by lakes. Modeled after inferred patterns of motion associated with the formation of the Dead Sea. The pull-apart occupied by the lake basin (central hole) and the master strike–slip faults (sense of motion indicated by arrows) are embedded in a valley delimited by normal faults. Modified from Garfunkel (1997).

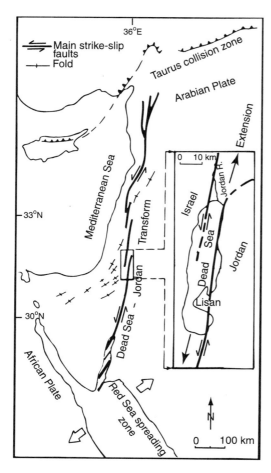

Figure 2.13. The Dead Sea and its tectonic setting relative to regional plate boundaries. Modified from Gardosh et al. (1997).

no significant topographical closure. In the Late Miocene subsidence rates within the Dead Sea basin appear to have accelerated, and the region was inundated by marine flooding and subsequent infilling by evaporites. Continued rapid subsidence but reduced sedimentation rates from the Late Pliocene onward resulted in the closure of the basin's marine connection and its conversion into a lake (Stein, 2001). Since the Late Pliocene the basin has subsided rapidly, accumulating almost 4000 m of sediments in some areas (Gardosh et al., 1997). Older lake basinal deposits toward the south reflect their progressive transport by the transform fault system (Zak and Freund, 1981; Garfunkel and Ben-Avraham, 1996).

The San Andreas transform fault system evolved during the Late Oligocene following the cessation of

plate convergence along the western margin of North America (Atwater and Molnar, 1973). This fault system directly generated many well-studied transtensional lake basins, such as Clear Lake and the Ridge Basin paleolake (Link and Osborne, 1978; Hearn et al., 1988). Another consequence of the cessation of plate convergence as the San Andreas Fault System evolved was the formation of a highly extended zone of North American continental lithosphere, the modern Basin and Range of the western United States. The initiation of motion along the San Andreas appears to have relieved intraplate stresses within North America, allowing extension to begin in the Basin and Range region during the Miocene (Dickinson, 1981). A vast province of topographical closure and lake, marsh, and playa development occurred in the basins formed by this extension.

The Basin and Range lakes could be grouped with rift systems, because both are the product of extension and thinning of the crust, generating subsidence. The formation of half-grabens as the primary units of basin development is also common to both systems, as are the complications induced by extensive volcanic damming for lake formation. However, the tectonics and geomorphology of the Basin and Range differ in significant ways from rifts, and these differences have important consequences for lake basin evolution. The modern Basin and Range is a broad zone of normal faulting and diffuse extension, in contrast to the single or few axial valleys of rifts (Wernicke et al., 1993). Also, in contrast to rift valleys, the degree of crustal extension in the Basin and Range is quite large, typically 10–50% but locally up to 100%. Drainage interconnections around relatively short, uplifted blocks with low drainage divides have allowed very different types of lake systems to evolve in the Basin and Range region (Leeder and Jackson, 1993). At times when lake levels are low, individual basins may be hydrologically isolated as marshes or playas. But during high lake stands, intervening fault blocks of the Basin and Range are often too small to prevent lake interconnection and extremely broad, moderately deep, island-filled lakes formed, such as Late Pleistocene Lakes Bonneville and Lahontan.

Lakes associated with continental collisions involve major continental collisions, as well as the "suturing" of smaller crustal fragments or "microplates" onto each other or onto larger continents. Continental collisions produce a complex array of compression, transform motion, and extension of the crust as one continental plate indents or overrides the other, resulting in a wide variety of basin types and lakes, including foreland basins, transtensional pull-apart basins, and rifts (Anadón et al., 1988b).

The Tertiary collision of India with Eurasia has produced a large number of basins housing lakes. These lake systems can be grouped geographically according to the nature of tectonic processes across the vast region impacted by this collision. On the Tibetan Plateau, large areas of interior drainage with numerous lakes have formed straddling faults and relict lows in topography related to the suturing of Mesozoic microcontinental fragments. In the humid belt south of the Himalayas, sedimentation rates are so high and sediment delivery through large rivers is so efficient that topographical closure and tectonic lake formation does not occur. But in the more arid regions of the Tibetan Plateau and central Asia, lakes have formed within intermontane piggyback basins and in foredeep depressions. East of the direct zone of collision, major strike–slip faults developed to accommodate the movement and rotation of large crustal blocks forced eastward by the collision. Numerous lakes in Mongolia, as well as the early phases of Lake Issyk-Kul, developed in pull-apart basins associated with these faults. Even farther from the zone of collision, in eastern China and Siberia, movement of crustal blocks has induced large-scale continental rifting, best exemplified by Lake Baikal. Because of river diversion these deep rift lakes and paleolakes receive comparatively low sediment supplies despite their subhumid climate, in contrast to the Himalayan foreland where the extraordinary volumes of sediment generated under humid conditions overwhelm receiving basins and prevent the topographical closure required for lake formation.

Cratonic basin lakes form in the low interior regions of continents as a result of subsidence mechanisms that are still poorly understood. They may be related to broad-scale subsidence of continents as they migrate across topographical irregularities of the earth's geoid (Russell and Gurnis, 1994). Alternatively, this type of subsidence may occur between elevated domal uplifts formed by plumes of upwelling material within the earth's mantle, as in the Lake Chad Basin, or associated with failed rifting events (Hartley and Allen, 1994). Cratonic basins are extremely long-lived, and undergo slow but persistent subsidence, often lasting many tens of millions of years. Through this history both marine and lake phases may alternate, depending on the relative elevation of the basin with respect to sea level and its potential for marine inundation. Cratonic basin lakes are characteristically broad and shallow, a consequence of both the low relief of these older, central continental regions (or *cratons*, hence their name) and the limited moisture supplies typically available in continental interiors. Although these basins may persist for geologically long intervals, their stratigraphical records indicate that they may be totally dry for lengthy periods. During these times the basin center may become covered by eolian sands, or may experience no net deposition, or be actively deflated by wind erosion. Consequently, the lake deposits of cratonic basins are typically very thin and stratigraphically discontinuous.

The evolution of Lake Eyre in central Australia illustrates this type of cratonic basin history (Russell and Gurnis, 1994; Alley, 1998). This playa lake presently occupies the topographical low of an enormous, arid to semiarid region of interior drainage (1.3 million km^2) of east central Australia. During the Tertiary, lacustrine sedimentation occurred during several discrete intervals. A Late Paleocene–Early Eocene interval was marked by the presence of shallow swamps associated with braided-river-system floodplains. During the Late Oligocene–Miocene, extensive, shallow alkaline lakes developed under climatic conditions considerably wetter than at present. Aridification of central Australia occurred during the Late Tertiary, and by the Pliocene the basin was filled by playas. Larger freshwater and saline lakes formed in several areas within the basin during the Middle Pleistocene, and the modern playa was excavated by wind deflation between 60,000 and 50,000 years ago (Magee and Miller, 1998). The aggregate thickness of all Tertiary lake deposits in the Lake Eyre Basin is at most 400 m, illustrating both the low rates of subsidence typical of this type of lake basin and the limited and highly episodic accumulation history of its sediments.

Fluvial Lakes

Fluvial processes are the third most important category of lake-forming mechanisms. Fluvial lakes make up 8% of the world's total lake area, but only about 0.3% of total world lake volume (Meybeck, 1995). Fluvial lakes are almost invariably quite shallow (Herdendorf, 1990). Among the world's largest lakes (> 500 km^2) of purely fluvial origin, none are deeper than 20 m, and even the deepest plunge pool lakes are only about 50 m deep. Nevertheless, fluvially formed lakes are extremely common elements of alluvial fan, floodplain, and delta geomorphology, and are probably the second most common class, after glacial lakes, in terms of numbers of individual lakes. Although individual lakes are relatively short-lived features of fluvial systems, they have provided important paleolimnological records of watershed and regional-scale paleohydrology and paleoclimate change, making a detailed examination of their origins important for our purposes.

Flowing water produces topographical closure either by excavating topographical depressions or by depositing sedimentary dams. Erosional lakes form by bedrock scour beneath waterfalls and cascades (*plunge pool lakes*), or during catastrophic flood discharges of upstream lakes. Plunge pools may undergo active excavation over periods of thousands of years, forming roughly circular basins of as much as 50 m depth. Spectacular examples of erosional lakes formed during catastrophic flood discharge occur in the Channeled Scablands region of eastern Washington State, the product of multiple flooding events during the Pleistocene from upstream, ice-dammed Lake Missoula (Bretz, 1930; V.R. Baker and Numedal, 1978) (figure 2.14). Although some of these flood events clearly occurred during the Middle Pleistocene, the best-recorded events are Late Pleistocene in age, between 18–13,000 ^{14}C years ago. Peak discharges of up to 2×10^6 m^3 sec^{-1} close to the flood's breakout point from Paleolake Missoula and velocities of over 20 m sec^{-1} caused the formation of extraordinary erosional and depositional features across a large region. Among these features are numerous lakes, many of which were formed almost instantaneously geologically speaking, from turbulence, bedrock plucking, and the undercutting of giant potholes (V.R. Baker, 1973).

These erosional examples notwithstanding, the most common lake-forming mechanisms associated with rivers involve the formation of topographical closure as a result of deposition. Aggradation, the vertical buildup of sediment from depositional processes, occurs unevenly within river valleys, accounting for the considerable variety of settings in which standing water can be ponded. Levees, scroll bars, paleochannel margins, crevasse splay deposits, and debris dams can all create topographical closure on a floodplain. Additionally, seasonal inundation in many large river systems creates temporary floodplain lakes, often of immense size, for example on the Amazon floodplain. Aggradation rates can vary over several orders of magnitude even when measured within the same river system and over the same time interval (Bridge and Leeder, 1979). However, the extent of topographical relief

in the active portion of alluvial valleys is quite limited in comparison with the depth of scour or tectonic subsidence that can form in glacial or tectonic lakes. Deposition on floodplains is a locally self-limiting process, because temporary topographical highs typically undergo slower rates of aggradation relative to their surroundings. In this process local topographical highs may cause clastic sediments to be diverted to infill low areas. Therefore, maximum depths in fluvially formed lakes are shallow, and with the exception of lakes formed by competitive aggradation, are scaled to the dimensions of channel scour and differential aggradation within the river basin.

Floodplain lakes form by a combination of groundwater infiltration, flood-season inundation by rivers exceeding bankfull conditions, and through the formation of *crevasse splays*, where levees are broken during high water condition, resulting in widespread inundation of the local floodplain. Floodplain lakes are especially common adjacent to channels that remain stable over long time intervals. Anastomosing river systems consist of multichannel rivers, set in banks that are highly resistant to lateral erosion, often from being clay-rich or heavily vegetated. Shallow lakes are maintained on such floodplains for periods of up to several thousand years by a combination of groundwater infiltration, crevasse splay activity, and local runoff (figure 2.15) (D.G. Smith, 1983; Sippel et al., 1992).

Abandoned channel lakes, also referred to as cutoff or oxbow lakes, form where channels migrate through lateral erosion of bank materials (figure 2.16). As meander loops become extremely sinuous they are prone to abandonment by river flow, which can be channeled more directly down the valley slope. Infilling and shallowing of the resultant lake occurs most quickly early in the lake's history. Occasionally a paleochannel may be reactivated, and its stratigraphy may record this compound lacustrine/fluvial history. The combination of no differential subsidence between these small lakes and their surrounding floodplains and their rapid infill rates (typically ~ 1 cm yr^{-1}) limits their life spans to a few hundred to a few thousand years (Lewis and Lewin, 1983; Kozarski et al., 1988). Water depth, shape, and area of these lakes is a direct reflection of the original channel geometry, as modified by subsequent infilling. Abandoned channel lake systems also evolve as a result of climate change. In northern Europe, accelerated formation of abandoned channels, formation of narrower channel lakes, and faster infilling rates were all associated with the termination of glacia-

Figure 2.14. Lakes of the Channeled Scablands, Washington State, erosional basins formed by the catastrophic outbursts of Late Pleistocene, glacially dammed Lake Missoula. The Dry Falls Cataract (Sun Lakes State Park) lies "upstream" from Dry Falls Lake (upper left), Red Alkali Lake (middle right), and Green Lake (lower right).

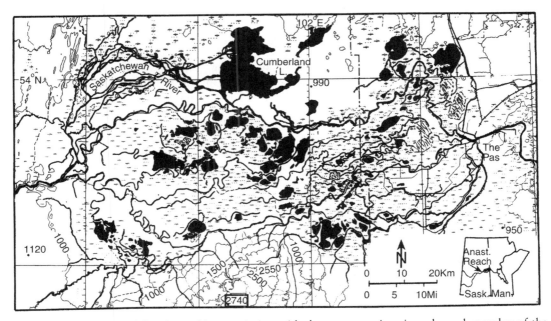

Figure 2.15. Floodplain lakes formed in association with the anastomosing river channel complex of the lower Saskatchewan River, Saskatchewan (Canada). Modified from D.G. Smith (1983).

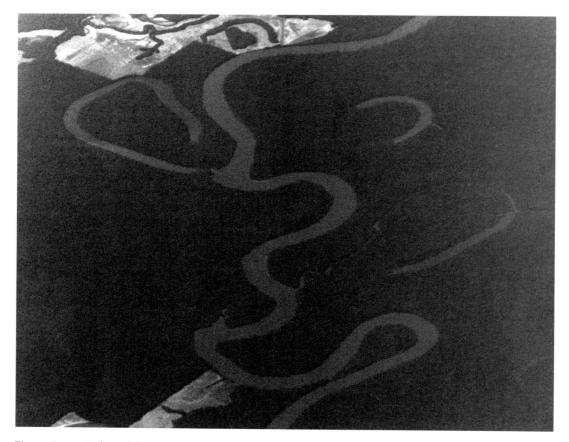

Figure 2.16. Oxbow lakes formed as meander cutoffs of the White River, Arkansas. Note the various stages of meander cutoff infill represented by the various small lakes.

tion in the Late Pleistocene and a transition to less erratic sediment discharge, higher precipitation and runoff, the establishment of vegetation cover, and greater channel stability (Bohncke et al., 1988; Starkel, 1991).

Lakes also formed in fluvial systems by *competitive fluvial aggradation*. Where rivers intersect there will inevitably be a difference in their relative rates of valley aggradation, caused by differences in their relative sedimentation rates and local variations in subsidence rate. Sometimes this causes the more rapidly aggrading system to block the smaller system, resulting in the formation of a lake. This process is particularly common in regions with large and mobile sediment supplies, for example around volcanoes. Unlike the lake systems formed on floodplains or in abandoned channels, the depth and persistence of these lakes is highly indeterminate. They can be much deeper than other fluvially formed lakes, depending on the size of the flow

obstructing dam, the relative aggradation rates, the depth of the valley being blocked, and local differences in subsidence rate. For example, prior to the twentieth century (when it was drained for farmland), the Tulare basin, in south central California, contained a large, shallow lake, up to 1600 km^2 at its outlet level area (Atwater et al., 1986) (figure 2.17). The lake overlies a tectonic depression but its outlet is blocked principally by the Kings River alluvial fan, draining the southern Sierra Nevada. Lakes have occupied the basin at various times during at least the past 600,000 years, and overspilled the outlet fan dam regularly. The size of these lakes seems to have been dependent on climate, as the rate of aggradation on the Kings River fan dam increased during periods of increased glacial outwash.

Exceptionally rapid aggradation of alluvium, creating "instant lakes," can occur when large volumes of volcanogenic sediment are deposited in

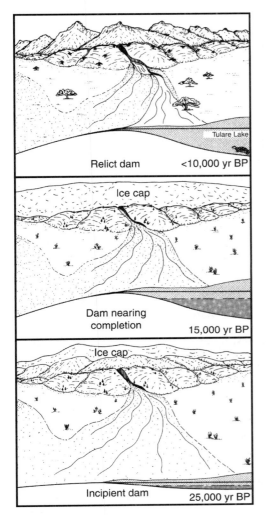

Figure 2.17. Schematic diagram of the Late Pleistocene–Early Holocene evolution of Tulare Lake, California and its outlet. From Atwater et al. (1986).

Lakes often form on the upper *delta plain* environments of major rivers. Broad-scale subsidence caused by pre-existing tectonic activity and/or sediment loading across the delta front can depress land surfaces below local base level. Because sedimentation adjacent to distributary channels of the river system may locally exceed this delta-wide subsidence, topographical closure can develop in numerous areas between both active or recently active levee systems. This allows both interdistributary lakes and brackish water bays to form. The depth and persistence of deltaic lakes is controlled by the interplay between sediment supply, from overbank flooding, crevasse-splay formation, avulsion, and organic sedimentation, and subsidence rate (Elliot, 1974). Most such lakes are quite shallow (< 10 m), although they may cover vast areas, as is the case of lakes in the Mississippi, Atchafalaya, Mackenzie, and Nile Deltas. In principle, such lakes might persist for very long intervals, associated with balanced subsidence and sedimentation. However, their intimate connection with active or recently active channels to provide topographical closure probably limits their normal life spans. Individual lake cycles of ancient delta plain lakes are rarely more than 30 m thick, and are usually much thinner (e.g., Fielding, 1984; Falkner and Fielding, 1993).

Lakes Formed Along Coastlines

Coastal lakes form a heterogeneous assemblage of water bodies adjacent to the ocean or larger lakes, formed by various, often unrelated processes. Holocene coastal lake formation and stabilization has primarily resulted from the consequences of major post-Pleistocene sea-level rise and shoreline migration, and isostatic rebound in deglaciated terrains, coupled with Late Holocene sea-level stability. Coastal lakes formed as glacial fjords or from delta plains have already been discussed. Here we will concentrate on lakes formed by drainage blockage behind coastal sand bodies, primarily barrier bars, dunes, and sand spits. Coastal lakes formed by such blockages comprise about 60,000 km^2, about 2.3% of total lake area, most of which is concentrated in a small number of larger (100–10,000 km^2) lakes (Meybeck, 1995). Most such lakes are shallow; the mean maximum depth of the largest 10 coastal lakes (> 500 km^2) is only 6 m (Herdendorf, 1990).

The proximity of coastal lakes to the ocean, coupled with their physical isolation from reworking of bottom deposits by marine storms, allows

a river valley adjacent to tributary streams following an eruption or debris flow event. In the Samalá River Valley of Guatemala, numerous small (all < 3 km^2 and up to 6.5 km long) dendritic lakes formed in the early 1900s following nearby eruptions and massive sediment discharge (Kuenzi et al., 1979). Extremely rapid aggradation allowed short-lived lakes of up to 15 m depth to form during this event. Paleolakes formed by similar mechanisms following volcanic eruptions are known from the Late Miocene Ellensburg Formation of central Washington (G.A. Smith, 1988).

them to record many coastal events that have been obliterated or time-averaged in the adjacent marine sedimentary record. For example, large tsunamis and their recurrence intervals can be recorded in normally quiet water coastal lake deposits by large washover sediment lobes (Bondevik et al., 1997). Furthermore, coastal lakes are the source of some of the most detailed records of coastal processes, isostatic rebound from deglaciation, and sea-level rise during the Holocene.

Repeated episodes of Pleistocene sea-level fall resulted in the deep incision of many coastal river valleys. It is these deepened valleys that form the axes along which many coastal lakes subsequently formed as sea level rose in the Late Pleistocene and Holocene. The ~120-m global (*eustatic*) sea-level rise of the Late Pleistocene–Early Holocene is thought to have caused the migration of large volumes of sediment toward the continents, particularly along coasts with broad continental shelves (Pethick, 1984). The deceleration of sea-level rise about 7000 yr BP caused the modern barrier bars, beaches, and islands to stop migrating landward, except in areas experiencing local subsidence. Slowing of sea-level rise also allowed sand spits to prograde in a single area for long enough to generate topographical closure of the previously overdeepened coastal river valleys, producing numerous lakes in the process.

As barrier bars, islands, and spits aggrade along coastlines they can produce topographical closure along their landward sides through a combination of marine and eolian processes. Shallow but often large lakes can form behind these features and may continue to migrate landward, or occasionally seaward as the barrier position moves. Numerous examples of these lake types exist, especially in the eastern United States, Europe, Australia, and southern Africa (Hutchinson, 1957; Timms, 1992). For example, a typical enclosed marsh and swamp area formed at Ramsay Bay, Queensland (Australia) through an episodic transgression since the Early Holocene (Pye and Rhodes, 1985). Eolian dune barriers were progressively drowned by sea-level rise in this area. Periods of dune formation associated with shoreface erosion eventually blocked the drainage behind the barrier, creating a large wetland. These types of lakes are usually quite shallow, although they may cover vast areas if the coastal plain surface was flat prior to inundation.

Where pre-existing river valleys graded to Pleistocene sea-level low stands, relatively deep valleys were incised into many coastlines. In most parts of the world these coastal river systems have been drowned as large estuaries during the Holocene sea-level rise. But in some areas where coastal sand transport or the development of marshes has been large relative to river outflow, barrier bars, dune systems, or peat deposits cut these river valleys off from their prior marine connections, forming relatively deep lakes in the process. This is the origin of the prominent and large coastal lakes found along the southwestern coast of Australia (Hodgkin and Hesp, 1998) and the eastern margin of southern Africa (South Africa and southern Mozambique) (Hill, 1975; Hobday, 1979) (figure 2.18a). Carbon-14 ages of sediments in these lakes indicate that they are probably Late Holocene in age, consistent with their formation following the stabilization of sea level in the Middle Holocene. Many of the larger ones, such as Lake Sibaya, are flat-bottomed but relatively steep-sided and deep (40 m), with water depths increasing along well-defined channels toward the ocean. Low rates of sediment input from tributary rivers entering these lakes allow them to retain their original fluvially derived shape. The extent of Holocene fill (mostly lake and marsh deposits) surrounding Lake Sibaya today demonstrates that this type of lake can be infilled rapidly (figure 2.18b).

Lakes Formed by Volcanic Activity

Lakes are prominent features of most volcanic terrains. Meybeck (1995) estimates that crater lakes, one major group of volcanically formed lakes, comprise an area of about 3150 km^2 globally. Volcanogenic lakes formed by other mechanisms probably comprise about another 3000 km^2, which means that volcanogenic lakes as a whole account for <1% of global lake area. Although this category is much less important numerically than the previous ones, lakes are often clustered in regions of modern or past volcanic activity, making them locally important.

For paleolimnology, the importance of volcanic lakes far outweighs their numerical abundance. Volcanogenic, and especially crater lakes are often very deep, particularly in humid climates. Many of the world's deepest lakes fall into this category, despite their small surface areas. Great depth combined with the wind-sheltering provided by steep crater walls combine to inhibit wind-driven mixing of these lakes. Where this has led to anoxia in bottom waters, these lakes can provide exceptionally resolved and long paleolimnological records.

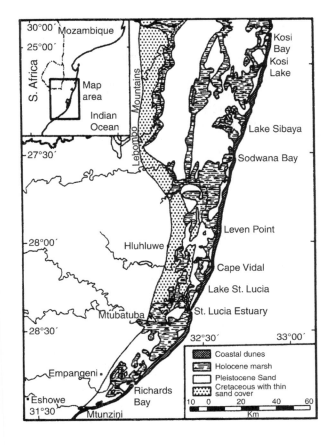

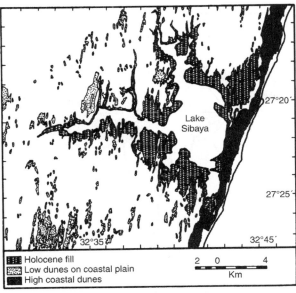

Figure 2.18. Lakes formed on the coastal plain of northeastern South Africa following Holocene sea-level stabilization. (a) Extent of coastal lakes and their relationship to Holocene marsh systems and massive coastal barrier dunes. (b) The rapid infilling of Pleistocene paleodrainages (indicated by the extent of Holocene lacustrine and marsh fill) and progressive shrinkage of Lake Sibaya. From Hobday (1979).

Two broad categories of topographical closure encompass almost all volcanic lakes: lakes formed in *craters*, and lakes formed where *volcanic dams* of various types impound pre-existing rivers. A third category, competitive aggradation by erupted and resedimented volcanogenic sediment in rivers, has been discussed in the previous section. Topographical closure produced by the above mechanisms is often coupled with tectonic, glacial, or fluvial processes, producing complex lake histories.

Crater lakes form in the topographical closure produced by volcanic crater formation. Because volcanism occurs in response to regional tectonic processes such as subduction or rifting, crater lakes tend to occur in geologically defined clusters. Craters may develop on a relatively small scale following single explosive eruptions. Such craters, known as *maars*, are circular or semicircular features, typically only a few hundred meters to a few kilometers in diameter (Büchel, 1993). Maars are normally excavated into the surrounding country rock, and are surrounded by a crater rim and walls of ash ejecta, which provides closure for subsequent lake formation (figure 2.19).

At the other extreme, crater formation may occur on top of large, stratified volcanoes, forming complex and very large *calderas* up to several tens of kilometers in diameter. Collectively, crater lakes are characterized by circular shapes and steep flanks, the result of a tendency of surficial rocks to collapse into the emptied magma chambers below the volcanic edifice following eruption. This produces a series of *ring fractures*, near-vertical faults that define the edges of the underlying cavity. Deeply collapsed calderas lying in humid climates are likely to be infilled by lakes that are very deep in comparison with their surface areas. By contrast, in arid or semihumid climates, small drainage basin areas lead to evaporitic conditions in closed basin caldera lakes, often strongly interacting with hydrothermal groundwater systems (Last, 1992; Kazanci et al., 1995).

Caldera collapse is often followed by resurgence of volcanic domes, either in the crater's interior or along the ring-fracture margins. As the crater fills with water these may form islands within the lake. In more complex calderas multiple generations of eruptions may produce lakes with several bathymetric depressions and islands. Over time crater lakes become integrated with local drainage patterns as the flanks of the enclosing volcano are eroded. This may result in a gradual decline in lake levels as the caldera walls are breached, or the catastrophic draining of a crater lake. In either case, the potential for a deep lake to develop or persist within the caldera will be diminished. For example, a 38-km^2 caldera lake at Aniakchak vol-

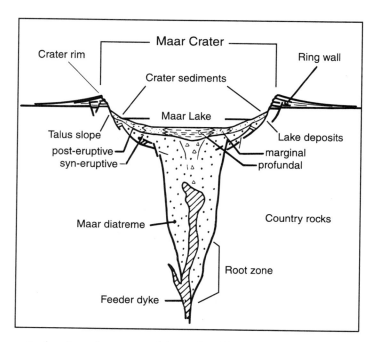

Figure 2.19. Schematic drawing of a maar and its enclosed lake basin. From Büchel (1993).

cano, Alaska drained catastrophically about 3400 years ago (Waythomas et al., 1996). Prior to that time the lake occupied more than half of the caldera floor and was 98 m deep, but only a small remnant lake persisted following the breaching event.

Lakes formed within calderas are commonly hydrologically closed basins, at least for surface water flow. Volcanoes form topographical highs, so drainage interconnections with upstream water bodies are unlikely. Similarly, in the case of small, explosively formed maars, the deposition of debris around the crater rim effectively seals off the lakes from external drainage inputs. Either case results in the low drainage basin to lake surface area ratios that typify crater lakes.

The combination of basin depth, limited drainage basin input, and ongoing subsidence makes volcanic crater lake basins potentially long-lived systems. Paleolimnological records spanning hundreds of thousands of years are not uncommon in crater lake deposits, although when interpreting these records we must recognize the great potential for interplay between the climate and geothermal systems affecting the host lake environments.

Maar lakes have been particularly important and fruitful subjects of paleolimnological research, far in excess of their abundance. Their small drainage areas make them particularly sensitive to local climate variations. Maars have been well studied geologically in western Germany, central Italy, Indonesia, and southern Australia, and many long cores have now been collected from these lakes documenting their histories (Negendanck and Zolitschka, 1993a). For example, in the Westeifel region of west central Germany, where maars were first described, an extensive cluster of maars have formed in connection with regional rifting (Büchel, 1993). Most of these are Middle–Late Pleistocene in age, although a few date from the Eocene (Mingram, 1998).

Although maar lakes have the potential for producing extremely well-resolved records, their histories in northern European regions that were glaciated during the Pleistocene are largely unknown. South of the Alps, maar lake records have been retrieved that provide this same type of resolution over intervals of hundreds of thousands of years (Niessen et al., 1993; Zolitschka and Negendanck, 1993). One maar, Lac du Bouchet, located in the Massif Central of south central France, provides a record that may span 800,000 years (Bonifay and Truze, 1991). Despite its small size (800-m diameter) and modest depth (currently 27 m), the lake is quite old, with each episode of comparative lake-level stability lasting several hun-dred thousand years. Slow sediment accumulation rates ($0.3\ mm\ yr^{-1}$) and a moderately deep crater basin have allowed the lake to persist through several Pleistocene climate cycles (Truze and Kelts, 1993).

Lakes in large collapsed calderas share many similarities with maars. Secondary eruptions, along ring fractures and/or as resurgent domes, may significantly modify the lake's evolution following the initial episode of caldera formation. Magmatic activity in the presence of a large lake establishes conditions for especially violent subsequent eruptions involving large quantities of steam (R.C.M. Smith, 1991). Crater Lake, Oregon is a well-studied small caldera lake, that despite its small size, about $8 \times 10\ km^2$, is the second deepest lake in North America (622 m) (Nelson et al., 1994) (figure 2.20). The lake formed during the Middle Holocene (6850 yr BP) atop Mt. Mazama, although Mt. Mazama itself had built up over a period prior to the caldera-forming eruption. Because of its geologically recent formation, its morphology is well preserved. The caldera itself formed from the collapse of the Mt. Mazama volcano along a circular ring-fracture system which lies well inside the present-day lake margin. A lake formed within 150 years of caldera formation and was filled over a 300-year period. Seismic reflection and core studies suggest that Crater Lake has undergone four stages of evolution, which may be typical of small caldera lakes. The first stage involved the initial caldera collapse, which defined the ring-fracture morphology and centers of subsequent deposition. During a second phase, subaerial sediments, landslides, and debris fans accumulated on the caldera floor and margin, prior to lake formation, filling the ring fractures in the process. A third, early lake-filling stage occurred while volcanic eruptions were continuing, resulting in high sedimentation rates. As infilling proceeded the crater walls were eroded, causing the lake to expand in area as its margins became shallower. This shallowing and lake enlargement process is also well documented for paleocaldera lakes (Larsen and Crossey, 1996). Following the cessation of eruptions, sedimentation was dominated by turbidite deposition.

Sedimentation in crater lakes continues until the crater is filled or breached. The rates of this infilling are highly dependent on local climate conditions as well as volcanic activity (Newhall et al., 1987). Because of their small watershed areas and sensitivity to local surficial precipitation, cyclical climatic change or cyclical volcanism may produce large lake-level fluctuations in calderas. In the case of extremely large and deep calderas like Lake Toba,

Figure 2.20. Crater Lake, Oregon. Note the steep flanks and limited watershed area typical of crater lakes.

Sumatra (Indonesia), a lake may persist for very long periods, although the recurrence of volcanic activity following a repose may completely reorganize or even destroy the pre-existing caldera lake (Chesner and Rose, 1991).

River valleys can be dammed by a variety of volcanic obstructions, resulting in the formation of *volcanic dam lakes*. Lava flows, domes, or volcanoes can all block off valleys or create topographical closure between adjacent volcanic edifices. When pre-existing rivers are suddenly blocked in this fashion they are converted into elongate or dendritic lakes. The maximum depth and sediment infill of such lakes is controlled by the volcanic dam sill height (McLeroy and Anderson, 1966; Malde, 1982; Haberyan and Hecky, 1987). For example, Paleolake Florrisant (Oligocene, Colorado) was a relatively shallow lake, formed when a low basaltic lava flow blocked a fairly broad valley. About 30 m of sediment fill accumulated behind the dam, approximating its spillway height.

Lakes that have formed through a combination of crater collapse, fluvial erosion and damming, and tectonic subsidence are common in volcanic terrains. Lake Taupo, New Zealand formed 250,000–300,000 years ago through the flooding of two large calderas and a number of adjacent grabens (R.C.M. Smith, 1991; Timms, 1992). Numerous Late Pleistocene and Holocene eruptions later modified the basin and resulted in dramatic fluctuations in lake level. Similarly, the topographically complex Yellowstone Lake (Wyoming) has evolved over the past 600,000 years through an interplay of volcanic eruptions, caldera collapse, river valley incision, and blockage by volcanic dams (Otis and Smith, 1977).

Lakes Formed by Wind

Wind generates topographical closure through erosion, referred to in this context as *deflation*, sand dune migration, or some combination of the two. Lakes formed entirely by deflation or deposition are mostly very small (< 1 km^2). However, secondary deflation of playas formed initially by other mechanisms can occur over vast areas. No census exists tabulating the number or areal extent of wind-formed lakes, but they are clearly a minor category on a global scale. Their concentration in certain areas, however, inflates their paleolimnological importance, particularly as recorders of regional hydroclimate, climate, and vegetation histories. Records from such lakes tend to be discontinuous, because of repeated episodes of deflation or sand dune migration.

Deflation basins are typically shallow circular or elliptical basins that develop in areas of intense wind scour over unvegetated landscapes. The depth of deflation basins is often correlated with the erodability of the substrate, which in turn is a function of both the cohesiveness and degree of fracturing of the material. Playas or other dry lake beds are ideal areas for deflation because they present a combination of both large, unvegetated expanses, coupled with easily transported sediment. Deep deflation hollows occasionally form in areas undergoing scour without secondary sediment infill (Pillmore, 1976). Deflation basins often display preferred orientations, normally related to prevailing wind direction at the time of excavation. In some areas, pre-existing topographical highs of easily eroded materials surrounded by more cohesive gravel pediments also control the orientation of deflation. Prominent areas of deflation basin lake development include the High Plains of the United States, the northern part of South Africa, and central and western Australia. Deflated playas are often rimmed by small, arcuate, shore-parallel dunes, referred to as *lunettes*, composed of eroded lake bed material, clay pellets, or secondarily eroded evaporite minerals. Lunettes can provide evidence for timing of deflation events, which might otherwise be difficult to constrain (Chen, 1995).

Deflation basins can persist for long intervals because of recurrent episodes of deflation, combined with eolian aggradation in surrounding areas (Holliday et al., 1996). Fills of 15 m or more are not uncommon in these lakes, but sediment accumulation is very discontinuous because of a combination of shallow basins and periodical re-deflation. These episodes of deflation impart a "memory" to the ongoing evolution of this type of lake (Bowler, 1983). Successive episodes of deepening of the deflated basin hole when the basin is dry will cause the lake to respond differently to the following periods of increased precipitation and groundwater discharge during periods of wetter climates.

The depth and extent of deflation in playas is often controlled by groundwater discharge. During intervals of high groundwater discharge flooding impedes deflation, whereas low water tables and reduced discharge onto the playa surface increase the likelihood of deflation. Because variations in groundwater discharge over long time intervals are linked to changes in precipitation and infiltration, deflation records can be used to infer regional or long-term patterns of climate change (Magee et al., 1995).

Dune migration and the deposition of eolian sands and silts create blockages in drainage networks, thereby creating lakes. The morphology and bathymetry of such lakes is constrained by the thickness of the mobile sand sheet generating barriers to flow, but in almost all cases, dune-blocked lakes are small and shallow, and therefore have short life expectancies. During periods when dunes are mobile, slumping and abundant sand supply increases the likelihood that many such basins will be quickly infilled as marshes, whereas dune sand stabilization allows open water conditions to persist for longer intervals (Stokes and Swinehart, 1997). In the Sand Hills of Nebraska a ~50,000 km² area of currently inactive dunes houses numerous small lakes in interdune depressions or behind dunes that cut off older alluvial valleys (figure 2.21). The Sand Hills sand sheet is generally 5–10 m thick, placing a cap on the depth of lakes in the region. Early workers thought that the Sand Hills and its lakes dated from arid intervals of the Pleistocene, but it is now clear that there were also phases of dune mobilization during the Holocene (Loope et al., 1995). Dune migration in the Sand Hills during these arid intervals blocked major river systems and raised the water table by up to 25 m, creating over 1000 lakes in interdune depressions in the process. The deepest lake basins lie directly upstream from the drainage network dams, which were created by the deposition of impermeable lake muds along the dune flanks. Dune-field lake chemistry in the Sand Hills is also a function of their position within dune fields, with the freshest water lakes occurring where the hydraulic gradient of the water table is steepest.

Lakes Formed by Solution

In regions of the world underlain by limestone, dolomite, or evaporite minerals lakes can form when ground surfaces collapse over dissolved subsurface cavern or fissure systems. Such *karst terrains* are important in certain regions (central Florida, the Yucatan Peninsula, the Balkans region, southern China), although the total area of lakes formed by this means is small, less than 1% of total global lake area. Most lakes occupying solution depressions are small, with only two solution lakes in the world > 500 km² (Herdendorf, 1990).

Dissolution of limestone, the most common type of karst parent rock, is controlled by the reaction of $CaCO_3$ with carbonic acid (from atmospheric CO_2) or other acids. Karst formation and the formation of sinkhole lakes can also occur in the absence of soil-generated carbonic acid, through dedolomitization caused by gypsum dissolution (Bischoff et al., 1994).

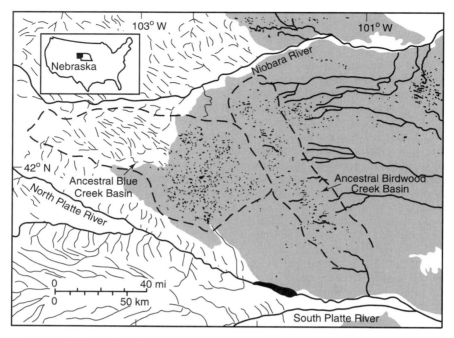

Figure 2.21. Map of the Sand Hills of Nebraska lakes district, showing the extent of blockage of paleo-drainage systems, resulting in reduced drainage network density in Sand Hills area (shaded) relative to surrounding Great Plains and the formation of numerous small lakes. From Loope et al. (1995).

Karstification occurs in a variety of settings where calcium carbonate dissolution is enhanced, both in upland and coastal environments. In central Florida upland karst forms from direct infiltration of undersaturated, slightly acidic groundwaters, and in freshwater/seawater mixing zones, which can promote dissolution (Upchurch and Randazzo, 1997). Because the latter process is focused at sea level, repeated fluctuations in sea level have led to multiple cycles of karst formation.

The surficial expression of karst includes simple funnel-shaped depressions, or elongated fissures and rectilinear depressions that follow fractures or fault systems. Karst lakes form when cover material over a developing fissure or cavern system can no longer support its own mass, leading to either gradual subsidence or catastrophic collapse, and the formation of a topographical depression. Where solution fissures lead directly to the surface, karst lakes may be very steep-sided or even overhanging. However, karstic bedrock is frequently mantled by other sediments that are not undergoing dissolution, but collapse from insufficient strength as the underlying bedrock deteriorates. In such cases, both the size and morphometry of karst lakes will be partly controlled by the nature of the cover material that collapses into the sinkhole. Where this material is cohesive (e.g., clays), a steep-sided lake basin may

result, whereas sandy or loose cover leads to a more gently sloping lake basin. The dissolution of bedrock is localized around areas of groundwater recharge at the tops of limestone aquifers, where there are local sources for soil acids and CO_2. Because of their subsurface groundwater connectivity, karst solution lakes are typically surficially closed depressions. Karst lakes formed either by the direct connection of fissure systems to the surface or collapse of overlying materials can be extremely deep for their size (Bischoff et al., 1994). However, karst formation and evaporite dissolution can also create more subtle depressions when they are extensive but deeply buried (Paine, 1994).

Karst lake morphology is strongly influenced by local physiography, fault and joint plane development, and sedimentary cover, that in turn influence groundwater recharge and dissolution. A good example of how these influences affect both the density and morphology of karst lakes comes from north central Florida (Schmidt, 1997) (figure 2.22). In the northeastern upland domain, a cover of siliciclastic sands overlying limestone bedrock is locally intact, and relatively few sinkholes exist because recharge to underlying limestones does not occur. At the edge of this siliciclastic sand cover (scarp domain), recharge is important but is

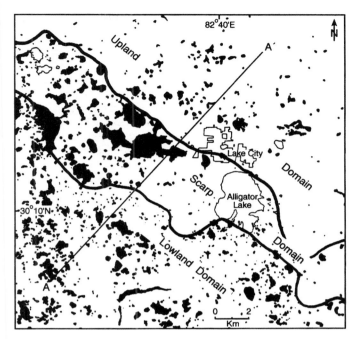

Figure 2.22. Distribution of karst lakes and sinkholes in the Lake City area of north central Florida. Karst lakes are relatively rare in the upland domain to the northeast, where a thick siliciclastic cover remains, inhibiting recharge and dissolution. In the scarp area, recharge is focused into narrow belts and larger deeper collapse zones and lakes form. The lowland areas, where siliciclastic cover has been completely stripped, are characterized by more diffuse karstification and numerous smaller lakes. Modified from Schmidt (1997).

localized to areas already stripped of cover. Larger and more complex sinkhole lakes develop under these conditions because sinkhole diameter is partly a function of cover thickness. Numerous but small sinkholes have formed in the lowland domain where the siliciclastic cover has been completely stripped away.

The combined potential for great depth and multicyclical episodes of karstification can lead to long persistence times for solution lakes (Brenner, 1994). Karst lake longevity is also enhanced in areas where karst formation acts in concert with tectonically induced subsidence, focusing dissolution along structurally controlled fracture systems (Juliá, 1980; Juliá and Bischoff, 1991).

Other Mechanisms for Lake Formation

Lakes formed by landslide dams are common in mountainous terrains. Although such lakes may be large and quite deep, they are typically short-lived, and therefore of little paleolimnological importance. Landslide damming produces dendritic lakes that may be quite deep, reflecting the back-filled river valley morphology behind the dam. Such lakes, however, are prone to catastrophic failure and rapid drawdown because of the unconsolidated and unstable nature of their topographical closure. Even when they do not fail immediately, landslide-dammed lakes are prone to rapid infilling because of their association with the high relief and sediment accumulation rates typical of mountainous areas. Most landslide lakes persist only for a few hundred to a few thousand years, and some have been known to drain in a single day (Hutchinson, 1957; Meybeck, 1995)!

Meteorite impact lakes are rare but of considerable paleolimnological interest because of their potential for housing long and highly resolved sedimentary records. Meteorite craters are produced by a combination of initial shock-wave compression, the deformation of surrounding rocks, and excavation by the impacting meteorite (Melosh, 1989). The basin structure of a meteorite impact crater lake is dependent on the size of the impacting body. Small impact craters (< 10-km diameter) show simple basin-like morphologies, formed during the small-scale collapse of the crater rim. At diameters of between 10 and 20 km, however, there is a transition to more complex craters, with wall terraces and central peaks, involving more complex and dramatic collapse events.

The recognition of meteorite impact crater lakes is often impeded by difficulties in finding evidence of an impact origin (Rondot, 1994). Generally such lakes have been recognized by their circular shape (although oblique impacts may not produce this), their occurrence outside of areas where other mechanisms for circular depression formation are readily apparent, and the preservation of meteorite or ejected glass (tektite) fragments (Hartung and Koeberl, 1994).

Large surface area to watershed area ratios and resulting slow sedimentation rates allow impact crater lakes to persist for long periods of time, in some cases for several million years. Most impact craters are steep-sided, which shelters their enclosed lakes from vigorous wind mixing, promoting the formation of anoxia and nonbioturbated sediments at depth. Taken together these factors make impact crater lakes favorable environments for long, highly resolved paleolimnological records, as in the case of Lake Bosumtwi, Ghana (W.B. Jones et al., 1981; Talbot et al., 1984; Koeberl et al., 1998) or Lake El'gygytgyn, Siberia (Brigham-Grette et al., 2001).

Artificial reservoirs are extremely important in terms of both their total volume and area. Although they are often considered separately from natural lakes, they are amenable to the same types of paleolimnological investigations. Large reservoirs are a phenomenon of the twentieth century, and because of their ongoing construction any compilation of total reservoir area or volume risks becoming obsolete. Fels and Keller (1973) estimated that reservoirs occupy a total area of $> 300,000 \, km^2$, or about 10% of total global lake area, a number that is certainly larger today. The recency of formation and well-documented history of most reservoirs allows their paleolimnological records to be used in ways that are difficult in natural systems, for example in calibrating relative sedimentation rate studies with upstream watershed erosion rates and in separating anthropogenic, slope, and climate components of erosion rates (Einsele and Hinderer, 1998).

The intentional backflooding of constricted and deeply incised river valleys in order to maximize reservoir capacity is responsible for characteristic features of reservoir morphology and bathymetry. Most are highly dendritic in shape and deepen consistently toward their spillways. Reservoirs differ

from natural lakes in that the lake's spillway elevation is almost always considerably below the maximum elevation of the dam. Because reservoirs are designed so that they will not overtop their retaining structures, they are hydrologically open systems but can vary considerably in both level and area. Their paleolimnological signals can reflect this unusual decoupling of lake-level variability and hydrology. The construction of reservoirs initiates the sudden development of deltas at the upstream confluence of the reservoir's tributaries and its slackwater zone. Sedimentation rates in reservoirs vary greatly, but in most cases they fill rapidly from a paleolimnological perspective, limiting their life spans from decades to perhaps thousands of years (Dendy et al., 1973; Einsele and Hinderer, 1997).

Lake Origins and the Quality of Paleolimnological Records

Modes of lake evolution need to be considered when a paleolimnological study is being contemplated to address a particular question. Not all lake sediment archives are created alike; some are better suited for addressing certain problems than others. Four critical factors need to be considered by the investigator (table 2.1):

1. *What is the required temporal duration of the record?* Lakes that undergo continuing subsidence through their history, or secondarily, those with low sediment accumulation rates, provide the longest stratigraphical records.

2. *What are the required levels of intersample time resolution?* Lakes with high sediment accumulation rates and/or low degrees of sediment bioturbation provide the highest resolution records.

3. *How critical is depositional continuity or stratigraphical completeness of a lake record to addressing the question at hand?* Lakes whose sedimentary records display high degrees of stratigraphical completeness are typically deep and/or rarely undergo large-scale erosional scour.

4. *Over what spatial scale should the archive integrate data collection?* Lakes with small watersheds and airsheds often archive local events, which may or may not be congruent with regional patterns of environmental change. Conversely, large lakes integrate signals from broader areas, but their records

may obscure local or short time scale processes.

Some processes produce numerous lakes across a geographical gradient, whose records can be used for comparative purposes, such as examining regional climate change across a latitudinal belt. Other classes are represented over intermediate-scale areas (throughout a mountain belt), whereas still others are much more restricted in area (most karstic regions), or are singular events (meteorite impact basins), making regional comparisons more difficult.

Not surprisingly, no one class of lake is optimal in all of these respects, and even for a single lake, considerable variation may exist in the quality of records from one area, water depth, or depositional setting to another. Some lake classes have severe limitations imposed on the quality of the records they can provide because of the nature of their history. Because of their rapid infilling and recent construction, reservoirs can only provide records of short temporal duration. Cratonic basins have sporadic histories of topographical closure and, despite their long-lived nature, are unsuitable for addressing questions where high levels of stratigraphical completeness are required. Large rift lakes provide exceptionally long records of continental climate and ecosystem change. However, they do so at the expense of local detail, since their signals are integrated over large watersheds and airsheds. Conversely, small crater lakes provide exceptional detail for climate change at the local scale and, where regionally abundant, provide comparative data sets. However, they do so over much shorter time scales than the longer-lived rift lakes.

Summary

Our conclusions from this chapter can be summarized as follows:

1. Lake basin evolution is controlled by factors that excavate a topographical enclosure at the earth's surface, and both climatological and depositional processes that serve to fill that enclosure with water and sediment.

2. Glacial lakes, formed by erosional and depositional impacts of glacial ice, are widespread features of the modern landscape, and provide a major container archive for Pleistocene earth history. Pre-Pleistocene glacial lake sediments are relatively uncommon.

3. Tectonic lakes form in a wide variety of extensional and compressional basins in the earth's crust. They are long-lived features, often filled with vast thicknesses of sediments. Because of their preservation potential, they provide our primary paleolimnological archives prior to the Quaternary.

4. Fluvial lakes form in topographical depressions on floodplains and deltas, through competitive aggradation, and, less commonly, through erosional processes. They are an important source of Quaternary paleolimnological records, although the shallow depths of these lakes limit their resolution and duration.

5. Volcanic lakes, especially crater lakes, are important sources of paleolimnological data because of their combination of long duration and high resolution records.

6. Coastal processes, wind, bedrock solution, landslides, dam construction, and meteorite impact represent secondary mechanisms of lake basin formation. However, all of these can be locally important sources of paleolimnological data.

7. Modes of lake formation and evolution must be considered carefully in judging whether a particular lake is appropriate for addressing a particular paleolimnological question.

3

The Physical Environment of Lakes

Before discussing paleolimnological archives, we need to consider those aspects of limnology that regulate how information is produced, transmitted, and filtered through the water column. Although many limnological processes leave behind sedimentary clues of their existence or intensity and are thus amenable to paleolimnological analysis, others leave little or no detectable trace. Our consideration of limnology here emphasizes the former. Throughout the next three chapters we will examine the properties of lakes, the implications of these properties for paleolimnology, and the types of physical, chemical, and biological information that can be transcribed into sedimentary archives.

Physical processes in lakes are of interest because they act as intermediary hydroclimate filters between external forcing events of interest, like climate, and the paleolimnological record. For example, understanding the hydrology of a lake is important because water inputs and outputs, which are often controlled by climate, determine lake levels, which in turn are recorded by ancient shoreline elevations, or indirectly by salinity indicators. Light and heat penetration regulate the distribution of organisms and the mixing of the water column, recorded by the distribution of various fossils, sediment types, and geochemical characteristics of sediments. Also, current and wave activity affect the transport of sedimentary particles and therefore the distribution of sediment types around a lake basin. Understanding these physical processes therefore provides us with a means of linking sedimentological, geochemical, and paleobiological records of lake deposits to the external environment.

cycle (figure 3.1). The lake components of this cycle include a series of inputs and outputs of water, which in combination with the morphometry of the lake basin, collectively determine the lake's level. Inputs include precipitation, surface runoff from rivers, and groundwater discharge into the lake. Outflows include surface outflow, evaporation, evapotranspiration losses from emergent aquatic plants, groundwater recharge, and hydration reactions with underlying sediments.

If water inputs and outputs for a lake are equal over a short time span, the lake surface elevation will remain constant. This is approximately the case in most lakes that are *surficially open basins*. Increasing precipitation + runoff + discharge has only a limited effect on lake level in open basins because this increase is matched by an increase in surficial outflow. This situation is approximated in many humid climate lakes. In contrast, if a lake is *surficially closed*, significant changes in lake level can occur as the ratio of inputs to outputs changes (Street-Perrott and Harrison, 1985). Surficially closed lakes behave like rain gauges, with lakes going up as inputs > outputs and down during the reverse situation. Such lakes are typical of relatively arid regions, where inputs are insufficient to cause the lake level to overflow the basin's spillway. To the extent that such increases and decreases in lake level relate to changes in precipitation, evaporation, or drainage diversion they may provide records of climate or anthropogenic change, and are therefore of interest to the paleolimnologist.

Light Penetration in Lakes

The Hydrological Cycle

Water enters and exits lakes through a variety of paths that comprise part of the earth's hydrological

The depth to which various wavelengths of light penetrate into lake water, and how that light is secondarily absorbed, is a major determinant of the distribution of organisms and heat in lakes,

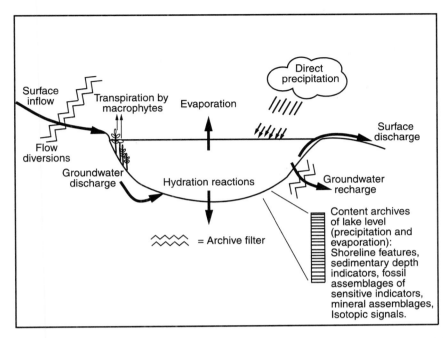

Figure 3.1. The hydrological cycle of lakes and relationship/filters to sediment archives of lake level, precipitation, and evaporation. The interpretable content archives of lake hydrology, as recorded in lake sediments (for example paleoshorelines, water depth indicators, etc.) are *filtered* through various agents modifying their information content, shown as zigzag lines in this and similar figures.

and further, of the sedimentary and fossil records of those organisms. Lakes can be divided into regions that are penetrated by sufficient light to allow photosynthesis to occur in the *photic zone*, and regions that lie below the region of photosynthesis, the *aphotic zone* (figure 3.2). The lake floor or *benthic* environment that lies within the photic zone is called the *littoral zone*, whereas the area of lake floor below that depth is referred to as the *profundal zone*. The term *sublittoral* is applied to those intermediate depths where photosynthesis occurs exclusively by benthic algae and cyanobacteria, whereas the littoral zone is restricted to the region where larger submerged plants (*macrophytes*) can or do occur.

There is no one depth at which light becomes a limiting factor to growth in a given lake. Rather, the growth ranges of photosynthesizing organisms depend on a host of factors. Light penetration varies because of differences in suspended sediment (*turbidity*), plankton, dissolved solid content of the water, and the orientation of the water body with respect to incoming solar radiation. Furthermore, different photosynthesizing organisms use light from different parts of the electromagnetic spectrum, depending on the types of

photosynthetic pigments they possess that allow them to photosynthesize under variable light conditions. Thus there is actually a photic zone for each species, which in total represent the region of active photosynthesis. This should serve as a caution to paleolimnologists searching for paleowater depth information from what are, in reality, complex paleolight-level indicators.

Heat, Mixing, and Stratification

Penetration of heat, turbulence, and water column mixing in lakes all merit the close attention of paleolimnologists for two reasons. First, a lake's content archives accumulate in response to variations in the duration of mixing events and water column stability, and therefore provide us with an interpretable record of mixing phenomena (lake hydroclimate), and often indirectly of climate. Second, the duration and intensity of mixing that occurs in a lake is a major, albeit indirect, determinant of the maximum resolution attainable from a lake's content archives. This is because mixing can both directly stir bottom sediments, and can introduce dissolved oxygen to the deep bottom waters of

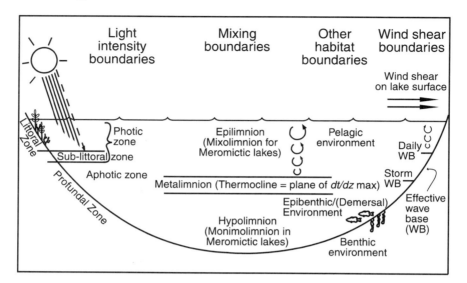

Figure 3.2. Zonation of the lake environment, illustrating terms used in this book. The vertical positions of boundaries related to light, vertical mixing, habitats, and wind shear are not fixed with respect to one another, and are shown here in a single diagram for convenience only.

a lake, allowing bioturbation of the sedimentary content archives by benthic organisms. The extent of sediment smearing by bioturbators places limits on the maximum resolution that can be obtained from a given lake's sedimentary record. Climate, lake morphology, the nature of influent rivers, and other factors influencing the mixing dynamics of a lake, all constrain the temporal scales at which we can interpret a given lake's history.

Heat enters lakes both as short wavelength solar radiation and as long wavelength radiation from geothermal sources (Ragotzkie, 1978). As solar radiation penetrates a lake, it is absorbed, resulting in heating of the lake water over a depth range of meters in turbid lakes to tens of meters in very clear lakes. The penetration of light, however, is exponentially reduced with depth. Thus, radiative heat is distributed very unequally within the surface waters of a lake and heat must be secondarily redistributed by other mechanisms in order for the deep water mass to be warmed. This occurs through wind mixing, density instabilities in the water column, and turbulence in influent water masses. Mixing occurs whenever the kinetic energy supplied by these external inputs is sufficient to counteract density contrasts between overlying, lighter water versus underlying denser water. Small-scale turbulence, set in motion by these external forces, drives mixing between adjacent water masses (Imboden and Wüest, 1995).

The potential for turbulent mixing to occur can be described by Richardson's number (R_i), a dimensionless number describing the ratio between the density gradient between adjacent water layers and the velocity of the turbulent *eddies* acting along the boundary between the layers:

$$R_i = g(d\rho/dz)/\rho(du/dz)^2 \qquad (3.1)$$

where g = acceleration due to gravity, ρ = density, z = depth, and u = velocity.

Turbulence will propagate and mixing will occur when $R_i < 0.25$. When turbulent mixing or density instabilities are insufficient to mix vertically adjacent water masses, the water column of a lake can become *stratified*. A stratified water mass is one with a stable or quasi-stable density configuration, in which the water column is at its lowest potential energy state. The development of stratification in a lake is influenced by both intrinsic properties of the lake, such as its morphometry, depth, fetch, and solute concentrations, and extrinsic factors, such as air temperature and solar radiation, wind speed and duration, air mass stability and humidity. In a wind-mixed lake of uniform salinity, thermal stratification is determined by a stratification index S (Holloway, 1980):

$$S = \mu^3 \rho C_\rho / zg\alpha(Q_0 - 2Q_1/cz) \qquad (3.2)$$

where μ = frictional velocity of air, ρ = density of water, C_ρ = specific heat of water at a constant pressure, z = water depth, g = gravitational constant, α = coefficient of thermal expansion, Q_0 = net surface heat input, Q_1 = solar radiation penetrating water column, and c = extinction coefficient.

Empirically, a transition from a well-mixed water mass to a stratified water mass occurs when the value of S drops below about 6700.

We can think of the *stability* of a lake as an indication of the amount of work required to break down its stratification and cause it to mix uniformly, without net heat loss or gain. One commonly used model of the stability of the mixed layer in a lake, the Wedderburn number (W), accounts for the ambient wind stress affecting the lake's surface and the lake water density structure (Thompson and Imberger, 1980). This model is applicable for lakes in which the wind stress is expected to run down the lake's length (its *fetch*) and where the wind persists for longer than a quarter of the internal seiche period, discussed in detail later in this chapter (Spigel and Coulter, 1996):

$$W = [(\rho_2 - \rho_1)/\rho]gh^2/\mu^2 L \qquad (3.3)$$

where ρ_2 = density of the *hypolimnion* (the deeper, stratified part of the water column), ρ_1 = density of the *epilimnion* (the upper, mixed part of the water column), ρ = reference density of water, g = acceleration due to gravity, h = depth of the epilimnion, μ = shear velocity in the water at the water surface, and L = length of the lake.

When W is large (generally $> L/4h$) stratification is strong and mixing is inhibited. Thus, the amount of energy required to destabilize a water body increases as the density contrast between overlying and underlying waters increases, and as the water volume that must be mixed increases. For example, in Lake Valencia, Venezuela, surface water temperatures rise each year between February and May as windiness, which causes evaporative cooling, declines (W.M. Lewis, 1983, 1984) (figure 3.3). Heat is transferred to deeper water by turbulent mixing, but in a subdued and delayed fashion. As the surface waters warm and the thermal contrast between surface and deep waters increases, the water mass becomes progressively more stable and remains stratified each year from about April to November. As the winds pick up again in November, evaporative cooling ensues, reducing the temperature contrast between surface and deep water and reducing water column stabi-

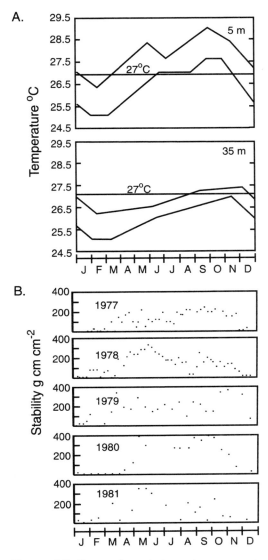

Figure 3.3. Seasonal variations in temperature range and stability of Lake Valencia, Venezuela. (A) Average monthly temperature variation over the five-year study period at 5 and 35-m depths. (B) Resultant water mass stability. From W.M. Lewis (1984).

lity. When this happens, the lake can once again mix uniformly.

Variations in the stability of lakes are correlated with latitude. High-latitude lakes experience large seasonal temperature fluctuations, but these variations occur at relatively low temperatures, at which density contrasts in freshwater are relatively small. In contrast, the temperature fluctuations in tropical lakes are much smaller, but occur in the temperature range where density increases rapidly per degree temperature rise. On a global basis this tradeoff between greater thermal variability and greater density difference per increment of temperature change results in lakes reaching their maximum stability at subtropical latitudes, between 20° and 30° from the equator (W.M. Lewis, 1987) (figure 3.4). Over time, changing thermal regimes have probably resulted in both N–S shifts in this latitudinal zonation and changes in the shape of the curve, since the absolute values of temperatures in latitudinal belts and temperature ranges are unlikely to have varied consistently.

Lake mixing varies greatly over a variety of spatial and temporal scales, and is affected by many types of surface and internal waves (Imboden and Wüest, 1995) (figure 3.5). Some of these wave types occur and dissipate over very short time intervals, from seconds to hours. This includes the setup of surface waves or Langmuir circulation, which involves counter-rotating vortices in the upper part of the water column. Others, such as storm mixing, large-scale internal waves, and annual stratification cycles, occur over much longer periods. From a paleolimnological viewpoint it is these latter phenomena that will concern us most, particularly those of seasonal duration or longer, since shorter-term events are unlikely to be archived by sedimentary indicators of mixing.

Seasonal and Longer-Scale Mixing Cycles

Over the course of a year a lake responds to heat exchange with its surrounding environment through changes in its thermal structure. Consider two contrasting lakes, Junius Pond #7, New York, and Lake Tanganyika, Africa. The former is a small, shallow, temperate, kettle lake [length = 200 m, maximum water depth (z_{max}) = 8 m], and the latter a large, deep, tropical, rift lake (length = 600 km, z_{max} = 1470 m) (Pendl and Stewart, 1986; Plisnier et al., 1999) (figures 3.6a and 3.7). Junius Pond #7 is ice covered in winter. A lack of wind-generated turbulence causes the lake to become stratified under the ice, with the densest water, near 4°C (the temperature of maximum density for freshwater), lying at the bottom of the lake and progressively cooler water nearer the surface. After the spring ice melt surface and deep-water masses mix rapidly, since the density contrasts between different layers are minimal. Through spring the surface waters of Junius Pond #7 become warmer. Initially, wind mixing can overcome the minor density differences between surface and deeper waters and the lake remains unstratified. But eventually wind-induced mixing becomes inadequate to overcome progressively higher density contrasts, and by May the lake begins to stratify. Mixing between the epilimnion and the hypolim-

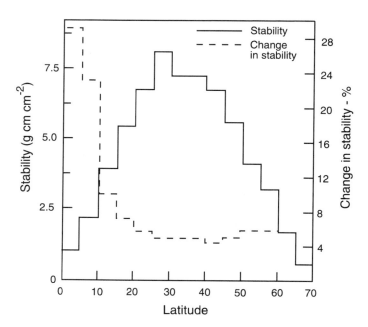

Figure 3.4. Latitudinal trend in average water column stabilities for freshwater lakes. From W.M. Lewis (1987).

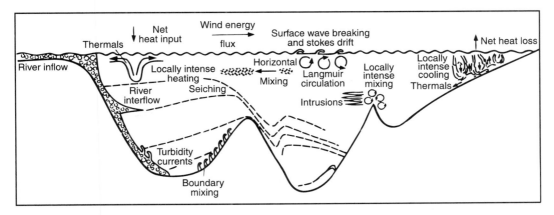

Figure 3.5. Schematic model of major mixing processes in lakes. Modified from Imboden and Wüest (1995).

nion is inhibited. A zone of strong temperature and density contrast, the *metalimnion*, develops between the two water masses. Within this zone a plane of highest temperature change can also be defined, termed the *thermocline*. As summer stratification intensifies, the thermal and density contrast between the epilimnion and hypolimnion increases. The thermocline also deepens as epilimnetic mixing progres-

sively warms its upper reaches. This stratification in turn inhibits mixing of gases, solutes, and nutrients across the thermocline, resulting in various forms of chemical stratification that can impart distinctive geochemical and fossil archives to the sediments. Surface waters respond quickly to declining air temperatures after mid-summer, although the thermocline continues to deepen because the total solar

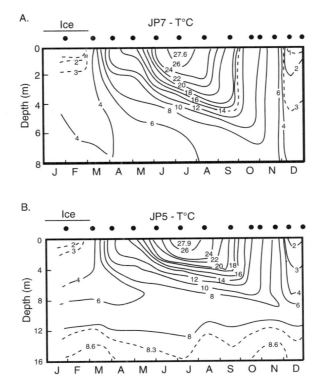

Figure 3.6. Seasonal thermal profile of the Junius Ponds, shallow, temperate lakes from New York. (A) Dimictic Junius Pond #7. (B) Meromictic Junius Pond #5. From Pendl and Stewart (1986).

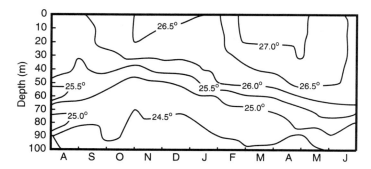

Figure 3.7. Seasonal thermal profile for the northern part of Lake Tanganyika, a deep, tropical, mero-mictic lake, near Kigoma, Tanzania (eastern side of the lake). The diagram shows a profile for the upper 100 m of the lake only. Deep, metalimnetic and monimolimnetic temperatures show little seasonal variation, and are approximately 23.3–23.5°C at 300 m, and 23.2–23.3°C at 1200 m. Derived from Plisnier et al. (1999).

input of heat still exceeds loss from surface evaporation. Thermal and density stratification decay in late September, as the lake surface begins to cool, promoting mixing for a second time in the annual cycle. This condition persists until the lake ices over and stratification is once again established.

The thickness of the epilimnion in Junius Pond #7 and other lakes is regulated by both the thermal structure of the water column and the wind energy available to counteract thermal stratification. Since mean wind stress across a lake surface increases with increasing fetch, there is a strong relationship between lake size or shape and the depth of the epilimnion (Ragotzkie, 1978; Patalas, 1984). Large lakes, or lakes that are elongated in the dominant wind direction, have deeper epilimnia than do small lakes or lakes oriented in directions opposed to the average wind flow. For temperate lakes this can be expressed as:

$$D_{th} = 4F^{0.5} \qquad (3.4)$$

where D_{th} = depth of summer thermocline (m) and F = maximum fetch (km).

Lakes with higher summer temperatures, generally at lower altitudes or latitudes, require greater inputs of wind energy to overcome the progressively higher density contrasts observed between warmer waters, and therefore tend to have shallower thermoclines. Paleolimnological comparisons of inferred mixing dynamics in space and time must account for these differences, particularly considering the fact that a lake's shape, size, and average epilimnetic temperature may change considerably over its history.

Hutchinson and Löeffler (1956) developed a useful classification of lakes based on seasonal patterns of mixing. Although this scheme has been criticized as not accounting for other time scales of mixing (Imboden and Wüest, 1995), it remains valuable for paleolimnologists for two reasons. First, it ignores very short-term events, unlikely to be resolved in sedimentary content archives. Second, complete mixing of a lake, down to the sediment–water interface, results in a qualitatively different sedimentary archive for paleolimnology versus partial mixing, as a result of the potential for bottom water ventilation and sediment bioturbation by benthic animals under conditions of complete mixing. In the Hutchinson and Löeffler model, a lake that completely mixes on a twice-annual basis, as a result of thermal overturn, like Junius Pond #7, is referred to as a *dimictic* lake. This type of mixing typifies many temperate water bodies.

This mixing regime contrasts strongly with that of tropical Lake Tanganyika (Plisnier et al., 1999) (figure 3.7). Here seasonal temperature differences from varying solar insolation are minimal, and no part of the lake ever drops to 4°C, so overturn of the entire water column from cooling does not occur. At the high ambient temperatures that prevail throughout the Lake Tanganyika water column, between 23 and 28°C, density contrasts for a given temperature difference are great. Because the lake is very deep, mechanical energy from wind is insufficient to overcome density stratification and a permanent thermocline exists. The lake is referred to as being *meromictic*, because it mixes only incompletely. Seasonal variation in windiness does cause the depth of mixing to vary throughout the year, but wind energy is never sufficient to overcome stratification below about 100–200 m.

Meromixis can persist over varying time scales and examples, such as the Dead Sea, are known where the condition has ended abruptly after persisting for centuries (Steinhorn, 1985). Various types of meromixis are recognized, depending on the cause of the density contrast between the regularly circulating upper water mass, the *mixolimnion*, and the poorly mixed deep water, the *monimolimnion*; these can sometimes be distinguished from sedimentary archives (Boyer, 1981). *Endogenic meromixis* occurs when salts are released from decomposing organic matter in the deep water mass of a lake, or as a result of solutes concentrating as brines during prolonged freezing of surface waters. *Ectogenic meromixis* results from an external input of saline water into an otherwise freshwater lake, or freshwater into a saline lake. Either situation results in a cap of light, relatively freshwater overlying a dense lower saline layer. Junius Pond #5, adjacent to the previously discussed Junius Pond #7, is a good example of such a meromictic lake (figure 3.6b) (Pendl and Stewart, 1986). Here meromixis is maintained by injection of saline waters from brines released from an underlying salt-bearing formation. Ponds #5 and #7 experience almost identical climate, ice-over seasonality, and surficial hydroclimate, and shallow water mixing and thermocline formation in the two lakes is similar. However, Pond #5 does not undergo complete overturn, as a result of its dense, saline bottom waters, which remain warm from brine discharge throughout the winter. It is instructive for paleolimnologists to consider the fact that very different sedimentary histories would be recorded by these two lakes, reflecting very different hydroclimates, despite their identical climate regimes.

A variety of other mixing regimes occur in lakes. *Monomictic lakes* mix once per year. In *warm monomictic lakes*, the lake cools sufficiently in winter to mix completely but the lake never freezes, and therefore no winter stratification occurs. This condition is commonly observed in lakes within the warm temperate zone and subtropics, and in saline lakes in mid-latitude arid environments that do not freeze readily. In *cold monomictic lakes*, typical of subpolar or alpine regions, water temperatures during the summer warm sufficiently to allow ice to break up but not to establish density stratification, and the water mass mixes freely throughout the summer. *Polymictic lakes* mix repeatedly throughout the year. Brief periods of stratification may occur, but this layering is broken up regularly and

both density stratification and the thermocline are weakly developed. This condition is most common in shallow lakes of the tropics and subtropics, particularly in areas with strong winds that can break up incipient stratification, and in arid regions with low humidity, enhancing evaporative cooling and mixing. *Oligomictic lakes* rarely and irregularly mix. They are found at low latitudes and altitudes, particularly in regions of low temperature variability and high humidity, the latter limiting the potential for evaporative cooling. Generally these are sheltered and/or deep, with mixing occurring after episodic cooling or strong wind events. *Amictic lakes* "never" mix, most commonly as a result of being permanently covered by ice. Most such lakes are restricted to polar regions or very high altitudes. In fact, recent studies of perennially frozen lakes in Antarctica have shown that even these lakes undergo slow circulation and convective mixing (L.G. Miller and Aiken, 1996; Tyler et al., 1998).

The presence, absence, and duration of vertical mixing is recorded in sediments because vertical mixing redistributes oxygen to the hypolimnion, and consequently to multicellular organisms, which can bioturbate sediments. Mixing also affects the primary production of phytoplankton in lakes by replenishing nutrients into the upper water column. Meromictic lakes are probably the easiest category of lakes to recognize from their hypolimnetic sedimentary archives, as they produce distinctive suites of indicators, in particular, very finely laminated sediments, formed in the absence of bioturbation by burrowing organisms (e.g., Valero-Garcés and Kelts, 1995). In most meromictic lakes laminations are generated by seasonal variations in runoff, nutrient availability, and upwelling in the mixolimnion, leading to changing patterns of primary productivity and/or calcium carbonate precipitation (e.g., Brunskill, 1969). Occasionally meromixis is also archived by changes in the abundance of fossils from animals feeding higher in the food web (Culver et al., 1981).

Other types of mixing regimes may also produce distinctive sedimentary signatures. Seasonal differences in the timing of maximum productivity of the lake's surface waters, relative to when the lake becomes stratified, might allow the distinction between warm and cold monomictic lakes from a careful examination of seasonally laminated sediments. Similarly, the presence of *authigenic* minerals (minerals formed in situ) whose formation is both temperature- and oxygen-dependent, or of wind-blown dust laminations, indicative of strong wind-mixing seasonality, may allow the interpretation of paleomixing patterns. In some cases, mixing patterns may be recognizable from the occurrence of seasonally diagnostic sediments that have been secondarily resuspended by mixing events at other times of the year (Hilton, 1985). In practice, however, distinguishing seasonal mixing types from paleolimnological records alone is difficult because of the similarities between sedimentary archives produced under very different seasonal mixing regimes.

Fluid Flow in Lakes

Currents and waves transport and redistribute sediment in lakes, and are major determinants of the lateral variation in particle size and sedimentary structures, sediment accumulation rates, and thickness of strata. Without an understanding of how physical transport mechanisms operate in lakes, it is impossible to accurately interpret the significance of either temporal or lateral changes in lacustrine sedimentary archives.

The transport of water and its suspended particles is affected by both its density and its dynamic viscosity, μ, defined as $\tau/(du/dt)$, or the shear stress, τ, required to produce a given rate of deformation in the fluid. Shear stresses act parallel to the surface of a fluid layer, by transferring momentum from one fluid to another. In water, shear stress may be applied from wind, or from an overlying water layer, in the process accelerating the adjacent water mass. The viscosity of water is strongly temperature-dependent. Freshwater in a tropical lake at about 25°C requires only about half the shear stress to accelerate it to a given velocity as water just above freezing.

Flowing water moves in one of two ways. When the streamlines of adjacent parcels of a fluid are moving smoothly and parallel with each other, the flow is referred to as being *laminar*. The velocity of a laminar flow is easily defined, because streamlines are moving unidirectionally. The contrasting state, *turbulent flow*, occurs when particles of water are no longer moving parallel to one another and streamlines become distorted. In turbulent flow, water particles are transferred in variable directions by eddies, and velocity can only be defined as an average for the entire water mass.

Two factors determine whether a fluid will move in a laminar or a turbulent fashion. *Inertial forces* are those involved in accelerating the fluid from a resting state, and are determined by the velocity of the flow (U), a reference length (L, typically the water depth, or for an object moving through the water, its size), and its density (ρ). *Viscous forces* are defined by the dynamic viscosity (μ), discussed above. The ratio of these two forces is a dimensionless number, called the *Reynolds number* (R_e):

$$R_e = UL\rho/\mu \qquad (3.5)$$

When R_e is small (typically < 500), flow will be laminar, whereas when R_e is large (> 2000), flow will be turbulent. Increasing flow velocity or the fluid's density both result in higher inertial forces, and increase the likelihood that the flow will be turbulent, whereas more viscous fluids, for example colder or more sediment-laden water, are more likely to move in a laminar fashion. In the open water of lakes, laminar flow is rare because the viscosity of water is so low. However, near a lake's sediment–water interface, viscous, sediment–water mixtures can move in a laminar fashion, and this is an important mechanism of sedimentation.

Surface and Internal Waves in Lakes

Waves are periodic oscillations that cause water particles to move in vertical orbits. Waves occur both at and below a lake's surface, and are active on many spatial and temporal scales, reflecting the variety of forces that generate their motion and restore the water particles to equilibrium (Imboden and Wüest, 1995). Waves may be *progressive*, where individual waveforms display forward motion, or *standing*, where waveforms do not undergo horizontal displacement. Understanding wave motion is important to paleolimnology for two reasons. First, waves are a form of turbulence and therefore serve to redistribute nutrients and gases through the water column, and their strength and distribution can determine regions and cycles of productivity or anoxia, both commonly recorded in lake sediments. Second, in shallow water waves can entrain and transport sediment, creating distinctive suites of sedimentary features that are interpretable from lake deposits.

Waves can be described by their characteristic *wavelength*, λ, or crest-to-crest distance, their height, H, the trough-to-crest vertical distance, and their period, T, the time required for movement of λ past a fixed point. Surficial progressive waves are primarily forced by wind energy, although the water particles themselves experience little net forward motion. Surface wave orbitals become progressively smaller with depth, dissipating at a depth of about $\lambda/2$. Where $\lambda/2 <$ water depth (z), the waves are referred to as *deep-water waves* and the orbitals are circular throughout their depth range. Deep-water wave velocity is simply λ/T, whereas deep-water orbital water particle velocity is $\pi H/T$. Waves that are active in areas of a lake where $\lambda/2 > z$ (local water depth) are referred to as *shallow-water waves*, because their orbitals impinge on the bottom, and can apply shear stress, entraining sediment in the process. The velocity of shallow-water waves is constrained by their interactions with the lake floor, proportional to $z^{0.5}$. Interaction with the lake floor causes the orbitals to flatten into horizontal ellipsoids. The velocity of water particles in these orbitals can be given as:

$$u = \pi HT \sin z(2\pi z/\lambda) \qquad (3.6)$$

The entrainment of a sediment grain of a given size by oscillatory waves is a function of this velocity, and inversely of the wave period (Komar and Miller, 1975). Wave velocity and shear stress on sediment is not constant throughout the wave's orbit. As the wave moves forward, it travels with a higher velocity for a shorter interval, whereas the return flow moves more slowly over a protracted period, a process that can result in a net separation of particles by size in shallow water, with larger particles concentrated landward and finer ones lakeward.

The size of λ and H for surface waves is controlled by the fetch over which wind energy builds the waves and the intensity of the wind stress creating the wave. For Lake Michigan, with a maximum fetch of about 480 km, this would translate into a wave base of about 60 m under a sustained storm wind of 20 m sec^{-1}, or about 30 m at 10 m sec^{-1}, assuming the wave base is $\lambda/2$. In contrast, a small lake, with a 2-km maximum fetch, would have maximum wave bases of only 1.5 m and 0.75 m under the same wind conditions. Internal waves, discussed below, can secondarily distribute solutes and mechanical energy to even greater depths. This implies that the use of sedimentological indicators of wave mixing for paleolimnological interpretation of water depth must be done with some understanding of probable wave dynamics and estimate of lake size in mind (e.g., Olsen, 1990).

Waves approaching shallow water maintain a constant period but their interactions with the bottom cause them to slow down and change their shape, becoming shorter and steeper-crested. Eventually, the waveform becomes unstable and breaks, creating surf and turbulence. Because a wave crest rarely approaches shallow water precisely parallel to depth contours, the wave will decelerate at different times along its crest, causing it to refract, becoming progressively more parallel to the coast. This refraction almost never results in

a complete reorientation of the net direction of wave attack on the coast. Therefore, shoaling waves can generate *longshore currents* running parallel to the coastline in the direction of net wave attack. In large lakes such currents can be important sediment transport processes, with velocities as high as 30 cm sec^{-1} (Freeman et al., 1972; Sly, 1978).

Coastal shallow-water wave and longshore current activity impart characteristic depositional signatures such as ripples and sand bars. The relatively small wavelengths of progressive surface waves in lakes, generally no more than a few tens of meters, imply that the depth range over which this type of shallow-water wave reworking can occur is highly restricted, even in large lakes. In most lakes this also translates into a narrow lateral range where such processes can occur. Therefore the presence of characteristic indicators in lake beds of shallow-water waves can be highly diagnostic for inferring water depth, and sometimes proximity to shoreline, at the time of deposition.

Seiches are standing waves whose wavelengths are comparable to the size of the lake basin (figure 3.8). They are associated with the occurrence of some forcing mechanism, normally steady winds blowing over the surface of a lake. This tilts both surface waters and internal density contrasts, causing surface waters to accumulate at the downwind end, and deep water to be drawn toward the surface at the windward end. In small and medium-sized lakes the thermocline at the downwind end becomes more sharply defined and weaker at the upwind end (Mortimer, 1974). Storm conditions, or the presence of persistent onshore winds, often coupled with low atmospheric pressure, causes a piling up of water at the coastline, and may result in a significant elevation of the water surface. This is counteracted by gravity, causing an underlying return flow of water in an offshore direction. In the shore zone, where shallow-water oscillatory waves are acting on the lake floor, this combination of lakeward-directed currents and waves, referred to as *combined flow*, generates very complex oscillatory turbulence during storms. This produces distinctive sedimentary features recognizable in lake deposits, discussed further in chapter 7.

When the wind stops, gravity acts to restore equilibrium in water surface elevations, both at the lake's surface and along internal density contrast surfaces. This causes the lake to oscillate around one or more nodes down the axis of the lake. In larger lakes, particularly those at higher latitudes, these oscillations also have a rotational component around the lake basin, driven by

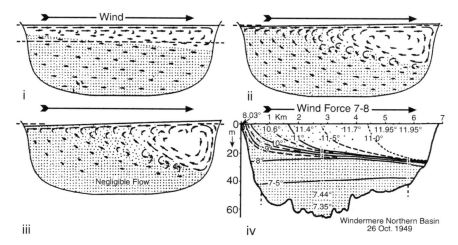

Figure 3.8. Models of seiche movement in lakes. A two-dimensional representation of the effect of wind on a stratified lake. (i)–(iii) Stages in the development of shear instability and deepening of the epilimnion at the downwind end of the thermocline. (iv) Temperature profile in the northern basin of Lake Windermere, U.K., after about 12 hours of relatively steady wind. For all panels the layer that was initially below the thermocline is shown stippled and the direction and relative speed of flow are shown by arrows. Modified from Mortimer (1974).

Coriolis effects (Saggio and Imberger, 1998). Vertical displacement of water increases away from the seiche nodes, as the epilimnetic water oscillates, causing both upwelling and downwelling to occur. Internal seiches that cause large amplitude oscillations on internal density discontinuities such as thermoclines can generate substantial mixing in lakes (Plisnier et al., 1999). These types of seiches can be archived in lake sediments through characteristic seasonal signals of nutrient regeneration and productivity in the upper water column as deep waters upwell to the surface.

The Coriolis Effect

Although water particles on the earth's surface are accelerated in a straight line by motive forces, that motion, as a result of gravity, is constrained to occur on a sphere. For example, consider a particle of water in the Northern Hemisphere, set in motion in a northerly direction. That particle also has a component of momentum directed toward the east, as a result of the earth's rotation. The velocity of that motion is a function of latitude and decreases in both directions away from the equator. From the reference point of a stationary observer on earth, the particle that is accelerated due "north" will be deflected toward the right or clockwise in the Northern Hemisphere (counterclockwise in the

Southern Hemisphere), as its eastward motion progressively exceeds that of the earth beneath it. A particle moving due "south" will be deflected clockwise, as it progressively lags behind the earth's faster motion at lower latitudes.

These deflections, referred to as Coriolis (or geostrophic) effects, become significant factors in lake circulation and wave formation when the period over which the particle motion occurs is large relative to the period over which the particle would rotate around the basin. A lake's latitude is also important, because the rate of decrease in the earth's orbital velocity increases with latitude, causing the Coriolis effect to increase toward the poles:

$$f = 2\Omega \sin \varphi \qquad (3.7)$$

where f = Coriolis effect, Ω = angular velocity of the earth = 7.29×10^{-5} rad sec^{-1}, and φ = latitude. Coriolis-driven circulation includes several phenomena: the deflection of incoming rivers entering lakes, causing currents to hug the right shoreline in the Northern Hemisphere; the rotation of progressive long-wavelength waves as oscillatory cells in the centers of large lakes (*Poincaré waves*) and seiches; and the transformation of this rotational momentum into shoreline-parallel waves (*Kelvin waves*) and currents along coastlines (Saggio and Imberger, 1998). In all cases these waves and currents generate turbulence and mixing, and have the potential to entrain particles, thereby directing patterns and rates of sediment accumulation.

Currents in Lakes

Currents are unidirectional flows at or below a lake's surface that lack the periodic oscillations of waves. In lakes currents tend to be transitory, developing after unique forcing events or seasonally, as a result of wind, from momentum supplied by waves, or gravity. The latter is driven by the through-flow of rivers in lakes with relatively small volumes, or by density instabilities between an overlying heavier fluid and an underlying lighter one. Currents, like waves, are subject to Coriolis effects, and in some lakes these are highly predictable (Endoh et al., 1995). Stratification enhances horizontal current transport and mixing by channeling momentum within confined depth zones (Imboden and Wüest, 1995). In medium and large-sized lakes, surface currents are often sufficient to transport clay and silt-sized particles over considerable distances (Sly, 1978). However, sand-sized particles are rarely transported significant distances in this way.

Lakes whose volumes are small relative to their inflows and outflows are strongly influenced by through-flowing currents, the strength of which often vary seasonally. River inflows are also a primary source of density-driven currents that descend to a depth at which they are in equilibrium with the water column. Density contrasts between ambient lake water and an inflow may result from temperature, salinity, or suspended load differences. As the flow enters the lake it may float on the ambient lake water, an *overflow* current, descend to some intermediate depth as an *interflow*, or if denser than all lake water, descend to the bottom of the lake as an *underflow* (figure 3.9). Seasonal or even diurnal

differences in temperature and sediment load can result in an influent current changing its flow position, resulting in very different sediment distribution patterns (Rea et al., 1981; Carmack et al., 1986; Weirich, 1986; Okamoto et al., 1995).

Overflows occur when influent rivers have low sediment loads, or are entering saline lakes, and often develop into interflows as they advance (N.D. Smith and Ashley, 1985). Interflows develop in density-stratified lakes, where the influent sediment/water plume sinks until it reaches a depth of equivalent density. In the process it generates considerable turbulence and mixing, thereby enlarging its volume. Upon reaching its equilibrium depth, the plume will normally spread horizontally. Currents of 2-15 cm sec^{-1} are typical of interflows, although higher velocities are recorded near inflows (N.D. Smith and Ashley, 1985). Both overflows and interflows are frequently observed to undergo deflection by Coriolis forcing, greatly influencing the ultimate distribution of river-borne sediment. Underflows occur when sediment loads are high and the river is entering a freshwater lake. The relatively high density of underflows compared with the ambient lake water limits the amount of mixing between the two water masses. The velocities of underflow currents are quite variable, but can be high, occasionally over 100 cm sec^{-1} (Lambert, 1982; Weirich, 1984). As a result of their high velocities and relative coherence, underflows may travel for tens or even hundreds of kilometers with sufficient velocity to transport sand (Grover and Howard, 1938; Pickrill and Irwin, 1982).

Currents caused by the incorporation of dense suspensions of sediment particles as part of the fluid are termed *gravity flows*, and are important mechanisms for sediment transport in many lakes. Although they commonly form in association with influent rivers, gravity flows can also be triggered by other factors, like earthquakes or storms, in

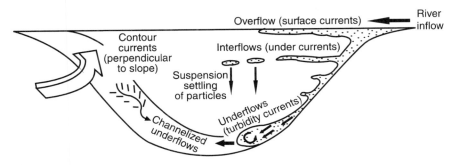

Figure 3.9. Density-driven currents in lakes, showing the variety of interactions that may occur as water of varying density enters a lake. Adapted from Sturm and Matter (1978).

any sloping area where unconsolidated sediment occurs.

The most common gravity flows in lakes are *turbidity flows*, initiated where sediment-charged underflows move down a sloping surface, or through the partial resuspension of unstable soft sediment on a slope. The resultant sediment/water mixture experiences little frictional resistance to flow and can spread over large areas. Turbidity flows have a fairly predictable anatomy. A sediment-rich lobate *head*, often with more than one lobe, leads the flow in a highly turbulent fashion. The velocity of this head is a function of the difference in density between the flow and the ambient lake water, gravity, the size of the head, and inversely, of the ambient lake water density. The head is followed by a progressively more dilute *body*, and ultimately a low-velocity *tail*, whose passage over a given place may be delayed by hours or days relative to the head. Turbidity flows in lakes are frequently erosive and may produce extensive sublacustrine channel systems, particularly along steep lake bottoms (Giovanoli, 1990; Scholz et al., 1993). Where these flows settle, they can form extensive lobate deposits, called sublacustrine fans.

Debris flows are another type of gravity flow that often leave deposits along the flanks of steep-sided lakes. Debris flows are extremely dense, water-saturated flows that exhibit plastic flow behavior and have sufficient internal strength to keep very large particles, even boulders, aloft. Whereas turbidity flows are highly turbulent, debris flows move in a laminar fashion, with low Reynolds numbers as a result of their very high viscosity, depositing massive, unstratified layers of debris.

Summary

Our conclusions from this chapter can be summarized as follows:

1. Physical processes in lakes, such as the distribution of light and heat, the mixing of the water column and development of stratification, and the motion of waves and currents, all strongly affect both the distribution and quality of sedimentary archives for paleolimnology. Most of these processes have strong linkages to external climate or watershed processes.

2. Water inputs and outputs, which are largely regulated by climate, determine whether a lake will be surficially open or closed. Paleolake levels provide an archive of these hydrological changes.

3. Light penetration determines the extent of the photic zone in a lake and the entire range of paleobiological archives directly or indirectly connected to benthic photosynthesis.

4. Vertical mixing and stratification of a lake's water column regulate the distribution of dissolved gases, especially oxygen, and nutrients on various temporal cycles. Those of seasonal or longer duration can be archived in sediments by differences in bioturbation and surface water productivity signals.

5. Waves affect lakes by generating turbulence, which can redistribute gases or nutrients and, in shallow water, locally rework bottom sediments.

6. Currents, driven by wind, waves, or gravity, redistribute suspended sediments over longer distances in lakes, at the surface, at intermediate depths, and at the lake floor. Both current and wave-induced motion of sediment-bearing water parcels are influenced by Coriolis effects.

4

The Chemical Environment of Lakes

Understanding lake chemistry is critical for correctly interpreting the geochemical archives of lake deposits. Elemental and isotopic distributions in lakes are closely linked to external climatic and watershed processes. Solute concentrations regulate the distribution of organisms, and the precipitation or dissolution of mineral phases. Both fossils and minerals leave sedimentary archives, and when we can interpret aspects of ancient water chemistry from these records we may be able to reconstruct paleoclimate or prior human activity around the lake. Interpretation of isotopic records likewise requires an initial understanding of their behavior in lakes and the links between this behavior and external factors such as rainfall or nutrient discharge.

Oxygen

Oxygen (O_2) is the second most abundant component of the atmosphere ($\sim 21\%$) after nitrogen ($\sim 78\%$). Paleolimnological indicators of oxygenation at the sediment–water interface may provide clues as to the nature and frequency of water-column mixing, water depth, the lake's trophic condition, and possibly climate. Thus, it is important to understand how oxygen is generated and consumed in lakes to properly interpret the oxygenation archives and filters (figure 4.1).

Oxygen is dissolved in lakes directly from air, and as a byproduct of photosynthesis by autotrophic organisms (multicellular plants, algae, and photosynthetic bacteria), greatly simplified as:

$$6CO_2 + 6H_2O + \text{sunlight} \rightarrow C_6H_{12}O_6 + 6O_2 \tag{4.1}$$

Photosynthesis is performed by both organisms on the lake floor (*phytobenthos*), and by floating algae and bacteria (*phytoplankton*). Molecular oxygen is lost from the water column through respiration, secondary oxidation of organic or inorganic compounds in the water column or sediments, or directly to the atmosphere following supersaturation.

Oxygen concentrations in the water column are a function of productivity, respiration, temperature, and mixing, all of which vary both diurnally and seasonally within a lake, and because of climatic or morphometric differences between lakes. To understand this variation, it is useful to return to our examples at Junius Ponds #5 and #7 (figure 4.2). In dimictic Junius Pond #7, the water column just below the ice is well oxygenated during the winter, but is anoxic at the lake floor. Because the lake is density-stratified during winter, the small quantities of oxygen that are generated from near-surface photosynthesis are not mixed into the hypolimnion. As the ice melts in the early spring and the lake mixes, oxygen is transmitted to the bottom. However, this mixing is increasingly inhibited as surface waters warm and stratification sets in. Despite complex interactions in the rate and location of maximum primary production in the epilimnion, during the summer months the bottom waters remain anoxic. Surface cooling and the deepening of the thermocline in late summer and fall is accompanied by a reintroduction of oxygen to deeper water, and by early November the second period of mixing reventilates the bottom of the lake with oxygen. The cycle is completed as the lake freezes over and stratification sets in again.

In meromictic Junius Pond #5 near-surface waters undergo a similar seasonal cycle. High concentrations of oxygen in surface waters during the winter are related to low temperatures (greater solubility) and low respiration, and decline as surface waters warm. However, permanent density stratification prevents mixing, and the monimolimnion remains anoxic throughout the year.

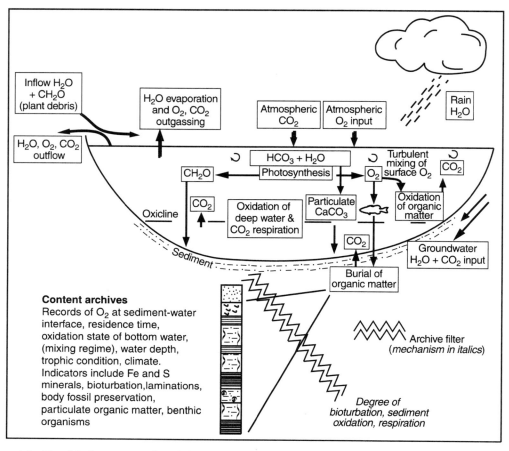

Figure 4.1. Simplified oxygen cycle in lakes and relationship/filters to sediment content archives of oxygenation at the sediment–water interface. Note that isotopic signals of oxygen and their filters are considered elsewhere. Information filter notation as in figure 3.1. Modeled after Talbot (1993).

Oxygen concentrations are also strongly influenced by differences in productivity or *trophic state*, which in turn are regulated by nutrient flux and light. In *eutrophic lakes*, those exhibiting high productivity and nutrient levels, O_2 is produced in large quantities in surface waters and readily mixed through the epilimnion. However, in the hypolimnion, oxygen regeneration cannot compensate for various types of consumption, and this zone becomes depleted in O_2 with the onset of summer stratification. Under conditions of extreme productivity and high water temperatures, even very shallow lakes can become anoxic throughout most of the water column, particularly where emergent plants are abundant, sheltering the water surface. In contrast, *oligotrophic lakes*, with low nutrient and productivity levels, may have oxygen profiles regulated primarily by physical mechanisms, particularly the lake's thermal profile.

This causes oxygen depletion in warmer surface waters and increasing O_2 concentrations with depth. Temporal changes in climatic factors, such as air temperature, or wind speed, as well as changes in nutrient inputs can cause lakes to undergo significant changes in their oxygenation profiles, even over short time periods. Oxygen profiles of small lakes can react very quickly to such changes, but even very large lakes can show pronounced changes over geologically brief intervals (Lehman et al., 1998).

Differences in morphometry and climate regulate oxygenation through their intermediate effects on mixing. Deep or sheltered tropical lakes are prone to meromixis or oligomixis, and typically display near-permanent deep-water anoxia. In contrast, elongate lakes exposed to substantial wind mixing are more likely to have oxygenated deep waters under comparable climatic regimes. Lakes

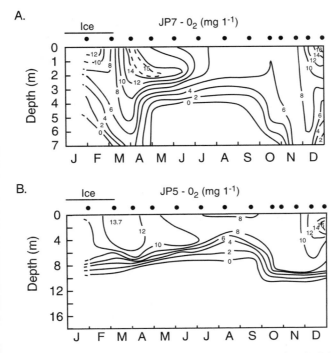

Figure 4.2. Seasonal variation in oxygen concentrations at the Junius Ponds. (A) Dimictic Junius Pond #7. (B) Meromictic Junius Pond #5. From Pendl and Stewart (1986).

with increased oxygen concentrations with depth during the summer are overwhelmingly restricted to very low productivity settings, for example at high altitudes. They may, however, retain dissolved oxygen throughout the water column during their period of ice cover, as a result of high solubility coupled with extremely low respiration rates. Because of their long stratification seasons, weaker geostrophic circulation, and high rates of respiration, tropical and subtropical lakes are subject to oxygen depletion about three times faster than comparable lakes in the temperate zone and as much as five times faster than in subpolar regions (W.M. Lewis, 1987).

Oxygen concentrations at the sediment–water interface vary greatly between lakes in response to hydroclimate. Oxygen penetration into the sediment both affects and is affected by bioturbation, and varies seasonally in lakes where the intensity of bioturbation is linked to seasonal population growth of burrowing organisms (Rippey and Jewson, 1982). Some climate-driven differences in oxygenation depths may exist where climate is linked to organic productivity. Lake muds in eutrophic lakes are typically oxygenated to depths of < 3 cm, both under tropical and temperate conditions (e.g., Rippey and Jewson, 1982; Cohen,

1984). In contrast, in oligotrophic lakes, with low rates of organic matter burial, bioturbation depths of 5–10 cm are common (Robbins and Edgington, 1975; Johansen and Robbins, 1977; Martin et al., 1998).

Reduction and Oxidation (Redox) Reactions and the Behavior of Fe and Mn

Reduction and oxidation reactions are those involving the transfer of electrons. An ion is oxidized when it loses an electron and reduced when it gains an electron. Electron transfer must occur involving a pair of ions, so an electron donor, known as the reducing agent (i.e., causing the reduction of the recipient) must interact with a recipient, the oxidizing agent. If an electrode is placed in lake water, it will capture free electrons, generating an electrical potential (voltage) in the process. The size of this potential depends on the capacity of the water to oxidize (positive potential) or reduce (negative potential). The measure of this voltage potential, standardized to a neutral pH, is referred to as E_H. Oxygen is only one of a number of important oxidizing agents present in lakes (for example,

Fe^{3+} or Mn^{4+}) and reduction can occur under lake conditions where free oxygen is present, if an appropriate reducing agent is available (Hamilton-Taylor and Davison, 1995).

The potential E_H regulates the solubility of numerous ions, and important ionic transformations occur as E_H thresholds are passed. Specific E_H values for redox transformations are also dependent on pH. The most important of these changes for the paleolimnologist are transformations that occur involving Fe, Mn, C, S, and N. Many other elements, while not directly involved in redox reactions, are nevertheless affected by their outcome, as they become adsorbed to, or coprecipitate with, Fe minerals, or are released from them upon dissolution. These include phosphorus, as well as a number of pollutant metals (e.g., Pb, Cu, Cd, Co, Cr, and Ni) of interest to paleolimnologists (e.g., Boyle, 2001a). Our discussion here will concentrate on Fe and Mn, since the redox behaviors of these elements are well understood.

The redox transformation of iron can be expressed as:

$$Fe^{2+}(\text{ferrous iron}) \Leftrightarrow Fe^{3+}(\text{ferric iron}) + e^- \quad (4.2)$$

Ferric iron is the insoluble, stable state of iron under oxidizing conditions, at high E_H. Under more reducing conditions iron undergoes a transition to soluble ferrous (Fe^{2+}) iron. Iron enters most lakes as iron oxide or hydroxide particles, which settle through the water column until they encounter a strong reducing agent, either anoxic water or organic matter. Ferrous iron can be quite abundant in anoxic, hypolimnetic waters of lakes, particularly near the sediment–water interface (Hamilton-Taylor and Davison, 1995) (figure 4.3). Therefore, factors that govern oxygenation of deep waters, such as mixing frequency or trophic state, also regulate the abundance of soluble iron. Meromictic or oligomictic lakes in particular tend to have high dissolved iron concentrations in their bottom waters, although this can be strongly affected by the availability of sulfur. At very low E_H values, sulfate (SO_4^{2-}) is reduced, allowing H_2S to form. If sufficient sulfur is available, highly insoluble iron sulfides, especially pyrite, can form in the water column, reducing the quantity of dissolved iron in the hypolimnion. Pyrite crystals formed within the water column are smaller and less variable in size than those formed during burial diagenesis, because of their short growth interval while suspended (Suits and Wilkin, 1998). This provides a means of discriminating pyrite formed in a reducing water column from pyrite formed under reducing burial conditions.

Seasonal mixing of Fe^{2+}-rich water with oxygenated water of the epilimnion results in the reoxidation of Fe^{2+} to Fe^{3+}, causing iron to reprecipitate. This secondary oxidation also occurs at the sediment–water interface, as sediments are ventilated by mixing, mediated by iron-oxidizing bacteria. It would seem likely that following burial in sediment below the depth that oxygen can reach through bioturbation, iron would be quickly reduced, with only insoluble Fe^{2+}-bearing minerals remaining.

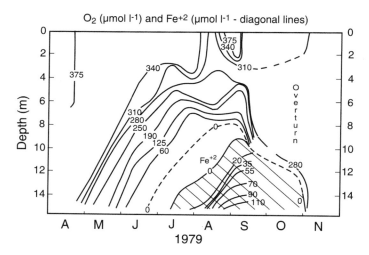

Figure 4.3. Seasonal variation with depth of dissolved O_2 and dissolved iron (Fe^{2+}—diagonal lines) at temperate, dimictic, Esthwaite Water, UK, illustrating solubility of iron as a function of redox condition. From Sholkovitz (1985).

However, this does not seem to be the case. Dissolution rates of Fe^{3+} oxides, while initially quite rapid under anoxic conditions, slow considerably after burial (Wersin et al., 1991). Iron oxides may persist in otherwise reducing sediments for as long as 1000 years.

Manganese redox reactions are similar to those of iron, except that they proceed more slowly, allowing manganese to diffuse over greater areas of the water column as it is being oxidized (Sholkovitz, 1985). Reduced Mn^{2+} is oxidized to Mn^{4+} under somewhat more oxidizing conditions than iron.

The extent to which Fe, Mn, or coprecipitated trace metals can be used to determine paleoredox conditions is complicated by the degree to which these various elements cycle back and forth across the sediment–water interface or undergo postburial diagenetic alteration. In lakes undergoing a transition from oligotrophic to eutrophic conditions, there is normally a major acceleration in the recycling of Fe from sediments to the water column. The response to this change recorded in subsequent sedimentation depends on both the external inputs of Fe as well as the amount being released from preexisting sediment.

In principle, the precipitation of Fe and Mn oxides might be used as an indication of episodic oxygenation of normally anoxic (Fe- and Mn-rich) deep waters, since these elements are concentrated by dissolution at depth. This could be used as a signal of oligomixis or the episodic movement of the metalimnion in a meromictic lake. Seasonally stratified lakes can also produce cyclical patterns of oxidized iron deposition, although subsequent bioturbation and sediment mixing may obliterate this signal. In lakes where the history of anoxia and trophic status is well known, it is often possible to interpret sedimentary records of Fe, Mn, and trace metals as a result of the interplay between surface oxidation state, organic matter input, and metal behavior (Schaller et al., 1997). Patterns of abrupt increases in sedimentary Fe have been observed in some lakes at the onset of eutrophication (Mackereth, 1966; Tracey et al., 1996). However, interpretation of metal cycling histories from Fe and other redox-sensitive metal profiles in lake sediments where there is no a priori knowledge of lake trophic or pH history should be done carefully, with consideration of other independent indicators of oxidation state.

Inorganic and Organic Carbon

Carbon occurs in various inorganic and organic forms in lakes, as both particulate and dissolved matter, and moves readily between these states through numerous reactions (figure 4.4). The carbon cycle of lakes expresses itself in sedimentary archives as a potential indicator of the rates of organic and inorganic carbon fixation in the water column, both of which may be related to organic productivity, as well as the paleo-pH of the lake.

Carbon dioxide comprises about 0.035% of the atmosphere, and is increasing by approximately 0.4% of that amount per year, largely as a result of the combustion of fossil fuels and net atmospheric CO_2 accumulation (Siegenthaler and Sarmiento, 1993). Despite its relatively small concentration in the atmosphere, carbon dioxide's high solubility in water makes it and its byproducts important components of lake water. Carbon dioxide plays a central role in photosynthesis, as well as in other key aspects of the carbon cycle of lakes. There is increasing interest in the global role that lakes may play as both sinks or reservoirs of carbon, because of the suspected relationship between atmospheric CO_2 content and climate change (e.g., Kling et al., 1991; Cole et al., 1994; W.E. Dean and Gorham, 1998). Furthermore, the ability of a lake to consume atmospheric carbon is closely linked to its internal biological processes (D.E. Schindler et al., 1997). Understanding the potential of paleolimnological archives to accurately record changes in atmospheric CO_2 requires an understanding of how carbon cycles through lake systems.

Upon dissolution in lake water, a small proportion of CO_2 hydrates to form carbonic acid:

$$H_2O + CO_2 \Leftrightarrow H_2CO_3 \qquad (4.3)$$

Two subsequent dissociations may occur, each of which yields a proton (H^+):

$$H_2CO_3 \Leftrightarrow H^+ + HCO_3^- \qquad (4.4)$$

and

$$HCO_3^- \Leftrightarrow H^+ + CO_3^{2-} \qquad (4.5)$$

Dissociation is pH-dependent, with carbon dioxide and carbonic acid dominating at low pH, bicarbonate at neutral pH, and carbonate at high pH (figure 4.5). Release or acceptance of protons in the reversible reactions above forms the basis for the primary buffering system (capacity to accept OH^- or H^+ ions without changing pH) in lakes where a buffer is readily available. *Alkalinity* is the measure of the buffering capacity of water, generally the total quantity of bases such as bicarbonate, carbonate, or hydroxide available to react with a strong acid. For most lakes, alkalinity is ultimately determined by bedrock composition and increases in areas underlain by limestones or other

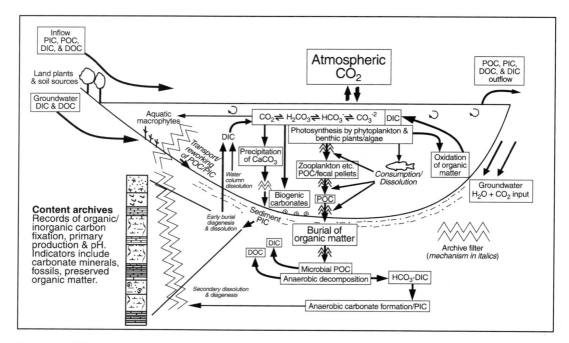

Figure 4.4. The inorganic and organic carbon cycle in lakes, illustrating major interactions, pathways, and relationship/filters to sediment content archives of organic and inorganic carbon fixation rates (primary and secondary productivity?) and pH. Here, DIC, DOC, PIC, POC, and OM are dissolved inorganic carbon, dissolved organic carbon, particulate inorganic carbon, particulate organic carbon, and organic matter, respectively. Isotopic archives are considered elsewhere. Modeled after Kelts (1988).

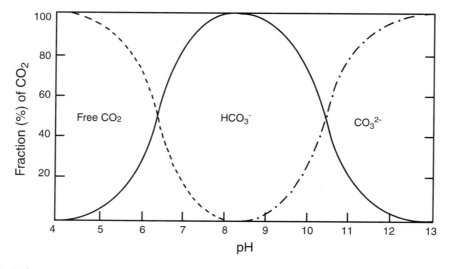

Figure 4.5. Relative proportions of CO_2 and its derivatives in relation to pH. Modified from Wetzel (1983).

carbonate rocks. Bacteria and plant decay in soils releases CO_2 into surface runoff, which when converted to carbonic acid can dissolve available Ca- or Mg-carbonate minerals present in the bedrock or soil, forming Ca- or Mg-bicarbonates:

$$CaCO_3 + H_2O + CO_2 \rightarrow Ca(HCO_3)_2 \qquad (4.6)$$

Subsequent dissolution produces Ca^{2+} and bicarbonate ions:

$$Ca(HCO_3)_2 \rightarrow Ca^{2+} + 2HCO_3^- \qquad (4.7)$$

In lake waters, reaction (4.6) is readily reversible and may be expressed as an equilibrium relationship. The direction in which the reaction is driven depends on the availability of CO_2. Processes that result in the addition of CO_2 result in the dissolution of $CaCO_3$ until equilibrium is re-established with $Ca(HCO_3)_2$, whereas removal of CO_2 has the opposite effect, causing precipitation of $CaCO_3$. Two primary mechanisms for causing removal of CO_2 and subsequent $CaCO_3$ precipitation are photosynthesis and CO_2 degassing, for example where supersaturated spring waters are discharged. In some lakes, temperature or salinity increases may also cause loss of CO_2 through lowered solubility and calcite precipitation. All of these mechanisms can be important sources for the accumulation of calcium carbonate minerals in lake sediments. In open water this may occur through the precipitation of microscopic crystals of calcite or aragonite. Lake waters often remain supersaturated with respect to calcium carbonate, and massive precipitation may not occur until a triggering mechanism develops, such as the presence of abundant nucleating surfaces provided by phytoplankton. However, increasing primary productivity in a lake does not translate into a simple increase in calcium carbonate accumulation, and in fact may result in just the opposite. Respiration and decomposition of organic matter are the primary means for adding CO_2 and causing $CaCO_3$ dissolution in lakes. Because increasing the deposition rate of organic matter through the hypolimnion results in an increase in CO_2 production, eventually accumulated calcite will be dissolved (W.E. Dean, 1999).

Poorly buffered lakes are subject to acidification whenever dissolution of reactants releases abundant H^+ ions. This occurs during CO_2 dissolution, forming carbonic acid, or in the case of input of common industrial pollutants, such as H_2S, which when oxidized converts to sulfuric acid. Acidification of lakes has profound biological consequences for lake ecosystems, frequently recorded in their fossil records.

Much of the carbon content of a lake is also in the form of organic carbon molecules including a wide variety of lipids, carbohydrates, and proteins, as well as humic substances formed *diagenetically* from the partial breakdown of organic compounds (Meyers and Ishiwatari, 1993). The origin, abundance, and cycling of this organic carbon load affects a variety of lake processes, for example benthic organism distribution or sediment redox reactions of interest to the paleolimnologist.

Carbon enters lakes in a variety of ways in addition to *autochthonous* (internal) dissolution of CO_2 and uptake through aquatic photosynthesis. Dissolved and particulate carbon, in organic and inorganic forms, also enters lakes from *allochthonous* or external sources. Allochthonous particulate and dissolved organic carbon (POC and DOC) is derived from the physical breakdown and chemical decay of terrestrial organisms. Dissolved organic matter in lakes is derived from the breakdown of autotrophic organisms within the lake, from allochthonous plant matter, and from excretion of organic matter by zooplankton (Søndegaard et al., 1995; Wetzel, 2000). The DOC concentrations from autochthonous sources often display complex seasonal variation in the epilimnia of temperate lakes, since reduced winter bacterial activity, increased zooplankton grazing, and increased summer photosynthesis can lead to increased DOC. The potential for DOC to affect paleolimnological records rests on its potential to contribute acidity to lakes and its relationship with water transparency, particularly in nonturbid, oligotrophic lakes (Rasmussen et al., 1989; Fee et al., 1996). Conditions that cause lowered DOC (for example, reductions in runoff or terrestrial vegetative cover, or increases in atmospheric CO_2) can also result in increased transparency and an expansion of the photic zone.

Particulate organic carbon only accounts for about 10% of the total organic carbon load of an average lake. However, its importance for paleolimnologists is disproportionately large, because different groups of organisms within a watershed or lake produce different organic compounds. Proportional differences between these compounds provide important clues to past community structure and change (Meyers and Ishiwatari, 1993). In most lakes, except very oligotrophic boreal lakes and small ponds in forests, the flux of autochthonous particulate organic carbon to lake sediment exceeds that of allochthonous POC.

Several factors affect the proportion and types of autochthonous POC in lake sediments. In shallow lakes, a high proportion of POC reaches the bottom of the lake, and is incorporated into sediments. In contrast, in deep lakes most POC is dissolved in the epilimnion. Therefore, paleolimnological interpretations of POC fluxes cannot be viewed as simple indications of primary production, but must also account for the probable residence time of POC within the epilimnion, which in turn is affected by particle size and density, and water viscosity. This helps explain the low total organic carbon concentrations observed in sediments from some deep, meromictic, tropical lakes (Katz, 1990). Even though primary production may be high, dissolution of organic matter prior to settling on the bottom as well as during early diagenesis is also high, limiting accumulation rates. In well-mixed lakes POC is remobilized and oxidatively altered by bioturbation and resuspension. In Lake Michigan, only about 2% of the organic carbon fixed in the epilimnion actually accumulates in the sediments (Eadie and Meyers, 1992).

Solute and Particulate Sources

Solutes and particulates enter lakes through rivers, groundwater discharge, gaseous exchange with the atmosphere, air-borne particulate matter, and precipitation. The relative importance of these sources depends on the location of a lake, its regional climatic setting and, particularly for the industrial era, its proximity to human activity. Rock weathering, including mechanical degradation and biochemical decay processes, provides the bulk of dissolved solids and particulates to most lakes. Over geological time scales, rates of weathering must ultimately be determined by the availability of fresh bedrock, which can only be generated through renewed tectonic uplift. On shorter time scales weathering rates and intensities are governed by bedrock type, climatic forcing, and human activities, all of which affect vegetation and soil types and the potential for solute and particulate transport away from the site of weathering.

Weathering occurs through the aqueous reaction of organic or atmospheric acids with minerals, releasing a combination of alteration product particulates and dissolved ions. For example, weathering of the common silicate feldspar, albite, in the presence of soil-borne acids, yields particulate clay (kaolinite) and dissolved Si and Na:

$$\underbrace{2NaAlSi_3O_8}_{\text{albite}} + 2H^+ + 9H_2O \rightarrow \underbrace{Al_2Si_2O_5(OH_4)}_{\text{kaolinite}}$$
$$+ 2Na^+ + \underbrace{4H_4SiO_4}_{\text{silicilic acid}} \qquad (4.8)$$

The rate at which this and similar hydrolysis weathering reactions actually occur in nature is greatly accelerated by vegetation.

Bedrock composition determines the solutes generated during weathering. For example, in regions of limestone ($CaCO_3$) bedrock, Ca^{2+} and HCO_3^- ions will usually dominate surface and groundwater chemistry. In addition to solutes, the weathering of iron-bearing silicate minerals generates insoluble iron oxides or hydroxides under atmospheric conditions, often bound to other particulate matter. Relatively nonreactive minerals, such as quartz, are released during weathering as particles, and transported mechanically.

Very soluble products of weathering, particularly the monovalent metallic ions, are easily flushed out of their sites of formation. Therefore sodium and potassium salts generally only accumulate in soils under extremely arid conditions, with very limited surface or groundwater flow. In contrast, the common, less-soluble divalent cations often accumulate in soils of semiarid to semihumid regions. In very humid, tropical climates all soluble cations tend to be leached from the soil and carried off by groundwater, leaving behind as residues very insoluble byproducts of weathering, such as Fe-hydroxides or kaolinite.

In addition to weathering, industrial or agricultural runoff also supply high concentrations of nutrients (N and P) and metals to lakes in many parts of the world today. Atmospheric precipitation is also important, particularly for sea spray entering coastal lakes, and in regions subject to industrially produced acidic precipitation. Atmospheric particulates include dust, charred particles in regions with extensive forest burning, coal use, or cooking fires, and industrially produced soot around major urban areas.

Major Ions, Salinity, and Residence Times

Global compilations of inland water composition show that a relatively small number of ions comprise the bulk of the dissolved solid content of dilute rivers that enter lakes. These include the cations Na^+, K^+, Ca^{2+}, and Mg^{2+}, and the anions Cl^-, HCO_3^-, CO_3^{2-}, and SO_4^{2-}. Na^+, K^+, and Cl^- are derived primarily from the oceans (Drever, 1997). Salt is carried into the atmosphere from the ocean surface as sea spray and carried over the continents, where it can be redeposited as dust or dissolved in precipitation. In the interiors of continents this marine source is supplemented or replaced by direct weathering of Na^+- and K^+-bearing silicate minerals, especially feldspars. Under pre-industrial conditions the primary source of SO_4^{2-} was also oceanic, or locally through the solution of marine sulfate deposits, but discharges from the burning of sulfur-bearing fossil fuels are now of equal or greater importance for many areas. Mg^{2+} and Ca^{2+} are derived from a combination of marine sources and bedrock weathering, especially of limestone, dolomite, or mafic igneous rocks. HCO_3^- and CO_3^{2-} are also derived from the weathering of limestones.

Salinity is the total concentration of all dissolved ions, expressed as milligrams per liter. Total salinity in lakes is controlled by the net accumulation of salts from all sources, minus salts lost through outflow or mineralization. As a general rule, hydrologically open lakes do not build up to extremely high salinities, except for unusual circumstances where lakes directly overlie pre-existing evaporite deposits. Although moderately saline, open basin lakes of considerable size do exist (e.g., Lake Manitoba, Canada; Last, 1982), they are also normally fed by relatively saline influents, which in turn form under strongly evaporative conditions. In most open-basin lakes, outflow over time eliminates solutes in approximate equilibrium with average dissolved solid concentrations in the lake water. Under such circumstances most dissolved solids will remain in the lake water for relatively brief intervals of time, determined in large part by the size of the lake and its through-flow rate. In closed basins, dissolved solid loads can build up over time, increasing salinity. Changing salinity in closed-basin lakes is of particular interest to paleolimnologists because hydrological closure, the buildup of salinity, and the deposition of salinity-sensitive fossils or evaporite minerals are all closely linked to precipitation + runoff: evaporation ratios, and ultimately, climate.

For any solute, it is possible to define a *residence time* (R), an average duration that any given molecule of that substance will remain in the water column:

$$R = m/r \qquad (4.9)$$

where m = mass of solute in solution and r = rate of removal from lake (by sedimentation, evaporation, outflow, etc.) (mass/unit time).

R is only meaningful under near equilibrium conditions, that is, when there is an approximate balance between inputs and outputs over the time interval of interest. Long residence times are characteristic of substances that occur in great abundance and/or are retained in the water column for

long periods owing to their lack of reactivity in important limnological and metabolic processes. Dissolved ions of this type are referred to as being *conservative* when they concentrate in the water mass through long intervals of evaporation and authigenic mineral accumulation without themselves being lost from solution. In contrast, short residence time substances cycle through the lake system quickly. Lake volume, and the inflow and outflow rates of water also play an important role in defining residence times. For example, consider the residence time of H_2O, the primary constituent of all lake waters, given a very large lake holding a large mass of water. If the inflows and outflows are small relative to the lake volume, R_{H_2O} will be very long, perhaps decades or centuries. A similar-sized lake with much larger inflowing and outflowing streams will have a much shorter R_{H_2O}, on the order of days to months.

All of the major ions are essential elements for metabolism and growth of organisms, representing alternative pathways for their consumption aside from the inorganic processes of hydrological outflow or mineral precipitation. Several (Cl^-, Na^+, K^+) almost always occur in abundances exceeding those required for growth, even under quite dilute conditions. In contrast, Ca^{2+} and $HCO_3^- + CO_3^{2-}$ are required in abundance by organisms that produce calcium carbonate skeletons, and under certain circumstances, even in relatively freshwater, low availability of these ions becomes a *limiting factor* for growth for paleolimnologically important organisms such as mollusks or ostracodes. Likewise, low abundances of SO_4^{2-} will serve to limit the productivity of various sulfate-reducing bacteria inhabiting the sediment, whose former presence is recorded by the production of diagnostic sulfide minerals and sedimentary concretions. Far more attention has been paid in paleolimnology to the behavior of major ions under highly concentrated conditions than in dilute freshwater, simply because the depositional signals of saline lakes (evaporite mineral suites, diagnostic fossils) are much more readily interpreted.

Evaporative Concentration and Other Fractionation Mechanisms for Major Ions

The geochemical and biological conservatism of Cl^-, Na^+, and K^+ make them excellent comparators of what happens to freshwater as it is evaporatively concentrated into a *brine*, and as numerous types of salts are deposited from that brine. The bewildering variety of evaporite salts found in lake deposits begs for simplifying models to explain the order of their precipitation. Fortunately, evaporation leads to selective and partially predictable sequences of evaporite mineral precipitation, dependent on the original elemental composition of the dilute water, the solubilities of the ionic species present, and the kinetics of precipitation rates.

The concept of tracing *brine evolution* derives from studies of the fate of spring and lake waters during evaporative concentration (Garrels and Mackenzie, 1967). Both the concentration (salinity) and the composition of rivers and the closed-basin lakes into which they flow differ dramatically (Hardie and Eugster, 1970; Eugster and Hardie, 1978; Eugster and Jones, 1979). Triangular compositional plots can be used to display the relative abundances of major anions and cations in dilute natural inflow waters (figure 4.6A)

and the concentrated brines of their associated closed-basin lakes (figure 4.6B). For readers unfamiliar with these diagrams, a data point falling precisely on the SO_4^{2-} corner indicates water in which sulfate is the only major ion present. Alternatively, a point falling midway between the Mg^{2+} and Ca^{2+} corners indicates water in which 50% of the major cations present are Ca^{2+}, 50% are Mg^{2+}, and Na^+ and K^+ are absent. In this data set, dilute influent waters are in most cases dominated by CO_3^{2-}, HCO_3^-, and Ca^{2+}. Sulfate-dominated rivers are restricted to those flowing off sediments (chiefly sulfate-bearing marine deposits), and there are no chloride-dominated rivers (typical of coastal areas). In contrast, the example's lake brines are dominated by Cl^-, SO_4^{2-}, Na^+, and K^+, and Ca^{2+}, Mg^{2+}, CO_3^{2-}, and HCO_3^- are all minor constituents or entirely absent. How has this happened?

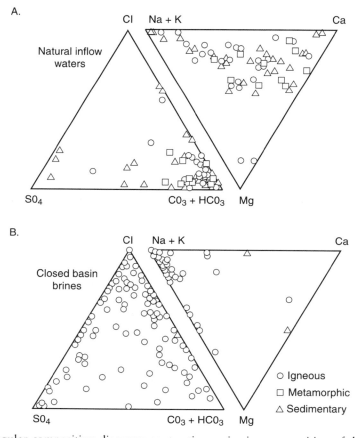

Figure 4.6. Triangular composition diagrams contrasting major ion composition of dilute inflow waters (A) with continental closed-basin brines (B). Symbols indicate drainage basin lithology. The predominance of igneous bedrock watersheds surrounding closed basins is an artifact of sampling. From Eugster and Jones (1979).

In most natural waters, the early stages of mineral precipitation involve the formation of relatively insoluble calcium and magnesium carbonates. Because equilibrium precipitation of $(Ca,Mg)CO_3$ dictates the consumption of equivalent molar proportions of $Ca^{2+} + Mg^{2+}$ and HCO_3^-, the initial proportional concentrations of these ions will regulate their relative depletion rates from the solution as precipitation proceeds. Consider the precipitation of calcite from a freshwater precursor in which SO_4^{2-}, Mg^{2+} and Na^+ are proportionately abundant, HCO_3^- is more abundant than Ca^{2+}, and Cl^- is a minor constituent (figure 4.7). Once all available calcium is consumed in the formation of calcite, the remaining brine will still have residual alkalinity $(CO_3^{2-} + HCO_3^-)$ and, as a result of evaporative enrichment of the remaining brine, the alkalinity will be much higher than the dilute precursor. Other, more abundant and more conservative ions, such as SO_4^{2-}, Mg^{2+}, or Na^+, are not consumed in precipitation of salts until extremely

high concentrations are reached, when highly soluble Na–Mg sulfates begin to form. This accounts for the plateau in concentrations for these ions. Within the range of concentrations illustrated, neither K^+ nor Cl^- attain sufficient concentrations to become supersaturated, and they behave conservatively as evaporative concentration rises.

Early calcium carbonate precipitation is extremely common in continental waters, frequently occurring prior to any significant evaporative concentration. Initial water composition coupled with early calcite precipitation determines the future possible pathways or branches along which subsequent *brine evolution* can occur. In cases where $HCO_3^- + CO_3^{2-} \gg Ca^{2+}$, the brine will be Ca + Mg-free and alkaline. Such a brine cannot yield Ca-bearing evaporite minerals unless there is some subsequent input of Ca-rich fluid. In figure 4.8 this *alkaline* pathway is labeled I. Different initial proportions of $HCO_3^-:Ca^{2+}$ will yield very different brines and potential evaporite minerals. An excess of initial Ca will produce a *hard-water* pathway, labeled II. Subsequent branching points in the Hardie–Eugster model represent alternative pathways based on the consumption and relative availability of progressively more soluble salts. Ultimately, highly concentrated brines from these various pathways will yield very different suites of evaporite minerals.

The brine evolution model provides an extremely useful model for understanding brine evolution and interpreting evaporite histories in lake deposits. Nevertheless, paleolimnologists should be aware of its limitations. The common uptake of cations by lake-floor clay minerals violates some of the relatively simple branch point assumptions of equivalent molar uptake of divalent cations and anions. Other authigenic Na and Mg silicates also form in lakes and may act as intermediate sinks for cations, particularly in alkaline environments (e.g., Darragi and Tardy, 1987). Secondary dissolution of lake margin salt crusts may reintroduce salts into the residual brine in very different proportions than the salts were originally precipitated, owing to localized effects of rainfall or runoff. In cases where high partial pressures of CO_2 occur in groundwater, the precipitation of early-forming calcite may be suppressed, leading to a different set of early pathways than those indicated by Hardie and Eugster (Herczeg and Lyons, 1991). The activity of sulfate-reducing bacteria, particularly in meromictic lakes or organic-rich muds, accelerates the loss of this anion from the water column faster than predicted by inorganic thermodynamics alone. Sulfate reduction not only eliminates SO_4^{2-} faster

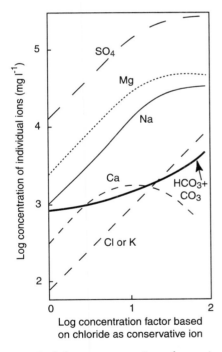

Figure 4.7. Solute concentrations for major ions (SO_4, Mg, etc.) as functions of total concentration, using Cl^- as the conservative tracer of concentration in an Mg–SO_4 pathway (Type IIIB brine of figure 4.8). Note the depletion of Ca and eventual plateauing of other ions with increasing concentration. Adapted from Nesbitt (1974) and Eugster and Hardie (1978).

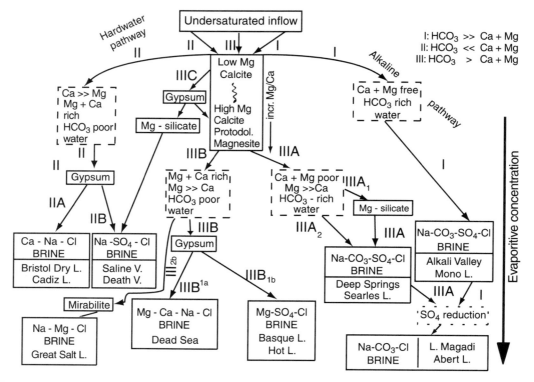

Figure 4.8. Brine evolution pathways, based on the original composition of influent waters that undergo subsequent evaporative concentration. From Eugster and Hardie (1978).

than expected, but it also introduces a new source of alkalinity:

$$8SO_4^{2-} + 2Fe_2O_3 + 15C + 7H_2O \rightarrow 4FeS_2$$
$$+ 14HCO_3^- + CO_3^{2-} \quad (4.10)$$

The extent to which this process can alter sulfate and carbonate concentrations is dependent not only on sulfate availability and reducing conditions, but also a ready source of organic carbon (Lyons et al., 1994). Finally, the very solubility of lacustrine evaporites makes them extremely susceptible to diagenetic alteration and secondary precipitation under very different groundwater conditions, obscuring the original depositional signal.

Brine evolution leaves a paleolimnological record both directly through the formation of evaporite minerals, and indirectly through its impact on species distributions and ecological communities (preserved as body fossils, organic residues characteristic of certain species, or isotopic variation). Because of the limitations evident in interpreting evaporites, fossil data are often invaluable in understanding lake chemical history. Many readily preserved species of organisms respond not merely

to total salinity of a water body, but also to its composition.

Nutrients

Nutrients are substances critical to the wide array of metabolic processes in organisms. Numerous elements and simple compounds are critical for the growth of lacustrine autotrophs. However, only a few of these substances are present in such small concentrations in lake waters, relative to their requirements, as to potentially limit the growth of cells and populations. The most important of these *limiting nutrients* in lakes are P, N, Si, and Fe. Phosphorus and nitrogen are essential for an extraordinary array of functions, such as formation of nucleic acids, energy transfer in photosynthesis, and protein synthesis. Silicon (in the form of silica, or SiO_2) is of particular importance to the growth of diatoms and other siliceous algae. Because of their critical role in the metabolism and reproduction of lacustrine organisms, nutrients cycle rapidly through lake systems, passing through food webs and between organic and inorganic states on a regular basis.

An increasing supply of a limiting nutrient (a *nutrient load*) can have a fertilizing effect on autotrophs, causing extremely rapid growth in populations. This supply may come from sources internal to the lake, for example in the upwelling of hypolimnetic waters into the photic zone, or externally, from the atmosphere, or as river-borne solutes and particles. Strictly speaking, nutrient shortages act to limit the growth of particular species assimilating that nutrient, since each species has unique requirements. Over time, the species composition of a lake may change in response to varying availability of nutrients. It has been known for a long time that phosphorus is the predominant limiting nutrient for productivity in mesic temperate lakes (Dillon and Rigler, 1974; D.W. Schindler, 1974; Vollenweider, 1976). Phosphorus may serve as an "ultimate" regulator of productivity, because all organisms require it and because of its relative scarcity. However, lakes in the tropics, subtropics, and arid regions of the world frequently show stronger dependence of productivity on nitrogen content, and at times on silica or iron as well (L.A. Baker et al., 1981; Cowell and Dawes, 1991; Lebo et al., 1992). Complex seasonal or longer-term variations in nutrient depletion patterns have been observed in

many warm-water lakes, and these changes in turn appear to restructure phytoplankton communities. These observations illustrate the fallacy of thinking of a single nutrient as being uniformly "limiting" for a given lake.

Nutrient fluxes and their effects on lake communities are of interest to the paleolimnologist because they leave both direct records, as sedimented minerals or in organic matter, and indirect records, through changes in fossil algal assemblages, reflecting changing nutrient loads. Accurate interpretation of these records requires a careful consideration of the sources of nutrients, and their behavior in various parts of the water column and surficial lake sediments.

Phosphorus

Phosphorus is derived naturally from the weathering of phosphate-bearing igneous and sedimentary rocks (figure 4.9). Phosphorus makes up only about 0.1% of the earth's crust, mostly in the form of apatite minerals, such as $Ca_5(PO_4)_3OH$ or $Ca_5(PO_4)_3F$. Weathering releases these minerals, which may be carried directly to lakes, or secondarily concentrated in sediments, as phosphorite deposits and then reweathered. Also of considerable

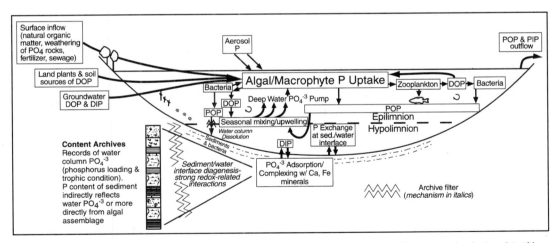

Figure 4.9. The phosphorus cycle in lakes, illustrating major interaction pathways and relationship/filters to sediment content archives of phosphorus loading (and perhaps productivity). Note that the highly reactive nature of phosphates at the sediment–water interface make the direct interpretation of P-concentrations in sediment extremely difficult, except in cases of extremely high loadings of P (generally anthropogenic), or in deep meromictic lakes, where bottom water accumulation of P is effectively decoupled from surface processes. However, alternatives such as diatom analysis may provide this information indirectly. Here POP, PIP, DOP, and DIP refer to particulate organic phosphorus, particulate inorganic (mineralized) phosphorus, dissolved organic phosphorus (in various molecular and colloidal forms), and dissolved inorganic phosphorus, respectively.

interest are anthropogenic sources of phosphates, especially those derived from the application of agricultural fertilizers and from urban and agricultural sewage.

Most dissolved phosphorus in lakes occurs as orthophosphate (PO_4^{3-}). Soluble phosphate also occurs as low molecular weight compounds, and as high molecular weight colloids. These compounds are produced both as excretions from cells, or through decay of organic matter in the water column and sediment, which releases soluble reactive phosphorus (SRP) into the water. Particulate phosphates include both organic forms and those bound in mineralized phases. The latter include both *detrital* phosphate mineral grains derived from weathering and *authigenic* mineral phosphates formed in the lake, mostly Ca- and Fe-phosphates. Phosphates are readily adsorbed to particulate iron hydroxides and $CaCO_3$, of considerable importance for its interactions at the sediment–water interface.

Phosphate can become depleted in the epilimnion during the growing season because of high demand by autotrophs, which remove available SRP during growth. In more productive lakes, especially during stratification, there is a tendency toward increasing phosphate concentrations with depth. At these times surface water primary production may reduce levels of available phosphates to negligible amounts in the upper water column. Hypolimnetic or mixolimnetic waters frequently act as zones of phosphate accumulation, both from settling organic matter that is no longer within the zone of algal assimilation, or from the flux of SRP from lake sediments. These same deep waters can periodically "pump" phosphate back to the epilimnion, through complete or partial mixing (Hecky et al., 1996).

The trend toward increased SRP with depth is well exemplified by a wide variety of temperate lakes, especially those of northeastern North America and northern Europe during the summer, but is often a permanent feature of meromictic tropical lakes as well. The trend becomes less marked during periods of lowered nutrient demand in surficial waters, for example, in winter under ice, as light, rather than nutrients, becomes a limiting factor for growth, and largely disappears during whole-lake mixing. In less productive, oligotrophic lakes, phosphorus concentrations also show less vertical or seasonal variation.

Under different conditions of seasonal mixing, or during periods of climate change, other patterns of seasonal variation in phosphorus may occur. In warm monomictic lakes, primary productivity may

be quite high during winter months, depleting SRP from the water column even as the lake is mixing (Porter et al., 1996). In such lakes, seasonal stream discharge events may play a much more important role in determining annual phosphorus cycles. Changing climatic conditions may lead to very complex responses in phosphorus concentrations and seasonality, however, since a number of factors (temperature, runoff, community structure) all have an effect on nutrient uptake and may vary independently (D.W. Schindler et al., 1996).

Phosphorus loading is of particular concern in many parts of the world, where heavy phosphorus discharges from agricultural, sewage, and industrial sources have led to serious lake eutrophication problems. Given the tremendous interest in problems of anthropogenic lake eutrophication, as well as the longer-term effects of climate change on lake productivity, it is critical that paleolimnologists develop methods of assessing water column P dynamics from sedimentary archives. Unfortunately, obtaining such records from phosphorus profiles in sediments alone is problematical. Lake-bottom sediments can act as both sources or sinks for phosphorus relative to the overlying lake waters, depending on redox and pH conditions, and sedimentary Fe concentrations (Baccini, 1985; Jensen et al., 1992; Eckert et al., 1997; Olila and Reddy, 1997). Phosphates are readily adsorbed onto Fe hydroxides. Under reducing conditions, Fe^{3+} is reduced, releasing adsorbed phosphate into pore waters. Therefore, meromictic and oligomictic lake sediments, or those of other lakes during stratification seasons, can discharge phosphorus to the deep waters of a lake. Higher pH also leads to a release of phosphate and possibly reprecipitation as P-minerals, or adsorption onto $CaCO_3$. Microbial activity in the sediment also affects SRP cycling because microbial uptake of phosphorus prevents SRP from being released into overlying waters under oxic conditions, but allows its release under anoxic ones. Phosphorus can move from pore waters into the water column through mixing by bioturbation, as well as by diffusion. However, two opposing effects complicate the net impact of bioturbators. On the one hand, burrowing ventilates the sediment, resulting in more microbial activity and more rapid breakdown of organic matter, releasing more mineralized P into pore waters. This same ventilation, however, increases the precipitation of Fe, thereby increasing adsorption of P on newly oxidized sediment around the burrows (Hansen et al., 1998). Because of these various interactions, long-term phosphorus accumulation in sediments only occurs in proportion to P abundance or loading in the water column under special circumstances.

Nitrogen

Nitrogen, as N_2, is the most abundant component of the atmosphere. Nitrogen is also an essential nutrient for amino and nucleic acid synthesis and many other functions in all living organisms. However, the strong covalent bonding of gaseous N_2 makes it relatively inert, and it must be biologically reduced or fixed by various bacteria and blue–green algae (cyanobacteria) from N_2 into the NH_3 form, which can be assimilated by other organisms. Nitrogen fixation occurs terrestrially and in water, and therefore both watershed and lake sources of fixed nitrogen are archived in sediments.

Following fixation, nitrogen can be passed from phytoplankton through the food chain, primarily to zooplankton, and then to higher levels (figure 4.10). At each stage, nitrogenous waste is created from dead cells and waste products, in both particulate and dissolved forms. Particulates may either dissolve during settling or be sedimented on the lake floor. Fixed nitrogen also enters lakes from external sources, such as rivers, precipitation, and aerosols. Some of the N-bearing particulate organic matter is sedimented, and although no longer involved in limnological processes, nevertheless figures in analysis and interpretation of nitrogen content and isotopes in paleolimnological studies. The relative importance to a lake's N-budget of external sources

versus internal fixation varies greatly. Similarly, N-fixation rates vary over several orders of magnitude between lakes and can change rapidly as nitrogen-fixing cyanobacteria become more or less abundant (D.W. Schindler, 1977; Howarth et al., 1988).

Nitrogen undergoes a number of important reactions within lakes with implications for paleolimnological interpretation. As bacteria break down organic matter they generate ammonia (NH_4^+) as a byproduct, the principal form of N-uptake by algae in most lakes. Nitrogen also undergoes reactions to more oxidized states, referred to as *nitrification*, and to more reduced states, called *denitrification*. Nitrification involves the oxidation of ammonia (NH_3 gas or aqueous NH_4^+), through a series of intermediary, bacterially mediated steps to nitrate (NO_3):

$$NH_4^+ + 2O_2 \rightarrow NO_3^- + H_2O + 2H^+ \quad (4.11)$$

Nitrate can subsequently be reduced to ammonia and reassimilated by plants and algae. For this reason ammonia concentrations in the photic zone are low in most lakes. An alternative fate for NO_3^- is *denitrification*, the reduction of nitrate to various N gases, primarily N_2, resulting in the loss of useable nitrogen from lakes (Seitzinger, 1988). Bacteria accomplish this as they oxidize organic matter, generally under low oxygen or anoxic conditions. In well-mixed lakes sediments are also a major source of nitrogen. Some of this is carried to the upper water column, in the forms of NH_4^+ and NO_3^-, where it can be secondarily fixed or assimilated

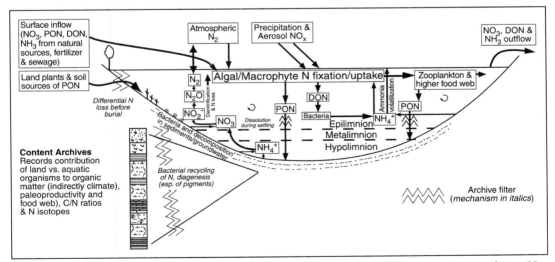

Figure 4.10. The nitrogen cycle in lakes, showing major interactions and filters to content archives. Here PON and DON are particulate organic nitrogen and dissolved organic nitrogen, respectively. Modeled after Hecky et al. (1996).

by phytoplankton or higher plants. An alternative fate is for nitrate to be subsequently reduced to N_2 in the water column along the pathway $NO_2 \rightarrow NO \rightarrow N_2O \rightarrow N_2$, and then either lost into the atmosphere, or fixed again by cyanobacteria.

Ammonification is a particularly important process in deep water in the nitrogen cycle of stratified lakes. In highly alkaline lakes, much of the ammonia produced in this way is volatilized and lost into the atmosphere (Talbot and Johannessen, 1992). In contrast, nitrate is a more common sedimentary byproduct within the oxidizing zone of lakes.

Nitrogen signals enter the paleolimnological record in ways that give us clues to both terrestrial watershed conditions and internal N-cycling. Nitrogen concentrations differ considerably between terrestrial versus lacustrine plants, particularly when expressed as C/N ratios (Premuzic et al., 1982; Meyers and Ishiwatari, 1993). Terrestrial plants with woody tissue and cellulose have relatively low concentrations of N (high C/N ratios), whereas N-concentrations in phytoplankton (which lack cellulose, but are rich in N-compounds) are much higher. Therefore, an important application of C/N ratio studies in lake deposits is the determination of the relative importance of contributions of terrestrial versus lacustrine (primarily phytoplankton) organic matter. The precise cutoff values of C/N for differentiating terrestrial from lake-derived organic matter are controversial. There is a general consensus that values < 10 indicate a phytoplanktonic origin. Some authors suggest that values > 20 clearly indicate a dominantly terrestrial origin, or accumulation in very unproductive, oligotrophic lakes (W.E. Dean et al., 1993; Meyers and Ishiwatari, 1995). However, phytoplankton growing in extremely N-deficient environments can have C/N ratios > 14.6 (Hecky et al., 1993). Furthermore, burial of organic matter is frequently followed by rapid loss of N, elevating C/N ratios by up to 30–40%. C/N ratios > 20 have been reported for sediments from the central parts of Lake Victoria, organic matter that is certainly derived from phytoplankton sources (Talbot and Lærdal, 2000). Information filters affecting this interpretation arise because organic matter from terrestrial and lake sources is not deposited uniformly following transport and settling. Terrestrial material may be concentrated or phytoplankton dissolved prior to settling.

A dominance of N-fixation as a means of supplying nitrogen to a lake ecosystem may be archived in lake sediments indirectly by the abundance of chemical fossils such as cellular pigments indicative of cyanobacteria (Leavitt et al., 1994b). The nitrogen-stable isotopic composition of sedimented organic matter in lakes is also affected by processes in the nitrogen cycle, discussed later in this chapter (Talbot, 2001).

Silica

Silicon in lakes is derived from the weathering of silicate minerals, the most common class of minerals in the earth's lithosphere. Dissolved silicon forms during the hydrolysis of silicates, especially the dissolution of feldspar minerals, to release silicic acid [H_4SiO_4, see equation (4.8)], and is carried into lakes either by surface flow or groundwater. Silicon in this form may be converted in lakes to particulate form through the growth of opaline silica (SiO_2)-secreting organisms, primarily diatoms, or through precipitation of authigenic silicate minerals. Particulate silicates also enter lakes in abundance as mineral weathering products like quartz, feldspars, or micas. Studies of these minerals are useful to paleolimnologists investigating the *provenance* (geological source of origin) and weathering histories of terrigenous components. Finally, silica can adsorb onto organic matter, clay mineral, and metallic hydroxide particles.

Silicon is an essential nutrient for the growth of diatoms and several other groups of lacustrine organisms, and its consumption, depletion, and cycling in lakes are closely tied to mixing and the productivity cycles of diatoms. Depletion of silica by diatom growth may be accompanied by changes in the phytoplankton flora of a lake, toward diatom species with lowered requirements for silica, or away from diatoms altogether (Kilham, 1971). As diatoms die, they sink and/or redissolve in the water column. Dissolution rates in the water column decline as the concentration approaches saturation, so most dissolution occurs near the surface and decreases toward the lake floor and in the sediments. A lack of secondary consumption below the mixed layer causes most meromictic lakes to have relatively high concentrations of silica in their monimolimnia, and their sediments are effective sinks for silica accumulation (Michard et al., 1994; Hecky et al., 1996). In large, deep lakes, the vast majority of settling diatoms are redissolved, and this silica will become available for renewed consumption by diatoms if it is retained in, or mixed into, the upper water column (e.g., T.C. Johnson and Hecky, 1988; Shafer and Armstrong, 1988).

At low or intermediate pH conditions, the accumulation of silica in sediments is overwhelmingly biogenic, because of the relatively low solubility of silicic acid. At high pH (> 9), however, inorganic precipitation of authigenic silicates assumes much greater importance. Under such conditions, typical of alkaline, closed-basin lakes, silicic acid dissociates, forming different species of silica such as $H_3SiO_4^-$ and $H_2SiO_4^{2-}$, whose solubility rises rapidly with increasing pH (Eugster and Jones, 1979). Slight declines in pH, or cooling temperatures around thermal springs in such silica-charged waters, can then lead to rapid precipitation of authigenic silicate minerals (B.F. Jones et al., 1967; Renaut and Tiercelin, 1994).

Stable Isotopes and Isotopic Fractionation

Isotopes are variants of elements that differ in their number of neutrons but that have identical numbers of protons and electrons. For example ^{12}C, the common form of carbon, has six neutrons and six protons, giving it a mass of 12, whereas the less abundant isotope ^{13}C contains seven neutrons and six protons. *Stable* isotopes are those which do not undergo radioactive decay over time, whereas *radiogenic* isotopes eventually decay to some other isotope or element, through spontaneous emission of α (= helium core) particles or β decay.

The different masses of two isotopes of the same element cause them to behave differently in any physicochemical or biochemical process that is affected by mass. For example, molecules of a substance containing the light isotope of an element can move more rapidly in a gaseous form than the same molecule containing the heavier isotope. Although isotopic variants can undergo the same chemical reactions because of their similar electrochemical properties, the rates at which reactions occur varies between isotopes. Physical processes such as evaporation, or biological ones like photosynthesis, can lead to the enrichment of one isotopic form over another in the product being formed and relative depletion of that same isotope in the elemental fraction left behind in the source material. For example, an element with two isotopes occurring in a liquid undergoing evaporation will display an enrichment of its lighter isotope in the gas, and a depletion in the remaining liquid, whereas the pattern of enrichment and depletion for the heavier isotope in the gas and liquid will be the reverse. This process of enrichment and depletion is referred to as *fractionation*. In addition to this type of physical fractionation, chemical fractionation occurs

because chemical bonds involving heavier isotopes are stronger than those involving lighter isotopes.

If a fixed amount of two isotopes undergoes a fractionation process, the preferential consumption of the lighter isotope, for example, causes the residual pool from which the isotopes are being drawn (referred to as the *substrate*) to become enriched in the heavy isotope. This means that subsequent fractionation is starting from a somewhat heavier isotopic pool, causing the new fractionation products to be slightly heavier than those formed earlier. Over time, as the substrate is completely consumed the average isotopic composition of the product approaches the original isotopic ratio. This process is called *Rayleigh distillation*. However, in the real world the isotopic products may be dispersed unequally, for example over some geographical range. Rayleigh distillation is a simplifying model that is not always attained precisely in nature, because from the observer's viewpoint, isotopic systems are rarely closed. However, it provides a useful conceptual model for understanding long-term trends in isotope depletion and enrichment processes.

In this section, we will consider processes of stable isotope fractionation affecting the lacustrine environment, both those occurring immediately within the lake and its surroundings. These processes are often strongly linked to environmental factors of great interest to the paleolimnologist, such as evaporation rates, terrestrial vegetation cover, or community productivity. For this reason, the ratios of important elemental isotopes (particularly of C, O, and N) comprise some of the most important content archives for paleolimnology.

Fractionation and δ Notation

The extent of fractionation in a physicochemical process can be expressed by comparison of the end products to the ratio of the two isotopes in a standard. For example, in the case of oxygen isotopes, standards include the ratio of $^{18}O/^{16}O$ in *Standard Mean Ocean Water* (SMOW) or their ratio in well-characterized reference materials. It is difficult to measure absolute quantities of isotopes in the laboratory; instead isotopic ratios are measured relative to materials with known ratios, and expressed as the extent to which an element in one sample has been isotopically fractionated with respect to the standard, through use of the δ value (‰ = per mil) :

$$\delta‰ = [(R_{sample} - R_{standard})/R_{standard}] \times 1000$$

$$(4.12)$$

Multiplication by 1000 is a convenience introduced because of the relatively small isotopic fractionation differences observed for most elements.

So, for example:

$$\delta^{18}O‰ = \{[(^{18}O/^{16}O)_{sample} - (^{18}O/^{16}O)_{standard}]/ \\ (^{18}O/^{16}O)_{standard}\} \times 1000 \qquad (4.13)$$

In addition to SMOW and (since its consumption) Vienna Standard Mean Ocean Water (V-SMOW), the other commonly used reference standard has been powdered $CaCO_3$ from the fossil *Belemnitella americana* from the Cretaceous Pee Dee Formation (hence, PDB), and its secondary standard V-PDB. SMOW values can be related to PDB values by the formula (Coplen et al., 1983):

$$\delta^{18}O_{V-SMOW} = 1.03091\delta^{18}O_{PDB} + 30.91 \qquad (4.14)$$

V-SMOW is used for hydrogen and oxygen isotopes, generally for water and other noncarbonate samples, and PDB is used in reference to both oxygen and carbon ($^{13}C/^{12}C$) isotopes in carbonate sediment or fossil samples. Stable isotope measurements of nitrogen ($^{15}N/^{14}N$) are referenced against the standard composition of air.

The isotopic composition of any sediment is a function of both the isotopic composition of the original parent material from which the substance was formed and the degree of fractionation involved in the formative process. For any formative reaction, whether in equilibrium or irreversible, an isotopic fractionation factor α can be defined by the proportion of an isotope in the reaction products. For example, during $CaCO_3$ precipitation from water, the dominant source of oxygen isotopes is the water itself. So we can define a fractionation factor describing this relationship between some calcium carbonate mineral formed and the host water as:

$$\alpha_{(CaCO_3-H_2O)} = (\delta^{18}O_{CaCO_3} + 1000)/ \\ (\delta^{18}O_{H_2O} + 1000) \qquad (4.15)$$

The α value can be thought of as an expression of the expected per mil (δ) difference between the reactant and the product. For example an α value can be defined for $^{18}O/^{16}O$ ratios during the evaporation of liquid water (w) to water vapor (v) in air:

$$\alpha_{w-v}{}^{18}O = (^{18}O_w/^{16}O_w)/(^{18}O_v/^{16}O_v) = 1.0115 \qquad (4.16)$$

The value 1.0115 implies that this reaction will result in an $\sim 11.5‰$ difference in the ratios of liquid and vapor with liquid water retaining a larger proportion of ^{18}O, assuming no other changes are involved.

Fractionation processes that occur during chemical reactions are temperature-dependent, a fact that suggested one of the first potential uses of stable isotopes in paleoenvironmental reconstruction, that of determining paleotemperatures from fossil material or authigenic minerals (Urey, 1947). However, a wide variety of both inorganic equilibrium processes and biochemical kinetic processes also cause fractionation.

Oxygen and Hydrogen Isotopes in the Hydrosphere

Oxygen has three stable isotopes, ^{16}O, ^{17}O, and ^{18}O, of which only ^{16}O and ^{18}O are normally measured (i.e., no additional information useful to paleolimnologists would be gained from analysis of ^{17}O). The two stable isotopes of hydrogen, ^1H and ^2H (generally labeled D, for deuterium) are discussed here in conjunction with oxygen isotopes because of their joint importance for understanding the hydrological cycle controls on stable isotopic fractionation in modern lakes. These hydrological relationships form the basis for some of the most common applications of oxygen isotopic analysis in paleolimnology: interpreting the hydrological history of precipitation sources, evaporation:precipitation ratios, and relative importance of surface or groundwater sources for paleolakes.

The most common paleolimnological analyses of oxygen stable isotopes use calcium carbonate from authigenic minerals or fossils, so it is useful to concentrate our discussion on fractionation processes for calcium carbonate minerals. Fractionation of oxygen during the formation of calcium carbonate in water is primarily dependent on temperature. Empirically:

$$1000 \ln \alpha_{CaCO_3-water} = a \times 10^6 T^{-2} - b \qquad (4.17)$$

where $a = 2.559$ for aragonite and 2.78 for calcite, T = temperature (C), and $b = 0.715$ for aragonite and 2.89 for calcite (Friedman and O'Neil, 1977; Grossman and Ku, 1986).

Using this relationship it would be possible to calculate a reaction temperature from a given calcite sample, if the starting isotopic composition of the water were known, but in paleolimnology this is almost never the case. Lake waters are highly variable in $\delta^{18}O$, in contrast with the oceans. Thus, the isotopic composition of a carbonate sample is a function of both the ambient temperature at the time of its formation and the prior isotopic history of the host water.

To understand this history, we must look outside the lake, to the sources of water feeding it, from atmospheric precipitation, surface runoff, and groundwater. The primary fractionation mechanisms for oxygen and hydrogen isotopes in surficial waters are evaporation and precipitation. Because the oceans are by far the largest reservoir of water on the planet, they are also the primary source of atmospheric moisture. Furthermore, evaporation is most intense at low latitudes, and warm air can hold more moisture, making the equatorial oceans the dominant source of this atmospheric water. Following its evaporative enrichment of ^{16}O and ^{1}H, water vapor is transported to higher latitudes or toward the interiors of continents and higher altitudes, where it undergoes a process that approximates Rayleigh distillation. The condensation process favors the heavier isotopes (^{18}O and D), and the removal of liquid water from the air mass as rainfall leaves the residual water vapor depleted in these components (i.e., more negative

in $\delta^{18}O$ and δD ratios) (Dansgard, 1964; Grootes, 1993). Subsequent precipitation will be even more negative through this process, as is shown in a transect of isotopic composition of precipitation inland from the Pacific Ocean toward the Canadian Rocky Mountains (figure 4.11). As a result, precipitation in both polar regions and the central regions of continents contains substantially less ^{18}O and D than equatorial and coastal precipitation. On a global level, there is a strong correlation between surface air temperatures and $\delta^{18}O$ (Rozanski et al., 1993).

Although reduced amounts of moisture are transported to far inland or polar regions, the extreme distillation processes involved lead to precipitation highly depleted in ^{18}O. These general trends are reflected in a strong global covariance between $\delta^{18}O$ and δD in precipitation, referred to as the *Meteoric Water Line* (Craig, 1961), that can be defined as approximately:

$$\delta D = 8\delta^{18}O + 10 \qquad (4.18)$$

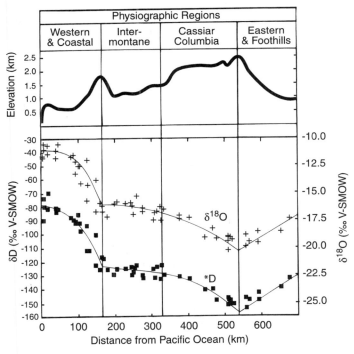

Figure 4.11. The evolution of $\delta^{18}O$ and δD in precipitation across the high-relief continental margin of the western Canadian Cordillera, from the Pacific Ocean (Western & Coastal) to the interior plains of Alberta (Eastern & Foothills). Decreasing temperature with distance and altitude along the trajectory from the Pacific coast toward the east drives rainout and depletion of $\delta^{18}O$ and δD, leaving residual atmospheric vapor (and precipitation produced from that vapor) isotopically light (increasingly negative $\delta^{18}O$ and δD). A similar trend from eastern sources of moisture moving west reaches its limit at the eastern end of the profile. From Yonge et al. (1989).

On the regional scale, the slope and intercept of this equation vary, depending on evaporation and precipitation conditions. Arid regions generally display shallower slopes than do more humid areas (B.F. Jones et al., 1994). Isotopic composition of precipitation in a given area is not constant, however, since the sources and conditions of precipitation vary throughout the year. To the extent that such source changes are predictable they produce characteristic seasonal changes in isotopic composition of precipitation (Rasmussen, 1968; Merlivat and Jouzel, 1979; Gat et al., 1994). In areas undergoing systematic changes in the source of precipitation over longer time periods, shifts occur in the isotopic composition of rainfall that may be evident in paleolimnological records (McKenzie and Hollander; 1993).

Aside from precipitation that falls directly on a lake's surface, the $\delta^{18}O$ and δD of water in lakes are affected by three important processes or inputs:

1. *Isotopic fractionation that occurs from the time that meteoric water falls in a drainage basin until it reaches the lake.* This results from evaporation over the days to months required for surface runoff to reach a lake. The extent of fractionation that occurs during this stage is determined by such factors as permeability of soils, local geomorphic gradients, and relative humidity (e.g., Dowd et al., 1992).

2. *Subsequent fractionation of the water mass in the lake* (Gat, 1981). This fractionation involves evaporation from the lake's surface, and is strongly dependent on the residence time of water in the lake, and interactions between evaporating water and the humidity already present in the air mass above the lake's surface. Such processes can occur over much longer time periods (and frequently result in much more significant fractionation) than #1. In lakes that are connected as a series of downstream basins, this can result in progressive fractionation in each lake and river step (Yang et al., 1996).

3. *Groundwater inputs.* Sometimes groundwater is a major component of the lake's isotopic budget (Krabbenhoft et al., 1990). Some groundwater entering a lake is of meteoric origin, basically reflecting the same origins and processes as #1 above, but deeper groundwater sources, like hydrothermal springs, may impart water with isotopic compositions completely unrelated to the recent history of the surficial watershed or its climate.

Fractionation resulting from evaporation leaves the residual runoff or lake water enriched in ^{18}O and D. Because of the longer residence times of water in lakes versus overland flow, it is the evaporative history of a lake itself that typically has the greatest postprecipitation impact on the isotopic composition of the lake water. In hydrologically open lakes, especially in climate regimes where evaporation is limited relative to surficial outflow, the isotopic composition of lake waters may be quite similar to that of its source water inflow (Stuiver, 1968; Talbot, 1990). Under these conditions, the isotopic composition of lake water may be primarily a signal of precipitation temperature (Yurtsever, 1975).

Evaporative enrichment is most evident in the isotopic composition of closed-basin lakes, particularly those found in deserts, or at low latitudes where evaporation rates are high (Talbot, 1990). Under these conditions annual–decadal changes in lake hydrology are often reflected in oxygen isotope trends for lake water (Benson et al., 1996). However, even in open-basin lakes the isotopic composition of lake water can be highly modified by evaporation over time if the proportion of water lost through the lake's outlet is small relative to evaporative loss. The isotopic composition of a lake is also strongly influenced by the residence time of the water mass itself, because the water mass over time can become increasingly fractionated through evaporation.

Archiving Oxygen Isotopic Change

During calcite and aragonite formation there is a series of oxygen isotope fractionations between carbonate, dissolved CO_2, and water. Because the exchange of oxygen atoms is very rapid and water is an overwhelmingly larger reservoir of oxygen than dissolved CO_2 for this process, the oxygen isotope composition of water must regulate the oxygen isotope composition of authigenic carbonates. If we could assume that the $\delta^{18}O_{calcite}$ that we measured, for example in a shell, formed in direct equilibrium with $\delta^{18}O_{water}$, we would have a tool for inferring paleo $\delta^{18}O_{water}$, and from that, information on local water residence time, precipitation/ evaporation ratio conditions and perhaps, paleotemperature. This assumption underlies most applications of oxygen isotope studies to understanding climate from lake deposits.

To what extent is this assumption reasonable? For some carbonates, such as "inorganically" formed precipitates of the upper water column, or carbonates formed inorganically at the lake floor, the assumption seems sound. Carbonates formed by benthic microbial activity also appear to form in near-equilibrium to lake water for oxygen isotopes (Chafetz et al., 1991). However, the relationship between carbonate isotopic composition and water isotopic composition for benthic invertebrates, like ostracode crustaceans and mollusks, is somewhat more complex. Isotopic analyses of live-collected ostracodes and bivalves from modern lakes are often consistent with the measured range of temperatures and $\delta^{18}O_{water}$ from which they were collected (figure 4.12) (Dettman et al., 1995). However, in experimental studies, geochemists have observed systematic offsets between the observed $\delta^{18}O_{shell}$ and what would be predicted on the basis of starting $\delta^{18}O_{water}$ and water temperature alone. These so-called *vital effects* appear to be species-specific, and may relate to differences in oxygen uptake kinetics in the construction of shell material (McConnaughey, 1989a,b; Von Grafenstein et al., 1999). The second part of the assumption, that $\delta^{18}O_{lake\ water}$ primarily reflects atmospheric/lake interactions, must also be treated cautiously, given the possible influence of local groundwater

sources, particularly on benthic carbonates (Palmer et al., 1998).

Hydrogen Isotopes

Because of the large proportional mass difference between ^{1}H and D, these isotopes are subject to extremely large degrees of fractionation. The δD composition of H-bearing compounds is dependent upon temperature effects on fractionation, coupled with the original isotopic composition of the hydrogen source, which for the purposes of paleolimnology is invariably water. In principle, δD_{water} is calculable from $\delta D_{algal\ cell\ walls}$ (Meyers and Lallier-Vergès, 1999). Given the limited range of temperature-related fractionation effects that exist in lakes, this provides a tool for inferring atmospheric precipitation sources and in situ fractionation of lake waters through evaporation, with greater precision than using oxygen isotopes. To date, attempts to make use of hydrogen isotopes in paleolimnology have been limited by the uncertainty as to whether sample materials have been effectively closed to OH exchange with surrounding pore waters. However, some very promising results have been obtained on a variety of organic and inorganic sediments, especially lipids from phytoplankton, terrestrially derived plant cellulose, and

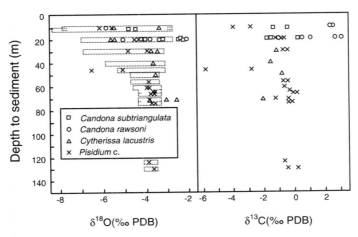

Figure 4.12. $\delta^{18}O$ measurements on living ostracodes and bivalves (*Pisidium* spp.) from Lake Huron (symboled data points), and their relationship with predicted $\delta^{18}O$ ranges for low-magnesian calcite formed in equilibrium with measured temperature and $\delta^{18}O_{water}$ ranges (outlined bars). Most data points fall within the equilibrium range, although the precise environmental conditions of T and $\delta^{18}O$ that any given shell formed in are unknown. From Dettman et al. (1995).

insect skeletons (Edwards, 1993; Schimmelmann et al., 1993; Krishnamurthy et al., 1995).

Carbon Isotopes

Two common stable isotopes of carbon exist on earth, ^{12}C and ^{13}C, the former much more common. The complexity of carbon pathways in lakes (figure 4.6) does not allow for the type of simplifying assumptions about fractionation pathways that we have seen applied for oxygen isotopes. Carbon enters lakes in many forms, and is also preserved as both organic matter and as inorganic carbonate minerals. Whereas both forms can be the source of isotopic information, that information records different sets of processes. Unlike oxygen, no single carbon reservoir dominates a lake's carbon budget to the extent that its isotopic properties will be mirrored by the lake as a whole.

A variety of important fractionation processes are known to affect carbon isotopes in lakes (figure 4.13). Prior to its accumulation as organic matter entering lakes, the isotopic composition of terrestrially derived plant debris already reflects strong fractionation mechanisms associated with photosynthesis. During photosynthesis terrestrial plants preferentially uptake ^{12}C from the atmosphere, causing terrestrial organic matter to have negative $\delta^{13}C$ values compared to the V-PDB standard. Several photosynthetic pathways exist, which differ in their diffusion rates of CO_2 to the chloroplasts, and in their use of different enzymes to drive carboxylation reactions. These differences also result in different degrees of C-isotope fractionation. The C_3 photosynthetic pathway is typical of most trees, shrubs, and temperate–cold climate grasses, and yields $\delta^{13}C$ values of about -25 to $-32‰$ (Cerling and Quade, 1993). The C_4 photosynthetic pathway (also referred to as the Hatch–Slack pathway) is typical of tropical climate grasses and sedges, and produces $\delta^{13}C$ values of -10 to $-14‰$. It is less energy-efficient than the C_3 pathway, but uses less water, and is therefore favored in water-limited conditions with high evapotranspiration rates, and in conditions of limited atmospheric CO_2 availability (Street-Perrott et al., 1997). A third type of photosynthesis, involving crassulacean acid metabolism (CAM), is used by a variety of succulent plants, generally in arid and semiarid regions. It produces $\delta^{13}C$ values of -10 to $-20‰$. The differences in C-isotope fractionation produced by these mechanisms

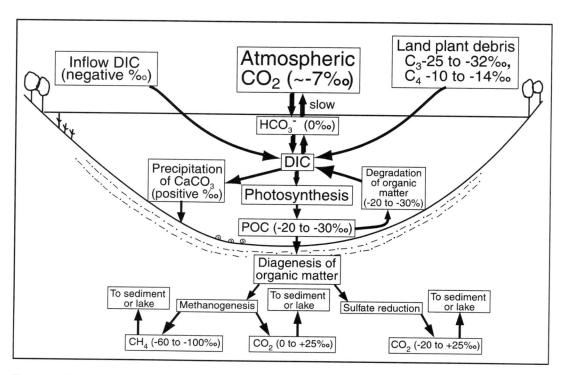

Figure 4.13. Major fractionation pathways for carbon isotopes in lakes. Modeled after Talbot (1993).

(especially the C_3 vs. C_4 contrast) are sufficient to allow the precursor plant type to be identified from isotopic analysis of organic matter. This is of particular interest to paleolimnologists because of the broad climatic significance that can be attached to organic matter primarily generated by C_3 plants (cooler and wetter, or higher CO_2 availability) versus C_4 plants (warmer and drier, and/or lower CO_2).

In addition to external inputs of organic matter, photosynthesis and organic production also occurs in the lake itself, using dissolved CO_2 or, to a lesser extent, HCO_3^-. This also results in strong fractionation (less so when HCO_3^- serves as a source of carbon), the formation of ^{13}C-depleted organic matter from phytoplankton, and a residual ^{13}C-enriched dissolved inorganic carbon pool from which authigenic carbonates may form. The extent of this photosynthetic fractionation and its impact on carbon isotopes is dependent on both the CO_2 concentration of the lake water, and the rate of photosynthesis (Herczeg and Fairbanks, 1987; Hinga et al., 1994).

Values of $\delta^{13}C$ in organic matter derived from phytoplankton are very negative (extremes of -12 to $-40‰$, but more typically -25 to $-30‰$), and cannot normally be easily distinguished from C_3-derived organic matter based on C-isotopes alone (Meyers and Ishiwatari, 1995). However, in alkaline lakes, where HCO_3^- is a more important source of carbon for photosynthesis, the reduced extent of fractionation and more positive carbon isotopic values of organic matter ($\sim -12‰$) may make phytoplankton distinguishable from carbon from C_3 terrestrial plants (Hollander, 1989; Talbot and Johannessen, 1992).

The effects of this production-driven fractionation on water column $\delta^{13}C$ can be seen in figure 4.14 (McKenzie, 1985). Lake Greifen, a small temperate-climate lake in northeastern Switzerland, experiences a winter and early spring period of low primary production, when the ^{13}C isotopic composition of DIC in the water column is relatively uniform. With the onset of late spring and summer algal blooms, the production of organic matter with very negative $\delta^{13}C$ causes surface water DIC to be depleted in ^{12}C, leading to more positive $\delta^{13}C$ values. As the lake undergoes summer stratification, a strong difference develops between the isotopic composition of the epilimnion, where primary production is continuing, and the hypolimnion, dominated by respiration. In the epilimnion ^{13}C-enriched DIC is incorporated in calcite, whose production rate is tied to epilimnetic productivity, producing a potential inorganic carbon archive of increasing primary production. Simultaneously,

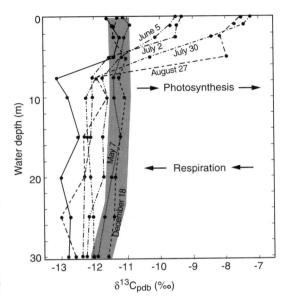

Figure 4.14. Monthly depth variation in $\delta^{13}C_{DIC}$ in Lake Greifen, Switzerland, showing effects of productivity in the epilimnion and organic matter degradation and respiration in deep water. The shaded areas represent the range of $\delta^{13}C_{DIC}$ found between December and May. The ^{13}C increase in surface waters due to photosynthesis and the ^{13}C depletion in deep water resulting from the respiration of the sinking organic matter are depicted for the summer months from May to September. From McKenzie (1985).

much of the settling particulate organic matter dissolves in the hypolimnion, releasing an excess of ^{12}C back into the DIC pool (i.e., producing more negative $\delta^{13}C$ values).

Over long time periods removal of ^{12}C from the surface waters of a lake can result in a trend toward heavier $\delta^{13}C$ values in lacustrine organic matter (Hinga et al., 1994; Talbot and Lærdal, 2000). This could happen as a result of a long-term increase in productivity in a meromictic or poorly mixed lake. In well-mixed lakes, however, the regular upwelling of previously deposited and isotopically light organic matter may prevent such trends.

As with oxygen isotopes, the long residence times of closed basins can also lead to isotopic enrichment of ^{13}C in the DIC pool (Talbot and Johannessen, 1992). This complicates the discrimination of the effects of productivity changes versus aridity from the analysis of $\delta^{13}C_{carbonate}$ alone, since both will favor the enrichment of ^{13}C. For this reason many geochemists advocate, where possible, the combined analysis of $\delta^{13}C$ from both the organic and inorganic fractions of sediment and its interpretation in conjunction with other analyses.

Variable mixing with surface waters and additional fractionation processes occurring on or near the lake floor complicate the interpretation of carbon isotopes in calcite produced in benthic environments. In well-mixed, shallow water, algal carbonates are likely to grow in equilibrium with lake water DIC. But benthic invertebrates grow under the influence of seasonal inputs of both dissolving particulate organic matter and calcite, and are subject to vital effects in carbon assimilation. The carbon fraction of skeletal carbonate in mollusks and ostracodes may be partially derived from food carbon, not necessarily in equilibrium with the local DIC pool (McKenzie, 1985; Tanaka et al., 1986; Von Grafenstein et al., 1999). And sediment diagenesis, including methanogenesis and sulfate reduction, can produce strongly fractionated DIC pools near the sediment–water interface, making the interpretation of C-isotopes from carbonate cements particularly complex (e.g., Talbot and Kelts, 1986).

Covariance of Oxygen and Carbon Isotopes

Carbon and oxygen isotopic data can both be derived from carbonate sediment samples from a lake, and it is possible to plot the results of such analyses for a single lake to examine long-term trends in such data. Geochemists have noted the common tendency for such oxygen and carbon isotopic values to covary along defined trends (figure 4.15). Talbot (1990) has noted that $\delta^{18}O$–$\delta^{13}C$ covariance is particularly characteristic of closed basin lakes, and has argued that it might be possible to recognize hydrologically closed lakes paleolimnologically on the basis of such trends. A single lake may be characterized by a relatively invariant slope to this trend, possibly the result of the morphological characteristics of the basin, for example, how broad or narrow the lake's evaporative surface is for a given lake volume. Open basins, in contrast, seem to show little variation in oxygen isotopes, reflecting their limited evaporative fractionation, coupled with variable ranges for carbon isotopes, the latter a consequence of the extent of seasonal productivity fluctuations.

Talbot's model of covariance sparked much controversy over the meaning of these trends, in part

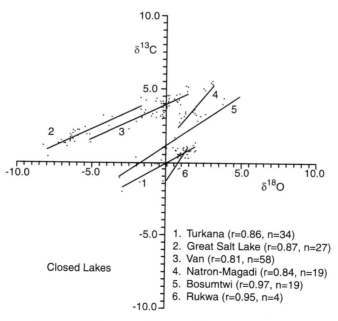

Figure 4.15. Isotopic covariance trends between $\delta^{18}O$ and $\delta^{13}C$ for several closed-basin lakes. From Talbot (1990).

because the mechanism for sustaining such covariance is uncertain, and a number of exceptions to the closed lake/open lake generalization have been noted. In open-basin lakes of the upper Midwest (United States), covariance is established by a combination of seasonal difference in precipitation sources and productivity (Drummond et al., 1995). In contrast, at Mono Lake (California), a hyperalkaline, closed-basin, covariance is evident over long time scales (> 5000 years), but is less apparent at the decadal to century time scale (Li and Ku, 1997). Under conditions where extremely high DIC levels exist (as in hyperalkaline systems), it may be very difficult to modify the existing $\delta^{13}C$ of the water mass simply by increasing stream discharge and raising water levels. Given our present state of knowledge it is probably prudent to use stable isotope covariance trends as a possible indicator, but not a foolproof one, of basin closure.

Nitrogen Isotopes

Two stable isotopes of nitrogen, ^{14}N and ^{15}N, occur, the former making up 99.6% of the total. Like carbon, organic and inorganic nitrogen is introduced to lakes from a variety of terrestrial and atmospheric sources. Over the last two centuries fossil fuel consumption, fertilizer application, and fuel wood burning/deforestation have greatly increased this flux worldwide. All of these sources have initial isotopic ratios that may be archived directly in lake deposits by sedimented terrestrial organic matter, or further fractionated by lacustrine processes. Rain and snow have dissolved $N-\delta^{15}N$ of ~ -5 to $+ 5‰$, relative to atmospheric N_2, depending on the sources of N in the precipitation (Talbot, 2001). Typical $\delta^{15}N$ compositions for C_3 land plants are $\sim +1‰$, whereas cyanobacterial phytoplankton have values of ~ 0, and other phytoplankton (non-N_2 assimilators) are typically between +2 to + 14‰. Changes in the relative proportions of these components are often reflected in sedimentary profiles of plant material from terrestrial versus aquatic sources (Pang and Nriagu, 1977; Meyers and Ishiwatari, 1995; Meyers et al., 1998). Soil organic matter is generally somewhat enriched in $\delta^{15}N$ ($\sim +5‰$) relative to average C_3 plants.

The numerous state changes that nitrogen undergoes in lakes generate numerous points in the lacustrine nitrogen cycle where fractionation can occur (table 4.1). Most of these processes, like the preferential assimilation of ^{14}N during the uptake of NH_4^+ or NO_3^- and the subsequent fixation of N in algal cells, lead to an enrichment of ^{14}N in the reaction product, and an accumulation of increasing quantities of ^{15}N in the residual dissolved inorganic nitrogen pool.

Because fixed nitrogen is required for growth in most groups of phytoplankton, strong fractionation often accompanies high rates of primary productiv-

Table 4.1. Major lacustrine processes of nitrogen isotope fractionation discussed in this chapter. Modified from Talbot (2001), after Collister and Hayes (1991)

Reaction Type	Predominant Location in Water Column	Type of Process (I, inorganic, equilibrium factors; B, biochemical, kinetic factors)	Fractionation Factor $(\alpha)^a$
Nitrogen dissolution	Air/lake interface	I	1.00085
Ammonium assimilation/ cellular N uptake	Eukaryotic algal cells within the epilimnion	B	0.993–1.013
Nitrate assimilation/cellular N uptake	Eukaryotic algal cells within the epilimnion	B	1.011–1.023
Direct nitrogen (N_2) fixation	Prokaryotic algal cells (primarily cyanobacteria) within the epilimnion	B	0.996–1.0024
Bacterial denitrification	Bacterial process, primarily in the anoxic zone or anaerobic sediments	B	1.02
Nitrification	Bacterial process, aerobic, but often at low O_2 concentrations	B	1.02
Ammonia volatilization	From sediments and water column	I	1.034

[a]Value greater than 1.0 indicates an enrichment of ^{14}N in the reaction product.

ity. In dimictic lakes this occurs as nutrients are recirculated into the epilimnion during the spring or early summer, causing $\delta^{15}N$ of organic matter, primarily phytoplankton remains, to decline. Nitrogen isotopes are also fractionated during consumption by animals, resulting in a $3 - 4‰\delta^{15}N$ enrichment at each trophic level (Minigawa and Wada, 1984; Peterson and Fry, 1987). Both seasonal bursts of primary productivity, and the dominance of settling animal remains when primary productivity declines, are evident from settling organic matter in lakes collected in sediment traps (Hodell and Schelske, 1998) (figure 4.16). These changes in N-isotopes provide a potential paleoarchive of productivity and food-web structure. There are, however, several very important caveats to this general relationship between productivity and nitrogen isotope fractionation. Little or no fractionation occurs during nitrogen fixation by cyanobacteria. Switches to cyanobacteria-dominated phytoplankton communities are often marked by declining ^{15}N concentrations, but in nitrogen-limited systems an isotopic trend toward more positive $\delta^{15}N$ values cannot be expected to directly result from high productivity alone. Also, decay processes lead toward isotopic enrichment of ^{15}N in organic matter, and when these processes are very important they may counteract the seasonal influence of primary production in the upper water column (Ostrom et al., 1998).

Other processes also affect nitrogen isotopic ratios in lakes, and potentially, sedimented organic matter. Nitrification and denitrification reactions that affect the state of dissolved inorganic nitrogen in the water column cause an enrichment of ^{15}N in the dissolved inorganic nitrogen, particularly for waters within the hypolimnion or monimolimnion (Yoshioka et al., 1988). The volatilization of ammonia involves strong fractionation and may serve (along with denitrification) as a major pathway for loss of ^{14}N from a lake. Because ammonia volatilization is correlated with pH (i.e., rate increases with increasing pH), alkaline lake systems are particularly prone to isotopic modification through this mechanism (Talbot and Johannessen, 1992).

Over longer time periods, fractionation can produce an overall ^{15}N-enriched source of nitrogen from which nutrients can be drawn by algae and bacteria. As a result, long-term trends toward more positive $\delta^{15}N$ values in lacustrine organic matter are often interpreted as indicators of a rise in lacustrine productivity, especially when accompanied by other consistent indicators of the same. This is most evident in lakes with an expanding nutrient supply, for example under conditions of anthropogenic discharge of nutrients or as a result of long-term increases in terrestrial productivity and nutrient discharge to a lake.

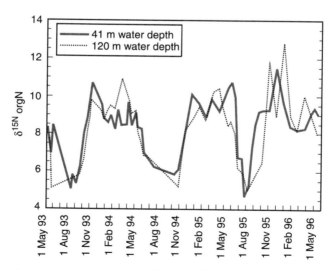

Figure 4.16. Nitrogen isotopic composition of sediment collected in two sediment traps over a three-year period from the eastern (Rochester) basin of Lake Ontario. Both traps were moored in 145 m water depth, with one placed 41 m below the surface and the other 25 m above the bottom (120 m). The data show an enrichment in ^{14}N (more negative $\delta^{15}N$) in organic detritus settling during summer, following the spring bloom, and a rise in $\delta^{15}N$ as heterotrophs become more common in fall and winter. From Hodell and Schelske (1998).

Summary

Our conclusions from this chapter can be summarized as follows:

1. Geochemical archives form a major source of information about lake history for paleolimnologists. Understanding gas, solute, and isotopic behavior, and their relationship to both internal limnological processes, and external climate and watershed characteristics, is essential for accurate interpretation of geochemical archives.

2. Oxygen content and redox conditions in the water column are closely tied to lake mixing, stratification cycles, and productivity. Oxygenation conditions affect the distribution of benthic organisms, bioturbation, and the formation or dissolution of redox-sensitive, authigenic minerals.

3. Carbon cycles through lakes in complex pathways, involving particulate and dissolved forms, and organic and inorganic forms. Carbon is primarily preserved in sediments as particulate organic matter and carbonate minerals, which collectively provide information about productivity and pH conditions in the lake, and external inputs of organic matter from the watershed.

4. Major ions enter lakes from weathering or atmospheric deposition. Their residence times are determined by lake volume, and the balance between inflow and outflow rates, both functions of climate. Major ion composition and concentration of dilute inflow plays a major role in determining subsequent brine evolution and the types of authigenic minerals formed in saline lakes.

5. The availability and cycling of commonly limiting nutrients (N, P, Si, and Fe) is determined by watershed characteristics and internal mixing processes, and in turn regulates the primary productivity of lakes. Nutrient availability over the past two centuries has been strongly influenced by anthropogenic loading, particularly of P and N. Records of nutrient loads and cycling can come from fossils of sensitive organisms, or sediment geochemistry.

6. Stable isotopes of oxygen, hydrogen, carbon, and nitrogen provide some of the most informative sources of geochemical information. Fractionation processes for O and H are closely linked to regional climate parameters, especially the characteristics of local precipitation, and residence time and evaporation/precipitation ratios of the lake waters. The C and N isotopes of lake sediments record both relative proportions of terrestrial to lacustrine organic matter and the internal cycling of these elements as a response to photosynthesis, secondary consumption, and respiration.

5

The Biological Environment of Lakes

Biological processes form the basis for a rich source of information for paleolimnologists. Populations of organisms are sensitive to variations in their external environment, and this sensitivity can be recorded as proportional changes in fossil abundances, evolutionary change, or extinction. Variations in lake temperature or water chemistry below the threshold of geochemical archives would normally go unrecorded in lake deposits were it not for fossils capable of registering these changes. Biotic systems are also the most complex components of lake systems, involving numerous species, their interactions with each other, and with their external environment. As a result, the interpretation of lacustrine fossil records is rarely straightforward, and must be viewed in the context of complex ecological dynamics, unfolding against a background of environmental and evolutionary change. In this chapter we will consider the biotic structure of lakes from a paleolimnological perspective, focusing on organisms and ecological interactions likely to be preserved in a lake's fossil record.

Lake Habitat Zonation

A transect running downslope and offshore from the shoreline will almost invariably reveal a change in habitat and lake organisms (see figure 3.2). In the shallow, littoral zone, high rates of photosynthesis can normally be supported, as light is not a limiting factor for growth. A high diversity of autotrophic and *heterotrophic* (consuming) organisms is encountered here. Near the shoreline, a fringe of emergent or submerged macrophytes is often present, either attached to the substrate, or floating nearshore. These plants form a substrate for many attached (*epiphytic*) or crawling organisms. On wave-swept, rocky, or sandy coasts macrophytes may be absent, but abundant algae or photosyn-

thetic bacteria may be present, attached to rock surfaces (*epilithic*), or adhering to sand grains.

In the sublittoral zone, light penetration is reduced, and large macrophytic plants are absent, but lower levels of benthic primary production may persist from algal or bacterial growth. Although algae are frequently found below the photic zone, because of circulation or settling, they are not photosynthesizing under such conditions. In the aphotic, profundal zone food resources are provided exclusively through secondary productivity, consumption of settling detritus (or the organisms that feed on such detritus), and microbial food resources.

The photosynthetic gradient between the littoral and profundal zones indirectly regulates heterotroph communities, for example by limiting the range of herbivores. But benthic habitats and communities are structured by other factors besides light availability, such as wave action. Waves and currents regulate substrates, changing food resource availability, shelters, and even reproductive success by affecting particle settling. Effective wave base sets a physical upper limit on the settling of fine-grained particles, which are important food resources for benthic detritus feeders. The specific depths at which transitions in light availability, wave-induced turbulence, and particle size change occur in the benthic environment of lakes vary with lake size, climate regime, and water clarity. Even in a single lake, and certainly over the history of any lake, the position of these transitions can be expected to change, imposing a challenge on the paleolimnologist trying to interpret biotic indicators that may be strongly associated with such gradients.

Moving up and away from our lake floor transect we encounter the open-water or *pelagic* environment of the lake. Floating, or weakly motile, plankton and mobile *nekton* (primarily fish) dominate this region. Although the habitat structure of this open water is less obvious to a casual observer, considerable variation in species abundance and diversity also occurs in this environment, driven by such factors as currents, nutrient availability, and open-water predation. Limnologists normally distinguish between primary producers in the plankton (*phytoplankton*) and secondary consumer animals (*zooplankton*), although some single-celled planktonic organisms can function in both roles. Much of the biomass in the pelagic environment of lakes is also made up of extremely small *picoplankton*, primarily bacteria, which consume dead organic matter in suspension or solution. Distinctions between benthic, planktonic, and nektonic organisms are to some extent artificial, since many organisms routinely move between these realms during their lifetimes.

A Synopsis of the Principal Groups of Lacustrine Organisms

It is useful to briefly consider those groups of lacustrine organisms that are commonly preserved as fossils, along with a few other groups of central importance to lacustrine ecology. I have divided the discussion by major habitat zones, discussing first those groups predominant in the plankton, then the nekton, and finally in the littoral zone and benthos. In keeping with common practice in modern systematic biology, I will use several terms to describe the evolutionary context of taxonomic names used in this book. A *monophyletic* group is one in which all members of the taxon share a common ancestor that is also a member of the taxon (i.e., has the unique, shared-derived characteristics of that group). Living ray-finned fish form a monophyletic taxon, since all living ray-finned fish and only ray-finned fish are descended from a common ancestor that was also a ray-finned fish. Monophyletic groups of organisms, regardless of taxonomic rank, are referred to as *clades*. A *paraphyletic* taxon is one in which only descendants of a common member of the taxon are included in the taxon, but in which not all descendants are included. "Bony fish" would be an example of a paraphyletic taxon, since the common ancestor of all bony fish was also the ancestor of tetrapods such as mammals. Finally, *polyphyletic* refers to a group whose common ancestor was not a member of the taxon (i.e., the group includes neither all nor only members of a unique clade).

Major Groups of Planktonic Organisms

Planktonic Algae (Phytoplankton)

Phytoplanktonic communities comprise a wide variety of groups of both single-celled and colonial algae and bacteria (Canter-Lund and Lund, 1995; Van den Hoek et al., 1995). "Algae" are an artificial, polyphyletic grouping of organisms, united by functional and ecological similarity rather than by evolutionary ancestry (Kumar and Rzhetsky, 1996; Lipscomb, 1996).

Major groups of phytoplankton are differentiated ecologically, and to some extent evolutionarily, by differences in their dominant *pigments*, used to absorb visible light energy during photosynthesis (table 5.1). All blue–green bacteria and algae use the pigment chlorophyll *a*, but many other types of chlorophylls or carotenoid pigments occur in more restricted groups of algae. Chlorophyll *b* is restricted to the chlorophytes, euglenophytes, and higher plants. Even when algal cells are completely degraded in sediments, these pigments or their distinctive byproducts may be preserved, thereby allowing paleolimnologists to infer major components of fossil algal communities.

The *Blue–Green Algae (Cyanobacteria)* are a monophyletic group of asexual autotrophs (figure 5.1A–C). *Prokaryotic* organisms like cyanobacteria lack a well-defined nucleus, mitochondria, golgi bodies, and chloroplasts, differentiating them from all *eukaryotic* algae. Unlike other photosynthesizing bacteria, cyanobacteria undergo oxygen-producing photosynthesis and utilize chlorophyll as a photosynthetic pigment. About 2000 species of cyanobacteria have been described, mostly from inland waters. Numerous unicellular, colonial, and filamentous species of cyanobacteria occur in the plankton, but they are common benthic autotrophs as well. Individual cyanobacterial cells are very small (up to a few micrometers in length), but colonies may be much larger (up to a few millimeters, and much larger for cyanobacterial mats). Cyanobacteria differ from other phytoplankton in their possession of specialized cells referred to as *heterocysts*, which are capable of fixing molecular nitrogen (N_2), allowing cyanobacteria to dominate phytoplankton communities when fixed nitrogen is scarce. Cyanobacteria often dominate the phytoplankton of highly eutrophic lakes. Although cyanobacterial cells are rarely preserved as lacustrine fossils, their pigment derivatives are distinctive, allowing their former abundance to be inferred from sediments.

Table 5.1 Important pigments for major algal and embryophyte (land plant) groups. X = pigment is important, x = pigment is present, +/− = pigment occurs rarely, * = found in dinophytes with brown endosymbiotic algae, ** = found in Oscillatoriaceae only. Adapted from Van Den Hoek et al. (1995) with additional information from Leavitt (1993)

	Cyanobacteria (Blue–Green Algae)	Dinophyta (Dinoflagellates)*	Bacillariophyceae (Diatoms)	Chrysophyceae (Golden-Brown Algae)	Cryptophyta	Euglenophyta	Chlorophyta (Green Algae)	Embryophyta
Chlorophylls								
chlorophyll *a*	X	X	X	X	X	X	X	X
chlorophyll *b*						X	X	X
chlorophyll *c1*		X	X	X				
chlorophyll *c2*			X	X	X			
chlorophyll *c3*								
Phycobilins								
phycocyanin	X				X			
allophycocyanin	X							
phycoerythrin	X				X			
phycobilisomes	X							
Carotenes								
α-carotene		x*		x	X		+/-	x
β-carotene	X	X	X	X	X	X	X	X
Xanthophylls								
zeaxanthin	X						+/-	
echinenone	X							
canthaxanthanin	X							
myxoxanthophyll	X							
oscillaxanthin**	X							
lutein						x	X	x
violaxanthin				x			X	
fucoxanthin		X*	X	X				
diatoxanthin		X*	X					
diadinoxanthin		X	X	x		+/-		
alloxanthin					X	x		
dinoxanthin		x						
peridinin		X						
neoxanthin		x		x		X	X	

98

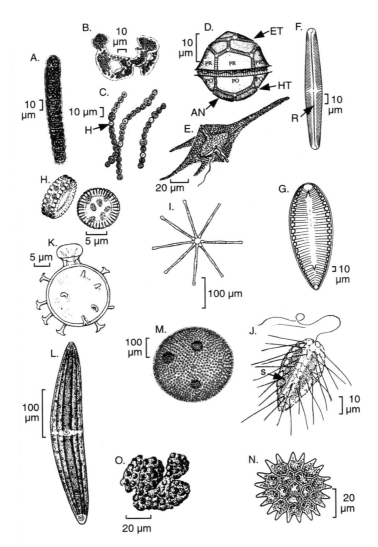

Figure 5.1. Representative species of major phytoplankton clades discussed in this book. (A) *Oscillatoria* sp. (Oscillatoriales, Cyanobacteria). (B) *Microcystis aeruginosa* (Chroococcales, Cyanobacteria). (C) *Anabaena* sp. (Nostocales, Cyanobacteria), note nitrogen-fixing heterocyst cells (H). (D) *Peridinium cinctum* (Peridiniales, Dinophyta dinoflagellate). (E) *Ceratium hirundinella* (Peridiniales, Dinophyta dinoflagellate). (F) *Navicula oblonga* (Pennales, Bacillariophyceae pennate diatom), note raphe (R). (G) *Surirella capronii* (Pennales, Bacillariophyceae pennate diatom). (H) *Cyclotella* sp. (Centrales, Bacillariophyceae centric diatom). (I) *Asterionella* (Pennales, Bacillariophyceae colonial pennate diatom). (J) *Mallomonas* (Mallomonadales, Chrysophyceae golden-brown algae), note fossilizable siliceous scales (S). (K) Fossilizable statospore from *Ochromonas fragilis* (Ochromonadales, Chrysophyceae). (L) *Closterium lunula* (Desmidales, Chlorophyta desmid green alga). (M) *Volvox globator* (Volvocales, Chlorophyta colonial green alga), with three daughter colonies. (N) *Pediastrum* sp. (Chlorococcales, Chlorophyta). (O) *Botryococcus braunii* (Chlorophyta). From Van den Hoek et al. (1995) (A,B,D,F,G,L,M); From Burgiss and Morris (1987) (C,N); From Cole (1979) (E,H,J); From W.D. Williams (1983) (I,O); From Hutchinson (1967) (K).

Dinoflagellates (*Dinophyta*) are small (10–300 μm), unicellular organisms bearing two dissimilar flagella (figure 5.1D,E). About 200 freshwater species are known, some of which are photosynthesizing autotrophs and some of which are heterotrophs (Van den Hoek et al., 1995). Most species are surrounded by a thick, often spinose, cell wall. Dinoflagellates undergo a complex life cycle, including seasonal changes in the cell-wall morphology of the motile stage, and the formation of very thick-walled, fossilizable, resting cysts. Pigments in dinoflagellates are dominated by golden-brown xanthophylls, particularly peridinin, which is unique to this group. Chlorophyll *a* and *c* are also present (though not *b*), but are generally masked by the dominant xanthophylls. Unfortunately, fossil pigments from dinoflagellates tend to be difficult to identify with precision (Leavitt et al., 1989).

Diatoms (*Bacillariophyceae*) are a diverse group of unicellular or colonial algae with thousands of described species, differentiated largely based on their distinctive, siliceous, cell walls (figure 5.1F–I). Individual cells are moderately large (typically 50–400 μm). Two overlapping *valves* form a *frustule*, a case enclosing the organism. Systematists recognize two major groups of diatoms, the radially symmetrical *centric* diatoms and the axially symmetrical *pennate* diatoms. Pennate species are subdivided longitudinally by a long slit or *raphe*, through which cytoplasmic structures extend, allowing these diatoms to glide along the substrate (Van den Hoek et al., 1995).

All diatoms possess chlorophyll *a* and c_2, plus β-carotene. The xanthophylls fucoxanthin, diatoxanthin, and diadinoxanthin dominate accessory pigments, collectively imparting a brownish color to many species. Diatoms may be planktonic, benthic, or may move seasonally between the plankton and benthos (*meroplankton*), with the majority of planktonic species being centrics. Collectively they inhabit a wide range of water bodies, in terms of nutrient availability and water chemistry. Benthic species live on the surfaces of rocks (*epilithic*), macrophytes (*epiphytic*), or soft substrates (*epipelic*). Some benthic diatoms are also facultative heterotrophs, and can live well below the sediment–water interface, or below the photic zone.

Diatoms are arguably the most important group of algae for paleolimnology. Their paleoecology has been extensively studied, largely as a result of their excellent preservation potential and abundance as fossils. The fact that many species have highly predictable distribution patterns related to water chemistry characteristics (for example nutrient concentrations or pH) makes them extremely useful as paleoenvironmental indicators.

The *Golden-Brown Algae* (*Chrysophyceae*) are a group of small (5–40 μm), usually flagellate, eukaryotic algae (figure 5.1J,K). Most species are unicellular or colonial, although a few are multicellular (Van den Hoek et al., 1995). The majority of chrysophyte species are planktonic, although benthic taxa are abundant in some environments (Douglas and Smol, 1995). The photosynthetic pigments of chrysophytes are similar to those of the diatoms, with a dominance of the xanthophyll fucoxanthin (Smol, 1988). Chrysophytes undergo alternating life stages involving both resting and motile forms. The distinctive spherical resting stages (*statospores*—figure 5.1K) possess easily fossilized siliceous skeletons. In some groups a siliceous (occasionally calcareous) armor of scales and bristles is present in the motile stage as well. Chrysophytes form a moderately diverse group (~1000 species) of mostly unicellular algae of interest to paleolimnologists because of their preservability and species-specific sensitivity to variations in water chemistry.

The *Green Algae* (*Chlorophyta*) are a very large (over 8000 species recognized) and morphologically diverse group of unicellular and colonial algae (figure 5.1L–O). Unicellular forms are small (a few micrometers), but colonial, filamentous, and coccoid varieties may be macroscopic. Molecular genetic evidence shows that the group is paraphyletic, with some green algae actually forming a common evolutionary lineage with higher plants, distinct from other green algae (McCourt, 1995; Kumar and Rzhetsky, 1996).

Green algae possess chlorophyll *a* and *b* as primary pigments. The presence of several xanthophyll carotenoids, notably the commonly preserved lutein, allows their pigments to be recognized in lake sediments (Leavitt, 1993). Some chlorophytes, such as the genera *Pediastrum* (figure 5.1N) and *Botryococcus* (figure 5.1O), have large quantities of silica in their cell walls and as a result are readily preserved as fossils.

Zooplankton

Zooplankton are a heterogeneous assemblage of mostly small heterotrophs, which are readily transported by currents. Feeding strategies vary greatly among major groups and species, and include phytoplankton grazers, particulate filter feeders and microcarnivores. Ecologically, the most important zooplankton in lakes are unicellular ciliates and amoebans, rotifers, and arthropods (mostly crustaceans and insects). However, only a limited number of these zooplankters are commonly preserved as fossils in lake beds.

Rotifers (*Rotatoria*) are small (0.1–1 mm) animals found almost exclusively in freshwater (figure 5.2A,B). Rotifers are suspension feeders, moving particles into their mouths through the use of cilia. Rotifers are found in a wide diversity of aquatic and semiaquatic habitats (Pennak, 1978). In many lakes they are the most abundant zooplankters, although species diversity is probably higher among benthic forms. A rotifer's body is surrounded by a cuticular covering, the lorica or shell, which may be preserved as a fossil, particularly in acidic peat bogs (Frey, 1964; Warner, 1990c). Rotifer eggs are also occasionally fossilized, although both loricae and eggs,

being quite fragile, are rarely encountered in pre-Quaternary lake sediments.

The *arthropods* are an enormous, and possibly paraphyletic, group of organisms, characterized by body segmentation (secondarily lost in some groups), and the possession of an exoskeleton, which is periodically shed or molted during the animal's life. Pieces of this exoskeleton are what most commonly become preserved from these animals. Differences in the durability of the exoskeleton, particularly between different groups of crustaceans and insects, account for major differences in preservation potential as fossils.

Microscopic and mesoscopic *crustaceans* are probably the most conspicuous zooplankters in

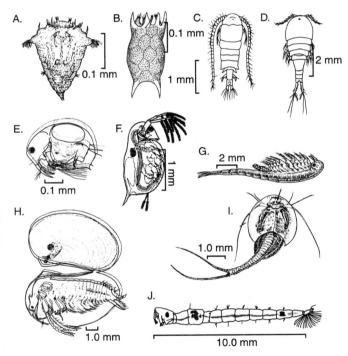

Figure 5.2. Representative species of major zooplankton clades discussed in this book. (A) *Synchaeta* sp. (Synchaetidae, Rotifera rotifer). (B) Fossilizable lorica of *Keratella* sp. (Brachionidae, Rotifera rotifer). (C) *Diaptomus* sp. (Diaptomidae, Copepoda, Crustacea calanoid copepod). (D) *Cyclops strenuus* (Cyclopidae, Copepoda, Crustacea cyclopoid copepod). (E) *Bosmina longirostris* (Bosminidae, Cladocera, Branchiopoda, Crustacea bosminid cladoceran). (F) *Daphnia pulex* (Daphniidae, Cladocera, Branchiopoda, Crustacea daphniid cladoceran). (G) *Artemia* sp. (Anostraca, Branchiopoda, Crustacea brine shrimp). (H) *Cyzicus californicus* (Conchostraca, Branchiopoda, Crustacea clam shrimp). (I) *Triops longicaudatus* (Notostraca, Branchiopoda, Crustacea tadpole shrimp). (J) *Chaoborus* sp. (Chaoboridae, Diptera, Insecta midge larva). From Cole (1979) (A); From Nicholas (1998) (B); From Burgiss and Morris (1987) (C,D,G); From Dodson and Frey (1991) (E,H,I); From Sorrano et al. (1993) (F,J).

most lakes. In freshwater "permanent" lakes, cladocerans and copepods are dominant groups, whereas fairy shrimp (anostracans) are common in salt lakes and temporary ponds.

Copepods are a large group of free-living and parasitic crustaceans (figure 5.2C,D). Free-living species are mesoscopic (typically 0.5–3 mm), with cylindrical bodies, prominent antennae, and head segments. Approximately 1000 freshwater species are known, living in both planktonic and benthic environments. Copepods are extremely important components of the planktonic food webs in most lakes (Williamson, 1991; Wetzel, 2000). Despite their importance in modern lakes, copepods are relatively unimportant for paleolimnologists, because of their poor preservation potential. Body segments are weakly skeletonized, and pre-Quaternary fossils of copepods are extremely rare. The only copepod fossils with any significant likelihood of preservation are their fecal pellets, and the relatively durable male spermatophores (Warner, 1990c).

Cladocerans (water fleas) are a diverse group of mesoscopic crustaceans (mostly 0.2–2 mm) that occur throughout the pelagic and littoral zones of most lakes (figure 5.2E,F). They, along with the *anostracans, notostracans (tadpole shrimp), conchostracans (clam shrimp)*, and some other groups, are all members of the dominantly nonmarine clade *Branchiopoda*, crustaceans characterized by the possession of flattened, leaf-like appendages.

Although cladocerans as a whole occur in a wide array of aquatic environments, individual species can be quite selective in terms of habitat. Most cladoceran species are selective filter feeders, separating algal and detrital particles by size. Some are also herbivorous scrapers or microcarnivores. Members of two families, the Bosminidae (figure 5.2E), and the Daphniidae (figure 5.2F), are important zooplankters, and can be differentiated at the family level based on the shape and arrangement of the antennae. Like copepods, cladocera are extremely important parts of pelagic and littoral food webs, and can undergo enormous seasonal or spatial variation in population densities, because of changing predation or food availability. But in contrast to copepods, cladocera are frequently preserved as fossils, because of their more robust, chitinous exoskeleton (Frey, 1986; Whiteside and Swindoll, 1988; Hann, 1990). Numerous durable skeletal parts, such as head shields, carapaces, claws, and ephippia (brood pouches), are produced during molting by a single individual through its lifetime.

Other branchiopods are mostly restricted to temporary and/or saline ponds, and predominantly occur in warm, semiarid climates today (Kerfoot and Lynch, 1987; Frank, 1988; Dodson and Frey, 1991). The relatively large size of most of these groups of branchiopods, and their slow speeds, make them easy prey for fish, partly explaining their modern rarity in fish-bearing permanent lakes (Kerfoot and Lynch, 1987).

Anostracans are 1–10 cm long, shrimp-like, filter feeders, with long, and often elaborate, antennae (figure 5.2G). Conchostracans are more ovate branchiopods (2–16 mm in length), largely enclosed within bivalved carapaces, the valve opening oriented downward as the animal moves (figure 5.2H). Most species are weakly swimming, near-bottom detritus feeders, but some are burrowers. Notostracans are tadpole-shaped detritus feeders and predators, 1–6 cm in length, with prominent shield-like carapaces covering the anterior portion of their body (figure 5.2I).

Conchostracans are common as fossils. In Cretaceous to Recent lake deposits they are generally associated with similar habitats to today (temporary pools or saline ponds) (Gray, 1988b). However, older occurrences of these clades appear to be more widespread, occurring in larger and freshwater lake deposits. Anostracans are rare as fossils, represented almost exclusively by their relatively durable eggs.

Insects are represented in the zooplankton of lakes primarily by dipteran (fly) larvae of the family Chaoboridae. Members of the genus *Chaoborus* (the phantom midges) are important zooplankters in a wide variety of lakes (figure 5.2J). Most chaoborids are episodic zooplankters, undergoing daily migrations from the lake floor to the surface during their fourth molt stage (Ward, 1992). In contrast, nonmigratory or partially migratory *Chaoborus* species are dominant in fishless lakes (Borkent, 1981). While in the upper pelagic zone, chaoborids themselves are voracious predators, foraging on a variety of microcrustaceans and other zooplankton. Chaoborids are commonly represented as fossils in lake beds, most commonly by their distinctive chitinous mandibles (Uutala, 1990).

Major Groups of Nektonic Organisms

Most nektonic animals in lakes are bony fish. Other common nektonic groups [larger crustaceans, agnathan "fish" (e.g., lamprey), cartilaginous fish, and all classes of tetrapods] are greatly outnumbered, both in terms of diversity and biomass, by bony fish in most lakes. In some lakes, larger swimming crustaceans (atyid or mysid shrimp) can also be numerically abundant.

There are approximately 10,000 species of fresh-water fish in the world, accounting for perhaps 40% of all species of fish globally (Lundberg et al., 2000). Fish display an extraordinary diversity of body morphologies, often strongly related to habitat and trophic specializations. Although neither "fish" nor "bony fish" form monophyletic groups, the vast majority of freshwater fish do fall into a well-defined, monophyletic group, the ray-finned fish (Actinopterygii), characterized by relatively flattened fins comprising numerous spine-like bones.

Ray-finned fish inhabit a wide variety of pelagic, littoral, benthic, and demersal (i.e., swimming just off the bottom) lacustrine habitats. Many species occupy the littoral zone as juveniles, or for breeding, and move to open water as adults. Other species of "lacustrine" fish are migratory, moving into or out of lakes into rivers or the sea at various life stages, generally for spawning.

Fish feed in an extraordinarily wide variety of ways, a fact that has profound consequences for lacustrine food webs. Herbivorous grazers, browsers, pelagic filter feeders, deposit feeders, and higher level carnivores are all represented, and in some lakes considerable feeding specialization is observed by individual species. Fish species assemblages are also strongly structured by foraging requirements, and many species have quite narrow requirements, limiting their ability to disperse to neighboring environments. Limitations to dispersal across land or aquatic habitat barriers have resulted in spectacular speciation and endemism (restriction of a species to a single lake) among many lacustrine fish clades, and also subjects these same clades to high rates of extinction.

Bones, teeth, and scales of ray-finned fish are commonly preserved as fossils in lake deposits. Because fish bones are weakly articulated in life, these skeletal elements generally disintegrate into individual pieces during burial, making accurate identification of fossils more difficult. Within meromictic lakes, however, complete or nearly complete skeletons of fish are quite common, and under these types of preservation conditions a great deal of paleolimnological information can be gained from fish fossils (Elder and Smith, 1988).

Amphibians, aquatic reptiles, birds, and mammals are common nektonic or nearshore animals in many lakes. Although their fossils are relatively common in lacustrine deposits, they have been little utilized by paleolimnologists, and therefore a discussion of them is beyond the scope of this book.

Major Groups of Littoral and Benthic Organisms

All of the groups of planktonic and nektonic organisms previously discussed above are also represented in the littoral and profundal zones of lakes. Additionally, a number of clades commonly represented in the fossil records of lakes are largely or completely restricted to benthic habitats. Benthic organisms require more durable body parts than planktonic organisms of similar size, because of their continuous contact with rigid and potentially damaging surfaces. Not surprisingly, benthic organisms are better represented in the lacustrine fossil record than plankton, and we will concentrate here on benthic groups that are readily fossilized. Nevertheless, there are many common, soft-bodied, benthic organisms, such as nematodes, which are almost completely unrepresented as fossils.

Cyanobacteria, as mats and mounds, are common littoral and sublittoral organisms in many lakes. Often these mats are actually complex microbial communities, including diatoms, chrysophytes, fungi, and sponges, as well as cyanobacteria (e.g., Winsborough and Seeler, 1984; Winsborough and Golubic, 1987; Bertrand-Sarfati et al., 1994; Cohen et al., 1997a). In hardwater and moderately alkaline lakes, mat and mound formation is accompanied by calcite precipitation, forming so-called *microbialites*. This can occur either indirectly, by decreased solubility of $CaCO_3$ associated with photosynthesis, or directly, through calcite sheath formation around the cyanobacteria (Pentecost and Riding, 1986). Microbialites can form as layered carbonate–algal tissue buildups (*stromatolites*), or *thrombolites*, when the buildup is unlayered. Microbialites are common features in the fossil record of lakes (Osborne et al., 1982; Casanova and Nury, 1989), generally preserved as a calcareous buildup, and occasionally incorporating microbial fossils.

Several groups of littoral chlorophytes merit discussion here, because of their importance in structuring littoral communities, and their potential for fossilization. Species of *Cladophora* are conspicuous, green algae, typically growing as filamentous masses on rocky substrates in eutrophic lakes and ponds (Van den Hoek et al., 1995). The *stoneworts* (Charales or *charophytes*) are distinctive, macroscopic algae, more closely related to mosses and vascular plants than to other chlorophytes (McCourt, 1995) (figure 5.3A). Common genera like *Chara*, *Nitella*, and *Lamprothamnium* often form extensive "meadows" in the littoral zones of hardwater lakes and coastal lagoons. Individual

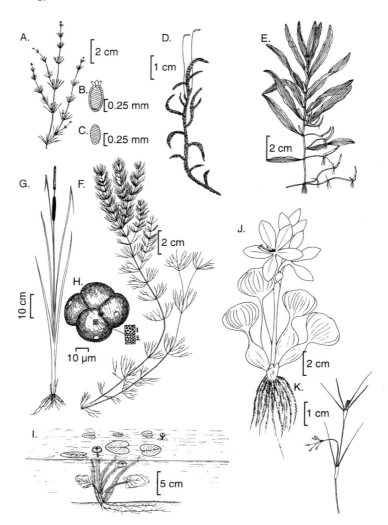

Figure 5.3. Representative species of major macrophyte clades and ecological groups discussed in the text. (A) *Chara contraria* (Charales stoneworts). (B) Intact gyrogonite (oogonium) of *Chara contraria*. (C) *Chara* gyrogonite. (D) *Scorpidium scorpoides* (Amblystegiaceae, Bryophyta aquatic moss). (E) *Potamogeton alpinus* (Potamogetonaceae, Angiosperm submerged flowering plant). (F) *Ceratophyllum demersum* (Ceratophyllaceae, Angiosperm submerged flowering plant). (G) *Typha angustifolia* (Typhaceae, Angiosperm cat-tail, an emergent flowering plant). (H) Pollen grain from *Typha latifolia*. (I) *Nuphar lutea* (Nymphaceae, Angiosperm water lily, floating, rooted flowering plant). (J) *Eichornia crassipes* (Pontederiaceae, Angiosperm water hyacinth, a floating, unattached flowering plant). (K) *Ruppia maritima* [Ruppinaceae, Angiosperm salt-tolerant (halophytic) flowering plant]. From Hutchinson (1975) (A–G, I–K); From Erdtman (1943) (H).

plants may be up to a meter or more in height. Charophytes are of particular significance to paleolimnologists because of their excellent fossilization potential. A calcified envelope, known as a *gyrogonite*, surrounds the reproductive zygote of the plant. These small (~ 0.5 mm length) football-shaped objects (figure 5.3B,C), in addition to calcareous encrustations and molds formed around the plant's axes and branches, are commonly found in littoral lacustrine limestone deposits (Soulié-Märsche, 1979).

Aquatic mosses are common, though not particularly diverse plants in many lakes (figure 5.3D). Mosses are simple embryophytic plants, lacking the vascular tissues of other, nonalgal macrophytes, which probably form a monophyletic group (Garbary et al., 1993; Capesius, 1995; Hedderson et al., 1996). Early mosses were apparently amphibious or terrestrial plants that have secondarily invaded freshwater, perhaps on several occasions. Aquatic mosses are very common in cool to cold-water lakes, and in regions where the groundwater table is extremely shallow.

Fossil moss remains are common in Quaternary lake sediments, represented primarily by leafy stems, and occasionally by the asexual sporophytes (Dickson, 1986; Janssens, 1990). Fossil assemblages typically include a mixture of aquatic, semi-aquatic, and terrestrial species.

Numerous *vascular plants* inhabit the littoral zone of lakes. Most of these are flowering plants (angiosperms), but in some habitats club mosses (lycopsids), horsetails (sphenopsids, esp. *Equisetum*), an assortment of "fern-like" plants, including tiny floating species of *Azolla*, as well as true ferns (primarily *Ceratopteris*), and conifers (e.g., *Taxodium*) may also be important. Ecologically, it is useful to subdivide macrophytes into four categories, based on their style of attachment to the substrate and mode of reproduction (Den Hartog and Van Der Velde, 1988; Wetzel, 2000):

1. Submerged plants that undergo primarily asexual, vegetative reproduction, such as the cosmopolitan genera *Potamogeton* (figure 5.3E) and *Ceratophyllum* (figure 5.3F). Some submerged species that occur in temporary ponds are capable of flowering and reproducing sexually when water levels decline.
2. *Emergent macrophytes*, that are rooted underwater, but whose leaves and reproductive structures extend above water on erect stalks, such as cat-tails (*Typha*) (figure 5.3G,H) or sedges (e.g., *Cyperus papyrus*, the papyrus). These plants reproduce either sexually, or through vegetative growth.
3. Plants that are rooted, but which have floating leaves and reproductive structures, and which primarily reproduce sexually, like the water lilies *Nymphaea* and *Nuphar* (figure 5.3I).
4. Unattached plants, with free-floating leaves and inflorescences, such as the water hyacinth (*Eichornia crassipes*) (figure 5.3J) or duck-weeds (*Wolffia* and *Lemna*).

This variation in growth form structures the littoral environment, and strongly influences local water circulation, chemistry, and the distribution of littoral animals. Although disagreement exists as to the direct nutritive importance of macrophytes for grazing organisms, there is no question as to their importance as substrates for epiphytic organisms, which are important fish and invertebrate food. Macrophyte distributions are regulated by water chemistry, particularly nutrient availability in water or sediments, and salinity. Changes in nutrient loading can dramatically alter the structure of macrophyte communities (Wetzel, 2000). Groundwater flow is also important in determining macrophyte distribution patterns (Verhoeven et al., 1988). Growth of most freshwater angiosperm species is inhibited by high salinities (the genus *Ruppia*, figure 5.3K, is a notable exception), and the species found in coastal and estuarine marshes are generally quite distinct from, and much less diverse than, those of inland waters.

Macrophytes are commonly preserved as lacustrine fossils, as fruits, seeds, spores, pollen (figure 5.3H), or emergent plant leaves, which often have large amounts of lignin, or are silicified (Collinson, 1988). An absence of a cuticle around leaves and stems in most submerged macrophytes reduces the probability of preservation of these body parts. The primary exception to this generalization occurs in acidic bogs, where macrofossil preservation of leaves and stems is enhanced (Grosse-Brauckmann, 1986).

Fungi are important components of lacustrine ecosystems, in both benthic and planktonic habitats. Fungi form a distinctive clade of heterotrophic thallophytes, which reproduce with sexual or asexual spores, and it is these spores that are most often preserved as fossils. Most aquatic fungi are saprophytes, playing an extraordinarily important ecological role in lakes by decomposing animal and plant remains, both within the sediment and as floating organic matter. Mycorrhizal fungi (commensal fungi on plants) are rare in aquatic environments, although their transported fossil spores may be preserved in lake beds. Some fungi are specialized animal or plant parasites, or dung decomposers. These close relationships provide a means for paleolimnologists to determine past distribution patterns of host organisms, which are frequently not as well preserved in lake sediments as the fungi themselves. Fungal spores are abundant in lake sediments and potentially very informative for paleolimnology. However, both perceived and real difficulties in taxonomy, identification, and relating ecological

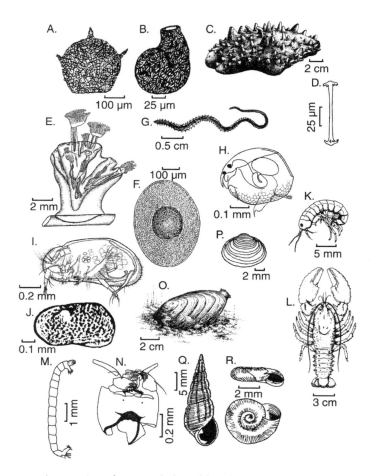

Figure 5.4. Representative species of major clades of benthic protists and metazoans of importance to paleolimnology. (A) *Difflugia corona* (Difflugiidae, Sarcodina testate amoeban). (B) *Lesquereusia spiralis* (Difflugiidae, Sarcodina testate amoeban). (C) Typical growth morphology for benthic, encrusting sponges (Porifera) (note: species displayed is a marine sponge). (D) Spicules from *Corvomeyenia everetti* (Metaniidae, Porifera). (E) Colony of *Lophophus* sp. (Plumatellidae, Phylactolaemata, Bryozoa colonial bryozoan, showing extended and retracted tentacular lophophores for filter feeding). (F) Fossilizable statoblast (dormant colonial structure of phylactolaemate bryozoans) from *Hyalinella punctata*. (G) *Tubifex* sp. (Tubificidae, Oligochaeta, Annelida segmented worm). (H) *Chydorus brevilabris* (Chydoridae, Crustacea benthic cladoceran). (I) *Candona suburbana* (Cyprididae, Cypridoidea, Ostracoda, Crustacea ostracod, showing internal view). (J) *Limnocythere africana* showing exterior of valve (Limnocytheridae, Cytheridoidea, Ostracoda). (K) *Austrochiltonia* sp. (Amphipoda, Crustacea amphipod). (L) *Euastacus armatus* (Parastacidae, Decapoda, Crustacea crayfish). (M) *Chironomus* sp. (Chironomidae, Diptera, Insecta midge). (N) Head capsule of a chironomid *Tanytarsus* sp. (showing the commonly fossilized portions of a midge larva). (O) *Anodonta* sp. (Unionidae, Unionacea, Bivalvia, Mollusca swan mussel, shown in life habit with siphon extended). (P) *Sphaerium striatinum* (Sphaeriidae, Corbiculaceae, Bivalvia, Mollusca pea clam). (Q) *Melanoides tuberculata* (Thiaridae, Prosobranchia, Gastropoda, Mollusca prosobranch snail). (R) *Gyraulus deflectus* (Planorbidae, Pulmonata, Gastropoda, Mollusca pulmonate snail, showing bottom and side views of shell). From Taylor and Sanders (1991) (A,B); From Bergquist (1998) (C); From Frost (1991) (D); From Brusca and Brusca (1990) (E); From Wood (1991) (F); From W.D. Williams (1983) (G,K,L); From Dodson and Frey (1991) (H); From Delorme (1991) (I); From Carbonel et al. (1988) (J); From Hillsenhoff (1991) (M); From Hoffman (1986a) (N); From Burgiss and Morris (1987) (O); From McMahon (1991) (P); From Brown (1991) (Q,R).

data derived from culturing experiments to geologically relevant conditions have inhibited their widespread study in paleolimnology (Van Geel, 1986; Sherwood-Pike, 1988; Pirozynski, 1990).

An enormous diversity of unicellular ciliate, flagellate, and amoeboid heterotrophs (referred to here as *protists*) occur in the littoral and benthic zones of lakes. These organisms occupy almost every conceivable trophic niche, and can be found in an extraordinarily broad range of habitats, including many extreme environments (highly acidic or permanently frozen lakes) where metazoans are excluded. An absence of durable body parts results in very infrequent preservation of fossils for most heterotrophic protists. Only two benthic groups are abundantly represented as fossils. The *testate amoebans* (also referred to as thecamoebans or arcellaceans) are microscopic (50–300 μm) unicellular, ameoboid protists, whose bodies are enclosed in a simple, sack-like shell or test (figure 5.4A,B). This shell, which is the only part of the organism normally preserved as a fossil, is composed of organic material, siliceous, or calcareous plates, or agglutinated particles that the organism incorporates from the environment (such as diatom frustules, sand grains, etc.) (Medioli and Scott, 1988; Warner, 1990b). Testate amoebans are common in terrestrial, wetland, and freshwater lake environments, and include both benthic and planktonic species.

A second group of amoeboid protists, the *foraminifera*, is a predominantly marine group of protists, occasionally represented in lakes, mostly in coastal regions, where they are readily introduced by birds, or during intermittent connections with the sea (De Deckker, 1988a). However, saline lakes even thousands of kilometers from the sea have been colonized by foraminiferans, particularly when those lakes lie along bird flyways (Patterson et al., 1997). Most lacustrine foraminiferans are benthic species, and are typically species that also occur in estuarine environments. Calcified or agglutinated foraminiferan tests are commonly found as fossils in saline lake deposits (Anadón, 1989).

Sponges (*Porifera*) are simple, multicellular, heterotrophs, which have traditionally been classified as multicellular animals, although there is considerable evidence that they represent an independent radiation from an independent unicellular ancestor (figure 5.4C). Sponges have cellular differentiation but no organs or true tissues. Adult sponges are benthic filter feeders, but gametes and larvae are dispersed in the plankton. All freshwater sponges belong to the Class Demospongiae, most in a single family, the Spongillidae. Spongillid sponges have a skeletal meshwork comprised of siliceous *spicules*, needle or barbell-shaped structures a few tens of micrometers in length (figure 5.4D). They are also characterized by the presence of *gemmules*, spherical structures formed for asexual reproduction. Adult spongillids may be encrusting or erect, branching organisms, usually only a few centimeters or less in height, although larger species are known.

Sponges occur in a wide variety of substrate and water chemistry conditions. Most sponges thrive under conditions of moderate turbulence, required to promote filtration. Heavy siltation eliminates many species, but some can take refuge by growing underneath rocks or logs (Harrison, 1988). Sponges are common lacustrine fossils, normally preserved as disarticulated spicules and gemmules (Harrison, 1990).

Bryozoans are small, colonial animals that occur on a variety of substrates in lakes (figure 5.4E). Although they are not diverse (about 50 freshwater species are known), bryozoans are common in lakes, often dominating the surfaces of rocks or macrophytes, and are important food sources in some lakes for bottom-feeding invertebrates and fish. (Bushnell, 1974). Their abundance seems to be related to the extent of the littoral zone and phytoplankton biomass (Crisman et al., 1986). Some studies have shown a general preference among bryozoan species for alkaline conditions (Bushnell, 1966, 1968; Ricciardi and Reiswig, 1994), but in other areas this relationship has been disputed (Crisman et al., 1986).

Bryozoans possess a cilia-bearing feeding apparatus, the lophophore, which filters algae, protists, detritus, or small metazoans from the water column. Freshwater bryozoans, most of which are members of the Class Phylactolaemata, are entirely soft-bodied as adults, and are preserved as fossils almost exclusively by asexual *statoblasts*, small reproductive structures that form as buds on their colonies (Kuc, 1973; Warner, 1990c; Francis, 1997) (figure 5.4F). Two sclerotized valves surround these statoblasts, their rigidity allowing them to be easily dispersed to form new colonies.

Annelids, or segmented worms, form a major group of primarily benthic invertebrates (figure 5.4G). Most freshwater species fall into one of two major clades, the oligochaetes and leeches (Hirudineans) (Brinkhurst and Gelder, 1991; Davies, 1991). Most oligochaete species are infaunal worms a few millimeters to a few centimeters in length, and form shallow burrows in sandy or muddy substrates, where they feed on detritus, bacteria, and algae (Brinkhurst and Gelder, 1991). This burrowing is manifest in lake sediments through extensive bioturbation, and occasionally by discretely preserved burrows, evident in the x-radiographs of sediments. The quantity and quality of sedimented organic detritus significantly affects oligochaete species distribution (Brinkhurst and Cook, 1974). Oligochaetes produce sclerotized "cocoons" around their eggs, the shape of which is species-specific, and which may occasionally be preserved as fossils (Bonacina et al., 1986).

Crustaceans are abundant and diverse members of littoral and benthic communities in lakes. Two major groups, the branchiopods (especially the cladocerans) and copepods, have already been discussed in relation to the zooplankton and are only briefly considered here. Most littoral and benthic cladocerans are members of the family Chydoridae, and in many lakes these animals represent the bulk of cladoceran diversity, though their ecology remains much less well known than their planktonic counterparts (Dodson and Frey, 1991). The majority of chydorid species crawl on or around macrophytes, but many species also occur on sandy, muddy, or rocky substrates (figure 5.4H). Chydorids can be extremely abundant, and in many lakes are the dominant benthic animals in terms of sheer abundance of individuals (over 10^6 per square meter in some instances). Many species show strong affinities for particular water chemistries or substrate types. Chydorid species feed in numerous ways, as herbivorous scrapers, ectoparasites, filter feeders, carnivores, or scavengers. As with bosminids, the skeletal parts of chydorids are readily fossilized in Quaternary lake beds. Copepods are also abundant and diverse benthic crustaceans, living both on the surface and burrowing in the upper few centimeters of the sediment. Like planktonic copepods, benthic species are rarely fossilized.

Ostracodes are one of the most important groups of benthic crustaceans in lakes and ponds, and because of their excellent fossil record, of great importance to paleolimnology (Delorme, 1990) (figure 5.4I,J). They are a diverse group (> 1000 species), with representatives in a wide range of temporary and permanent aquatic environments, and semiaquatic settings. Most freshwater species fall into one of two clades, the Cypridoidea (figure 5.4I) and the Cytheroidea (figure 5.4J), determined by both carapace and anatomical differences. Ostracodes are small (generally 0.5–3 mm) crustaceans, whose bodies are enclosed in bivalved calcite carapaces (Delorme, 1991). The overwhelming majority of freshwater species are epibenthic or infaunal, swimming or crawling just above the sediment–water interface, or moving interstitially between sediment grains.

As with the branchiopods, predation on ostracodes, especially by fish, has played an important selective role in their evolution and ecology. Species in permanent freshwater lakes are typically smaller than those of temporary pools or ephemeral fishless lakes, and particularly small species dominate the faunas of long-lived tectonic lakes, which possess numerous endemic predators (De Deckker, 1983; Martens, 1988, 1994). Lacustrine ostracodes have a rich fossil record; in many lake deposits they are the most abundant fossilized invertebrates. Normally only their calcite carapaces (which often disarticulate into isolated valves) are preserved.

Other important freshwater crustacean groups include the *pericarid crustaceans* (the amphipods, figure 5.4K, isopods, and mysids, or oppossum shrimp), and the *decapods* (crayfish, crabs, and true shrimp, figure 5.4L). Collectively these organisms occupy a diverse array of benthic, planktonic, and parasitic habitats. As a result of the relatively weak sclerotization of their skeletons, pericarids have poor lacustrine fossil records, known mostly from exceptionally preserved fossils, or from crawling or burrowing trace fossils. Decapods are much more heavily skeletonized than pericarids, and as a result are moderately common as lacustrine fossils. To date, however, they have received little attention from paleolimnologists, except for evolutionary studies.

Benthic *insects* are an abundant and diverse group of freshwater organisms, often accounting for the vast majority of metazoan species diversity and biomass in lakes (Ward, 1992). Species that are aquatic at some or all phases of life occur among 13 orders of insects, representing independent invasions of freshwater (table 5.2) (Hillsenhoff, 1991; Ward, 1992).

Insect communities in lakes display strong habitat zonation, primarily related to vegetation and substrate. Most aquatic insects are benthic or epibenthic, with the vast majority occurring within the littoral zone; species diversity declines regularly with depth (Ward, 1992). Well-vegetated littoral zones are frequented by numerous species of Odonata (dragonflies), Hemiptera (water bugs), and Coleoptera (beetles). In well-vegetated ponds fish predation appears to reduce insect diversity. As with the branchiopods, many insect species are restricted to or are much more abundant in temporary ponds, often aestivating or laying desiccation-resistant eggs to maintain the population through periods of drought. The insect fauna of rocky littoral areas is often similar to that found in stream

Table 5.2. Aquatic association and habitats of insect orders containing aquatic or semiaquatic representatives in lakes and ponds. Adapted from Ward (1992).

Order	Aquatic Association	Aquatic Life Stage
Collembola (wingless hexapods)	2	A, L
Ephemeroptera (mayflies)	1	L
Odonata (dragonflies)	1	L
Hemiptera (water bugs)	2	A, L
Orthoptera (grasshoppers, crickets, etc.)	3	A, L
Plecoptera (stoneflies)	1	L
Coleoptera (beetles)	2	A, L, P
Diptera (true flies)	2	L, P
Hymenoptera (aquatic wasps)	2	A, L, P
Lepidoptera (moths, butterflies)	2	L, P
Megaloptera (alderflies, dobsonflies, fishflies)	1	L
Neuroptera (spongillaflies)	2	L
Trichoptera (caddisflies)	1	L, P

Aquatic associations: 1 = all species are aquatic at some life stage, 2 = primarily terrestrial group, with some aquatic representatives, 3 = primarily terrestrial group, with some semiaquatic representatives, no truly aquatic species. Aquatic life stages: A = adult, L = larvae and nymphs, P = pupae.

beds, with abundant Trichoptera (caddisflies). Usually the only insects commonly found in the profundal zone of lakes are dipterans (true flies), especially chironomid and chaoborid fly larvae, although in large oligotrophic lakes mayflies, stoneflies, and caddisflies also occur in deep water. A few taxa, like *Chaoborus*, discussed earlier, are partly planktonic, or live at the air–water interface (*pleustonic*). A somewhat larger number are strong nektonic swimmers, particularly among the Hemiptera and Coleoptera.

Insect diets are extremely variable (Roback, 1974). Algal grazing is most common among Ephemeroptera (mayflies) and Lepidoptera (aquatic moths). Herbivores include some stoneflies, lepidopterans, and beetles, the latter among the few groups that directly consume living macrophytes. Many insects, including most dragonflies, water bugs, and some water beetles, stoneflies, and dipterans are active predators. Most caddisflies and dipterans are omnivorous.

Insects are common fossils in lake beds, preserved as sclerotized skeletal parts (wing coverings, mandibles and other head parts, and appendages). However, most preserved insect fossils in lake deposits are terrestrial, rather than aquatic, species. Some aquatic insect groups such as the Trichoptera are both common as lake fossils and potentially very informative for paleoecological purposes, but have received relatively little attention by paleolimnologists to date (N.E. Williams, 1988a,b). However, one aquatic insect order, the Diptera, is

of particular importance to both limnology and paleolimnology and merits special attention.

Dipterans make up over half of all aquatic insect species, occurring in a wide variety of habitats (Ward, 1992). The clade includes such familiar animals as horse flies, midges, mosquitoes, and the previously discussed zooplanktonic chaoborids. The *Chironomidae* (*midges* or *lake flies*) are undoubtedly the most diverse group of aquatic dipterans, and account for over 15% of all described aquatic insects (figure 5.4M). Chironomid larvae have elongate (~ 1-3 cm), segmented bodies, with a pair (sometimes more) of fleshy appendages at the anterior end. Most species live in the sediment in loosely fitting tubes, although some are free-living. The animal spends most of its life growing through a series of larval molts or *instars*, eventually molting as a pupa, and then rising to the lake surface, where it emerges to breed as a short-lived, nonfeeding adult.

Chironomids are extraordinarily important members of the benthic fauna of most lakes. They feed in a wide variety of ways (as herbivores, detritivores, carnivores, and web-constructing filter feeders), and in turn are important parts of the diet of numerous fish and invertebrates. Midges occur in all depth and vegetation zones, and can be found in both permanent and temporary lakes. However, they are of greatest ecological importance in the profundal zone, where they may dominate the zoobenthos, numerically, by species diversity and in biomass (Hillsenhoff, 1991). Chironomid fossils are common in lake beds, and have proven exceptionally useful for paleolimnologists. They are generally preserved by their heavily sclerotized larval head capsule and feeding apparatus (figure 5.4N, Hoffman, 1988; I.R. Walker et al., 1991a).

The *mollusks* are a large clade of animals, united by possession of a muscular foot and a fleshy *mantle*, an organ that produces their calcareous/proteinaceous skeleton. Approximately 5000 species of freshwater mollusks are known, all of which fall into two major subclades, the gastropods and bivalves (Van Bruggen, 1992). Both groups are diverse and conspicuous macroinvertebrates (1–150 mm), found in a wide variety of inland water habitats (Dillon, 2000). The Gastropoda also includes a group (the Pulmonates) with amphibious and fully terrestrial species, many of which are known as fossils from lake beds.

Bivalves (clams and mussels) possess two external skeletal valves, within which the animal can completely retract (McMahon, 1991). About 2000 species of nonmarine bivalves have been described, the majority of which fall into one of two clades, the subclass Paleoheterodonta, which includes most of the freshwater "mussels" (Deaton and Greenberg, 1991) (figure 5.4O), and the diverse and primarily lacustrine, heterodont "pea-clams" (figure 5.4P, the Pisidiidae or Sphaeriidae of some authors).

Bivalves occur in a wide variety of nonmarine aquatic habitats, in greatest diversity in nonturbid rivers, but also in lakes and ponds. A large number of species of bivalves are capable of resisting desiccation for long periods, up to several months for species of pisiids and corbiculids, and even for years among some unionid mussels. Some species of freshwater clams can therefore inhabit water bodies that dry up seasonally, such as small ponds or flood-plain lakes (Burky, 1983; Burky et al., 1985). Larvae of the freshwater paleoheterodonts (various families) are parasitic, primarily on fish, and use this mechanism to disperse between suitable habitats. Most freshwater bivalves are filter feeders, but some species are capable of deposit feeding. Fossil lacustrine bivalves are very well represented in the fossil record, preserved by their shells, and also by their burrows.

The *Gastropoda* (snails and slugs) are mollusks that possess a univalved, coiled shell (secondarily lost in slugs, or with very open coiling in freshwater limpets) composed of layered proteinaceous material and calcium carbonate, a distinct head, a rasping feeding organ called a *radula*, and a muscular foot for locomotion. Two major groups of gastropods are represented in freshwater. The

Prosobranchia, a polyphyletic group of aquatic snails, are almost entirely aquatic, and possess a gill and an *operculum*, a rigid trap door that is used to protect the animal when it withdraws inside its shell (figure 5.4Q). The Pulmonata are primitively terrestrial mollusks, but have reinvaded freshwater. They have a lung (formed within the mantle cavity) and lack an operculum (figure 5.4R).

Several thousand species of freshwater gastropods have been described, all of which are benthic animals. They can be found crawling over rocky, macrophytic, or soft substrate surfaces, inhabiting crevices, or living infaunally, in burrows or interstitially between sediment grains. Freshwater gastropods include herbivores, deposit feeders, scavengers, and filter feeders (Dillon, 2000). Most species feed on epiphytic and epilithic algae; fewer species consume macrophytes directly, presumably because of their lower nutritive content (Russell-Hunter, 1978; K.M. Brown, 1991). Gastropods are common and conspicuous fossils in lake deposits, represented by their shells, and less commonly, their opercula.

Controls on Species Distribution

Paleolimnological studies of fossils are often conducted because of the presumed sensitivity of organisms to changes in lake hydroclimate. The ability to infer abiotic factors such as past temperature, water depth, light penetration, or water chemistry from fossils allows paleolimnologists to make statements concerning climate change, change in nutrient loading, or changes in watershed characteristics. Biotic factors such as predation also regulate species distributions, and inferences about their past importance may also be gleaned from the fossil record of lakes.

In some cases organisms may respond quite directly to an environmental variable of interest. For example, a species of fish may have very specific thermal requirements for successful maturation of its eggs. If we then find that species represented as a fossil, we might infer a range of paleotemperatures present at the site and time of fossil accumulation, assuming the fossil organism we are examining had the same physiological requirements as the modern population from which we draw our inference.

More commonly, environmental variables operate collectively to impact an organism or community. Physiological experiments might determine that a particular quantity of dissolved oxygen is required by a species of invertebrate of interest to a paleolimnologist. However, a higher concentration may have been necessary to maintain the increased metabolic rate this species would have experienced if it were living at a higher temperature than that at which experimental data was collected. An organism's abundance may be correlated with several environmental variables, and it may not always be possible to determine which variables were responsible for the change observed.

Perhaps even more common, and unfortunately more complex, is the situation in which an environmental variable indirectly affects a species' distribution pattern. Numerous studies have shown particular benthic invertebrates to have specific water depth ranges. Paleowater depth is a variable of great interest to paleolimnologists, since changes in water depth over time may allow us to establish a hydrological budget for a lake, often a major step in understanding regional climate for the time of deposition. However, benthic organisms rarely respond to water depth per se. More commonly their distributions are linked to water depth through some intermediate steps, such as water column algal concentration or sediment turbidity. A benthic grazing invertebrate dependent on epiphytic algae growing on macrophytes may well show a consistently measurable depth distribution in a given lake, but it has additional factors separating its distribution pattern from the original variable of interest. Are the appropriate epiphytes available for foraging? Is competition or predation excluding the invertebrate from parts of its potential range? The implication is that our attempts to interpret past environmental variables from fossil species occurrences must be tempered by a clear understanding of how those variables individually and collectively affect both species and species interactions.

Finally, the sensitivity that organisms display to environmental variables is subject to evolution over time. Generalizations that we make in paleolimnology about the sensitivity of a particular species or clade to some variable of interest must be tempered with the understanding that our ability to make these inferences weakens as we go back farther in time.

Abiotic Factors

Light

Light is perhaps the single most critical abiotic factor structuring lake ecosystems. Light availability controls the vertical extent of photosynthesis within the water column, thereby determining the size of the littoral zone and the volume of open water in which primary production can occur. Light is also required by predatory animals that hunt visually, and as a cue for vertical migration among many zooplankton species.

The transmission of light through water is dependent on the specific wavelength of light involved, as well as the color and clarity of the water. Because photosynthetic pigments absorb light energy with varying efficiencies at differing wavelengths, selection occurs for autotrophs with pigment combinations that can optimally use available light. Colored or turbid waters have different light transmittance characteristics than clear water and may therefore favor different groups of autotrophs. Many cyanobacteria and diatom species, as well as some mosses, have pigment combinations that allow them to grow under lower light conditions than most chlorophytes or angiosperms (Wetzel, 1988; Hill, 1996; Van den Hoek et al., 1995; Lampert and Sommer, 1997).

Adaptations to low light conditions notwithstanding, there is a general pattern of decreasing species diversity of autotrophs with decreased light (increasing depth or turbidity), as fewer can continue to photosynthesize under those conditions. Most macrophytes and phytoplankton are restricted to depths where the incident light is approximately 1% or greater than that at the surface. At about this depth gross photosynthesis is matched by respiration (referred to as the *compensation depth*). For submerged macrophytic angiosperms this depth may be anywhere from less than a meter (in turbid water) to as much as 12 m (Wetzel, 1988, 2000; Lampert and Sommer, 1997), but cyanobacteria and aquatic mosses, with their low light pigment combinations, have been found as deep as 122 m in ultraoligotrophic lakes (Hutchinson, 1975; Spence, 1982). Light intensity also structures autotrophic communities at the water surface. In the shallow, littoral zone shading effects of dense masses of emergent macrophytes reduce the light available on the lake floor, often eliminating benthic competitors in the process. Algal growth can also be inhibited under the excessive light levels (especially UV levels) found near the water surface, causing the depth of maximum photosynthesis to occur below the surface (Hill, 1996). In addition to the direct controls light exerts on the extent of the littoral and photic zones, light also indirectly controls the distribution of herbivores. Both the abundance and diversity of many groups of invertebrates reach their maxima within the littoral zone (Crisman, 1978; Jónasson, 1978; Paterson, 1993; James et al., 1998). Grazing selectivity further limits the distribution of herbivorous species to particular light levels (Lodge, 1985, 1986). In paleolimnological records the extent of the littoral zone and aspects of its internal structure may be recorded as changes in fossil macrophyte, benthic diatom, or herbivore abundance, differences in carbon: nitrogen ratios, differences in preserved pigment concentrations, or differences in stromatolite morphology.

Temperature

The most direct effects of water temperature on aquatic organisms are related to changes in metabolic rates. Over the normal range of temperatures tolerated by freshwater organisms (0–30°C) there is frequently a strong relationship between increased temperature and increased productivity. However, all organisms have ultimate thermal limits beyond which metabolism ceases and the organism will die. Few organisms can tolerate temperatures in excess of 50°C, and only prokaryotes are found in lakes above ~ 60°C (Melack, 1988). A more restricted set of temperatures circumscribes viable populations because of limitations on successful reproduction. A still more limited temperature range allows optimal growth, what is probably recorded when populations are abundantly preserved as fossils. As with other factors, the effects of temperature can rarely be isolated because of their interactions with such variables as water chemistry, especially oxygen solubility, as well as the effect of temperature on interacting species. Also, organisms such as insects, with partly terrestrial/partly aquatic life histories, are influenced by a much broader range of air temperatures, which may actually exert a greater influence on their distribution than water temperatures per se.

Temperature ranges for various types of thermal controls are still poorly known for most aquatic organisms. However, some generalizations of value to paleolimnologists can be drawn. The dominant class of periphytic algae commonly shifts from diatoms (< 20°C) to chlorophytes (15–30°C) to cyanobacteria (> 30°C) with increasing temperature. These trends are reflected in both species diversity patterns and overall biomass (DeNicola, 1996). Extremely cold-water environments are also characterized by specialized assemblages of chlorophytes (Melack, 1988). A variety of heat-loving (*thermophilous*) cyanobacteria occur in lakes and ponds surrounding geothermal springs, and may be useful paleoecological indicators of such environments when found as fossil stromatolites (Brock, 1978; Renaut and Tiercelin, 1994). Metazoans as a group have much more limited temperature ranges than algal or cyanobacterial groups. Individual species distributions may be more circumscribed, and their occurrence as fossils is correspondingly more informative. Insect species' distributions, especially for midges, are often strongly linked to temperature, particularly through the controls temperature exerts on periods of hatching and fecundity (Rossaro, 1992; Ward, 1992; I.R. Walker et al., 1997). Restricted temperature ranges are also known from other paleolimnologically important groups, such as ostracodes and diatoms (Carbonel et al., 1988; Vinson and Rushforth, 1989; Cox, 1993; DeNicola, 1996).

Dissolved Oxygen

The importance of dissolved oxygen concentrations in regulating the distribution of aquatic organisms has been recognized for almost a century, since the pioneering work of Thienemann on chironomids in European lakes (Thienemann, 1913, 1922). With few exceptions metazoans require oxygen for aerobic respiration. Numerous vertebrates and invertebrates can survive in low oxygen (*dysaerobic*) environments, but very few species can survive for prolonged periods with no oxygen. Profundal macroinvertebrate biomass normally declines rapidly below the epilimnion in lakes that display clinograde oxygen profiles (Dinsmore et al., 1999). Some oligochaete and chironomid species survive under dysaerobic conditions by reducing respiration rates to very low levels, or by using specialized blood, gill, and water-pumping adaptations (Brinkhurst and Gelder, 1991; Heinis et al., 1994). Fish species inhabiting dysaerobic environments often have enlarged gill surface areas (Greenwood, 1961; Beadle, 1981). In some organisms, low oxygen levels in the hypolimnion elicit behavioral responses, such as migration into shallower oxygenated water, or the formation of resting phases that have extremely low respiration rates (Rabette and Lair, 1998). Amphibious organisms, such as pulmonate gastropods or adult water bugs, breathe through lungs or trachea, and such animals may forage in dysaerobic or anaerobic water, revisiting the surface periodically. This limits their depth distribution to the upper few meters of a lake (Harman, 1974; Ward, 1992).

It has long been known that significant differences in community structure exist between waters with low and high oxygen concentrations (Thienemann, 1920; Saether, 1975). Chironomids, along with tubificid oligochaetes, often dominate the benthos of eutrophic lakes with low bottom water oxygen concentrations, and the two groups may be the only metazoans present under conditions of extremely high biological oxygen demand, surviving even through periods of anoxia (Roback, 1974). Long-term studies of chironomid communities have been made to determine their systematic responses to deoxygenation. A 60-year record of chironomid communities and oxygen concentrations in Lake Geneva indicates a progressive loss of deep-water species as cultural eutrophication and deep-water anoxia increased (Lodz-Crozet and Lachavanne, 1994) (figure 5.5). In shallow water where nutrient loads increased but oxygen concentrations remained adequate for chironomids, the response has been an increase in both species abundance and diversity. However, paleolimnologists need to interpret diversity trends in profundal chironomid communities with caution, as hypolimnetic aeration experiments have in some cases resulted in a decline in midge species diversity (Dinsmore and Prepas, 1997). The interactive relationships between water temperature, dissolved oxygen (DO), and productivity can also result in surprising responses to total DO. Under conditions of either high DO and low water temperature, or warm temperatures but low DO, invertebrate biomass and secondary productivity can be depressed (figure 5.6). Maximum abundances of invertebrates under such a gradient will occur at intermediate levels of DO content.

Oxygen concentration effects have also been investigated for other paleolimnologically important clades. Variations in oxygen requirements for ostracode (Delorme, 1982; Newrkla, 1985), mollusk (McMahon, 1991; Mouthon, 1992; Bechara, 1996), and fish species (Tonn and Magnuson, 1982; Rahel, 1984) all provide potential tools for the paleolimnologist investigating past O_2 availability.

pH

Since the early twentieth century it has been known that many organisms are sensitive to changes in lake acidity (Kolbe, 1927; Hustedt, 1939). This relationship has attracted considerable attention because of the potential ecological effects of acidic precipitation on lakes. Paleolimnologists have been interested in this relationship because of the potential for lacustrine fossils to record changes in lake communities over a protracted period of acidification.

Extreme acidity affects many physiological processes in algae, macrophytes, and animals, mostly in ways that reduce viability. Algal species are probably most strongly impacted by the indirect effects of low pH, such as decreased dissolved inorganic carbon and increased solubility of dissolved inorganic nitrogen and toxic metals such as aluminum, mercury, and cadmium (Planas, 1996). Aquatic

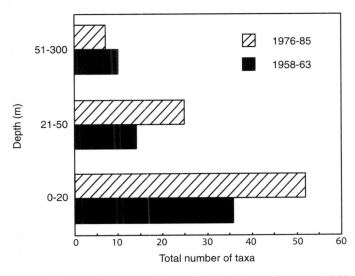

Figure 5.5. Bathymetric distribution of chironomid taxa in Lake Geneva between 1958–63 and 1976–85. Note consistent differences between littoral and profundal species richness, and the decline in species richness over time in the profundal zone, probably the result of decreased oxygenation of deep water. From Lodz-Crozet and Lachavanne (1994).

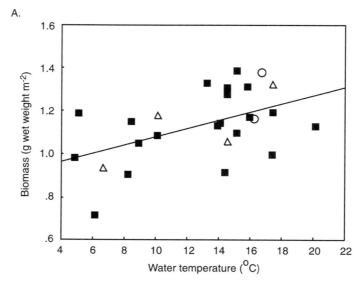

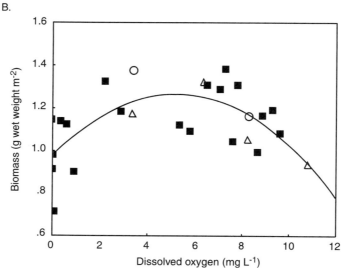

Figure 5.6. Relationship between profundal macroinvertebrate biomass (PMB) and (A) water temperature and (B) hypolimnetic dissolved oxygen concentration for 26 boreal lakes in northern Alberta. Solid squares, lakes with unimpacted catchments; open triangles, lakes with catchments impacted by forest fire; open circles, lakes with catchments impacted by timber harvesting. From Dinsmore et al. (1999).

metazoans suffer reduced oxygen diffusion across gill membranes at low pH, and a variety of complications with osmoregulation (Jobling, 1995). Community-wide effects of low pH often result from elimination of *keystone* species, such as predatory fish or larger benthic macroinvertebrates. Conversely, increases in littoral, filamentous,

green algae under acidic conditions may have profound effects on nutrient cycling within lakes and food quality for heterotrophs (D.W. Schindler et al., 1985; Turner et al., 1995). Another lakewide consequence of increased acidity is increased transparency, which in turn affects a wide variety of lake processes, such as redistribution of heat or extent of the littoral zone (Grahn et al., 1974; D.W. Schindler et al., 1985). Experimentally manipulated lakes show a wide range of consequences of acidification, most of which can be archived by paleolimnological records (table 5.3). An intolerance of acidity by algal species that normally inhabit neutral waters is reflected by a marked decline in species diversity during the course of these experiments (figure 5.7).

At less extreme conditions (circum-neutral pH) the effects of acidity are mostly a consequence of the relationship between acidity and carbon-species dissociation. Some species of algae or macrophytes can only take up inorganic carbon in specific forms, which may become unavailable as pH rises or falls.

Several paleolimnologically important clades show species-specific sensitivity to pH, and their assemblages can characterize the pH of modern or ancient lake waters quite accurately. Diatoms are particularly useful in this regard, as there are species assemblages spanning a very wide range of pH conditions (e.g., Gasse and Tekaia, 1983; Battarbee, 1984; Charles et al., 1989; Dixit et al., 1993). Chrysophytes are generally restricted to low–intermediate pH conditions, related to their obligate use of inorganic carbon for photosynthesis in the form of CO_2, which is unavailable at high pH (Royackers, 1986; Sandgren, 1988). Many chrysophyte species have specific ranges of lake acidity over which they are common (Duff and Smol, 1991). Testate amoebae also appear to be most common in mildly acidic (and oligotrophic) environments. Gastropods and fish are both eliminated from lakes at extremes of pH (Rahel, 1984; Lodge et al., 1987) and there is normally a marked decline in overall species diversity under these conditions.

Salinity and Solute Composition

The ability of aquatic organisms to osmoregulate sets limits on their physiological tolerance to changes in salinity or water composition. A wide range of metabolic activities in multicellular organisms relies on cells being able to maintain gradients of ionic concentration between their interior and exterior. In dilute freshwater, metazoans must maintain body fluids that are more concentrated, or *hypertonic*, relative to surrounding water, requiring a retention of salts. In contrast, organisms inhabiting hypersaline waters must be *hypotonic* with respect to ambient lake water, involving excretion of salt and water retention. Because salinity is highly variable in most saline lakes, their biotas are frequently exposed to both hypertonic and hypotonic conditions, complicating adaptations to this environment. Some algae, like the chlorophyte

Dunaliella, or vascular plants, such as the macrophyte *Ruppia* (figure 5.3K), can survive by producing hypertonic intracellular fluids (Melack, 1988). Organisms that thrive under conditions of elevated salinity are called *halophiles*, and are of great interest to paleolimnologists because of their hydroclimatic significance when found as fossils.

Some organisms are more sensitive to external salinity changes than others and inhabit waters with relatively narrow ranges of dissolved solutes. Organisms that are poor osmoregulators in principle can make excellent indicators of paleochemical change in lake waters, because they can only be abundant in lakes with a narrow range of solute concentrations and compositions (Frey, 1993). In contrast, organisms that are good osmoregulators or osmoconformers typically occur over a wide range of salinities. Their presence per se is usually uninformative with regards to salinity, although species assemblages that are *exclusively* composed of such species are usually indicative of lakes that are either currently saline or that undergo dramatic fluctuations in salinity (W.D. Williams, 1998). Only a few organisms, such as the brine shrimp *Artemia salina*, have been shown to be physiologically restricted to high salinity lakes (D'Agostino, 1980; Bos et al., 1999) (figures 5.2G and 5.8). Groups of paleolimnologically important organisms with species known to be sensitive to modest salinity changes include the diatoms (Fritz, 1990; Cumming and Smol, 1993a), chrysophytes (Cumming et al., 1993), cladocerans (Frey, 1993), and ostracodes (Carbonel et al., 1988).

Under conditions of elevated salinity, species distribution is determined by both composition and concentration of dissolved solids. For example, some ostracodes are restricted to particular portions of the brine pathway "space" outlined in figure 4.8, being exclusively alkaline or hardwater species (Forester, 1986) (figure 5.9). Because of the phenomenon of $Ca^{2+} + Mg^{2+}$ depletion, lakes that are both highly alkaline and highly saline present special problems for aquatic organisms, since both Ca and Mg are requisite for growth in almost all metazoans. Maximum "salinity" thresholds under such conditions are likely to represent conditions where reduced Ca and Mg inhibits survival. This is most notable among organisms such as bivalves, which must secrete robust calcareous skeletons. Hyperalkalinity also affects lake ecosystems indirectly because of the rapid volatilization of ammonia that occurs under highly alkaline conditions. This condition favors the growth of cyanobacteria, which can fix atmospheric nitrogen, over other algae.

Table 5.3. Trends in aquatic ecosystems at the Experimental Lakes Area, Ontario, Canada that have been stressed by experimental acidification or eutrophication and suggested possible paleolimnological signals. "R-strategists" are "weedy" species that can undergo rapid reproduction rates, frequently have short life spans, and colonize new habitats quickly. Modified from D.W. Schindler (1990)

Response Type	Acidified Lakes (Lakes 223 and 302S)	Eutrophied Lakes (Lakes 227 and 226NE)	Possible Manifestation in Paleolimnological Archives
Energetics and Nutrient Cycling			
Community production/ respiration ratio	Increases	Increases	Organic matter flux rates, N-isotopes
Fate of unused production	No change in exported or unused primary production; decreased utilization of allochthonous inputs	Exported and unused production increases	Sedimentary P-profiles (meromictic or highly eutrophied systems only); accumulation of terrestrial organic matter; changing C/N ratios?
Rate of Nutrient Cycling	Minor disruption to carbon and nitrogen cycle; nitrogen cycle may decrease slightly	Increased rates of P, N, and C cycling	N-isotopes
Rate of nutrient loss	Nitrogen losses due to denitrification and outflow increase, but lost proportion of incoming N decreases; sulfur losses to sedimentation and outflow are proportional to inputs	Outflow and sedimentation losses of P, N, and C are proportional to inputs, therefore higher	P and S profiles? TOC profiles; N-isotopes
Community and Population Structure			
Proportion of R-strategists	Increase in R-strategists among zooplankton; decrease in R-strategists among fish	Increase in R-strategists among zooplankton	Flux of zooplankton fossil remains (esp. cladocerans, rotifers); fish remains in meromictic lakes
Size of organisms	Average size of phytoplankton increases, zooplankton and chironomids decrease; fish sizes increase	Phytoplankton increase, zooplankton decrease	Direct measurement of fossils in high-resolution deposits
Life spans of organisms	Life spans of fish and benthic crustaceans decrease	Life spans of crustacean zooplankton decrease	Estimation of average life span from skeletal parts of organisms with growth bands/rings (e.g., bivalves, fish otoliths) or size for species with determinate growth patterns with age
Species Diversity and Dominance	Decreases at all trophic levels examined	Decreases at all trophic levels examined	Direct estimates from well-preserved fossil groups (e.g., cladocerans, diatoms, ostracodes); fossil pigments

In inland saline lakes there is strong inverse correlation between salinity and both species richness and diversity, and many clades are eliminated at high salinities (Timms, 1998; W.D. Williams, 1998). As with pH, salinity effects at the community level are often indirect, often manifesting themselves through profound changes in biological interactions following the elimination of some key species (Wurtsbaugh, 1991). Diversity changes that have been observed to occur over time as a salt lake

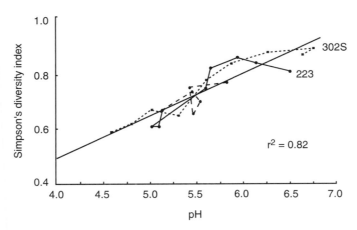

Figure 5.7. The relationship between pH and phytoplankton species diversity in experimentally acidified lakes, in the Experimental Lakes Area, Ontario, Canada. The dotted section for Lake 223 denotes the period of recovery (increasing pH). From D.W. Schindler (1990).

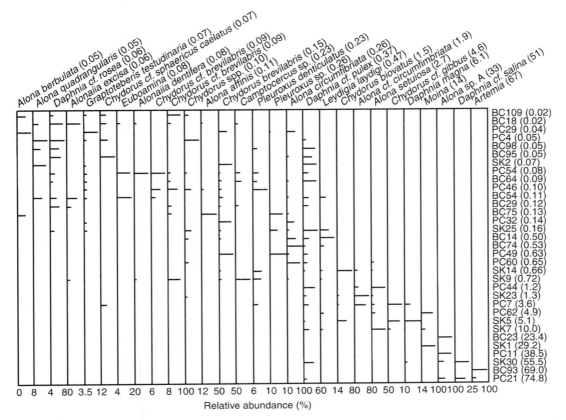

Figure 5.8. Relative abundance of microcrustacean (Cladoceran and Anostracan) remains from a series of lakes on the Interior Plateau of British Columbia, Canada, illustrating relationship of species distribution with salinity. Lakes are arranged in order (top to bottom) of increasing salinity. Values in brackets for lakes are salinity in grams per liter. Values in brackets after species names are the weighted average optimum salinity for that species. From Bos et al. (1999).

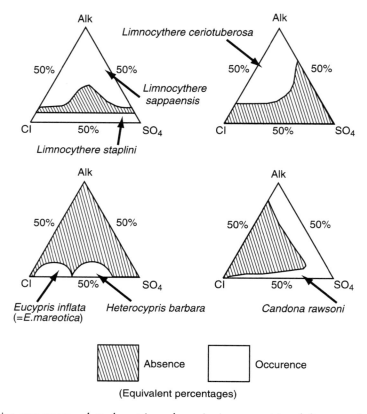

Figure 5.9. Species occurrences plotted on triangular anionic compositional diagrams for several common ostracode species found in North America and Europe. *Limnocythere ceriotuberosa*, *L. sappaensis*, *L. staplini*, and *Candona rawsoni* are all widespread North American species. *Eucypris inflata* (*E. mareatica*) and *Heterocypris barbara* are Spanish species. From Palacios-Fest et al. (1994), after data from Baltanás et al. (1990), and Forester (1986).

changes in salinity have important implications for paleolimnologists (figure 5.10). As a closed-basin lake rises and salinity falls, increasing diversity is usually most evident at the lake margins. In the central part of the basin saline groundwater may continue to accumulate and/or the lake may be chemically stratified, maintaining the original low diversity. A signal of changing surface inflow and average salinity may not be apparent under such circumstances if the paleolimnological signals being interpreted only come from the center of the basin.

The trend toward decreasing species diversity with increasing salinity in continental saline lakes is in sharp contrast with the species diversity gradient from freshwater to brackish/estuarine/coastal lakes to fully marine conditions. Here diversity is highest at the marine end of the spectrum, generally somewhat lower in fully freshwater, and much lower in the intermediate salinities at the fresh-

water–marine interface, where conditions are highly unstable.

Nutrients

Anyone who has casually observed a seasonal or longer-term buildup of large quantities of "pond scum" around the edges of urban lakes has seen qualitative evidence of the important role nutrients have in structuring lacustrine communities. Nutrient input establishes the "potential productivity of lakes," whereas the realized productivity of a lake results from the interplay of abiotic inputs and biotic interactions (Kitchell and Carpenter, 1993). The role nutrients play in algal and macrophyte communities has been quantitatively studied through a combination of empirical observations of lakes in regions experiencing changes in nutrient loading, manipulations of whole lakes or parts of

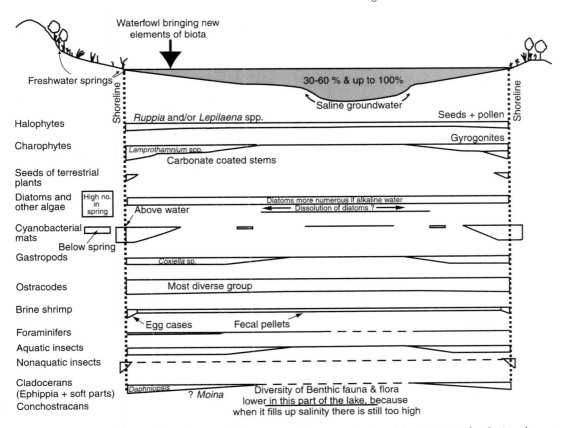

Figure 5.10. Typical spatial distribution of biota across the floor of saline lakes in Australia during the wet season, when lakes are full. Note that all members of the biota shown here do not necessarily occur together in the same lake; this depends on numerous factors such as water salinity and ionic composition, temperature, water type (ephemeral vs. permanent), water turbidity, presence of aquatic vegetation, water depth, and so on. If a lake such as the one shown here dries annually, the salinity would still be too high ($\gg 10\%$) for many organisms, such as gastropods, halophytes, and charophytes, to hatch or start growing when water first fills its deepest depression. Not until the lake fills further does salinity drop (e.g., between 3 and 6%) and the biota become more diversified. From De Deckker (1988a).

lakes (mesocosms) through nutrient addition experiments, laboratory experiments, and long-term studies using paleolimnology.

Limnologists have long recognized that particular algal groups are associated in a general way with either low or elevated levels of nutrient input. Oligotrophic lakes have algal communities that are commonly dominated by desmids (green algae) or chrysophytes, whereas eutrophic lakes are more commonly dominated by assemblages of diatoms and cyanobacteria. The abundance of chrysophytes in oligotrophic lakes probably results from their competitive advantages over other phytoplankton under conditions of low nutrient availability. In contrast, under eutrophic conditions chrysophytes are both outcompeted by other phytoplankton and heavily grazed by zooplankton, greatly reducing their abundance (Smol, 1995a; Van den Hoek et al., 1995).

Normally, limnologists are more concerned with the effect that supplies of specific nutrients have on autotrophs, rather than on a simplistic classification of trophic state. Marine phytoplankton biomass contains the major nutrients in a cellular $C : N : P$ ratio of $106 : 16 : 1$ (Redfield, 1958). This *Redfield ratio* represents a bulk average of many species, and individual freshwater taxa may deviate from this ratio. Nevertheless, it represents a useful mean from which to start our consideration of nutrient limitations. In many situations nutrients are not present in excessive supply (i.e., more abundant than can be consumed by the individuals present).

In such cases a phytoplankton community that conforms to the Redfield ratio will be phosphorus-limited when the $N : P$ cellular ratio is much greater than $16 : 1$ (conventionally $> 20 : 1$) and will be N-limited when $N : P$ is much less than $16 : 1$ (conventionally $< 10 : 1$). Silica, which is not considered in the Redfield ratio, is a primary nutrient for diatoms, and its depletion can cause diatom species to decline.

Specific $N : P$ ratios select for species that grow optimally under those conditions, thereby creating a mechanism for community change to occur as some species gain competitive advantages over others (Sommer, 1990). Nevertheless, some generalizations about nutrient effects on lakewide productivity can be made with reference to seasonality, specific geographical regions, or lake types. Lakes in the north-temperate regions of the world respond more strongly in terms of increased productivity (i.e., grams C fixed per square meter per day) to the addition of phosphorus than any other nutrient (D.W. Schindler, 1974; Borchardt, 1996; Wetzel, 2000). In warmer climates results of enrichment experiments are more equivocal, with some productivity increases in some lakes responding most strongly to P additions, some to N or Si, and some to multiple nutrients, depending on the season or algal community present. Many lakes are characterized by a great degree of annual or aperiodic variability in phytoplankton composition, and in some cases these community changes can be linked to changing nutrient availability (Leland and Berkas, 1998).

Phosphorus enrichment can result not only in major increases in primary productivity and eutrophication, but also in major changes at the species level in algal communities, forming the basis for most paleolimnological studies of nutrient loading, in particular using diatoms (table 5.3 and figure 5.11). Extreme eutrophication of lakes is often correlated with declining phytoplankton species diversity, and in lakes that have been experimentally enriched with nutrients a decline in diversity often occurs (Cottingham and Carpenter, 1998). Conversely, in lakes that have seen significant declines in phosphorus input, for example from sewage diversion, total diversity is often observed to increase over time (Willén, 1992; Ruggiu et al., 1998) (figure 5.12). Phosphorus-induced eutrophication affects a variety of limnological variables, such as biological oxygen demand and available dissolved oxygen. As a result, other lake organisms frequently respond to nutrient enrichment, and may serve as indirect paleolimnological indicators of changing trophic status and phosphorus enrichment

(e.g., Hoffman, 1996; Findlay et al., 1998; Belis et al., 1999; Brenner et al., 1999).

Water Turbulence and Substrate

Changes in benthic communities in lakes are commonly associated with substrate boundaries. The relationship may be direct, as in the case where an organism requires a specific particle size for foraging, or indirect, for example when both a soft substrate texture and an organism's distribution are regulated by some third variable such as turbulence.

Lake floor particle size is closely linked to turbulence, wave energy, and proximity to shoreline. In most cases the relative effects of these linked variables on benthic communities are extremely difficult to disentangle. Distinct compositional changes in the benthos are often associated with the environmental transition from surf zone to deeper water (Jónasson, 1978). Sandy substrates within the surf zone of large lakes or within the discharge areas of deltas present a mobile substrate that is difficult for most organisms to colonize and diversity tends to be low in these areas, dominated by interstitial microinvertebrates (Fuller, 1974). Cobble and rock surfaces in the surf zone provide attachment surfaces for epilithic diatoms, cyanobacteria, and green algae, and their variable levels of turbulence promote correspondingly higher diversity among bryozoans, sponges, crawling insects, crustaceans, and mollusks (Cooper, 1984; Burkholder, 1996).

In small lakes or the sheltered coastlines of large lakes macrophytes may be abundant, providing attachment surfaces for epiphytic algae, forage for algal and macrophyte grazers (including many gastropods), and forage and cover for many fish species. The importance of such refuges appears to increase with increasing nearshore turbulence (Fischer and Eckmann, 1997). Some macrophytes, for example certain species of *Chara*, require a combination of light and anaerobic sediment (generated by macrophyte debris) in order to germinate, and are therefore restricted to a narrow range of littoral habitats (Hutchinson, 1975). This region is usually marked by high species diversity, notably for our purposes among several readily fossilized clades, including the gastropods and chydorid cladocerans (Crisman, 1978; Michel, 1994; James et al., 1998).

Deeper parts of the littoral zone are often characterized by an abundance of filter-feeding invertebrates, particularly bivalves, and an increase in deposit feeders. Below normal wave base terrigenous particle size tends to fine significantly, and

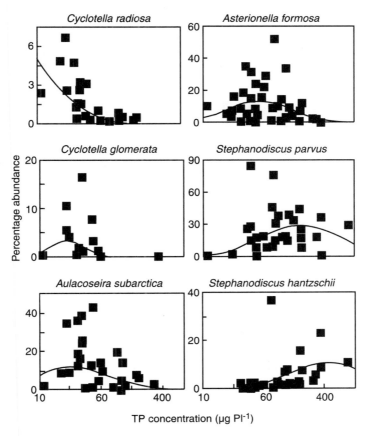

Figure 5.11. Species distribution of six common planktonic diatoms in surficial sediments of Northern Irish lakes, which illustrates their changing abundance along a phosphorus gradient. The vertical axes are relative abundances (of total diatom count); the horizontal axis is total phosphorus (TP) on a log scale; only positive occurrences are shown. The weighted-average optimum for each species is indicated by the vertical dash on the upper horizontal axis and the Gaussian log-fit curves fitted to the abundances are indicated. From N.J. Anderson (1997a).

detritus feeding dominates (Kilgour and Mackie, 1988; Machena and Kautsky, 1988). Invertebrate community structure in detritus feeding-dominated habitats is strongly influenced by the quality and quantity of particulate food available, which in turn is linked to particle size and current transport mechanisms (Cohen, 1984; R.K. Johnson and Wiederholm, 1989). Species diversity is generally low in profundal lake environments, dominated by chironomids, chaoborids, ostracodes, oligochaetes, and bivalves.

Substrate, turbulence, light gradients, and oxygen gradients frequently operate in concert to produce complex zonation patterns in lakes, in which it is difficult if not impossible to disentangle the "ultimate" causes from the community pattern. These interactions should caution us against overeager searches for "causality" in explaining modern lake ecosystem structure, and all the more so in paleolimnology.

Habitat Heterogeneity

Habitat heterogeneity introduces structural variation into the external environment inhabited by organisms. These differences in food resources, shelter, and breeding opportunities can be a source of local advantages for different species. For example, a rock on an otherwise sandy lake floor provides protection for small animals from predation, and attachment points for sessile organisms, conditions that would not exist in that place in the rock's

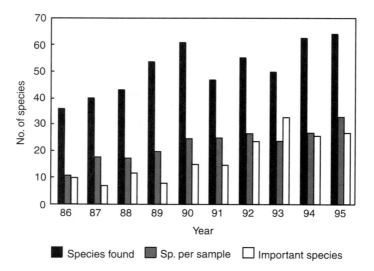

Figure 5.12. Number of differently categorized phytoplankton species in Lake Maggiore (Italy) through a period of declining nutrient input and improving water quality (oligotrophication) from 1986 to 1995. Species found (solid bars) = total number of identified species; Sp. Per sample = annual average number of species per sample; Important species (solid white bars) = number of species which, at least on one occasion, make up part of the dominant species group, those making up the most abundant 80% of the total biovolume in a series of ranked species. From Ruggiu et al. (1998).

absence. Both theory and observation suggest that heterogeneous environments can lead to increased species diversity for both competitors and commensal or symbiotic organisms. This occurs because numerous habitat patches confer advantages to differing species in small areas, allowing many to coexist over a larger area (Mittlebach, 1988; Benson and Magnuson, 1992; Eklöv, 1997). Habitat heterogeneity can occur at any spatial scale and may change over time as well, leading to temporal change in species diversity that is detectable by paleolimnologists. It is easiest to visualize habitat heterogeneity in terms of benthic habitats, but even open-water environments can be heterogeneous, the result of discrete patches of water with different physical or chemical characteristics.

Macrophytes and littoral rocky substrates probably provide the strongest habitat-structuring elements in the benthic environment of lakes, and most lakes display declining offshore gradients of species diversity that are related to one or both of these sources of heterogeneity. Macrophytes directly determine the extent of herbivores and animals dependent on littoral plant cover (Rossier et al., 1996). They provide refuges from predation (Hershey, 1985), and reduce aggregation of prey

species, particularly when prey occur at low densities (Hosn and Downing, 1994). Furthermore, the increased surface area of macrophytes can inflate the number of epiphytic organisms that can inhabit surfaces, and provide a greater spectrum of light, nutrient, and temperature levels for them to exploit (Carpenter and Lodge, 1986; Sher-Kaul et al., 1995). All of these causes for higher diversity in macrophytic habitats also hold for rocky substrates (Schmude et al., 1998). Conversely, factors that reduce this heterogeneity, such as anthropogenic siltation, also result in declining species diversity (Alin et al., 1999b).

Biotic Factors

Aquatic ecologists have assembled a large body of data suggesting that biotic factors, such as grazing, predation, competition, and symbiotic interactions, are at least as important in regulating ecosystem structure and function in lakes as the "abiotic" factors discussed above. Many of these interactions have such complex feedback loops that their interpretation over short-term studies remains obscure even for experimental manipulations (e.g., Carpenter, 1988). So why should paleolimnologists worry about them? I can think of three important reasons. First, some interactions do leave clear archived signals in fossil or biogeochemical remains for the interpretation of paleolimnologists (e.g., Leavitt et al., 1994a). Second, the theoretical understanding of these interactions often suggests that they are played out over long periods of time (perhaps centuries or more in some cases), but experimental and field observational ecology rarely if ever provides a framework for investigating phenomena over that time duration. Finally, many interpretations of aquatic ecological data have been based on assumptions about prior lake history that may or may not be accurate. By understanding the questions ecologists are trying to ask and providing relevant data, paleolimnologists can be important contributors in testing or falsifying hypotheses and resolving important debates in biology.

Probably the most important biological interactions in lakes are those involved in *food webs*, networks of predators, herbivores, autotrophs, and detritus feeders linked by their pathways of consumption and energy flow (figure 5.13). Arrows on this diagram show the direction of consumption, who is eating whom. Notice that alternative food resource pathways exist for various consumers. Planktivorous fish may feed on either smaller planktivorous invertebrates, in this case chaoborid fly larvae, or on large herbivores, such as the *Daphnia* cladoceran, depending on preference or availability.

How are population sizes of members of this or any food web regulated? One viewpoint holds that

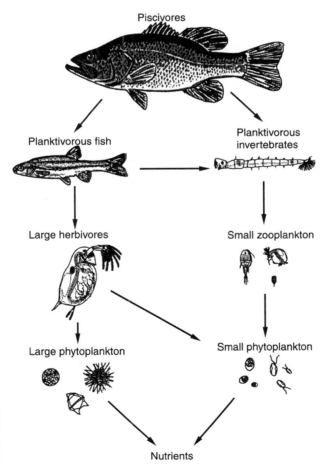

Figure 5.13. Major interactions of the trophic cascade observed in studies of experimental lakes from Wisconsin (United States) (see chapter 1 for additional discussion). Nutrients (mainly inorganic phosphorus and nitrogen) are provided by inputs from the watershed and recycling by animals. From Carpenter and Kitchell (1993).

the dynamics of population growth and community formation is controlled by the external input of nutrients or light, allowing food to be created and consumed at progressively higher trophic levels. Put in more nutrients and there will be more primary producers, allowing more herbivores, and so on. This *bottom-up* notion views the biological outcome (community structure) as resulting largely from abiotic forcing, with biotic interactions largely confined to competition at each trophic level for the food resources from below. An opposing viewpoint sees the community structure as resulting from *top-down* effects, predation and consumption. Put in more piscivorous (fish-eating) fish and the planktivorous fish species being consumed will decline in abundance, causing a subsequent *increase* in the large herbivores, and so on. Control in the most extreme version of this viewpoint is strictly biotic, and even nutrient availability might be seen as exclusively the result of organisms if essentially all nutrients are recycled in the lake. More commonly, proponents of the trophic cascade hypothesis (chapter 1 and figure 1.7) see predation and selective consumption as regulators of community structure operating within a context of available nutrients and energy (Kitchell and Carpenter, 1993).

Not surprisingly, the real world of food-web controls comprises elements of both top-down and bottom-up effects, a result also supported by theory and modeling (e.g., Bartell et al., 1988). Experimental manipulations of ecosystems within single lakes, for example adding or eliminating piscivorous fish, tend to support the trophic cascade model, whereas regional comparisons of lakes experiencing different levels of nutrient input provide evidence for bottom-up effects as dominant. Some authors have tried to synthesize the two viewpoints (McQueen et al., 1989), or suggest that the differences in observations of relatively short-term lake manipulation experiments versus regional patterns of lake community structure (presumably developed over centuries or millennia) are a result of differences in temporal scale of observation (Lampert and Sommer, 1997). Paleolimnological archives of community structure that are both highly resolved and that cover long time intervals have an important role to play in resolving this scaling issue. At the same time, it is important for paleolimnologists interested in these problems to realize that complex interactions in food webs (directed both from and to consumers) often lead to unexpected outcomes in community structure, productivity, and diversity, and in some cases to entirely different community states. Although food-web hypotheses are in principal amenable to

"long-term" testing and perhaps falsification using paleolimnological methods, paleolimnologists should enter this zone with their eyes wide open to the limits of their prior understanding of the system they are studying.

Species Diversity

Paleolimnologists commonly record the diversity of species found in a fossil assemblage, and try to make paleoecological interpretations from these data. But why do some lakes have many or few species? And when and under what circumstances did they acquire those species? These represent arguably the broadest questions about the controls on species distribution in lakes, and answering them normally requires a comparative or biogeographical approach, using data covering large spatial scales and very long temporal scales. As we have already seen, a number of abiotic and biotic variables, such as salinity, pH, temperature, or grazing, have local effects on diversity, and it is tempting (and perhaps often justified) for paleolimnologists to interpret diversity trends over time strictly on the basis of changes in such variables. However, this list misses some important variables that are related to species diversity in lakes, but that operate over much longer time scales or larger areas than the types of processes we have discussed so far. We also need to consider variables such as dispersal processes for organisms and lake history and stability. These long-term factors are evidenced in the common relationship between species diversity and lake area.

Species diversity is strongly correlated with habitat area for many, although not all, lacustrine ecosystems (Barbour and Brown, 1974; Eadie and Keast, 1984; Brönmark, 1985; Cantrell, 1988). As a group, larger lakes may house more species than smaller lakes for several reasons. First, large lakes on average are more persistent habitats than small lakes, and their populations are therefore less frequently subject to local extinction. Second, large lakes encompass a greater range of habitats on average than small lakes. Third, large lakes may be located closer, as a whole, to more centers of species origin than small lakes. And finally, large lakes may be more connected to a larger number of potential immigrant species pools than small lakes. However, it has been known for many years that phytoplankton and zooplankton diversity show little relationship with lake area (Ruttner, 1952; Järnefelt, 1956). In fact some of the largest lakes, Baikal, Tanganyika, and Malawi, which house extremely diverse benthic invertebrate and fish faunas, have remarkably simple and species-poor planktonic communities (Dumont, 1994). Clearly other biogeographical or historical factors are at work besides area.

Several important candidates are known. First, the biota of lakes can be thought of as a "sample" of a regional species pool (Koskenniemi, 1994; Neill, 1994). The pool's cumulative diversity is the result of regional historical processes operating on long time scales (figure 5.14). The degree to which any particular lake is able to sample its pool depends on two things:

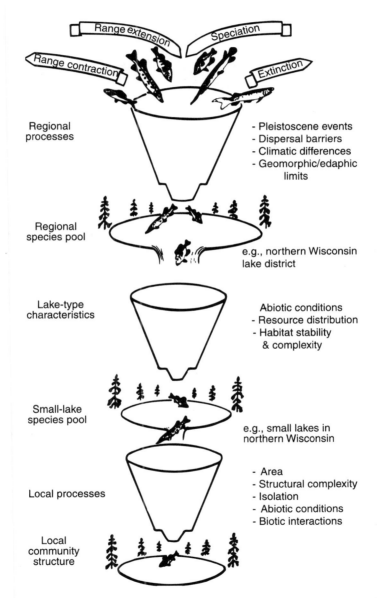

Figure 5.14. Conceptual model of the origin and maintenance of lacustrine species assemblages, illustrating the interplay of regional and local processes, and how local lakes sample a broader regional species pool. The specific illustration is for fish in Jude Lake, a small forest lake in Wisconsin, but the point of how local species assemblages originate is general for all organisms, lake types, and regions. From Tonn et al. (1990).

1. How interconnected the lake is to the entire pool of potentially colonizing species, from the perspective of the organism; and

2. How appropriate the lake is environmentally for any given member of the pool, that is the abiotic and biotic factors we have already considered.

Some species, particularly wind- or bird-dispersed insects or algae, may move between lakes more readily than others, for example profundal macro-invertebrates or fish. A high degree of interconnectedness may reduce the difficulty of colonization, making a series of lakes appear to the organisms in question like one large lake. In this case area may be unimportant as a determinant of diversity relative to factors like water chemistry or species interactions. Lake history is also important. Lakes are well known for their ability to isolate populations genetically from other lakes, and a long history of in situ speciation can result in a lake having far more diversity than would be predicted by lake area alone. Conversely, lakes in regions that have been very recently glaciated may have simply had insufficient time for colonization, let alone speciation, to occur.

Lakes that have persisted for long periods without undergoing extreme fluctuations in salinity, temperature, or pH are particularly likely to accumulate locally evolved species beyond what might be expected based on area alone, and consequently have higher than average species diversity for their area (Martens et al., 1994). Most lakes, however, are geologically ephemeral, or undergo major changes in abiotic conditions over their history, even over time scales of as little as decades to centuries. Their biological communities may be subject to major perturbations as a result of frequent and drastic change, or even complete annihilation if the lake dries completely.

Summary

Our conclusions from this chapter can be summarized as follows:

1. Biological processes in lakes, archived for paleolimnologists by fossils, form one of the richest sources of information for the paleolimnologist. Interpreting these archives requires a basic understanding of the relevant abiotic and biotic controls on species distribution, diversity, and community structure.

2. Lake habitats are structured by gradients in light, turbulence, and proximity to the lake floor. Very different communities of organisms occupy the habitats generated by this structure.

3. Planktonic and nektonic organisms, including various preservable groups of algae, rotifers, crustaceans, insects, and fish, occupy open-water habitats, and their fossils record paleolimnological conditions at the surface or at intermediate depths of the water column.

4. Benthic organisms, including various algae, vascular plants, protists, sponges, bryozoans, annelids, crustaceans, insects, and mollusks, provide paleolimnologists with records of conditions at the lake floor, and differentiate gradients across the lake bottom in light penetration, turbulence, and oxygen availability.

5. Abiotic factors, such as light, dissolved oxygen, pH, salinity, nutrient availability, water turbulence, substrate, as well as the degree of local heterogeneity in all of these factors, are major determinants in the species composition of a part of a lake. To the extent that individual species are sensitive to changes in particular factors, they can be used to infer past limnological conditions.

6. Species interactions, such as predation, are also important regulators of species composition in lakes, and can sometimes be inferred from paleorecords. Abiotic and biotic variables, operating on various temporal and spatial scales, must all be considered by a paleolimnologist when they try to interpret the significance of fossil diversity or species composition data.

6

Age Determination in Lake Deposits

It is almost impossible to overemphasize the importance of good chronological control to paleolimnology. Age control allows us to determine rates of processes and fluxes of materials, and to test hypotheses of linkage between archives and hypothesized external controls of those archives. Geologists differentiate between relative age versus absolute dating methods. Relative age determinations are based on the concepts of superposition (older sediments are on the bottom, in the absence of tectonic disturbance) and lithological correlation. In contrast, absolute dating methods are done without necessary reference to other analyses or locations, to produce an age determination (i.e., 100,000 yr before present). Some methods, such as paleomagnetics, amino acid racemization, and biostratigraphy, lie in a gray area between these two, providing absolute dates or age ranges in certain circumstances and relative age constraints in others. In this book, I will refer to the general study of both relative and absolute age determination as *geochronology*, and use the term *geochronometry* to refer to absolute dating.

Relative Age Determination and Correlation in Lake Deposits

Lithological correlation involves matching similar lithologies between outcrop or core localities, allowing a network of age relationships to be established between various sites. This can be done at any scale, from within a lake to intercontinental, although lithostratigraphical correlations based on core or outcrop observations are most commonly useful only at a local, intrabasinal level. Correlation within basins is often achieved using reflection seismic stratigraphy. Depositional or unconformity surfaces can normally be recognized on seismic lines that extend over the scale of individual sub-basins

to entire lakes (Nelson et al., 1994; Lezzar et al., 1996; Van Rensbergen et al., 1998). When dated cores are obtained or outcrops studied along these seismic lines, a correlation network can be established, with probable ages attached to specific seismic horizons. Intrabasinal correlation can also be done by correlating distinctive patterns of change in features such as magnetic intensity, patterns of stable isotopic change in sediments, or biostratigraphical markers, that may be consistent across a lake basin. Sometimes, relative correlations can be made between lakes. The character of lake deposits can change over broad regions, when some forcing phenomenon, such as climate change, causes a regional change in sedimentation style, causing, for example, a regional shift toward slower sedimentation rates or increased organic carbon content in the sediments.

The most common approach to feature correlation between two cores or outcrop sequences of data is simple "wiggle matching," visually aligning the most prominent peaks and troughs in whatever variable is being correlated. This seat of the pants approach, however, only makes use of some of the data potentially available for correlation purposes. More sophisticated numerical techniques of relative correlation also exist, which attempt to correlate all points within two stratigraphical sections, by minimizing the dissimilarity between all parts of the records (Thompson and Clark, 1989).

Highly distinctive or unusual *marker horizons*, possessing unique lithological characteristics, such as volcanic ashes, can sometimes be used for cross-correlation within a lake basin or between adjacent lakes. Ultimately, however, the utility of relative correlation methods is limited in lake deposits by their high degree of lithological heterogeneity. This limitation, coupled with the rapid lake-level fluctuations observed in many lakes, makes lithostratigraphical correlation based on general sediment type difficult at best. As a result, absolute age determinations are an essential element of most paleolim-nological investigations.

Age Models and Display of Age-Modeled Data

Before we consider the various techniques available for geochronometry in lake deposits, it is useful to consider some general issues about absolute age data. These are concerns that face all paleolimnologists, which transcend analytical technique.

Does the age make sense? The first encounter that a paleolimnologist has with a disagreement between absolute ages and the law of superposition is always disconcerting (figure 6.1). We naturally expect to obtain ages that are younger in stratigraphically higher deposits, but sometimes this doesn't happen. There are several possible explanations for this. First, the material being dated may not be the same age as the stratigraphical horizon (the age of deposition). This can result in absolute ages younger or older than the age of deposition. In many cases, it is not immediately evident which specific date is disordered. When a single date is wildly out of accord with a series of other dates, both above and below, it is usually considered most parsimonious to reject the one, rather than the many. Paleolimnologists should keep in mind, however, that parsimony does not always lead us to truth. Second, all absolute ages are accompanied by uncertainty, usually expressed as a $\pm x$ yr notation. This uncertainty reflects the inherent statistical uncertainty in most measurement techniques. In radiometric dating methods, an uncertainty is introduced by the variation in particle counting statistics, or the accuracy of the measurement itself. These are usually expressed as ± 1 or 2 standard deviations (1 or 2σ) around the mean date. When the uncertainty is large in the ages of two stratigraphically adjacent, dated samples, the distribution of probable ages may actually overlap. A disordering may then come about simply from the "true" ages falling in the overlapping tails of the two distributions, so that the apparently younger sample (i.e., the sample with the younger mean derived from counting statistics) is in fact older. Third, there may be an error in the age determination. As much as we value geochronometric data, we must also cling to its validity lightly. Mistakes happen, because of laboratory problems, inadequacies of techniques, or mislabeling of samples. And fourth, superposition may be violated itself, from slumping in a core, or for tectonic reasons of thrust faulting, overturn of beds, and so on. All of these possibilities

should remain in the near recesses of the paleolimnologist's brain as any ages, even apparently good ones, are being considered. Careful observation of outcrop/core stratigraphy, handling, and labeling of samples, and choice of methodologies, all go a long way toward alleviating some of these problems, but most workers will face them eventually regardless of quality control.

How disordered age data should be handled is a matter of debate. There is almost always a strong temptation among earth scientists to exclude age data points that "don't fit" (figure 6.1B,C). Such exclusion is widely practiced, although of questionable scientific legitimacy. At a minimum, "bad" dates should be reported, and shown in the initial stratigraphical presentation of the data. Also, whenever dates are excluded there should be a good reason for their exclusion, which is unrelated to the age estimate itself, for example evidence of contamination. The same threshold of sample acceptability should then be applied to all samples to make sure they also fit the criteria.

Commonly, absolute age inconsistencies exist, but not to a degree that we want to completely disregard the significance of the data. This can be done by developing a general *age model* for the lake beds being studied. The purpose of an age model is to provide age estimates for intermediate stratigraphical levels, for which no direct date is available. The simplest way to do this, when age data is ordered, is through point-to-point interpolation (figure 6.1B,C). This approach is commonly used, particularly when chronological data is limited. However, there are two important weaknesses with this approach. First, there is a strong likelihood of error in the interpolated ages if few data points are involved. Also, this type of model assumes that sedimentation rate changes exactly coincided with data points. There is no particular reason to think this should be true. An alternative is to fit some form of regression to the data set. This produces a smoothing function that attempts to account for the totality of the data and its trends. Some age models attempt to smooth mass accumulation rates ($g\,cm^{-2}\,yr^{-1}$) as opposed to thickness ($mm\,yr^{-1}$) (Bischoff et al., 1997b). Mass accumulation rate models are also more appropriate than thickness models for very young sediments, where water content and compaction are highly variable. Linear, least squares regression models assume a constant sedimentation rate (figure 6.1D). These are generally used when an initial inspection of the data suggests this is a reasonable assumption. Note however that in the case shown, the simple linear regression model suggests a future age for

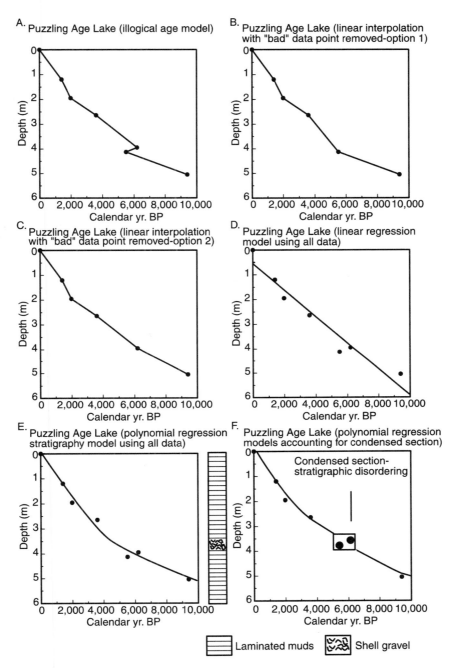

Figure 6.1. Development of an age model for a hypothetical stratigraphical section (in this case, a core). In this scenario there is no tectonic disruption or slumping of beds that would violate the law of superposition. (A) Plot of raw age data versus depth, showing age reversal. This age model incorporates all data points in a point-to-point interpolation, but that does not make sense, since it requires that older sediments lie on top of younger ones. (B,C) Point-to-point interpolations that address the disordered age dates by the rejection of one or the other contradictory points. Age dates from intermediate horizons are interpolated by the line connecting adjacent points, and changes in sedimentation rate are assumed to correspond with dated horizons. (D) A least-squares regression fit that incorporates all data points, and assumes no change in sedimentation rate. (E) A polynomial regression fitted to the data, that accommodates changes in sediment accumulation rate. (F) A possible explanation for the stratigraphical disorder observed in the radiometric dates, given additional stratigraphical information.

the uppermost sediments which cannot be true. More complex logarithmic or polynomial equations are used to fit the data when the pattern suggests a significant change in sedimentation rates, as is the case in figure 6.1E, and these can also be constrained to pass through the origin.

Careful analysis of lithological and chronometric data often indicates that significant unconformities are present in lacustrine deposits. Where lithological evidence for hiatuses exists it should be incorporated into an age model (figure 6.1F). Unconformity surfaces, or condensed sections, should be used by the investigator to break the overall age model from a continuous data set (with one regression fit) into a series of submodels. Studies of cores taken in the deepest part of a lake, which cover a few centuries or less, commonly ignore the possibility of significant erosional unconformities or hiatuses. However, the likelihood of such interruptions increases as the duration of the study interval increases. In any situation where major variations in lake level or circulation are likely to have occurred over the study period, the lithostratigraphy of the core or outcrop needs to be carefully assessed for major gaps in the stratigraphical record, especially when these coincide with abrupt changes in apparent sedimentation rate.

Once an age model has been established, it is possible to display any other indicator data as a time series plot (figure 6.2). In most cases the original data set is shown in this way (figure 6.2A). However, some indicator records are extremely spiky, or display erratic changes that obscure important temporal trends. When this occurs it is a common practice in paleolimnology to smooth a record, producing an integrated, lower-resolution record from an originally higher-resolution data set. There are various ways to do this. One common technique is referred to as a *running average* (figure 6.2B). In this technique a mean value is determined for a series of data points that includes a *central point* and n data points (typically 2 or 3) that lie stratigraphically adjacent to (both above and below) the central point. This mean value ($\Sigma_{data}/[2n + 1]$) is then plotted against the model age for the central point. The calculation is then repeated, with each data point serving as a central point and the $2n + 1$ data points to be averaged moved accordingly. Using this method a smoothed record is obtained, with no data plotted for the first and last n data points in the series. Another simple method is to take the means for stratigraphically adjacent sets of samples and display these means against their respective mean model ages for the samples (figure

6.2C). This approach (sometimes referred to as an *interval means record*) is particularly useful when the objective is to compare data from a high-resolution record against a lower-resolution record.

One final caution about age models needs to be noted here. In cases where independent age control is unavailable, or sparse, it is common practice to use some well-established pattern of change in an indicator as a chronological tool. This is the basis of the use of the marine oxygen isotope record as a chronological record in the ocean; the oxygen isotope changes do not record time per se, but are so consistent between localities that their peaks and troughs can reasonably be correlated with dated events. Although use of indicator records in this way may be necessary in the absence of other age information, it can also lead the investigator into a trap of circular reasoning as to the timing or origin of paleolimnological events.

Geochronometry

There are four broad classes of geochronometric techniques available to the paleolimnologist to obtain absolute age information: *radiometric dating*, *dating by radiation damage*, *amino acid racemization dating*, and *event correlation methods*. Many methods have been developed within each class, and are summarized in several general reviews (Faure, 1986; Geyh and Schleicher, 1990; Rutter and Catto, 1995).

Radiometric Dating Using Radiogenic Isotopes

The use of unstable (radioactive) isotopes for geochronometry relies on the fact that these isotopes are transformed into daughter isotopes, with the emission of various particles at known rates. Four types of decay are known:

1. α Decay, or the release of a ^{4}He nucleus from the parent isotope with n neutrons, resulting in a daughter product with $n - 2$ neutrons.
2. β Decay, or the transformation of one neutron to a proton in the parent isotope, resulting in a daughter product with an additional proton and one fewer neutrons. An electron β^- is emitted in the process.
3. β^+ Decay and electron capture, involves the capture of an electron by the parent nucleus, causing a proton to be converted to a neu-

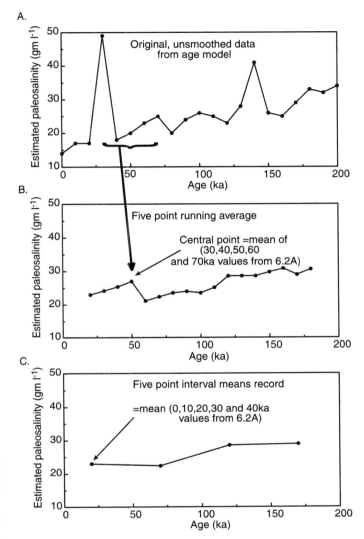

Figure 6.2. Approaches to illustrating time series data. (A) An originally "spiky" data set is smoothed, through the use of running averages. (B,C) The originally higher resolution data set has been intentionally "unresolved" to make it comparable with some other, lower resolution data set.

tron, or the conversion of a proton to a neutron, with release of a positron (e^+). In either case the atomic mass remains constant but the atomic number of the daughter decreases by one.

4. Spontaneous nuclear fission involves the fragmentation of heavy isotopes into daughter products of approximately equal size.

Radioactive decay occurs in three contexts of interest to paleolimnologists. *Cosmogenic isotopes* are produced through interaction of cosmic rays with previously existing atoms, either in the atmosphere or on earth. Second, a number of metallic element isotopes decay at extremely slow rates and are present in residual quantities in the earth's crust. Third, in addition to these naturally occurring radioactive decay processes, geochronologists can also study the decay of artificial radioisotopes, released into the atmosphere by nuclear bombs or reactor accidents.

All processes of radioactive decay proceed at rates that vary strictly in proportion to the relative abundance of radioactive nuclei that are present in the material. A decay constant (λ) can be defined as the proportion of atoms decaying per unit time. A *half-life* (τ), the time required for half of the original nuclei to decay, can be expressed as:

$$\tau = \ln 2/\lambda = 0.693/\lambda \qquad (6.1)$$

Variations in the half-lives of different radio-isotopes provide a range of time scales over which radiometric dating techniques can be applied (table 6.1).

Cosmogenic Isotopes and ^{14}C Dating

Cosmic rays from various sources continuously bombard the earth's atmosphere and land surface with highly energetic particles. When these cosmic ray particles collide with an atomic nucleus they can cause the original nucleus to be converted into a new nucleus. Some of these conversions result in the formation of radioisotopes that are otherwise rare in the earth's environment. The decay of these newly formed cosmogenic isotopes can be exploited as a dating tool. By far the most important cosmogenic isotope for paleolimnologists is ^{14}C.

Radiocarbon dating is probably the single most widely used geochronometric tool in paleolimnology today, and has occupied this position of importance almost from the development of the technique (Libby et al., 1949). Its time period of application (past ~ 50 ka) provides an excellent match to the age of the vast majority of extant lakes. Furthermore, the material needed to obtain a radiocarbon date, primarily organic matter and fossils, is abundant in the lacustrine environment, simplifying the task of finding appropriate samples for dating. The widespread use of radiocarbon dating has led to a much greater understanding of both the controls and limitations of the technique than exists in other radiometric dating methods, as well as a greater development of conventions used for analyzing and interpreting radiocarbon data (Arnold, 1995; Litherland and Beukens, 1995).

Radiocarbon dating relies on the upper atmospheric production of radioactive ^{14}C and its subsequent decay to ^{14}N. Carbon-14 is produced continuously in the earth's upper atmosphere, as cosmic radiation bombards ^{14}N, causing neutron emission and conversion to ^{14}C (figure 6.3). After its production, radiocarbon is oxidized to $^{14}CO_2$. At this point it is available for consumption in photosynthesis by plants or CO_2 exchange with water bodies to produce ^{14}C-bearing carbonate minerals. As with stable isotopes, an isotopic equilibrium is established between the atmosphere and plant tissues. Equilibrium values are not constant

for all plants, however; for example there are differences between C_3 and C_4 plants.

Radiocarbon decays with a comparatively short half-life (5730 yr) back to ^{14}N, and it is this decay that establishes the clock for radiocarbon dating. Two distinct methods have been developed for calculating the amount of decay that has already occurred. *Conventional radiocarbon dating*, or beta counting, involves counting the decay events (β-particle emissions) that occur over some interval of time. For example, modern carbon (i.e., equilibrated with the modern atmosphere) produces about 15 β-particle emissions per minute per gram. This value will decline as the sample ages and loses ^{14}C. Because the decay rate is directly related to the concentration of ^{14}C remaining in the sample, a radiocarbon age can be calculated. *Accelerator mass spectrometry* (also referred to as AMS or dating by atom counting) is a more recently developed technique than conventional counting (Bennett et al., 1977; Muller, 1977). It relies on the direct counting of ^{14}C atoms in a sample by first separating ^{14}C from ^{14}N, and then destroying other carbon isotopes in a tandem mass accelerator. In both conventional and AMS radiocarbon dating, results are presented in radiocarbon "years" relative to 1950. The actual calibration samples for this "zero" date (the inception of radiocarbon dating) are 1850 wood, which grew prior to the advent of the massive influx of ^{14}C-free carbon into the atmosphere from fossil fuels. Radiocarbon "year" data must be calibrated by various means discussed below to obtain a calendrical age estimate. Therefore it is important when presenting radiocarbon data that it be clear whether the data is calibrated or uncalibrated.

Sources of Error and Correction Factors in Radiocarbon Measurement

Stratigraphical mixing and contamination form the most common and serious problems for radiocarbon dating. These range from relatively obvious, and correctable, problems with modern contaminants (roots, fibers, etc.) becoming admixed in a sample, to more intractable problems of stratigraphical disordering. A radiocarbon date is an estimate of the age of the sample material itself, which may or may not be contemporaneous with the surrounding deposits, or the age of deposition. Older carbon-bearing material is commonly reworked into younger deposits through secondary erosion and transportation. Less commonly, younger particles can be worked into older deposits through bioturbation. These types of problems can often be avoided through careful examination of the stratigraphical context of the material being considered for dating. Oligotrophic lakes, whose sediments have low organic carbon concentrations, present special problems for ^{14}C dating. Because of the limited amount of identifiable organic remains that are normally preserved in these types of lake deposits, it is frequently necessary to obtain ages in oligotrophic lakes from bulk, total organic carbon samples (Abbott and Stafford, 1996; Colman et al., 1996). Bulk samples obtained in this way are prone to incorporate carbon from various sources that are not contemporaneous with sedimentation (terrestrial soils, or reworked sediment), and often yield anomalously old ages.

More intractable problems arise when natural sources of old carbon (^{14}C-depleted) influence the ambient CO_2 pool. Unlike the atmosphere, which is relatively well mixed, lakes are commonly stratified and subject to nonequilibrium and local sources of CO_2 from runoff and groundwater. *Reservoir effects* arise when these older carbon sources, such as bedrock limestone, or old soils, result in significant lowering of the initial $^{14}C/$ ^{12}C ratio relative to the global mean atmospheric ratio at the time of formation (Broecker and Walton, 1959; Stuiver and Pollach, 1977; Benson, 1993). For example, inorganic carbon runoff from Paleozoic limestone bedrock contains no radiocarbon, and will cause an erroneously old "age" for a carbonate deposited from this water in a downstream lake (e.g., Vance and Tekla, 1998). The date will reflect some mix in the lake water of the old ^{14}C-free carbonate and the current atmospheric $^{14}CO_2$ input. Offsets caused by such old carbon sources are particularly complex in large lakes, where variation in bedrock outcrop input of ^{14}C-depleted carbon, and subsequent complex internal mixing, can introduce significant spatial variation in the ^{14}C isotopic composition of carbonate (Rea and Colman, 1995; Moore et al., 1998).

This problem, known as the *hard-water effect*, is most simply addressed by avoiding the dating of materials that are likely to have assimilated HCO_3^- from the lake itself, such as authigenic carbonates, lacustrine invertebrate fossils, or aquatic plants and algae. Unfortunately, no simple criterion exists for determining the severity of the hard-water effect in a lake (Aravena et al., 1992). Even lakes with very similar drainage chemistries can exhibit varying levels of this problem, depending on the photosynthetic pathways and lake residence times involved in the internal cycling of carbon.

Reservoir effects also exist in the atmosphere, although efficient mixing makes this less of a problem here. Differences in ^{14}C content exist between the northern and southern hemisphere atmospheres that need to be taken into account in very high-resolution studies. A more serious reservoir effect problem arises in lakes with long residence times. The monimolimnion of a meromictic lake may retain DIC for very long periods of time, exceeding 10^3 yr in some lakes (e.g., Talma et al., 1997). When this carbon is incorporated into carbon-bearing deposits, either through direct precipitation or from gradual leakage into the epilimnion and incorporation into shallow-water carbonate, it will again cause an erroneously old age estimate.

No simple correction factor can be universally applied to these problems. In some regions the magnitude of the old carbon input is reasonably well understood, and a calendar year correction can be added into the calculation of an age (Olsson, 1986). More commonly, the amount of old carbon entering the system is unknown and/or variable, making accurate corrections difficult or impossible. When primary lacustrine carbon sources must be used, investigators should attempt to obtain samples of autochthonous organic carbon of known age, but collected prior to the beginning of nuclear testing (explained below), before attempting to interpret radiocarbon data. A better solution, when possible, is to avoid the use of autochthonous carbon sources entirely in younger (i.e., Mid–Late Holocene) samples, where reservoir effects can amount to a significant proportion of the sample age, or in any case where very high-precision dating is required. For these samples, terrestrial plant fossils or charcoal are preferable sources of carbon (R.E. Taylor, 1987). These sources of course are also subject to other errors of reworking. As a result, the best option is to try and use very delicate materials, such as whole terrestrial leaves, which are unlikely to survive long periods of transport and reworking.

A correction factor is normally added to radiocarbon age data to account for the fact that not all datable materials have the same starting $^{14}C/^{12}C$ equilibrium value (Broecker and Olson, 1959). The fractionation offset in $^{14}C/^{12}C$ from atmospheric equilibrium is mirrored by a proportional offset in the $^{13}C/^{12}C$ ratio, which can easily be measured using a mass spectrometer. The resulting δC correction is usually on the order of 100 yr or less, and therefore of great concern only for quite young materials, or when very high-resolution results are required.

Table 6.1. Isotopic systems of interest to geochronometry and paleolimnology

Isotope System	Parent Radionuclide	Half-life (yr)	Appropriate Materials for Analysis	Time Scale over which Technique is Potentially Useful (AOE = age of earth, effectively unlimited upper end)	Optimal Resolution (statistical counting precision ±1σ)	General References
^{14}C	^{14}C	5730	Organic materials (wood, charcoal, leaves, bone, bulk organic matter, soil, humus, pollen) and organic or inorganic carbonates	$1 \sim 50 \times 10^3$ yr[a]	$\sim 0.5\%$ of age[b]	Olsson (1986), Évin (1987), Arnold (1995), Litherland and Beukens (1995)
^{129}c	^{129}I	15.7×10^6	Organic-rich sediments	$3–80 \times 10^6$ yr (also as a bomb tracer)	?[c]	Fabryka-Martin et al. (1985, 1987), Fehn et al. (1987)
$^{230}Th/^{234}U$	^{238}U and ^{234}U[d]	$^{234}U = 2.45 \times 10^5$ $^{230}Th = 7.52 \times 10^4$	Carbonates, evaporites, carbonate-rich muds, possibly diatomites	$10^3 – 5 \times 10^5$ yr[e]	$2–15\%$[f]	Cherdyntsev (1971), Edwards et al. (1986/87), Lalou (1987), Schwarcz (1989), Blackwell and Schwarcz (1995)
^{210}Pb	^{210}Pb[g]	22.3	Fine-grained lacustrine muds	1–150 yr	~ 1yr	Olsson (1986)
(UTh)/He	Numerous alpha particle emitters	Variable, depending on numerous parent isotopes	Zircon or apatite grains in tephras	$> 3 \times 10^5$ yr[h]. Upper limit will depend on U content (lower for very high U minerals)	Better than 2% of age	Kohn et al. (2000)
^{230}Th unsupported	^{230}Th	7.52×10^4	Clay-rich mudrocks	$3 \times 10^3 – 3 \times 10^5$ yr	Semiquantitative	Bernat and Allègre (1974), Geyh and Schleicher (1990)
$^{228}Th/^{232}Th$	^{232}Th and ^{228}Ra	5.76 yr for ^{228}Ra and 1.91 yr for ^{228}Th[i]	Fine-grained lacustrine muds	0.1–15 yr	Months (approx.)	Koide et al. (1973)

134

Isotope	Half-life (yr)	Material dated	Source/event	Resolution	References
^{137}Cs	30	Fine-grained lacustrine muds	Post-1952 (thermonuclear weapons testing), peak fallout 1963. Secondary peak associated with Chernobyl accident	NA (age marker)	Pennington et al. (1973), Erten et al. (1985), Crusius and Anderson (1995)
^{239}Pu	2.4×10^4	Fine-grained lacustrine muds	Postnuclear weapons testing. Like ^{137}Cs	NA (age marker)	Wan et al. (1987), Crusius and Anderson (1995)
^{240}Pu	6.6×10^3	Fine-grained lacustrine muds	Postnuclear weapons testing. Like ^{137}Cs	NA (age marker)	Wan et al. (1987), Crusius and Anderson (1995)
^{90}Sr	28	Fine-grained lacustrine muds (carbonates)	Postnuclear weapons testing. Like ^{137}Cs	NA (age marker)	Wan et al. (1987)
$^{40}K/^{39}Ar$	1.250×10^9	K-bearing minerals (e.g., leucite, K-feldspars, muscovite, biotite), occasionally volcanic glass. Primarily from lava or tephras	0.5×10^6 yr – AOE for $^{40}K/^{40}Ar$; 10^4–10^5 yr – AOE for $^{40}Ar/^{39}Ar$ [i]	1% of age	Piboule et al. (1987), McDougall (1995)

[a]Useful for ultramodern samples (post-1950) and prior to 300 yr BP. Intermediate ^{14}C ages subject to multiple age interpretations.

[b]For prebomb age dating. Expressed as ^{14}C yr.

[c]Still in experimental stages of development as a paleolimnological dating tool.

[d]Via two very short-lived intermediates, ^{234}Th (24.1 days) and ^{234}Pa (1.18 min), form as intermediates between ^{238}U and ^{234}U. See figure 6.7.

[e]5×10^5 yr with TIMS. For conventional α spectroscopy, about 3.3×10^5 yr.

[f]Resolution decreases with increasing age.

[g]Via two very short-lived isotopes, ^{210}Bi and ^{210}Po.

[h]Probably an order of magnitude shorter may ultimately be possible (A. Gleadow, personal communication, 2000).

[i]Through numerous, short-lived intermediaries.

[j]Lower end of age range highly dependent on K-mineral analyzed and age determinations generally of lower precision.

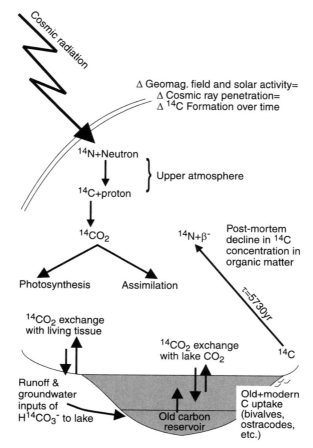

Figure 6.3. Pathways of ^{14}C formation, uptake, accumulation in a lake, and decay.

Atmospheric Variability in ^{14}C Production and Concentration

Since the 1950s it has been known that the upper atmospheric production of ^{14}C does not occur at a constant rate. Radiocarbon dating of wood samples that had been directly dated using tree-ring counts initially demonstrated that systematic offsets exist between radiocarbon ages and calendrical ages, which must be the result of changes in ^{14}C production (de Vries, 1958; Olsson, 1970; Stuiver and Kra, 1986). The pattern of these offsets suggests they result primarily from changes in the strength of the earth's magnetic field, allowing greater or lesser amounts of cosmic radiation to enter the atmosphere. Secondary effects related to sunspot cycles, or changes in cosmic ray flux, are superimposed on the longer-term changes (Damon et al., 1978, 1989; Suess, 1986).

Based on the relationship between the radiocarbon apparent age record and the tree-ring ages obtained from the same samples, calibration curves have been developed, which allow an investigator to obtain a calendar age from a radiocarbon age (Stuiver et al., 1993). Recently, varve-counted lake sediments and annually layered cave deposits (speleothems) have extended the range of these calibration curves back tens of thousands of years (figure 6.4) (e.g., Kitagawa and van der Plicht, 1998). Unfortunately, the calibrations do not always give unique solutions in the calculation of an absolute age; because of the wiggles in the radiocarbon production curve more than one calendrical age may correspond to a single radiocarbon age.

Ultramodern (Nuclear Testing) ^{14}C

With the advent of frequent atmospheric nuclear weapons testing, starting in the late 1940s and 1950s, the quantity of radiocarbon in the earth's atmosphere increased dramatically. The absolute increases have varied between location, depending

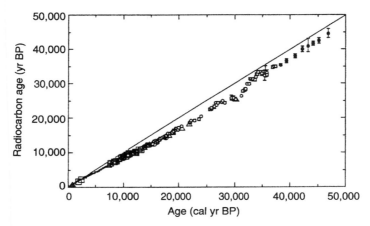

Figure 6.4. Radiocarbon calibration back to 45,000 yr BP, reconstructed from annually laminated sediments of Lake Sugetsu, Japan. Approximately 250 terrestrial macrofossil ^{14}C age dates were plotted against a floating varve chronology from 8830 to about 45,000 calendar (varve) years. The floating chronology is fixed in time by curve matching with the pre-existing annual tree-ring record of ^{14}C variability. Deviations from a 1:1 correlation of ^{14}C age with varve age are caused by fluctuations in the ^{14}C production rate in the upper atmosphere. Intervals of higher ^{14}C production result in ^{14}C ages that are systematically too young relative to true calendar age. Modified from Kitigawa and van der Plicht (1998).

on proximity to nuclear testing and prevailing atmospheric circulation patterns, but were commonly over 120% "modern" (i.e., calibrated 1950) levels at the height of nuclear testing. Since the partial nuclear test ban treaty prohibiting atmospheric tests went into effect in the 1960s, atmospheric ^{14}C has declined gradually. *Ultramodern* [14]C exceeding 100% of modern carbon has occasionally been used as a dating tool for very high-resolution studies of lake deposits that postdate the 1950s. Doing so, however, requires that the bomb-related atmospheric ^{14}C curve be known for the area where the lake is located, and only a few regions of the world are well characterized in this way.

Uranium-Series Dating

Uranium-series dating is a term applied to a group of radiometric dating methods that rely on the radioactive decay series of ^{238}U, ^{232}Th, and ^{235}U, all of which ultimately decay to stable lead isotopes (figure 6.5). Most U-series dating methods in lake deposits are applied to uranium-rich sediments, typically carbonates, or less commonly evaporites. However, samples for ^{210}Pb dating can be taken from a wide variety of fine-grained sediments. The relevant half-lives exploited by the various U-series

techniques make them useful over a wide range of ages (days to $> 10^6$ yr), although most paleolimnological applications fall in the 10^1 to 5×10^5 yr range.

During its initial weathering from silicate rocks, uranium is normally oxidized into a soluble U^{6+} form, which readily complexes with bicarbonate or phosphate ions. Although seasonal variability exists in the supply of U isotopes to lake sediments, in most lakes this appears to be time-averaged on the time scales of interest for U-series dating (Blake et al., 1998). Longer-term variation in U-series isotope supply however may occur as a result of differences in climate and weathering (Edgington et al., 1996). Thorium isotopes, which are the daughter products of ^{232}U, ^{235}U, and ^{238}U, are uniformly insoluble during weathering. They become separated from uranium, accumulating instead with clays, hydroxides, or organic matter. Other mechanisms further down the decay chains also serve to chemically separate precursor from daughter isotopes, for example in the formation of ^{222}Rn gas from ^{226}Ra.

When uranium is incorporated into a crystalline carbonate or phosphate its daughter isotopes can begin to accumulate. These daughter products will increase in proportion until an equilibrium concentration of each is formed, when its formation rate is matched by subsequent radioactive decay. This con-

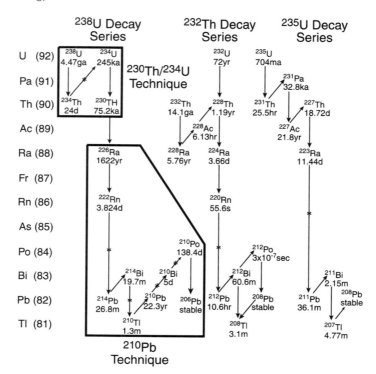

Figure 6.5. The uranium and thorium radioactive decay series. Major radioisotopes are shown, along with their half-lives (ga = 10^9 years, ma = 10^6 years, ka = 10^3 years). The illustrated decay series has been simplified to eliminate some of the very short-lived isotopes of little importance to geochronology (simplified portions indicated by *). Vertical lines signify α decay pathways, whereas diagonal lines are β decay paths. The portions of the ^{238}U decay series of relevance to ^{230}Th/^{234}U dating and ^{210}Pb dating are enclosed in boxes.

dition is referred to as *secular equilibrium*. Most U-series dating techniques are only effective during this period of disequilibrium, which varies between isotopes depending on their respective half-lives. Some U-series dating methods rely on the initial complete isolation of the daughter isotope from the parent. In these cases, for example ^{210}Pb, the daughter product will completely disappear with time, if the parent/daughter separation has been complete and no parent isotope is present. This daughter isotope activity that exists without being replenished by further decay of its isotopic precursor is referred to as being *unsupported*. This is in contrast to a *supported* background production of the daughter isotope, when the daughter has not been completely isolated from the precursor.

An alternative possibility exists for dating lake deposits when the parent isotope is sedimentologically isolated from the daughter (^{230}Th/^{234}U). Isotopic dating then rests on the comparison of remaining parent and daughter isotope concentrations, with the implicit assumption that the sample

has behaved as a "closed system" (i.e., there has been no leakage in or out of the relevant isotopes). Unfortunately, many sedimentary geochemical systems only serve to partially isolate the parent from the daughter isotope, and/or are not truly closed to isotopic exchange. Lake sediments present particular problems to the U-series geochemist, because they may incorporate isotopes from a variety of sources, and lose them through pore-water exchange. Therefore, assumptions of isotopic isolation and absence of exchange are normally tested through a variety of comparative isotopic techniques, followed by the application of correction factors. In the oceans, initial concentrations of parent and daughter isotopes are reasonably well known and constant, facilitating the correction of the assumption of parent/daughter isolation. This is not the case in lakes, where isotopic concentrations can be quite variable over space and time.

There are a large number of U-series dating techniques that, in principle, could be used in dating lacustrine sediments, or have been used occasionally (Blackwell and Schwarcz, 1995). In practice, however, only two of these have been commonly used to date; ^{230}Th/^{234}U and ^{210}Pb, while a third (U–Th)/He, is likely to become an important tool in the near future.

^{230}Th/^{234}U

The ^{230}Th/^{234}U dating system is used to determine the time since uranium incorporation into a sample. Following the precipitation of the various isotopes of U present, equilibrium will be approached between the extremely long half-life parent isotope ^{238}U and its much shorter-lived daughter isotopes ^{234}U, and subsequently ^{230}Th. The age since uranium deposition is then determined by the degree of disequilibrium that exists in the sample between ^{234}U and ^{238}U, and between ^{230}Th and ^{234}U. It can be calculated as:

$$^{230}\text{Th}/^{234}\text{U} = (\lambda_{230}/[\lambda_{230} - \lambda_{234}])$$
$$\times (1 - [1/(^{234}\text{U}/^{238}\text{U})])(1 - e^{[\lambda_{234} - \lambda_{230}]t})$$
$$+ \,^{238}\text{U}/^{234}\text{U}(1 - e^{-\lambda_{230}t}) \qquad (6.2)$$

where ^{230}Th/^{234}U and ^{238}U/^{234}U = measured isotopic activity ratios, λ_{230} = ^{230}Th decay constant (or λ_{234}, the ^{234}U decay constant), and t = sample age.

The equation can be solved iteratively, or by graphical comparison of the ^{230}Th/^{234}U versus ^{234}U/^{238}U activity ratios (figure 6.6). The presence

of initial Th is detected and corrected by analysis for the extremely long half-life ^{232}Th, which is not fractionated relative to ^{230}Th. The two Th isotopes are therefore assumed to have been present in constant isotopic proportion at the time of deposition, and an initial ^{230}Th is calculated from the measured ^{232}Th (Bischoff and Fitzpatrick, 1991; Luo and Ku, 1991).

The ^{230}Th/^{234}U method is applicable for samples in the age range of 10^3 to 5×10^5 yr (Edwards et al., 1986/87), making it a valuable tool for extending the dating range of Quaternary lake sediments beyond the radiocarbon time scale. It can also compliment ^{14}C dating, especially where hard-water effects are a problem for radiocarbon interpretation. A variety of lake sediment types have been used as sample material for ^{230}Th/^{234}U, including stromatolites, tufas, organic-rich carbonate muds, and evaporites (e.g., Bischoff et al., 1985; Hillaire-Marcel et al., 1986; Lao and Benson, 1988; Ku et al., 1998). In cases where both radiocarbon and ^{230}Th/^{234}U dates are available, it is often possible to cross-check results between the two methods (Israelson et al., 1997).

(U–Th)/He

Although the idea of (U–Th)/He dating was conceived by Rutherford in the early twentieth century, its implementation as an effective dating tool was impractical until quite recently. Helium is produced throughout the ^{238}U, ^{232}Th, and ^{235}U decay chains.

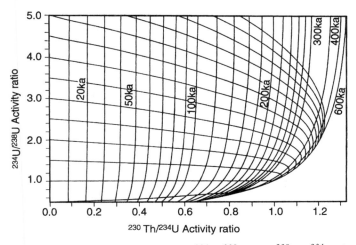

Figure 6.6. Diagram for evaluating U/Th ages from ^{234}U/^{238}U and ^{230}Th/^{234}U data. The relationship between ^{234}U/^{238}U and ^{230}Th/^{234}U for closed systems initially free of ^{230}Th. The subhorizontal curves are the decay paths for samples initially deposited with ^{234}U/^{238}U ratios equal to the intercepts at ^{230}Th/^{234}U = 0. The subvertical curves are isochrons (ka). From Blackwell and Schwarcz (1995).

Measurement of He entrapped within zircon, apatite, or other mineral grains, coupled with the measurement of the various parent isotope alpha particle emitters, provides a means of estimating the elapsed time since the mineral became a closed system, trapping He (e.g., Kohn et al., 2000). Early studies of the (UTh)/He system on tephras suggest that it can provide a high-precision dating tool ($\pm 2\%$ or better). The lower age range of this technique, currently as low as 3×10^5 yr, but with potential to perhaps as low as 3×10^4 yr, will make it extremely well suited to dating Early–Middle Pleistocene tephras, which are often difficult to date by other means.

^{210}Pb

The ^{210}Pb dating method is probably the most widely used technique for high-resolution dating of very young (approximately last 100–150 yr) lake beds. Because of its short half-life (21.6 yr), the majority of ^{210}Pb applications have addressed questions of rapid ecological change and human impacts on lake and watershed ecosystems (e.g., Brenner et al., 1999). Lead-210 dating was developed in the 1960s, and first applied to lake deposits in the early 1970s (Goldberg, 1963; Krishnaswami et al., 1971). The method relies on decay along the ^{238}U decay chain. An intermediate isotope, ^{226}Ra, exists in soils and exposed bedrock in trace amounts. As it decays (1622 yr half-life), ^{226}Ra is transformed into the gas ^{222}Rn, which moves into the atmosphere. In turn, ^{222}Rn decays rapidly (3.8 days), through a series of very short-lived isotopes to ^{210}Pb, which is quickly redeposited as unsupported ^{210}Pb from the atmosphere onto land or water. Redeposited lead is carried with Fe and Mn oxides and organic matter into lakes, where it can become sedimented, undergoing its own decay to ^{210}Bi. For the purposes of ^{210}Pb dating, the residence time of this lead prior to its deposition is generally assumed to be very brief. This is probably an oversimplification. Most lakes are subject to *sediment focusing*, whereby fine-grained particles over time are winnowed into deeper-water environments (Likens and Davis, 1975; Hilton, 1985). Because ^{210}Pb is carried with fine particulates, this can result in anomalously large accumulations of ^{210}Pb in deep water (Crusius and Anderson, 1995). Many lakes have shallow areas where sediment is likely to be trapped prior to final burial. As a result, the transport residence time of fine particulates may cause a ^{210}Pb age model to significantly overestimate the time of deposition, and underestimate sediment accumulation rates. After

burial, sedimented ^{210}Pb decays through several very short-lived radioisotopes to stable ^{206}Pb.

The decay of low-energy beta particles of ^{210}Pb to ^{210}Bi is difficult to measure, and therefore ^{210}Pb is generally not measured directly. Instead, the subsequent higher energy decays that culminate in the formation of stable ^{206}Pb are measured. Because these half-lives are all very short, the activities of the intermediate isotopes (^{210}Bi and ^{210}Po) are assumed to be in equilibrium with the ^{210}Pb present, and are used to estimate ^{210}Pb concentration. Lead-210 is measured either by alpha counting of ^{210}Po $\Rightarrow$ ^{206}Pb, or by gamma counting of ^{210}Bi $\Rightarrow$ ^{210}Po (Schell et al., 1973; Schell, 1977; Joshi, 1987). Alpha counting has the advantages of using a smaller sample and greater precision, but the technique destroys the sample. Gamma counting allows the sample to be retained for further analysis, and allows simultaneous analysis of other radioisotopes of interest to paleolimnologists, such as ^{137}Cs and ^{226}Ra.

In addition to the supply of unsupported ^{210}Pb that is deposited from the atmosphere, there is also a background of supported ^{210}Pb that is present in all samples from the in situ equilibrium decay of its precursor isotopes. This is evident in a typical ^{210}Pb profile as an asymptote of low, relatively constant values at the bottom of a core. Because of the very brief half-lives of the intermediate isotopes formed prior to ^{210}Pb, they are likely to be retained in previously deposited sediments until conversion to ^{210}Pb. If ^{226}Ra is assumed to be in equilibrium with supported ^{210}Pb, a measurement of mean ^{226}Ra can be used to infer total supported ^{210}Pb. This amount can be subtracted from total ^{210}Pb to calculate the unsupported ^{210}Pb. A more expensive, but also more precise way of doing this is to measure ^{226}Ra along with every ^{210}Pb sample, using gamma emission counting (Schelske et al., 1994). Estimating supported ^{210}Pb from ^{226}Ra on a level-by-level basis allows variable inputs of radium and supported lead activity to be incorporated into the age model.

Lead-210 dating is complicated by the fact that the concentration of ^{210}Pb at the sediment–water interface is a function of both the flux rate of the unsupported lead itself, as well as the background sedimentation rate. At least one of these must be known, or assumed, in order to transform ^{210}Pb activity data into an age. It must also be assumed that no significant ^{210}Pb migration or exchange occurs with pore waters. Two methods have been devised to accomplish this. In the *constant initial concentration* (CIC) model (sometimes referred to as the constant activity model), the dilution of ^{210}Pb through the period of deposition is assumed to be constant. Thus, the initial activity of unsupported lead per gram dry weight is assumed to have remained unchanged throughout the deposition of the core (Goldberg, 1963; Krishnaswami and Lal, 1978; Robbins, 1978). This model requires that an excess reservoir of ^{210}Pb be available in the water column at all times for scavenging by settling particles. It can be useful in lakes where sedimentation rates have been relatively constant through the study period. However, since many lakes have been subject to increasing sediment flux from human activities, this assumption is frequently unwarranted. Even in lakes that have experienced little or no change in watershed erosion rates, differences in local sedimentation rates can result as sediment is secondarily winnowed and moved to other areas. An alternative, *constant rate of supply* (CRS) model is applicable in a broader range of situations. This assumes that the absolute flux rate of ^{210}Pb to the sediment–water interface remains constant, regardless of background sedimentation, such that a higher rate of background sedimentation will lead to a lower ^{210}Pb concentration (Goldberg, 1963; Appleby and Oldfield, 1977). Where independent means of dating very young lake sediments exist, the CRS model has been found to yield more accurate ages than CIC, especially when the residence time of the lake water is long enough to allow all available ^{210}Pb in the water column to be sedimented (Appleby et al., 1979; Binford et al., 1993; Blais et al., 1995). Most lakes undergo changes in sediment accumulation rates. In any lake that is undergoing an acceleration of sediment accumulation rates, the CIC model will consistently underestimate the age of sediment prior to the rate increase, therefore also overestimating the sediment accumulation rate during the period of increase.

Commonly, there is an exponential decline from the surface downcore in the activity of unsupported ^{210}Pb, reflecting the radioactive decay of the unsupported ^{210}Pb against a background of relatively constant sedimentation (figure 6.7). These types of profiles provide the most straightforward age information, and where this is the case CIC and CRS models yield equivalent age models. Deviations from this type of profile most often take the forms of flattening of the profile near the surface (figure 6.8), or sharp breaks in the overall exponential trend. Flattening can be caused by several mechanisms; bioturbation of the upper sediments, dramatic increases in background sedimentation rate, diagenetic mobilization of lead into pore waters and out of the sediment,

and changes in lead supply rates (Binford et al., 1993). Because bioturbation is commonly observed in lake sediments, a number of models have been developed to accommodate its effects in the interpretation of ^{210}Pb profiles (Goldberg, 1963; Robbins, 1978; Appleby and Oldfield, 1983). Increasing sedimentation rates can lead to very complex profiles, even resulting in declines in ^{210}Pb at the surface, if sedimentation rates are increasing faster than a doubling per half-life.

Artificial Radioisotopes from Nuclear Weapons Testing and Atmospheric Emissions

Atmospheric testing of nuclear weapons during the 1950s and early 1960s resulted in the production and fallout of a variety of artificial radioisotopes. The peak fallout deposition of bomb-produced radionuclides occurred in 1963 and declined rapidly after the cessation of large-scale atmospheric testing, with a secondary peak following the Chernobyl nuclear accident (Hardy, 1977; Cambray et al., 1987).

Following deposition in soils or water, fallout is transported by surface or groundwater to lakes, and ultimately sedimented in lakes or the oceans. Isotope geochemists have used the occurrence of several of these isotopes, especially ^{137}Cs, ^{239}Pu, ^{240}Pu, and ^{90}Sr, as geochronometric time markers in lake deposits (Wan et al., 1987; Appleby et al., 1993, 1995). Their application differs from natural radioisotopes as geochronometers, in the sense that researchers are generally looking for a downcore spike in the radioactivity profiles of these isotopes, that can be correlated with the 1963 period of maximum fallout accumulation (figure 6.8). Most fallout tracer studies have concentrated on ^{137}Cs, because it is one of the easier artificial radionuclides to measure (Crusius and Anderson, 1995). The interpretation of ^{137}Cs profiles is complicated by the relative diagenetic mobility of this isotope (R.B. Davis et al., 1984; Brenner et al., 1999). Unlike ^{210}Pb, which becomes tightly bound to sedimented particles, ^{137}Cs can be quite mobile after deposition. Bioturbation and sediment focusing can also redistribute ^{137}Cs (Leavitt and Carpenter, 1989). These processes result in artifacts, such as the presence of appreciable ^{137}Cs activity in both pre- and postatmospheric testing aged sediments, and the absence of early 1960s ^{137}Cs peaks in some young cores. Because of these problems, some authors advocate the replacement of ^{137}Cs investigations by other, less mobile, fallout tracers,

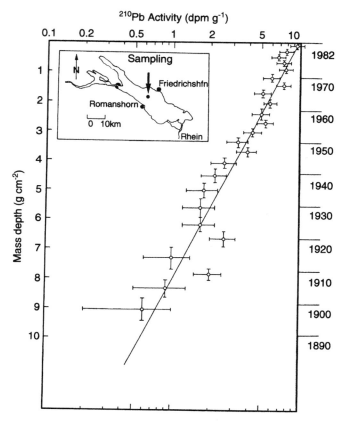

Figure 6.7. Lead-210 profile and least-squares age model for a short core from Lake Constance, Switzerland (inset map). The daughter product of ^{210}Pb, ^{210}Bi, was measured by gamma ray spectroscopy. Supported ^{210}Pb was estimated from constant activity levels at the deepest portion of the core, supplemented by occasional upcore measurements of ^{226}Ra. The core shows a progressive downsection decline in ^{210}Pb activity. Very similar results of sedimentation rates and ages were obtained using CIC and CRS models, supporting the notion that sediment accumulation rates remained relatively constant through the core. Note the use of mass depth in place of an absolute vertical depth scale, reflecting the need to correct for highly variable water content near the sediment–water interface. Also note the large standard errors in the lower part of the core, typical of variable counting statistics at low ^{210}Pb activities. From von Gunten and Moser (1993).

such as ^{239}Pu and ^{240}Pu (Crusius and Anderson, 1995).

K/Ar and Ar/Ar Dating

The radiometric dating methods discussed up to this point are all restricted in use to relatively young (Middle Pleistocene–Recent) lake sediments. However, potassium/argon and ^{40}argon/^{39}argon dating methods are useful in both Pleistocene and pre-Pleistocene lake sediments. They allow the direct dating of interbedded lava flows and

volcanic ashes (tephras), and authigenic K-silicate minerals.

Radiogenic decay of the long-lived radioisotope ^{40}K produces the noble gas ^{40}Ar. Prior to crystallization, any ^{40}Ar produced by such decay at the earth's surface (for example in lava) equilibrates with atmospheric argon. After the lava cools, any subsequent ^{40}Ar formed is trapped within the crystalline lattice of the K-bearing mineral from which it was derived. Thus, the age information provided is one of time since crystallization (cooling in the case of lava). Argon will continue to be lost until the mineral cools below a "closure temperature,"

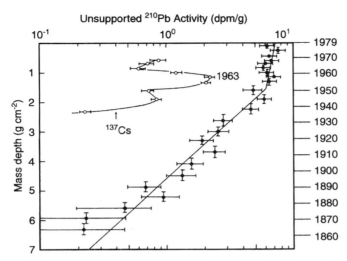

Figure 6.8. Unsupported ^{210}Pb activities and age model and ^{137}Cs activity profile for a core from Lake Zürich, Switzerland collected in the late 1970s. The ^{210}Pb profile shows a plateau in the upper 6 cm (1.5 g cm^{-2}). The preferred explanation for the plateau in this case is postdepositional mobilization of ^{210}Pb, although similar patterns are also observed in bioturbated sediments. The ^{137}Cs curve has a well-defined peak, which is consistent with the ^{210}Pb age model to represent the 1963 peak in artificial radionuclide fallout. From von Gunten and Moser (1993).

which varies between about 150 and 500°C. Because this occurs quickly at the earth's surface, the cooling temperature is generally considered equivalent to the age of emplacement. If the rock is reheated at some later time, however, some proportion of the Ar produced through radiogenic decay will be lost, creating a potential error in the interpretation of the "age" of the sample.

The potassium/argon dating method requires the measurement of the quantity of ^{40}K and ^{40}Ar present in a sample (McDougall and Harrison, 1988), using multiple subsamples for different parts of the analysis. Under optimal circumstances, when ^{40}Ar > 10% total Ar, the radiogenic argon concentrations in the sample can be calculated to a precision of 0.5% standard deviation, allowing a K/Ar age precision of ~ 0.7%. Because of the extremely long half-life of ^{40}K, the dating of relatively young (Quaternary) samples generally requires some combination of high potassium minerals (for example K-feldspars) and relatively large sample sizes, to insure the availability of sufficient argon for analysis (McDougall, 1995).

The ^{40}Ar/^{39}Ar technique is a more recently developed variation on K/Ar dating (Merrihue and Turner, 1966) that makes all measurements on a single, carefully chosen crystal, allowing the analysis of smaller quantities of optimal material. It also allows for higher precision dates to be determined. It involves irradiating the sample in a

nuclear reactor, to convert a proportion of the ^{39}K present to ^{39}Ar. The rate of this conversion is known, determined by the characteristics of the irradiation treatment. Then the argon is released from the sample and purified, and then measured isotopically for both ^{40}Ar and ^{39}Ar. In nature, the ^{40}K/^{39}K is essentially a constant; therefore the measurement of the quantity of both argon isotopes allows determination of the ^{40}K/^{40}Ar ratio, and hence the sample's age.

The ^{40}Ar/^{39}Ar method also allows the investigator to determine if the sample was subject to any reheating after formation, which would release Ar from the sample, resulting in an erroneously young age in a conventional K/Ar analysis. Normally, the heating of a ^{40}Ar/^{39}Ar sample proceeds in an incremental fashion, releasing more and more argon as the temperature increases (figure 6.9). Modern methods of this *step heating* allow a series of apparent ages to be calculated from the sample. Argon diffuses outward sequentially from the exterior of a crystal to the interior, both as a result of secondary reheating during its geological history, and in the step-heating process. Thus, by determining the difference in Ar isotopic ratios from the edge to the interior of the mineral grain, the step-heating process allows the argon loss history to be determined.

The most common application of ^{40}K/^{40}Ar and ^{40}Ar/^{39}Ar dating methods to lake sediments comes from the analysis of K-bearing minerals (e.g., potas-

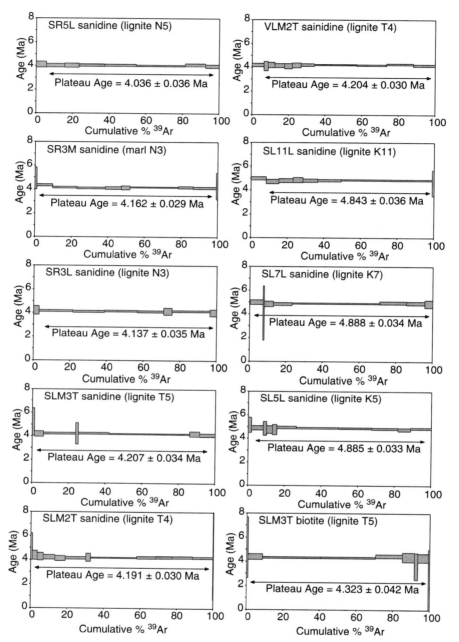

Figure 6.9. Application of $^{40}Ar/^{39}Ar$ geochronology to a series of tephras from the Lower Pliocene Ptolemais lake deposits of northwest Greece. The tephras occur interspersed through a highly cyclical sequence of organic-rich sediments (lignites) and carbonate muds. Laser incremental heating spectra are for sanidine (K-feldspar) and biotite samples from nine of the tephras (SLM3T) dated on two separate samples. Error bars are 2σ statistical precision. Incremental heating results in release of increasing amounts of ^{39}Ar. Variations in calculated age in this step-heating method are interpreted as evidence of significant reheating during the thermal history of the crystal (for example during burial diagenesis). The consistency of age dates and their similarity to their plateau ages suggests there has been relatively little reheating, which would reset the $^{40}Ar/^{39}Ar$ clock. Note the high degree of statistical precision in the dates ($\pm 0.61\%$ of age). From Steenbrink et al. (1999).

sium feldspars, muscovite, biotite, etc.) in tephras. Other sources, such as authigenic K-bearing minerals in sediments, are used less commonly. Tephras can provide a ready source of K-rich minerals, which normally cool rapidly below their closure temperature after eruption. The principal concern in analysis of volcanic ash for dating is the question of whether the sample has been significantly reworked by sedimentary processes of transportation and resedimentation. When this has occurred there is a potential for calculated K/Ar dates to be considerably older than the time of deposition. Chemical alteration, particularly with glassy tephras, is also a potential source of error, as this can lead to the escape of radiogenic argon, even at very low temperatures. As a result of these concerns, considerable effort normally goes into the microscopic inspection of tephras prior to their analysis, to insure that the sample appears to be unreworked and unaltered.

Radiation-Effect Techniques

In addition to direct dating of radioisotope decay, it is also possible to date minerals by the indirect effects of radioactivity and external radiation sources. A class of radiation-effect dating techniques exploits the fact that the effects of many of these processes on minerals accumulate at constant or measurable rates. Two groups of these techniques, fission track dating and luminescence dating, are relevant to paleolimnology (table 6.2).

Fission Track Dating

As the byproducts of nuclear fission pass through a mineral grain they can produce zones of intense disruption of the crystal, known as fission tracks. The radioactive isotope ^{238}U produces spontaneous fission products at a sufficiently high, and known, rate, such that a count of these tracks forms the basis of geochronometry in minerals, such as apatite or zircon, with sufficiently high uranium concentrations. This is done by counting the number of tracks that cross a polished surface over a measured area. The number of tracks produced is a function of both the age and the uranium content of the sample. The latter is generally determined by artificially inducing the production of new fission tracks in the sample by irradiation in a nuclear reactor. The density of newly formed tracks in the sample is then compared with a sample of known and uni-

form uranium concentration irradiated under the same conditions.

The potential for successful fission track dating is determined by whether the sample contains sufficient uranium to produce a statistically significant number of tracks, and whether the sample has been exposed to postdepositional heating, which anneals previously formed tracks. After their formation tracks can disappear or *anneal* over time, as they are subject to elevated temperatures; track counting from such samples can lead to erroneously low age estimates (Gleadow, 1980; Naeser et al., 1982; Kohn et al., 1992). Various techniques are available to determine the extent of annealing and attempt to correct for this problem (Storzer and Poupeau, 1973; Burchart et al., 1975; Westgate, 1989).

Analysis of a large number of single zircon grains is usually the best approach to obtaining an age of a tephra, and makes identification of reworked outlier grains possible. Because of their small numbers of tracks, younger, Quaternary samples generally require a larger number of individual counts in order to obtain adequate counting statistics. Fission track chronology is generally applicable for tephras as young as about 10^5 yr. Although it is theoretically possible to count tracks in samples younger than this, the time required and statistical imprecision usually make the method impractical under these circumstances.

The primary applications of fission track geochronology of relevance to paleolimnology have been in the dating of tephras, especially in conjunction with tephrochronology and K/Ar or Ar/Ar dating methods (e.g., Gleadow, 1980). However, the importance of fission track methods for absolute dating purposes may decline in future years, for a number of reasons. First, large uncertainties are normally associated with fission track age estimates, limiting their use for high-precision dating. Also, $^{40}Ar/^{39}Ar$, (UTh)/He, and other radiometric dating methods can often be applied to the same U-rich samples as can fission track methods, and provide higher precision dating than fission tracks.

Luminescence Dating Techniques

Luminescence techniques of absolute dating (thermoluminescence and optically stimulated luminescence) exploit the buildup of stored charges in sediment grains from an effectively constant, low-level flux of ionizing radiation from environmental nuclear decays and from soil-penetrating cosmic rays (Wintle and Huntley, 1980; Aitken, 1998). Sedimentary particles can store some portion of

Table 6.2. Nonradiometric geochronological methods for lake sediments

Method	Appropriate Materials in Lake Deposits	Time Scale over which Technique is Potentially Useful	Optimal Resolution	General References
Radiation-Effect Techniques				
Fission-track dating	Volcanic tephras (zircons, volcanic glass)	10^5–10^9 yr (possible but generally impractical for younger sediments	3–15% of age[a]	Walter (1989), Westgate and Naeser (1995)
Luminescence dating[b]	Fine-grained quartz or feldspars in lake muds	10^2–10^6 yr	10–30% of age[c]	Aitken (1994, 1998), Berger (1994, 1995)
Chemical Dating				
Amino acid racemization	Mollusk and ostracode shell, ratite eggshell, organic-rich sediment	10^1–10^6 yr[d]	10–20% of age	D.S. Kaufman and Miller (1992), Rutter and Blackwell (1995), Goodfriend et al. (2000), Wehmiller and Miller (2000)
Event Correlation Techniques				
Varve chronology	Fine-grained muds and silts, diatomites, organic-rich sediments or carbonates	Any age[d]	Subannual to annual	O'Sullivan (1983), Saarnisto (1986), R.Y. Anderson and Dean (1988), Boyle (1993)
Tephrochronology	Volcanic ash (tephra) beds—occasionally disseminated ash particles	Any age[e]	Indeterminate	Self and Sparks (1981), Einarsson (1986), Westgate and Naeser (1995)
Magnetostratigraphy	Fine-grained sediments, volcanics	10^2–10^5 yr[f]; 7×10^5–10^8 yr[g]	10^2 yr[f]; 10^5–10^7	Thompson and Oldfield (1986), Verosub (1988), Barendregt (1995), King and Peck (2001)
Biostratigraphy	Lacustrine fossils (esp. diatoms, charophytes, ostracodes, and mollusks) and terrestrial fossils entombed in lake beds (esp. pollen)	10^6–4×10^8 yr[h]	5×10^5 yr[h]	Hedberg (1976), Colin and Lethiers (1988)

[a] Relative resolution decreases with decreasing age. Statistical precision of $> \pm 10\%$ is the norm.
[b] Including a variety of thermoluminescence and optically stimulated luminescence techniques.
[c] Resolution decreases significantly in samples $> 10^5$ yr.
[d] For floating chronologies. Fixed chronologies currently limited to last 45 ka.
[e] Older or lithified ashes may be impossible to fingerprint geochemically as a result of diagenetic alteration.
[f] For secular variation records.
[g] For polarity reversal records. Resolution varies with duration of polarity reversals.
[h] For evolutionary events (species or lineage ranges, assemblage ranges, and concurrent ranges). Paleoecological zones of maximal abundance are of variable durations.

this energy for geologically significant periods of time (up to 10^6 yr), as electrons displaced from their original atomic configuration into crystalline lattice defects or impurities. This accumulated energy has been likened to a dosimeter (radiation) badge, in that the accumulated radiation can be "read out" since the time that electron displacement energy began to accumulate (Berger, 1995). In the case of luminescence dating methods, the relevant "start date" is the last exposure to sunlight or significant heating; sunlight empties light-sensitive traps, whereas heat empties all electron traps. The calculation of an age using luminescence techniques involves artificially emptying electron traps through either heating, to produce thermoluminescence (TL), or through photonic stimulation using various strong light sources to produce optically stimulated luminescence (OSL). A flux of photons is emitted from the sample, which can be measured using very sensitive photon detector systems. The measured TL must be corrected for various artifacts, such as any residual TL retained from prior episodes of burial. The artifact-corrected flux of photons can be scaled or converted to an equivalent-dose value (D_E). The D_E is then divided by the dose per unit time for the materials being analyzed, to calculate a TL or OSL age. Several methods have been developed to estimate D_E, that vary in their applicability depending on the TL or OSL properties of the minerals and depositional conditions at the time of burial. The age range of materials that can be dated using luminescence techniques may extend back to 10^6 yr, although large uncertainties (± 20–30%) generally accompany dates greater than 10^5 yr.

Typically, TL dating is performed on source materials such as quartz, feldspar, zircon, calcite, or volcanic glass. Ambient radiation dose rates are determined by the energy supplied by radiogenic isotopes in the neighborhood of the sample. This is inferred by measurement of the concentrations of radiation sources (U, Th, K, and sometimes Rb) within the sample. This is a potential source of inaccuracy, as radioactive decay can dose a sediment grain at distances of up to 30 cm. As a result, luminescence measurements are sometimes supplemented by field measurements of background radiation in the vicinity of the sample site. Radiation rate estimates also make assumptions about other factors that will secondarily affect dosing, such as sediment water content, or soil-penetrating cosmic rays.

An important question in TL dating of lake sediments is to what degree the TL signal is actually "zeroed" at the time of deposition (Berger, 1990). Coarser bedload particles may actually reside in the water column for only a short time. These particles will inherit some amount of relict TL signal from prior burial intervals that is not erased by their brief exposure to sunlight. Berger (1990) and Berger and Easterbrook (1993) showed that the adjacent coarse and fine laminae within a single glaciolacustrine rhythmite had different initial TL signals. The rapidly deposited, coarser silt particles retained a significant residual TL signal, and therefore provided TL apparent ages that were too old. This contrasted with the finer clay particles, which would have remained in the water column for a much longer period, during which time their prior TL signal was zeroed out. In the case of OSL, since the signal is more rapidly zeroed than TL's light-sensitive component, there is greater potential for accurate application to lake sediments (Berger and Doran, 2001).

Although luminescence dating has been most extensively applied to eolian deposits and buried soils, a number of studies have also used TL/OSL techniques to date Quaternary lake and peat marsh deposits (Berger and Easterbrook, 1993; Balescu and Lamothe, 1994; Berger and Eyles, 1994; Lang, 1994; Krause et al., 1997; Rogalev et al., 1997; Doran et al., 1999; Berger and Anderson, 2000). The greatest value of TL/OSL dating to paleolimnology is likely to be in the postmagnetic reversal polarity time scale, preradiocarbon age range (~ 730–45 ka), where few alternative dating tools exist. Luminescence dating also has great potential for dating organic-poor lake sediments (high latitude, high altitude, and arid, playa-lake environments), where the ^{14}C dating option may not exist (Dutkiewicz and Prescott, 1997; Krause et al., 1997; Berger and Doran, 2001). However, these older sediments have also proven more difficult for TL workers to date and/or obtain stratigraphically coherent results (Berger, 1995). One way to demonstrate the likely value of older TL dates may come from dating a continuous sequence of sediments that spans from the radiocarbon time range into older lake beds, and cross-dating the younger sediments using ^{14}C (Berger and Anderson, 2000) (figure 6.10).

Amino Acid Racemization

Amino acid dating techniques are based on the transformation (racemization) of amino acids from their form in living organisms, to a different form after death (Abelson, 1954; Hare and Abelson, 1966; Hare and Mitterer, 1967). The

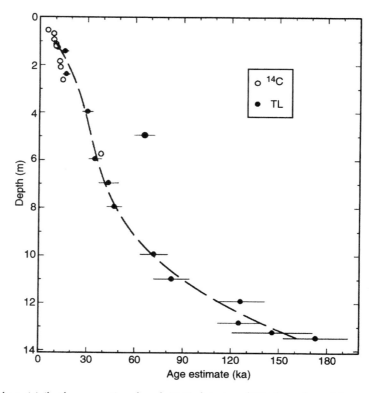

Figure 6.10. Carbon-14 (both conventional and AMS dates) and TL dates for a 14 m core from Squirrel Lake, northern Alaska. The two sequences of dates are relatively, though not perfectly, concordant. Four sedimentological zones were described from the core: (I) 14–11.2 m below surface, massive sandy mud; (II) 11.2–5.5 m, monotonous, crudely laminated muds; (IIa) 5.5–3.6 m, compositionally similar to II but lacking laminations; and (III) 3.6–0 m, silty, sandy mud, but with more organic matter. The sediment types are indicative of suspension settling or slow rainout of particles from the upper water column. This type of depositional mechanism is most effective for allowing time for the TL signal to undergo natural zeroing after its prior burial and erosion history, critical for obtaining reliable TL dates. The TL age series shows the potential for using the technique to extend age dating in lake sediments continuously beyond the ^{14}C range. Modified from Berger and Anderson (2000).

extent of this transformation for an individual amino acid varies with both time and temperature. Therefore, if the racemization rate is known, amino acids can be used to date organic materials by either assuming some burial temperature, or assuming that the average burial temperature history of the sample is the same as the history of the calibration material. Alternatively, amino acids can be used to determine burial temperatures if some independent age control exists for the sample. The potential for variations in racemization rate is great, leading to erroneous age estimates in many early studies. As a result, many authorities in the field recommend caution in interpreting amino acid data in terms of absolute ages unless they are calibrated for specific taxa or geographical settings (Rutter and Blackwell,

1995). In principle, samples of anywhere from 10^1 to 10^6 yr can be dated using amino acid racemization, although most absolute dating applications so far have been in a more restricted (10^3–10^5 yr) range. Also, because different amino acids racemize at different rates, it is in principle possible to use several amino acid systems to extend the time range over which dates are obtained (table 6.2).

Amino acids, the relatively simple building blocks of proteins, occur in all living organisms, and also in sedimented or aqueous organic matter. Most amino acids contain an asymmetric carbon atom core (occasionally two carbon atoms are involved), around which the other parts of the molecule can be distributed in mirror image fashion, either in a dextral (D) or levral (L) orientation.

L-Amino acids are the dominant forms in living organisms; these can convert to D-amino acids after death, as proteins degrade. In inorganic amino acids L and D forms occur in approximately equal proportions, or in other equilibrium mixtures for amino acids that contain more than one asymmetric carbon atom. Postmortem racemization (or epimerization, for amino acids with more than one asymmetrical carbon) then involves the transformation of the pool of individual amino acids from exclusively L forms to an equilibrium mixture of L and D forms. The rate of this transformation depends on the specific amino acid involved, its position and environment within a protein, the temperature and pH of the burial environment, the taxa involved, and a number of other factors (Rutter and Blackwell, 1995; Wehmiller and Miller, 2000). The differences between amino acids and the effect of temperature on reaction rates are particularly noteworthy; for typical amino acids, racemization rates approximately double for every 4°C temperature increase, implying that slight errors in estimated burial temperature can lead to significant errors in modeled age (Bada and Schroeder, 1975; Bada, 1984; Miller and Brigham-Grette, 1989).

Until quite recently, either ion exchange chromatography or gas chromatography were used to separate amino acids, the former technique requiring considerably less sample material and pretreatment. Recent advances in the use of reverse phase liquid chromatography have now reduced the sample sizes required to the microgram range, for example permitting the analysis of single or small numbers of ostracode valves (D.S. Kaufman and Manley, 1998).

The potential of amino acid racemization dating to paleolimnology comes from both the range of materials that can be dated and the age range over which the technique is applicable. Several amino acids are routinely used for geochronological purposes. Isoleucine and leucine are probably the most commonly applied; because of their slow rates of racemization they are appropriate for materials up to 10^6 yr in age. Other, more rapidly racemizing amino acids, such as aspartic acid, are more useful when higher resolution but shorter duration records are being investigated (Goodfriend, 1992; Oviatt and Miller, 1997; D.S. Kaufman, 2000).

Amino acids are ubiquitous in most types of organic matter, creating a large potential for the application of amino acid dating in lake deposits. In practice, however, analysis of lacustrine materials to date has been more limited. Bulk sediments have been analyzed for amino acids, and sometimes yield consistent and interpretable age results (e.g.,

Blunt et al., 1981; Blunt and Kvenvolden, 1988). However, there are several problems with obtaining amino acid data from bulk sediment that to date have limited its utility. As with bulk ^{14}C analyses, amino acids within a sample are likely to have been derived from various sources, not necessarily of the same age. Older sediments will introduce high D/L ratio molecules (Kessels and Dungworth, 1980; Dungworth, 1982). These different sources are also likely to have been subject to varying racemization rates, because of their variable molecular environment during preservation, and because of taxonomic differences in proteins. In the future it may become possible to resolve these problems, as our ability to identify and extract specific classes of molecules of known origin improves.

More consistent results have been obtained using ostracode and mollusk shells (McCoy, 1987; Oviatt et al., 1987; Bouchard et al., 1998; Magee and Miller, 1998; D.S. Kaufman, 2000). Proteins involved in the biomineralization of these animals occur as thin membranes around their shells, and are often preserved with their fossils (D.S. Kaufman and Miller, 1992). The most ideal benthic invertebrate fossils are those derived from depositional environments where temperature can be assumed to have remained relatively constant over the burial history. This condition is met in the hypolimnion of temperate or high-latitude lakes (where a uniform 4°C history can be assumed). For the Holocene, temperature constancy is probably also a reasonable assumption for tropical lakes. However, it is important to note that the vast majority of amino acid geochronology studies of lakes have focused on fossil mollusks from exposed deposits, where complex postburial temperature histories are likely to have occurred. With the advent of techniques that use smaller sample sizes, the application of amino acids to core studies, for example using ostracodes, is likely to increase (Oviatt et al., 1999).

Some of the most intriguing applications of amino acids of relevance to paleolimnology actually involve the use of terrestrial fossils. The fossilized eggshells of flightless, ratite birds (ostriches, emus, etc.) are sometimes found in Pleistocene lake margin deposits. These eggshells have proven to be some of the best recorders of the racemization signal, in the sense that they behave as almost completely closed systems (no diagenetic exchange with surrounding sediment) and provide very consistent age information (Brooks et al., 1990; G.H. Miller et al., 1991).

Amino acid data is generally expressed as D/L ratios. Once these ratios have been calculated for a variety of amino acids and samples the data can be used in several ways. Univariate plots of individual

D/L ratio data can be compared between cores to establish probable correlations. Downcore plots generally show an increase in D/L with depth for all amino acids (Blunt and Kvenvolden, 1988). A common approach among amino acid geochronologists is to group sample populations with similar D/L ratios into *aminozones*, stratigraphical intervals that are of roughly contemporaneous age (McCoy, 1987) (figure 6.11).

If the goal of the research is to use the D/L ratio data for absolute dating, then either the racemization curve must be calibrated against some independent age model, for example from ^{14}C data (e.g., Magee and Miller, 1998), or the effects of the various factors that impact racemization rates must be assumed to be relatively invariant. As a result, absolute ages from amino acid racemization must be treated cautiously unless they are calibrated, in which case the age uncertainty is proportional to the uncertainty in the calibration.

Event Correlation Techniques

Unusual earth history events can be correlated in paleolimnological records by the characteristic signatures they leave behind in lake beds. The most widely used event correlation techniques are those involving the identification and counting of annually layered sediments (varve chronology), the interpretation of the history of volcanic ash

deposits (tephrochronology), earth magnetism (paleomagnetism), and organism evolution and extinction (biostratigraphy) (table 6.2).

Varve Chronology

Annually layered lake sediments are referred to as *varves* (figure 1.2h). Where laminated sediments can be shown to be annual deposits they provide an independent and very high-resolution means of dating Late Quaternary lake beds and their contents (O'Sullivan, 1983). Furthermore, in both Quaternary and pre-Quaternary deposits, the metronome of varve deposition provides a means of determining rates of a variety of lacustrine, watershed, and paleoclimatic processes, even when the absolute age of those deposits is unknown (R.Y. Anderson and Dean, 1988).

The history of varve chronologies extends back into the nineteenth century, well prior to the published use of the term (Heer, 1865). Most early studies, while recognizing the annual nature of many lake deposits, did not take full advantage of the potential of this archive, specifically the opportunity for paleolimnologists to establish independent, annual chronologies, against which the history of events could be compared, and against which other dating techniques could be refined. The earliest explicit use of this approach was the work of De Geer (1912). He proposed that by putting together a series of overlapping varve counts from

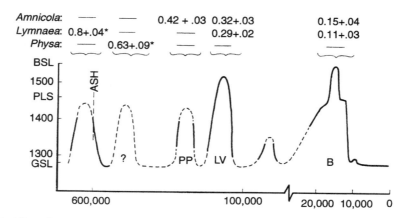

Figure 6.11. D-Alloisoleucine/L-isoleucine data for fossils of three gastropod genera (*Amnicola, Lymnaea,* and *Physa*) taken from deposits of the Bonneville Basin, Utah. The D/L ratios form tight clusters (aminozones) that are readily differentiated from one another, and that increase with age. The data are plotted against estimated lake-level fluctuations through a series of lake cycles. More recent studies by Oviatt et al. (1992, 1999) have suggested absolute age ranges for some of these cycles: B = Bonneville, 28–12 ka; LV = Little Valley, 150 ka; PP = Pokes Point, 417 ± 55 ka; Lava Creek (marked "ash"), ~ 620 ka. From McCoy (1987).

a number of lakes in Scandinavia, it would be possible to determine the timing of the retreat of the Fennoscandian Ice Sheet. Since De Geer's early work, an extensive network of over a thousand varve records has been established throughout Scandinavia (e.g., Cato, 1987; Björck et al., 1992; Wohlfarth et al., 1993; Holmquist and Wohlfarth, 1998). Similar varve chronologies exist for other parts of the world as well (Antevs, 1922; Verosub, 1979a,b; Ridge and Larsen, 1990).

In the simplest case, varves are counted from the sediment–water interface, or some other surface of known age, down or up section to some feature of interest, yielding an age in varve years. For cores, this normally involves some type of resin impregnation of continuous vertical cross-sections, followed by careful imaging of the varves to insure all are counted (e.g., Landmann et al., 1996). Outcrops can be treated in the same way when the sediments are finely laminated, or simply measured in outcrop when the varves are relatively thick. The equivalency of varve years to calendar years is normally based on some independent evidence from sedimentology or geochronology that the rhythmic beds are in fact annual deposits. Using this approach, the timing of specific earth or lake history events can be dated with very high precision (e.g., Leonard, 1995). Single core varve chronologies now exist from a number of locations around the world that extend back to the Late Pleistocene (Landmann et al., 1996; Kitagawa and van der Plicht, 1998) (figure 6.12). These chronologies provide some of the most highly resolved records of lake history available, providing direct year-by-year data on changes in annual sediment accumulation rate, sediment composition, grain size and fossil accumulation, and by extension the timing of regionally and globally important paleoclimate events.

Errors accrue in single-core varve counts for several reasons. First, there is the possibility that the layers that are being counted are not in fact annual deposits, and that indistinct varves may be miscounted. This is best evaluated using sedimentological and independent dating criteria, as discussed in chapter 7. Second, there may be uncertainty about the age of the datum from which the varves are being counted. Normally this is the surface layer of sediments in a lake. Because the uppermost sediments in extant lakes are very susceptible to disturbance during coring, some varves may be lost or amalgamated during coring operations. To a great extent, the use of appropriate methods for retrieving undisturbed cores (freeze cores, box cores, or multicores) alleviates this problem. The final, and most difficult, problem occurs when intervals of an otherwise varved record are interrupted by massive, slumped, or otherwise unvarved sediments, or by erosional unconformities. This problem is relatively intractable when only a single core is available.

With multiple cores, however, depositional artifacts like the ones mentioned above can often be overcome, by piecing together correlated, overlapping records, and making corrections for unconformities in specific core records (Hajdas et al., 1995).

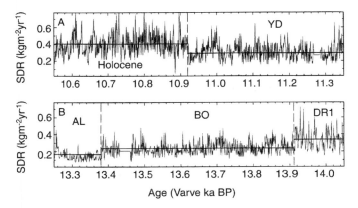

Figure 6.12. An annually resolved varve chronology of sediment accumulation rates from Lake Van, eastern Turkey for two intervals in the Late Pleistocene and Holocene. Horizontal lines are means for the individual chronological intervals (European Quaternary stages). DR1 = Oldest Dryas, BO = Bølling, AL = Allerød, YD = Younger Dryas. Changes in sediment deposition rates probably reflect variable weathering and sediment transport rates into the lake under different climate regimes. From Landmann et al. (1996).

This is the principle behind the European and North American varve networks (figure 6.13). This generally involves matching marker horizons, such as distinctive tephras, or patterns of individual varve thickness from one record to another, or from one record to a regional average record. Thickness matches are most commonly performed by simple visual pattern matching. Using these methods, *floating varve records*, which are not directly tied to the present, can be dated, allowing the extension of a composite varve record into older time periods. In sediments where varves are not evident from lithology alone, it may also be possible to count other microscopic features in a sectioned core, such as cyclical variation in microfossil species abundances (Card, 1997). The labor-intensive nature of such counting however probably restricts this application to the analysis of relatively short core intervals.

It bears emphasis that varve chronologies pieced together from multiple records and multiple lakes are often subject to miscorrelations. This is because most correlations are based on a limited, visual analysis of the zones of correlation. When more sophisticated statistical analyses of correlations have been performed on these correlations they are sometimes found to be in error, even for intensively studied varve sequences such as those of Sweden (Holmquist and Wohlfarth, 1998).

The chronological methods discussed above can also be applied to older (e.g., pre-Late Pleistocene) varved lake sediments (R.Y. Anderson and Dean, 1988). These varved sequences provide floating chronologies only, since there is no way to establish an absolute age estimate for any given varve horizon at an annual level of precision, as can be done with more recent varves. However, these records still allow a paleolimnologist to determine levels of interannual variation in the entire range of indicators present in a local record. In exceptional circumstances, individual pre-Quaternary varve sequences can even be correlated between distantly separated outcrops.

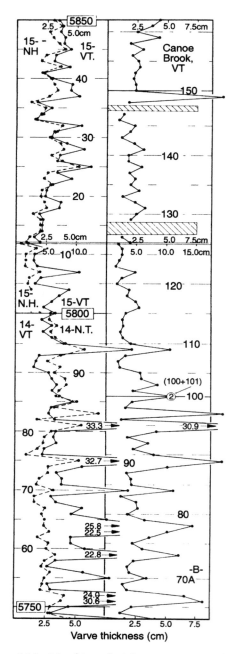

Figure 6.13. Matching glacial varve stratigraphies from several sites in the northeastern United States separated by over 100 km. Sediment thicknesses versus varve count years are plotted for paleolakes Albany (dashed line, 14-NY, from eastern New York), Ashuelot (dashed line, 15-NH, from southwestern New Hampshire), and Hitchcock (solid lines, 14-VT, 15-VT, and Canoe Brook site, all in Vermont). From Ridge and Larsen (1990).

Tephrochronology

Explosive volcanic eruptions can distribute *tephra* over extremely large distances, sometimes up to several thousand kilometers (figure 6.14). When individual tephras are dated and can be identified in different areas, they can serve as correlation tools. This provides the possibility of obtaining age control without recourse to expensive and repeated radiometric dating of the same tephra in different

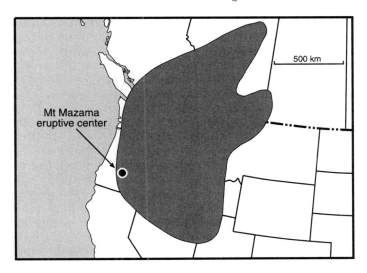

Figure 6.14. Distribution of the Mt. Mazama (Oregon) tephra in western North America. The source of the tephra is located at Crater Lake, a collapsed caldera that formed ~ 6600 yr BP. Approximately 40 km^3 of tephra was released in this eruption, with the furthest currently recognized accumulations in central Alberta (1550 km from the source). Note the asymmetry in distribution of ash around Crater Lake, reflecting prevailing winds at the time of the eruption. Most lakes in the region blanketed by Mazama ash have a recognizable tephra layer from this eruption. Adapted from Westgate and Naeser (1995).

locations. Even in the absence of direct dates, accurate correlations between tephra horizons in multiple locations can allow a researcher to develop a relative age scheme for the deposits under study. Lakes that lie in or near volcanically active regions of the world often have many ash horizons interbedded with their sediments, and are therefore suitable for obtaining a tephrochronological record.

Correlation of two tephras requires an analysis of the physical and chemical characteristics of both, to insure they are in fact derived from the same eruptive event (Izett, 1981; Westgate and Gorton, 1981; Sarna-Wojcicki et al., 1984). In practice, this requires the characterization of several features of the tephra, such as its color and gross physical attributes, the shape of the volcanic glass shards, the shape and size of crystals in the shards, the major and minor element geochemistry of the tephra, and the degree of alteration of the tephra. Most tephra analyses are done on recognizable volcanic ash beds. However, volcanic glass shards may occur as disseminated particles in otherwise nondescript lake beds. Particularly in lakes that are located long distances from eruptive centers, the collection of volcanic particles from these types of deposits may be critical for dating purposes. Increasing attention is being given to methods for separating these isolated glass shards (e.g., Turney, 1998).

Whether or not an unknown tephra can be matched by its physical and geochemical characterization is largely dependent on how well the regional "pool" of tephras is known. In some regions, such as the western United States, Iceland, and Kenya, detailed records exist of the geochemistry and physical properties of numerous tephras. This increases the likelihood that an unknown tephra can be matched to one that has been previously described (Naesser et al., 1981; Sarna-Wojcicki et al., 1984; F.H. Brown et al., 1985; Einarsson, 1986). Correlation networks between multiple lake basins in the western United States have allowed both the identification of matching tephras and the age-bracketing of previously unknown ones (Izett, 1981; Westgate and Gorton, 1981; Sarna-Wojcicki et al., 1988; Negrini et al., 2000) (figure 6.15). This involves the progressive refinement of age relationships between tephras, dated stratigraphical horizons, and frequently, identifiable paleomagnetic events (Sarna-Wojcicki et al., 1984, 1997).

Secondary reworking of the volcanoclastic particles produced during an eruption sometimes complicates tephra correlation. Although an eruptive cycle may be geologically brief, the transport and redeposition of water-lain tephra can sometimes continue long after eruptive activity ceases. This can cause a tephra to become mixed with unrelated particles (lake muds, diatoms, etc.). When this

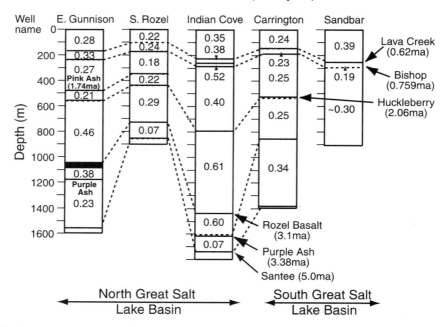

Figure 6.15. Age estimates (ma) and sediment accumulation rates for five deep wells at the Great Salt Lake, Utah, based on tephra correlation. The upper three tephras (Lava Creek, Bishop, and Huckleberry) had been dated independently, prior to this study, and are widespread throughout the western United States. These known tephras were matched geochemically to the unknown ones of the Great Salt Lake to establish an absolute age model. Higher sediment accumulation rates are observed in sites that lie in close proximity to major border faults and active subsidence. Modified from Kowalewska and Cohen (1998), after work by Moutoux (1995).

occurs, the tephra can take on very different physical characteristics from the pure air-fall material derived from the original eruption. Stratigraphically, this type of volcanic ash mixture may also have a significantly younger age of deposition than its eruption.

Paleomagnetism and Rock-Magnetism Records

Paleomagnetic dating and correlation methods take advantage of the fact that the earth's magnetic field varies over time, and this variation is recorded in magnetically susceptible minerals (Butler, 1992; Jacobs, 1994). The earth's magnetic field can be described at any given time and place by three vector components. *Inclination* is the angular deviation of the magnetic field vector measured down from the horizontal in a vertical plane containing the field vector. *Declination* is the angular difference between geographical north and the horizontal trace of the field vector (magnetic north). *Intensity*

is the strength of the magnetic field, which generally increases with proximity to the magnetic poles.

The magnetic field has two states, *normal* and *reversed*. In the normal (i.e., present-day) state, the north end of a freely mobile compass needle orients itself toward the north and down. During periods of reversed magnetic polarity, a needle at the same locality would point upward and toward magnetic south. *Magnetic reversals* between the two states occur aperiodically, generally on time scales of 10^5 to 10^6 yr, and last for at least 10^4 yr (Barbetti and McElhinny, 1976). Transitions between the two states take on the order of a few thousand years. Paleomagnetists have recognized two types of polarity intervals (figure 6.16). *Chrons* are longer intervals of predominantly normal or predominantly reversed magnetism. These are punctuated by brief intervals of the opposing polarity, referred to as *subchrons*. Following many years of careful study of very well dated stratigraphical sequences, a highly resolved history of chrons and subchrons is now available for the past 5 ma (Cande and Kent, 1995). A somewhat lower-resolution record of

Age (Ma)

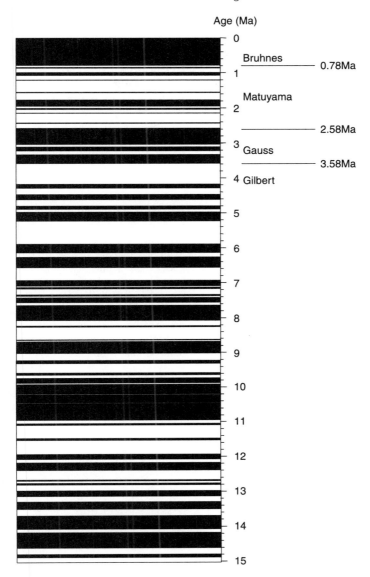

Figure 6.16. Geomagnetic polarity time scale for the past 15 ma. Normal (i.e., like present-day) intervals (chrons and subchrons) by convention are shown in black, reversed polarity intervals in white. Important Late Tertiary intervals of predominantly normal (Bruhnes, Gauss) or reversed (Matuyama, Gilbert) magnetism are labeled on the right, along with the current best estimates of their boundary ages. Derived from data in Cande and Kent (1995) and King and Peck (2001).

polarity reversals at the chron level encompasses the past 160 ma, based on the record of magnetic anomalies on the sea floor (Heirtzler et al., 1968). In addition to these reversals, shorter-term variability in the magnetic vector occurs, on time scales of 10^1 to 10^4 yr. *Geomagnetic excursions* involve major migrations of the dipole, approaching a polarity reversal but ultimately returning to the prior polarity regime (King and Peck, 2001). *Secular variation* involves migration of the local magnetic vector that is evident over a relatively large area. The source and geographical scale of both geomagnetic excursions and secular variation are controversial; they may be global, or they may be derived from the nondipole components of earth magnetism. Both polarity and secular variation records are in a continuous state of being improved, as newer and more complete stratigraphical sections become available from which paleomagnetists can obtain improved age estimates of particular geomagnetic events. Many of these improvements have come from the very high-resolution paleomagnetic records of lake deposits (e.g., Lund, 1996; Sarna-Wojcicki et al., 1997).

Changes in magnetic polarity and secular variation can be preserved as *remanent magnetism* in both sediments and volcanic rocks as they accumulate in a lake basin. Individual magnetically susceptible (Fe and Ti-bearing) mineral grains contain variable numbers of zones of uniform magnetic properties. In the case of volcanic rocks, *thermoremanent magnetization* (TRM) occurs as the rock cools below several hundred degrees Celsius. The magnetization of an oriented sample of this rock can be measured, to determine the inclination and declination of the magnetic field at the time it formed.

Volcanic rocks are absent from most lacustrine sediments, and where present are generally stratigraphically discontinuous. However, *detrital remanent magnetization* (DRM) occurs when sedimentary magnetic carriers take on the ambient magnetic field characteristics. This type of magnetization can be preserved and measured in lacustrine sediments, especially finer-grained detrital magnetic minerals, which will be oriented primarily by the ambient magnetic field. The extent of such magnetization is also a function of the minerals involved. Both DRM and TRM are susceptible to various types of secondary magnetic overprints. These overprints may have developed at times when the ambient magnetic field was quite different from the time of deposition; altering the total remanent magnetism of a sample. Various demagnetization techniques are used to remove these secondary effects, in the process yielding a record of the original remanent magnetization.

Sediments and volcanic rock record several types of magnetic archives of interest to paleolimnologists. At the longest time scales, a record of polarity reversals may be preserved in lake deposits, and when they are accurately identified, can provide absolute age information (e.g., Liddicoat et al., 1980; Glen and Coe, 1997; King and Peck, 2001). At moderate to high latitudes (> 30°) inclination records alone are generally adequate to characterize a sample as normal versus reversed polarity. This is extremely useful for paleomagnetism studies from cores, as it means that azimuthally oriented cores are not required (King and Peck, 2001). At lower latitudes, however, declination data is needed to

determine polarity, which does require sample orientation information.

In very long cores from modern lakes or ancient lake beds the recognition of reversals can be extremely useful as a geochronometric tool. However, the binary nature of magnetic reversals (normal or reversed) implies that some independent means must be available to infer which polarity transition is being observed in a particular sequence of lake beds. In practice this is usually done by cross-correlation with some other source of age information, such as biostratigraphical or radiometric dates. For example, if a tephra in an apparently continuous lacustrine stratigraphical sequence were dated at 0.80 ma, it would be reasonable to assume that a persistent magnetic reversal from reversed to normal polarity, lying in a stratigraphical sequence 3 m above this tephra, was the Bruhnes–Matuyama Polarity Chron boundary. Because this boundary has been globally dated at 0.78 ma, that same age could be attributed to sediments at the reversal in the local stratigraphical section. This type of reasoning of course depends on there being no unrecognized major hiatuses or unconformities in the section, and on the assumption that all reversals at about that time interval have been previously recorded. Even if a reversal in a local section cannot be pinned to a particular reversal in the geomagnetic time scale, however, its recognition may prove useful. A distinctive pattern of stratigraphical thicknesses between reversals might be recognized from several locations within a basin, thereby providing a means of relative correlation in otherwise lithologically homogeneous sediments. An extremely long lacustrine record of paleomagnetism from the sediments of Tule Lake, northern California, provides an excellent illustration of how paleomagnetic reversals and excursions can be used in conjunction with radiometric dating and tephra correlation, for absolute dating control in lake deposits (Rieck et al., 1992) (figure 6.17).

The concepts of absolute and relative paleomagnetic dating can, in principle, be applied to both geomagnetic excursions and particular patterns of secular variation. The greater frequency of excursion and secular variation events also makes their recognition more relevant to most paleolimnological studies, those documenting time periods of 10^3 to 10^5 yr BP. Unique patterns of change in declination, inclination, or intensity might be recognized between core localities within a lake basin or between lake basins (Peck et al., 1994; Negrini et al., 2000) (figure 6.18). Several major geomagnetic excursions have been recognized for the past 200 ka alone, and some of these are recorded at

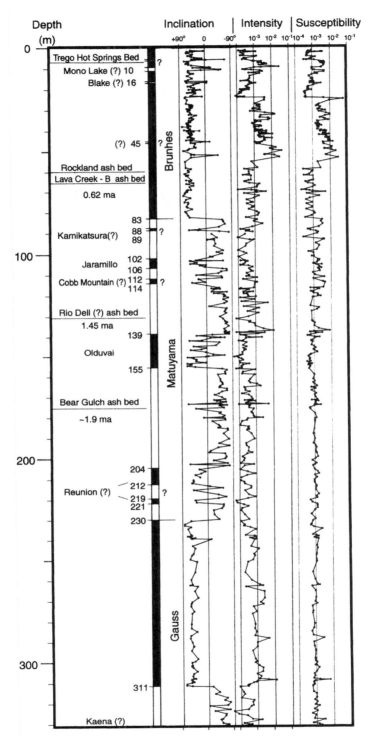

Figure 6.17. An example of combining magnetostratigraphy with tephrochronology from a Plio-Pleistocene core taken at Tule Lake, northern California. Composite stratigraphical plot of remanent magnetic inclination, intensity, and bulk magnetic susceptibility data on the right. An interpretation of these records relative to the geomagnetic time scale is shown in the middle. Horizons of major polarity changes and excursions are shown in meters. Tephra correlations used to constrain the magnetic correlations are also shown, along with their estimated ages. From Rieck et al. (1992).

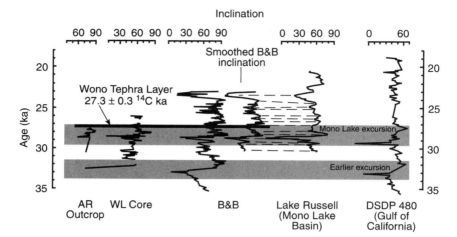

Figure 6.18. High-resolution correlation of paleomagnetic inclination features between several sites in western North America. Three intrabasinal sites at Summer Lake, Oregon (Paleolake Chewaucan—AR Outcrop, WL Core, and B&B Core) are compared with sites in the Mono Lake Basin (Paleolake Russell) and the Gulf of California (Mexico). A radiocarbon age on the Wono Tephra provides absolute age control for part of this record. Major correlations are shown by dashed lines. A widely recognized geomagnetic event, the Mono Lake Excursion, is highlighted, along with another prominent, earlier excursion. From Negrini et al. (2000).

multiple localities. Although some secular variation events may reflect relatively local events, others have been correlated over distances of thousands of kilometers. This suggests that even some short-lived geomagnetic events may be recording perturbations in the dipole component of earth magnetism (King et al., 1983). Particularly intriguing global-scale correlations have been recorded in magnetic intensity records. Wiggle-matching between the intensity records of Lake Baikal (Siberia, Russia) and the Mediterranean Sea have demonstrated extremely long distance (8000 km) correlations with a resolution of as short as 7000 yr (Peck et al., 1996).

Another approach to using magnetic properties of sediments for correlation purposes comes from the study of various rock-magnetic properties of sediments. *Magnetic susceptibility* (often given as κ) is the ratio of the induced magnetization of sediments to the strength of the inducing laboratory field. Magnetic susceptibility is a function of the mineralogy, concentration, and, to a lesser degree, grain shape or size of the magnetic minerals present. In many lake sediments, κ is closely correlated with total magnetite or titanomagnetite content (e.g., Rosenbaum et al., 1996). The concentration and mean grain size of magnetic minerals in lake sediments varies in response to a number of variables,

including weathering intensity, dilution of terrestrially derived magnetic minerals by organic and chemical sediments, and dissolution or alteration of magnetic minerals (Thompson and Oldfield, 1986). For example, the high-susceptibility mineral magnetite can be altered to low-susceptibility hematite during weathering. Decreasing the proportion of magnetite in sediment will also have the effect of decreasing κ. Dissolution of magnetite is known to occur through bacterial oxidation under conditions of high productivity and organic carbon flux (Karlin, 1990; Leslie et al., 1990; Snowball, 1993). Less importantly, κ values can be affected by magnetic mineral grain size, with somewhat higher κ values resulting from smaller grain size. Several test parameters have been devised by paleomagnetists to differentiate between these possible causes of change in κ (Thompson and Oldfield, 1986; Verosub and Roberts, 1995).

Intrabasinal, regional, and perhaps even global correlations may be possible through the interpretation of rock-magnetic data from lake sediments (e.g., Thouveny et al., 1990; Negrini et al., 2000). At the within-lake basin level, similar patterns of magnetic susceptibility change can be expected to occur in different parts of the lake, as long as the controlling factors (magnetic mineral composition of bedrock and climatically induced weathering) are reasonably constant. This is most likely to be the case in relatively small lakes lying on a uniform type of bedrock. A more exciting possibility is the use of rock-magnetism records in lake sediments for global correlations, based on the relationship of magnetic mineral changes to profound climate change. Several studies from Lake Baikal and western North American pluvial lake sediments have found that higher magnetic susceptibility is associated with high lake stands and glacial intervals (pluvial periods), whereas warmer and drier interglacial conditions are associated with lower susceptibility (Peck et al., 1994; Rosenbaum et al., 1996; Negrini et al., 2000). Various explanations have been offered for this correlation, and it seems unlikely that any one mechanism is valid for all cases.

Biostratigraphical Correlation and Age Determination Using Lacustrine Fossils

Biostratigraphy is the science of correlation and age determination of strata using fossils. Biostratigraphers determine the chronostratigraphical range of their embedded fossil organisms, from the time of their evolution to their extinction. This can be done in both a relative or absolute time sense. Biostratigraphical units (biozones) are defined on the basis of the fossil content of a sequence of rocks, that allows them to be differentiated from the sediments above (younger) or below (older). Biozones can be defined by the interval over which a specific fossil species or lineage occurs, by the interval over which an assemblage of species co-occur, by an interval of stratigraphical overlap between two species, or by a zone of maximal abundance for a species.

Frequently, the biostratigrapher is also interested in determining absolute age information from the fossils. To accomplish this, the fossil ranges initially must be related to some independent source of age control, for example a ^{39}Ar/^{40}Ar – dated tephra, or a known paleomagnetic reversal. Once a *first appearance datum* (FAD) or *last appearance datum* (LAD) are securely tied to such age controls at one locality, they can be used to estimate the absolute age of the same species FAD and LAD at another site. However, FADs and LADs should not be thought of as precisely equivalent in time (*coeval*). There are inherent and variable lags in how long it takes for a species to reach the site under investigation from its area of origin, and how long after its arrival the first fossil of that species was preserved. There is also uncertainty in how likely

it is that a biostratigrapher will find a fossil close to this earliest preserved occurrence. A similar set of concerns operates at the LAD/extinction end of the time period for the fossil organism in question.

The most useful fossils for biostratigraphy are those with short temporal ranges, which evolved and went extinct quickly. Fossils of this type are commonly found in long-lived, tectonic lake basins. Fossil organisms with broad geographical range can be correlated over greater distances, improving their utility. Most lacustrine organisms have subcontinental ranges and would seem to be good candidates for interbasinal correlation. Unfortunately, the same factors that promote rapid speciation (physical barriers to dispersal and genetic isolation) in some clades of lacustrine organisms also limit their geographical ranges. As a result, the most rapidly evolving clades tend to be restricted to single paleolake basins, and correlation and dating using them is therefore restricted in area. Extremely large but long-lived lake basins provide the best combination of circumstances for developing useful biostratigraphical schemes in lakes, and it is also in these settings that most research has been focused (e.g., Carvalho-Cunha and Alves-Moura, 1979; Colin and Lethiers, 1988).

Despite these limitations, biostratigraphical schemes have been widely developed in lakes, using diatoms, charophytes, ostracodes, and mollusks. Lake deposits also frequently incorporate the fossils of terrestrial organisms that are biostratigraphically useful, especially pollen and small mammals. As a result of variation in the rates of both evolutionary and ecological change, it is impossible to pin down a precise range of resolution over which biostratigraphy is applicable. In principle, lake deposits of any age that contain fossils are amenable to biostratigraphical analysis, with time resolutions of anywhere from $\sim 2 \times 10^5$ to 10^7 yr (table 6.3). In part, the biostratigraphical resolution is dependent on the degree of study that the fossils in a region have received. With increased study, biostratigraphical zones can often be subdivided, producing a more refined, higher-resolution biostratigraphy. However, some of the differences evident in table 6.3 probably reflect real differences in evolutionary rates. The rift lake basins of the South Atlantic (Cretaceous) and East Africa (Tertiary), the Pannonian Basin of Central Europe, and the Tertiary Chinese basins are notable for their brief biostratigraphical zones, a consequence of rapid speciation and extinction. In contrast, some of the much longer duration zones are likely to be a reflection of the frequently great longevity of lacustrine species. In practice, lacustrine biostratigraphy using

Table 6.3. Commonly used organisms in pre-Quaternary lacustrine biostratigraphy and approximate time resolution attained. Note that biostratigraphical resolutions are commonly improved with additional research. Only lake-dwelling organisms are listed (as opposed to fossils, such as pollen, small mammals, etc., that may be found in lake deposits). The list is not exhaustive and a number of biostratigraphical schemes also exist as proprietary commercial information with petroleum companies

Taxonomic Group and Region	Time Interval	Resolution (by Stratigraphical Interval)	Approximate Chronometric resolution[a]	References
Ostracodes				
West Africa and Brazil	Late Jurassic–Early Cretaceous	Subage	1 ma in best studied sections	Krömmelbein (1966), Grekoff and Krommelbein (1967), Moura (1972, 1988), Carvalho-Cunha and Alves-Moura (1979), Tambareau (1982), Grosdidier et al. (1996), Bate (1999)
United States (west)	Jurassic	Age	—	Schudack (1998)
Northern Europe (possibly intercontinental)	Late Jurassic–Early Cretaceous	Subage	2.5 ma	F.W. Anderson (1973)
China	Triassic–Late Cretaceous	Sub-Epoch	8–30 ma	Gou and Cao (1983), Hao et al. (1983), Y. Li (1984), Qiping and Whatley (1990)
China[b]	Tertiary	Subage	5–10 ma	Defu et al. (1988)
Caribbean (Greater Antilles)[c]	Neogene	Epoch (approx.)	3–8 ma	van den Bold (1976, 1983, 1990)
Southern/Central Europe (Paratethyan Basins)	Neogene	Subepoch	2–8 ma	Carbonel (1978), Van Harten (1990), Cipollari et al. (1999), Gliozzi (1999)
Dinoflagellates				
Central Europe (Pannonian Basin)	Neogene	Subepoch	0.3–3 ma	Magyar et al. (1999)
Diatoms				
United States (west)	Neogene	Subepoch–age	3–15 ma	Van Landingham (1964a, 1985), Bradbury and Krebs (1982), Krebs et al. (1987), Krebs and Bradbury (1995)
Charophytes				
Europe	Jurassic–Cretaceous	Subage	1–2 ma	Feist and Schudack (1991)
Europe	Cretaceous–Paleogene	Subage	1–2 ma	Feist-Castel (1975), Feist-Castel and Columbo (1983), Martin-Closas and Grambast-Fessard (1986), Riveline (1986), Galbrun et al. (1993)
United States (west)	Triassic	Epoch	5–22 ma	Kietzke (1989)[d]
United States (west)	Jurassic	Subage	0.7–2 ma	Schudack (1995, 1999)[d]
United States (west)	Cretaceous	Age	5–10 ma	Peck (1957), Soulié-Märsche (1994)
Mollusks				
China	Jurassic	Epoch	19–24 ma	Yu et al. (1993)
East Africa	Neogene	Subage	0.5–1.5 ma	Pickford et al. (1991), Williamson (1982)
Central Europe (Pannonian Basin)	Neogene	Subepoch	0.2–2 ma	Magyar et al. (1999)
W. North America	Paleogene	Subage	0.2–3 ma	Fouch et al. (1979), Hartman and Roth (1998)

[a]Estimated either by range of absolute age determinations, when known, or as mean estimated duration of biostratigraphical zones, by dividing total time duration by number of zones recognized.
[b]Gastropods also zoned at comparable resolution.
[c]Fresh to brackish water marine species.
[d]Zonation based on a combination of charophytes and ostracodes.

species, lineage, or assemblage ranges is most commonly applied to Early Pleistocene or pre-Quaternary sediments, where resolutions of > 1 ma are acceptable. In Mid–Late Quaternary lacustrine biostratigraphy, identification of zones of maximal abundance of some species, rather than FADs and LADs, is a more common practice. The presumption in correlating abundance peaks between cores is that the environmental conditions that promoted these peaks were coeval. Given the variability of lacustrine environments that we have already discussed, this is a risky proposition, and is generally only useful at a local, intrabasinal scale. An important exception, however, comes in the application of local abundance peaks related to human disturbance, or intentional and accidental species introductions, in very recent paleolimnological records (e.g., Ogden, 1966; McAndrews, 1968, 1988; Blais et al., 1995).

Summary

Our conclusions from this chapter can be summarized as follows:

1. Developing an accurate chronology should be a primary goal in all paleolimnological studies. This involves both the establishment of relative age relationships, through principles of superposition and correlation, and, to the extent possible, geochronometric age models.
2. Paleolimnologists have a variety of techniques at their disposal for determining absolute ages of lake deposits, including methods based on radiometric dating, dating by radiation damage, amino acid racemization (a form of chemical dating), and event correlation methods. Which technique is appropriate for a given study or lake deposit will vary depending on age, desired resolution, lithology, and expense.

3. Radiometric techniques, relying on radiogenic isotope decay, form the bulwark of most paleolimnological age modeling. For Quaternary lake deposits, ^{14}C and (for very young sediments) ^{210}Pb are the most commonly applied radiometric dating methods. Artificial radionuclides from nuclear explosions or reactor accidents also provide useful age constraints for very young (post-1950s) lake deposits. In older lake deposits, radiogenic systems with longer half-lives are required, especially K/Ar and Ar/Ar dating.
4. Radiation techniques, especially fission track dating and various luminescence dating methods, rely on the indirect and time-dependent effects that radioactivity and other radiation sources have on minerals. Luminescence techniques are particularly exciting for paleolimnologists because they hold potential for dating Quaternary materials such as carbonate and organic-poor sediments, which are not generally amenable to other dating methods.
5. Amino acid racemization dating relies on the clock-like transformation in symmetry of amino acids. It provides a means of absolute dating for organic material of Quaternary age, especially when calibrated by independent age controls.
6. Event correlation techniques include varve counting, tephrochronology (the identification and correlation of volcanic ashes or tephras), correlation of paleomagnetic records, and biostratigraphy. All provide potential sources of absolute age control (and in the case of varve counts, absolute age intervals) whenever they can be calibrated to an independent (generally radiometric) chronology. All are potentially useful methods for both Quaternary and pre-Quaternary-aged lake deposits.

7

Sedimentological Archives in Lake Deposits

Lake sediments are both repositories and sources of information about lake history. Depositional products tell us about the mechanisms of transport or accumulation of important geochemical and fossil archives, but important clues about that history are imbedded in the pattern of sedimentation itself. Geologists have recognized this fact since the earliest paleolimnological studies. Although he would certainly not have called himself a paleolimnologist, Charles Lyell's (1830) classic studies and interpretation of the depositional environments of the Eocene Paris Basin set the tone for a time-honored approach to the study of ancient lake deposits. Lyell recognized that understanding the physical, chemical, and biological attributes of lakes that affect sedimentation, obtained through modern observation, must be applied to a four-dimensional (spatial plus time) analysis of sedimentary deposits and depositional history. However, not everything we need to know or every process we need to invoke will necessarily arise from our short-term observations of modern lakes. Events that are unlikely to occur in the course of a brief, several-year experiment or period of monitoring may become virtual certainties over the long history of some lakes and may leave a sedimentary archive of which we have little prior understanding from modern studies (Dott, 1983).

Furthermore, the sedimentary response that we observe to some external forcing event may differ depending on the time scale over which we observe the response (Dearing, 1991). Consider a hill slope that is undergoing accelerated erosion, and that is producing an accumulation of sediment in a downstream channel as a result of land-clearing activities. Initially there may be no response in terms of sedimentation rate in the downstream lake; all of the sediment is being held in temporary storage. This process may occur over time scales of a few decades. At some later time a triggering event, perhaps a series of abnormally high rainfall and discharge

years, causes this sediment to be released to the lake, now at an accelerated rate. This becomes a sedimentary response that the paleolimnologist can record. But, over geological time scales of millennia or longer, the original process may be modified, and new ones may gain in importance. The same location may experience a long-term decline in sediment accumulation rate as a result of reduced subsidence rates. There is an element of spatial scale dependency in this too; large lakes and/or large catchments will experience complex feedbacks with greater time delays, because of their stronger hydroclimate filters. For all of these reasons paleolimnologists must do more than apply theory gained from modern observation. They must also incorporate the element of time, the probability of unlikely events, and the probability of delayed responses to processes into their analysis of lake history.

Commonly Used Sedimentological Archives of Lake History

The physical properties of sediments provide a variety of indicator records for paleolimnologists. Several excellent reference books are available that elaborate on "how to" aspects of these archives (e.g., Håkanson and Jansson, 1983; Berglund, 1986; Miskovsky, 1987; Last and Smol, 2001). Here I will only briefly review some of the more important indicators.

Descriptive Sedimentology of Cores and Outcrop

Detailed stratigraphical logs of outcrops or cores are essential elements of any paleolimnological study. These include basic descriptions of *lithofacies*, the visible characteristics of color, grain size, particle composition, bedding, contact relationships, sedimentary structures, and macrofossils. The importance of these data cannot be overestimated; although they are technically some of the simplest data to collect, they are also frequently some of the most informative. Accurate interpretation of most geochronological, geochemical, and paleoecological data relies on sound core or outcrop descriptions, and interpretations using more "sophisticated" analytical techniques than the qualitative assessments of the human eye can go wildly astray when supporting core and outcrop observations are not available.

Descriptive sedimentology provides direct information on major changes in depositional environment, such as the vertical relationship of the sample at the time of deposition to lake surface level, wave base level, and the oxicline. Large-scale facies changes usually provide the most direct evidence of major hiatuses in deposition, or patterns of lake infilling. On a gross scale, grain size analysis provides an indication of proximity of the sample site to sources of sediment, and the power of transporting and reworking waves and currents at the depositional site. Because most lake sediments are fine-grained, this is usually accomplished in unconsolidated sediments through one of several granulometric techniques specific to small particles, such as pipette analysis or the use of laser particle size analyzers. For consolidated or lithified lake sediments, particle size is often estimated from microscopic analysis.

Water Content

The analysis of water content is extremely important for unconsolidated lake beds. Although water content and porosity are of limited interest to most paleolimnologists in their own right, water content data are essential for the calculation of particle or geochemical flux rates, and for correctly interpreting geochemical trends that may be influenced by pore fluid diagenesis.

Petrographical Examination and X-ray Diffraction

Usually an early part of a physical description of a core or lake bed outcrop involves a microscopic examination of the samples collected, using some combination of binocular microscope, petrographic microscope, and scanning electron microscope (SEM) techniques. Light microscopy allows the texture and composition of clastic particles and fossils

to be quantified, and for coarser-grained materials, mineralogy to be determined. Some types of fine-grained particles are highly diagnostic of specific origins of formation and can be counted petrographically. For example, the incomplete combustion of oil and gas, by automobiles or from industrial emissions, produces microscopic *spheroidal carbonaceous particles*, which can be counted in lake sediments as an indicator of atmospheric hydrocarbon emissions. SEM observation extends the size range of particles that can be identified and measured. With the coupling of a backscattered electron detector and energy dispersive spectrometer (EDS), SEM study allows semiquantitative analyses of elemental composition to be obtained on the grains under study.

X-ray diffraction (XRD) techniques are the most direct and commonly used methods in paleolimnology to determine the mineralogical composition of fine-grained, inorganic sediments. Many minerals provide quite specific information about geochemical and thermal environments at the time of formation, but because of their small size, may be impossible to identify microscopically. Although some of these minerals may be identifiable using SEM and related techniques, this rarely provides the type of bulk analysis of the sample that can be obtained with XRD.

Physical Properties

Extremely valuable data can be obtained from *physical property* measurements of lake sediments. These data are essentially geophysical, but are obtained directly on cores or subsamples from outcrops. The purpose of using these techniques is to rapidly obtain an indirect measure of some facies characteristic of interest; large numbers of measurements can be made in a very brief period in comparison to labor-intensive techniques like petrography. Some of the most commonly obtained measurements are *magnetic susceptibility* and *gamma-ray attenuation*. Both of these measurements can be made on discrete samples, or in pass-through data loggers. The latter allow a core to be "logged" semicontinuously and very rapidly for the relevant geophysical parameters. Down-hole logging, as practiced by the petroleum industry in open wells and boreholes, can be used to accomplish many of the same tasks. Magnetic susceptibility is a function of the composition and concentration of magnetic minerals and, to a lesser extent, the grain size of the magnetic carriers (Thompson and Oldfield, 1986; Verosub and Roberts, 1995). Because magnetic minerals in lake sediments are primarily detrital grains, their abundance, and therefore the total susceptibility of a sample, can be diagnostic of the relative contribution of lithogenic versus authigenic +biogenic contributions to the sediment. Furthermore, watershed changes in dominant detrital magnetic minerals can result in substantive changes in total sample susceptibility. Because the common, high susceptibility mineral magnetite can be reduced to pyrite under anoxic conditions, magnetic susceptibility measurements are also a function of mixing conditions in a lake (Karlin and Levi, 1983). Gamma-ray attenuation is a function of the bulk density of sediment; it varies depending on water content and lithology.

Major Factors Controlling Sedimentation in Lakes

Six major factors directly control the broad-scale differences we observe in lake sedimentation patterns (figure 7.1). These factors are interrelated in complex ways; most are ultimately linked to regional climate, geology, and for the past few millennia, human activities. One of the more exciting challenges of paleolimnology is to understand the linkages or feedback loops between these factors, and how they interact to create unique depositional signals. These interaction loops also indicate the need for caution in paleolimnological interpretations that lean too heavily toward a single causative explanation for an observed pattern of deposition.

Watershed Geology

The topographical relief, bedrock composition, and underlying structure of a lake's watershed all play major roles in determining both the rate of sediment input and the lake's sedimentary geochemistry. Topographical relief is a first-order control on mass-wasting rates, resulting in high sediment accumulation rates in lakes downstream from mountainous terrain. Watershed geology influences the texture of available sediments for deposition based on both the initial composition and fabric of the bedrock. For example, easily dissolved and corroded limestone bedrock rarely produces the type of sandy beaches that typify the coastlines of lakes whose watersheds are underlain by more resistant materials such as coarse-grained igneous rocks or quartzites. The geological characteristics of the watersheds of small lakes tend to be relatively uniform, whereas in larger lakes with more varied

watershed terrain, these relationships become more complex, reflecting different bedrock, surficial geology, and orographic/climatic variation within the basin (Osborne et al., 1985; Kalindekafe et al., 1996). The range of dissolved solutes and buffering capacity of lakes are tied to bedrock composition, thereby determining the types of authigenic minerals that can form in a lake.

Because of the chemical instability of their mineral assemblages, some types of bedrock, such as fine-grained basalts, produce very little sand, weathering directly from the outcrop or cobbles to altered muds. Prior metamorphism or jointing in rocks also produces characteristic grain shapes that are available for subsequent deposition in a lake. Coupled with watershed microclimate, bedrock geology regulates the rate of surface versus subsurface runoff and groundwater discharge to lakes, by influencing such factors as the permeability and porosity of land surfaces and their local erodability (Reid and Frostick, 1986). Bedrock relief, composition, and texture also influence the response of watersheds and sediment discharge rates to land clearing.

Water chemistry is directly influenced by watershed geology. Aerosol and dust contributions notwithstanding, the watershed provides the most direct source of chemical input to a lake. Alkalinity and pH are directly influenced by the composition of watershed bedrock, leaving depositional signals directly through their controls on calcium carbonate precipitation, and indirectly through their influence on organisms. In closed-basin, saline lakes, the watershed geology normally determines the types of brine pathways and therefore the evaporites that can form.

Watershed Climate

Local precipitation within a lake's watershed affects vegetation cover, soil development, and watershed erosion rates. All of these factors influence the rate of sediment delivery to a lake and the chemistry of particulate matter and dissolved solids entering the lake. Particulate compositional differences related to climate change are most evident in the analysis of elemental concentrations and clay minerals deposited in lakes (e.g., Yemane et al., 1996; Yuretich et al., 1999). Lakes in cold and dry climates often have clays dominated by minerals, like chlorite, with abundant easily removed cations. These ions are stripped under wetter and warmer conditions of increased weathering, leaving a residue of cation-poor clays, like kaolinite.

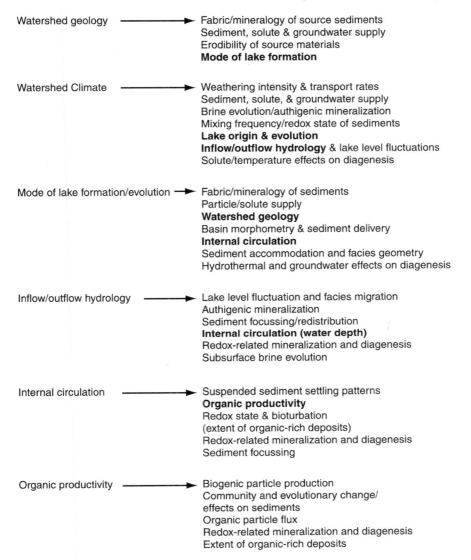

Figure 7.1. Major factors affecting the broad-scale differences observed in lake sedimentation patterns, facies development, and facies geometry. Note that even these "major" factors are not independent of one another, but instead affect each other (shown in bold) through a complex web of interactions.

In extremely arid climates downstream delivery of sediments will usually be slower than under intermediate conditions of precipitation. Arid watersheds may simply have insufficient means of delivering sediment to a lake because of the infrequency of overland flow. As a result, sediment accumulation in the downstream basin may be quite slow, dominated by wind-blown dust and chemical precipitates, rather than clastic particles. Most data suggests that sediment transport increases significantly as mean annual precipitation reaches the semiarid, 300–400 mm yr^{-1} level. Conflicting data

exist, however, as to the effect of high levels of effective precipitation and mean annual runoff on sediment loading to lakes (Langbein and Schumm, 1958; Knighton, 1984). Because of the seasonality of vegetation growth in the humid tropics, peak discharges and the degree of seasonality of discharge may have a greater influence on sediment yields to lakes than mean annual precipitation alone (Walling and Kleo, 1979). Furthermore, on a short time scale, sediment yields within channels or from overland flow cannot be directly translated into sediment accumulation in a downstream lake.

In many watersheds, storage within channels, deltas, and coastal plains partially decouples the short-term climatic or anthropogenic effects of sediment erosion from immediate lacustrine sediment accumulation (Dearing and Foster, 1993).

Temperature also plays a role in determining sediment delivery patterns to a lake through its influence on soil, vegetation, and for cold climates, ice development. In temperate to tropical regimes, warmer mean annual temperatures result in higher rates of biochemical weathering, and higher sedimentation rates. Higher discharge and lacustrine sedimentation rates are associated with higher mean annual temperatures, longer runoff seasons, and higher mean annual precipitation in most cool temperate and boreal regions (Itkonen and Salonen, 1994). However, relationships between temperature and sediment delivery are much more complex in cold, glaciated regions. Glacial scour is an effective mechanism for generating fine sediment, but much of this material is bound for long periods in ice, unavailable for deposition in a downstream lake. The delivery of this fine sediment (*glacial rock flour*) to a lake therefore does not have a simple relationship to temperature. Sedimentation rate effects are closely linked to the time scale of measurement (Leonard, 1986, 1997). Over diurnal and seasonal scales, ice extent and the velocity of glacial movement are more or less constants and sediment discharge to lakes is closely linked to temperature; more sediment is delivered when temperatures, melting, and discharge rates are higher (figure 7.2). For short-duration paleolimnological studies this relationship may be exploited through examination of annual lamination thickness patterns. On these short time scales temperature also affects the density relationships between inflow and lake water, determining the formation of overflows or underflows, and therefore the spatial distribution pattern of sediment being delivered to a lake (Weirich, 1986). However, at longer time scales other factors come into play. Leonard (1986) found that in glacial-fed Hector Lake (Canadian Rocky Mountains) sedimentation rates increased during Late Holocene glacial advances as a result of long-term increase in bedrock erosion rates (figure 7.3). However, high sedimentation rates were also evident immediately following the periods of glacial advance, when a large quantity of sediment, previously in ice-bound storage, became available for transport to the downstream lake. In contrast, the maximum warming period, a few centuries later, was marked by slower lacustrine sedimentation rates, as previously stored glacial debris had been used up.

Mode of Lake Formation and Evolution

The processes responsible for forming and maintaining a lake basin invariably have strong effects on its sedimentary infill. The reasons for this are easy to understand when you consider that in order to produce and maintain the topographical depression that contains the lake, either material was physically excavated or the margins of the depression were relatively elevated. In either case a large mass of erodable material on the lake margin owes its origin to the same processes that caused the depression to form initially.

Many of the sedimentological differences between lakes that are related to lake origin result from the differences in both subaerial and subaqueous slope associated with basin-forming processes, reflected in the morphometry of the basin. For example, deposits from gravity flows are characteristic of the steep lake margins associated with lakes formed by volcanic caldera collapse, overdeepening of glacial rock basins, or along major border faults in tectonically formed rift basins (e.g., Pickrill and Irwin, 1983; Tiercelin et al., 1994). In contrast they are relatively unimportant components of floodplain or wind deflation basins, since these mechanisms rarely produce slopes sufficient to promote gravity flows. In tectonic lakes bulk sediment composition is often linked to specific basin structural contexts (Soreghan and Cohen, 1996). In tectonic and collapsed caldera basins, long-term sediment accumulation rates are controlled by underlying rates of fault motion and subsidence, which in turn vary systematically across the lake floor.

Inflow and Outflow Hydrology

The connections with upstream water sources and downstream discharge place a first-order constraint on depositional patterns in lakes. Most obvious is the question of whether a lake is hydrologically open or closed at its surface. Presence or absence of a spillway determines the extent to which lake level can fluctuate in a basin, and therefore the extent to which deposits that are linked to a particular environmental zone of the lake are likely to migrate spatially over time. Whether a lake is closed or open can and does change over time, depending largely on climatic conditions but also occasionally on upstream changes in river flow direction (stream piracy, etc.). Lakes with surficial outlets undergo relatively small changes in surface elevation, because increased upstream discharge to the lake is compensated by increased outflow. As a result both their surface area and the lake floor bound-

A.

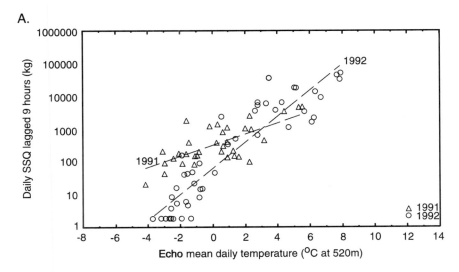

B.

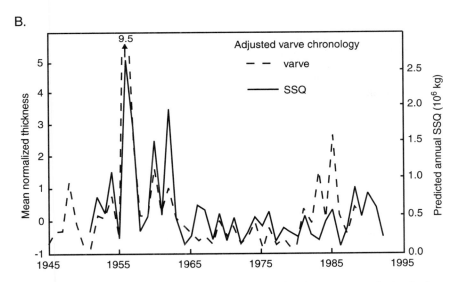

Figure 7.2. Climatic effects on streamflow sediment discharge and sediment accumulation in the watershed of Lake C2, a high Arctic lake basin of northern Ellesmere Island, Canada. (A) Strongly correlated relationship between mean daily temperature at the nearby Echo weather station and measured daily suspended sediment discharge (SSQ). This relationship can be used to predict annual lacustrine sediment accumulation based on long-term temperature records. (B) A comparison between predicted annual sediment discharge (SSQ) into Lake C2 and adjusted annual sediment accumulation (mean normalized thickness) over a 50-year period shows that temperature effects account for over 60% of the variance in interannual sediment accumulation. From Hardy et al. (1996).

aries between environmental zones with differing sedimentary processes (for example, above or below shallow water wave base) are relatively stationary. In contrast, closed lakes undergo much more spatial and temporal variation in area, envir-

onmental boundaries, and local sediment accumulation rates (figure 7.4).

Variations in lake level have numerous sedimentary consequences. Deltas in closed basins undergo large and erratic changes in position in response to

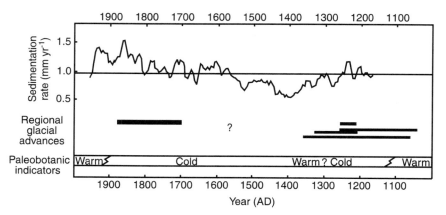

Figure 7.3. Composite sedimentation rate curve for glacial-fed Hector Lake (Canadian Rocky Mountains) compared with periods of glacial advance and climate (paleobotanic) indicators over the past 900 years. Note correlation between periods of increased sedimentation rates and both Late Holocene glacial advances (resulting from long-term increase in bedrock erosion rates), and periods immediately following glacial advances, when large quantities of previously ice-bound sediment were discharged. From Leonard (1986).

rapid lake-level fluctuations (e.g., Kroonenberg et al., 1997). During times of low lake levels, deposits that have previously accumulated along the margins of a lake may be eroded and redeposited in the deeper parts of the basin, leaving behind characteristic erosional surfaces in their wake. In large lakes, wave action in the surf zone where waves break can also be erosive, admittedly on a smaller scale than

what occurs as lakes fall. As lake levels rise a path of eroded and reworked deposits may also develop.

An important exception to the link between basin closure and rapidity/frequency of water depth and shoreline fluctuations occurs in proglacial lakes and their postglacial successors (figure 2.9). Spillway blockage by ice can cause lakes to rapidly rise, until a new outlet spillway level is

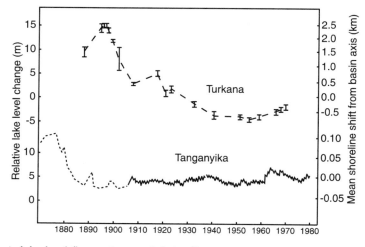

Figure 7.4. Historic lake-level fluctuations and their effects on mean shoreline positions in Lakes Turkana (a closed basin) and Tanganyika (an open basin since the late nineteenth century). Both lakes are in the African rift valley. Short-term lake-level fluctuations in L. Turkana are greater than in L. Tanganyika; these translate into order-of-magnitude larger shoreline positional shifts for Turkana as a result of its much lower lake floor topographical gradient. Modified from Cohen (1989a).

reached. When ice blockage is removed from the lower outlet the same lake can fall catastrophically, as we saw in chapter 2. Following deglaciation relative lake levels can continue to change rapidly as a result of isostatic rebound. This produces equivalent facies shifts, as for example in the adjustment of shoreline beach-ridge deposits (Thompson, 1992).

Apart from the largely physical consequences of lake closure, there are also profound chemical and biological ones that occur in lakes that are closed because of insufficient water inflow, and that indirectly affect sedimentation. Hydrological closure allows ionic concentration to rise over time in lakes, producing evaporite deposits from concentrated brines, a distinctive depositional signature. Except in those relatively rare circumstances where saline groundwater discharge causes hydrologically open lakes to become saline, low levels of salinity and an absence of evaporite deposits characterize most open-basin lakes. The accumulation of dissolved solids also determines what organisms can potentially invade the lake, and secondarily affects depositional processes (extent of bioturbation, accumulation of shelly deposits, etc.).

Up to this point we have primarily considered hydrology in terms of surface discharge. However, groundwater discharge and recharge are also important regulators of lake level. This is most evident in closed basins, where groundwater input can make up a very large proportion of the total influent water budget. In arid or semiarid climates, the rate of groundwater flow into a lake basin can determine whether a lake persists as a body of water through dry seasons or longer droughts, thereby dampening lateral shifts in environmental boundaries, or water chemical changes over time. Likewise, subsurface outflow, particularly in highly fractured volcanic bedrock terrains, can cause a surficially "closed" basin lake to behave like an open basin, maintaining relatively low salinities and limiting lake level instability (Darling et al., 1990).

Most aspects of inflow and outflow hydrology of concern to paleolimnologists are either directly or indirectly linked to climate. In regions with low precipitation to evaporation ratios (P/E), inflow is normally insufficient to infill depressions, and there is always some minimum threshold of P/E required to maintain a lake's spillway flow. In North America almost all closed-basin lakes are confined to the arid, western interior part of the country, where rainfall is low and potential evapotranspiration high. Open basins occur in the east, in the extreme west, and at high elevations in the interior, all areas of high P/E ratios. Groundwater discharge

is also often closely linked to climate through rates of surficial groundwater recharge. In most parts of the world, the dominant source of shallow groundwater that discharges into lakes is of recent, meteoric origin. Therefore, while groundwater discharge may dampen seasonal fluctuations in lake inflow, it is also sensitive to long-term changes in precipitation and evaporation.

Internal Water Circulation

Internal circulation and mixing affects the redox state of sediments, their resuspension following deposition and subsequent transport, and the accumulation of organic matter in sediments. Amictic and meromictic lakes provide paleolimnological records that can be more highly resolved in time than those of monomictic or dimictic lakes, as a result of the lack of resuspension and/or bioturbation in the former lake types. Under conditions of no resuspension or seasonal turnover, sedimentary signals can be preserved that resolve monthly events, perhaps even shorter durations under exceptional circumstances. In contrast, resuspension from bottom current activity reduces the potential resolution of sedimentary records (Håkanson, 1982; Fillipi et al., 1998). Such mixing normally blurs seasonal signals and moves sediment, especially fine particulates, long distances from its area of origin (Bloesch and Uelinger, 1986). In conditions of either extremely vigorous circulation or slow accumulation, resuspension can cause mixing of deposits accumulated over decades or longer. Fine particulate matter can also be maintained in suspension for long periods by other internal hydrodynamic mechanisms, such as internal waves (Lee and Hawley, 1998). Internal circulation also affects diagenetic alteration of sediments after deposition, again primarily through impacts on redox state. Methanogenesis, a common occurrence in permanently stratified lakes, can lead to the formation of minerals such as siderite or dolomite (Talbot and Kelts, 1986).

Horizontal transport causes physical mixing of sediments from different watersheds, and allows sediment to be carried to particular areas of a lake. When coupled with vertical mixing (gravitational flows) and resuspension, this generates the phenomenon of *sediment focusing*, an increase in sediment accumulation rate in a particular part of a lake basin (M.B. Davis and Ford, 1982). Focusing into the deepest part of a lake basin is a common attribute of morphologically simple and small lakes, but patterns of accumulation become more complex as lake size and shape complexity increase. In such lakes the spatial variation in rates of sediment accumulation can often be linked to some combination of location of sediment supply, opportunity for particle settling, and possibility of resuspension (e.g., Ludlam, 1984).

Organic Productivity

The sedimentary consequences of changing organic productivity are seen most directly in the accumulation rates of "organic" sediments. These include combustible organic matter, the skeletal remains of organisms, such as biogenic silica, and the calcium carbonate precipitation commonly associated with photosynthesis. Increases in primary production rate within a lake are commonly associated with changes in the amount of organic debris that is sedimented to the bottom of the lake, unconsumed by dissolution, heterotrophic organisms or

bacterial decay, or in isotopic changes in organic sediments. For example, sediment trap and core data for open-water sites in Lake Ontario (United States/Canada) show that the accumulation rates of organic carbon and fine, particulate calcium carbonate are strongly correlated with nutrient loading, and these relationships are also mirrored in carbon stable isotope records (figure 7.5) (Schelske and Hodell, 1991; Hodell et al., 1998). However, numerous differences in the efficiency of heterotrophic consumption and solubility of different types of organic matter exist that prevent organic flux rates from being interpreted as direct proxies for productivity (Hicks et al., 1994). On short time scales the rates of organic sediment accumulation are also affected by the rate at which phytoplankton debris settles to the lake floor, a function of water depth, vertical circulation, particle aggregation, and the formation of fecal pellets (Haberyan, 1985; Pilskaln and Johnson, 1991). Particle degradation, resuspension, and grazing all impose limits on our

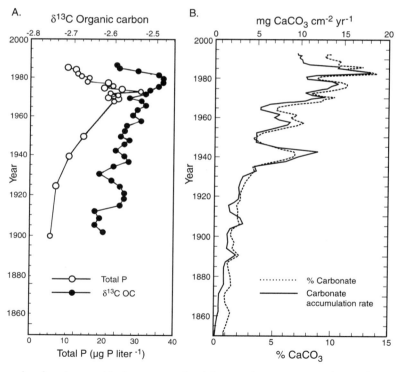

Figure 7.5. Records of carbon stable isotopes and calcium carbonate accumulation from Lake Ontario, showing their strong relationship with total phosphorus loading and increasing productivity in the lake during the twentieth century. (A) Historical relationship between $\delta^{13}C$ of organic C in sediments and total P in Lake Ontario water. Total P data were obtained by computer simulation for 1900–1970 and by lake sampling for 1968–1982. (B) Comparison of sediment carbonate concentration and flux rates for a core from the eastern basin of Lake Ontario over the same time interval. From Schelske and Hodell (1991) (A); Adapted from Hodell et al. (1998) (B).

ability to resolve short-term productivity changes from accumulation rate data alone (Zohary et al., 1998).

Facies Models for Lacustrine Depositional Systems

Facies models are descriptive, graphic, or mathematical simplifications that relate some set of transportational, depositional, and diagenetic processes to a deposit. For paleolimnologists they serve as tools to help us conceptualize what happened in a lake to create a particular sedimentary sequence. The heterogeneity of lakes in time and space makes the process of simplification daunting. A model developed in one lake may not be suitable for understanding a related process in another, because of subtle differences in process that the investigator has overlooked. This is a particular challenge for the paleolimnologist, since we generally do not know the physical and chemical conditions of the lake whose deposits we are studying; quite the opposite, we are usually trying to interpret prior conditions from the deposits.

Sedimentologists study the facies characteristics of lacustrine environments at varying scales (table 7.1). At the microenvironmental scale, different environments of deposition are characterized by variation in composition, texture, and small-scale sedimentary structure reflecting local or short-term processes. Study of these variations is most important when questions about annual to century-scale processes are relevant, for example involving recent human impacts to lake ecosystems or short-term climate change. At larger, macrostratigraphical scales, facies variation is expressed in the stacking pattern and three-dimensional geometry of sedimentary deposits. These features are controlled by lake infilling by river-borne sediment, major changes in lake level, or tectonic or volcanic activity, disrupting drainage patterns or changing relative lake level. Compositional differences in lake sediments related to sediment inputs from different watersheds or lakewide changes in productivity also become more important at larger scales.

Seismic Facies and Sequence Stratigraphy

A sedimentary body can be characterized not only by its physical or chemical attributes but also by its geophysical properties. Differences in seismic reflectivity are probably the most widely used geo-

physical facies characteristics in paleolimnology. Seismic stratigraphers have found that differences in the seismic response of sediments allow differentiation of both facies type (e.g., fine vs. coarse-grained, well-bedded vs. chaotically bedded) and larger scale geometry of deposits. Differences in the lateral continuity of reflections are some of the most important characteristics separating seismic facies. Some reflections can be traced over long distances, representative of depositional processes that are uniform over large areas. This would be the case for example in the central part of a lake experiencing uniform sedimentation of suspended materials. Discontinuous reflections, on the other hand, are indicative of more abrupt or frequent lateral facies transitions, for example within channels. The amplitude of reflections is controlled by the contrasts in the seismic responses of vertically adjacent facies. A vertically uniform deposit will produce only low-amplitude reflections, whereas major vertical differences in deposits will produce high-amplitude reflections. The geometry of sedimentary reflections also gives clues to their physical attributes. Dipping reflectors are characteristic of inclined bedding surfaces, such as one often finds at the front of deltas. At a larger scale the shape of a deposit may give clues to its origin. For example, a package of reflections that radiate outward, forming a diverging wedge away from the sediment source, would be indicative of a sublacustrine fan. The manner in which reflectors terminate can also be indicative of depositional setting (figure 7.6) (Mitchum et al., 1977; Miall, 2000). Erosional truncation, such as one would find on the edge of a channel, or following a period of erosion and subsequent deposition, is marked by the termination of reflections of one type and the superposition of a contrasting type. Lapout terminations reflect the depositional limits of a body of sediment, beyond which no sediment of some particular origin was ever deposited. Several different types of these lapouts are recognized, each suggestive of different environmental conditions of formation. On the outer edge of an expanding body of sediment, for example a prograding river delta or sublacustrine fan, reflections may downlap onto the pre-existing lake floor. When sediment builds up over a previously formed erosional surface (for example, as a channel fills up, or during a major lake-level rise following a protracted period of low lake stand and erosion) a series of onlapping reflectors form. Toplap is observed where sediment accumulation terminates at the upper edge of a deposit, for example on the landward side of a delta.

Table 7.1. Facies characteristics for different lacustrine depositional environments and their potential for providing paleolimnological records

Depositional System	Textures/ Lithologies	Small-scale Bedforms and Sedimentary Structures	Large-scale Geometry, Seismic Facies Characteristics, and Stratigraphical Architecture	Typical Fossils	Potential for Paleolimnological Records (high resolution, short time scale, 10^0 to 10^2 yr)	Potential for Paleolimnological Records (low resolution, long time scale, $>10^3$ yr)
Deltas		*Large-scale coarsening upward cycles*	*Prograding clinoforms or broad aprons (often stacked), toplaps and downlaps, extensive normal faulting*			
Topset-distributary channels	G, S	Fining up cycles, trough X-bedded and Rip X-laminated S w/scoured lower surfaces	Elongate lenses and channels, erosive into lower unit, discontinuous	Only robust fossils, often abraded V	Poor–Fair	Fair (generally short-lived)
Topset-subaerial levees	S, Si	Thin-thick beds, Lam, climbing Rip Lam, R, Bur	Discontinuous lenses, convex up	Fossils rare	Poor–Fair	Fair (generally short-lived)
Topset-delta floodplain	Si, Cl, Mic Nod, P	Thin beds, massive or mottled (bioturb), R, Bur, Nodular paleosols	Aggradational sheets	V, Pl, Ins, Pol	Poor–Fair	Good
Topset-deltaic ponds	Si, Cl, P, Gy, Carb	Thin Si beds, Rip/ climbing Rip Lam S and Si, Bur	Lenses	Pl, Ins Ost, Clad, Chir, Mol	Fair–Good	Fair (generally short-lived)
Delta mouth bars and radial bars	S, G, minor Si	Trough X-bedded and planar bedded S, G; Rip X-Lam S, flat bedded G, thin mud drapes but large-scale coarsening up seq, imbrication	Lenses	Rare Mol (abraded)	Poor	Poor
Delta slope channels	S, G	Thick to massive beds, scour surfaces	Lenses with incised lower surfaces, often flanked by convex-up levee lobes	Rare Mol (abraded)	Poor	Poor

172

Environment	Lithology	Sedimentary structures	Bedding geometry	Fossils/biota		
Delta front/slope surfaces	S, Si, Cl where flocculated	Inclined rip (near top), wave rip, climbing rip and parallel lam. Bedding (often draped over rip) (slumps)	Prograding clinoforms or broad aprons and toplaps and downlaps, moderate-high-amplitude reflections	Pl, Ost, Mol, Clad, Chir, Pol, Dia, Ins, Sp, Dis Fish	Fair–Good	Good
Prodelta	Pelletized Si, Cl, FS and VFS, Mic lenses	Rhythmites, thin graded beds, Lam	Moderately continuous sheets, low-moderate amplitude reflections	Pl, Ost, Mol, Clad, Chir, Pol, Dia, Sp, Dis Fish	Good	V. Good
Shallow lake or littoral environments					*Fair–Good*	*Depends on persistence (generally low in small, shallow lakes)*
Open water, oxygenated lakes or ponds	Si, Cl, VFS and FS, Gy (reddish, gray, green or light brown)	Lam, thin–thick Beds, Bur, oscillatory Rip	Sheets and lenses (variable, depending on water body size)	Pl, Ost, Mol, Clad, Chir, Pol, Dia, Sp, Dis Fish	Fair–Good	Fair
Marshes	Si, Cl, VFS and FS, occas. Carb (Mic), P	Bur, R, massive (bioturbated) muds and peats	Sheets (generally thin)	Pl, Ost, Mol, Clad, Chir, Pol, Dia, Sp, Dis Fish	Fair	Fair
Acidic Bogs	P, Si, Cl, VFS		Lenses (sometimes convex-up)	Br, Pol, Pl, Clad, Ins	Good	Fair–Good
Beaches of larger lakes			Often progradational, entire complex may be sheet like, but individual facies elements more discontinuous, like rectangular wedges, individual bars are often convex upward			
Lower shoreface	VFS, FS, Si	Coarsening-up cycles or mud/sand interbeds, pinch and swell Lam, thin graded beds, wave/current Rip, hummocky X-beds, scour surfaces, minor Bur	Laterally discontinuous thin sheets or stringers	Ost, Mol, Chir, Dis Fish, (Pol, Dia, Sp in silty units)	Poor–Fair	Fair

(continued)

Table 7.1. (*continued*)

Depositional System	Textures/ Lithologies	Small-scale Bedforms and Sedimentary Structures	Large-scale Geometry, Seismic Facies Characteristics, and Stratigraphical Architecture	Typical Fossils	Potential for Paleolimnological Records (high resolution, short time scale, 10^0 to 10^2 yr)	Potential for Paleolimnological Records (low resolution, long time scale, >10^3 yr)
Upper shoreface	S, G	Current and wave Rip, med to large trough and hummocky X-beds, scour surfaces, flat beds. Gentle lakeward dip	Laterally discontinuous thin sheets or stringers	Fossils rare and abraded Mol, Dis Fish, V	Poor	Poor
Foreshore and subaerial bar surfaces	S, G, B, shell lags	Flat bed Lam, low angle X-beds	Laterally discontinuous thin sheets, stringers and lenses	Fossils rare and abraded Mol, V	Poor	Poor
Backshore/washover fans	FS	Flat bed, X-beds, Bur, Mudcr, R. Often landward dip	Sheets and lobes	Fossils rare and abraded Mol, V	Fair	Fair
Backshore lagoons and sand/mudflats	Si, Cl, FS, P, Gy	Massive to unevenly Lam, Mudcr, Bur, R. occas. Coarsening-up cycles, Nod	Sheets	Strom, Pl, Ins, Dia, Pol, Ost, Mol, Char, conchostracans	Fair	Fair
Nearshore carbonates						
Beachrock	S, G, Carb cement, Gyp	Flat bed and low angle X-bed. Gentle lakeward dip	Laterally discontinuous thin sheets or stringers	Mol, Ost	Poor	Fair (episodic indicators of lake level)
Marl benches	Si, Mic, S, G (shell or coated grains)	Thin-thick beds, massive, Bur, coarsening-up cycles	Sheets or occas. prograding clinoforms	Mol, Ost, Char, Clad, Chir, Sp, Pol, Dia, Ins	Fair–Good	Good
Microbial bioherms/spring travertines	Carb (occas. AS near hydrothermal sources)	Lam or massive, cemented boundstone, interbeds of S/Si/shell	Convex-up mounds, banks or discontinuous lenses, often assoc. w/ faults, slope breaks and groundwater discharge interfaces	(As bound clasts) Mol, Ost, Char	Fair (Good within laminated stromatolitic zones. Poor in thrombolitic buildups)	Fair–Good (depends on continuity)

(continued)

Ooid shoals	S, some G and shell	Thin–thick X-beds and flat beds, wave and current Rip	Discontinuous lenses, occas. Prograding clinoforms and downlap	Fossils rare (abraded, coated Ost and Mol)	Poor	Fair (episodic indicators of lake level)
Protected, low-energy coasts & embayments	FS, S and Mic, Gy, biogenic Carb	Thin-bedded Lam, Bur, Nod, brecciated and soil horizons		Mol, Ost, Char, Clad, Chir, Sp, Pol, Dia, Ins, some V	Good	Fair
Sublacustrine fans			Fan-like or wedge-shaped geometry w/ divergent reflections, both aggradational and progradational forms depending on lake type		*Poor*	*Good for lake level, sed. discharge or tectonic cyclicity*
Canyons	G, B, S, some Si and C	Unsorted debris flows, fining-upward turbidites, massive sands (grain flow deposits), slumps	Deeply incised surfaces, onlapping bases and convex-up or planar upper surfaces, elongate channels and lenses, high-mod amplitude reflections w/internal terminations	Fossils uncommon, Rare Mol in coarse units, Dia, Pol, Ost in fines	Poor	Poor
Constructional channels	S, G, minor Si	Thick Turbs and fining-upward sequences	Lenses and elongate channels, concave-up lower surfaces, convex-up at top, high-mod amplitude, discontinuous refl.	Fossils uncommon, Rare Mol in coarse units, Dia, Pol, Ost in fines	Poor	Poor
Levees	Si, VFS and FS	Thick mud and sand interbeds, occas. Si and S Lam, Bur, soft sed. deformation structures	Convex-up lenses	Dia, Pol, minor Ost, Ins, Dis Fish	Poor	Poor–Fair
Interchannel & overbank environment	Si, Cl, occas. FS	Thin–thick homogen. muds, Bur, thin-med Turbs, occas. debris flows and slumps	Sheets, broad lobes, flat lenses, low amplitude, discont. and parallel reflections	Dia, Pol, minor Ost, Ins, Dis Fish	Fair	Fair–Good
Lake floor & pelagic Wave/current reworked platforms and benches	FS, occas. Si, shells	Thin–thick beds, current ripples and ripple X-lamination, shell/gravel lags	Tabular w/occas. clinoforms, high-amplitude reflections	Mol, Ost, Dis Fish, Pl (abraded and less delicate parts)	Fair	Fair–Good

Table 7.1. (*continued*)

Depositional System	Textures/ Lithologies	Small-scale Bedforms and Sedimentary Structures	Large-scale Geometry, Seismic Facies Characteristics, and Stratigraphical Architecture	Typical Fossils	Potential for Paleolimnological Records (high resolution, short time scale, 10^0 to 10^2 yr)	Potential for Paleolimnological Records (low resolution, long time scale, $>10^3$ yr)
Open water adjacent to steep slopes	S, G, B	Thick Turbs, unsorted or crudely tabular debris flows, slumps massive, occas. Strom	Chaotic lobes, hummocky, low-continuity, low-amplitude reflections	Fossils rare	Poor	Poor
Open water of shallow basins (subject to resuspension) or shallow slopes of deep basins	Si, Cl, Mic, Diat (light colored), Gy occas. VFS (gray or brown, occas. red)	Wispy Lam, thin–thick tabular beds, common scour and Bur, some contorted Lam, thin Turb	Sheet drape deposits, variable amplitude reflections, laterally continuous, often onlapping sequences	Mol, Ost, Ins, Dia, Pol, Clad, Chir, Sp, Dis Fish, Pl	Fair–Good	Excellent
Deep basin floors below wave base, limited resuspension but regularly oxygenated	Cl, Si, Mic, Diat, Gy (gray or brown)	Thin–thick beds, massive, Bur, thick rhythmites, Teph	Sheet drape deposits, variable-amplitude reflections, laterally continuous, often onlapping sequences	As above, but lower density/diversity of spp	Good	Excellent
Meromictic or amictic lake floors	Cl, Si, Mic, Diat, Sa (dark brown, green or black, reflecting high TOC)	Thin rhythmites and microlaminites (combinations of Org, Diat, Sili, Carb), Lam, Turb, Teph	Sheets, drape pre-existing topography, variable-amplitude reflections, laterally continuous	Dia, Pol, Ins, Art Fish, Pl (well preserved), little or no benthic inverts except w/indications of transport	Excellent	Excellent
Saline lakes/playas						
Dry mudflats	Si, Cl, FS (red and gray)	Thin–thick bedded muds and channels, Mudcr, some R, pedogenic cracks and cements	Sheets w/imbedded sand channels	Fossils rare (minor V, P and Pol)	Poor	Fair

Environment	Characteristic textures/lithologies	Characteristic bedding features	Geometry	Characteristic fossils		
Wet mudflats	Interbedded Sili (dominant) and Evaps, Si and Cl, AS, Org	Lam, thin–thick bedded or massive, cyclical bedding, Mudcr, solution collapse features, Evap Nod, isolated crystals, beds and mineral casts, occas. S lenses	Sheets	Br eggs and fecal pellets, bacterial mats and Stroms, clam and tadpole shrimp, occas. Char and Ost, Pol, Dia, occas. Ins	Fair	Good
Ephemeral lake (salina, salt pan)	Evap and Sili Si and Cl, AS, Org, minor FS	Thick Evap beds, tepees and salt crystallization displacement and solution collapse features, occas. mud, tephra and eolian sand horizons	Sheets	Bacterial mats and Stroms, Br eggs and fecal pellets, clam and tadpole shrimp, occas. Char and Ost, Pol, Dia, occas. Ins	Good	Good
Deep basin stratified saline lake	Cl, Si, Mic, Diat, Sa, Evap, occas As, Org, Sa (dark gray, green or black)	Thin rhythmites, microlaminites, Turb, Teph, Nod	Sheets, drape pre-existing topography	Br eggs and fecal pellets, bacterial mats, Pol, Dia, Pl, Ins, occas. Mol, no bioturbation below chemocline	Excellent	Excellent

Codes: (1) Characteristic textures/lithologies: AS = authigenic silicates (zeolites, feldspars, chert, etc.), B = boulder/cobble, Carb = carbonate, Cl = clay, CS = coarse sand, Evap = undifferentiated evaporites, FS = fine sand, G = gravel, Gy = gyttja, Gyp = gypsum + anhydrite, Ha = halite, Mic = micrite (carbonate mud), MS = medium sand, Org = organic – rich, P = peat/lignite, S = sand, Sa = sapropels, Si = silt, Sili = siliciclastic, Teph = tephra/volcanic ash beds, VFS = very fine sand. (2) Characteristic bedding features: Bur = burrows and other trace fossils, Lam = laminations, Mudcr = mudcracks, Nod = nodules, Rip = rippled beds, X – bed = cross – bedding. (3) Characteristic fossils. Art Fish = articulated fish remains, Br = brine shrimp, Bry = mosses, Char = charophyte oogonia and encrustations, Chir = chironomids, Clad = cladocerans, Dia = diatoms/chrysophytes, Dis Fish = disarticulated fish bones, Ins = terrestrial insect remains, Mol = molluscs, Ost = ostracodes, Pl = plant debris (leaves, twigs, seeds), Pol = pollen and spores, R = root traces/root mats, Sp = sponge spicules, Strom = stromatolites/algal mats, V = terrestrial vertebrate remains.

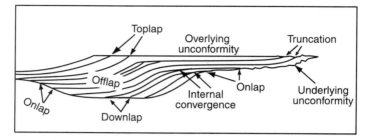

Figure 7.6. Terminology for stratigraphical terminations as developed by Mitchum et al. (1977) from seismic data. From Miall (1990).

Over the past 20 years macrostratigraphy has undergone a revolution in thinking about the relationship of facies to environmental conditions, and how to interpret that change. This revolution has come about largely as a result of our ever-improving ability to visualize the subsurface with reflection seismic techniques and the development of theory to provide a context for interpreting those images. We have come to realize that large bodies of sediment are normally bounded by surfaces of erosion or nondeposition. These might be created during periods when previously deposited sediments were stripped and redeposited at lower surface elevations. Such erosion comes about as a result of major changes in relative *base level*, the relative elevation of a receiving basin (in our case a lake) with respect to the watershed(s) from which the sediment is derived. Such a change might occur because lake level has gone down, because the land surface of the watershed has gone up through tectonic uplift or the lake basin down from subsidence, or for a variety of less important reasons. Alternatively, surfaces of nondeposition might form when lake level rises, moving the shoreline toward land and reducing or stopping the deposition of sediments in offshore locations. Deposits of sediment that are bound by such surfaces are referred to as *sedimentary sequences* and their study is called *sequence stratigraphy*. Several principals of sequence stratigraphy merit mention here:

1. Sequence boundaries occur at a variety of scales, reflecting both short-term adjustments in base level and longer-term, more profound events.
2. Shifts in base level affect not only erosional processes, but also the position and processes of deposition throughout the depositional system, from the upland to the lake basin. Sequence stratigraphers refer to this interdependent body of environments as a *systems tract*.
3. The style of deposition throughout a systems tract is dependent on the status of relative base level. Whether a lake is at a high or low stand, and whether its level is rising or falling, affects how sediments are distributed.

Although the significance of erosional or nondepositional surfaces was evident to earth scientists prior to the ready availability of abundant seismic data, the large-scale geometry of sedimentary sequences could not be fully appreciated without the large geographical scope of information provided by seismic data. Sequence stratigraphy was originally developed for application to marine sediments, particularly those of the continental shelves. However the general principles of sequence stratigraphy are also applicable to lacustrine settings (Bohacs et al., 2000). Lake basins do differ in some profound ways from the ocean, most notably in that:

1. Lakes can undergo frequent and large magnitude elevation changes; and
2. In contrast with the World Ocean, there is no globally consistent pattern of *eustasy* or absolute level. Individual lakes fluctuate independently, although there may be similarity in change over broad areas as a result of similar climate or tectonic interconnections.

The study of stratigraphical sequences provides paleolimnologists with a means of assessing the magnitude and timing of major shifts in relative lake level, and in some cases differentiating the interactive effects of changes in sediment supply, subsidence, and climate on the infilling of lake basins.

Deltas

Deltas are the primary sites of siliciclastic (i.e., terrestrially derived silicate minerals) sediment introduction into lakes. As such they provide the most immediate link in a lake to processes occurring within the watershed, and can be important for understanding human impacts on lake ecosystems. Deltas are also important zones of sediment accumulation, particularly in lakes undergoing active coastal subsidence, and therefore can be the source of long paleolimnological records.

Depositional Processes of Deltas

The deceleration of flow that occurs as a river reaches standing water causes sediment to begin to settle. Depending on the slope and extent of the coastal plain over which the delta is developing, this deceleration may begin well before the river actually reaches the lake. Therefore, delta deposition affects upstream portions of the systems tract, including fluvial and terrestrial overbank processes as well as truly lacustrine ones. These terrestrial or fluvial deposits, if present, normally form horizontally deposited or channelized bodies of sediment, referred to as deltaic topsets. They are heterogeneous deposits, including very coarse to very fine-grained clastic sediments, organic-rich peats and muds, and even occasionally carbonate deposits. This diversity of lithology reflects the diversity of depositional environments that may occur in this area, including fluvial channels and bars, lateral levees, overbank floodplains, swamps and delta-plain lakes. Soil formation is a common process throughout the delta plain, and preserved paleosols are often diagnostic criteria for these deposits. Fossils associated with these deposits are normally from upland organisms (terrestrial plants, insects, or vertebrates) or those of organisms inhabiting shallow aquatic environments, rather than open lakes.

Inflowing currents from rivers introduce particulate sediment both as sands and gravels along the bottom and as suspended material. As a river approaches a lake, decelerating flow normally causes considerable quantities of sediment to be deposited, often resulting in the breakup of a main channel into multiple distributor channels. Each channel, where it enters the lake, can form a separate sedimentary body or lobe, although these may merge and lose their distinctiveness from subsequent reworking. At the channel mouth the remaining bedload may be either deposited as elevated river mouth bars, transported laterally by longshore current activity, or downslope through some combination of inertial and gravity flow mechanisms.

The fate of the suspended load is controlled by the density relationship between the inflowing river water/sediment mixture and the lake water. This relationship determines whether sediment will spread along the lake's surface, descend to some intermediate depth, or plunge to the lake floor. Other factors of importance in determining the distribution of suspended load include the inertia of the river itself, the effectiveness of current activity within the lake, and the underlying slope and maximum depth of the lake floor (Bates, 1953; L.D. Wright, 1977; Postma, 1990).

Overflows (chapter 3) can transport fine-grained sediment considerable distances into the lake, in the process separating it from the zone of bedload deposition. This is the case, for example, in most saline lake deltas. Turbulent interactions between the lake floor and the overlying river water are minimized in these conditions. The coarse bedload tends to remain close to the channel path, forming river mouth bars, and finger-like bodies of sand. Fine-grained sediments deposited outside of the confines of the channels during flood events tend to stabilize the channel by forming cohesive channel banks that are difficult to erode. Under these conditions, delta lobes of fine-grained sediment can rapidly build out or *prograde*, as both subaerial and subaqueous deltas, over long distances (figure 7.7).

Given their grain size, surprisingly large quantities of clay-sized particles are deposited within a few kilometers of the river outlet (e.g., Hyne et al., 1979). The principal cause of this unexpectedly rapid settling is *flocculation* of clay minerals. Dissolved salts and organic matter, both readily available where a river enters a saline lake, enhance the aggregation of clays into larger particles. Some types of organic compounds also induce flocculation, although others may have the reverse effect. Clay flocculation rates also vary between clay minerals, causing kaolinite to be deposited closer to the river mouth on entry into saline lakes than illite. Smectites are comparatively unaffected by these flocculating mechanisms and therefore can be transported longer distances. Fine particles that are not rapidly flocculated can be dispersed over large areas of the lake, eventually to be deposited as hemipelagic sediments.

In situations where the river and lake waters are of approximately the same density, thorough mixing occurs near the river mouth, resulting in very rapid deposition of sediment near the shoreline.

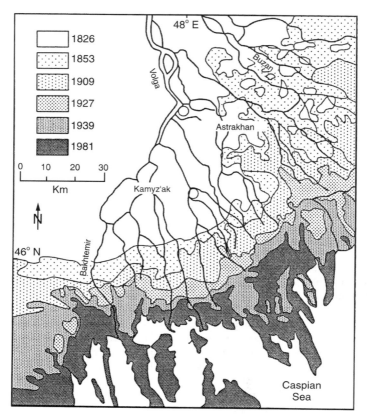

Figure 7.7. The Volga River delta, where it enters the Caspian Sea, illustrating its rapid progradation over the past 170 years. From Kroonenberg et al. (1997).

This situation prevails in many freshwater lakes, particularly when the inflowing river does not have an extremely high sediment load. As sediment piles up near a channel mouth it forces subsequent discharge to radiate over a broad angle, producing many radiating sand or gravel bars. If the receiving lake is relatively deep these bars will be noticeably inclined, referred to as *foresets* by sedimentologists. Where these slopes are steep the foresets are subject to frequent avalanching and formation of grain flows and debris flows. The abundant reworking of these deposits limits their utility for fine-scale interpretation of event history. Fine-grained particles can remain in suspension in interflows for long periods. Both internal waves and the Coriolis effect can transport these sediments at velocities of 2–15 cm sec^{-1}, redistributing them as broad blankets of mud. The lower salinity typical of interflow situations makes flocculation relatively unimportant as a sedimentation mechanism, although fine particles may still become aggregated through zooplankton fecal pellet formation.

Variations in the geometry of deltas seem to be strongly controlled by the dominant texture of their bedload. In gravel-dominated rivers, or other rivers with high proportions of bedload to suspended load, foresets tend to be very steep, reflecting the limited ability for secondary mechanisms such as long-shore currents and waves to rework the sediment (e.g., Nemec, 1990; J.D.L. White, 1992). This is the most common origin of *Gilbert-type* deltas, named after G.K. Gilbert, who first recognized this type of construction in the deltaic deposits of Pleistocene Lake Bonneville, Utah. Such deltas are particularly common in glacial lakes. In an aerial view of a Gilbert delta, horizontal *topset* beds are deposited at or close to the lake surface. These give way lakeward to steeply inclined foresets, dipping down the delta surface into deep water. In large lakes, these foresets can be huge; in Lake Bonneville, foresets with relief of over 70 m and dips of up to 30° have been observed (figure 7.8) (D.G. Smith and Jol, 1992). This environment is characterized by frequent slope failure and various

Figure 7.8. Gilbert-type delta steeply dipping foresets from Late Pleistocene delta deposits where they entered Lake Bonneville, Utah.

types of gravity flows (Nemec, 1990). At the base of this slope much finer *bottomset* muds and sands accumulate. In all deltaic environments these bottomset prodeltaic deposits provide the most interpretable, high-resolution record of delta history, since deposition here is much more continuous and reworking much less frequent than in the higher regions of the delta system.

Sandier bedload systems tend to produce much more gently inclined deltas. Waves and nearshore currents can rework bedload more effectively under such circumstances, resulting in more dispersion of sediment. Such reworking is most effective in large lakes, where waves and currents are vigorous (e.g., T.C. Johnson et al., 1995; Scholz, 1995).

Underflows occur primarily when extremely sediment-rich (or occasionally saline) water enters a freshwater lake, or where cold water enters a warm lake, for example in tropical lakes surrounded by high mountains. They are most characteristic of flood conditions, when sediment discharge from rivers is normally highest. Frictional interaction with the lake floor is extreme in this circumstance, causing erosion and channelization, and reducing the importance of secondary mechan-

isms for redistributing sediment, such as Coriolis circulation. In this case, little sediment accumulates near the shoreline and the delta will not prograde. Under these circumstances sediment tends to accumulate in or near channels, carried through a combination of inertial and gravitational flow mechanisms into deep water. Semicontinuous downslope flow of sediment-rich water is probably the main mode of fine sediment transport, resulting in blankets of silt or mud in the more basinward (distal) portions of the delta. Such flows can travel for over 100 km (Grover and Howard, 1938)! This type of deposition is often highly seasonal, increasing during flooding and waning at other times of the year. The deposits that form under these circumstances may be distinctly laminated, reflecting the dominant flood seasonality. In many lakes this results in an annual cycle of lamina formation, but numerous examples are also known where flood-related laminations form more than once a year, or only every few years during extreme flows (Sturm and Matter, 1978).

Local oversteepening or sudden changes in underflow inertia can trigger secondary gravity flows of deltaic sediments (e.g., Pickrill and Irwin,

1983). Turbidity flow, grain flow, and debris flow deposition can all occur under these conditions, although the former is probably the dominant mechanism of transport. The most characteristic deposits of lacustrine turbidity flows are silty or sandy graded deposits, beds that show an upward decrease in grain size reflecting waning flow conditions of the passing turbidity current. Small-scale ripple lamination and cross-bedding, and in sandier turbidites, planar laminations, are also commonly observed, particularly in deposits from larger lakes. Grain flows produce relatively massive or structureless sandy beds, often with a variety of fluid escape structures, whereas subaqueous debris flows are chaotically bedded deposits, with larger fragments jumbled in a matrix of finer material.

Load-related subsidence is a hallmark feature of all larger deltas. As deltas grow they can locally depress the crust, creating new accommodation space for additional sediment to accumulate. This is a particularly important process for deepening areas adjacent to the sites of active deposition. Sediment loading often results in sedimentary deformation features as well. This occurs most commonly when coarse-grained topset and foreset deposits prograde over relatively fluid and light prodeltaic muds, causing the underlying sediments to rise as diapirs through the overlying sands.

In some lakes alluvial fans discharge directly into standing water, without any intervening river course. This type of deposit, referred to as a *fan delta*, is typical of lakes bounded by steep slopes, for example along fault escarpments in rift valley lakes, or in calderas. Fan deltas may develop in either deep or shallow lake conditions, with differing consequences for sedimentation (Massari and Colella, 1988). In deep lakes fan deltas act as feeder systems, providing sediment to deep water through underflow mechanisms, but not undergoing significant progradation (Cohen, 1990). Their deposits often give way abruptly to pelagic, fine-grained sediments settling from suspension. In shallow lakes, fan deltas prograde as a broad apron of sediment over the lake floor, with much more gradual facies transitions.

Fossil content in deltaic deposits is quite variable. Generally the region immediately adjacent to the outflows of rivers (channels, river mouth bars) presents an extremely harsh environment for organisms, mostly because of their extremely high concentrations of suspended sediment and bedload. Consequently, fossilizable lacustrine organisms are rare in the upper reaches of most deltas, regardless of their mode of origin. Fossils found in nearshore parts of deltas, for example on delta foresets, tend

to be from older abraded material that has been reworked, or the remains of terrestrial or fluvial organisms. Burrows and other trace fossils are similarly restricted in these environments. In prodeltaic environments, however, fossil remains are normally much more abundant, including mollusks, ostracodes, cladocerans, chironomids, and fish, as well as transported terrestrial insect remains.

Vertical Facies Sequences

As deltas in lakes accumulate they take on distinctive patterns of facies accumulation that are not apparent from a two-dimensional analysis of surface sediment facies alone. Sedimentologists refer to these features as the delta's internal geometry. Lithofacies at any given location vary as a result of changes in absolute lake level, tectonic activity, progradation of the delta (causing a site to become shallower over time), or the migration of delta channels and lobes. Delta structure can be inferred through a combination of core, outcrop, and geophysical studies (seismic or radar stratigraphy), allowing larger-scale facies models to be developed and three-dimensional aspects of geometry to be interpreted.

At a macroscale, deltaic deposits commonly display overall coarsening-upward patterns during the construction of individual progradational lobes (figure 7.9) (Farquharson, 1982). This results from an environmental transition of any given location to shallower water conditions, where simple suspension settling of clays and silts is replaced by more energetic mechanisms of deposition. Multiple cycles of coarsening-upward sediment packages become stacked on top of one another as a result of subsidence, migration of delta lobes, or lake-level fluctuations.

A relatively simple case of facies geometry resulting from delta progradation can be visualized using Gilbert-type deltas as examples (figure 7.10). Steep-fronted Gilbert deltas can accumulate from material cascading down a steep bedrock surface (figure 7.10A). Once a fan builds up to approximately the elevation of its river input, delta progradation can begin, with foresets accumulating as steeply dipping lobes of sediment, periodically transported downslope by gravity flows. As the foresets built outward, topset stream deposits also prograded, forming a blanket of horizontally deposited sediments at approximately lake surface level on top of the foresets. The foresets eventually began to accumulate over flat-lying lake floor muds. The net result of this deposition is often evident in seismic or ground-penetrating radar profiles of Gilbert deltas

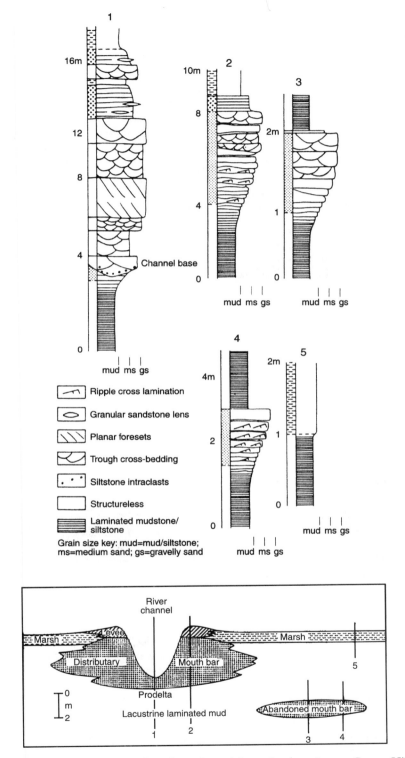

Figure 7.9. Lacustrine deltaic stratigraphy, from Late Mesozoic deposits at Camp Hill, Antarctic Peninsula. (A) Representative measured sections, showing typical coarsening-upward sequences observed in individual delta lobes. (B) Cross-sectional interpretation of paleoenvironmental setting for the five measured sections illustrated. From Farquharson (1982).

A.

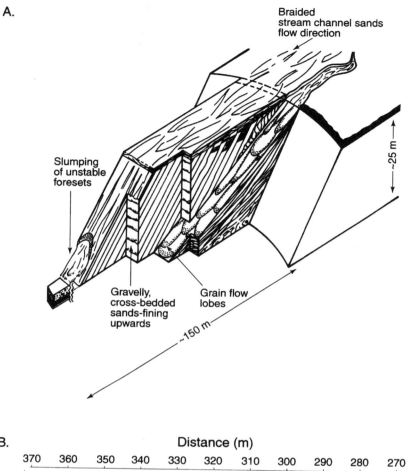

Braided
stream channel sands
flow direction

Slumping
of unstable
foresets

Gravelly,
cross-bedded
sands-fining
upwards

Grain flow
lobes

~25 m

~150 m

B.

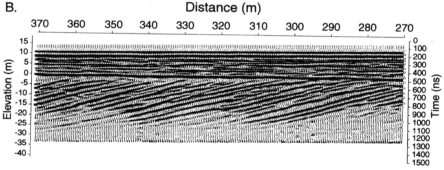

Figure 7.10. Facies geometry in a Gilbert delta. (A) Schematic block diagram showing the internal features of a Gilbert delta building out over a sublacustrine fan core, from the Pliocene Hopi Buttes Maar Crater Lake deposits, Arizona. The underlying sublacustrine fan was built up by elongate grain-flow lobes. Once the fan aggraded to the lake surface, this style of deposition was replaced by the prograding delta system, building a braided stream out into the lake. Foresets from the prograding delta underwent periodic slumping on the steep surface. (B) Ground-penetrating radar cross-section from a Gilbert-type delta entering Peyto Lake, Alberta, Canada. Note the geometry of poorly reflective bottomsets, overlain by steeply inclined prograding foresets, in turn overlain by near horizontal topsets. From White (1992) (A); From Smith and Jol (1997) (B).

(figure 7.10B). More complex delta geometries arise when lake-level fluctuations and tectonic activity are superimposed on progradation (figure 7.11).

The stacking patterns that result from deltas formed under varying lake-level positions can be quite complex. Also, deposits and systems tracts formed when lakes were at relatively high stands might be quite different from those at low stands. This results in a series of depositional sequences being formed, each representing a period of delta development and bounded by erosional unconformities that develop when lake level falls (Lemons

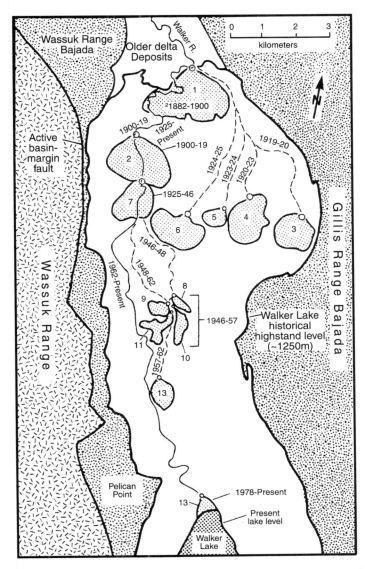

Figure 7.11. Planview distribution of the Walker River deltas and channel tracts in the northern Walker Lake basin, showing period in which each tract and delta was active. The delta series developed as the lake contracted over the past ~120 years, with successive deltas shrinking in size as the water and sediment delivery to the lake declined. Progressive tilting of the floor of the half-graben basin on which the lake and deltas developed also influenced stream and delta lobe positions. From Blair and McPherson (1994).

and Chan, 1999). In seismic profiles from Lake Malawi, Africa, low lake stand paleodeltas can be readily seen as inclined groups of strata (figure 7.12). Some low-stand lobes of these deltas may be up to hundreds of meters below modern lake level, indicating the magnitude of lake-level fluctuations that have occurred in this lake during the Pleistocene. Dating such deposits, coupled with determining their elevation, would then give a precise chronology of lake-level change.

On a microscale, vertical changes at a single location often reflect short-term variation in river/lake density interactions. Many river outflows alternate between over-, inter-, and underflow conditions on a seasonal, or even diurnal basis, depending on their rate of discharge, temperature, and suspended sediment concentrations. The most obvious manifestation of seasonal to multiannual change in sediment discharge is the rhythmic deposition of fine sediments (typically silts and clays) deposited as thin beds or laminations. Because of the more limited effects of shallow-water waves and bioturbation under deeper-water conditions, these couplets are most evident in the more distal, prodelta portions of a delta system. The limited amount of bioturbation that occurs in low-productivity glacial lakes or meromictic lakes also makes these settings more favorable for the preservation of such laminations. In lakes where texturally distinct couplets can be shown to form

on an annual basis they are referred to as *varves* (De Geer, 1912). The term varve has unfortunately been applied to many types of laminated or thin-bedded lake sediments, not all of which formed on an annual cycle. Varves can form in lakes for a variety of reasons related to seasonal changes in inflow, productivity, overturn, or some combination of these. In prodeltaic (and sometimes in lower foreset) settings episodic flooding events are the primary cause of varving and the relative thickness of silt versus clay layers is normally related to proximity to the river mouth (Ashley, 1975; Pickrill and Irwin, 1983) (figure 7.13). For example, in proglacial lakes, high spring and summer discharge results in silt layers that thicken toward the river mouth. These silty layers are frequently graded, reflecting their origin from underflows or turbidity flows (De Geer, 1912; Gustavson, 1975). During winter months, ice cover allows a more uniform blanket of clays to be deposited from suspension, and silt discharge from rivers is reduced. Because most glacial lakes have low salinities, clays deposited in them are rarely flocculated, instead settling as oriented particles that readily transform into laminae. Varve thicknesses may vary over decadal to century time scales as a result of changing sediment discharge from rivers, or more locally, from changes in discharge positions as distributary channels migrate or avulse (Lamoureux, 1999). Differentiating between these causes is an important chal-

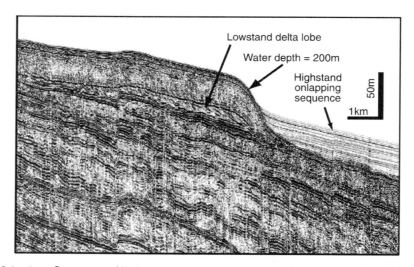

Figure 7.12. Seismic reflection profile from the Dwangwa River Delta region, Lake Malawi, Malawi. Buried low lake stand paleodeltas can be readily seen as inclined groups of strata, indicating the existence of lake levels that were hundreds of meters below present during the Pleistocene. From Scholz (1995).

Figure 7.13. Late Pleistocene glaciolacustrine varves from the Severn River area, northern Ontario, Canada.

lenge for paleolimnology, as varve records of sedimentation rates are increasingly being used as direct paleoclimate indicators.

Littoral Depositional Environments

Lacustrine beaches and shallow littoral environments receive siliciclastic sediments from several sources. River-derived sediments can be redistributed along coastlines away from their original deltaic source, especially in large lakes, with strong secondary transport mechanisms. In some cases, such as where low-gradient fan deltas are prograding only a lake margin, the sedimentological distinction between lake and coastal plain may be very subtle. This is especially true of closed-basin lakes that undergo significant seasonal lake-level fluctuations. Alternatively, on exposed coastlines subject to vigorous reworking by coastal currents, a wide variety of distinctive geomorphic features can form, including spits, barrier ridges, and so on. Many of these are recognizable in well-preserved ancient lake shorelines (figure 7.14). Local alluvial sources and

coastal erosion processes provide sediment away from major river deltas, particularly in lakes surrounded by large amounts of unconsolidated materials. Wind-derived sediment is important in arid regions subject to wind deflation. Even in humid climates, however, wind can also secondarily rework coastal sands around large lakes, generating large dune fields in the process.

Depositional Processes in the Lacustrine Coastal Zone

Lacustrine beaches are usually not areas of long-term sediment storage, acting instead as temporary or episodic vessels for sediments, which are subject to frequent erosion. As such their deposits are generally less informative for paleolimnological purposes than deeper water zones or subsiding deltas. Beaches may be quite useful for recording episodic events such as high lake stands, since they accurately mark paleolake level. In conditions where lakeshores are prograding they may also give relatively long (thousands of years) records. But in

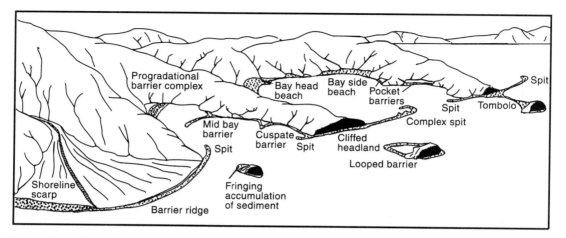

Figure 7.14. Shoreline depositional and erosional features found in many large lakes with high-energy coastlines. From Adams and Wesnousky (1998).

small lakes, where coastal sediment delivery is minimal, the records provided by beach deposits almost invariably are patchy.

The coastal sand belt associated with beaches can be subdivided into several subenvironments, based on nature of physical wave energy and exposure (figure 7.15). A *shoreface* zone occupies the deeper-water region of the beach, where shallow-water waves begin to "feel bottom." The lower part of the shoreface will only be exposed to wave attack during storm events. A narrow foreshore marks the shoreline of most lakes. Beach berms, elevated areas of coarse particle accumulation, are usually small along lakeshores. However in larger lakes subject to wave washover during storms, a backshore environment may develop. Depending on local water table configuration these areas may be dry, occupied by marshes, or open-water lagoons. Back beach environments are also areas of eolian dune formation around some lakes.

Within the zone of coastal sediment transport wind-driven waves are the most conspicuous agent for redistributing coarse-grained sand and gravel; tides in even the largest lakes are minor features and create only insignificant currents. The transition from offshore to shoreface to beach face, where breaking waves run out onto the beach, is very narrow in most lakes, as a result of limited fetch and the shallow depth of storm wave base (Sly, 1978). Longshore currents are generally much weaker than in coastal marine environments, and coastal pileup of water, causing rip currents, is rare. As a result the belt of sandy coastal sedimentation away from deltaic sources is quite restricted and facies transitions to "offshore" fine-grained

sedimentation occur rapidly. In lakes with steep, offshore profiles, longshore transport is limited to small pocket beaches, beyond which sand is carried into deep water by gravity flows (Osborne et al., 1985; Soreghan and Cohen, 1993). Where coastal waves and longshore transport energies are adequate to move sand they create a mix of planar and ripple bedding, producing planar and cross-stratified sands in the foreshore and upper shoreface environments (figure 7.16). Lacustrine fossils tend to be rare and/or abraded in these deposits. Larger bedforms, such as megaripples, or large sandbars, are primarily features of exposed coastlines in larger lakes. In very large lakes with gentle topographical gradients, combined flow during the storm reworking of shoreface environments can produce types of hummocky or swaley cross-stratification more commonly associated with marine continental shelf deposits (Eyles and Williams, 1992; Buatois and Mángano, 1994).

In small lakes, or protected embayments of large ones, the "coastal belt" may be only meters wide, and muddy shorelines are typical features of small ponds. This is expressed sedimentologically as thin-bedded, laminar to massive (bioturbated) silt and silty sand. These deposits are much more likely to contain fossil invertebrates and aquatic macrophytes. Pond and embayment deposits are often overwhelmingly dominated by aquatic plant debris, particularly in areas away from siliciclastic sediment input. Where macrophytes occur in wave-swept larger lakes they also shelter the lake floor from wave energy, acting as silt traps, even under relatively vigorous wave conditions. In a big lake this zone can be very extensive, covering tens or

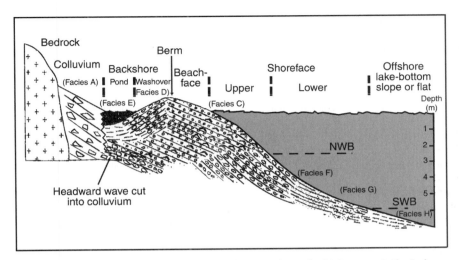

Figure 7.15. Generalized facies model for shoreline deposits from the Pleistocene Lake Lahontan shoreline (Churchill Butte area, Nevada). The model shows depositional patterns perpendicular to shoreline. Abbreviations: NWB, normal wave base; SWB, storm wave base. Water depths are approximate. Facies: A, unsorted, unstratified muddy–bouldery gravels; (no B); C, lakeward-dipping pebbly gravels; D, landward-dipping sandy or pebbly gravels; E, mixed gravel, sand, mud, and diatomite; F, lakeward-dipping, poorly sorted, pebbly gravel; G, burrowed and rippled, granular sands; H, horizontal, laminated silt and thick-bedded mud and clay. From Blair (1999).

Figure 7.16. Foreshore deposits (gently inclined, lakeward-dipping sands) from the Pliocene Bouse Formation, California. The sands are composed of highly abraded fragments of barnacles, which grew on nearby rocky and vegetated substrates in a saline lake.

hundreds of square kilometers. Away from macro-phyte beds, large, shallow lakes may experience considerable wave reworking in their central areas. However, mechanisms for suspending and transporting sand to far offshore areas usually do not persist for long periods, and wide reaches of macrophytes generally rim the coastal areas of such lakes. As a result, fine, organic-rich sediments or wind-blown dust dominates even very shallow water offshore sedimentation in these lakes.

The deposits of ponds, bogs, and small lakes have a long and sometimes confusing history of description and terminology. The unconsolidated, organic-rich sediments of shallow lakes were originally classified on the basis of their organic content. *Gyttja* is a gray or dark brown sediment, containing humic material, plant fragments, terrigenous sediment, diatom, and other algal and zooplankton remains. Gyttja has a textured appearance and feel when wet, a result of having been transformed as a mass of fecal pellets by chironomids or other benthos. As a result these sediments are also sometimes referred to as *copropels*. Gyttjas form under oxygenated conditions. *Dy* is a type of gyttja formed under acidic (typically bog) conditions, with high concentrations of acidic humic materials. It is brown and fibrous in appearance, from abundant peat particles. *Sapropels* are glossy black (when fresh, from reduced FeS) and generally finer-grained organic sediments. They contain abundant CH_4 and H_2S and other products of microbial reduction. Sapropels require reducing conditions to retain their appearance and composition, although this can develop seasonally from low oxygen conditions, or the rapid accumulation of organic-rich sediment, without permanent anoxia.

Fossil content and preservation within nearshore depositional environments is highly variable. In well-developed, coarse-grained beach deposits fossils are rare, as a result of poor habitat conditions and secondary abrasion and dissolution. However, in finer-grained nearshore deposits (gyttjas or their lithified equivalents), fossil preservation is often excellent and the biota diverse, reflecting the diversity and habitat heterogeneity of this zone. Macrophytes, pollen, benthic algae, insects, crustaceans, and mollusks are all commonly found as fossils in the deposits of marshes, ponds, and small lakes, or in protected embayments of large lakes.

Vertical Facies Sequences

Because of their position at the lake's edge and the narrowness of the coastal zone, slight lake level falls quickly lead to exposure of beach deposits. When lake level remains relatively constant beach environments cannot aggrade vertically to accommodate new sediment, and beach deposition can only continue though progradation. For these reasons ancient lacustrine beach sand deposits are generally thin. In small or low-gradient lakes individual packages of planar and cross-bedded sands are typically only a meter in thickness or less. In these conditions coastal sediments frequently show signs of erosion and soil process modification; the formation of various types of paleosols, pedogenetic alteration, diagenetic mineralization, and nearsurface cementation (Calvo et al., 1989; Cohen et al., 2000a). Rapid vertical and lateral facies transitions are typical of this environment (Calvo et al., 1989).

In somewhat deeper water environments or during a relative lake-level rise, accommodation space is greater and thicker sand bodies can accumulate. This is most important in large, gently sloping lakes, where storms are adequate to rework sediment relatively long distances offshore. Under these conditions prograding beach complexes, with cross-stratified sands, can develop in water depths of as much as 10 m, producing sand bodies of comparable thickness (e.g., Castle, 1990). These types of lake margins are also ideal for producing vertical sequences of hummocky or swaley cross-stratified sands, formed during large storm events (figure 7.17) (Eyles and Williams, 1992). In relatively young (Late Pleistocene and Holocene) lake deposits it is often possible to recognize former strand lines of lakes from the sequential deposition of *beach ridges*, which incorporate sediments from upper shoreface, foreshore, and dune environments (figure 7.18). Sequential beach ridges develop over brief intervals while lake levels are falling from a temporary high stand (Hunter et al., 1990; Thompson, 1992).

Shallow-Water Carbonate Depositional Environments in Lakes

Many lakes provide the necessary environmental conditions for the accumulation of calcium carbonate deposits, notably high rates of photosynthesis and relatively alkaline water chemistries. As a result, lime muds and sands are common features of lake deposits. *Marls* are carbonate muds produced with only subordinate amounts of clastic material. Carbonate deposits sometimes form near rivers, if some transport or sorting mechanism exists to separate carbonate from noncarbonate debris. Where both siliciclastic and carbonate sediment sources exist in subequal proportions, mixed carbonate/clastic deposits will accumulate.

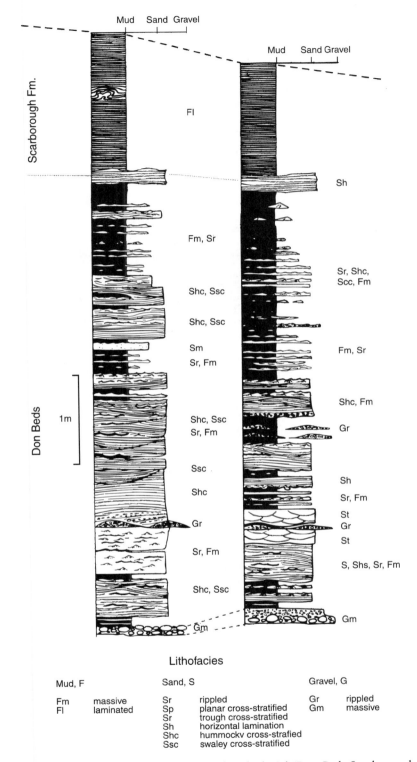

Mud Sand Gravel

Mud Sand Gravel

Scarborough Fm.

Fl

Sh

Fm, Sr

Sr, Shc,
Scc, Fm

Shc, Ssc

Shc, Ssc

Fm, Sr

Sm
Sr, Fm

Shc, Fm

Don Beds

1m

Shc, Ssc
Sr, Fm

Gr

Ssc

Sh

Shc

Sr, Fm

Gr

St
Gr

Sr, Fm

St

S, Shs, Sr, Fm

Shc, Ssc

Gm

Gm

Lithofacies

Mud, F		Sand, S		Gravel, G	
Fm	massive	Sr	rippled	Gr	rippled
Fl	laminated	Sp	planar cross-stratified	Gm	massive
		Sr	trough cross-stratified		
		Sh	horizontal lamination		
		Shc	hummocky cross-strafied		
		Ssc	swaley cross-stratified		

Figure 7.17. Stratigraphical sequence at the last interglacial–glacial (Don Beds-Scarborough F) transition exposed in the Don Valley, Toronto, Ontario, Canada. In the Pleistocene Don Beds of ancestral (early Wisconsin-aged) Lake Ontario, a sequence of lake margin and shoreface sands and gravels gives way vertically to finer muddy sediments, reflecting progressively deeper water conditions. From Eyles and Williams (1992).

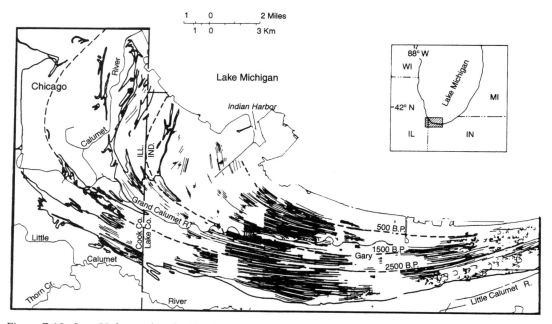

Figure 7.18. Late Holocene beach ridge formation at the south end of Lake Michigan, Illinois and Indiana. A series of over 100 such ridges, accumulated during overall coastal progradation, provide a detailed record of lake level history during the late Holocene. Ridges are cored by foreshore deposits and overlain and underlain by dune and upper shoreface deposits respectively. Palustrine (swamp) and lacustrine sediments occur in the swales between beach ridges. From Thompson (1992).

Deposition of Carbonates

Carbonate sediments behave differently from siliciclastic sediments, both during and after deposition, with implications for their paleolimnological interpretation. Carbonate minerals are relatively reactive under varying lake and diagenetic conditions, and mineral dissolution and replacement is common. These processes are particularly important in highly concentrated brines, near shorelines, and in areas of groundwater–sediment interaction. The exposure of carbonate sediments through falling lake level causes them to develop characteristic diagenetic and soil-forming features, including the formation of carbonate nodules, calcite-filled cracks, dissolution spaces (microkarst), and secondary cementation around roots (Plaziat and Freytet, 1978; Mount and Cohen, 1984; Wright et al., 1988; Platt, 1992). When such features can be recognized they provide valuable paleolimnological evidence of lake-level change. Also, coarse-textured carbonate sediments, from shells or coated grains for example, can form in situ in lakes, and need not indicate anything about hydrodynamic conditions at the time of their emplacement.

A number of different carbonate minerals form in lakes (table 7.2). The most common of these minerals are low (0–5%)-Mg calcite, high-Mg calcite, and aragonite. Specific carbonate mineral assemblages provide information about lake water chemical and temperature conditions and brine evolution at the time of precipitation. Brine evolution normally causes Mg concentration to rise during evaporative concentration, and favors carbonates that form preferentially under high Mg conditions, such as high-Mg calcite, aragonite, or perhaps dolomite. In alkaline pathway lakes ($Ca^{2+} + Mg^{2+} \ll CO_3^{2-} + HCO_3^-$), depletion of all Ca and very highly concentrated water promotes the formation of Na-carbonates such as trona. Other minerals such as nahcolite ($NaHCO_3$) appear to require high partial pressures of CO_2, generated in sediment pore waters by bacterial decay. The formation of some carbonates is sensitive to temperature variation as well (Shearman et al., 1989).

Shallow-water carbonates form a distinctive suite of facies types because of the secondary physical transport, reworking, and sorting processes that can affect their textures, structures, fabric, and large-scale facies geometry. Carbonate sedi-

Table 7.2. Important lacustrine carbonate and bicarbonate minerals and their occurrences in lakes

Carbonate Mineral	Formula	Typical Occurrence in Lakes	References
Low Magnesian Calcite	$Ca_{0.95-1.00}Mg_{0.05}CO_3$	Precipitated mud, shell, coated grains, springs, playa crusts, cements	Müller et al. (1972), Eugster and Hardie (1978), Popp and Wilkinson (1983)
High Magnesian Calcite	$Ca_{0.7-0.95}Mg_{0.05-0.3}CO_3$	Precipitated mud, coated grains, stromatolites, cements	Müller et al. (1972), Kelts and Hsü (1978), Cohen and Thouin (1987)
Aragonite	$CaCO_3$	Precipitated mud (high Mg/Ca_{water}), coated grains, mollusk shells	Kelts and Hsü (1978), Popp and Wilkinson (1983)
Monohydrocalcite	$CaCO_3 \cdot H_2O$	Precipitated mud (unstable)	Kelts and Hsü (1978)
Ikaite	$CaCO_3 \cdot 6H_2O$	Springs under cold conditions	Shearman et al. (1989)
Dolomite	$Ca_{0.90-1.1}Mg_{0.90-1.1}(CO_3)_2$	Diagenetic alteration, high Mg/Ca lake waters or brines, efflorescent crusts, primary muds?	W.E. Dean and Gorham (1976), Eugster and Hardie (1978), De Deckker and Last (1988), Janaway and Parnell (1989), Platt (1992), Komor (1994), Arenas et al. (1997)
Huntite	$CaMg_3(CO_3)_4$	Diagenetic alteration on playas-high Mg concentrations	Kelts and Hsü (1978)
Magnesite	$MgCO_3$	Diagenetic alteration on playas-high Mg concentrations	Kelts and Hsü (1978)
Hydromagnesite	$Mg_4(OH)_2(CO_3)_3 \cdot 3H_2O$	Low-temperature hydrothermal springs, stromatolites	Schmid (1987), Braithwaite and Zedef (1994)
Nesquehonite	$MgCO_3 \cdot 3H_2O$	Diagenetic alteration on playas-high Mg concentrations	Kelts and Hsü (1978)
Siderite	$FeCO_3$	Diagenetic alteration and nodules	Kelts and Hsü (1978)
Dawsonite	$NaAlCO_3(OH)_2$	Alkaline brines	Eugster and Hardie (1978)
Natron	$Na_2CO_3 \cdot 10H_2O$	Alkaline brines, crusts (cooler conditions than trona)	Eugster and Hardie (1978), Eugster (1980)
Thermonatrite	$Na_2CO_3 \cdot H_2O$	Efflorescent crusts of alkaline playas, early diagenesis in nodules. High temperature (e.g., springs) or low pCO_2	Bradley and Eugster (1969), Eugster and Hardie (1978), Renaut and Tiercelin (1994)
Nahcolite	$NaHCO_3$	Muds, diagenetic alteration (bacterial decay under alkaline, high pCO_2 conditions, >10 × atmospheric pCO_2) in nodules and concretions in alkaline lakes	Eugster and Hardie (1978), Eugster (1980)
Gaylussite	$Na_2CO_3 \cdot CaCO_3 5H_2O$	Interstitial brine interactions (alkaline pore waters and freshwater) in alkaline playas	Eugster and Hardie (1978)
Pirssonite	$Na_2CO_3 \cdot CaCO_3 \cdot 2H_2O$	Interstitial brine interactions (alkaline pore waters and freshwater) in alkaline playas	Eugster and Hardie (1978)
Trona	$NaHCO_3 \cdot Na_2CO_3 \cdot 2H_2O$	Precipitated muds in warm conditions (>20°C), interstitial (bacterial decay under alkaline conditions), efflorescent crusts	Bradley and Eugster (1969), Eugster and Hardie (1978), Eugster (1980)
Burkeite	$Na_2CO_3 \cdot 2Na_2SO_4$	Interstitial interactions under alkaline conditions	Eugster and Hardie (1978)

ment is produced primarily within the photic zone, through the formation of a number of distinctive types of particles. The most ubiquitous of these are lime muds, formed dominantly by direct calcite precipitation, and secondarily by degradation of larger particles. The precipitation of clay and silt-sized calcite or aragonite particles occurs seasonally as a result of photosynthesis-induced precipitation, for example around bacteria, or charophyte stems. Other mechanisms also exist for precipitation of carbonate mud that are independent of their position relative to the photic zone, for example adjacent to spring discharge (rapid degassing sites for CO_2), or where Ca^{2+}-bearing rivers or groundwater meet highly alkaline lake water. Calcium carbonate muds are common in shallow water along a variety of low-energy lake margins, for example in small lakes and ponds, among macrophytes, or, in large lakes, along protected or gentle-gradient coastlines.

Sand and gravel-sized carbonate particles are also produced by several mechanisms in lakes. In lakes occupying limestone bedrock basins, eroded carbonate clasts can be transported into the lake. Bioclastic sediments, primarily fragments from ostracode and mollusk shells or charophyte oogonia and stem coatings are a major source of in situ carbonate grains. Another source is from the formation of *coated grains*, concentrically layered particles of calcite or aragonite (occasionally other minerals as well). The most characteristic types of coated grain in lakes are *ooids*, sand-sized particles with spherical or other rounded shapes. Ooids gain their shape through a combination of accumulation of calcite or aragonite around some nucleus, coupled with continuous rolling to expose new surfaces to growth. Internally they display a wide variety of crystalline fabrics, some of which have paleoenvironmental significance (Popp and Wilkinson, 1983). As a result they form on wave-swept beaches or platforms in shallow water, although they may be secondarily transported and deposited in deeper water through gravitational flow. Larger coated grains, called *oncoids*, nucleate around gravel or cobble-sized objects like shells or rocks. Calcite precipitation on most oncoids is mediated by cyanobacteria; thus they can be thought of as a special type of stromatolite (Freytet and Plet, 1996; Verrecchia et al., 1997). These objects are generally too large to roll continuously and their coatings are therefore more irregular (F.G. Jones and Wilkinson, 1978).

Lithified carbonate also forms cemented surfaces in lakes through the development of beachrock (nearshore pavements formed by carbonate cementation around existing sand grains, algal bioherms (*stromatolites* and related features), and spring-deposited tufas. Beachrock pavements provide excellent paleoshoreline indicators. They comprise variably cemented (friable to highly indurated) sand and gravel surfaces that dip gently in a lakeward direction (Cohen and Thouin, 1987; Last and De Deckker, 1990).

Carbonate *bioherms* are buildups formed primarily by cyanobacteria in lakes, although some are also formed by green algae (Riding, 1979; Rouchy et al., 1996) (figure 7.19). They develop into a variety of shapes, from simple encrustations on existing surfaces, to extremely large "reefs," with relief of as much as 18 m off the lake floor (e.g., Braithwaite and Zedef, 1994; Rouchy et al., 1996). Internally these structures have extremely variable and often complex fabrics, comprising layered zones of accretionary growth, clotted regions of dense carbonate, or cavities that can become secondarily filled with debris. Tufas are porous carbonate spring mound deposits. They most commonly form as thickly layered deposits with numerous spongy cavities, and often incorporate surrounding particulate matter (Pedley, 1990). The formation of tufas involves a combination of microbial calcite precipitation and inorganic degassing. They often build up into spectacular columns, many meters tall (figure 7.20).

The type of carbonate sediment that forms in any nearshore setting is strongly influenced by lake floor gradient and local wave energy conditions (Platt and Wright, 1991). In lakes with low gradients and low energy conditions ramps of carbonate muds occur, often with only subtle facies changes associated with lake water depth. This type of accumulation is characteristic of shallow playas and carbonate-producing marsh and swamp (palustrine) environments. Shorelines in playa lakes of this type may be marked by the presence of stromatolites, and abundant soil carbonates indicative of exposure. In palustrine settings, charophyte, ostracode, and gastropod remains may occur throughout the lake deposit, with little or no indication of a central, deep-water basin (Platt, 1989). Slump features and other indications of gravity flows are also rare in these settings.

Low-gradient, but higher energy conditions promote the formation of carbonate ramps, with extensive ooid shoals and algal bioherms. These are encountered in many large but shallow lakes, such as the modern Great Salt Lake (Utah) (Sandberg, 1978). Ooid and other carbonate sand deposits in these settings form as planar or lenticular beds, in

(A)

(B)

Figure 7.19. Fossil and living stromatolite bioherms. (A) Miocene stromatolites from the Chalk Hills Fm., Idaho. (B) Modern stromatolite bioherm from Lake Tanganyika , growing on the edge of a slope break, associated with normal faulting (Burundi coast, 15-m water depth, fish are about 75 mm for scale).

Figure 7.20. Spring-fed carbonate tufa towers of Mono Lake, California.

some cases up to several meters in thickness (Camoin et al., 1997).

In lakes with low-energy shorelines but steep gradients, calcareous benches can form at or around the slope break of the lake floor. Peats, marls, and oncoids are characteristic of shallow-water tops of these benches (figure 7.21A). Charophytes, which can grow even under relatively low light levels (chapter 5), may comprise a larger proportion of sediments in slightly deeper water, along the bench slope. In the lower littoral or pro-fundal zone gastropod and ostracode shell frag-ments, along with suspension-settled mud, become the dominant carbonate particles. This facies model is most characteristic of small but relatively deep hardwater ponds and lakes, or in protected embay-ments of larger lakes (e.g., Murphy and Wilkinson, 1980).

A.

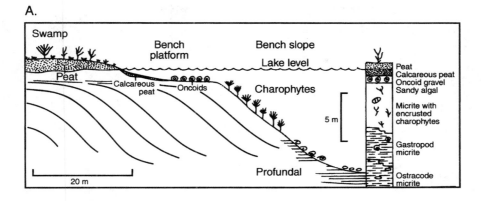

B.

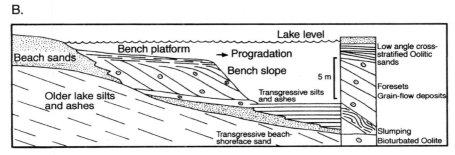

Figure 7.21. Contrasting lake margins in carbonate lakes. (A) Low-energy, bench-type carbonate lake margin and a vertical stratigraphical profile developed during progradation, based on observations from Lake Littlefield, Michigan. (B) High-energy, bench-type lake margin: ancient example and idealized stratigraphical sequence from the Pliocene Shoofly Oolite, Glenns Ferry Formation, Idaho. Modified from Platt and Wright (1991), modeled after Murphy and Wilkinson (1980) (A); Modified from Platt and Wright (1991), modeled after Swirydczuk et al. (1980) (B).

High-energy conditions coupled with steep gradients promote the formation of the carbonate facies types depicted in figure 7.21B. These are characteristic of large lakes, particularly those with steep coastlines. Under well-agitated conditions mobile oolitic sands may form large benches, slumping or flowing into deep water at their margins (Swirydczuk et al., 1979; Platt and Wright, 1991) (figure 7.22). Where strong slope breaks occur over bedrock, stromatolites can accumulate, often forming extensive bioherms and reefs. Another characteristic feature of high-energy lacustrine carbonate deposits is the formation of winnowed shell deposits. These "lags" and shell bars form when bivalve or gastropod shells are concentrated by storm activity, at or above wave base. On large platforms these shell lags can develop into vast horizontal accumulations of shell material (Cohen, 1989b; Abrahno and Warme, 1990).

Vertical Facies Sequences

On a relatively small scale, the progradation of a shoreline along a carbonate depositional lake margin will produce different vertical sequences of deposits depending in large part on the type of lake margin and energy conditions that prevailed during deposition (figure 7.23). The buildup of benches and ramps over older deep-water deposits produces a variety of vertical sequences, all of which coarsen upward. This coarsening occurs over a vertical distance that is proportionate with the available accommodation space for sediments; in the absence of significant lake level change or subsidence this approximates original water depth at the base of the ramp or shoal (e.g., Swirydczuk et al., 1979).

The superimposition of lake-level fluctuation and tectonic subsidence or uplift creates stratigra-

Figure 7.22. Oolitic foresets from lake-margin bench deposits of the Shoofly Oolite, Glenns Ferry Formation (Pliocene), Snake River Plain, Idaho.

phical sequence boundaries in lacustrine carbonates in a manner analogous to deltas. Declining relative lake levels (regressions) cause erosion of previously deposited lake margin carbonates and the rapid formation of carbonate soils and other carbonate exposure features. Lowering lake levels elicit other responses as well, such as changes in carbonate mineral assemblages, reflecting more concentrated water chemistry, or the formation of shell lags. Alternatively, rapid lake-level rise or subsidence can cause lake margin carbonate-producing areas to be "drowned," replacing relatively coarse-grained bench and ramp deposits with deeper-water carbonate muds, although the potential of carbonate deposits to "keep up" with a transgression is probably greater than along siliciclastic

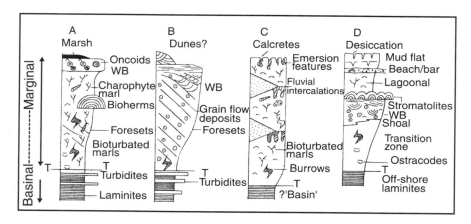

Figure 7.23. Facies models for lacustrine carbonates showing typical shallowing upward, progradational "regressive" sequences for: (A) low-energy bench margin; (B) high-energy bench margin; (C) low-energy ramp margin; (D) high-energy ramp margin. T, depositional transition from below to above the thermocline; W, transition to above wave base. From Platt and Wright (1991).

coasts, because they are not dependent on local rivers supplying sediments. Carbonate production may shift to new locations further from the basin center. Also, the trapping of siliciclastic sediments in drowning river valleys may allow the expansion of new carbonate production areas in parts of the lake that previously were inundated by river deltas. Closed-basin lakes are particularly prone to repeated episodes of progradation during periods of relatively constant lake level superimposed on cycles of relative lake-level rise and fall (e.g., Platt, 1994).

Sublacustrine Fans

In deep lakes with steep margins the formation of large prograding deltas is often precluded, as sediment is carried through various types of gravity flows directly into deep water. This also occurs offshore from prograding deltas, where complex "stair-step" type topography occurs, for example as a ramping surface gives way to a steeply dipping fault plane. Under either circumstance sediment may be transported as unconfined sheets of mud and sand that cascade downward as gravity flows. Mass movements of debris and grain flows can result in the aggradation of chaotic deposits of sand, gravel, and boulders at the foot of steep slopes, for example below crater walls or the bedrock exposures of faults. Frequently, however, the movement of sediment into deep water is facilitated by the occurrence of sublacustrine channels or bedrock canyons. Because of their steeper slopes such channels act to focus sediment delivery to specific areas of the lake floor.

Some bedrock channels are clearly relict river channels, formed during prior low lake stands. However, active scour from gravity flows and faults or tectonically formed ramps can also produce similar features. In all of these cases, where confined or partially confined sediment/water gravity flows moving through a relatively steep channel are discharged onto a relatively flat lake floor, a sublacustrine fan can develop. Lake types that particularly favor the formation of such fans include steep-sided rift basins, back-arc basins, pull-apart basins, overdeepened glacial rock basins, and calderas.

Sublacustrine fans superficially resemble alluvial fans and some types of deltas, in their lobate or radiating geometry, and in some smaller-scale aspects of their sedimentology. All three consist of complexes of channelized and overbank deposits, and typically show general trends toward finer particle size in more distal portions of the sediment body. In many lake deposit sequences sublacustrine fans grade into, or are replaced over time by deltas, as deep and steep lake margins are transformed to gentler ones through sediment infill. However, important differences also exist, which allow these types of depositional systems to be distinguished (Soreghan et al., 1999; Wells et al., 1999).

The most proximal deposits associated with fan systems are those that accumulate in incised canyons. Such canyons may be up to tens (in exceptional circumstances hundreds) of meters deep and become partly infilled by stacks of sands and gravels, derived from either rivers or shallow water portions of the lake (Harris et al., 1994; Soreghan et al., 1999). Major channels that emerge from these canyons are some of the most prominent features of large fans. Channel geometry in the largest lakes can be quite complex, with individual channels splitting and rejoining over many kilometers (figure 7.24) (Tiercelin et al., 1992). Individual channels are often flanked by levees, which when present demonstrate that they are constructional features (figure 7.25) (Soreghan et al., 1999). The channels are filled primarily by graded turbidites, normally somewhat finer grained than the canyon fill deposits. When lithified, these channel deposits often preserve current indicator features such as ripple marks, tool marks, and flute casts, indicating the direction of paleoflow. Soft-sediment deformation features are common here. Levees surrounding the channels are generally composed of finer-grained, laminated, or massive muds.

Away from the channels deposition occurs both episodically, when the nearby channels overtop their banks, or as a background of more continuous settling of suspended hemipelagic mud. These interchannel or overbank deposits are finer grained, more laterally continuous, and less frequently deformed than the channel fill. They consist of repetitive sequences of fine-grained turbidites and hemipelagic mud, which are laterally continuous over large parts of the fan surface (figure 1.2i). At a large (kilometer) scale they often form lens-shaped or radiating bodies of mud, forms that can be observed in seismic profiles (e.g., Lezzar et al., 1996; Soreghan et al., 1999). In extremely distal settings (far from their original source), deepwater lacustrine turbidites become very muddy (siltstones and shales when lithified), with only occasional thin fine sands.

Fossil content in sublacustrine fan deposits tends to be limited. Some gravity flows introduce shelly fossils, especially ostracodes and mollusks, chironomids, and benthic diatoms into deep water but these occurrences are sporadic. More commonly, plank-

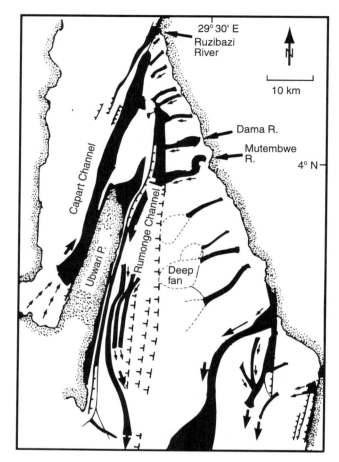

Figure 7.24. Complex, modern channel, and sublacustrine fan systems of the northern part of Lake Tanganyika. The orientations of the numerous, large SSW-directed channels are controlled by underlying structure of the northern faults, sub-basins and structural highs, particularly the uplifted Ubwari Peninsula. Adapted from Tiercelin et al. (1992).

tonic or other suspension-settled fossils (diatoms, other algae, and pollen) are found in the finer hemipelagic deposits that are interbedded with the fan.

Microscale vertical facies variation in sublacustrine fans is created by the episodic nature of sedimentation across their surfaces. Periodic channel flooding and turbidity flows, interspersed with background hemipelagic sedimentation, create laminated rhythmites similar to those of prodeltaic environments. At a larger scale, channel instability may lead to the movement of areas of coarse, in-channel versus fine, out-of-channel deposition. At the largest scale, coarsening and fining of sublacustrine fan deposits reflects major variations in sediment supply. Depending on specific lake type this may be driven by glacial/interglacial cycles (Back et al., 1998; Van Rensbergen et al., 1998, 1999), fluctuating lake levels (Scholz et al., 1998), or changes

in tectonic activity (Blair and Bilodeau, 1988; Tiercelin et al., 1992; Magnavita and Da Silva, 1995; Scholz et al., 1998; Soreghan et al., 1999). Internal unconformity surfaces can also be generated by changes in sediment discharge, and may be readily visible in seismic reflection profiles (Back et al., 1998).

Open Water (Pelagic and Hemipelagic) Deposition

In the central parts of a lake basin, away from the direct influence of river deltas or the growth of sublacustrine fans, sedimentation is dominated by the accumulation of materials settling from suspension. These include four important categories of sediments. Hemipelagic muds are transported far

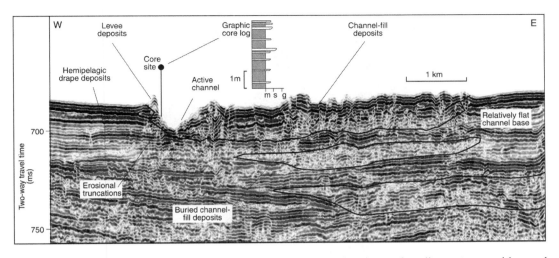

Figure 7.25. Single-channel seismic profile and simplified graphical core log illustrating muddy–sandy channel-fill deposits, from the north-central part of Lake Malawi, offshore from the South Rukuru River, Malawi. The western end of this seismic line lies approximately 12 km to the northeast of the present-day S. Rukuru River mouth. The channel fill deposits are recognizable in this shore-perpendicular profile by their irregular or hummocky upper surfaces; note that the modern active channel lies on the western margin of these deposits. Similar but older channel-fill deposits can be seen in the subsurface. From Soreghan et al. (1999).

offshore by surface or midwater currents. These sediments are derived from a combination of river or shoreline erosion processes. Organically derived particles are primarily skeletal "oozes," or partially degraded organic matter, mostly from phytoplankton. Chemically precipitated sediments include carbonates, Fe + Mn oxides, and hydroxides, and to a lesser degree evaporite or other authigenic minerals. Finally, wind-blown particles (terrigenous dust, pollen, and volcanogenic tephra) are important in some lakes. All of these types of deposits are characteristic of the central parts of most large lakes, or small lakes where river input is limited, for example in craters or solution lakes. In large but shallow lakes, whose central lake floor lies within the photic zone, organic debris from macrophytes or benthic algae can be added to this list. Small lakes and ponds often lack a true pelagic depositional zone, since most or all of the lake floor comes under the direct influence of coastal or prodeltaic depositional processes.

Pelagic sediments are dominantly silt and clay-sized particles. Sand-sized particles, if present, are primarily derived from eolian origins, the occasional large turbidite, or from fecal pellet aggregates. Because these areas are remote from river inputs they typically experience slow rates of sediment accumulation. These types of sediments provide the highest degree of stratigraphical resolution

obtainable from lake deposits, and are therefore frequent coring or sampling targets for studies aimed at very high-resolution reconstructions.

The types of facies that ultimately accumulate in these open-water environments depend not only on their proximity to terrestrial sediment sources but also on three important secondary effects: travel time of sediment to the lake floor, which affects secondary dissolution, sediment resuspension, and bioturbation. All serve to mix sediments that were originally formed or deposited at different times, thereby blurring, to various extents, the potential for a high-resolution record. In situations where mixing of particles of various settling age is limited, laminated sediments can be formed, preserving a record of events on as short as a seasonal time scale.

Laminations and Varves in Open Water

Close examination of laminated lake sediments normally reveals that they comprise monotonous repetitions of a limited number of sediment types. These repetitive clusters are referred to as *couplets* when comprising two facies, for example carbonate laminae interbedded with organic-rich laminae (figures 1.2h and 7.26). Occasionally the pattern of repetition may be more complex, involving three (*triplets*) or more repetitive sediment types.

Figure 7.26. Rhythmically laminated, organic-rich shales from the Toca Limestone (Lower Cretaceous), offshore Congo. Alternating units of organic-rich (dark-colored) and organic-poor muds in this lacustrine shale are typical of deep water, hemipelagic deposition below the thermocline or within the permanently anoxic zone of a meromictic lake. The repetitive bedding is suggestive of annual depositional cycles, although in this unit the case for annual varving is only circumstantial.

Sediment traps are used to collect settling sediment from the water column at regular intervals over a period of months to years. These devices have been instrumental in helping sedimentologists understand the mode, rate, and seasonality of formation of different laminae types. Through the use of such trapping it has become evident that in many lakes the major lacustrine cycles of overturn, complete or partial mixing, productivity blooms, and runoff can be differentiated by the types of material settling through the water column (R.Y. Anderson and Dean, 1988). Generally, though not always, these changes are played out over an annual cycle. This provides a mechanism for creating an annually varved record of lake processes and sedimentation in open-water depositional environments.

Several seasonally variable components of sedimentation stand out as particularly important for varve formation. Variation in clastic sediment from seasonal runoff is important for producing varving in the prodeltaic environment. This signal often transmits to more distal, offshore settings as well, particularly where sediment is transported as overflows or interflows. Sometimes, fine-grained to medium-grained silts are carried up to several kilometers offshore by seasonal river flooding (R.Y. Anderson and Dean, 1988). In temperate, boreal, or subpolar climates runoff seasonality is associated with snow melting and spring discharge. In the

tropics and subtropics, runoff seasonality may be controlled by monsoonal circulation or other factors determining the timing of local rainy seasons. In contrast to thermal seasonality, the wet–dry seasonality of the tropics and subtropics may alternate once or more than once in a year, depending on regional patterns of precipitation.

Overturn events and the partial or complete disruption of stratification can create annual or biannual periodicity in the formation of autochthonous Fe and Mn oxides (Anthony, 1977). If Fe and Mn have accumulated in large quantities in the hypolimnion, sudden precipitation events can accompany seasonal overturn as these metals are oxidized, with the frequency of cyclical deposition depending on the seasonality of stratification. In temperate dimictic lakes this type of precipitation event can occur twice a year, whereas different patterns can develop in other climate regimes.

Increased production of precipitated carbonate mud is caused by seasonal increases in the photosynthetic uptake of CO_2. In the Temperate Zone this normally occurs during the spring, and can result in massive precipitation of calcite or aragonite over the course of days or weeks. Seasonal precipitation of evaporite minerals can also be driven by factors such as periodic supersaturation related to temperature or evaporation thresholds being exceeded.

Other biogenically formed materials can also produce sediment on a seasonal basis. Diatom blooms are related to overturn and nutrient generation seasonality. Dominance of one or a small number of species under specific nutrient conditions can lead to a seasonal succession of blooms and accumulation that may be evident in the fossil microstratigraphical sequence of varves (Glenn and Kelts, 1991). Finally, in the absence of pelletization the settling of extremely fine-grained or light material is inhibited. In lakes that experience ice-over conditions, the deposition of fine organic detritus increases during winter months.

Before varves can archive seasonal records, several conditions have to be met (Kelts and Hsü, 1978):

1. There must be a seasonally variable flux of mineral and organic matter from epilimnion, settling through the water column.
2. Particle settling must occur fast enough to transmit the seasonal signal to the lake floor. This is problematical because most suspended materials in open water are initially extremely small (micrometers to tens of micrometers) and will remain suspended with only the slightest current. The formation of zooplankton fecal pellets facilitates settling, since pellets may be hundreds of micrometers in diameter, and can settle to a lake floor in hours or days (N.D. Smith and Syvitski, 1982; Haberyan, 1985; Pilskaln and Johnson, 1991). The frequency of pelletization, therefore, is particularly important in determining whether deposits of very deep lakes will be laminated. Another factor affected by settling rate is the potential for secondary dissolution, particularly when the settling particles are carbonates or volatile organic matter.
3. The water column must be sufficiently stratified to prevent large-scale resuspension and mixing of previously settled materials.
4. Bioturbation must be minimized for laminae to be preserved. This can occur because of permanent or partial stratification and anoxia, or because food supplies are too limited and sedimentation is too fast to allow subannual mixing by burrowers (e.g., Cohen, 1984).
5. There must be minimal bottom currents to preclude resuspension.
6. Gas formation during microbial degradation of organic matter must occur slowly enough so as not to destroy laminae.

Annually laminated lake sediments provided material for study in some of the classic facies investigations of the late nineteenth and early twentieth century (De Geer, 1882, 1912; Nipkow, 1920; W.H. Bradley, 1929). Work by De Geer and some of his contemporaries first established that many rhythmically bedded deposits were deposited on an annual cycle in glacial lakes. Nipkow (1920) showed that varves could also form in nonglacial systems. Based on comparisons with Nipkow's work on nonglacial varves from Swiss lakes, W.H. Bradley (1929) used the accumulation of laminites in the Eocene Green River Formation (western United States) to develop a model of laminae formation under meromictic conditions. This model continues, in various guises, to be applied today in the interpretation of laminites. In the Bradley model seasonal phytoplankton bloom production alternated with periods of hemipelagic mud settling in a meromictic lake, to produce the laminations characteristic of many oil shales. Ironically, Bradley's general application of a meromictic lake model to the Green River Formation itself was subsequently challenged by several authors, who reinterpreted at least some of Bradley's varved intervals (the Wilkin's Peak Member) to represent deposition in a relatively shallow playa lake (e.g., Eugster and Hardie, 1975; Smoot, 1978). During the 1970s and 1980s this debate generated considerable controversy, but probably the true history of the Green River Formation falls somewhere between these two models, with intervals of both "permanent" stratification and intervals of playa lake development.

Varves in modern Lake Zurich below 50 m depth form as triplets (Kelts and Hsü, 1978), consisting of:

I. A mesh of organic sludge (algal filaments, FeS, and clay-sized hemipelagic particles).
II. Diatom frustules (mostly cold-water species) plus subordinate hemipelagic and organic matter.
III. Relatively pure calcite (only minor diatoms), with decreasing calcite crystal size upward.

This sequence forms as fine phytoplankton-derived organic particles settle during late autumn and winter and reduced sulfides accumulate under stratified conditions (I), followed by spring diatom blooms, after overturn mixing and accelerated spring production (II), followed by the production of fine, micritic calcite (III) (figure 7.27).

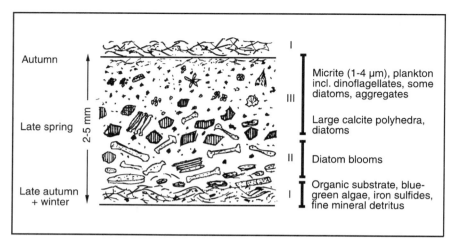

Figure 7.27. Schematic representation of a typical nonglacial varve. Units I and part of II form the dark part of the varve, and unit III forms the light part. From Kelts and Hsü (1978).

In small lakes, varve formation can probably occur without aggregation in fecal pellets (Brunskill, 1969). However, in larger lakes with more vigorous epilimnetic currents, or in deep lakes, seasonal varve formation probably requires aggregation to allow particles to reach the hypolimnion within the same season that they are formed. This emphasizes the fact that whereas a varve couplet or triplet represents an annual cycle of deposition, a single lamina does not represent a half or third of a year. Deposition of the individual components of a varve may occur over only a matter of days or weeks, with little or no deposition for much of the remainder of the year (Glenn and Kelts, 1991).

Tropical lakes have different seasonal cycles from temperate lakes that can lead to different patterns of varve formation than the temperate examples above. In meromictic Lake Malawi, Africa, Pilskaln and Johnson (1991) have shown that light, diatom-rich laminae represent deposition during the dry windy season (April–October), when mixing and blooms occur. Dark laminae are formed from runoff during the wet (but calm) season, when terrigenous input is high but productivity is low.

Some pulses of open-water sedimentation occur at frequencies other than those controlled by annual cycles. As a result paleolimnologists must exercise caution in automatically assuming that laminated sediments are true varves (J.M. Dean et al., 1999). In some river systems (particularly in arid regions), runoff is inadequate to produce terrigenous laminae

on an annual basis, and may represent extreme flood years. In coastal lakes, laminations can form after episodic incursions of seawater, causing massive mud flocculation and settling (e.g., Hodgson, 1999). Also, laminations are known to form in circumstances where oxygenation is periodically, or even permanently, high. Varved sequences can be preserved in dimictic lakes if sedimentation rates are very high, the case in many proglacial lakes, if bottom waters are sufficiently protected from turbulence, common in small or deep lakes, or in lakes where spring bioturbation is limited. The latter is most likely to occur under oligotrophic or high-latitude conditions.

Preservation of Organic-Rich Sapropels and Laminites

The accumulation of organic-rich sapropels in open-lake conditions involves an interplay between rates of organic primary production and transfer of organic matter to the lake floor, input of carbonate or clastic sediments that dilute organic matter, and secondary degradation by microbial decay and diagenesis. This interplay is of considerable interest to paleolimnologists concerned with the formation of hydrocarbon deposits, since many important petroleum source rocks occur in sapropelic lake sediments (Fleet et al., 1988; Katz, 1990). Sapropelic sediments can be transformed into *oil shales* when they become lithified.

Organic sediments can be derived from both autochthonous and allochthonous sources (Kelts, 1988). In the central portions of large lakes the primary source of organic matter is autochthonous, from mid-lake production. This can consist of epilimnetic phytoplankton, bacteria growing at chemocline, or benthic microbial mats (Valero-Garcés and Kelts, 1995). In small lake basins organic matter may represent more of a mix of both autochthonous and allochthonous organic matter sources. Alkaline lakes are particularly prone to accumulation of laminated organic sediments because alkaline lakes take up CO_2 rapidly as carbon is fixed by photosynthesis, allowing very high biomass levels to accumulate. Even in productive lakes, however, only a small proportion of organic productivity is actually sedimented (Niessen, 1987). Even after sedimentation there is continued loss of organic content in sediments because of bacterially mediated diagenesis. Organic matter undergoes microbial decomposition in sediment through a variety of processes. In the aerobic zone this involves oxidation, whereas anaerobic processes involve reduction of manganese, nitrate, iron, sulfate, and carbonate. All result in the formation of simpler organic compounds, and the formation of various hydrocarbon compounds. In some lakes part of this organic matter is also secondarily transformed as bacterial mats on the lake floor. Pyrite commonly forms in sulfate-rich lakes under meromictic conditions, or where organic accumulation rates are extremely high (e.g., Kelts, 1988; Loftus and Greensmith, 1988).

Nonlaminated Sediments of the Mid-lake Environment

When the lake floor lies within the aerobic zone of the water column for long periods of the year, bioturbation tends to destroy seasonal or episodically formed laminae. Vigorous bottom currents and resuspension have the same effect. Such sediments may be organic-rich or -poor, depending on the rate of organic matter accumulation and dilution by other forms of sedimentation. The central basins of large, shallow lakes, or lakes with strong bottom current systems are commonly underlain by homogeneous, thin to thick-bedded muds. Bedding discontinuities here are more likely to represent very strong, episodic resuspension events such as storms, rather than seasonal deposits.

In very large lakes suspended material is redistributed by internal wave activity, and depocenters may reflect the dominant orientation of Coriolis-related flow. Large lake sediments often remain in suspension for long periods, allowing their broad distribution, and preventing the archiving of seasonal signals (e.g., Harrsch and Rea, 1982). Even where suspension intervals are shorter, vigorous bottom currents, resuspension, and bioturbation still insure that sediments are well-churned, homogeneous muds (e.g., Henderson and Last, 1998). Open-water lake deposits display characteristic, large-scale facies geometries, most evident from reflection seismic survey data. In meromictic lakes, very deep lakes or lakes with simple deep-water topography and limited bottom currents, comparatively uniform and widespread deposition across large areas leads to a uniform seismic appearance of "sheet drape" type deposits (Eyles et al., 1991; Lezzar et al., 1996; Van Rensbergen et al., 1999). More complex fill patterns are typical of lakes with strong bottom currents or relict topography, such as the Laurentian Great Lakes (Dobson et al., 1995). In these situations, deposits (and therefore seismic reflections) are more discontinuous laterally, and show pronounced thickening and thinning related to local sediment focusing or scour conditions.

Vertical Facies Sequences in Offshore Deposits

Facies changes over time in mid-lake depositional environments are driven by limnological changes at a wide variety of time scales. For much of the early part of the twentieth century paleolimnologists focused on an ontogenetic or successional view of lake change, in which variation was seen as more or less unidirectional. For example, many small ponds in the North Temperate Zone formed by late glacial processes have undergone more or less predictable stages of limnological and depositional change as they filled in and warmed during the Holocene. This often involves a transition from varved lake muds and silts to gyttjas through stratigraphical sections measured in meters. At the opposite end of the scale, longer-lived, tectonic lakes also show ontogenetic sequential development related to diminution of subsidence and their progressive infilling. These types of shallowing-upward sequences can occur over thousands of meters of strata.

Superimposed on these *lifetime* trends of lakes, however, are cyclical changes in environments of deposition. The interpretation of repetitive change in lake history has assumed increasing importance among paleolimnologists in the past 30 years. This has occurred because of the recognition that various phenomena that force lake conditions to vary, such as the annual cycle, solar output, or variations in the earth's orbital parameters, have the potential to create rhythmic depositional patterns in lakes. R.Y.

Anderson (1986) suggested the sedimentary impacts of cyclical changes in lake hydroclimate are replicated at different time scales in the stratigraphical record; in other words they are fractal. He referred to this phenomenon as the *varve microcosm*. For example, consider a series of varve couplets consisting of alternating carbonate and evaporite laminae. In our hypothetical scenario the carbonate laminae form during a spring season of enhanced productivity, while the evaporites form during periods of supersaturation for the minerals present. This might be the result of enhanced evaporative concentration during a dry season. This annual cycle in deposition is generally the strongest component of cyclical bedding, because in lakes throughout most parts of the world seasonal variations in temperature or precipitation exceed in magnitude changes over longer time periods. However the components of this annual varve cycle are commonly duplicated over longer time series as rhythmic repetitions of one dominant component or the other, depending on longer-term climate cycles.

In the simplest extension of the varve microcosm, cyclical change in seasonal runoff or temperature might favor the accumulation of carbonates over evaporites, resulting in a trend toward thicker carbonate parts of the couplets for some period of time, and thinner carbonates at others. More commonly, the longer-term cycles will be depositionally "played out" by something other than an exact mirror of the seasonal cycles. For instance, there may be thresholds of sediment availability, productivity, saturation, or dissolution below which no carbonate or evaporite forms. If the lake becomes more dilute as a result of a long-term trend, seasonal evaporation or cooling may be insufficient to deposit any of the evaporite minerals, and deposition will become entirely dominated by carbonates. Amalgamation by bioturbators can mix two sediment types, so that laminae couplets are no longer evident, but the overall proportion of one mineral product versus the other changes in the homogenized beds.

Long-term hydroclimate cyclicity is manifest through a variety of depositional trends, which provide a linkage between varve pattern and climate. Laminae thickness patterns may change as a result of decadal–centurial scale trends in rainfall or temperature (Kelts and Hsü, 1978). Over longer (10^3 years) time intervals sedimentation responses incorporate the lag times inherent in changing vegetation cover and erosional patterns (R.Y. Anderson and Dean, 1988). Long-term mineralogical contrasts are also a common result of changing aridity. Cycles of low-Mg calcite-dominated carbonate deposition, formed in relatively freshwater, alternating with high-Mg calcite, aragonite, or even dolomite, formed in more saline water, are an expected outcome of cyclical changes in salinity. At much longer time scales (10^4 to 10^5 years) mid-lake depositional cycles frequently involve large lake-level fluctuations, or complete or near-complete desiccation of lake basins, transforming a deep-lake depositional environment into a shallow lake, lake margin, mudflat, or soil. These types of cyclical lake-level fluctuations are commonly expressed as deposits of laminated, black or brown mudstones (oil shales), alternating with nonlaminated, massive, or thicker-bedded, mudstones and marls, which are lighter in color (figure 7.28).

Summary

Our conclusions from this chapter can be summarized as follows:

1. *Lithofacies* of lacustrine sediments can be described in terms of outcrop, core, and microscope-scale lithology, mineralogy, and geophysical properties.
2. Lacustrine sedimentation is regulated by the interactions of watershed geology, climate, mode of lake formation, inflow and outflow hydrology, internal water circulation, and organic productivity. Variations in these factors within or between lakes create the extreme variability observed in lithofacies of lake deposits.
3. *Facies models* are useful simplifications of the characteristic features observed in lake deposits from different subenvironments, incorporating changes in depositional systems over time as deposits prograde or aggrade. These models can incorporate outcrop, core, or geophysical observations.
4. Lacustrine deltas are the major entry points of sediments into lakes, and therefore important zones of accumulation. Variations in deltaic sedimentation rates and quality of potential archives depend on distance from the original river mouth and the nature of sediment/water inflow to the lake.
5. Lacustrine coastal environments are areas of short-term sediment storage and vigorous sediment reworking, and usually provide only low-resolution paleolimnological archives. However, shoreline indicators are

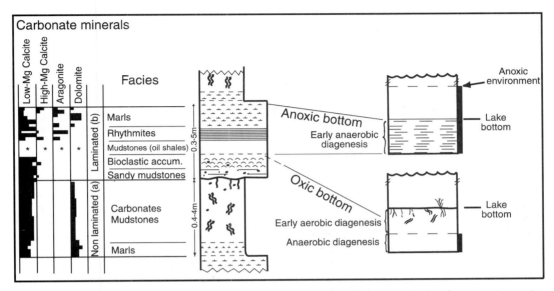

Figure 7.28. A model of oxic/anoxic depositional cycles from the Miocene Rubielos de Mora Formation, Spain. Variable redox conditions at and below the sediment/water interface related to changing lake levels resulted in the deposition of a suite of lithofacies that repeats itself numerous times over a stratigraphical interval of at least 200 m. From Anadón et al. (1988a).

useful markers of episodic high lake stands. The organic-rich deposits of the central parts of small lakes can provide good records of local watershed and climate processes.

6. Carbonate deposits are common in lakes, and yield important information about biological activity, productivity, and water depth.

7. Sublacustrine fans form in deep-water environments, where sediment builds out over the lake floor. They can provide informa-

tion about long-term climate and tectonic changes affecting lake basins.

8. Open lake and pelagic deposits provide the most continuous and highest resolution records available from lakes, and are often *varved*, recording subannual events. Variations in settling rates, dissolution, and lake-bottom conditions affect the quality of these records. High-resolution lake records often display cyclical facies changes on various scales, indicative of fluctuations in productivity, lake level, and climate.

8

Facies Models at the Lake Basin Scale

Understanding the historical evolution of sedimentation in a lake requires not only a grounding in facies interpretation but also an understanding of the larger-scale, lakewide linkages between deposition and those factors influencing sedimentation. The facies models we examined in chapter 7 can be linked to understand the differences in deposits between lake basins. Basin-scale facies models focus on the major interactions between climate or tectonic/volcanic activity and sedimentation, attempting to explain why particular facies types develop in particular areas or at particular times in a lake's history. Here I will focus on a few examples from the most intensively studied depositional settings, including lake types defined by mode of origin and evolution (rifts, glacial lakes, etc.) as well as saline lakes and playas, which share chemical and climatic attributes.

Rift Lakes

Large-scale facies modeling in rift lakes has been driven by a need to understand the occurrence of hydrocarbons in ancient rifts (Lambiase, 1990; Katz, 2001). This in turn spurred a rapid accumulation of seismic reflection and facies data in the East African rift lakes and Lake Baikal (Russia) during the 1980s and 1990s, as well as attempts to synthesize these data and integrate them into general models. As we saw in chapter 2, the evolution of rift basins involves the development of asymmetric half-grabens and, in larger lake systems, the linkage of these half-grabens in a linear chain. As rift basins age, progressive deformation will eventually cause extensive deformation on both sides of the basin, transforming them into asymmetric full grabens, as seen in Lake Baikal today. This pattern of tectonic development has consequences for geomorphology, sediment delivery rates and locations, and sediment

composition, that also vary depending on whether the lake basin is relatively full (high-stand conditions) or empty (low-stand) (Rosendahl et al., 1986; Cohen, 1990; Scholz and Rosendahl, 1990; Tiercelin et al., 1992; Soreghan and Cohen, 1996). Large-scale depositional patterns in a rift lake therefore represent an interplay between tectonic and climatic forces, factors that operate on somewhat different time scales.

Sedimentation within a half-graben basin can be subdivided into four zones (figure 8.1):

(1) *Escarpment margin* sedimentation comprises all depositional belts along the major border faults bounding a half-graben. Major border faults typically produce substantial topographical relief along the lakeshore, with the highest rift-shoulder mountains and the deepest topographical basins often in close proximity. It would seem that such relief should generate abundant sediment that could accumulate immediately below these mountainous areas. In fact, much of the erosion associated with such uplifts is initially directed away (backshed) from the rift valley (Frostick and Reid, 1987). Because many rift lakes comprise multiple basins, sediments backshed from the rift highlands may in fact be redirected into the lake in an adjacent half-graben, on hinged margins (figure 2.10), or at axial margins and accommodation zones. The escarpment margin area experiences the highest long-term rate of subsidence within a rift basin. It is also a focal area for sediment accumulation, primarily through the accumulation of fan deltas (in shallow lakes), sublacustrine channels, and fans. Piles of sediment up to several kilometers thick can accumulate adjacent to major border faults. Very high subsidence rates here promote the vertical stacking of sublacustrine fans, rather than their outward progradation (Gawthorpe et al., 1994; Scholz, 1995; Nelson et al., 1999). Because of their proximity to major faults and high topographical relief, the areas closest to escarpment mar-

A.

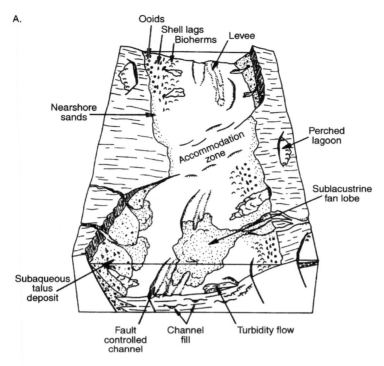

B.

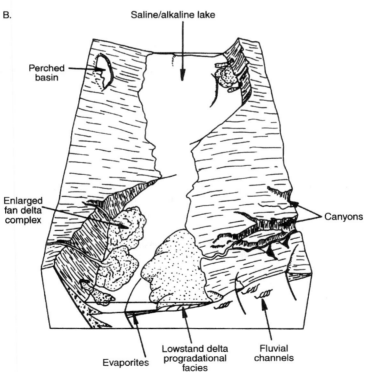

Figure 8.1. Idealized systems-tract models (not to scale) of rift-basin lake deposits and structure in an alternating half-graben framework. An escarpment margin is shown on the left front and hinged margin on the right front of each diagram. (A) Lake high-stand model. (B) Lake low-stand model. Note the more extensive development of coarse-grained facies in mid-lake areas during low-stand conditions. Modified from Scholz and Rosendahl (1990).

gins are commonly sites of accumulation of very coarse debris, conglomerates and breccias, often with enormous blocks. This material is emplaced by rock falls and sublacustrine debris flows. When lake stands are high, little sediment can accumulate in shallow water; most material is transported through gravitational flows to deep-water sites of accumulation (figure 8.1A). Sediments entering a rift lake along hinged margins are often secondarily transported by sublacustrine channels across the basin and into deep areas at the base of escarpments, and the positions of these channels and their deposits are also controlled by local faulting patterns (Lezzar et al., 1996). Local bathymetric highs, which are largely created by the orientation or intersection pattern of secondary faults, focus sedimentation more narrowly into discrete depositional zones along the deep lake floor (Lezzar et al., 2002). Only narrow and discontinuous belts of littoral deposits develop along escarpment margins and their potential for long-term preservation in the stratigraphical record is limited by secondary erosion. When lake stands are moderately low, escarpment margins are zones of fan-delta formation (figure 8.1B). If lake level drops below the point where the border fault enters the basin the lake margin will normally be rimmed by alluvial fans and sand or mud flats (Tiercelin and Vincens, 1987).

(2) Sedimentation along *hinged margins* (also called *shoaling* or *flexural* margins) comprises deposition along the opposing side of a rift basin from its major border fault. In contrast to the steep topographical gradients of the escarpment margin, hinged margins represent broad and often gently sloping ramps, across which river deltas can prograde and broad littoral and sublittoral zones can develop. Large, backshed river systems often enter rifts along these ramps. Accompanying this gentler topographical gradient is a more gradual facies transition from nearshore to offshore deposits. Extensive shallow-water carbonate deposition is more likely to occur here than on the opposing escarpment margin. Sediment accumulation on hinged margins is more sensitive to slight changes in lake level and wind or current-driven resuspension, and depositional sequences are punctuated by more frequent and more profound unconformities than at the base of escarpment margins. Reduced clastic sediment input to the lake also occurs under high lake stand conditions, primarily from sediment being temporarily stored on deltas. Such conditions are ideal for the accumulation of littoral and sublittoral carbonate deposits. At times of low lake stands an increase in sediment discharge allows

the progradation of broad delta systems toward the lake's center (Scholz, 1995). As rift basins age and full graben systems develop, the hinged margin increasingly comes to resemble the escarpment margin, with reduced lateral inflow and steeper flanks (Scholz et al., 1998). As this occurs sedimentation is first ponded, and then blocked altogether from entry along much of the lake margin, and sediment input becomes concentrated at the axial margins and accommodation zones.

(3) The *axial margins* of rift lakes located at the narrow ends of the basins are commonly sites of major river inflows, entering the lake on low gradient ramps. These rivers also build prograding deltas, with broad deltaic plains and prodelta environments, similar to those of hinged margins. At high lake stands they are also the sources of major sublacustrine channel systems that can traverse the length of a half-graben lake segment.

(4) *Accommodation zones* are the intersection areas between adjacent half-grabens. The intersection of half-graben segments is sometimes accompanied by strike–slip faulting, local compression, or local extension. These different types of interactions and fault intersections generate a variety of lake-floor morphologies, some steep-sided and some gently sloping, which serve to direct sediment transport and accumulation. Accommodation zones are also important sites for river inflow, which often follow major basin-bounding faults. As noted above, the importance of these types of deltas increases as the rift ages; in Lake Baikal the ancient accommodation zone delta of the Selenga River has provided a stable entry point to the lake for millions of years (Scholz et al., 1998).

Away from lake margins, deep rift lake sediments are dominated by fine-grained facies, clastic muds, sapropelic muds, diatom oozes, and under appropriate water chemistry conditions, fine-grained carbonates. Local water chemistry, mixing regimes, and productivity all exert controls on variations in lithology. In deep, tropical rift basins, meromixis promotes the formation of laminated rhythmites, whereas more thickly bedded muds typically floor well-mixed rift lakes at higher latitudes. Basin floor suspension deposits are dissected by distal sublacustrine channel systems. These channels frequently follow pre-existing structural features such as local faults, which cause sediment to be distributed axially along the basin's length (Lezzar et al., 1996; Nelson et al., 1999).

At low lake stands, basin floor deposits are also muddy, but may contain a wide range of evaporites if precipitation/evaporation ratios are low and the lake is saline. Evaporite accumulation is normally focused toward the deeper parts of fault-bounded basins, and can result in very thick accumulations of bedded salts in the depocenters of such lakes.

Climate effects are superimposed on the structural control of facies development in rifts. In boreal and subpolar regions, lower evapotranspiration rates make the hydrological closure of a lake basin difficult to achieve (Colman, 1998). Two lines of paleolimnological evidence have been commonly used in rifts to infer major eustatic lake level fluctuations driven by precipitation balance changes:

1. The widespread occurrence of shoreline terraces both well above and well below modern lake level; and
2. The development of stacked, prograding delta packages, evident on seismic lines, at depths considerably below modern lake level.

Both of these lines of evidence must be evaluated carefully, since tectonic adjustments can also affect the elevations of shorelines and deltas. Sedimentation in rift lakes with a strongly positive water balance is more likely to be affected by thermal change than by changes in precipitation/evaporation ratios, expressed by such changes as productivity responses (biogenic silica) or sediment yield from glacial erosion (Colman et al., 1995; Back et al., 1998; Karabanov et al., 1998).

Tropical and subtropical rift lakes are poised to react more strongly to precipitation changes than higher latitude ones, because of the much higher potential evaporation rates encountered at low latitudes. In these regions major lake-level excursions of hundreds of meters are possible, and have been documented from a number of rift lakes (e.g., Scholz and Rosendahl, 1988; Scholz et al., 1998). The depositional response of a rift lake to such fluctuations seems to be regulated primarily by the slope of the lake margin (Cohen et al., 1997b). The *Capart Cycle* (figure 8.2), a depositional model for cyclicity observed primarily in seismic records in the high-relief Lake Tanganyika basin, can be contrasted with the *Van Houten Cycle* of the low-relief Triassic Newark Basin (figure 1.11). Based on their estimated average duration, the Van Houten cycles of the Newark Basin and the Capart cycles of the Tanganyika Basin are both thought to be driven by Milankovitch-scale climate cyclicity, manifest by increasing or decreasing precipitation/evaporation ratios. Van Houten cycles are markedly thinner than Capart cycles (10 m vs. 50 m average thick-

ness). Deposition at high lake stands is similar in both the Van Houten and Capart cycles; both are marked by the accumulation of sapropelic laminites and other sheet-drape type deposits. Depositional differences related to slope are best expressed during low or rising lake level intervals. In the gently sloping Newark basins, the formation of laterally continuous paleosols and evaporites mark low stands. In the steeply sloping Tanganyika Basin, low stands are marked by deeply incised erosion surfaces, overlain by coarse sedimentary infill. As lake level begins to rise these valley fills give way to the formation of sublacustrine fans. Very similar depositional sequences to the Capart cycles have been observed in Lake Malawi. Both field observations and computer modeling show that a transformation from Van Houten to Capart-type cyclicity can be generated through an increase in local subsidence rate and the formation of higher local relief (Anadón et al., 1991; Cohen et al., 1993).

At the longest time scales, the three-staged history of rift basin evolution, discussed in chapter 2, is mirrored by a three-part depositional history (Lambiase, 1990; Schlische and Olsen, 1990; Magnavita and Da Silva, 1995; Bohacs et al., 2000) (figure 8.3). In the early synrift phase (1 on figure 8.3), the development of small and localized block faults creates only minimal barriers to throughgoing sediment transport. River systems, broad swamps, or shallow lakes fill rifts at this stage, and sedimentation easily keeps pace with subsidence, overfilling the basins and preventing deep topographical basins from developing. As subsidence continues and the zone of subsidence becomes enlarged the rift basin enters a second synrift phase. The rate of sediment delivery from uplifted rift escarpments to the lake basin declines in proportion to the rate of subsidence. During this second phase (2) the lake can deepen quickly as sedimentation no longer keeps pace with subsidence. Rift lakes can remain in this state for very long intervals, as long as the balance of subsidence to sedimentation remains high. Eventually all rifts either cease to subside or break out into incipient ocean basins, at which point they cease being lakes, as in the Red Sea. In the former case, declining subsidence results in the progressive infilling of the basin. This is often accomplished through sequential infilling of a half-graben with deltaic and fluvial deposits (3 and 4), in which axial sediment input progrades over one basin, spilling into the next and so on, until the rift is completely infilled. Often this final postrift phase is marked by extensive fluvial deposition (5), as the rift is transformed from a lake to a river valley.

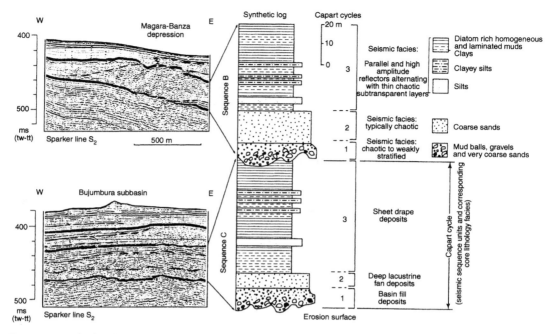

Figure 8.2. Idealized Capart Cycles from northern Lake Tanganyika. Facies interpretations relating core and high-resolution seismic data, average cycle, and facies thicknesses are based on Tiercelin and Mondeguer (1991), Baltzer (1991), Bouroullec et al. (1991, 1992), and Lezzar et al. (1996). Each cycle consists of an erosion surface, developed at low lake stand, followed by infilling as the lake level rises. Initial coarse-grained basin-fill, mostly restricted to channel systems, gives way to more extensive deep lacustrine fan deposits. Cycles are capped by very fine-grained (laminated diatomaceous ooze) sheet-drape deposits that extend over much of the lake basin, that form when the lake level is high. Compare this cyclicity from a steep-sided rift lake with that of a gently sloping rift basin, shown in figure 1.11. Modified from Cohen et al. (1997).

Foreland Basins

Sedimentation and facies analysis in foreland basin lakes has a long history of study. As with rift lakes this has been driven largely by a search for depositional models useful in the exploration for hydrocarbons. Unfortunately there are relatively few studies of sedimentological processes in modern foreland basin lakes in comparison with rifts; therefore much of our knowledge is inferential and based on studies of ancient deposits, or based on modern lakes with different origins. The gigantic Early Tertiary foreland basin lakes of western North America simply have no direct counterparts today.

Lacustrine sedimentation in foreland basins shares some similarities with rift basins, in that both basin types can produce thick and asymmetric piles of sediment, whose facies characteristics are controlled by both climatic and tectonic factors. Like rifts, foreland basin lake sedimentation at the largest scale is commonly characterized by a three-phase history, which can be repeated in the case of episodic tectonic activity (Steidtmann et al., 1983; Yuretich, 1989; Liro and Pardus, 1990; Sladen, 1994). In the case of foreland basin lakes this involves an initial phase of shallow, temporary lake deposition, related to the initiation of thrust development and basin subsidence. A second phase of maximum subsidence is marked by deposits of lacustrine muds from very large and relatively stable lakes. A final phase involving the termination of flexure is recorded by basin infilling deposits related to delta progradation and the elimination of lakes.

Characteristic Features of Foreland Basin Lakes

Several important differences exist, which differentiate foreland basin lakes from rift lakes. These differences have implications for facies development

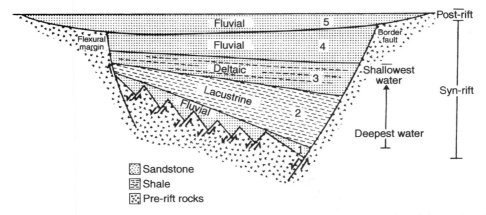

Figure 8.3. Idealized section of a half-graben containing a sequence of fluvial and lacustrine deposits formed during synrift and postrift periods. From Lambiase (1990).

and geometry, and for the lake's sedimentological response to climatic forcing.

Proximity to Border Faults

In rift lakes major border faults are often coincident with or in close proximity to the shoreline. This can put bedrock surfaces adjacent to the lake (the *coincident margins* of Donovan, 1975), allowing for the deposition of very coarse materials from rock fall, subaqueous debris flows or other gravitational flows directly within the lake. This type of sedimentation is much less likely in foreland basin lakes, where propagating thrusts or basement-cored uplifts are generally surrounded by broad aprons of alluvial fan, fluvial, and alluvial plain deposits, and removed from direct contact with the lake basin (*noncoincident margins*). In most foreland basins coarse sediments are actually trapped by rapid subsidence near the thrust front, and only fine-grained materials (sand and mud) accumulate along more distal topographical lows, where lakes can form (Heller et al., 1988). Somewhat surprisingly, large volumes of coarse sediments probably only move out toward the basin center as tectonism slows and the subsiding region adjacent to the thrust front oversteepens from accumulated sediment. For example, in the Oligocene–Miocene Molasse Basin (Switzerland), fine-grained alluvial and lacustrine deposition occurred to the northeast of the propagating thrust front of the rising Alpine mountain chain (Platt and Keller, 1992). Deposits from alluvial fans and river systems flanked the thrust front. An elongate lake ($\sim$ 200 km long $\times$ 20 km wide)

formed in the topographical low of this basin, but this was far from the major thrusts, generally over 50 km away from the rising mountain front (figure 8.4).

Although foreland basin lakes are normally well removed from major erosional bedrock surfaces, there are at least two circumstances where this generalization breaks down. First, lakes can sometimes form as piggyback basins, on the surfaces of advancing thrust sheets, and may subsequently feed into the foredeep, through overflow or seepage (figures 2.9 and 8.5) (Salvany et al., 1994). These piggyback lakes are normally much smaller and more ephemeral than the ones formed in foredeep settings. Also, foreland basins are occasionally deeply flooded by subsequent uplifts that block prior drainage systems in more distal parts of the basin. This can cause a transgression of the lake basin margins onto the margins of a pre-existing foreland uplift. At this point the lake basin technically no longer owes its origin to its foreland basin precursor. However, its deposits will typically onlap the eroded thrust belt, and it may be difficult to distinguish the secondary cause of topographical closure without very careful study.

Facies Migration

In cases where the front of a fold–thrust belt is advancing, both the lake and its facies belts will migrate, shortening the lifetimes of lakes, and making their geometry less static than in rifts. The extremely thick, stacked sequences of deltaic or fan deposits, commonly observed in rift lakes, cannot form under such circumstances.

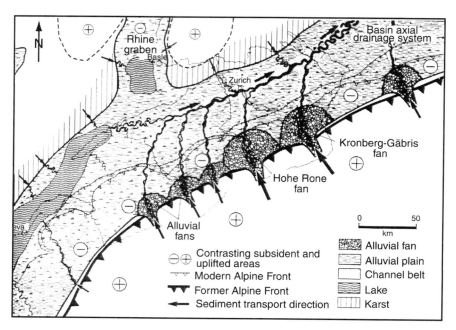

Figure 8.4. Paleogeography of the Swiss Molasse Basin at the Chattian-Aquitainian (latest Oligocene–earliest Miocene). Notice the position of the large foreland basin lakes, which formed well to the northwest of the advancing Alpine thrusts of that time. From Platt and Keller (1992).

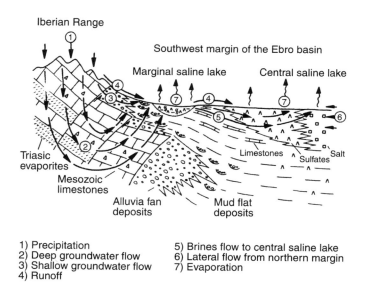

Figure 8.5. Sedimentary and paleohydrological model for the development of small piggyback lake basins on the advancing thrust front of the Ebro Basin (Lower Miocene), Spain. The small piggyback lakes were fed by groundwater discharge and runoff derived from the thrust sheet. These lakes acted as evaporator ponds, partially concentrating brines before discharging them to a larger foreland basin lake in the central part of the basin. From Salvany et al. (1994).

Basin Shape

Foreland basin lakes are more variable in shape than rift lakes. They are often circulate or ovate, with gentler bottom topography (as a result of more rapid, sediment infill relative to rifts). As a result lateral facies transitions in foreland lake basins tend to occur more gradually than in rifts. Also, as noted in chapter 2, many foreland basin "lakes" are actually broad shallow swamp systems, formed on overfilled depositional basins that are expressed topographically as ramps rather than as major depressions. These swamps are usually very shallow, with sedimentation derived from a mixture of peats from rotting vegetation and minor inputs of sand, silt, and clay.

Watershed Area and Sediment Supply

Foreland basin lakes tend to have larger watersheds than rift lakes, relative to lake basin size. Foreland basin lake margins rarely have the types of very short and steep feeder drainages that are so characteristic of rift escarpments. As a result, sediment supply in foreland basin lakes is generally greater than

in rifts. This fact, coupled with the observations of foreland basin shape, explains why the persistence of very deep topographical basins over long periods is much less likely in foreland basins than in rifts.

Water Depth and Facies Types

Because of their lower topographical relief, lake margin deposition in foreland basins is much more likely to be influenced by wave, current, and bioturbation-related processes than in rifts, and much less likely to be dominated by gravity flow deposition. Also as a result of low relief, these shallow-water processes in foreland lakes can operate over broad areas. A good example of the extent of wave-influenced deposition in foreland basins comes from the Douglas Creek Member of the Eocene Green River Formation, Utah (Castle, 1990) (figure 8.6). Green River deposits in the Uinta Basin formed in a large (80-km across), but shallow (20–30 m z_{max}) lake, surrounded by basement-cored uplifts. Lake margin sediments consist of about 100 m of alternating fluvial–deltaic and wave-dominated lacustrine shoreline deposits, which were derived from

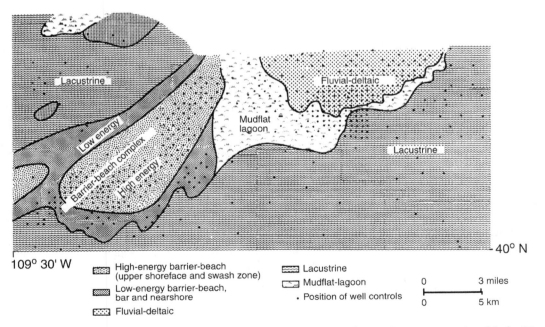

High-energy barrier-beach
(upper shoreface and swash zone)
Low-energy barrier-beach,
bar and nearshore
Fluvial-deltaic

Lacustrine
Mudflat-lagoon
• Position of well controls

0   3 miles

0   5 km

Figure 8.6. Paleoenvironmental map of nearshore deposits from the northeastern margin of Lake Uinta (Unit 1, Upper Douglas Creek Mbr., Lower Green River Formation, Eocene of northeastern Utah). A laterally extensive system of beach, mudflat, and deltaic deposits formed along this low-gradient lake margin, building out into a large but shallow lake. From Castle (1990).

the Uinta uplift, well to the north of the lake. Facies analyses indicate that the northern margin of this lake was rimmed by laterally extensive barrier beach systems and mudflats.

Sediment Supply and Climate

The high sediment supply available to foreland basin lakes allows them to become overfilled with sediment much more readily than rifts in humid climates, when long-term runoff and sediment discharge are high. The through-flowing swamps that form under these conditions in the more distal regions of foreland basins are represented in the stratigraphical record by peats, coals, palustrine limestones, fluvial deposits, and paleosols, with only rare indications of open-water conditions (figure 8.7). This is in strong contrast to rifts, whose more limited sediment supplies allow them to become deep lakes under high precipitation/evaporation conditions. The largest lake phases of foreland basins (as opposed to swamps) tend to occur when these lakes are saline, receiving less water but also less sediment input. (e.g., Bohacs et al., 2000). As in rifts, short-term climatically induced variability in foreland basin lake deposition is superimposed on long-term tectonic signals. This is expressed as cyclical deposits indicative of more saline versus less saline conditions, or lesser versus greater clastic sediment input (e.g., Drummond et al., 1996). In contrast, the complicating effects of climate on sediment supply in shallow lakes make lake-level indicators per se more difficult to interpret as paleoclimate signals in foreland basin lakes in comparison with topographically deeper rift lakes.

Glacial Lakes

The development of facies models for glacial lakes dates back to some of the earliest work in lake sedimentology (Jopling, 1975). The unique features of glacial lake deposits, particularly the association of unstratified or poorly stratified, coarse-grained tills and rhythmically bedded muds, were recognized early in the nineteenth century, well before their relationship with glacial activity was fully appreciated. Since the time of Louis Agassiz, geologists have recognized the potential for using the suite of characteristic glacial lake deposits to interpret climate change during the Pleistocene, and during the twentieth century this potential has been extended to pre-Quaternary glacial lake deposits as well.

Several features of glacial ice and its interaction with both lake water and lake sediments warrant our consideration. These include:

1. The potential for voluminous, seasonal discharge of meltwater, with high suspended sediment concentrations.
2. The potential for deposition of very coarse debris into mid-lake locations, through a variety of ice-related mechanisms, such as discharge from subglacial or englacial tunnels, calving of debris-laden glacial fronts, or melting of glacially-derived icebergs.
3. The potential deformation of lake deposits by glacial movement (glaciotectonics).
4. The potential for extremely rapid lake-level changes and catastrophic outbursts from the destruction of ice dams.
5. The potential for secondary lake-level fluctuations from postglacial isostatic rebound in lakes adjacent to the position of former ice sheets.

All of these factors are relevant to sedimentation and facies analysis in ice-contact lakes, those lying directly adjacent to glacial ice. From a sedimentological point of view the most important types of contact lakes are proglacial lakes, lying in front of an advancing or retreating glacier or ice sheet. Lakes of this type were extremely common along the margins of the major Pleistocene ice sheets. Noncontact, distal meltwater lakes are fed at least in part by glacial meltwater, but have no direct physical connection with glacial ice, and are therefore affected by fewer of the factors listed above. In cases where the proportion of the lake's water and sediment budget that is contributed by glacial runoff is relatively small, noncontact lakes may have few facies characteristics that distinguish them from other nonglacial lake types. In considering glacial lake facies models it is useful to keep in mind that many more studies of lake process sedimentology in modern distal meltwater lakes have been made than proglacial lakes, and recognize the potential bias this may create for the development of specifically "glacial" lake facies models.

Meltwater Discharge

Both seasonal and long-term melting of glacial ice release extremely large quantities of boulders, gravel, sand, and suspended silt, which can be discharged into a downstream lake. Deposition in both

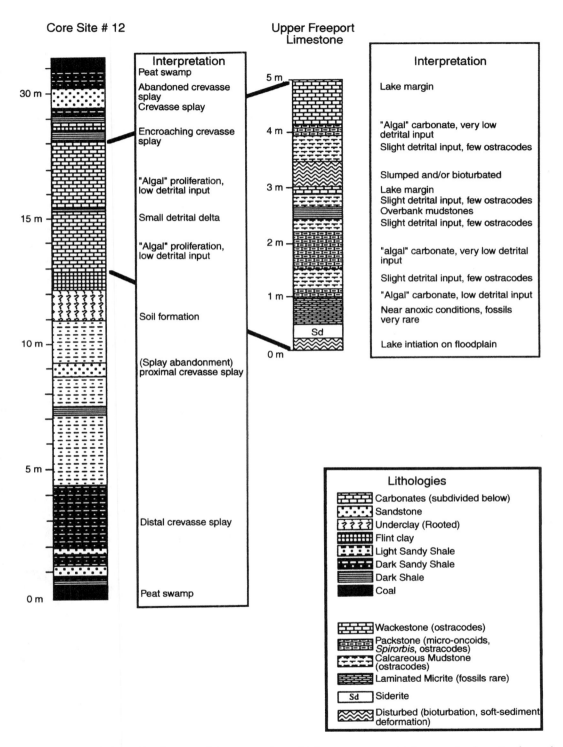

Figure 8.7. Lithological column and interpretation of lake, swamp, and lake margin deposits from the Allegheny Group (Mid-Pennsylvanian of western Pennsylvania). These short-lived lake and swamp deposits formed within the Appalachian foreland basin under relatively humid conditions. Details of the lacustrine Upper Freeport Limestone are shown on the right. From Weedman (1994).

the nearshore, deltaic, and offshore parts of a glacially fed lake all respond to variation in the discharge and inflow velocity of meltwater streams (Gustavson, 1975).

Meltwater discharge into glacial lakes has been investigated in a number of lakes through continuous monitoring of stream gauging stations and in lake sediment/water discharge. The latter is usually measured by vertical and lateral profiles of temperature, current flow, conductivity, and turbidity. Meltwater streams that are discharging directly from glacial ice frequently have suspended sediment concentrations in the range of $200-1000\,\mathrm{mg\,l^{-1}}$, sufficiently high to produce underflows where the stream enters a lake (Gustavson, 1975; N.D. Smith et al., 1982). Underflows are probably responsible for most sedimentation in proglacial lakes. However, where upstream ponds or outwash plains intercept the meltwater stream, much of the suspended sediment may settle out prior to its entry into a glacial lake, producing less concentrated interflows or overflows at the lake edge. Diurnal and seasonal temperature differences in meltwater also have strong effects on the relative density of the stream to the lake.

Both flow conditions and suspended sediment concentrations can change rapidly in meltwater streams, producing very dynamic interactions at the stream/lake interface. The strongest variation occurs seasonally. Very high sediment discharge occurs during spring and summer melting, and this translates into rapid deposition of coarser-grained particles or flocculated fines during the warmer times of the year. In contrast winter sediment discharge is normally much smaller. Under winter ice cover, deposition may be reduced to only the settling of fine-grained clay that still remains in suspension from the previous summer. Years of higher than average temperatures, with greater glacial melting normally yield greater amounts of sediment (figure 7.2) (Perkins and Sims, 1983; Hardy et al., 1996). The seasonal contrast between summer and winter discharge can be either accentuated or diminished, depending on differences in mean winter and summer temperatures (Pickrill and Irwin, 1983).

Diurnal variation in sediment discharge also occurs in meltwater, often producing very complicated inflow patterns into downstream lakes (Weirich, 1986). Inflows can change over time scales of hours or less between being overflows, interflows, or underflows, because of the complex interaction of diurnal temperature changes in the meltwater stream and sediment load. As meltwater warms above 4°C these two factors work in oppos-

ing directions with respect to fluid density causing very abrupt changes in flow conditions. The same conditions that apply to diurnal fluctuations in flow conditions can also operate over seasonal or multi-year time scales (e.g., Carmack et al., 1979; N.D. Smith et al., 1982). Alternations between overflow and interflow conditions can therefore produce rhythmites at various time scales, from daily to annual or even longer (P.G. Johnson, 1997).

Onshore/Offshore Depositional Trends Related to Meltwater Discharge

The temporal variability of glacial meltwater discharge is felt throughout the depositional system of a lake, from proximal deltaic environments to offshore muds (N.D. Smith and Ashley, 1985). Glacial lake deltas and beaches display relatively few features that clearly differentiate them from other deltas. Both beaches and the topsets of glacial lake deltas are typically gravelly. Delta topsets are often organized as bars, with imbricated pebbles or cobbles from high flow rates (e.g., D.G. Smith, 1994). The high proportion of bedload gravel and boulders that meltwater streams typically transport produces a steep upper delta front profile, the classic Gilbert delta (Bujalesky et al., 1997; D.G. Smith and Jol, 1997). The topset gravels usually occupy channels that erode into underlying upper delta foresets. Deltaic sedimentation in proglacial environments is frequently rhythmic, regardless of absolute grain size, reflecting their connection with the highly episodic flow of glacial meltwater discharge. Upper foreset deposits are commonly organized into thick (up to 1 m), rhythmically bedded (fining upward), sands and gravels. Lower down on the foreset surface these give way to finer, sandy and silty beds with lower dip angles (< 10°). The sands are commonly organized as climbing ripple drift stratification, reflecting very rapid piling of sediment. Silty layers often drape over these sands. Lower still, the deltaic rhythmites become even finer-grained, interbedded with ripple-bedded fine sands.

Such steep-fronted deltas are very prone to slope failure; slumps, compressional hummocks, and irregular lake-floor mounds are common in glacial lake deltaic deposits. The sand and gravel that dominates the proximal deposits of meltwater deltas are easily redistributed by coastal wave action, giving these deltas a relatively smooth shape in plan view, as opposed to the birdsfoot geometry more typical of muddier deltas. Thus, where these rivers prograde, they tend to do so over broader fronts, rather than as very discrete lobes. The degree to which glacial lake deltas can prograde is also strongly linked to the lake's origin and morphometry. Deeply excavated rock basins experience relatively slow rates of progradation, and meltwater streams entering some overdeepened glacial lake basins may not produce prograding deltas at all, instead feeding deep water sublacustrine fans. In contrast, meltwater streams entering shallow kettles can prograde quickly.

Suspended sediments are deposited in very different ways in glacial lakes, depending on three important factors:

1. Whether they are periodic flows (seasonal meltwater discharge) or aperiodic (turbidity currents from catastrophic melting or slumping).
2. If periodic, whether they enter the lake as overflows, interflows, or underflows.
3. Whether or not the lake becomes well stratified during the summer, generally a function of water depth and fetch.

Continuous underflows and aperiodic turbidity currents are strongly affected by lake-floor topography. In some ways the deposits of the two types of flows are quite similar. Both thicken toward their source, pond upstream from obstructions,

and infill local depressions (Ashley, 1975; Gustavson, 1975; N.D. Smith et al., 1982). Also, both types of current transport suspended materials onto the distal parts of deltas, and below the epilimnion both will produce rhythmites. Aperiodic, surging, turbidity currents are normally deposited in minutes to hours. In contrast, underflows are deposited over a protracted period. The high flow months for meltwater, generally the spring and summer, are the times of coarser sediment deposition, giving way to much finer particle settling during winter periods of ice cover.

Differentiating the deposits from these two types of flows is important for both paleolimnological reconstruction and geochronology by varve counting. Not surprisingly, considerable efforts have been made by several researchers in identifying key facies indicators to do so (N.D. Smith and Ashley, 1985; Teller, 1987) (figure 8.8). Although glacial lake rhythmites are often interpreted as annual events, numerous studies have shown that many glacial "varves" are in fact either aperiodic deposits, or deposits with other than annual periodicity (e.g., Lambert and Hsü, 1979; Pedersen and Noe-Nygaard, 1995). The most sensible approach seems to be to refer to deposits simply as rhythmites until either independent geochronology or detailed sedimentological studies demonstrate them to be annually deposited cycles.

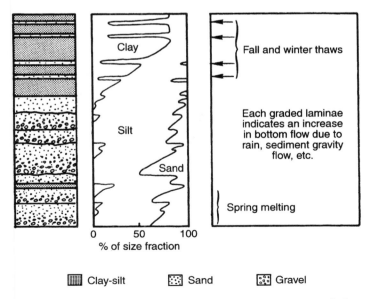

Figure 8.8. Schematic representation of grain-size variation in an annual varve cycle from a proglacial lake. From Teller (1987).

Because they constitute single events of waning flow, surge rhythmites generally display gradual grain size fining upward. Quantities of finer and coarser sediment in a turbidity flow are generally proportional to one another, resulting in a fairly constant ratio of silt to clay couplet thickness regardless of the total thickness of the rhythmite. Also, the rapidity with which the turbidite settles insures that it is very unlikely to be bioturbated along internal bedding surfaces. Annual rhythmites generally have sharp contacts between the coarse and fine parts of the couplet, reflecting the very different mechanisms of settling between summer and winter. On close inspection, the coarser silty layer may be made up of multiple fining-upward units, reflecting periods of greater and lesser warm season flow. Also, at any given site the thickness of silty layers tends to vary considerably from year to year, but winter clay settling and, therefore, thickness remains relatively constant. This reflects the fact that under summer stratified conditions clay is largely trapped within the epilimnion, mostly exiting the lake from overflows to the lake's spillway. Thus at the start of winter a fairly uniform concentration of clay will remain to settle under ice cover (M.D. Johnson et al., 1999). Also, because of the longer time available for bioturbation, burrows are common within single silt layers.

Seasonal overflows and interflows are less affected by bottom topography, but do show settling trends (thickness and grain size) driven by Coriolis circulation (N.D. Smith et al.; 1982). Coriolis and down-lake effects seem to be accentuated in deeper, summer-stratified lakes, rather than in shallow, unstratified lakes. Also, fines are more efficiently dispersed across the entire lake surface in stratified lakes. Both overflows and interflows tend to produce thinner deposits than underflows, consistent with their lower suspended sediment concentrations. In some lakes, where meltwater inflows have low suspended sediment concentrations, the deposits of overflows form thickly-bedded or even massive silts (Syverson, 1998). However, care must be taken when ascribing deposits to underflow or over/interflow origins based on obvious lamination or rhythmic bedding. Detailed petrographical studies of apparently "massive" clays sometimes show them to be subtly laminated, their structure obscured by internal load deformation produced by very rapid sedimentation (Van Der Meer and Warren, 1997).

In the more proximal parts of the prodelta environment, rhythmically bedded sands, silts, and clays display more complex facies characteristics, because of the greater mix of suspension and traction load

deposition (Lemoine and Teller, 1994). Sandy parts of these couplets are frequently ripple-laminated, reflecting moderate velocity traction currents active at the time of deposition. Where detailed chronologies have been determined, it can be shown that these rhythmites can be deposited very quickly; at a given location tens of centimeters can be deposited over a single year (Gustavson et al., 1975).

Controversy exists as to how much clay settling can occur during the period of the year when the lake is ice-free and turbulent. Fabric analyses of clay laminae from both laboratory settling studies and Pleistocene rhythmites show that clay particles are frequently aggregated, either forming during periods of high sediment concentration when particle interactions were most frequent, or by fecal pelletization (N.D. Smith and Syvitski, 1982; O'Brien and Pietraszek-Mattner, 1998). Textural gradations between the silty and clayey portions of some rhythmites suggest that clay settling may be derived from underflows during their waning stages of flow, as well as by settling from the epilimnion.

Deposition of Coarse-Grained Sediments in Glacial Lakes

Various types of sublacustrine fans form in association with meltwater discharge. Some are deep-water aggradational bodies, formed by the accumulation of sand or gravel at the base of steep slopes (Back et al., 1998). More unusual deposits form from outwash within or in front of ice tunnels, near the fronts of glaciers (Rust and Romanelli, 1975; Kaszycki, 1985; Clayton and Attig, 1989; McCabe and Cofaigh, 1994). Because of the difficulty inherent in studying processes of sedimentation within and below glaciers, almost everything we know about this type of deposition is inferred from ancient sediments. Where meltwater is discharged from ice tunnels the outflow is thought to form a powerful, turbulent jet, capable of producing stratified sands, gravels, and cobbles. These deposits are superficially similar to fluvial deposits; normally they are organized as cross-stratified gravel and sand bars, separated by numerous channels. In some cases these deposits develop as prograding foresets that are secondarily scoured or deformed by basal glacier movement. Their subglacial origin is usually inferred from their pattern of complex interbedding with fine-grained lacustrine sediments, and frequent evidence for ice-loading deformation.

Irregularly bedded lenses of very heterogeneous mixtures of particle sizes are commonly found interbedded with glacial lake deposits. These *diamicts* comprise various proportions of mud, sand, gravel, and cobbles, with larger particles imbedded in finer matrix. They are commonly thought to be unstratified, but most have at least some crude internal layering. The descriptive term diamict is preferable for sediments of unknown origin to the commonly used *till*, which is strictly speaking a genetic term describing sediments known to have been directly deposited by glacial ice, without being disaggregated or secondarily transported (Eyles et al., 1983).

Diamict facies are strongly related to the thermal conditions under which the glacier forms. In more temperate and/or humid settings *wet-based glaciers* occur, where ice is moving by gliding across the underlying substrate. This allows particles caught in between to be readily reoriented and/or carved into distinctive clast shapes. In this environment meltwater channels are common, resedimenting some particles through fluid flow. This is in distinction to the *dry-based glaciers* of arid-polar environments. Here ice is frozen to its rock base, and moves much more slowly through internal deformation alone.

When dry-based glaciers recede, generally through sublimation rather than through melting, they leave behind piles of internal debris, more or less in situ. Successive piles of higher and higher debris are lowered on top of one another, resulting in crude stratification. Reworking by englacial or subglacial water is, by definition, rare in these circumstances. Such diamicts are probably less common in the environments where proglacial lakes form than the wet-based glacial diamicts. The latter have more evidence for internal structure, suggesting some form of reorientation of clasts at the base of glaciers or fluid flow, and frequently are interbedded with cross-bedded or well-layered deposits indicative of channelized flow in englacial or subglacial tunnels. Deposits of this type are commonly interbedded with lacustrine rhythmites or turbidites, particularly at the bases of stratigraphical sequences formed during glacial recessions.

Some diamicts are produced by sediment gravity flows, showing internal structure, oriented pebble fabrics and soft sediment deformation consistent with downslope transport (Clark and Rudolf, 1990). Large debris flows can be triggered by oversteepening of delta or fan surfaces, or when bedrock surfaces are destabilized during extremely rapid lake-level rise, for example when glacial ice blocks a former lake outlet (Eyles, 1987). Other mechanisms for producing these textures include the accumulation of ice-rafted debris, as sediment-laden icebergs melt in the middle of a lake or as deposits from the soles of glaciers as they override a lake's floor.

Deformation of Lake Deposits by Glaciotectonic Movement

When glaciers advance over pre-existing lakes and lake beds they can create a variety of extraordinary sediment scour, resuspension, and deformation events. The magnitude of disturbance generally correlates with the rate at which the glacier is advancing; where glacial surging occurs the effects can be catastrophic. In a few instances the results of these events have been observed in modern proglacial lakes, where glacial surges at rates of > 10 m per day occur. This creates extremely deformed lake beds, causes very high sedimentation rates in front of the advancing ice, intensifies iceberg activity, and creates debris flows from sediment shed to the lake floor (N.D. Smith, 1990). Where Pleistocene proglacial lake beds are well exposed, evidence for glaciotectonics is frequently documented, especially the occurrence of lake-floor reverse faults and slumps (Jørgensen, 1982; Sadolin et al., 1997).

Effects of Extremely Rapid Lake-Level Fluctuations

Rapid lake-level fluctuations are a hallmark feature of proglacial lakes (figure 2.3). Since the nineteenth century histories of rapid lake-level fluctuations in Quaternary lakes of North America and Europe have been inferred from the positions and elevations of erosion surfaces, shoreline ridges, beaches, wave cut benches or cliffs, delta topsets, and river terraces (Taylor, 1990). At any given location rapidly changing lake levels were recognized by the occurrence of alternating sequences of glacially deposited stony diamict and finer-grained and uniformly bedded or laminated lake muds. A tradition of naming paleolakes based on periods of uniform outlet position developed out of these observations.

Proglacial lake-level fluctuations result from four interconnected causes (Hansel et al., 1985):

1. Advance/retreat of ice margins, blocking or opening outlets, and differentially downwarping the earth's surface.
2. Downcutting of outlets through unconsolidated tills or lake beds.
3. Major changes in the volume of water entering the lake.

4. Differential isostatic rebound. This last mechanism is notably slower than the other three in generating lake-level change, and produces a characteristic exponential slowing of lake-level rise as isostatic equilibrium is approached following deglaciation (Coakley and Karrow, 1994).

Since these factors operate together they lead to very complex lake-level histories, changes in the area and depocenters of lakes, and changes in the position of shoreline and deltaic deposits. This history, coupled with episodic erosion from subsequent glacial advances, accounts for the very discontinuous facies belts observed in many proglacial lake bed sequences. *Forced regressions*, where abrupt lake-level fall imposes basinward shifts in facies belts, are observed in the stratigraphy of many Late Pleistocene lakes (e.g., Pedersen and Noe-Nygaard, 1995). Lake-level fall also results in the development of unconformity surfaces. Unlike arid basin lakes, a proglacial lake can experience a major lake-level fall and continue to discharge water through an outlet. This allows major erosion surfaces to develop even in low basinal parts of the

lake floor, and may explain the presence of extensive erosion surfaces evident in seismic profiles from late glacial lake deposits of the Lake Huron Basin (Lewis et al., 1994). All of these events lead to very complex stratigraphical relationships. In many proglacial lake deposits it has been very difficult to disentangle which factors have been most important at any given time interval in causing lake levels to change rapidly. A number of controversies about the history of various lakes have arisen out of this confusion (Kehew, 1993).

The frequency of lake-level oscillations in proglacial lakes is in large part related to the size and shape of their glacial dams. Three types of ice-dams exist, each with a characteristic style of outflow and lake level stability (figure 8.9). Outlet flooding from ice-dammed lakes is predominantly subglacial or englacial, resulting in catastrophic flood events that might persist until the lake drained. However, the degree of drainage and flooding caused by these types of outflows can span a range of volumes (Clarke et al., 1984; V.R. Baker and Bunker, 1985). On the small end, outflow tunnels can anneal through collapse or ice movement, preventing the lake from draining entirely. At the opposite

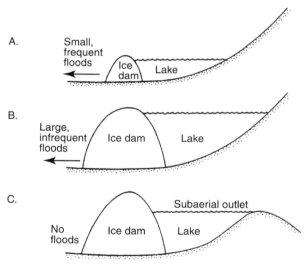

Figure 8.9. Schematic diagrams of ice-dammed lakes. (A) Low ice dams are self-dumping, releasing small floods and partly to entirely draining whenever water volume behind the dam rises sufficiently to create buoyant lifting of the dam and subglacial tunneling. These lakes dump frequently but produce small floods. (B) High ice dams are also self-dumping but impound deeper lakes. They dump infrequently but with larger magnitude lake-level fluctuations and produce larger downstream floods. (C) Subaerial outlet lakes form when the ice dams only raises the lake level to a secondary bedrock spillway, preventing subglacial dumping from occurring. These types of lake have relatively more stable lake levels, as long as the glacier continues to block the valley outlet. However, glacial recession will still cause the lake to drain catastrophically. From Atwater (1986), after Thorarinsson (1939).

end of the spectrum, an ice dam can rupture entirely, resulting in near total drainage of the lake (O'Connor and Baker, 1992).

Sedimentary consequences of rapid lake-level fluctuations come about both within the fluctuating lake and downstream, when catastrophic flooding occurs, depositing debris from upstream sources. A lake that lies downstream from a rapidly discharging proglacial lake will act as a pond for deposits from the upstream flood. By piecing together records of upstream lake-level fluctuations and downstream floodwater deposition, geomorphologists and paleolimnologists have attempted to reconstruct records of some of the major proglacial lakes of North America and Europe (e.g., Teller and Mahnic, 1988). Here again, controversy reigns. The origin of the spectacular deposits and erosional landforms in the Channeled Scablands region of Washington State has been a subject of research and controversy throughout the twentieth century (Bretz, 1923; V.R. Baker and Bunker, 1985) (figure 2.14). Although it seems firmly established that the primary cause of these features was from a cataclysmic outburst or outbursts from upstream proglacial lake sources, there has been an almost unending swirl of debate as to how many floods were involved and what their sources are. Most recent authors attribute the flooding to one or more outbursts of upstream Lake Missoula (Waitt, 1984; Atwater, 1986; V.R. Baker and Bunker, 1985; O'Connor and Baker, 1992; G.A. Smith, 1993). O'Connor and Baker (1992) estimated the peak discharge of the largest Lake Missoula outburst flood at $> 17 \pm 3$ million $m^3 \, sec^{-1}$, but there were probably numerous smaller flood episodes as well. Some of the sediment carried by these Lake Missoula floods was resedimented in downstream lake basins. However, there is uncertainty as to the sources of this sediment and the number of flood events that occurred (Waitt, 1984; Shaw et al., 1999). Downstream lake deposits comprise sequences of couplets (interpreted by some as varves), which are interrupted at regular intervals by graded beds. The graded units, interpreted as individual flood deposits, are on the order of 0.5–1 m thick, equal to tens of couplets, and fine upward from sand to clay. Atwater (1986) proposed that these deposits were emplaced by flows in which water movement forced the traction of sediment on the lake floor. The fact that the deposits are less frequently recorded at higher elevations along side arms of one of these downstream paleolakes was interpreted by Atwater to signal waning flow as the pulse of sediment moved upslope. The thickness of the graded units is proportional to the number of varves separating each graded bed from the next one underlying it. Atwater interpreted this to indicate that Lake Missoula was self-dumping about every 30–50 years.

Implicit in the "fixed number of floods" hypothesis are the assumptions that varved sequences can be correlated from one outcrop belt to another and that individual rhythmites correspond to single, discrete flood events. However, the spatial variability and lake-level fluctuations inherent in proglacial lakes makes their deposits very difficult to correlate. It is possible that rhythmite flood-deposit sequences resulted from transient pulses of floodwaters during single events, and that rhythmites observed in different areas formed at different times, or record multiple sources of flooding (V.R. Baker and Bunker, 1985; G.A. Smith, 1993). Most likely the history of outflow deposits from Lake Missoula will continue to be debated for years to come.

Temporal Change in Glacial Lake Deposits

Temporal trends in glacial lake sediment supply and deposition are driven on short time scales by seasonal melting, and on longer time scales by the advance and retreat of glaciers. On short time scales, this is responsible for the formation of rhythmic varves, and the cycles of changing mean annual temperature and glacial advance and retreat that are archived in varve thickness patterns. However, buffers that modulate intermediate to long-term signals of climate can obscure this record. At the decadal scale, upstream ponding of small lakes, for example in depressions formed by melting ice, can briefly interrupt sediment delivery to a lake that previously had been in the direct path of a meltwater stream.

At longer time scales, variation in sedimentation rates are influenced by the interaction of two factors:

1. Changes in sediment supply created by the erosive action of glacial ice; and
2. Changes in temporary storage of that sediment within the glacial system (Leonard, 1986, 1997).

As a result, changes in sedimentation rates are not simply a function of warm versus cold climate. On the scale of the entire Pleistocene deglaciation, or pre-Pleistocene deglaciation events, mean sedimentation rates seem to decline with warming temperatures, as unconsolidated glacial sediment supplies were exhausted (N.D. Smith and Ashley, 1985;

Sen and Banerji, 1991; Van Rensbergen et al., 1999). In the early phases of glacial retreat, however, sedimentation rates can be extraordinarily high, as large volumes of meltwater and sediment are focused into downstream lakes (Eyles et al., 1991).

The longest-term trends in glacial lake history are those related to deglaciation and glacial/interglacial cyclicity. Most of what we know about facies transitions at these time scales comes from observations of the transition out of the last glacial interval (figure 8.10). Evidence for multiple cycles of glaciation and deglaciation within single lake basins is relatively rare, for the simple reason that renewed glacial activity tends to destroy much of the evidence of intervening lake conditions!

Fining upward sequences are commonly produced in the deltaic deposits of proglacial lakes during deglaciation. The most straightforward explanation for this trend is a retreat of ice, the source of coarse sediment, from the delta margin (Shaw, 1975). In some cases changes in weathering regimes and increasing production of clays may also be responsible for this pattern.

Combined seismic and coring studies provide a three-dimensional view of proglacial lake facies geometry during deglaciation. Lake Annecy (France) is a glacial overdeepened lake (structurally controlled erosion) on the flank of the Outer Alps (Nicoud and Manault, 2001). Seismic surveys and coring at this lake have revealed a ~ 150 m thick section of late glacial and postglacial sediments.

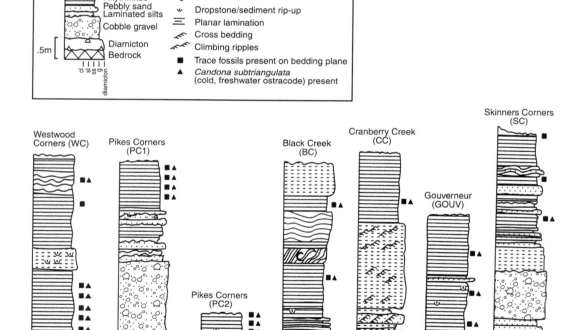

Figure 8.10. Stratigraphical sections from late Pleistocene–earliest Holocene deposits of the western Adirondack borderland, northern New York, illustrating typical deglaciation sequences in proglacial lakes. The most common transitions observed are (from older to younger): (1) diamicts and/or cobbly gravels with striated clasts (interpreted as subglacial accumulations), (2) better stratified sands and gravels, interpreted as proximal outwash sediments, and (3) organic-poor lake deposits, often rhythmites, with abundant dropstones, and laminated or unlaminated muds with higher organic content and more abundant fossils. From Pair et al. (1994).

These sediments comprise five depositional units that can be recognized in seismic facies and in cores, and that record the sedimentary response of this lake to deglaciation (Van Rensbergen et al., 1998; Beck et al., 2001) (figure 8.11).

Facies changes that accompany the transition from proglacial to postglacial conditions are partly driven by changes to warmer climate, longer ice-free intervals, and increased terrestrial input of organic matter. All of these changes promote increased primary productivity, reflected in increased fluxes of total organic carbon, authigenic calcite,

and biogenic silica (Willemse and Törnqvist, 1999). Lake morphometry plays an important role in determining both the rapidity and extent to which these changes occur. Indications of increased productivity following deglaciation are diachronous between small and large lakes, with small lakes more immediately responsive to warming conditions (e.g., Moscariello et al., 1998). Shallow ponds frequently display profound facies changes from proglacial inorganic silts and clays to postglacial gyttjas or peats (Zernitskaya, 1997). Similar diachroneity is observed in core records from

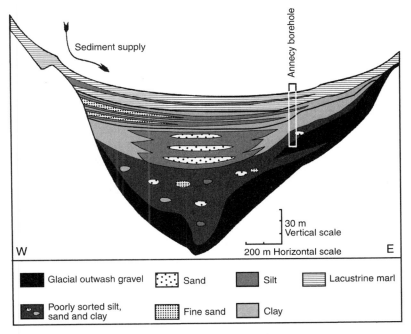

Figure 8.11. Generalized lithological cross-section of glacial and postglacial deposits of perialpine Lake Annecy, France, based on seismic and borehole data. From oldest to youngest, the deposits consist of: (1) Chaotic/irregularly stratified boulder and cobble deposits with irregular seismic geometry. These are interpreted as deposits from subglacial melting and development of subaqueous outwash fans. During this phase, glacial retreat had commenced and Lake Annecy began to expand. (2) Largely reflection-free unit, basin filling unit that ponds in the axial part of the lake basin. These are poorly sorted sand, silt, and clay units, interpreted to have been deposited as outwash by rapid meltwater discharge from glaciers at both axes of the basin. (3) Divergent reflectors that onlap the lake margin and pond in the axial deep areas. This seismic signature is associated with the appearance of rhythmites in cores. It is interpreted to represent the further retreat of glaciers and development of meltwater streams that were feeding underflow currents. (4) Progradational unit with reflectors diverging toward the northwest margin of the lake and downlapping toward the southeast. These deposits are well-bedded clay/silt rhythmites, that accumulated as a prograding, sublacustrine fan. Their thicker geometry along the northwest is assumed to result from Coriolis forcing of overflows or interflows from a major northern axial river system. Glaciers had largely retreated to beyond the watershed boundary by this time. (5) Low amplitude sheet drape of organic-rich calcareous marl. These thinner deposits are Holocene accumulations, formed under conditions of considerably warmer climate, increased terrestrial vegetative cover, and a change in depositional regime and productivity that allowed carbonates to accumulate. From Van Rensbergen et al. (1998).

lakes at varying altitudes, where the sedimentological effects of postglacial warming are first evidenced at lower elevations (Seltzer et al., 1995).

Peculiar effects on late glacial and postglacial lake stratigraphy are associated with the melting of "dead-ice" in kettle-hole lakes. In Denmark during the late glacial period numerous lakes formed behind stagnant ice or in kettle depressions produced by collapsing and melting ice (Brehmer, 1988; Sten et al., 1996). Large masses of dead ice dominated the landscapes of southern Scandinavia for up to 7000 years after the ice retreated. As a result, sedimentation across this surface was marked by very irregular patterns of infill. Because of the circuitous pattern of water flow around these ice obstacles and highly variable water velocities around remnant ice there is little in the way of systematic "proximal to distal" fining patterns in sedimentation away from the ice-sheet fronts. Gravel-filled ponds lie adjacent to mud-filled ones, in no particular order.

The melting of large ice blocks in periglacial environments often requires very long periods of time, particularly if they are partially or wholly thermally insulated by rock and debris cover. Thus the formation of a lake basin itself, as well as its initial stratigraphical infill of diamict debris and lake muds, may be greatly delayed after the retreat of regional glacial ice. Where volumes of ice are very large, for example in front of continental ice sheets, this lag time can exceed 1000 years (Warner et al., 1991). Melting also generates ongoing soft sediment deformation during the early phases of kettle-lake formation (e.g., P.G. Johnson, 1997).

Playas and Saline Lakes

The interaction of runoff, groundwater input, and climate conditions produces unusual suites of facies in saline lake environments. These lakes are most characteristic of arid and semiarid environments. Their deposits comprise not only evaporite minerals but also a range of clastic sediment types, carbonates, and other authigenic minerals that accumulate under arid conditions. A smaller number of saline lakes occur in semihumid coastal environments, where saline ocean-fed groundwater percolates through coastal ridges and dunes, and some are also found in areas of high but extremely seasonal rainfall.

The terms saline playa, salina, continental sabkha, and salt pan are often used interchangeably; all refer to basins that are intermittently filled by evaporating, concentrated brines. Most saline lakes and playas occupy topographically closed basins (Hardie et al., 1978). These are commonly, though by no means always, fault bounded; in some parts of the world, such as Australia and Canada, other mechanisms for creating topographical closure are more important than local faulting. A much smaller number of saline lakes are open basins, where the salinity is derived from groundwater discharge rather than evaporation of lake water.

The permanence or intermittence of inflow determines whether a lake will "permanently" hold water, or will only fill seasonally. Sources of inflow can be either surface runoff or groundwater, and the importance of the latter cannot be overemphasized in these systems, particularly for smaller saline lakes. Most playas and saline lakes act as areas of groundwater discharge, and are frequently surrounded by numerous springs and spring deposits. A smaller number, for example, on the High Plains of western Texas, United States, are recharge playas, filling with water seasonally when runoff is high, but emptying during dry seasons. In contrast to discharge playas, recharge playas remain fresh throughout their seasonal existence, as water is lost primarily through subsurface flow rather than evaporation.

A typical closed-basin, saline lake occupying a fault-bounded basin displays a wide range of facies and environments between the basin margin and center (figure 8.12). Saline lakes and intermittent playas are commonly surrounded by a variety of peripheral deposits, especially coarse-grained alluvial fans and sand flats, the latter being aprons of fine sand. Although these deposits are not strictly part of the lake or playa, they form sources from which surficial sediment, soil carbonates, or gypsum, surface crusts and cements can be derived and transported into the lake basin. These sediments are frequently interbedded with more basinal lake deposits. In coastal or deflation basin lakes, surrounding or interbedded sediments are generally derived from aeolian dunes.

Alluvial fans, sand flats, and sometimes more basinal parts of a salt lake system are common areas for spring discharge. These springs are of significance for lake sedimentologists because they create distinctive sediment facies types, and are a major source of water and sediment for many salt lakes (figure 8.5). Where spring discharge has high concentrations of dissolved CO_2, the loss of pressure at the earth's surface causes rapid outgassing and calcium carbonate precipitation of spring-mound tufas. Cemented pavements, partially bro-

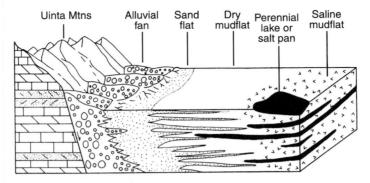

Figure 8.12. Schematic drawing showing the distribution of subenvironments in a saline lake, modeled after the Wilkins Peak Member of the Eocene Green River Formation. From Smoot (1978).

ken by crystallization and upward flow of groundwater, are also common at discharge interfaces (Warren, 1982). Small lakes or ponds that form sites of preliminary solute concentration prior to runoff into the main playa or lake basin commonly surround spring mounds.

The outer margin of a playa or saline lake is commonly rimmed by an area of seasonal mud accumulation, formed in pools of standing and drying water (Smoot, 1978; Renaut et al., 1994). On the outer edges of a playa, these *mudflat* environments are generally dry below the surface during dry seasons or periods of low surface runoff; the most important defining feature of the dry mudflat environment is the fact that its subsurface sediments are not permanently saturated with brines. Dry mudflats are flooded sporadically by surface runoff and then dry through a combination of evaporation and runoff to topographically lower areas. Deposition during these flooding events consists of millimeter-scale, graded silt and clay laminations from decelerating sheet wash and settling in temporary ponds. Relatively insoluble micritic muds are also deposited here. The low topographical gradient across these flats precludes the deposition of most coarser, eroded material; larger fragments are dominantly intraclasts, formed in situ from crusts or cracked surfaces. However channels often dissect the surfaces of playas, particularly near their outer edges, where coarser sand accumulates. During dry intervals, mudflat surfaces are covered by abundant desiccation features; polygonal cracks, crusts, and mud chips.

A transition from dry to saline mudflats occurs where more or less permanent groundwater brines saturate the sediment directly below the surface. Groundwater brine recharge allows crystals to continue to grow below the surface, forming crystal clusters or nodules. This is frequently accompanied by a compositional gradient across mudflats from purely siliciclastic to mixed siliciclastic and carbonate/evaporite muds to almost pure carbonate/evaporite muds, as well as subsurface microbial gradients, affecting redox conditions.

Seasonally wet playas often develop springs at the intersection of relatively dense lake brines and lighter groundwater discharge. This *Ghyben–Herzberg Interface* forms an important area for evaporative pumping, the deposition of various metals from solution, the formation of stromatolites, and diagenetic minerals (figure 8.13) (De Deckker, 1988b). Some characteristic minerals of saline lakes or playas only form where interfaces of very different composition fluids exist. Gaylussite, for example, requires the interaction of previously formed calcite (supplied by more dilute lake margin groundwaters) with Na and CO_3-rich brines:

$$CaCO_3 + 2Na^+ + CO_3^{2-} + 5H_2O \rightarrow \text{gaylussite}$$
$$(Na_2CO_3 \cdot CaCO_3 \cdot 5H_2O) \qquad (8.1)$$

Thick, efflorescent crusts commonly cover the surface deposits of wet mudflats. These crusts are formed by complete drying of shallow groundwater brine films, or by surficial movement of mineral dust by wind followed by dissolution and reprecipitation during rainfall (Eugster and Hardie, 1978; Hardie et al., 1978; Smoot and Castens-Seidell, 1994) (figure 8.14). These crusts can accumulate up to a meter thick, and are composed of a fine crystalline mass of a single or few minerals (for example, halite or trona). After several years of accumulation they can become very hard from repeated episodes of crystal growth and cementation. Polygonal cracking also often breaks the

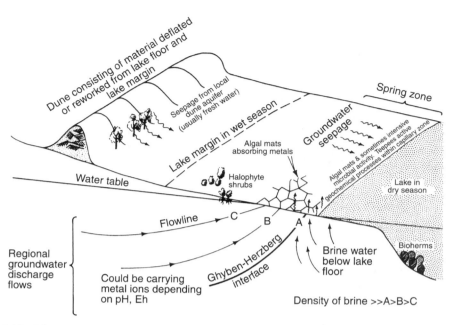

Figure 8.13. Diagram representing a typical saline lake affected by the brine-water pool below it. In this particular case, a Ghyben–Herzberg interface operates between regional groundwater and the brine pool, thus forcing both waters to converge toward the lake surface near the lake shore, in an area characterized by seepage, springs, and the growth of algal mats. From De Deckker (1988b).

Figure 8.14. Polygonally cracked halite deposits from a modern playa system, the Salar de Uyuni, Bolivia.

crust surfaces from repeated crystal growth and displacement. Internally they have a distinctive type of irregular layering and vertical tubes or cavities, which are caused by solution and collapse during the continual fluctuations in the position of the water table with respect to the playa surface. The facies transition from dry to wet mudflats is commonly marked by a change toward more massive or thickly bedded mixed salt and mud beds, and fewer well-laminated muds and silts.

A gradational transition toward longer periods of standing or subsurface water exists moving down toward the center of a playa or saline lake system. A downslope increase in evaporite deposition and decrease in siliciclastic deposits and mudchips normally accompanies the change from mudflat to longer-lasting saline lake environments (Schubel and Lowenstein, 1997). On the topographically higher areas, where efflorescent crusts are forming, any secondary wetting, for example from dilute rainfall, will tend to redissolve the most soluble minerals. These solutes are then carried downslope to topographical lows and reprecipitate. This progressive accumulation of more concentrated and evolved brines toward the center leads to a strong geochemical zonation on the playa or lake surface, and this is reflected in evaporite facies zonation.

In closed basins with a strong perennial flow of surface or groundwater perennial lakes may exist. Lakes that are only fed year-round by groundwater are generally smaller, often along major faults, or in coastal zones, where marine-derived aquifers form a large, local source of water (e.g., Warren, 1982). Larger saline lakes can have the same suite of shoreline depositional environments that occur in nonsaline lakes. However, the composition of sedimentary particles in these saline lake deposits frequently will include evaporites or carbonates as well as siliciclastic sediments (Eardley, 1938; Neev and Emery, 1967). Depending on salinity and alkalinity levels, macrophytes may be able to grow in these settings, affecting both the bioturbation of sediments and the accumulation of coarse-grained organic sediments. Stromatolites are also common features of coastal saline lake environments. Coarser-grained evaporite particles (individual crystals or aggregates and coated grains) often accumulate near the shoreline and may show signs of current or wave reworking. In older lithified lake deposits however these structures are easily obliterated by subsequent crystal dissolution and precipitation (Rosen and Warren, 1990; Salvany et al., 1994).

Finely laminated muds are the most pervasive characteristic of offshore zones of shallow saline lakes (e.g., Last and Vance, 1997). These laminites are often organic-rich. Occasionally current or wave-transported deposits, such as turbidites or intraclasts from the lake margin, interrupt these laminites.

Sometimes significant wind-derived sediment (clastic or evaporite sand horizons) is also blown into the central lake basin, identifiable by its good size sorting (e.g., Magee, 1991). In some shallow saline lakes coated grains or fecal pellet deposits are also common in offshore settings. Also in shallow and/or less saline lakes, muds may be secondarily bioturbated or wave-mixed into more thickly bedded muds. The diversity of fossil species is almost invariably low in saline lakes or playas. However there are often very large numbers of one or a few fossil species (for example brine shrimp eggs or fecal pellets, halotolerant ostracodes, terrestrial insects, halophytes such as *Ruppia*, charophytes, and root traces on margins). In some nonalkaline, saline lakes, mollusks may also be common (e.g., Truc, 1978; De Deckker, 1988a). At higher salinities, or in progressively deeper water, an elimination of benthic invertebrates results in a reduction in bioturbation and the preservation of laminites and horizontally oriented clay particles (Bowler and Teller, 1986).

Mineral Composition and Crystallization in Playas and Shallow Saline Lakes

Which evaporite minerals will form in a saline lake depends on original dilute inflow composition (often from multiple and different sources), extent of secondary fractionation during evaporite crystallization, and diagenetic interactions. As a brine evolves, progressively longer residence times provide the opportunity for both more concentrated and more variable fluids to form in different parts of the lake or playa system. Different evaporite minerals require particular conditions for formation (Eugster and Hardie, 1978). Halite can form simply from the removal of water, whereas Na-carbonate formation is sensitive to pCO_2 and temperature, with different mineral assemblages forming under different conditions. Some common evaporite minerals, for example gaylussite, only seem to form diagenetically, through interaction of brine and previously formed evaporites, or between both of these and surficial fresher water. Because of the variability in all of these processes, it is not uncommon to find ten or more discrete mineral phases within a single lake system. Suites of evaporite minerals that are found within a single playa or saline lake primarily reflect the average composition of the dilute inflows, so that certain minerals predictably co-occur or are generally mutually exclusive (Eugster and Hardie, 1978). For example a variety of Na-carbonates are normally found together in a single lake, precipitating from alkaline brines or forming diagenetically from alkaline brine/freshwater/organic material interactions. In contrast, Na-carbonates are rarely found with Ca-sulfates, since the brine pathways on which they form are quite distinct. Where major changes in evaporite mineral assemblages do occur within a single lake, in either space or time, they usually reflect very different dilute, inflow sources (e.g., figure 8.15).

Evaporite crystal morphology provides clues about precise conditions of formation. Crystal morphology is influenced by a wide variety of factors such as nucleation rate, pH, and other aspects of water chemistry, degree of supersaturation, presence of organic matter, temperature, and growth rate (Magee, 1991). For example, gypsum forming from a pure supersaturated, aqueous solution forms needle-like prismatic crystals. This might occur when crystals are forming within the water column. Very rapid growth promotes a stubbier shape to these prismatic crystals and their eventual replacement by lenticular growth forms, whereas low temperatures counter this trend. When other ions such

as Na^+, or organic matter are present in large quantities, they tend to complex on the crystal faces, interfering with growth, causing the growth of more pyramidal or disc-shaped crystals, flattened in the long (c-axis) direction. Gypsum crystals growing within muds, from the evaporation of capillary groundwater, are normally of these shapes.

In older, lithified lake sediments, highly soluble evaporite minerals are often gone, complicating the interpretation of paleobrine pathways or mineral zonation. If the original mineral had a distinctive shape or habit, and that shape is retained by a replacement mineral, a *pseudomorph*, it may however be possible to infer from the replaced mineral's chemistry the original composition of the playa waters.

In addition to evaporite minerals, numerous authigenic Na- or K-silicate minerals, such as clays, zeolites, K-feldspar, and chert, are also commonly found in saline lakes. These minerals form through interactions between brines and pre-existing silicates, such as volcanic glass, clays, or diatoms. In order for silicates to be highly reactive a combination of high pH and fine grain size (large surface area for reaction) is required (Hay, 1966; Sheppard and Gude, 1968). Clay minerals form authigenically in saline lakes from the alteration of detrital siliciclastic precursors, for example other clays:

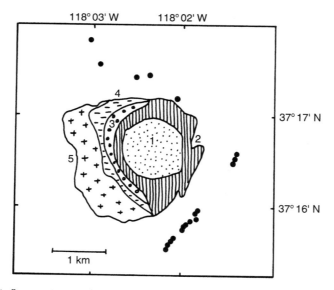

Figure 8.15. Major inflow springs and mineral zonation of the Deep Springs playa. Circles represent spring sources: (1) burkeite; (2) thenardite; (3) gaylussite; (4) dolomite; (5) calcite and/or aragonite. From Eugster and Hardie (1978), after B.F. Jones (1965).

Detrital Al smectite + Mg + SiO_2 + Na →

$$MgAl \; smectite \qquad (8.2)$$

Zeolites can be thought of as the chemical equivalent of hydrated feldspars; their open framework structure gives them substantial capacity for ionic exchange, particularly for Ca, K, and Na. This most commonly occurs from the reaction of tephras with alkaline brines. Some studies have suggested that direct precipitation of zeolites can occur from lake water, particularly where zeolite beds are thick and no obvious source of readily dissolved silica is apparent (Hay, 1966; Gall and Hyde, 1989). In most saline lakes, however, this mechanism is precluded by low solubility of aluminum. Zeolite transformation from tephra can occur over geologically short intervals in alkaline brines (< 10^3 years) (Renaut, 1993). Clay alteration can also be a source of zeolite (analcime) formation.

Other silicate minerals can also form from interaction of Na-rich alkaline brines and tephras (Eugster, 1980). Magadiite [$NaSi_7O_{13}(OH)_3$ $3H_2O$] is a precursor mineral for chert, originally described from alkaline Lake Magadi, Kenya (Eugster, 1967). Beds of magadiite can be traced into chert units over short distances, supporting the intermediate role of this mineral in bedded lacustrine chert formation. This may happen by leaching of sodium by dilute runoff (at the lake margin) or by spontaneous crystallization of quartz from magadiite (Hay, 1968).

The facies changes that occur between the topographically higher and lower parts of a playa or shallow saline lake basin often produce a characteristic *bull's-eye* depositional pattern. This pattern reflects the increasing concentrations and changing composition of surface and groundwater brines, and the greater distance from siliciclastic sediment sources (figure 8.15). Sometimes this zonation is expressed primarily through evaporite minerals, but in alkaline lakes, where fine-grained silica sources are readily available, it may also involve zeolites or other silicates (e.g., Sheppard and Gude, 1968). Alteration zones are only rarely perfectly concentric, because of variations in local conditions. The most common deviations from the bull's eye pattern occur where larger influent streams enter a lake, producing a cone of fresher water, siliciclastic, and biogenic (macrophytes and carbonates) sedimentation extending far into the basin. Large spring mounds emanating from the lake floor, or variation in detrital mineral composition can also cause the bull's eye pattern to break down.

Vertical Facies Transitions in Saline Lake Deposits

Vertical facies changes in saline lake deposits reflect both seasonal change or saline lake expansion and contraction over time, the latter connected to such factors as long-term climate change, and for coastal lakes, changes in relative sea level. Seasonal cycles in deposition are common in saline lakes from areas of highly continental climates, with high summer temperatures and very low winter temperatures. This type of seasonality is favorable for the alternating precipitation of evaporites created by evaporative concentration of some salts in the summer and the freeze out precipitation of others in winter (Renaut and Stead, 1994).

Coastal lakes in many arid and semiarid parts of the world have been infilled by regressive evaporite sequences over the past 6000 years, since the stabilization of eustatic sea level in the Mid-Holocene. (e.g., Warren, 1994). This produces characteristic sequences of sea water-related evaporites as the lake infills and dries, sometimes including dolomites. Recharge playas frequently display patterns of cyclical sedimentation, with lacustrine muds, eolian sands, and paleosols reflecting alternating wet–dry episodes and the rise and fall of the water table (Hovorka, 1997). The basin floors in playas are frequently deflated by low-stand wind erosion, and remaining lake-floor sediments simultaneously subject to pedogenic alteration, providing further evidence of arid climate or low water table conditions.

Variability in both the location and type of hydrothermal sedimentation is a common pattern in the cyclical histories of salt lakes. Commonly this is linked to climate cyclicity (figure 8.16). For example, in alkaline lakes silica precipitation would be favored during arid intervals, when groundwater pH is higher, silica is more soluble, and calcium is depleted. Conversely, during periods of more humid climate, influent hydrothermal fluids will be more dilute, with lower pH and higher Ca^{2+} concentrations.

In some watersheds, chains of lakes are connected during humid periods when "upstream" lakes overflow into topographically lower basins. At such times the upstream lakes remain relatively fresh, whereas during more arid intervals one or more of the lakes may close, turning it and its downstream successors into closed-basin saline lakes or playas. Such lake chains are particularly informative to paleolimnologists because they illustrate the different responses and precipitation

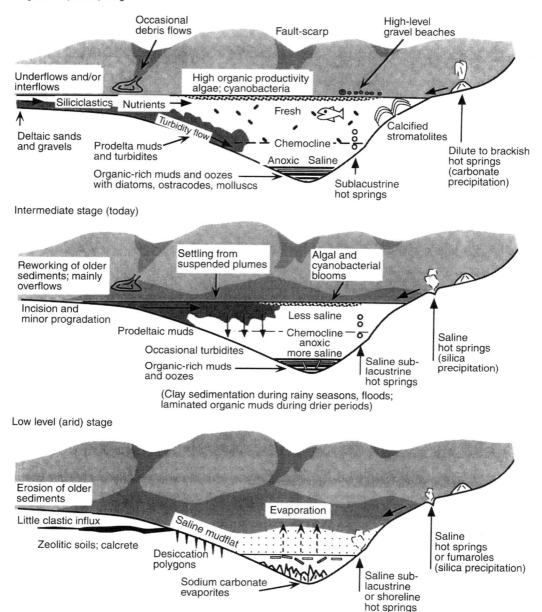

Figure 8.16. Depositional model for saline Lake Bogoria, Kenya, showing the main changes in the style of sedimentation according to prevailing climate and lake level. From Renaut and Tiercelin (1994).

thresholds that can exist from one lake to its downstream neighbor. The best-studied examples of these types of lake chains are the United States Basin and Range lakes between present-day Mono Lake and Death Valley (figure 8.17). Long core records from a number of these lakes provide a Late Pliocene–Recent record of lake level and overflow events, based on carbonate and evaporite mineralogy. Cyclical aspects of these sedimentation records are driven by glaciation and runoff from the

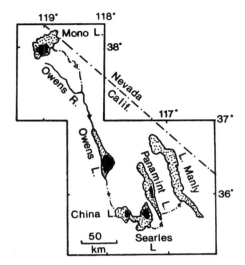

Figure 8.17. Chain of lakes in southeastern California that show linked but variable responses to climatic cycles. These lakes have been the source of a number of important paleolimnological records from long cores. Dots show the extent of lakes at high stands, and dashed lines paleodrainage connections, but all of the lakes are closed basins today. From Eugster and Hardie (1978).

Sierra Nevada, possibly superimposed on a longer-term aridification related to a developing rain shadow behind the uplifting mountains.

Owens Lake (the first lake below the uppermost Mono Lake basin) did not experience evaporite deposition over the past 800 ka, but has undergone cyclical deposition of $CaCO_3$ deposition during closed-basin conditions, alternating with siliciclastic deposition during pulses of glaciation and increased runoff in the nearby Sierra Nevada (Bischoff et al., 1997a). At downstream Searles Lake, alternating mud (with calcite, aragonite, and dolomite) and evaporite (halite, plus various Na-carbonate, sulfate, and borate minerals) deposition resulted from alternating wet–dry climate conditions (G.I. Smith et al., 1980). Here, in contrast to Owens Lake, upstream preconcentration insured that carbonate precipitation only occurs during the relatively wet phases. The lowest basin in the lake chain, Death Valley, California (Lake Manly), never overflowed. Its cyclical fluctuations in evaporite textures and composition reflect an alternation between mudflat and ephemeral lake deposits (i.e., variable water table elevation), with perennial salt lake conditions only developing during the wettest episodes (figure 8.18) (J. Li et al., 1996, 1997; Lowenstein et al., 1999).

Pulses of tectonism can also drive facies cycles in evaporites. This type of evaporite facies geometry can be visualized from the stratigraphy of the western Ebro Basin, Spain (figure 8.19). A large saline lake/playa system developed in this foreland basin foredeep during the Oligocene and Miocene (Anadón, 1994; Salvany and Orti, 1994). Vertical cycles of evaporites to siliciclastic mudstones in this case reflect progradation of marginal alluvial fans onto the playa surface. Over time the basin depocenter migrated toward the northwest, reflected in the progressive offset of the position of lake expansion.

Deep-Water Evaporites

In hypersaline lakes direct precipitation of evaporite minerals can occur in open-water conditions. Much of our understanding of this process comes from studies of the Dead Sea and smaller Canadian prairie lakes. In the Dead Sea, water is saturated to slightly supersaturated with respect to NaCl, and halite has been precipitating in open waters of the lake since the 1970s (Gavrieli, 1997; Stiller et al., 1997). The current phase of halite precipitation is driven by two primary factors. Industrially evaporated brines are returned to the lake today supersaturated with respect to halite. Until the late 1970s, the lake was meromictic and surface waters somewhat more dilute, which prevented surface water halite precipitation. However, long-term water diversion from the lake's tributary, the Jordan River, resulted in lowered lake levels, eventually allowing the lake to overturn and surficial

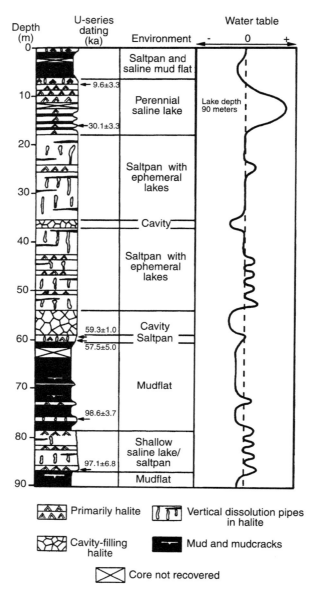

Figure 8.18. Stratigraphical column for a core from Death Valley, California, showing lithofacies, age control, and interpreted water table fluctuations for the past 100 ka. From Li et al. (1996).

halite crystallization to occur. This type of periodic overturn event is probably very important in the long-term brine evolution of deep-water saline lakes. In the absence of mixing salt accumulation in deep water is probably slowed, except in those cases where groundwater input accounts for the bulk of the solute inflow. Therefore, overturn events are required to introduce more dissolved solids from influent sources into the evolving deep-water brine, with each overturn event creating the possibility of raising the salinity of the hypolimnion (Anati et al., 1987). Rates of documented halite precipitation in the Dead Sea can be very fast; during the 1980s the sedimentation rate of halite was 3–6 cm yr^{-1}. Dead Sea halite forms both coarse (> 2 mm) and fine (0.1–0.4 mm) crys-

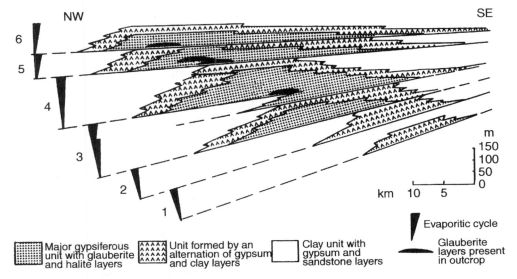

Figure 8.19. Stratigraphy and facies geometry of the Lerín Gypsum Formation, Ebro Basin, Spain, subdivided into evaporite–detrital cycles. Dashed lines represent stratigraphical sequence boundaries, followed by maximum saline lake expansion, and then general progradation of alluvial fan deposits (clay units) inward from the basin margins. Cyclicity and westward facies migration is regulated by the tectonic development of this part of the foreland basin. From Salvany and Orti (1994).

tals; there is no evident correlation of crystal size with depth. Precipitation events appear to be triggered by periodic cooling of the epilimnion, when halite solubility declines. Halite crystals form both as floating particles at the surface in calm weather, and through spontaneous crystallization on the lake floor.

At mid–high latitudes, evaporite precipitation can also occur seasonally from the reduced solubilities of some minerals at lower temperatures. This so-called *freeze-out* effect can result in rapid deposition of various Na and Na + Mg sulfate salts in the deeper parts of the lake basins during the wintertime (Last, 1994).

The monimolimnion of a "permanently" stratified saline lake might seem like a place where sedimentation would be extremely uniform. Detailed studies of such lakes, however, have shown that evaporite facies in deep-water environments can be quite variable (Valero-Garcés and Kelts, 1995). When precipitation is seasonal or episodic, deep-water evaporites are commonly laminated or thin-bedded, often alternating with carbonate or clastic muds. In the Dead Sea, sediments are often laminated, consisting of mud/evaporite couplets, although these laminations appear to represent episodic pulses of clastic sediment input punctuating a background of evaporite precipitation, rather than annual cycles. Deep-water deposits often contain thicker beds as well, derived from reworked shallow-water evaporites and carbonates.

Crater Lakes

Facies characteristics of maars and caldera lakes are defined primarily by the mode of origin and geomorphology of these basins. Crater lakes almost invariably have relatively small watersheds, measured in square kilometers or tens of square kilometers. It is this feature that makes them instantly attractive to certain types of paleolimnological investigations, because:

1. They do not integrate environmental (especially climatic) archives from a large catchment (Negendanck and Zolitschka, 1993a); and
2. They normally have a very limited range of lithologies within their watersheds.

However, these generalizations mask a considerable variation in size and complexity that exists among crater lakes. Very small explosion maars have relatively simple basinal morphologies (figure 2.19). The explosive maar formation process occurs over a period of weeks to months in single eruption maars (Wood, 1980). The effects of subsequent volcanism on lakes that form in these craters is primarily limited to the deposition of tephras from other adjacent maar explosions within the local or regional volcanic field. Posteruptive sedimentation in these types of monogenetic craters is controlled almost exclusively by the external forcing factors of climate, and for recent centuries, human activity (Bonifay, 1991; Bonifay and Truze, 1991; Negendanck and Zolitschka, 1993b).

Controls on sedimentation in larger calderas is considerably more complex. These systems typically experience multiple episodes of volcanism, faulting, and resurgence of the lake floor and related deformation, in addition to extrinsic climatic controls. Secondary eruptions and faulting create complex bathymetry, forming moats and isolated basins, which focuses sedimentation. Ongoing hydrothermal activity is a common feature of calderas, affecting lacustrine sedimentation long after an eruptive cycle ends. Although caldera lakes are still relatively small compared with many glacial and tectonic lake types, they are normally larger than maars, and can integrate environmental and climate signals from a broader area. Furthermore, many calderas are not closed systems hydrologically or sedimentologically, and may be integrated into larger drainage systems by regional faulting, allowing for more complex sedimentation histories.

Deep-water depositional processes, such as suspension settling, gravity flow deposition, and slumping are dominant modes of lake-margin sedimentation in crater lakes. The steep flanks and small watersheds of both small and large craters limit the accumulation of sediment at the lake margin. This is particularly noticeable in humid climate conditions, where lakes are deep. At low lake stands, littoral deposits are better developed, especially when the lake is reduced to only a portion of the relatively flat-lying basin floor. The watershed morphology and area of craters also tends to limit sedimentation rates after eruptions cease and easily eroded debris is washed into the lake; sedimentation rates of 0.1–2 mm yr^{-1} are typical for both large and small crater lakes.

Sedimentation in Small Maar Lakes

Maar sedimentation is strongly influenced by external climate variation, through its influence on mixing regimes, organic productivity, and dust influx. Following an initial influx of explosion-related rubble, the contribution of volcanoclastic debris is usually fairly constant and low, although slumping of crater wall debris may continue for some time. Some structural controls on sedimentation are expressed in maars, particularly in determining the locations of depocenters (Büchel and Pirrung, 1993). However these volcanotectonic influences are minor in comparison with their effects on larger caldera sedimentation.

The steep walls and small size of small crater lakes make them highly susceptible to permanent or semipermanent stratification. This tendency toward oligomixis or meromixis is accentuated in humid climates, where craters are filled more deeply, in warm climates lacking seasonal overturn, and/or in areas receiving heavy anthropogenic loading of nutrients. In arid or semiarid climates, maar sedimentation is dominated by the development of soils and playas, with mudflat, carbonate, evaporite, and eolian deposits surrounded by alluvial aprons from maar wall debris (J.D.L. White, 1990; Narcisi and Anselmi, 1998).

Maar sedimentation typically consists of relatively simple accumulations of four facies types:

1. Narrow aprons (typically < 100 m wide) of coarse rubble from crater weathering (Büchel and Pirrung, 1993);
2. Siliciclastic and volcanoclastic turbidites, derived from the margins of the crater;
3. Episodically occurring volcanic ash horizons (tephras); and
4. Fine-grained gyttjas and sapropels, consisting of diatomaceous, siliciclastic, organic, and carbonate debris, which are often annually laminated.

Because of their small size and closed drainages, the posteruptive histories of maar lakes are often strongly modulated by climatic forcing. For example, in European maar lakes, sedimentary infill during the colder Late Pleistocene was dominated by fine-grained turbidites, deposited under holomictic conditions. These lakes became meromictic during the warmer Holocene, generating the deposition of annually laminated sediments, and in a number of lakes, the development of siderite varves. In some of these lakes, such as Holzmaar (Germany), varve sequences are remarkably continuous, and can be counted for geochronometry back to 13,000 yr BP (Zolitschka, 1991). These long varved sequences display cyclical variations in the thickness of individual varves, which is thought to be under the control of various climate cycles. However, the controls are complex, and undoubtedly several different cyclical processes are driving their accumulation rates. Longer records of maar deposition from non-glaciated terrains in southern Europe also display cyclical sedimentation, probably driven by pulses of siliciclastic dust deposition during cold and arid intervals (Zolitschka and Negendanck, 1993, 1996; Narcisi and Anselmi, 1998) (figure 8.20).

Sedimentation in Larger Calderas

Facies patterns in large calderas are more complex than in simple craters primarily because of the more complex topography of their faulted and domed lake floors (Nelson et al., 1994). As with maars, sedimentation on the margins of caldera lakes is dominated by volcanoclastic debris. In the case of calderas, this often consists of debris avalanches derived from the rubbly walls of the caldera at the base of the steep caldera wall slopes. These are often interbedded with turbidites. In areas that are proximal to the crater wall turbidites are thick, sandy, and gravelly graded beds, whereas on the basin plain they are thinner and muddier. Diatomaceous oozes and other pelagic sediments accumulate very slowly in areas removed from terrigenous input. Secondary volcanic features of the lake floor, such as lava domes and flows, form after

the collapse of large calderas. These structures isolate sub-basins, and create structural highs and platforms that are isolated from terrigenous sediments derived from the watershed. (figures 8.21 and 8.22). They also cause very complex sediment focusing patterns to occur on the lake floor. Hydrothermal alteration is commonly observed in caldera lake sediments, manifest through the formation of hydrothermal deposits around ring fractures and faults (e.g., Larsen and Crossey, 1996), and the deformation of sediments by rising, hydrothermally generated gas, so-called fluid escape "pockmarks" (Pickrill, 1993). In stratigraphical sequences, evidence for this type of alteration typically declines over time, as the caldera becomes quiescent.

Long-term depositional history in calderas is regulated by the interplay of declining or renewed volcanism, with climatically induced fluctuations in lake level. A good example of this interplay comes from an ancient caldera lake of the Oligocene Creede Formation, Colorado (Larsen and Crossey, 1996). The Creede Formation represents the infill sequence of a large (12-km diameter) resurgent caldera, some though not all of which are lake deposits. After the collapse of the caldera, its central floor underwent resurgence forming a broad moat between the resurgent dome and the caldera walls. Lava, secondary volcanic domes, tuffs, and lacus-

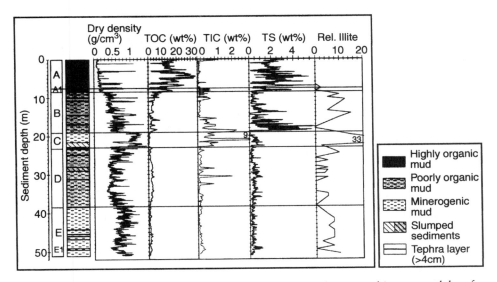

Figure 8.20. Stratigraphical column for a core from Lago Grande di Monticchio, a maar lake of southern Italy, showing lithostratigraphy, dry density, total organic carbon (TOC), total inorganic carbon (TIC), total sulfur (TS), and relative abundance of illite. The Pleistocene/Holocene transition is marked by a major decrease in dry density (mineral matter), increase in TOC and TS, and decline in illite, brought into the lake basin during arid glacial intervals as eolian dust. From Zolitschka and Negendanck (1996).

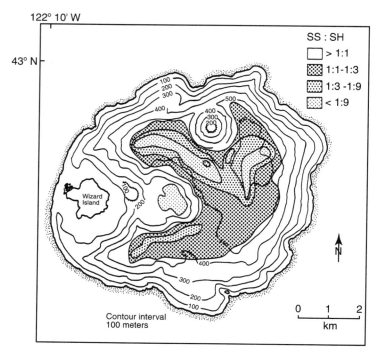

Figure 8.21. Bathymetry and areal distribution of sandstone/shale (SS : SH) ratios in Crater Lake, Oregon, calculated from original sand:mud ratios of 0.5–2.0 m gravity cores. Conversion to rock ratios shown here assumes that unconsolidated muds would be reduced to one-third of their original thickness in transformation to shale. The sub-basin centers and central platform areas are areas of finer sediment accumulation. Modified from Nelson et al. (1994).

trine sediments filled this moat. It is not known how long this Creede Lake persisted. The deposits are perhaps as much as 700 m thick, but radiometric dating suggests that the lake lasted less than 600 ka.

Deposition in the Creede Lake underwent changes driven by lake-level fluctuations and volcanic cyclicity. Following its initial filling, the lake was high for some period of time, resulting in deposition of turbidites across the lake's moat areas, and the progradation of deep-water fans (figure 8.23). Away from areas of sublacustrine fan aggradation, deposition was dominated by the seasonal accumulation of rhythmites, in this case, carbonate/fine clastic couplets. Travertines formed along lake margin fractures and faults at times when lake levels were high. Following its early high stand, the Creede Lake underwent a slow but continuous lake-level fall. During the low stand interval, littoral and ephemeral lake deposits, fan deltas, and alluvial fans extended into the basin. Incision and canyon cutting into caldera walls also occurred at this time.

Volcanism affected sedimentation in the Creede Lake in ways typical of large calderas,

by influencing both sedimentation rates and types of accumulation. Lake deposits formed during early eruptive activity are deposited in abrupt contact over older deposits. During this early phase, relatively massive fallout of volcanic tuffs occurs, producing very rapid sediment accumulation. This gives way to a posteruption phase with an increasing deposition of ash turbidites and suspension-derived rhythmites and increasing proportions of organic matter. During this period sedimentation rates begin to drop off quickly. These deposits in turn give way to relatively slowly accumulating intereruption deposits, primarily tuffaceous turbidites and rhythmites. In calderas that experience multiple cycles of eruptive activity, this entire sequence might be repeated numerous times.

Summary

Our conclusions from this chapter can be summarized as follows:

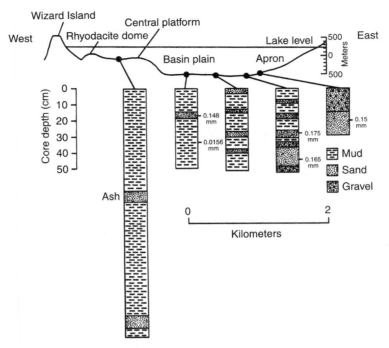

Figure 8.22. East–west transect of core lithologies from Crater Lake, Oregon. Median grain size diameter data are shown for sandy units. From Nelson et al. (1994).

1. Sufficient data exist for certain lake types to make generalized facies models, which describe deposition on lakewide scales, under varying climatic and tectonic conditions.

2. *Rift lake deposits* are strongly asymmetrical, reflecting the underlying structural differences between escarpment, hinged, and axial lake margins, and accommodation zones. These areas respond very differently to major lake-level fluctuations induced by climate change or changes in subsidence rates.

3. *Foreland basin lake deposits* share some similarities with rift deposits, particularly at the scale of larger-scale geometry. Major differences between the two mostly relate to the greater distance between fault margins and lake shores typical of foreland basin lakes, basin and watershed morphology, and the generally much shallower depths encountered in foreland basin lakes.

4. The most characteristic features of *glacial lake deposits* relate to the episodic nature of meltwater sediment discharge, and particularly how this affects the formation of overflow, interflow, or underflow deposits. Glacial lakes can deposit very coarse sediments in mid-lake locations by a variety of processes. Glacial lake deposits frequently show signs of secondary deformation from glacial movement. Lake levels in ice-contact lakes are subject to extremely rapid fluctuations, shifting the location of littoral deposits, and obliterating older deposits. Cyclicity is evident in glacial lake stratigraphy at various scales, from annual varving to long-term glaciation/deglaciation cycles.

5. *Playa lake deposition* is strongly regulated by the position of water table and groundwater inputs. The specific mineralogy of saline lakes is tied to bedrock chemistry and brine evolution, and varies within a basin and over time, as a result of lake-level or groundwater level fluctuations, most of which are linked to climate. Deep, saline lakes are commonly density stratified and meromictic, producing well-laminated deposits of evaporites, carbonates, diatomites, and terrigenous sediments.

6. *Deposition in maars and calderas* is strongly influenced by the restricted watershed areas

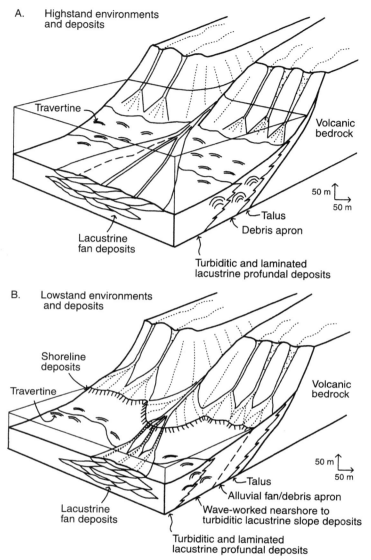

Figure 8.23. Distribution of depositional environments in the Oligocene Creede Formation caldera "Lake Creede" (Colorado) during (A) high-stand, and (B) low-stand lake conditions. From Larsen and Crossey (1996).

and steep marginal slopes of these lake types. Maars accumulate volcanoclastic sediment from very small areas, and their sedimentary records are therefore very responsive to local climate or anthropogenic changes. Caldera sedimentation is more complex, the result of basin segmentation, the existence of central lake topographical highs, and variable sediment focusing. Long-term stratigraphical changes in calderas reflect the interplay between episodic volcanic resurgence, declining volcanic and hydrothermal activity, and variable precipitation/evaporation ratios and lake-level change.

9

Geochemical Archives in Lake Deposits

As we saw in chapter 4, the isotopic, elemental, or molecular constituents of a lake and its sediments reflect both external chemical inputs and the lake's internal biogeochemical cycles. Lake sediment geochemistry is the product of interactions between these external inputs from watershed geology, groundwater, vegetation, and the airshed, and internal lake processes. Both external and internal inputs are heavily influenced by climate, and for the past few thousand years, human activities. With careful consideration of the various information filters affecting their records, geochemists can greatly broaden the scope of questions that can be addressed using paleolimnology.

It is of critical importance when interpreting chemical data, that it be placed in the context of other sedimentological or paleontological archives. With modern, automated techniques, it is possible to amass large amounts of geochemical data in a relatively short time, data that can be compiled into deceptively "simple" geochemical profiles. Perhaps more than any other types of indicator, geochemical profiles are often interpreted as standalone records, without reference to petrographical, or even gross lithofacies information. Although it is tempting to read chemical stratigraphies as a direct record of inputs from a watershed or airshed, the signals are blurred by the whole host of messy, internal processes that we have already encountered in the hydroclimate filter: lag and residence time effects, reworking, particle and redox focusing, organismal uptake, and bioturbation. Lake deposits integrate local changes in source conditions, background sedimentation rates, and geochemical focusing processes (Engstrom and Swain, 1986). As a result, different locations within a lake may provide different geochemical histories, and interpretations of an integrated lake history must take into account these internal variations and their probable causes. This is always harder to do with paleolake deposits, where the original basin morphometry and hydrology is obscure.

In this chapter we will also consider postdepositional information filters that affect geochemical archives, in particular bioturbation and diagenesis. Because many geochemical components of interest to paleolimnologists are bound to fine-grained particles, they can be readily mixed by bioturbation. Both bioturbation and geochemical diffusion-related mixing are complex phenomena; unraveling their consequences often requires a mathematical modeling approach, particularly for interpreting very recent lake deposits (Matisoff, 1982; Robbins, 1986). Bioturbators do not homogenize sediments uniformly (often larger particles are concentrated at the bottoms of burrowed zones). Most commonly the effects of sediment mixing are to reduce peak concentrations of geochemical components, broaden both peaks and valleys of distribution, and cause the maximum values of specific geochemical components to appear deeper, and thus apparently earlier in the stratigraphical record than actually occurred (Leavitt and Carpenter, 1989). The severity of blurring depends on the question being asked. For very fine-scale work, such as a study of twentieth century pollution history in a lake with slow sedimentation rates, bioturbation might pose an insurmountable problem, and paleolimnological techniques may be inappropriate for obtaining a record of geochemical fluxes. In contrast, the interpretation of long-term (thousands of years) cyclicity in lake deposit chemistry may be hardly affected at all by bioturbation effects, since long-term maximal and minimal concentrations of components are only minimally affected by bioturbation.

Diagenetic alteration of lake sediment chemistry begins prior to and continues long after deposition. The sedimentary geochemical environment is subject to relatively rapid changes induced by fluctuations in water column stratification, lake level, and changes in the flux of particles. As a result, many compounds, once incorporated in the sediment, have complex ionic exchange, degradation, and diffusion histories, with rates of these processes varying over time. As with bioturbation, this complexity increases in areas of slow sediment accumulation.

A Note on Data Collection and Expression

In raw form, chemical stratigraphical data are expressed as a mass fraction of sedimented materials (mass percent, ppm, etc.). Normally, these are expressed against dry sediment weight, to eliminate the problem of highly variable water content of lake sediments. Sometimes other components considered irrelevant by the investigator are eliminated from the sum. For example, in studies concentrating on watershed inputs, results may be expressed as elemental concentration per gram of mineral matter to eliminate authigenic and biogenic components (Engstrom and Wright, 1984). For elemental components, the convention I will follow in this chapter is to refer to elements that comprise > 1% of sediment dry weight mass as *major elements*, those that comprise 0.1–1% (1–10 ppt) as *minor elements*, and those comprising less than 1 ppt as trace elements. As in chapter 4, isotopic data is expressed using the δ notation of relative heavy-isotope enrichment or depletion.

In lake sediments for which bulk sediment accumulation rates are known, it is also possible to calculate a flux rate of a chemical component, for example, expressed as $mg\,cm^{-2}\,yr^{-1}$. This has the great advantage of allowing direct comparisons of input rates throughout the record, without the confounding variable of background sedimentation. In some studies, fluxes have been shown to be increasing, even as total concentrations remain constant or decline. The ability to measure such fluxes is of particular importance in studies of recent paleolimnological changes in lakes, where increases in background sedimentation from anthropogenically linked erosion may dilute the elements or compounds of interest (e.g., Findlay et al., 1998). Unfortunately, the imprecision of dating techniques, except in very young sediments, often makes this impossible. Furthermore, even when accumulation rates can be calculated, they are sometimes misleading, since their input rate may or may not be linked to the mechanisms responsible for changing bulk accumulation rates. For this reason, many authors present both concentration and flux data

for comparison (Boyle, 2001b). Chemical data are also expressed in the form of elemental, or molecular ratios (or equivalently, as crossplots of two components) to examine patterns of variance and covariance, and less frequently, using multivariate techniques, to simultaneously compare suites of geochemical components.

Major and Trace Elements

Comparative studies show that a limited number of elements comprise the vast majority of lake sediments, reflecting the typical compositional variability of the continental crust, and the relative reactivity or solubility of crustal materials as they enter lake water.

Silicon and Aluminum

Because they are relatively unreactive and insoluble at most soil E_H and pH conditions, minerogenic Si and Al tend to become concentrated in soil profiles as weathering intensifies. Therefore, increases in terrigenous Si and Al in lake cores have been used as indicators of warmer and/or wetter climatic conditions, and overall increased weathering intensity. Fe and Mn delivery to downstream lakes also increases proportionally during periods of warmer climate and soil stabilization, although the redox environment of the lake complicates their interpretation in lake sediment profiles. Si is also an important authigenic constituent of lake sediments, as biogenic silica, or as precipitates under highly alkaline conditions. This necessitates the separation of terrestrially derived (minerogenic) from biogenic Si.

The interpretation of Al profiles is also complicated by its increased solubility under acidic conditions. This has suggested the possible application of Al profiles in sediments from poorly buffered lakes that have undergone recent acidification. Unfortunately, the results of acidification on lake accumulation of Al are complex, and often contradictory (Boyle, 1994). The diagenetic reactions and mineralization that Al can or does undergo shortly after burial are still poorly understood. Furthermore, Al concentrations in modern acidified lakes and lake sediments show no simple relationship to pH (Schafran and Driscoll, 1987; J.R. White and Gubala, 1990; Boyle, 1994). In some lakes, declines in Al probably come about because of a more rapid export of Al from the watershed. Al increases in the sediment profiles of some anthropogenically acidified lakes after the acid inputs stop and the lake pH rises (Fritz and Carlson, 1982). However, if high dissolved Al loads enter a lake they may become supersaturated during seasonal increases in pH, resulting in increased precipitation rates. Fortunately, sequential extraction methods for aluminum now allow Al profiles to be deconstructed into Al content derived from direct precipitation within the water column versus that from eroded soils. This may ultimately make Al a more powerful tool in reconstructing watershed pH history.

Biogenic Silica (BSi)

A large part of the silica (SiO_2) deposited in most lakes is biogenic in origin, primarily derived from siliceous algae (diatoms, and to a lesser extent chrysophytes). In some nearshore or swamp deposits grass leaf phytoliths and sponges can also be important constituents of biogenic silica (Andrejko and Upchurch, 1977; Bates et al., 1995). As a result of its amorphous or cryptocrystalline structure, biogenic silica is chemically more reactive than most minerogenic silicates, and can be extracted through wet chemical digestion techniques that take advantage of its faster dissolution rate (DeMaster, 1981; Mortlock and Froelich, 1989).

Biogenic silica profiles are frequently interpreted by paleolimnologists as indicators of variation in overall productivity. More accurately, BSi represents the flux of siliceous skeletal matter from the epilimnion, minus the dissolution that occurs during settling and on the lake floor. Because this amount of dissolution is very large, and seasonally variable, the use of BSi as a productivity rests on two assumptions: that the accumulation of BSi is proportional to siliceous algal productivity (i.e., postmortem dissolution rates remain constant relative to production);, and that siliceous algal productivity accurately reflects total primary productivity. These assumptions receive most of their support from the correlation of BSi paleolimnological records with other paleoindicators or historical records of productivity change. For example, lake sediment records frequently show a direct relationship between BSi and TOC, and an inverse relationship between both of these variables and the soluble cations (MacDonald et al., 1993). Presumably this occurs because warmer and/or wetter climates lead not only to a decrease in unweathered (Na, K-rich) mineral grain input, but also an increase in productivity.

Diatom productivity is affected by nutrient availability, water temperature, turbidity, and for higher latitude lakes, duration of the ice-free period; various studies have interpreted BSi in terms of these variables. Over decadal–century time scales, rates of sedimentary accumulation of BSi may fluctuate substantially in response to the input of other nutrients and the varying growth and sedimentation rates of diatoms. In a series of papers, Schelske and his colleagues (Schelske and Stoermer, 1971, 1972; Schelske et al., 1986) argued that peak BSi concentrations in the sediments of several of the North American Great Lakes may be linked to human activities increasing phosphorus loading to the lakes. At the time, this proposal generated considerable controversy (Parker and Edgington, 1976; T.C. Johnson and Eisenreich, 1979; Shapiro and Swain, 1983), although subsequent studies have supported the general model (Schelske, 1988, 1991). Climatic explanations have generally been applied to BSi variability over longer (10^3–10^5 year) time scales. A 250-ka core record from Lake Baikal shows a strong relationship between global climate cycles (SPECMAP) and BSi concentration, with BSi concentration increasing during warmer interglacial intervals and decreasing during colder glacial periods (Colman et al., 1995, 1999) (figure 9.1).

In contrast to these apparently successful applications of BSi as a paleoproductivity indicator, independent experimental or physiological support for the BSi/productivity relationship is actually quite limited, and some paleolimnological studies have documented a lack of correlation between BSi and other productivity indicators (Engstrom and Wright, 1984). Diatoms are often outcompeted by other groups of phytoplankton, even as overall productivity rises. Changes in silica residence times or supply rates will have first-order consequences for BSi deposition that may completely mask total productivity relationships (T.C. Johnson et al., 1998). The numerous documented controls on diatom growth and Si supply suggest that biogenic silica data should be interpreted cautiously in terms of overall productivity and climate change, with attention primarily paid to major (i.e., order of magnitude) excursions in BSi concentration.

Sodium, Potassium, and Titanium

The conservative cations Na^+ and K^+ normally precipitate in authigenic lacustrine minerals only under highly saline conditions, or when they are sorbed onto fine particles (Engstrom and Wright, 1984). In freshwater lakes, sedimentary profiles of Na and K primarily reflect the mineralogical byproducts of soil weathering. Both Na and K are usually present

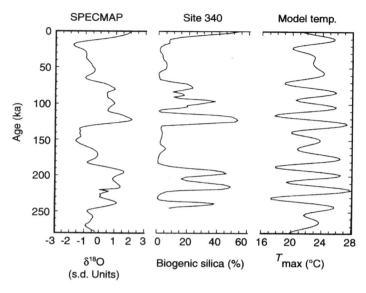

Figure 9.1. A comparison of the biogenic silica record from core site 340 at the Academician Ridge, Lake Baikal (Russia), with two paleoclimate models. The SPECMAP $\delta^{18}O$ deep-sea record is a nonlinear record with respect to orbital forcing, thought to predominantly reflect global ice volume at the poles. The model temperature curve is based on an energy balance model of maximum summer temperatures at 60N, 100E, and is driven by an assumption that linear changes in solar insolation will result from the composite of all orbital forcing mechanisms. The biogenic silica record shows a strong similarity to the SPECMAP record, and much less correspondence with the model temperatures, suggesting that temperature changes at Lake Baikal are well integrated into a nonlinear (with respect to orbital forcing) global climate system. See chapter 13 for further discussion of possible explanations for this pattern. From Colman et al. (1995).

in substantial quantities in crystalline or clastic sedimentary bedrock, and in watersheds with these types of bedrock most sedimented Na and K occurs bound in silicate mineral grains. Studies of northern European and North American lakes have demonstrated a relationship between weathering intensity and erosion, principally driven by climate, and the accumulation of these metals (Mackereth, 1966; Engstrom and Wright, 1984; Young and King, 1989). Na and K supply often increases during periods of rapid erosion and terrestrial sediment discharge, for example, during or just following glacial episodes, or during enhanced anthropogenic erosion, whereas vegetation, deep-soil profile formation, and soil stabilization normally reduce the flux of Na and K in mineral form. When this increased flux is not masked by proportionately greater sediment input from some other source, it will be recorded by higher Na and K concentrations in lake sediments. Ti is also released more from bedrock under heavy erosional regimes and has

been found to behave in a similar fashion to Na and K (R. Jones, 1984).

In hypersaline, closed-basin lakes, both Na and K, along with the dominant conservative anions, can be precipitated in authigenic minerals. Some attempts have been made to interpret Na and K profiles in saline lake sediments in terms of variable salinity (e.g., Valero-Garcés et al., 1996). However, even when authigenic Na and K minerals accumulate, these highly soluble cations readily diffuse through sediment pore waters, undergoing diagenetic alteration, and confounding their original concentration patterns. Furthermore, short core profiles frequently display increasing Na and K toward the sediment–water interface, even when other indicators do not suggest rising salinity, suggesting these trends result from diffusion gradients in the sediment, which may or may not be superimposed over historical trends in lake water composition (Haskell et al., 1996). For these reasons, geochemical interpretations of saline lake histories are more frequently made from mineralogical rather than elemental profiles.

Calcium, Magnesium, and Strontium

Calcium and magnesium are major bedrock constituents in many watersheds, and their concentrations are occasionally interpreted from silicate or organic fractions of lake sediments. In the detrital fraction these elements behave in much the same way as Na and K, increasing in proportion under conditions of deeper erosion and reduced soil stability. Of much greater importance are the applications of Ca, Mg, and Sr concentrations in authigenic carbonate minerals and skeletal carbonates from invertebrates (Chivas et al., 1983, 1986a,b; Valero-Garcés et al., 1996). Increasing salinity in hardwater lakes is generally accompanied by an increase in the Mg/Ca_{water} ratio, resulting in the mineralogical series (from low to high salinity) of low-Mg calcite ⇒ high-Mg calcite ⇒ aragonite ⇒ dolomite (Müller et al., 1972). However, the various carbonate minerals cannot accommodate Sr and Mg within their crystalline lattices equally; Sr is preferentially taken up by aragonite, whereas Mg is accommodated more readily by calcite. This implies that whereas both Sr/Ca and Mg/Ca ratios in authigenic carbonates may provide useful information concerning evaporative concentration, any interpretation of these profiles in carbonate sediments must be accompanied by mineralogical analysis.

To avoid some of the complications of varying carbonate mineralogy, geochemists turned to analyzing Mg/Ca and Sr/Ca profiles from ostracode shells, where the skeletal mineralogy is known to be uniformly low-Mg calcite (Chivas et al., 1983, 1986a,b). Ostracodes secrete their shells rapidly during molting; metals are acquired during this process, which lasts perhaps an hour (Turpen and Angell, 1971). Aside from Ca, Mg is the most abundant metal in the ostracode shell, typically comprising 0.1–1 mol% of total metals. Sr occurs in significant but generally smaller quantities (0.03–0.5%). Chivas et al. (1983) showed that partition coefficients could be calculated that relate the con-

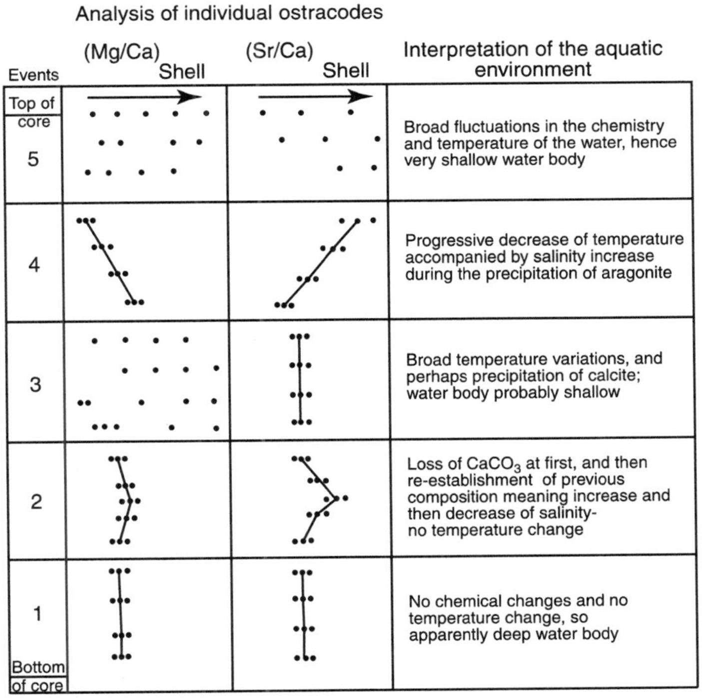

Analysis of individual ostracodes

Figure 9.2. Summary diagram showing the types of aquatic environments reconstructed from Mg/Ca and Sr/Ca analyses of individual ostracode valves from various horizons sampled from a hypothetical core. Mg/Ca_{shell} responds strongly to temperature variability but only weakly to salinity, whereas Sr/Ca_{shell} responds primarily to water chemistry. Dots represent analyses of individual valves and lines connect points of mean values for each horizon. Modified from De Deckker and Forester (1988).

centrations of each metal with respect to Ca in the ostracode shell to the free metal (uncomplexed) ratios of those same metals in the ambient environment:

$$K_p(Me) = (Me/Ca)_v / (Me/Ca)_w$$

where K_p = partition coefficient for Me, Me = either Mg^{2+} or Sr^{2+}, v = valve, and w = ambient water.

These partition coefficients are determined by culturing ostracodes or from field collections. Different species appear to have different K_p values, which seems to be evolutionarily controlled, since more closely related species have more similar K_p values. Based on their early work, Chivas and his colleagues proposed that the Ca–Mg–Sr system could be used as a combined paleosalinometer/ paleothermometer (figure 9.2). However, some complications to this application have become apparent in the years since these investigations took place. Most researchers agree that for Sr, metal partitioning is relatively insensitive to variation in temperature (Chivas et al., 1983, 1986a,b; De Deckker and Forester, 1988; Engstrom and Nelson, 1991; Holmes, 1992; Palacios-Fest et al., 1994; Palacios-Fest, 1996). As a result, Sr/Ca measurements are now routinely applied as a general indicator of Sr/Ca$_{water}$, which for many lakes corresponds with salinity. Complications arise in the interpretation of Sr/Ca data because the K_{Sr} value may not be a constant, even within a species (Wansard et al., 1998), and because sedimentary sinks for Sr, such as authigenic Sr-bearing minerals, affect available Sr for ostracode uptake (Haskell et al., 1996). Alkaline lake systems are particularly prone to depletion of Ca during early brine evolution and frequently deposit Sr minerals.

Unlike Sr/Ca, Mg/Ca ratios are temperature-sensitive. Early studies by Chivas et al. (1983) suggested that Mg/Ca$_{shell}$ was a function of both temperature and Mg/Ca$_{water}$, as would be predicted based on thermodynamic considerations alone. However, subsequent research suggests that biokinetic processes may play a much larger role in Mg uptake than was originally thought. Higher Mg uptake in juveniles (which molt very quickly) and differences in partition values between species all suggest that metabolic activity, which is temperature-dependent, is at least as important a determinant of Mg$_{shell}$ concentrations in ostracodes as water composition. In culturing and field studies, Mg/Ca$_{shell}$ is often highly correlated with ambient growth temperature, with little or no relation to ambient Mg/Ca$_{water}$ (Palacios-Fest and Dettman, 2001). Similarly, in paleolimnological studies of

ostracodes, where independent estimators of paleotemperature and paleosalinity are available, changes in Mg/Ca$_{shell}$ correspond closely to temperature trends, whereas Sr/Ca$_{shell}$ tracks ionic concentration (Cohen et al., 2000a). Mg uptake rates, as opposed to strictly free-metal ratios, are known to strongly influence both organic and inorganic carbonate composition, and uptake rates are often primarily temperature-dependent (Given and Wilkinson, 1985; Rosenthal and Katz, 1989; Bodergat et al., 1993). The variable results concerning Mg/Ca$_{shell}$ temperature sensitivity between authors suggest that the relative roles of thermodynamic and kinetic controls may be species-dependent.

Iron and Manganese

Mn and Fe are delivered to lakes as unaltered mineral grains, oxides, colloids, or organic complexes. Acidic or reduced conditions in some soils promote increased mobility of Fe and Mn, and some paleolimnological records suggest increased inputs of these elements during periods of acidic soil development, for example when coniferous forests dominate a watershed (Engstrom, 1983). Because the mobility of these elements greatly increases at redox boundaries, much effort has been put into trying to interpret Fe or Mn profiles in terms of redox history. Results, unfortunately, are contradictory as to how the profiles should be interpreted. The somewhat higher solubility of Mn versus Fe under dysaerobic conditions has been suggested by some researchers as a key to this interpretation, with rising Fe/Mn in a core being indicative of the onset of reducing conditions, especially when total Fe is low, followed by Fe/Mn$_{sediment}$ decline as anoxia becomes permanent (Kjensmo, 1968; Swain, 1984; Tracey et al., 1996). Elevated Fe/Mn can also be imparted by dysaerobic conditions within the watershed, in the waterlogged soils of wetlands, and secondarily transmitted to the lake through overland flow or groundwater (Pennington et al., 1972). In some paleolakes this has allowed the use of Fe/Mn ratio data as an indicator of paleowater-table elevation (Schütt, 1998).

Several other factors must be considered in the application of Fe/Mn ratio data, and suggest that these data be interpreted cautiously. First, not all stratified lakes display Fe/Mn patterns like those described above (Schaller et al., 1997). Diagenetic "stratigraphies" can develop in Fe and Mn for reasons that are only loosely connected to long-term lake history. Experimental studies of sediment Fe concentrations have shown that strong vertical trends can develop in as little as a few weeks in previously homogeneous sediments, from differential Fe mobility at or near the sediment–water interface (Carnignan and Flett, 1981). In carbonate sediments, Fe and Mn profiles will reflect mineralogy. Like Mg, Fe and Mn are preferentially taken up in calcite and excluded from aragonite. Variations in sulfate supply also complicate the interpretation of Fe/Mn ratios considerably.

Phosphorus

Given the tremendous interest in understanding the history of phosphate loading and eutrophication in lakes throughout the world, it would seem natural that obtaining and interpreting records of sedimented phosphorus would be important in paleolimnology. However, the diagenetic mobility and biological reactivity of phosphorus at the sediment–water interface works against this application, and only rarely does long-term phosphorus accumulation mirror phosphorus loading. The well-known interactions between P and both Fe and Ca often link the fate of sedimented P more to redox variability at or near the sediment–water interface than to P concentrations or loading in the lake (Engstrom and Wright, 1984; N.J. Anderson et al., 1993; N.J. Anderson and Rippey, 1994). Diatom communities are often sensitive to phosphorus loads, and can therefore provide a paleolimnological measure of epilimnetic P, independent of sedimented phosphorus (chapter 11). Such studies commonly demonstrate a lack of correlation between water column P history and long-term phosphorus sedimentation records. Where stratigraphical profiles of sedimentary phosphorus have been made, they also frequently do not correspond with other paleoindicators of nutrient loading, such as TOC, TIC, or isotopic signals, discussed later in this chapter (Engstrom and Wright, 1984; Schelske et al., 1988).

Where correlations between sedimentary phosphorus concentrations and water column phosphates do exist, they probably result from one of two situations. Extremely high loadings of phosphorus, to the point where phosphorus no longer limits productivity, may be recorded by high sedimentary phosphorus levels (Brenner et al., 1993; Brezonik and Engstrom, 1998). This is evident in P profiles from cores taken in Lake Okeechobee, Florida, a lake that at present experiences heavy P loading from agricultural fertilizers, but which prior to the twentieth century was extremely nutrient-

poor (figure 9.3). Phosphorus concentration trends in sediment may also prove useful in deep meromictic lakes, where deep-water phosphorus accumulation can become decoupled from surface water uptake and nutrient recycling.

Sulfur

Sulfur is supplied to lakes from a combination of atmospheric, stream flow, and groundwater sources, primarily in the forms of inorganic sulfates and organic S-bearing compounds. Interest in S as a paleolimnological tool has come about for several reasons. In very old sediments, where original environments of deposition are uncertain, sulfur concentrations are useful in differentiating between sediments of marine versus lacustrine origin. On very recent time scales, interest in S has been spurred by its role in acidic precipitation, its effect on poorly buffered lakes, and the potential for documenting anthropogenic effects on the sulfur loading from lake sediment records.

Settling organic matter (seston) plays a major role in the sedimentation of sulfur in lakes, especially oligotrophic ones, in contrast to the ocean, where sulfate is a much more important component of the sulfur cycle (Mitchell et al., 1990). This linkage of C and S in seston is reflected in covariant C/S profiles, primarily, though not exclusively, in oligotrophic lakes (B.B. Wolfe et al., 1996). Total S and sulfate concentrations in the oceans are considerably higher than most lakes (Berner, 1984; Berner and Raiswell, 1984; Davison, 1988). This is reflected in the difference between bulk carbon/sulfur (C/S) ratios in lake sediments (typically 40–120) versus 0.5–5 for marine sediments, and can be used to differentiate between marine and lacustrine deposits.

Sulfur sedimentation in lakes is also closely linked to the behavior and supply of carbon, iron, and phosphorus. We will concentrate on the formation and accumulation of sulfides here, because more is known about this process than other aspects of sulfur deposition. Dissolved sulfate can be converted to H_2S by sulfur-reducing bacteria, either within the water column in a stratified lake, or just below the bioturbated zone in the sediments of well-mixed lakes. The rate of this process is governed both by the availability of organic matter and sulfate, but in contrast to the oceans, it is almost always sulfate rather than TOC that is in short supply (Bates et al., 1995). Stable isotope studies of the $^{34}S/^{32}S$ ratio can provide valuable clues about sulfate availability to this process, supple-

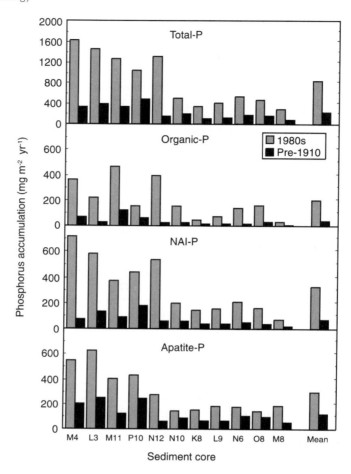

Figure 9.3. Accumulation rates of total P and major forms of phosphorus in a series of cores from Lake Okechobee, Florida, comparing the period prior to intensive agriculture (pre-1910). Accumulation rates based on ^{210}Pb-dating of all cores. The cores illustrate an $\sim 4 \times$ increase in phosphorus sedimentation, with most of that increase occurring since the 1940s. The timing of these increases correlates with a major expansion in channelization of inflow into the lake and livestock ranching in the watershed. From Brezonik and Engstrom (1998).

menting sedimentary sulfur concentration profiles in paleolimnological studies. ^{34}S/^{32}S fractionation occurs during bacterial sulfate reduction, with H_2S being depleted in ^{34}S relative to the sulfate precursor. Sulfides depleted in ^{34}S are formed when sulfate is effectively unlimited, whereas ^{34}S-enriched sulfide results when sulfate is in short supply.

Following its formation, hydrogen sulfide, both in the sediment and water column, can subsequently undergo several fates. It may react with available iron to form various Fe-disulfides, it can react with organic compounds, or it can be lost from the sediment through diffusion and secondary oxidation to sulfate or elemental sulfur. Fe–S inter-

actions are limited by their relative availability, and by the form of Fe present. Unlike sulfur, iron is rarely in short supply, except in extremely pure limestone bedrock terrains. However, the type of Fe available may determine the rate at which it can react to form disulfide iron minerals (pyrite, etc.), fine-grained Fe-oxides being more reactive than Fe-bearing silicate minerals.

In unstratified lakes the formation of Fe-sulfides generally occurs close to the site of sulfate reduction. This is not necessarily the case in stratified lakes, since H_2S will be dispersed through the water mass through advection. Once formed, FeS_2 is quite insoluble under anoxic conditions, and can accumulate in the sediment, which suggests the possibility of using insoluble iron sulfide concentrations as a paleoredox indicator. However, this potential is limited whenever secondary reoxidation has occurred.

The numerous diagenetic reactions that sulfur can undergo complicate the paleoenvironmental interpretation of S concentrations, and the study of $S_{sediment}$ is therefore best undertaken in conjunction with complimentary analyses of S-isotopes and the micrcoscopic examination of organic matter. Using this approach it is possible to use sulfur archives to interpret the history of water-table fluctuations, vegetation history, and changes in microbial communities (Bates et al., 1995). Very promising paleolimnological applications of sedimentary sulfur concentrations have also developed out of the study of lake acidification (Mitchell et al., 1988, 1990). The connection between atmospheric sulfate discharges and acidification of poorly buffered lakes has been well established for many years (Cogbill and Likens, 1974; Drabløs and Tolan, 1980; Likens and Butler, 1981). High sulfate loading from local sources, such as mine discharge, can also acidify lakes that have low alkalinity. A paleolimnological investigation of either atmospheric or local sulfate impacts using sedimentary sulfur concentrations makes two assumptions. First, the original sulfate loads in the lake water must be accurately portrayed in sedimentary sulfur, either as inorganic sulfides, organic S compounds, or some combination of the two. Second, sulfur in the sediment must be sufficiently immobile so as not to blur the record. The latter assumption is more probable for organically bound S than in inorganic sulfides. An additional concern in interpreting sedimentary sulfur profiles is the common covariance of S and TOC in lakes where most sulfur is bound in organic compounds. In such lakes covariant increases in sulfur content and TOC can be expected whenever seston flux increases, regardless of total atmospheric or runoff sulfate loading.

Even in lakes where sulfides are the dominant sedimented form of sulfur, and thus diagenetic mobility is a concern, there may be useful information about changing sulfate loading recorded in sediment geochemistry. It is reasonably well established from experimental lake acidification studies that increasing sulfate loading (as H_2SO_4) in the water column can accelerate sulfate reduction (Cook and Schindler, 1983). If adequate iron is present, this commonly leads to increased sulfide accumulation rates (D.W. Schindler, 1985). Sulfide concentrations in sediments of lakes exposed to large atmospheric sulfate loads have increased, often with other correlated effects on sediment

and water column geochemistry. For example, many years ago Hasler and Einsele (1948) predicted that increasing sulfate concentrations in lakes would have the effect of releasing iron-bound phosphate from the sediment, thereby increasing productivity. This combination of increasing sulfide concentrations and primary productivity caused by phosphate release occurred in Lake Bussjösjön, a small, highly eutrophic lake in southern Sweden (Olsson et al., 1997) (figures 9.4 and 9.5). Prior to the twentieth century, sulfate input to this lake was limited, and phosphates were sedimented as vivianite $[Fe_3(PO_4)_2 \cdot 8H_2O]$. With the advent of increased fossil fuel consumption in the region, however, sulfate reduction expanded greatly, and iron sulfides became a major sink for this increased S. As a result, the formation of ferrous phosphates declined, allowing P to be recycled back to the lake system. This in turn rapidly fertilized the lake to its present highly eutrophic condition, raising the rate of organic carbon deposition and other indicators of productivity in the process.

Trace Metals

A variety of elements that occur in very small quantities in lake sediments are routinely studied by paleolimnologists, either as geochemical tracers of sediment sources or, more commonly, as indicators of pollutant inputs. Especially in regions where long-term historical records are nonexistent or incomplete, paleolimnological techniques are often the only means of determining the sources and history of industrial pollutant discharges.

Most trace metals of interest to paleolimnologists enter lakes through multiple pathways, through bedrock erosion, as overland or groundwater discharges from mine tailings, industrial waste, or sewage, and as atmospheric emissions. Ideally, we would like to be able to distinguish the flux from all of these; not all may be of equal interest for a particular application. In some applied studies it may be critical to establish the time when a particular source, such as a mine, became an important input, and clearly differentiate this source from others. This is most easily accomplished when the source of metals to a lake is strictly atmospheric, with the watershed otherwise undisturbed, or where a watershed receives runoff from a very limited number of industrial sources. For these reasons the choice of what lake, or what part of a lake, to study using trace metal pollutants becomes critical; an inappropriately chosen site will always yield trace element profiles, but these pro-

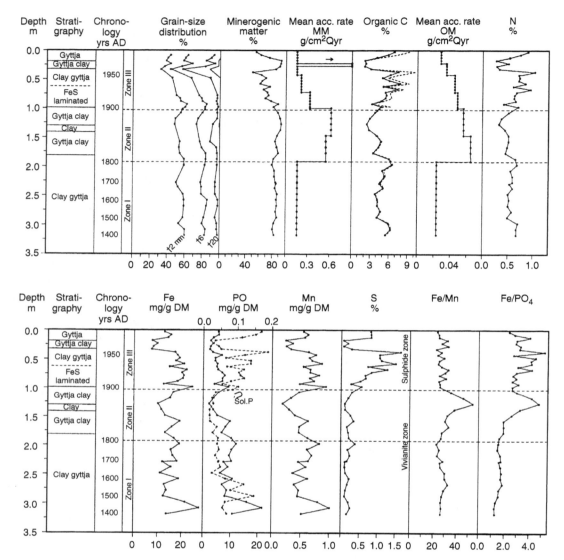

Figure 9.4. Chemical and physical stratigraphy for Bussjösjön, a hypereutrophic lake in southern Sweden. The core record demonstrates changes in chemical stratigraphy and sediment accumulation rates, indicative of high rates of nutrient sedimentation, the establishment of stratification, and the release of phosphates as Fe-phosphate (vivianite) accumulation ceased and Fe-sulfide accumulation began. These were related to changes in land use and nutrient loading associated with the expansion of farming in the early nineteenth century, and the introduction of commercial fertilizers in the late nineteenth century. MM = minerogenic matter; OM = organic matter; sol. P = H_2O_2−extractable phosphorus. Organic carbon is calculated in percentages of total dry mass (solid line) and on a carbonate-free basis (dashed line). From Olsson et al. (1997).

files may be largely uninterpretable in terms of the questions being asked. Many short-term trends in metal profiles in lakes result from very local phenomena, such as fires or road clearing. For this reason paleolimnological studies of regional phenomena, like changing atmospheric emissions of Pb or Hg, need to be investigated using records from multiple lake basins (Norton et al., 1992).

As with major element profiles, trace metal concentrations in sediments can be strongly affected by

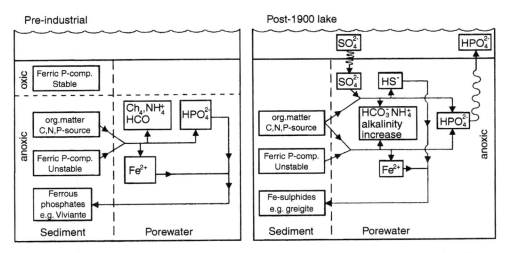

Figure 9.5. Simplified box model showing the cycling of Fe and P in the preindustrial and post-1900 sediments of Bussjösjön Lake, Sweden. In the preindustrial sediment, P was trapped in the oxic zone or as ferrous phosphates. A sulfate load to the anoxic post-1900 sediment–pore water system resulted in the removal of iron by FeS formation and release of P, fertilizing the lake. The generation of alkalinity connected with sulfate reduction may have supported precipitation of calcium carbonate. From Olsson et al. (1997).

watershed erosion, lake transport, and diagenetic phenomena that have little, if anything, to do with original pollutant fluxes from point sources or the atmosphere. Eroded soils often have naturally elevated concentrations of heavy metals, causing increasing metal flux to be strongly correlated with accelerated erosion, regardless of human inputs. To get around this problem it may be necessary to use an independent archive of soil erosion, with which the metal profiles of interest can be compared (Boyle et al., 1998).

Mine tailings discharges can cause elemental concentrations of a number of metals to rise dramatically in downstream lake sediments. Where the ore bodies are well underground, or in a separate watershed, it may be relatively simple to recognize the onset of such discharges. However, many ore bodies are located close to the earth's surface, and in these circumstances trace metal profiles should be interpreted cautiously, especially when elevated metal concentrations are observed in sediment profiles older than any conceivable mining activity (e.g., Sims and White, 1981).

Many trace metals such as Cu and Zn complex with organic compounds, and will naturally rise along with TOC and/or finer grain size, regardless of changes in anthropogenic flux. For this reason any interpretation of trace metal data should be accompanied by comparable grain size and TOC analyses. The concentrations of many environmen-

tally important trace metals are heavily affected by redox conditions, particularly Pb, Cu, Cd, Co, Cr, and Ni, causing their concentrations, and apparent flux rates, to rise, even in the absence of any "real" change in external input. The transition of a lake floor to anoxia, or short-term anoxia of the water column followed by a return to oxidizing conditions, can both result in short-term mobilization of some metals, which may be reflected in "spikes" in a metal profile (Schaller et al., 1997).

When the transport, sedimentation rate, and diagenetic complications discussed above are adequately understood, very useful information can be obtained from trace metal profiles of relevance to pollutant histories. The histories of lead and mercury discharges are particularly well documented from lake sediment studies. We will look briefly at studies of lead concentrations in lake sediments here, and return in chapter 12 to consider mercury.

Lead pollution in North America is a regional phenomenon, related to atmospheric Pb inputs related to smelting, coal burning, and combustion of leaded fuels. The latter source was probably dominant for most parts of North America prior to the rapid decline in Pb addition to fuels in the 1980s–1990s. Some of the first detailed paleolimnological studies of lead concentrations were made in the Seattle urban area, focusing on urban runoff and atmospheric lead input to Lake Washington (Barnes and Schell, 1973; Spyridakis and Barnes,

1978). These studies demonstrated the potential for not only calculating flux rate changes in Pb, but also for differentiating the sources of lead coming into the lake, as land use and industrialization patterns changed through the twentieth century.

A regional paleolimnological study of lakes in the eastern United States shows a long-term increase in sedimentary lead concentrations, starting in some cases by the early nineteenth century (Norton et al., 1992) (figure 9.6). The authors specifically chose remote lakes for the study, where there was no local industrial runoff; here atmospheric emissions can reasonably be assumed to represent the principal source of anthropogenic lead. In the early nineteenth century background lead concentrations were uniformly less than 100 ppm, and then rose steadily during the twentieth century in all lakes. The earlier rise of Pb concentrations in the northeastern United States versus Florida probably reflects the earlier industrialization of the former region. In lakes with rapid response times or low degrees of bioturbation, late twentieth century declines in Pb deposition are evident, which correlate with the decreasing use of leaded fuel.

Organic Matter in Lake Sediments

Because it may be derived from a variety of terrestrial and lake sources, organic matter (OM) has the potential for archiving information relevant to many questions about lacustrine and terrestrial ecosystem change. Terrestrially derived OM normally enters a lake as water-borne solutes or particulates, simplifying the interpretation of its sources to within the lake's watershed. Organic compounds and particles can be thought of as fossils, subject to reworking and alteration, like skeletal fossils (Meyers, 1997). Organic compounds vary considerably in the completeness of their records; there are orders of magnitude differences in the rate at which various organic geochemical "fossils" (different organic compounds) degrade, both during settling through the water column and after sedimentation. Decomposition in the water column is a particularly important part of this change for paleolimnologists to understand, since it is here that environmental conditions change most dynamically, and for many compounds this is the primary zone of geo-

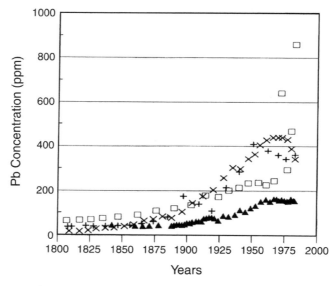

Figure 9.6. Lead concentrations in nineteenth and twentieth century lake sediments from four U.S. lakes. These lakes were chosen because of their minimally disturbed watersheds, to provide records of regional, atmospheric Pb flux. All cores are dated with ^{210}Pb profiles. Somewhat higher and earlier atmospheric Pb inputs are evident for the lakes from the northeastern United States, closer to industrial sources. Downturns in some late twentieth century records may also indicate declining usage of Pb-additives in gasoline after the 1970s. Mud Pond, Maine (□); Upper Wallface Pond, New York (×); McNearney Lake, Minnesota (+) and Lake Mary, Florida (▲). From Norton et al. (1992).

chemical alteration and therefore the major information filter (Meyers and Eadie, 1993). But this alteration itself can be the source of useful information, for example when diagenetic byproducts of organic matter breakdown are indicative of specific conditions of oxidation.

Because of the difficulty of separating organic components within lake sediments, the early studies of OM were done on bulk materials, and many studies are still done in this way. Bulk OM analyses have the advantage of simplicity and low cost in sample preparation. In offshore settings, where there is little question as to the extent of terrestrial OM input, bulk analyses are particularly cost- and time-effective. However, these advantages decline in nearshore and deltaic areas where mixing of terrestrial and lacustrine OM is substantial, or when the question being asked has to do with specific algal precursors of OM. Under these circumstances, paleolimnologists are increasingly analyzing specific compounds within OM. These methods allow organic geochemists to characterize such attributes as precise molecular composition and structure or isotopic composition of particular compounds, and from that determine their specific organic matter precursors and diagenetic pathways.

Bulk OM Indicators

Total Organic Carbon

Total organic carbon is one of the most common geochemical analyses in paleolimnology. TOC is estimated in several different ways; through the loss of mass during ignition at high temperature, commonly referred to as "loss on ignition" (LOI), chromatographically through the use of a CHN analyzer, or coulometrically. LOI is an extremely simple, and therefore widely used technique, and gives good results for organic-rich sediments, but is a progressively poorer estimator of actual organic carbon content at low OC concentrations, and its measurement is susceptible to mass and burn time-related artifacts (Heiri et al., 2001).

TOC is often taken as an indicator of productivity, particularly when TOC concentrations can be converted into fluxes. The underlying assumptions are that original productivity is quantitatively reflected in the amount of biomass that sinks to the lake floor, and that this biomass is proportionately degraded after burial. When comparing highly oligotrophic lakes with more productive systems, a qualitative record of productivity seems to be preserved in a wide range of lake mixing regimes, particularly in offshore settings, and where clastic input is limited or relatively constant. This qualitative use of TOC as a productivity indicator is borne out most dramatically by the numerous studies that demonstrate substantial changes in TOC concentrations associated with glacial advances (TOC decreases) or deglaciation (TOC increases) (e.g., Levesque et al., 1994). However, under conditions of moderate to high productivity other factors, such as mixing regimes, organic matter degradation and clastic dilution become more important regulators of TOC, and the interpretation of TOC profiles must be adjusted accordingly (Katz, 1990).

Once these complications have been accounted for, TOC profiles have been extremely useful for documenting long-term climatic cyclicity. This is most evident in lake systems that undergo dramatic swings in productivity or lake level (Beierle and Smith, 1998). For example, the sediments of glacially fed Owens Lake (California) show strong swings in TOC content over the past 800 ka, covariant with inorganic carbon (carbonate) for much of that period, and largely coincident with changes in global glacial/interglacial climate cycles (Bischoff et al., 1997a) (figure 9.7). In this case warmer, drier conditions led to decreased minerogenic sedimentation rates, and increased TOC concentrations. Under warm climate conditions, major swings in TOC profiles may reflect climate in very different ways, as intense evapotranspiration and drying expose large areas of lake floor to soil-forming conditions, which oxidizes and rapidly degrades previously formed lacustrine OM (Beuning et al., 1997).

C/N ratios

Nitrogen is sedimented in lakes from both terrestrial and lake-derived organic matter. Smaller proportions of N also occur in sediment adsorbed on clay minerals, and in the frustules of diatoms, although the importance of these for bulk N concentrations is unknown. As we saw in chapter 4, aquatic organic matter from phytoplankton is much richer in N, and has much lower C/N ratios, than terrestrial organic matter. Many studies have shown that the C/N ratio is a useful tool for distinguishing long-term transitions between terrestrial to algal-input dominance in lake sediment OM (Meyers and Ishiwatari, 1995). Such transitions might occur as a result of lake-level fluctuations, or the migration or progradation of a large river delta. Many coastal or closed-basin lakes show dra-

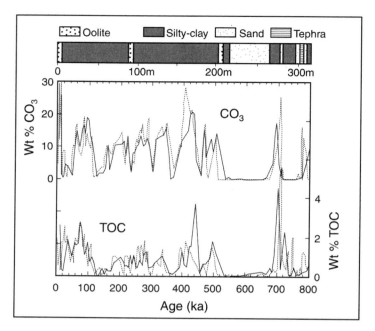

Figure 9.7. Lithostratigraphy, CO_3, and TOC records for the 800-ka core record from Owens Lake, California. Sample depth (below lithostratigraphy) and estimated age are both shown. Solid lines are from channel samples (homogenized strip samples covering 3-m intervals) and dotted lines are from point samples; the two records are in good agreement. Prior to 500 ka (220 m) environments of deposition at the Owens Lake Basin core site alternated between lacustrine and fluvial. Since 500 ka the record is entirely lacustrine. Values of CO_3 and TOC are very close to zero during the most recent glacial maximum at 17–25 ka, suggesting the lake was intensely overflowing during glacial advances. From Bischoff et al. (1997a).

matic declines in C/N ratios during the period when they are undergoing flooding and transgression, as terrestrial or nearshore habitats, with abundant vascular plant debris, are transformed into off-shore, algal-dominated environments.

Caution must be used in interpreting C/N profiles, because of the early diagenetic loss of N, which will elevate the C/N ratio, sometimes placing obviously lake-derived OM into the range of "terrestrial" values. The degree to which this occurs varies between lakes (Meyers and Benson, 1987). Selective degradation of N can sometimes be inferred from a lack of correlation between this ratio and other indicators of terrestrial organic input, for example carbon isotopes in organic matter.

Organic Hydrocarbon Analysis and Rock-Eval Pyrolysis

Hydrogen, like nitrogen, is present in different proportions in terrestrial versus aquatic OM, suggesting its use as an organic source material indicator for paleolimnology. Hydrogen is abundant in proteins and lipids, which make up a high proportion of algal cells. In contrast, the cellulose and lignin of terrestrial vascular plants contain little hydrogen. Organic H-compounds (hydrocarbons) are often quite volatile under the influence of oxygen. As a result, H concentration in various hydrocarbon compounds will be higher when OM is preserved under anoxic conditions, or in bacteria metabolizing under anoxic conditions.

Although direct elemental analysis of hydrogen is possible, H concentrations are more conventionally estimated indirectly, using rock-eval pyrolysis, a controlled heating process that releases hydrocarbon compounds of sequentially lower volatility in an inert atmosphere. (Espitalié et al., 1977; Tissot and Welte, 1984). A number of signals are produced in this process, including the amount of hydrocarbons present before heating (S_1), the amount of hydrocarbons released during pyrolysis (S_2), the amount of CO_2 liberated during pyrolysis (S_3), and TOC. From these measures, two important indicators can be derived. The Hydrogen Index (HI) is a measure of hydrocarbon concentration in organic matter relative to total organic carbon ($= S_2/TOC$, expressed as mg HC gas liberated during pyrolysis $g^{-1}C_{org}$) and is usually interpreted as a proxy for the H/C ratio in the organic matter present. The Oxygen Index (OI) approximates the amount of oxygen present per unit total organic carbon ($= S_3/TOC$, expressed as mg CO_2 $g^{-1}C_{org}$). Crossplots of HI and OI (referred to as *modified Van Krevelen diagrams*, figure 9.8) are used to differentiate the origins of hydrocarbon compounds, based on the relative values of the two indicators (Katz, 1988). Type I organic matter is rich in hydrocarbons from bacteria and/or the waxy cuticle surrounding vascular plant tissues. Type II organic matter is slightly lower in H but higher in O, and may be derived from aquatic algae, or from mixtures of aquatic and terrestrial organic matter. Type III organic matter is poor in H but is O-rich. It can be derived either from vascular land plants, or through oxidation of other organic matter. Aside from its utility in determining historical trends in the source of organic matter, HI has also been shown to be a powerful tool for distinguishing periods of lake floor exposure and soil formation (Talbot and Livingstone, 1989).

Compound-Specific Indicators

The analysis of specific organic compounds or *biomarkers* in lake sediments is predicated on the fact that molecular fossils are often indicative of specific organic origins, from particular groups of plants, animals, true bacteria, or methanogenic bacteria (Meyers and Ishiwatari, 1995). This specificity creates a potential for understanding quite detailed aspects of ecosystem history. Very exciting results have come out of this field, since the organisms that leave these geochemical fossil records, mostly algae and bacteria, typically leave no skeletal or undegraded body part fossils. At the same time, the differential volatility of organic compounds, both in the water column and during burial diagenesis, tells us that these compound-specific data must be interpreted carefully. Some organic materials, such as cellulose and lignin, undergo quite predictable rates of degradation (Meyers et al., 1980; Ishiwatari and Uzaki, 1987). When the loss rates of organic compounds are known, their proportions in sediments can be interpreted both in terms of diagenesis and original concentration. Unfortunately, low molecular weight compounds, many of which are of interest to paleolimnologists, are subject to rapid and variable depletion during early diagenetic microbial decay (Cranwell, 1984).

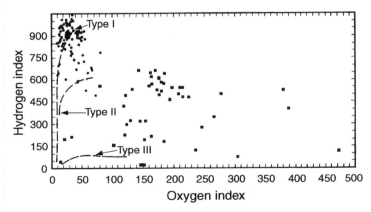

Figure 9.8. Modified van Krevelen diagram, showing a crossplot of Oxygen Index OI versus Hydrogen Index HI, for sediment samples from the East African Great Lakes (■) and Eocene lake deposits from the Green River Fm., western United States (Lake Uinta) (●). Type I (H-rich, O-poor) hydrocarbons are derived from bacteria or the waxy cuticle around vascular plant tissues. Type II are derived from aquatic algae or algal/terrestrial mixtures. Type III hydrocarbons (H-poor, O-rich) are derived from terrestrial plants, or from the oxidation of other organic matter. From Katz (1988).

Many organic geochemical studies of lake sediments have focused on lipids. Lipids comprise a large class of organic compounds that can be isolated by organic solvent extractions, and quantified using gas chromatography mass spectrometry (GCMS), which differentiates the abundances of individual hydrocarbon compounds by their carbon chain length (Meyers, 1997) (figure 9.9). In sediments, lipids represent a mix of original biologically produced substances, and geolipids, produced by diagenetic alteration. Lipid extraction techniques yield a variety of compounds, including hydrocarbons, carboxylic acids, alcohols, ketones, aldehydes, and other related compounds.

Lipids have proven particularly useful as paleolimnological indicators, because many are indicative of specific organic sources, and because they are less prone to diagenesis than other organic compounds. Lipid studies are also useful in paleolimnology for distinguishing hydrocarbons from natural versus anthropogenic sources, such as petroleum and its derivatives (Lipiatou et al., 1996). The straight-chained n-alkanes and n-alkanoic acids can be used as indicators because dominant molecular chain lengths vary between primary producers. Short chain lengths (C_{12}, C_{14}, and C_{16} n-alkanoic acids and C_{17} n-alkanes) are dominant in algae, although other plants produce them as well. In contrast, odd-numbered C_{27}, C_{29}, and C_{31} n-alkanes and C_{24}, C_{26}, and C_{28} n-alkanoic acids

are produced in the epicuticular waxes of terrestrial vascular plants. As a result of their usefulness for distinguishing organic source materials and their relatively low potential for microbial degradation, the n-alkanes are probably the most widely studied biomarkers in paleolimnology (Meyers and Ishiwatari, 1995).

The origins of a number of lipids are specific to particular groups of organisms, increasing their utility for paleolimnology (Leenheer and Meyers, 1983; Volkman, 1988). Some are limited to particular bacteria, cyanobacteria, chlorophytes, or seed-bearing terrestrial plants. Some biomarkers are indicative of specific alteration pathways related to diagenesis or metabolism. For example, the isoprenoid-saturated hydrocarbons are a common group of hydrocarbons in lake sediments that are indicative of diagenetic alteration of the chlorophyll a molecule.

The study of biomarkers in paleolimnology has proliferated since the 1970s, as geochemists have realized their potential for addressing questions about lake and watershed history. For example, lakes undergoing eutrophication have been commonly observed to display sedimentary profiles with an increasing dominance of short chain length n-alkanes, reflecting an increased contribution of organic matter from algal sources (Meyers and Teranes, 2001). Pristane/phytane ratios have also proven useful in characterizing paleolimnological

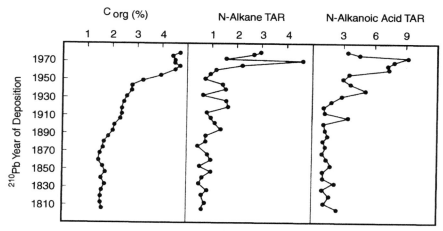

Figure 9.9. Organic geochemical stratigraphy (TOC and selected biomarkers) for a short core from eastern Lake Ontario. Chronostratigraphy based on ^{210}Pb-dating. The core records an increase in organic carbon concentrations and simultaneous increase in terrestrial organic indications, suggesting a greater contribution of terrestrially derived organic matter to the geolipid fraction. TAR = terrestrial/aquatic ratio, an index of the ratio between long-chain (generally terrestrially derived geolipids) versus short-chain (aquatic) geolipids. For n-alkanes this is calculated as $(C_{27} + C_{29} + C_{31})/(C_{15} + C_{17} + C_{19})$, and for n-alkanoic acids $(C_{24} + C_{26} + C_{28})/(C_{14} + C_{16} + C_{18})$. From Meyers (1997).

changes. Pristane and phytane are bacterially synthesized lipids. Phytane is particularly abundant in archaebacteria (methanogens) and various alkalophilic and halophilic bacteria, suggesting the use of the ratio pristane/phytane as an index of redox and/or salinity in paleolake deposits, with low values indicative of anoxia and/or high salinity (Didyk et al., 1978; Duncan and Hamilton, 1988; Mello and Maxwell, 1990).

A major focus of lipid biomarker studies of human impacts to lakes and their watersheds has been the effects of both increased runoff of terrestrial organic matter, and increased algal productivity in lakes undergoing eutrophication. Interpretation of biomarker results for this purpose is often complicated, however, because anthropogenic impacts can cause terrestrial inputs to either increase or decrease relative to aquatic inputs (Bourbonniere and Meyers, 1996a; Meyers, 1997).

Another important application of biomarkers to paleolimnology is in the recognition of combusted hydrocarbons. Polycyclic aromatic hydrocarbons (PAHs) are volatile organic compounds. They accumulate in lake sediments by one of several processes, through the combustion of fossil fuels, the combustion of other organic-rich compounds (for example in forest fires), through spills of fuel oil, or the diagenetic formation of aromatic compounds from previously buried organic matter (Meyers and Ishiwatari, 1995). The proportions of various types of PAHs give clues to their dominant source, because different PAH compounds form under different combustion temperature conditions.

Hydrocarbons derived from petroleum can also be differentiated from those of biological origin by other characteristics:

1. *Chain length*. Odd-number chain lengths typically dominate biological hydrocarbons. This is not the case with petroleum hydrocarbons.
2. *Molecular diversity*. This is much higher in petroleum hydrocarbons than biological ones. The large number of compounds found in petroleum frequently cannot be adequately separated by available analytical techniques (even by high-resolution capillary gas chromatography). A resulting "smeared" chromatogram pattern, termed an *unresolved complex mixture* (UCM), is characteristic of petroleum (figure 9.9) (Bourbonniere and Meyers, 1996b). In Lake Ontario, an increase in upcore concen-

trations of UCMs since the early twentieth century reflects an increase in the discharge of hydrocarbon fuels, from boats, oil spills, and nonpoint-source terrestrial runoff from paved roadways (figure 9.10).

Pigments

In chapter 5 we saw that photosynthetic bacteria, algae, and higher plants all produced pigments to collect light for photosynthesis. Despite organic degradation in the water column, some of these pigments are preserved in lake sediments, from which they can be isolated and studied by paleolimnologists. Effectively, sedimented pigments are biomarkers, although they are rarely referred to as such.

Pigments are extracted from sediments using various solvents. Early pigment studies were done using spectrophotometric techniques, thin layer and open column chromatography. Today, increasing use is being made of high-performance liquid chromatography (HPLC), which can differentiate more pigment derivatives, and isolate them from very complex mixtures (Leavitt et al., 1989). In the past few years, mass spectrometry techniques have also been applied to pigments, to identify the specific organic chemical composition and molecular ion characteristics of chromatogram peaks identified by HPLC (Hodgson et al., 1997). This method promises to both improve pigment identification and increase our understanding of pigment degradation pathways.

Early studies of algal pigments found systematic variation in total pigment concentrations, which were correlated with the trophic status of lakes; simply put, there appeared to be more pigments deposited in eutrophic lakes (Vallentyne, 1957, 1960; Gorham, 1960). This research suggested that the presence/absence of specific pigments, or their ratios, might serve as useful indicators of changing algal community structure, or trophic conditions, resulting from cultural nutrient loading, or deglaciation (Gorham et al., 1974; Guilizoni and Lami, 1988; Sanger, 1988). However, perhaps even more than most biomarkers, pigments are subject to rapid and differential microbial and diagenetic alteration, particularly in the presence of oxygen and elevated temperatures. The vast majority of pigments entering the water column are destroyed prior to burial. This includes both those formed in the water column by algae, and those from leaves of vascular plants, especially the chlorophyll compounds. Some pigments are systematically underre-

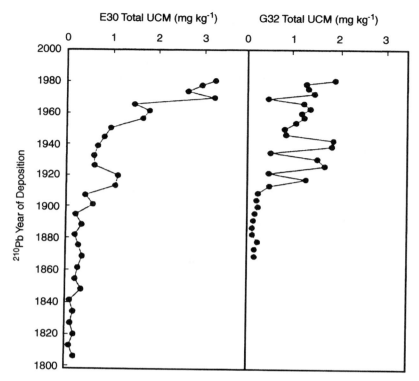

Figure 9.10. Changes in concentrations of unresolved complex mixtures (UCMs) of petroleum residue hydrocarbons at different intervals in the depositional records of two cores (E30 and G32) from the eastern (Rochester) Basin of Lake Ontario. Petroleum residues began to appear in the late nineteenth century, but are still relatively minor constituents of lake sediments in this region today. From Bourbonniere and Meyers (1996).

presented, for example those from dinoflagellates, diatoms, and chrysophytes (especially fucoxanthin and c-phorbins), relative to those from green algae (lutein, b-phorbins), cyanobacteria (zeaxanthin, a-phorbins), and cryptophytes (alloxanthin, χ-carotene, a-phorbins). Other pigments such as the bacterial carotenoids oscilloxanthin and myxoxanthophyll are quite resistant to degradation, and can be overrepresented relative to other pigments originally present. Pigments are also destroyed, or converted to predictable byproducts, during ingestion by zooplankton (Daley, 1973).

Leavitt (1993) recognized three major stages of pigment destruction each with its own time scale and rates of degradation:

1. *Sinking of algal cells.* This is the phase of most rapid degradation. As much as 99% of pigments are lost from photo-oxidation or microbial oxidation, herbivory, or enzy-

matic breakdown. Pigment preservation is promoted by low O_2, low light conditions, rapid deposition in fecal pellets, deep water burial environments, cold hypolimnetic waters, and limited ingestion of organic matter in the sediment by invertebrates (Carpenter et al., 1986; Sanger, 1988; Leavitt and Carpenter, 1990).

2. *Loss at sediment–water interface.* This process is slower than #1, and depends on the chemical and biological environment of the lake bottom environment.

3. *Slow postburial diagenesis.* This operates over millennia, to reduce total carotenoid and chlorophyll content. During the period shortly after burial, selective loss of certain pigments continues, especially those of diatoms (diadinoxanthin, fucoxanthin, Chl c). This is commonly reflected by downcore declines in pigment concentrations in the upper few centimeters of short cores.

Because of their rapid diagenesis, the study of pigments has primarily been applied to Late Quaternary lake sediments. Some pigments, however, degrade into characteristic byproducts, which themselves are much more stable. Studies of the proportions of these byproducts have been extended back into much older sediments (e.g., Kimble et al., 1974).

In recent years, pigment geochemists have tried in various ways to address the question of how serious a limitation pigment diagenesis places on paleolimnological interpretation. The results are mixed. Sediment trap and experimental studies in individual lakes suggest the problem is more severe than do regional surveys of lake algae and sedimented pigments. This discrepancy suggests that loss patterns are governed by the unique characteristics of individual lakes, their morphometry, stratification, and productivity. Losses of pigments probably don't vary greatly between years as long as these controlling factors remain relatively constant. Pigment studies are most justified when the duration of the study is short enough that these background variables do not change significantly (Leavitt, 1993). Conversely, long-term changes in lake morphometry or stratification regimes are likely to alter the taphonomic environment of pigment preservation sufficiently to raise questions as to the comparability of results.

Some early applications of fossil pigments to paleolimnology used trends in the ratio of chlorophyll derivatives to carotenoids as an indicator of organic matter source, or trophic state, and, indirectly from these, climate change, because chlorophyll derivatives are better preserved in terrestrial organic matter (woodland organic matter, peats, and swamps have chlorophyll/carotenoids > 1.0), whereas algal decay favors carotenoid preservation. (e.g., Gorham and Sanger, 1967; Sanger and Gorham, 1972). Similarly, the ratio of oscillaxanthin/myxoxanthophyll, both cyanobacterial carotenoids, was proposed as an indicator of phosphorus availability and, indirectly, lake stratification (Swain, 1985; Tett et al., 1985; Ganf et al., 1991; Sabater and Haworth, 1995).

However, the large amount of data collected on variable rates of pigment diagenesis in the past two decades points to caution in the application of ratio data. As a result, many researchers now prefer to interpret individual pigment profiles directly, without reference to a normalizing pigment to create a ratio (figure 9.11). Individual profiles for commonly preserved pigments provide extremely valuable information, when those pigments are diagnostic of particular algal groups (S.R. Brown et al., 1984; Leavitt, 1993; Hall et al., 1997). Oscillaxanthin, mentioned above, is only produced by Oscillatoriacean cyanobacteria, a group that often forms blooms during the onset of eutrophication (Griffiths et al., 1969; Züllig, 1981; Swain, 1985). Others include zeaxanthin (indicative of cyanobacteria), alloxanthin (indicative of cryptophytes), fucoxanthin (diatoms, chrysophytes, and dinoflagellates), diatoxanthin (diatoms and dinoflagellates with brown algal endosymbionts), and lutein, pheophytin b, and chlorophyll b (indicative of chlorophytes). Purple and green sulfur bacteria also have unique photosynthetic pigments, which have been used to discriminate changes in bacterial populations and interpret past changes in light intensity in lakes.

Detailed interpretations of ecosystem change are possible when profiles are compared from multiple pigments, characteristic of different algal and consumer groups. Leavitt et al. (1994c) studied the top-down effects of fish introductions on algal communities, as reflected by fossil pigments from three, low-productivity, alpine lakes in the Canadian Rocky Mountains (figure 9.11). Two of the lakes, Snowflake and Pipit, were originally fishless, but were stocked in the 1960s with several trout species, whereas Harrison Lake had native trout. The species introductions caused the disappearance of larger cladoceran zooplankton species and benthic amphipods, and allowed smaller copepod and rotifer species to flourish. By the 1980s trout had disappeared from Snowflake and Pipit Lakes. Larger invertebrates partially, but not completely, returned to their prefish introduction state. Fossil pigment profiles illustrate the contrasting ecological histories of the three lakes. Pigment concentrations were low and constant for all pigments in Snowflake and Pipit Lakes prior to the 1950s, followed by a rapid increase, which peaked about 1965. Pigments related to chlorophytes (lutein-zeanthinin and pheophytin b), diatoms (diatoxanthin), and cryptophytes (alloxanthin) all showed major increase at the time of trout stocking, which impacted nutrient cycling in those lakes, and declined after trout were eliminated. In unstocked Harrison Lake, changes over time were much less systematic.

Pigment data can also be used as an index of UV penetration, and indirectly, of paleodrought conditions (Leavitt et al., 1997, 1999). Lakes that have undergone acidification generally experience a loss of DOC, which is normally the primary attenuator of UV radiation in lakes. Net UV radiation also increases naturally during periods of decreased cloud cover (drought conditions). In either case

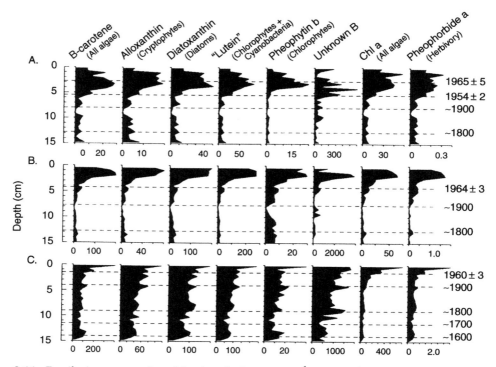

Figure 9.11. Fossil pigment stratigraphies (nmol pigment $gm^{-1}_{organicmatter}$) for three alpine lakes from the Canadian Rocky Mountains. Snowflake Lake (A) and Pipit Lake (B) were originally fishless, but stocked with various trout species in the 1960s, resulting in loss of zooplankton and abrupt changes in algal flora, discussed in text and reflected in pigment stratigraphies. Harrison Lake (C) was unstocked and shows only modest upcore changes in pigments, typical of diagenetic losses alone. From Leavitt et al. (1994c).

UV-absorbing pigments are produced by various benthic algal groups in response to this change in radiation, and increases in these pigments have been observed in core intervals corresponding to drought periods.

Pigments can provide information about the stratification history in lakes, using taxon-specific pigments from algal or bacterial clades that are characteristic of a particular stratification state (Hodgson et al., 1998). In some cases pigment alteration processes can also provide information about lake acidity (Guilizzoni et al., 1992). The conversion of chlorophyll to pheophytin is promoted by acidic conditions, and studies in European Alpine lakes have shown that progressive twentieth century acidification from atmospheric sources is recorded by increasing upcore concentrations of pheophytin.

Stable Isotopes

The stable isotope records of oxygen, carbon, and nitrogen have provided some of the most valuable information in the paleolimnological toolkit. Here, I will concentrate on what are arguably the two most important and thoroughly investigated archives of such information, sedimented carbonates (for O and C isotopes) and sedimented organic matter (for O, C, and N).

Carbonate Records of O and C Stable Isotopes

Calcium carbonate minerals provide a simultaneous record of the isotopic uptake of both C and O at the time of crystallization. Carbonate (primarily calcite and aragonite) isotope records have been successfully used to interpret the history of lakes and their climatic context since the 1960s (Oana and Deevey, 1960; Fontes and Gonfiantini, 1967; Stuiver, 1970). Since that time these records have been mined for information about paleoprecipitation, paleotemperatures, water residence time, runoff, paleogroundwater discharge, productivity, and lake mixing, to mention only a few of the more important applications. The relative importance attributed to these different factors usually varies with the climate regime relevant for the study lake. In cool water lakes, and those with very short residence times (flow-of-the-river lakes), oxygen isotopic records in carbonates may be good indicators of temperature, and are commonly interpreted as such in wet-temperate parts of North America and Europe. But in tropical and/or arid areas, where evaporation is more intense, residence times increase, and the relative importance of crystallization temperature as a determinant for overall patterns of oxygen isotopic variation declines. In deep lakes, with relatively low through-flow, or reduced temperature seasonality, the oxygen isotopic signal increasingly becomes one of residence time and paleohydrology.

A wide variety of sedimented carbonates may be available for any analysis, and in developing a research program or evaluating prior research work it is useful to consider the pros and cons of each for addressing particular questions in paleolimnology. Sources include precipitated carbonate muds (micrite) and grains, carbonate coatings, and skeletal materials. The nature and quality of record provided by each reflects the complexity of the isotopic pool in the source waters (for O) or carbon reservoir (for C) at the time of crystallization. Open-lake precipitates tend to reflect the broad average conditions of O and C prevailing for the lake as a whole; they provide an integrated archive that can reasonably be assumed to be in isotopic equilibrium with the lake water at the time of formation. Nearshore or benthic carbonates above a lake's chemocline may reflect more local conditions, for example from seepage, spring discharge, or runoff, as well as inputs from the average lake water. In some paleolakes changes in groundwater discharge appear to have been very important in determining isotopic compositional history (Hillaire-Marcel and Casanova, 1987).

For open-lake benthic carbonates, such as shells or ooids, the oxygen in a carbonate is likely to have accumulated in isotopic equilibrium with the lake waters at the time, because of the over-whelming influence of the water mass in determining O-isotope composition. Even early diagenetic precipitates occasionally provide information useful for limnological or paleoclimatic reconstruction (e.g., Yemane and Kelts, 1996). The same, however, cannot be said of the carbon isotopes in a benthic carbonate, which may reflect multiple sources (DIC_{water}, $DIC_{pore\ water}$, organic matter, or other older sediment sources of C). Micritic carbonate formed in the epilimnion is a source of C-isotope records that can reasonably be interpreted in terms of lake productivity in short residence time lakes, with increasing productivity reflected by higher ^{13}C concentrations in calcite (McKenzie, 1985). In lakes with long residence time, productivity plays a smaller role in determining $\delta^{13}C$, and C-isotope trends are more controlled by factors such as outgassing of ^{12}C-rich CO_2, variable rates of mixing between epilimnetic and hypolimnetic waters, and diagenetic changes associated with anoxia or elevated salinity (Talbot and Kelts, 1986, 1990).

The interpretation of the C-isotope record of benthic invertebrate shell carbonate is also complicated by the degree to which these organisms are influenced by the uptake of carbon from organic matter settling through the water column. When pronounced increases in productivity have led to the settling and decay of ^{13}C-depleted organic matter in deep water, this may be evidenced in a significant $\delta^{13}C$ difference in shallow-water micrite and benthic shell carbonate (Lister, 1988). Benthic carbonates formed by aquatic plants, especially charophytes, may also show systematic offsets in C-isotopic composition, relative to associated benthic invertebrates, related to the type of DIC being used by the plant (Hammarlund et al., 1997). The sediment–water interface is also an area where methanogenesis can release ^{13}C-enriched carbon (e.g., Curry et al., 1997). Finally, diagenetic carbonates from laminated sediments or concretions will reflect prevailing conditions below the sediment–water interface, including microbial activity, and different redox conditions than exist in the water column. During early diagenesis, these may be linked to the lake water mass, though through a strong microbial filter. However, later diagenesis can produce completely different compositions of carbonates than what was originally present in shallow pore waters. Because of these complexities, paleolimnologists should interpret stable carbon isotope records from carbonate much more cautiously than their associated oxygen isotope records.

Isotopic Records from Carbonate Muds (Micrite)

Carbonate muds that have been directly precipitated from a lake's water column provide one of the most reliable and coherent sources of oxygen and carbon isotope data for paleolimnological interpretation (Talbot and Kelts, 1990; McKenzie and Hollander, 1993). In most studies primary micrites are assumed to reflect temperature and isotopic composition in the mixed, upper water column. However, several factors must be kept in mind when interpreting isotopic records from micrite. First, the source of the sample should be clearly established; although authigenic micrites may be fairly easily distinguished in unconsolidated sediments this is much more difficult to do in limestones. Finely laminated, carbonate muds, from rhythmites or varves, provide some of the best sources for clean "inorganic" micrite muds, because they are less likely than thicker-bedded deposits to be mixed with abundant nearshore debris. Conversely, much of the micrite that is studied in isotopic investigations may in fact be finely broken fragments of ostracodes, *Chara* stems, or other benthic carbonates. In some settings carbonate muds may be overwhelmed by the input of detrital carbonates; isotopic records from lakes lying in dominantly carbonate bedrock terrains should be interpreted very cautiously. Micrites are commonly treated as inorganic precipitates, although, strictly speaking, this is frequently not the case; biological mediation is often involved in precipitation, and may induce systematic fractionation offsets unknown to the investigator. Finally, micrites should ideally be monomineralic. If multiple minerals are present between samples, investigators should be aware of their sample's composition; systematic isotopic fraction differences exist between different carbonate minerals (Tarutani et al., 1969; Romanek et al., 1992).

Other Inorganic Primary Carbonates

Benthic inorganic precipitates, such as coated grains, are occasionally used as sources of isotopic data for paleolimnological reconstruction. In low-resolution ($> 10^3$ years) studies, these may provide good sources of isotopic information. The possibility of extensive reworking of older carbonate grains into younger deposits makes these types of carbonates less suitable than laminites for high-resolution studies.

Benthic Microbial Carbonates (Stromatolites and Oncoids)

The laminar accretion of stromatolites and oncoids makes them excellent sources of isotopic information (Casanova and Hillaire-Marcel, 1993). Stromatolites that grew along paleoshorelines and in deeper water can provide detailed records of isotopic change during growth intervals, because the carbonate is effectively cemented and cannot be easily reworked, and those reworked fragments that are incorporated into the stromatolite are easily recognized through microscopic examination. Stromatolites "grow" episodically, whenever lake level conditions are appropriate, and a single stromatolite head may comprise several generations of growth bands, with individual bundles separated in age by many thousands of years. Individual growth couplets in stromatolites often form seasonally, allowing contrasts to be drawn between stable isotope compositions at different times of an annual cycle. Unlike varves, however, the accretionary pattern is usually quite discontinuous, between bundles of tens to hundreds of laminar couplets. As a result, stable isotope data from stromatolites are most useful for addressing questions that require the characterization of paleohydrological variability for discrete time intervals. Because stromatolites can grow at a range of depths below a lake's surface, their isotopic ratios can be used to study productivity and mixing at varying water depths (Casanova and Hillaire-Marcel, 1993).

Organically Precipitated Carbonates

Isotopic records from organically precipitated carbonates have been obtained from a wide variety of lacustrine fossils, including charophytes, ostracodes, mollusks, and so on. Most studies of these types of carbonates have focused on the interpretation of oxygen isotopes; the varied sources of DIC available to heterotrophic and/or benthic organisms makes interpretation of carbon isotope trends from these types of fossils very difficult. Because of uncertainties about species-specific isotopic fractionation effects, most authors try to use one or a very limited number of species for analysis throughout a stratigraphical sequence.

Ostracodes are probably the most commonly used biogenic carbonates, a result of their abundance, and the potential for combining shell isotopic studies with minor element analyses and/or faunal analyses on precisely the same samples (e.g., Curtis and Hodell, 1993; Palacios-Fest et al., 1993; Cohen et al., 2000a). Faunal analyses can be used to constrain the environmental conditions under which the ostracode grew, for example whether it was a benthic, burrowing, or swimming species, whether it inhabited warmer littoral or deeper, cooler profundal habitats, and whether it likely grew in the influence of groundwater discharge, thereby guiding the interpretation of the ostracode shell geochemistry. Curtis and Hodell (1993) made this type of multi-indicator study on Late Pleistocene and Holocene ostracode fossils from Lake Miragoane, a deep, warm-monomictic lake in Haiti (figure 9.12). The lake is presently open,

but probably loses most water from evaporation. Curtis and Hodell obtained isotopic and minor element records from *Candona* spp., benthic ostracodes whose shell chemistry would reflect bottom-water conditions.

Bivalves have also proven extremely useful as sources of oxygen isotopic data (Dettman and Lohmann, 1993, 2000; Wurster and Patterson, 2001). Their accretionary growth pattern allows them to be used to obtain records of interseasonal and interannual variability in $\delta^{18}O_{water}$. Monthly to seasonal growth increments are often evident in microscopic examination of bivalve shells. Annual cloudy layers or thickened conchiolin layers are

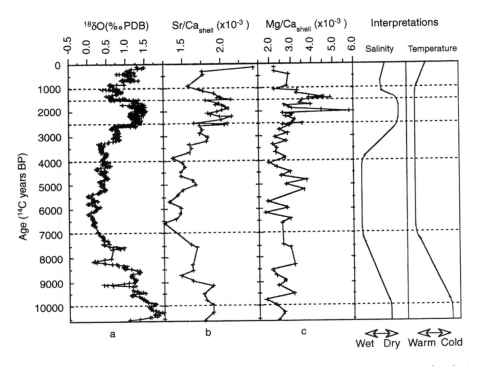

Figure 9.12. Summary of stable isotope and minor element records from ostracode fossils in Lake Miragoane, Haiti and the interpretations of those records. Sr/Ca and Mg/Ca records were interpreted based on studies from related North American *Candona* species. From 10.5–10 ka BP the lake was saline (high Sr/Ca) and cold with high evaporation/precipitation ratios (positive $\delta^{18}O$). The $\delta^{18}O$ of rainfall at this time was also heavier during the Late Pleistocene because of a global depletion of ^{16}O from the world's liquid water supply as a result of increased global ice volume. Mg/Ca remained constant, as a result of contrasting effects of low temperature and high salinity (also supported by pollen evidence). From 10–7 ka $\delta^{18}O$ and Sr/Ca decreased, indicating decreasing salinity and increasing temperature (global seawater $\delta^{18}O$ also became more negative as the ice caps melted). From 7–4 ka temperature and salinity were fairly constant and salinity was low (more mesic conditions are also supported by the pollen record). From 4–2.5 ka all variables increased, indicating increasing salinity and aridity (higher evaporation/precipitation). From 2.5–1.5 ka relatively constant conditions occurred, with positive $\delta^{18}O$ and high Sr/Ca indicating maximum aridity. All indicators decrease from 1.5–1 ka but then rise again in the past few hundred years. From Curtis and Hodell (1993).

often visible, providing an independent signal of annual growth increments, providing an independent means of interpreting the pattern of isotopic variation between growth layers. In temperate-climate bivalves, cloudy shell layers form during cold months, just before and after hibernation. Growth bands can be individually sampled by milling microscopic shavings off the shell in micrometer-range intervals, collecting the carbonate precipitated at different times of the year.

When analyzing bivalve growth bands, the investigator must try to determine whether the variability in carbonate $\delta^{18}O$ is dominated by changes in temperature or influent water $\delta^{18}O$ (Dettman and Lohmann, 1993) (figures 9.13 and 9.14). If the $\delta^{18}O$ composition of the ambient water is relatively constant, then temperature signals will dominate the annual isotopic cycle. This results in a record displaying more enriched $\delta^{18}O$ during the cooler months, with the greatest seasonal variability occurring in mid latitudes. At high latitudes, bivalves cease to grow during winter months (referred to as "biological shutdown") and part of the predicted temperature-induced cycle of isotopic change will be missing, as evidenced by the interpretation of seasonal cloudy layers and growth bands.

In lakes where water isotopic composition changes during the year, for example from variation in runoff or rapid evaporation, these effects will be superimposed on temperature, causing bivalve shell $\delta^{18}O$ to vary in more complex ways. Where seasonal oxygen isotope variation is greater than about 6‰, variability in $\delta^{18}O_{water}$ must be invoked because temperature differences that would explain this much $\delta^{18}O$ change are unrealistic for bivalve growth conditions. Furthermore, in the temperate zone, warm summer rain is almost always more positive in $\delta^{18}O$ than winter precipitation; and this trend, when recorded in shell carbonate, inverts the profile produced by crystallization temperature per se on precipitating aragonite. A larger seasonal contrast in summer versus winter precipitation will have the effect of dampening the resultant seasonal contrast in isotopic variation (figure 9.14). Areas like the Amazon basin, with no seasonal change in temperature, but strong seasonality in their source of precipitation, could also yield large seasonal variations in bivalve oxygen isotope records.

Subannual, isotopic records can also be obtained from otoliths, accretionary structures composed of aragonite that form in fish ears (Patterson et al., 1993; G.R. Smith and Patterson, 1994). Annual growth bands in otoliths may be > 1 mm yr^{-1}, but are themselves composed of seasonal bands

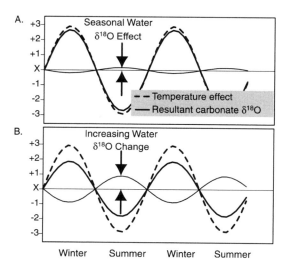

Figure 9.13. Simple model of seasonal change in shell $\delta^{18}O$ expressed as a deviation from a mean value (X) under two contrasting isotopic scenarios. The effects of temperature and water $\delta^{18}O$ are combined to produce carbonate $\delta^{18}O$. Temperature effects are identical in A and B. In model A, change in $\delta^{18}O_{shell}$ is driven primarily by a seasonal temperature effect. In model B, significant seasonal $\delta^{18}O_{water}$ variability is superimposed on temperature. The change in $\delta^{18}O_{water}$ could be caused by the size of the water body, increased summer evaporation at B, or greater seasonal precipitation at B. From Dettman and Lohmann (1993).

and even daily increments. Otoliths can often be identified to species. Otolith $\delta^{18}O$ data from a species with a narrow temperature range is most likely primarily controlled by water $\delta^{18}O$. In contrast, the isotopic record of otoliths from fish species that live in a wide range of temperatures is likely to be more strongly influenced by temperature. Simultaneous analysis of contemporary deep versus shallow-water species records has also proven useful for characterizing differences in residence time and temperature through the water column.

Applications of Oxygen and Carbon Isotope Data in Paleolimnology

Oxygen isotope data is used most frequently in paleolimnology as either an indicator of paleotemperature or general evaporative trends. In open-basin lakes with relatively short residence times and low evaporation rates, temperatures of carbo-

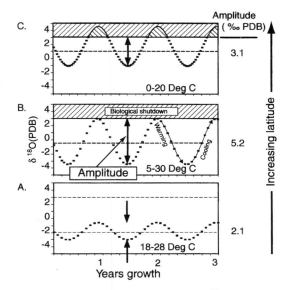

Figure 9.14. Schematic patterns of $\delta^{18}O$ variation in the growth of bivalves at different latitudes. $\delta^{18}O$ of the water is constant and temperature variation is based on a hypothetical transect from low latitude (A) to high latitude (C) sites. $\delta^{18}O$ recorded in shell carbonates (■); cross-hatched areas define cessation of growth resulting from cold weather shutdown. From Dettman and Lohmann (1993).

nate crystallization can sometimes be inferred quite precisely, if the average isotopic composition of the regional precipitation is well known. Early studies of the oxygen isotopic history of the Laurentian Great Lakes attributed most isotopic variability in terms of temperature (Fritz et al., 1975). However some isotopic excursions in open-basin, short residence-time lakes are simply too large to be realistically explained by temperature alone. In their study of the oxygen isotopic history of Lake Michigan, Colman et al. (1990) showed that other sources of isotopic variability, such as changes in runoff, in this case, the influx of Laurentide Ice cap meltwater, can be very important in open-basin lakes.

For long residence-time lakes, $\delta^{18}O$ trends often correlate with climate; lakes with low P/E ratios will over time display trends toward more positive $\delta^{18}O$ values (figure 9.12). The rate at which oxygen isotope composition responds to this forcing is largely a function of the lake volume; small lakes and lakes with small catchments are more sensitive to brief episodes of change in P/E, and their isotopic records will react more sensitively to such changes. Closed-basin lakes are particularly prone to

repeated cycles of more positive $\delta^{18}O$ values, reflecting drier times, and negative $\delta^{18}O$ values, from wetter times (Talbot and Kelts, 1990). Qualitative interpretations of isotopic trends are particularly valuable in pre-Quaternary lake deposits, where the original lake morphology and paleohydrology are unknown (e.g., Bellanca et al., 1992).

Comparisons of Deep Versus Shallow-Water Isotope Records

Differences between near-surface and deep water fractionation processes for carbon and oxygen isotope ratios can be exploited to interpret the history of lake stratification and productivity. As we saw earlier, oxygen isotopes in authigenic micrites normally record some combination of temperature, water input, and evaporative conditions at the lake's surface. In deep temperate or boreal, dimictic lakes, where bottom waters remain at or near 4°C year-round, it is possible to constrain the crystallization temperatures for benthic carbonates from mollusks or ostracodes. In these circumstances isotope stratigraphical changes can be interpreted in terms of the remaining variables alone. Carbon isotope offsets also exist between surface and deep waters. In shallow, well-mixed lakes this results primarily from differences in primary productivity, whereas early diagenetic reactions and methanogenesis are more important causes of shallow/deep differences in stratified lakes.

As a result of these differences, if a lake is well mixed, then benthic carbonates can be compared against the surface-water carbonates, to interpret offsets of shallow and deep-water records as a function of temperature for oxygen isotopes, and productivity for carbon isotopes (e.g., Lister, 1988). Some studies have taken this general approach because of its potential for differentiating productivity effects on isotopic composition from those driven by temperature (e.g., Fillipi et al., 1999). Particular care must be taken when making such comparisons to be certain that "shallow-water" carbonate silts are not in fact simply crushed ostracode shell fragments or detrital carbonates.

The observation that variation in $\delta^{18}O$ is correlated with variation in precipitation/evaporation ratios and residence time in closed-basin lakes suggests that $\delta^{18}O$ might be useful as an indicator of paleolake levels. In principle, if both isotopic and water volume inputs and outputs are known, it should be possible to model this relationship. In hydrologically simple, closed-basin lakes, where proportional water volumes change very rapidly (over a few years), it may be possible to infer lake levels more or less directly from O-isotope measurements. This approach was taken by H.C. Li et al. (1997) to investigate the isotopic and lake-level histories of Mono Lake, where a good correlation exists between isotopes and lake levels alone for the historical period (figure 9.15).

However, usually lake level cannot be "read" directly from an isotopic history. The balance of water volume and isotopic composition entering and exiting the lake, the residence time of water, the fractionation associated with evaporation, and the lake's volume all need to be accounted for in developing such a relationship. Furthermore, in larger lakes with multiple influent streams, compositional heterogeneity can be expected to exist in isotopic records from different areas. Composite records must be assembled to take this variability into account. For a hydrologically closed basin, where evaporation is the only output of water, this relationship can be expressed as:

$$\partial(V\delta_{lake})/\partial t = Q_{input}\delta_{input} - Q_{evaporation}\delta_{evaporation}$$

where V = lake volume, Q = amount of flux, δ = isotopic composition of each flux.

Ricketts and Anderson (1998) used this relationship to model the recent (late nineteenth to twentieth century) lake-level history of Lake Turkana (Kenya) (figure 9.16). They studied micritic carbonates from a number of well-dated freeze cores to obtain isotopic records over this time interval. They then combined and normalized these data to obtain a composite lakewide record of isotopic change. In doing so they obtained a reasonably good fit between the "predicted" lake-level history and what has actually been observed from historical lake-level records. To do this, however, required some a priori knowledge or assumptions about the isotopic composition and water volume fluxes into and out of the lake. For extant lakes or Quaternary lake deposits, where the paleohydrological interconnections are well understood, this is feasible, since reasonable assumptions can usually be made about these parameters. For more ancient lake deposits it is likely to be much more difficult.

In hydrologically complex lakes, oxygen isotope ratios may be largely or completely decoupled from lake level, usually as a result of upstream fractionation processes (e.g., Cohen et al., 1997a). Upstream lakes can provide water to downstream ones that is considerably more enriched in ^{18}O than other watershed runoff sources feeding the downstream lake. If the upstream lake is simultaneously providing an important water source for the overall hydrological balance of the lake, then a closure of the upstream lake, for example from drier climate, could lead to a simultaneous decline in the downstream lake level, coupled with more negative values in the downstream lake's $\delta^{18}O$ archive.

Oxygen isotopes are also effectively decoupled from lake level and most salinity indicators in closed basins during periods of rapid filling. A lake that has nearly or completely dried, and then

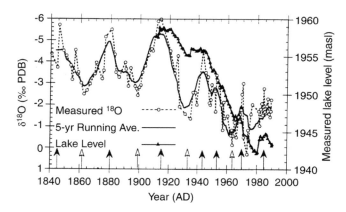

Figure 9.15. Comparison of measured $\delta^{18}O$ and their 5-year running averages in washed samples of a core from Mono Lake, California, with the historical measured Mono Lake level. Scales for the two ordinates are set arbitrarily to facilitate the comparison. $\delta^{18}O$ decrease corresponds to lake-level increase. Solid arrows indicate periods of high runoff and high lake stands. Open arrows indicate periods of low runoff and low lake stands. Note that the lake level continuously declined between 1941 and 1981, largely as a result of stream diversion, with secondary peaks denoting high runoff periods during this interval. From H.C. Li et al. (1997).

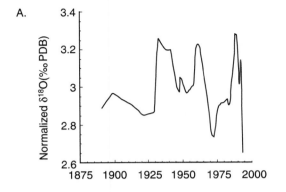

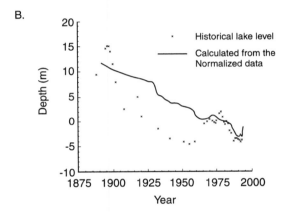

Figure 9.16. $\delta^{18}O$ freeze core record from closed basin, Lake Turkana, Kenya and comparisons between lake-level modeling and observed lake-level fluctuations. (A) Normalized and stacked $\delta^{18}O$ record from authigenic calcite in cores. The records from three cores with good chronologies were normalized ($[\delta^{18}O - \delta^{18}O_{mean}]/\delta^{18}O_{standard\ deviation}$) and interpolated to yearly values. This record served as the basis for isotopic inputs into a hydrological model of lake level (also requiring estimates of water inflow and evaporative output from the lake). (B) Water depth versus year lake-level record and model output results. The fit is remarkably good given that there was considerable spatial variability between freeze core records. From Ricketts and Anderson (1998).

rapidly begins to refill during a period of increased precipitation, will possess a $\delta^{18}O$ signature controlled by regional rainfall. Simultaneously, dissolution of previously precipitated efflorescent crusts on the lake bed may cause this same water mass to be quite saline. As the lake level continues to rise, the isotopic composition of precipitated carbonates

may become progressively more enriched in ^{18}O, while the lake becomes less saline (e.g., Chivas et al., 1993).

Oxygen Isotopic Records from Biogenic Si

Paleolimnologists frequently encounter lake sediments from which they would like to obtain an oxygen isotope record, but where carbonates are absent. In acidic or low-productivity systems, an alternative approach is to obtain oxygen isotopic data from biogenic silica, typically from diatom frustules (Leclerc and Labeyrie, 1987) although, in principle, from other sources such as sponge spicules as well. Analyses of oxygen isotopes from biogenic silica are much less commonly made than carbonate analyses, because of analytical and interpretive difficulties inherent in the technique. Oxygen is bound in biogenic silica from diatoms in two distinct fractions. The inner skeletal mass is relatively unhydrated, and its oxygen isotope composition reflects ambient conditions at the time of skeletonization. The outer, more porous skeletal layer contains abundant hydrated silica. This hydration can occur at any time after skeletal formation, up to and including the time of analysis. Special analytical techniques have been developed which can account for this hydration oxygen. Furthermore, because individual diatoms are very small, bulk samples must be used, necessitating special cleaning techniques to eliminate detrital silica. Oxygen isotopic fractionation in the formation of biogenic silica is also temperature-dependent, although this relationship differs from that of carbonates. When these difficulties are surmounted, however, very valuable $\delta^{18}O$ records can be obtained from carbonate-poor lakes, and these archives are interpretable for paleoclimate and paleohydrology studies in similar ways to carbonate records (Rietti-Shati et al., 1998; Barker et al., 2001; Leng et al., 2001).

Stable Isotopic Analyses of O, C, and N in Organic Matter

Isotopic analysis of O, C, and N are widely used by paleolimnologists to determine the history of internal lake processes of mixing, residence time, and whole-lake metabolism, to differentiate sources of organic matter, and more broadly, to infer paleoclimates. Because organic matter can come from both terrestrial and aquatic sources, differentiating provenance is critical for addressing most research questions. It is always desirable to have supporting and independent information on the origin of organic matter, since this will constrain the interpretation of the isotopic record. Increasingly, isotope geochemists are analyzing stable isotopes in organic matter from specific organic compound fractions, for example from biomarkers. This allows extremely detailed interpretations of origins for the various components of organic matter present (Hayes et al., 1987; Meyers, 1997). Analyses of compound-specific isotopes greatly enhance what can be learned from biomarker studies or bulk isotopic analyses of organic matter alone. Although several groups of organisms may synthesize the same biomarker, the way that isotopes of carbon are partitioned in the process of that synthesis may differ. For example, the isotopic differences between C3 and C4 photosynthetic pathways are reflected in analyses of individual, long-chain *n*-alkanes, just as they are in bulk organic matter. In some cases, the analysis of isotopic composition of specific biomarkers allows identification of source organic materials to quite specific levels, for example to specific genera of trees (Rieley et al., 1991).

When cellulose can be shown to be primarily derived from aquatic sources, analysis of its $\delta^{18}O$ can provide a record of lake water $\delta^{18}O$ (Epstein et al., 1977; Sternberg et al., 1984; Buhay and Betcher, 1998; Sauer et al., 2001). As with the analysis of $\delta^{18}O$ of biogenic silica, this is a useful approach when carbonates are absent. Oxygen isotopes show a strong but consistent +27 to 28‰ fractionation in aquatic cellulose, which appears to be relatively independent of water temperature (B.B. Wolfe et al., 1996). Carbon stable isotope analyses are performed far more commonly on lacustrine organic matter than either oxygen or nitrogen analyses. But increasingly C and N isotopes are being simultaneously analyzed in organic matter, especially in conjunction with other methods such as rock-eval pyrolysis, petrography, and C/N ratios.

Factors Responsible for Temporal Shifts in Carbon Isotopes

Temporal changes in carbon isotopes in organic matter result from a variety of internal and external processes:

(1) *Changes in dominant source of dissolved inorganic carbon.* In a stratified lake, the accumulation of a pool of carbon depleted in ^{13}C commonly occurs in deep water, as phytoplankton cells sink and degrade. If this water is recirculated to the surface, a pronounced decrease in $\delta^{13}C$ can occur. Conversely, a prolonged period of stratification will cause the available DIC pool in the epilimnion to become enriched in ^{13}C over time. Long-term variation in factors that alter the depth or intensity of upwelling can therefore be recorded as shifts between more negative and more positive $\delta^{13}C$. This would include such things as changes in overall windiness (windier periods reflected by deeper mixing and more recirculation of hypolimnion C), average water temperature (cooler water temperatures causing less stable stratification and deeper mixing), or changes in salinity-related stratification. This cause of shifts in carbon isotopes is most plausible in situations where a lake is known as, or suspected of, being episodically stratified, and in offshore settings, where terrestrial components of OM are unlikely to dominate the carbon isotope signal. Carbon isotope archives of upwelling events are likely to appear as spiky shifts toward lower $\delta^{13}C$ values in core OM records.

(2) *Changes in productivity/eutrophication.* Increased productivity speeds up the transfer of OM with relatively negative $\delta^{13}C$ values to the hypolimnion, thereby enriching the residual epilimnetic DIC in ^{13}C. This effect will be superimposed on mixing effects, either enhancing or suppressing the mixing signal (e.g., Schelske and Hodell, 1991).

(3) *Changes in metabolic pathways for carbon fixation.* In lakes that have undergone a major change in alkalinity, both benthic and planktonic primary production will undergo a switch in the dominant source of DIC for photosynthesis, with CO_2 favored at lower pH and HCO_3^- at higher pH. In a lake that is becoming progressively more alkaline, this will result in a trend toward more positive $\delta^{13}C$. This explanation is most plausible in lake sediments dominated by autochthonous organic matter, and where independent evidence for high alkalinity exists.

(4) *Changes in availability of dissolved CO_2.* Higher concentrations of CO_2 can be dissolved from the atmosphere into cool water than warmer water. For a given amount of productivity this larger pool of available CO_2 will not undergo as much fractionation as would be the case in warmer water. Thus, a major cooling event can cause a shift to lower $\delta^{13}C$ in organic matter. CO_2 availability can also be driven by changes in CO_2 concentrations in the atmosphere. For example, lower pCO_2 in the atmosphere during glacial periods would cause iso-

topic discrimination in plants that use dissolved CO_2, but not in terrestrial plants that extract CO_2 gas directly from the atmosphere (e.g., Meyers and Horie, 1993).

(5) *Changes of dominant vegetation within the watershed.* In some lake sediments, terrestrially derived plant debris is the dominant source of organic matter. This is likely to be the case in small lakes, and in the coastal or deltaic zones of large lakes. Under these circumstances, significant shifts in watershed vegetation, for example, from dominantly C3 to C4 photosynthetic pathways, can cause major changes in carbon isotopic composition in lake sediments. During a transition from a C3 to a C4 plant-dominated landscape, organic matter entering the lake would become more positive in $\delta^{13}C$; the reverse isotopic trend can be expected from a C4 to C3 transition. These trends can be interpreted more broadly in terms of paleoclimatic change prior to the introduction of agriculture, since C4 plants are favored under conditions of water stress, and, to a lesser extent, decreased available atmospheric CO_2 (e.g., Huang et al., 2001). Watersheds that have undergone conversion from forest cover to agricultural C4 grasses can also produce isotopic shifts in the $\delta^{13}C$ of lacustrine organic matter (Sackett et al., 1986). The likelihood that watershed vegetational changes are responsible for carbon isotope trends can be evaluated by examination of C/N ratios or rock-eval pyrolysis data. It is important to recognize that lag times may exist between the formation of terrestrial organic matter and its deposition in a lake, particularly in lakes with large watersheds, and for high-resolution studies these lags can be cause for concern (Hillaire-Marcel et al., 1989).

(6) *Diagenetic trends.* Some diagenetic processes also result in sustained shifts in carbon isotope ratios in organic matter. In the extremely carbon-rich sediments of a marsh, a loss of the more reactive components of organic matter, especially amino acids, that typically have more positive $\delta^{13}C$ values, has been observed (Spiker and Hatcher, 1984; Meyers and Ishiwatari, 1993). This results in a shift of several ‰ toward more negative bulk organic matter. In lake sediments with lower organic concentrations these shifts aren't observed. Isotopic changes in carbon that accompany major TOC changes should be interpreted in this light. Intense fractionation also accompanies methanogenesis; changes in carbon isotopes that accompany indications of redox shifts need to be interpreted in this light.

Factors Responsible for Temporal Shifts in Nitrogen Isotopes

Because aquatic algal tissue is much richer in nitrogen than terrestrial vascular plant tissue, both the bulk nitrogen content and nitrogen isotope ratios are dominantly controlled by lacustrine, autochthonous inputs of organic matter (Talbot, 2001). This imbalance in N sources is so great that lacustrine N isotopic signals will prevail even when deposited organic matter is not the dominant organic matter by mass. This implies that studies of paleolake processes using N isotopic ratios may not be compromised even when terrestrial organic matter input is substantial.

Despite the lack of terrestrial plant forcing factors driving N-isotopic ratios, there are still a multitude of possible aquatic, soil and diagenetic fractionation factors we have to contend with when explaining trends in $\delta^{15}N$ (table 4.1). This wide range of causes is both a curse and a blessing to the investigator, making it difficult to determine specific reasons for a given trend using N-isotopes alone. However, the power of nitrogen isotope data becomes evident when they are combined with other indicators. If a combination of independent indicators allows us to rule out most causes of fractionation, the remaining possibility may say something very specific about lake ecosystem history.

(1) *Changes in productivity/eutrophication.* Rapid burial of organic nitrogen, coupled with more intense demand for dissolved inorganic nitrogen by phytoplankton, will result in a progressive enrichment of ^{15}N in sedimented organic matter. Over time this will lead to a combination of rising N concentration in sediments and higher $\delta^{15}N$ values, commonly accompanied by a trend toward more positive $\delta^{13}C_{organic\ matter}$. Trends of this type have been observed in the recent sediments of lakes undergoing cultural eutrophication from fertilizers or sewage.

(2) *Changes in the volume of a lake's hypolimnion.* An increase in the proportion of a lake that is weakly mixed or permanently stratified promotes more voluminous denitrification. This can cause sedimented aquatic organic matter over time to become more positive in $\delta^{15}N$. Lakes that are undergoing transformation from seasonal mixing to oligomixis or meromixis may display this trend (e.g., Hecky et al., 1996).

(3) *Changes in lake water chemistry.* The volatilization of ammonia at high pH causes a strong enrichment in ^{15}N. This is most likely to occur in

closed-basin lakes under highly evaporative conditions, and thus could be used as an indirect indication of more arid climate.

(4) *Change in algal flora.* Especially from or toward N-fixing cyanobacteria. Nitrogen-fixers have access to the large reservoir of ^{15}N-depleted atmospheric N. Organic matter formed from them will be depleted in ^{15}N, whereas phytoplankton that assimilate ammonia or dissolved nitrate will produce ^{15}N-enriched organic matter.

(5) *Change in mixing regimes.* Stratification leads to more positive δ^{15}N values in the surface water dissolved inorganic nitrogen pool, and the production of ^{15}N-enriched organic matter from epilimnetic phytoplankton. Deep mixing episodes will have the reverse effect, returning ^{15}N-depleted nitrogen to surface waters. The effects of stratification, mixing, and algal flora on nitrogen isotopic ratios can be ambiguous because of their contrasting effects. Stratification often favors an increase in N-fixing cyanobacteria as well as decreased mixing. This demonstrates the need for other sources of information to resolve what has occurred, for example supplementary pigment or biomarker data to identify the algae involved.

(6) *DIN supply changes.* There is considerable debate about how the size of the internal DIN pool of a lake affects the nitrogen isotope composition of lacustrine organic matter. However, modification of the total DIN pool in a lake by external inputs should be recorded in the δ^{15}N. It has been shown that rapid rises in lake level or deforestation can release soil N to the lake, which is enriched in ^{15}N. This latter effect tends to be short-lived and produces a spiky increase-then-decrease pattern in δ^{15}N (e.g., Talbot and Lærdal, 2000). Changes in soil composition can also drive nitrogen isotopic ratios, especially in small lakes with abundant or very variable allochthonous inputs of soil (B.B. Wolfe et al., 1999). More intense biogeochemical cycling in soils, for example in more humid or warmer climatic conditions, leads to ^{15}N-enrichment, and this can be reflected in lake sediments if they contain large quantities of redeposited soil organic matter.

(7) *Diagenetic changes.* Nitrogen-rich compounds are very volatile and tend to decompose rapidly before and shortly after burial. Both theory and observation in marine settings suggest that nitrogen isotope ratios become more positive during early diagenesis, although studies of organic matter accumulating in lakes are inconclusive as to whether this effect is important or pervasive. Some evidence suggests that rapid burial reduces the likelihood of significant

diagenetic change in N-istopes (Hodell and Schelske, 1998).

Lake Bosumtwi (Ghana): An Example of Integrated Isotopic/Organic Matter Investigations

It should be evident from the preceding discussion that organic matter and stable isotope data are most powerful when presented as integrated studies, allowing trends or absolute values of one indicator to eliminate possible explanations of trends in another. A paleolimnological study of a 17-m core from meromictic Lake Bosumtwi illustrates this point, archiving a late Pleistocene–Holocene record of diagenetic and climate change for West Africa that relies heavily on various analyses of organic matter (figure 9.17) (Talbot and Kelts, 1986; Talbot and Johannessen, 1992). The C/N ratio, HI, and smear slide data all suggest that organic matter throughout the core contains a mixture of terrestrial and phytoplankton debris, consistent with the lake's relatively small size (8 km diameter). Four paleolimnological zones are evident in the core.

Zone 4. Prior to 25 ka, sediments are characterized by microlaminated diatomite, clastic units, and carbonates, with relatively low, but variable δ^{13}C and δ^{15}N values from organic matter, and very variable C/N ratios. Pollen data for this time suggests that forests covered the watershed (Maley, 1989). Nitrogen data, phytoplankton fossils, and sedimentology all suggest that the geochemical and sedimentological variability resulted from changes between a well-mixed (nutrient regeneration, replenishment of DIN pool, increasing diatom abundance) and a poorly mixed (cyanobacterial dominance) lake.

Zone 3. After 25 ka, rising δ^{13}C values indicate an increase in C4 grasses in the watershed, and generally more arid conditions. The lake's elevated alkalinity at this time is supported by (a) higher δ^{15}N, presumably from ammonia volatilization at high pH, and (b) the formation of diagenetic Mg-calcite and dolomites.

Zone 2. After 10 ka a dramatic shift occurs, to lower δ^{13}C and δ^{15}N values, indications of a rise in lake level and an increase in forest pollen. Lower δ^{15}N values suggest a switch occurred to a dominance of N-fixing cyanobacteria. These changes are all consistent with the establishment of greater water column stability and probably meromixis.

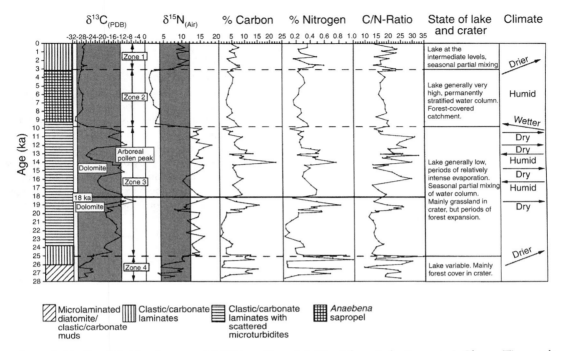

Figure 9.17. Simplified log for cores B6/B7 (total length 16.9 m) from Lake Bosumtwi, Ghana. Time scale is based on 13 conventional ^{14}C dates. Shaded areas show probable compositional ranges of terrestrial organic matter supplied to the lake. From Talbot and Johannessen (1992).

Zone 1. After 3.5 ka the core records a switch to more positive δ^{15}N, indicating the end of extreme stratification and a renewal of deep water, ^{15}N-enriched DIN sources. This also marked the end of algal domination by N-fixing cyanobacteria. Pollen remains show that forest cover persisted during this time, implying that the lake's watershed became windier rather than drier.

Summary

Our conclusions from this chapter can be summarized as follows:

1. Archives of isotopic, elemental, and molecular constituents of lake sediments are central components of most paleolimnological investigations. Correctly interpreting these archives depends on a good understanding of both the predepositional and diagenetic filters that affect lake sediment geochemistry.
2. In freshwater lakes, sediment profiles of the major elements Na, K, Si, and Al may pro-

vide information about watershed weathering intensity. Al also has potential as an indicator of lake pH history. Biogenic silica, mostly from diatoms, is often an indicator of paleoproductivity.
3. Ca, Mg, and Sr concentrations are most frequently analyzed in biogenic carbonates, especially in fossil ostracodes and mollusks, where elemental ratios can provide valuable information about paleotemperature and paleosalinity.
4. Fe, Mn, P, and S are all important and reactive elements in lake biogeochemical cycles, and archives of these elements would be of considerable interest to paleolimnologists, especially for understanding the history of anoxia, nutrient loading, and acid precipitation in watersheds. However, interpretation of sediment profiles for all of these elements is greatly complicated by their variable mobility under different redox conditions, and by their uptake in sediments by bacteria.
5. Trace metal profiles, especially for Pb and Hg, have been widely studied by paleolimnologists, to record the history of pollutant

inputs from watersheds and the atmosphere.

6. Bulk organic matter records (TOC, C/N, and rock-eval pyrolysis) provide inexpensive archives of the proportion of terrestrial to aquatic organic matter sedimented in a lake, and indirectly from this, of changes in lake productivity and watershed climate. Compound-specific analyses of specific biomarkers, especially lipids and pigments, provide more detailed records of organic inputs, allowing the investigator to more clearly define the provenance of organic matter, identify past sources of hydrocarbon pollutants, and interpret past trophic interactions.

7. $\delta^{18}O$ and $\delta^{13}C$ records from lacustrine carbonates are important sources of information about paleotemperature, water residence time, P/E ratios, and past lake productivity. Because isotopic fractionation is controlled by multiple factors, the accurate interpretation of lacustrine isotopic records relies heavily on a realistic understanding of lake hydrology and mixing.

8. $\delta^{13}C$ and $\delta^{15}N$ records from both bulk and compound-specific organic matter can greatly augment the analysis of OM provenance, and can provide very detailed information about internal nutrient cycling and paleolake productivity. However, for accurate interpretation, this information must be collected in conjunction with more routine organic geochemical information.

10

Paleoecological Archives in Lake Deposits I: Problems and Methods

Fossils provide some of the most detailed sources of information for environmental reconstruction available to the paleolimnologist. The use of lacustrine fossils to infer paleoenvironmental conditions is fundamentally based on inferences derived from modern correlations between the distribution of organisms and environmental variables, coupled with an understanding of *taphonomy*, the study of the fossilization process. No single group of organisms provides a comprehensive picture of lake ecosystems or environmental change, so it is always desirable to gather paleoecological records from multiple clades and habitats in a paleolimnological study. Analysis of multiple clades provides a means of establishing or testing ecological hypotheses that may not be possible from the study of one group alone. For example, many limnological processes affect the plankton, littoral organisms, and benthos in predictable sequences, and with predictable intensities.

The most comprehensive study of fossil data and data analysis will be meaningless if the fossils studied are misidentified. A good taxonomic framework is an essential element of paleoecological studies. Accurate identification of described species, and the curation of voucher specimens, photographs, and other descriptive materials of undescribed species is important, to insure the quality of a paleolimnologist's ongoing work, and to avoid future errors based on previously misidentified fossil specimens.

Special Considerations in Using Fossils for Lake Reconstruction

Using fossils to interpret lacustrine paleoenvironments requires not only an understanding of modern organism distributions, but also an understanding of four additional factors we did not con-

sider in chapter 5: (1) ecological causality and scale, (2) taphonomy and time-averaging, (3) historical contingency, and (4) evolutionary processes.

Ecological Causality and Scale

One of the most common uses of fossil data in lakes is to try and reconstruct changes in some physical or biological forcing process from changes in abundance or morphology of the fossil organisms affected by the process. We might be interested in reconstructing changes in nutrient flux to the lake, based on changes in the relative abundances of some fossil animals (figure 10.1). However, these animals actually responded to nutrient load only indirectly, through the effect of nutrient load on autotrophs and/or organic detritus. Now suppose that our ultimate objective is not really to understand productivity changes or nutrients, so much as to understand climate changes that may be driving productivity changes. This removes the organism another step away from the "cause" we are trying to investigate. Each of these steps involves a set of information filters, so with each step the connection becomes potentially more obscure between the proximal "cause" of some ecological change and the correlation we are actually interested in. Ideally, in an investigation of indirect causes of ecological change we would analyze multiple types of records (diatoms, cladocerans, etc.), since each may be subject to a different set of information filters.

Ecologists working over short time intervals commonly focus on proximate causes for community or morphological change in organisms. Such causal relationships are readily manipulated experimentally, or studied through monitoring, and are manifest over short time periods (figure 1.6). At their upper temporal limit, ecological studies overlap with the most highly resolved lake records. Therefore, questions at this scale may be of interest

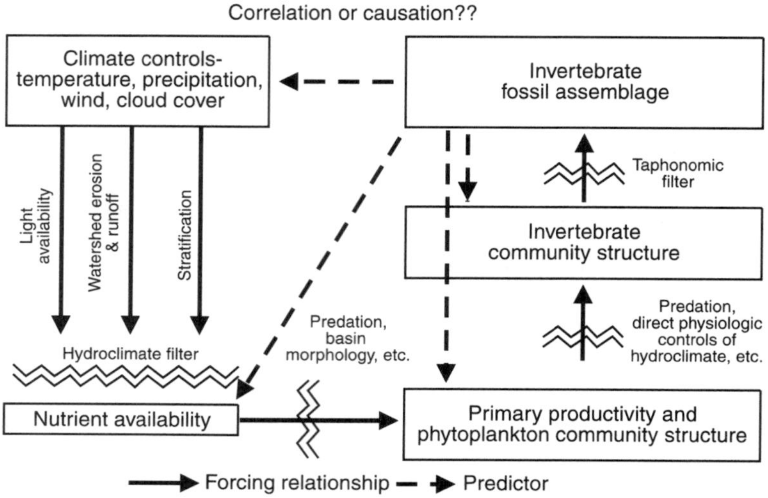

Correlation or causation??

Figure 10.1. Conceptual model illustrating direct and indirect forcing mechanisms controlling community composition and paleoenvironmental inference. The model illustrated shows only the more immediate indirect controls of climate and short time-scale environmental variables. Longer term, indirect controls on species assemblages (basin formation, evolutionary controls) are ignored here.

to a paleolimnologist as well. Returning to our hypothetical example, if our concern is to document the past history of changes in secondary productivity, then our study organisms might give us a rather direct picture of this history. However, many paleoecologists concern themselves with processes and interactions occurring over much longer time scales, where proximal causality becomes of lesser interest (Schoonmaker, 1998). If our question shifts to issues of watershed erosion or climate, then the inference derived from the fossils becomes more one of correlation than direct causation. In fact, many factors that have direct, physiological effects on organisms, such as light, nutrient, or oxygen availability, salinity, and major ion chemistry, are themselves indirectly related to climate. Over longer time scales, or larger spatial scales of analysis, these indirect causes of ecological change generally become more important targets of research. Climatic processes tend to dominate the interests of most paleolimnologists concerned with time scales of $\sim 10^3$–10^5 years, whereas at time scales > 10^5 years, questions are commonly directed at understanding both climatic and tectonic change. In other words, it is meaningless to ask the question, "What is controlling the distribution of organism X?" without reference to the scale of time and space over which the question is applied (N.J. Anderson, 1995a,b; Cohen, 2000).

To state that a cause of ecological change is indirect says nothing about the degree of correlation between the cause and the effect. As steps are added between the organisms and the process we wish to interpret, there are more filters that affect information transfer. In some cases these filters may blur the connection to the point where no useful information is provided from species assemblages about the process of interest. In other cases, indirect controls may be expressed quite precisely, to the extent that climate variables can be inferred directly and quantitatively from the fossil organisms themselves. But in almost all cases these filters reduce the temporal and spatial resolution at which we can ask an ecological question.

Determining whether a particular group of fossils can usefully archive some paleoenvironmental variables of interest requires sampling the organism's modern distribution across a broad range of environments. The implicit assumption here is that we can substitute a broad geographical scale of analysis for the long length of time that we cannot directly sample. A survey of organisms across a range of lakes experiencing varying precipitation and other climatic variables would give us some idea of how sensitive our target organisms are likely to be to these indirect causes of change. For older records, whose fossil species are extinct, the inference of causation or correlation must be less direct.

Taphonomy and Time-Averaging

Taphonomy is the study of processes that affects fossils from the time of death to the time of recovery by a paleontologist, including dissolution in the water column and after burial, transport, mixing of fossils prior to burial, physical abrasion, and compaction. Because these processes do not operate equally between species or depositional environments, taphonomic biases pervade the fossil record, and need to be carefully considered in any paleoecological investigation (Behrensmeyer et al., 1992).

A loss of information occurs prior to, during, and after burial, so most assemblages of fossils collected from lake beds are not some kind of snapshot of a lake community at some instant in the past (figure 10.2). Instead, the typical fossil assemblage includes individuals that were ultimately preserved together, despite being derived from different habitats or different times. A lacustrine fossil assemblage might comprise a mixture of the remains of benthic, planktonic, and nektonic organisms. Benthic organisms, such as chironomids or ostra-

codes, can be preserved in situ, close to the site where they lived, or reworked and transported by waves or currents. Plankton and fish settle through the water column, sometimes directly following death, or after some period of transport, settling in a depositional environment only indirectly linked to their original habitat.

Many important elements of lake communities, such as copepods, are rarely preserved as fossils. In other groups, such as fish, the disarticulation processes that accompany decay may make remaining body parts unidentifiable. Both the composition and size of organic remains play a major role in this bias during the early stages of taphonomic filtering. Some materials are highly inert and readily preserved, such as sporopollenin, the complex group of polymers that make up pollen grains. Others, such as the calcium carbonate skeletons of mollusks or ostracodes, are readily dissolved under

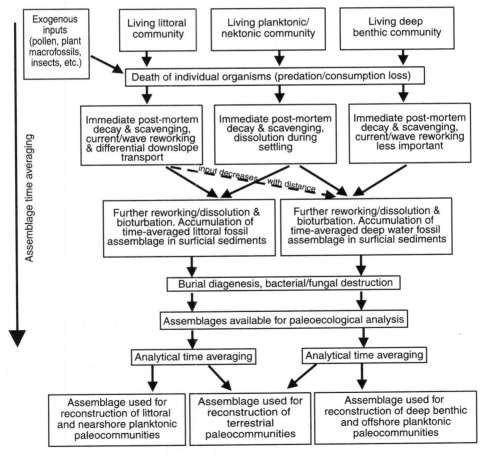

Figure 10.2. A conceptual model of the biological inputs and taphonomic processes affecting lacustrine fossil assemblages and the paleoecological inferences derived from those fossils.

acidic conditions. Nonmineralized cells (for example in nonsiliceous algae), "soft" animal tissues, and the weakly mineralized cuticle of some arthropods are only preserved under anoxic conditions. Normally, only a subset of species inhabiting a lake is actually preserved as fossils. However, under exceptional circumstances, the combination of such agents as rapid burial, anoxia, and blanketing by microbial mats can lead to the preservation of remarkably complete fossils of fish, insects, and other very delicate invertebrates, recording a broad array of a lake's biota (Stankiewicz et al., 1997; Briggs et al., 1998; Harding and Chant, 2000).

As a result of their relatively large surface area/volume ratios, small fossils are subject to relatively rapid chemical destruction. In deep lakes the overwhelming majority of biogenic silica produced by diatoms is redissolved during settling, prior to sedimentation (Eadie and Meyers, 1992). Physical processes are much more important for larger fossils, like mollusks, which are subject to significant inertial forces when they collide with other objects. Taphonomic degradation continues long after burial, through microbial attack, chemical dissolution, and compaction. The cumulative effects of these processes explain the fact that organisms with relatively weak skeletonization, such as cladocerans, are common in Late Quaternary sediments, but rarely encountered in pre-Quaternary lake beds.

Taphonomic history is also influenced by lake morphometry and depositional and burial environments. Variation in water depth, current and wave activity, and water chemistry, especially oxygen content, can lead to strong differences in preservation, even for a single species (Wilson, 1988). The central, deep-water regions of small lakes are subject to sediment focusing, which will artificially inflate the flux rate of microfossils relative to littoral environments, and can act as sinks for easily transported, upslope remains (Likens and Davis, 1975; M.B. Davis et al., 1984). In large lakes, where sedimentation is accentuated around prograding deltas, offshore focusing is less important. A greater proportion of settling algal fossils are dissolved within the water column in deep versus shallow lakes, and the proportion reaching the lake floor is increasingly dominated by those aggregated as fecal pellets. Fossils are almost invariably degraded more quickly under conditions of increased wave and current reworking, lake-floor oxygenation, bioturbation, and scavenging. Because all of these features of the lake-floor environment affect accumulating sediment as well as fossils, they also indirectly regulate the postburial degradation processes that affect organic remains.

For example, coarser and more organic-poor sediments generally promote more rapid skeletal degradation.

The physical transport and bioturbation processes that mix fossils together are actually advantageous for some types of paleolimnological questions. As the temporal resolution of our questions becomes lower we may not want an "instantaneous" picture of the lake's organisms, as this may not be sufficiently informative of the range of conditions we are interested in, or of the species present close to, but not precisely within, the limited sampling area. If we were examining the annual cycle of nutrient variability from diatoms in varved sediments, we would probably want our samples to contain the full range of species that occurred in the water column throughout the season, to provide some kind of average, rather than a snapshot of a phytoplankton patch that happened to occur over the burial site on one day. Physical taphonomic effects on fossils (sorting and concentration) also provide clues for environmental interpretation, such as wave reworking, the existence and prevailing direction of currents, or the effect of storms on the lake floor.

Fossil assemblages in all environments are *time-averaged*. In lakes, this mixing represents intervals ranging from months (in varved sediments) to $\sim 10^3$ years, with typical values in the 1–10-year range. For exogenic fossils, those transported from the watershed into the lake such as terrestrial pollen, there is also some additional interval between the formation of the organic remains and their arrival in the lake. This interval can vary considerably, depending on such things as the occurrence of temporary storage sites (e.g., soils), and the proximity of the organism when living to the lake. Some component of time-averaging results from the actual physical mixing of fossil particles at identical stratigraphical levels, whereas another fraction (*analytical time-averaging*) is an artifact of the need to incorporate some finite vertical stratigraphical range (time) within a single sample, even if this is only measured in millimeters (Fürsich and Aberhan, 1990).

In profundal environments the principal control on time-averaging is the tradeoff between depth of bioturbation and sediment accumulation rate. Minimal bioturbation and higher sediment accumulation rates both reduce time-averaging, as is evident from the subannual resolution preserved in varved sequences. In shallower water, physical transport, related to bottom slope and bottom current reworking, increases in importance and sediments are generally much more time-averaged

(Cohen, 1989b). Time-averaging therefore regulates the scale of questions a paleoecologist can appropriately investigate in a fossil sequence, and these scales will vary between organisms and burial environments. For these reasons, when trying to develop a fossil predictor of an environmental forcing process from modern organisms, paleoecologists normally do not compare instantaneous censuses of living organisms with the environmental variable of interest. Instead, they try and bypass some of the early taphonomic stages of dissolution, transport, and time-averaging, and compare the environmental variable with surface samples of "recently" dead fossils.

When time-averaging operates on a time scale shorter than a decade or so, it also has the virtue for paleolimnologists of providing a more complete picture of the local species "pool" than would be evidenced by an instantaneous census (Barker et al., 1990; Cummins, 1994; Alin et al., 1999a; Cohen, 2000). Although some of this is certainly attributable to environmental conditions at the collection site shifting over the time span of the time-averaging process, in part it is also a result of the inherently patchy distribution patterns of populations of any species in a lake. Although ecologists tend to think of patchiness in distribution as a spatial phenomenon, it is also a temporal one, with populations colonizing and disappearing from a locality in a lake over periods of time that can be integrated into a time-averaged sample. In other words, if time-averaging is not too severe, it may serve to provide a much better indication of the local species diversity in the vicinity of the core or outcrop site than an "instantaneous" sample of the same limited area would provide.

Historical Contingency

All ecological change in lakes occurs in a context of some pre-existing set of conditions that were present when the change began to impact the lake ecosystem. These conditions set boundaries on how a particular ecosystem response to a change will be played out. Consider two lakes, whose morphologies, bedrock geology, and hydroclimates are nearly identical. In one lake basin an earlier epidemic caused the local extinction of all fish species, whereas the other lake maintains a healthy population of several species of planktivorous and piscivorous fish. Now suppose that both are exposed to rapid influxes of nutrients driven by the same regional climate change event. Even though the nitrate and phosphate flux may be identical in the two systems, the response in terms of future ecosystem change will almost certainly differ, because the response is contingent on what came before, that is, the presence or absence of fish for some unique historical reason.

On short time scales, contingency can be incorporated into the experimental design of ecologists as they try to manipulate ecosystems and observe the outcomes of these manipulations. At time scales of years to a few decades, the range of contingent events that are likely to have profound effects on ecosystems is relatively predictable, and in many cases is known from historical observation. The types of contingent events that occur frequently are also likely to have relatively limited impacts across lake systems. Within single lakes their effects are more easily understood, either through manipulation or modeling, and their impacts can be readily incorporated into paleolimnological interpretations.

Over longer time scales, however, larger-scale but less frequent events gain greater importance for ecosystem history. A type of event that occurs once every 10^5 years is so rare on a human time scale that it is unlikely to have ever been observed, let alone experimentally mimicked in a realistic fashion. Furthermore, these types of rare events may have broad regional consequences. Returning to our fish-bearing versus fishless lakes example, if a single lake is fishless, it may only be a matter of a few years before some fish species naturally reinvade the lake from a downstream or upstream source. But if the elimination of fish were regional in extent, the differing consequences of the same regional climate change event would be played out over a much longer time interval. At the same time, the historical contingency upon which the original differences were based would, in hindsight, become harder to mimic or interpret correctly from ecological experiments alone, particularly as the underlying cause becomes too large to be encompassed in prior human experience. It is these large-scale and profound contingencies that confront lake paleoecologists with some of their most difficult, and in many ways most interesting, problems. An event that is statistically rare by the standards of human lifetimes might become almost a certainty over the duration of a long-lived lake.

Understanding the influence of rare events on future lake ecosystem development is not simply an academic exercise. If, as seems likely, we are living through an interval of global warming that is unprecedented in recorded human experience, we have to turn to high-resolution records of the past rare events, such as paleolimnological records of ecosystem change at the end of the last glacial period, to provide historical indications of how ecosystems have previously responded to such a rare event. At even longer time scales, processes such as episodes of continental rifting, or the evolution of land plants, may have provided nearly unique drivers of subsequent lacustrine biotic history, played out on continental or global scales, over millions of years.

Ecosystem Assembly and Evolutionary Processes

Lacustrine communities are the products of almost four billion years of evolution and ecosystem change. It is important for paleolimnologists to remind themselves that lacustrine organisms, and their interactions with other organisms and their physical environment, have changed greatly throughout this history. Even on time scales as short as 10^4 years, lacustrine paleoecologists are commonly confronted with assemblages of lacustrine fossils with no known modern analogs. These species assemblages may be composed of what appear to be entirely extant species, where the individual species, as far as we know from their modern distributions, inhabit very different environments. We might find an assemblage of lacustrine fossils that includes species that today are indicative of cold, Arctic lakes, co-occurring with warm temperate species. Assuming we could eliminate taphonomic mixing as the source of this puzzling assemblage, there are several plausible explanations to account for this finding. First, the modern biogeographical or ecological ranges of the species may be inadequately known, leading to erroneous interpretations about their ecology. This is likely for groups of organisms that are relatively poorly studied. Second, the modern range of the species may be circumstantial, contingent on historical events of dispersal pathways, disease, and so on, more than on actual ecological requirements imposed by diet or physiology. (e.g., M.B. Davis, 1986). When we use modern species' ranges as guides to past environmental conditions, even with organisms whose modern distributions are well known, we must always entertain the likelihood that these range boundaries do not circumscribe the full range of conditions the species may have inhabited in the past. Finally, there is no particular reason to believe that the evolution of changes in physiological tolerance of specific environmental conditions is, or should be, mirrored by morphological evolution evident in fossil remains.

A fossil species may appear superficially similar, or even identical, to a modern species, yet have different requirements in terms of habitat or other environmental variables. This problem is compounded by the fact that fossils only preserve a part of the organism's morphology; even if morphological features have evolved in concert with physiological adaptations, they may be invisible in fossil remains.

The implication of the above points is that our ability to infer environments and processes from fossils weakens with time. No specific time frame can be attached to the rate of this weakening, because the rates of change in ecosystem assembly and morphological/physiological evolution vary between environments and clades. The presence of a few nonanalog species does not automatically rule out interpreting a fossil assemblage, if the assemblage is dominated by species whose ecologies are well known. It is important, however, that paleolimnologists keep this principle in mind as they move between qualitative and quantitative interpretations of past conditions from fossil organisms. A quantitative comparison with modern species assemblages may be entirely justified for paleolimnological reconstructions of twentieth century lake deposits. We will be on much shakier grounds, however, when we try to use the same modern distributions to interpret Miocene lake beds.

The Paleoecological Toolkit

A discussion of sampling and processing methods for lacustrine fossils is beyond the scope of this book. Readers are referred to Last and Smol (2001) for excellent discussions of these methods. The analysis of fossil data is much more uniform between groups and warrants consideration here. Paleoecological data analysis involves two stages. First, the fossil data must be recorded, to describe what is present at what stratigraphical levels, in what condition, and so on. Second, the data must be interpreted, to develop some type of inference about paleolimnological conditions at the time the fossils were deposited.

Fossil Records and Presentation of Data

Fossil data can be recorded in a variety of ways. For organisms that are infrequently encountered in a core or outcrop, it may only be possible to record their *presence or absence* (figure 10.3a). Presence/absence tabulations are used when lithological or taphonomic conditions preclude accurate counting

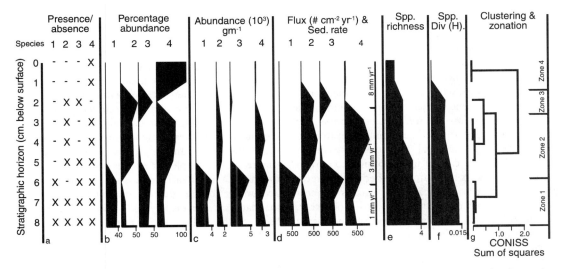

Figure 10.3. Presentation of paleoecological data, for a hypothetical data set consisting of information about four species (labeled 1–4) and eight stratigraphically ordered samples taken from a core. Species diversity (f) is the Shannon (*H*) index. Clustering (g) uses the constrained, incremental sum of squares method, with samples constrained to cluster with stratigraphically adjacent samples (Grimm, 1987). This allows the paleoecologist to distinguish fossil zones (h).

statistics, when comparisons of individual counts may not be very meaningful, and when paleoecological analysis is a secondary objective of the research. *Percentage data* are normally used when the above problems can be overcome, and are the standard display tool of paleolimnology (figure 10.3b). Here, the proportion of each species of interest relative to a sum is illustrated for each successive stratigraphical level. The sum may be of all organisms within the fossil group of interest, or of some subset of that group.

Percentage data are useful in drawing attention to relative changes in the species present. However, proportional change data cannot differentiate between the effects of dominance by some species and actual abundance changes over time. A species may be increasing in abundance through time, yet show a decline in percentage representation, because another species is increasing even more rapidly. To avoid this problem, abundance data are sometimes displayed relative to sediment (individuals per gram dry weight sediment) (figure 10.3c). In the scenario illustrated in figure 10.3, a major decline in the fossil abundance of all species occurs near the top of the core, suggesting that conditions for all four species were deteriorating, with species 1, 2, and 3 most heavily affected.

Relative abundance plots fail to accommodate the fact that abundance per unit sediment can change as a result of either absolute changes in

the flux of fossil remains, or changes in sedimentation rate and sediment focusing. When it is possible to accurately estimate sedimentation rate, this problem can be overcome by the display of flux rate data (accumulated individuals per unit area per year) (figure 10.3d). This is commonly done only in very recent cores, and must be done carefully, since even apparently "continuous" deposition may in fact be rather sporadic. Returning to our example, an upcore increase in sedimentation rates can be shown to be responsible for part of the core-top decline, through sedimentary dilution of fossils. However, these data show that most of the decline in the fossil accumulation rate of these species is real.

Various univariate statistics can be used to initially describe and synthesize fossil assemblage data. *Species richness* and *species diversity* statistics are used to provide some indication of the complexity of the species assemblage present at one stratigraphical level relative to other levels. As we saw in chapter 5, diversity commonly declines during episodes of environmental stress, and so may provide some useful information about changing paleolimnological conditions. Species richness is simply the number of species encountered in a sample, preferably of some standardized size (figure 10.3e). However, species richness does not account for the proportional differences between the species present. As a result, some paleolimnologists prefer to use one or more of the various *species diversity* statistics that have been developed by ecologists, to consider both richness and representation (figure 10.3f). If two fossil assemblages both have 10 species, but in one a single species makes up 99% of all individuals present, whereas in the other the species are equally represented, then the first assemblage is considered to be less diverse. Species richness and diversity statistics are both especially sensitive to assemblage and analytical time-averaging differences, and require careful interpretation to correct for these effects. If two samples of equal stratigraphical thickness are compared for species richness and diversity, but one accumulated over one year, and the other over 100 years (slower sedimentation rate), the latter could be expected to comprise more species, simply because it sampled a greater range of population patches and environments over a longer time interval.

Another important stage in the assessment of paleolimnological data is the comparison of similarity between samples. This is done primarily to identify horizons with very similar ecological characteristics and also to locate intervals of rapid change in fossil assemblages. Various multivariate techniques have been developed to allow these comparisons to be made, and to facilitate the identification of similar species assemblages (H.J.B. Birks and Gordon, 1985; Prentice, 1986; H.J.B. Birks, 1998). As with species richness and diversity indices, these methods are sensitive to the effects of time-averaging, and techniques that work well for low-resolution stratigraphical records, with highly time-averaged samples, may be inappropriate for very high-resolution records.

Cluster analysis techniques join samples into relatively homogeneous groups or clusters based on similarities in their assemblages, and display these data as tree-like diagrams, referred to as dendrograms (figure 10.3g). Groups of samples with low within-group variation relative to other samples are linked successively to more dissimilar groups. *Zonation* (figure 10.3h) involves constraining clusters to stratigraphically contiguous samples (Grimm, 1987), thereby allowing the investigator to identify stratigraphical intervals of similar fossil assemblages and periods of assemblage change.

Ordination techniques include a wide range of multivariate analytical methods applied to species assemblage and other paleoenvironmental data. Multivariate ordination techniques are used as exploratory, hypothesis-building tools, for simplifying the information present in complex fossil data sets into more readily interpretable and graphical form. This is made possible by the large degree of covariance that exists between variables in most paleoecological data matrices. Ordination methods reduce the number of dimensions within large, multidimensional data matrices (for example individual species abundances) to a small number of independent or uncorrelated axes of variation. They also maximize the dispersion between data points, in this case lake samples or individual species. These combined properties allow the similarity or dissimilarity between samples to be easily visualized (figure 10.4). Samples or species are ordered along the major axes of variation (usually only those axes accounting for the greatest amount of variance in the data set are displayed). Variation in sample "scores" (the values on the primary axes) are normally analyzed to try and infer their paleoenvironmental significance, the underlying causes of major patterns of covariance. Principal components analysis, correspondence analysis, canonical correspondence analysis, nonmetric multidimensional scaling, and detrended correspondence analysis are all examples of commonly used ordination techniques.

Types of Paleolimnological Inference

Paleoecological inferences made from lacustrine fossils can be qualitative, semiquantitative, or quantitative in nature. These terms are used to describe the type of inference obtained, as opposed to the counting or clustering procedures used to describe or display the fossil data. Qualitative inferences are general descriptions of the significance of a fossil or an assemblage, such as "the presence of *Artemia salina* (brine shrimp) eggs indicates high salinity conditions." Qualitative inferences rest on the occurrence of diagnostic "indicator" species, taxa that are known to be common under some specific type of environmental condition. Long-term trends in environmental change can frequently be inferred from trends in the abundance of some sensitive species or group of species. Qualitative inferences are unaccompanied by any form of statistical precision, and their use in explicit hypothesis testing is therefore limited (Charles and Smol, 1994). Qualitative inferences are the norm in the interpretation of fossils from old lake beds (i.e., pre-Middle Pleistocene), in which the presence of numerous extinct taxa or "no-analog" species assemblages makes developing explicit inferences difficult, if not impossible. In these cases environmental tolerances are either estimated by analogy with closely related living species, or indirectly from associated geochemical and sedimentological evidence. They are also the norm in cases where the variables controlling species abun-

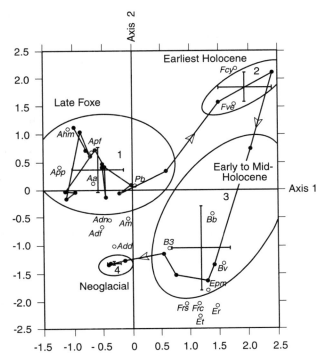

Figure 10.4. Illustration of a multivariate ordination of sample data sets from diatom frequency data from a group of Late Quaternary lake deposits from the Canadian Arctic. The method used here is correspondence analysis, one of several techniques commonly employed by paleoecologists to explore temporal or spatial trends in complex data sets. It allows visualization of the gross similarity between core samples (solid circles) and taxa (open circles). The diagram displays the first two (dimensionless) axes (those containing the greatest amount of information about the structure of the data set and its variability). Numbered ellipses delimit the diatom zones identified by stratigraphical constrained cluster analysis (see figure 10.3g), and the crosses represent two standard deviations around the mean of axis scores within each zone. Arrows indicate direction of lake ontogeny over time. From Wolfe (1994).

dances are only understood in a very general sense ("increases in species X indicate some combination of rising temperature, rising nutrient availability, and rising salinity").

Quantitative inferences take the form of "these diatoms indicate that at 1000 yr BP, Lake X had a pH of 6.5 ± 0.21σ. In this form, a paleoecological inference can be used statistically for hypothesis testing (Has the lake become significantly more acidic in the past 1000 years?). Quantitative inferences are largely limited to the study of lacustrine fossils in Mid-Pleistocene and younger sediments, where the issue of nonanalog assemblages is less problematical. The general approach has been to survey the modern distribution of the fossil group under study on a regional basis, in similar types of lakes to the one where the paleolimnological record is being obtained. A knowledge of the physicochemistry or climate conditions of the individual modern lakes is coupled with the respective species lists and abundance data to develop a mathematical model that "predicts" that set of past conditions most strongly controlling species distributions of the fossil group under study.

Paleoecological inferences sometimes fall between these two extremes and can be termed "semiquantitative." This term includes a range of approaches, such as the analysis of trends in groups of organisms that have been characterized based on their relationship to some environmental variable, the use of scaled, categorical, or weighted indices of some environmental variable, the determination of an environmental variable without any clear indication of the quality of the inference or with large associated uncertainties, or the application of models that are of uncertain relationship to the paleolake deposits in question. For example, "the proportion of alkaline-tolerant species increased over time" is a semiquantitative statement. An important virtue of semiquantitative inferences from fossils is that they are frequently attainable in pre-Pleistocene lake sediments.

Methods of Quantitative Inference from Fossils

All quantitative inference methods in paleoecology rely on mathematical models that relate a set of environmental variables (typically water chemistry and air or water temperature) to modern assemblages of organisms in a series of lakes. Data are collected from these lakes consisting of the environmental variables of interest, generally those thought by prior experience to be important regulators of species distribution, and the species abundances or indices. This is referred to as a *training set* (the terms calibration data set and reference set are also used interchangeably) (figure 10.5). The organisms tabulated in a training set are normally those found in the surface sediments of each lake, rather than a census of living populations, for the taphonomic reasons explained earlier (figure 10.5A). Training set data also provides data for a set of environmental variables relevant to the collection site. Ideally these are also time-averaged over the interval represented by the surface sample. Various types of regression models or *transfer func-*

tions can then be generated to calibrate the relationship between species and environmental variables in the training set. In the example shown, relevant to discussion below, this produces a normal species distribution curve (optimum and tolerance range) along each environmental variable gradient (figure 10.5B). Note in the example shown, as in reality, that optimal conditions and tolerance ranges of species vary greatly. Paleoecological analysis involves a calibration step from the model to solve the transfer function (an equation or system of equations) for the unknown environmental variable(s).

Of course, not all species or groups of species will be equally useful in reconstructing any given environmental variable of interest. Equally, some environmental variables may be unreflected in any measured species distributions. The value of a transfer function lies in its ability to accurately reflect the variable from the assemblage data. To evaluate this predictive ability, an inference derived from a modern data set (where the environmental variable can be measured) must be calibrated against the inferred (model-derived) values. In regression analysis the quality of the inferred versus measured relationship is expressed as a goodness-of-fit of the regression, described by the coefficient of determination (r^2), and the regression's 95% confidence intervals. In a useful transfer function, the scatter of data (residuals) away from the regression is small, indicative of a strong relationship

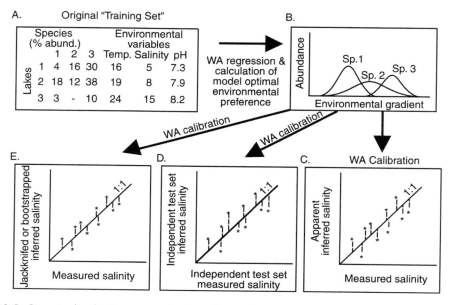

Figure 10.5. Steps in developing a quantitative paleoecological inference model. Modified from Fritz et al. (1999).

between the environmental variable and the species response. The calibration of a model in this way allows the paleolimnologist to analyze the error in the model and quantify its uncertainty. Several approaches exist for calibrating an inferred versus measured environmental variable once a transfer function is developed (H.J.B. Birks et al., 1990b). A simple method is to compare the transfer function values against the measured values for the original data set, generating *apparent inferred values* and associated errors (figure 10.5C). However, because the model data and the calibration data are identical in this case, the results (goodness-of-fit) can produce overly optimistic interpretations of the quality of the transfer function. The best approach would be to use a modern data set that is entirely independent (i.e., different lakes) from the original training set (figure 10.5D). This approach, however, is expensive, requiring much additional sampling, which could be used more profitably to improve the training set.

A compromise approach involves statistical sampling procedures known as *jackknifing* and *bootstrapping* (figure 10.5E), which allow all data collected to be used in model development, but provide quasi-independent test sets for model calibration. Jackknifing is an iterative procedure that involves the creation of a series of new training data sets from the original set of lakes, each of which leaves out one lake sample set. Each of these training sets is then used to independently derive a set of species response curves (the inference model). The measured environmental variable of interest from the one lake that was excluded from the model then forms an independent test of the inferred value from the model. The procedure is repeated until the species list of each individual lake data set has been excluded from the training set and its measured variable tested against the model output. The quality (uncertainty) of the overall model is then assessed by its performance in all of the individual tests. Bootstrapping involves the creation of many randomized training sets from the original training set. Each artificial set is the same size as the original set (i.e., same number of lakes), and is created by random sampling-with-replacement, meaning it may include one lake multiple times and others not at all. Transfer functions are generated from each bootstrap cycle and compared against the measured values for the test set of lakes left out of that cycle's training set. The quality of the transfer function is then assessed from the cumulative results of all bootstrap cycles.

The development and use of any transfer function makes several assumptions that paleolimnologists should bear in mind (Sachs et al., 1977; Charles and Smol, 1994):

1. The correct biological variables (species abundances, indices, etc.) have been chosen to predict the dependent variable of interest (water chemistry, climate, etc.). Various ordination techniques have been used to identify those environmental variables with particularly strong relationships to species distributions, which are therefore likely to be predictable with a high degree of accuracy from a fossil species data set. One of the most useful exploratory techniques, canonical correspondence analysis (CCA), makes realistic assumptions about the nature of ecological gradients along which species are distributed, and allows the investigator to simultaneously visualize the relationship between species and "controlling" environmental variables (ter Braak and Prentice, 1988; Walker et al., 1991a).

2. The environmental variables measured for the training set are representative of the conditions experienced by the organisms during the time-averaged interval when the training set fossils accumulated.

3. The training set data are typical and relevant to the fossil application intended, and cover the range of environmental variation expected to be encountered in the fossil lakes.

4. The observations (data matrices) are independent.

5. The random variation in the environmental variables is normally distributed.

6. In the solution of the transfer function, the independent (biological) variables have no random components.

The earliest transfer functions applied linear regression techniques to derive a relationship between a single environmental variable and some biotic index (e.g., Meriläinen, 1967). With the advent of accessible computerized solutions to multivariate statistical problems in the 1970s, it became practical to analyze paleoecological data sets for multiple environmental variables and complex biological assemblage matrices. Transfer functions accordingly shifted to the use of multiple regression analysis, principal components analysis, canonical correlation analysis, and several other multivariate techniques (e.g., Imbrie and Kipp, 1971; Webb and

Bryson, 1972; H.J.B. Birks et al., 1975; Delorme et al., 1977).

The statistical methods mentioned above all share an assumption of a linear response between the environmental variables of interest and the affected species' abundance, at least over the range of environmental variation relevant to the ecological conditions under analysis. In situations where the environmental gradient is short, and community change is limited along that gradient, such linear response models work very well as transfer functions (ter Braak and Prentice, 1988; H.J.B. Birks, 1998).

As the length of environmental gradients along which species are sampled increases, however, it becomes increasingly likely that species distributions will approximate some kind of bell-shaped (Gaussian distribution) curve along most environmental gradients of interest to paleolimnologists. This type of distribution is characterized by some optimal condition (the point on the environmental gradient where the species is most abundant), the response of the species to its optimal condition (the peakedness of the curve), and a tolerance range (the variance around the optimum in which the species remains common). In the late 1980s ter Braak (1987) discovered that for each of the major statistical methods mentioned above and used to relate environmental variables to biotic responses in a linear fashion, there was an equivalent method that assumed a unimodal response. The most widely used unimodal techniques are various forms of *weighted averaging* (WA) regression, all of which assume that the optimal environmental condition for each species present in an assemblage is represented by its maximal abundance (ter Braak and Van Dam, 1989; H.J.B. Birks et al., 1990b; ter Braak and Juggins, 1993). WA methods are powerful tools for developing quantitative inferences about past environmental conditions from fossil data. This is best evidenced by the goodness-of-fit seen in any number of comparisons between WA-inferred variables, derived from the training set data, and actual measured environmental variables (figure 10.6). WA regression/calibration models have now been developed for a number of taxonomic groups (diatoms, chironomids, etc.), to infer a variety of variables (pH, alkalinity, temperature, salinity, P, N, and substrate). Applications of WA models to paleoecological reconstruction have produced very detailed records of these variables, and also allow the investigator to calculate uncertainty associated with the inference (figure 10.7). At present this type of modeling is done on time scales extending back to the Late Pleistocene, but it seems

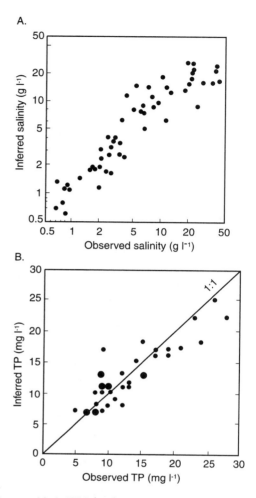

Figure 10.6. Weighted averaging-inferred lake water chemistry using diatom surface sediment training sets, showing the relationship between inferred values and observed data (analytical measurement). (A) Lake salinity ($g\,l^{-1}$) from the North American Great Plains (after Fritz et al., 1991). (B) Total phosphorus ($\mu g\,TP\,l^{-1}$) from British Columbian lakes (after Hall and Smol, 1992). These plots clearly illustrate the ability of simple weighted-averaging models to reconstruct water chemistry. From N.J. Anderson (1995b).

likely that for some evolutionarily conservative taxa, it will ultimately be possible to extend the approach even further into the past. It warrants emphasis that WA models, along with other transfer functions, are primarily valid for fossils from lake deposits within the study area from which the training set was derived. Although it is not

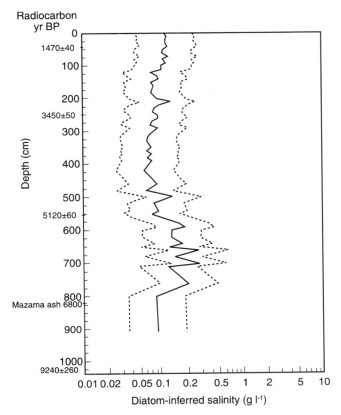

Figure 10.7. A plot of Holocene history of diatom-inferred salinity for Harris Lake, Saskatchewan. The salinity inferences were derived from a salinity transfer function developed from lakes in southern British Columbia and using the computer program WACALIB version 3.3 (Line et al., 1994). The major taxa encountered in the Harris Lake core were sufficiently represented in the surface sediment calibration data set, so modern analogs were not a problem. The inferred salinity values are mean bootstrap estimates and the errors (dotted lines) are estimated standard errors of prediction provided in the WACALIB output. From Wilson et al. (1997).

uncommon for WA models from one area to be applied to another, similar region, such applications should be treated cautiously, and the results considered semiquantitative, until the model can be validated more locally.

Summary

Our conclusions from this chapter can be summarized as follows:

1. Interpreting fossil archives from lake deposits requires understanding the linkage between the ecological cause and effect at varying temporal and spatial scales. The value of fossils to answer questions about lake or watershed history depends on both the time scales at which they respond to a forcing phenomenon, and the filters lying between the historical process and the fossil's accumulation.

2. Taphonomic processes that affect fossils after organisms die create filters that are very important for paleolimnologists to understand. Taphonomic processes skew fossil assemblages toward more durable body parts, and cause those assemblages to be time-averaged representations of past communities.

3. Our ability to interpret fossil assemblages in lake deposits weakens with age, a consequence of both changes in community struc-

ture through time and morphological evolution. Interpretation of past events must also allow for the fact that rare events become more likely to be important as the duration of an archive increases, and the importance of such historical contingencies in molding ecosystem history.

4. Various techniques have been developed to display historical changes in species abundance, diversity, and community structure, and to compare patterns between archives.

5. Inferences of paleoenvironmental change from fossils can be qualitative, semiquantitative, or quantitative. Quantitative methods involve the identification of environmental variables that strongly correlate with species assemblage changes, development of modern lake training set data, to link specific environmental conditions with species assemblages, and transfer functions, to calculate past environmental conditions from a specific fossil assemblage.

11

Paleoecological Archives in Lake Deposits 2: Records from Important Groups

The lacustrine fossil record comprises a mixture of *endogenic* fossils, such as cladocerans, derived from lakes, and *exogenic* fossils, such as insects or pollen, which are carried into lakes, by wind and water from surrounding areas. Our primary emphasis here will be on the endogenic fossil record of lakes; we will only briefly consider general aspects of the taphonomy and paleoecological significance of exogenic fossils for terrestrial plant and insect fossils.

Information about lake fossils varies greatly between groups. Some taxa, such as diatoms, are virtual workhorses of the field, with numerous investigators, and established methods of sampling, analysis, and interpretation. At the other extreme are organisms such as copepods, which, despite their importance in lacustrine ecosystems, are so poorly fossilized that they are unlikely to ever play a major role in paleolimnology. In between these extremes lie the majority of lacustrine organisms. Many relatively common groups have great potential for paleoecological interpretation, but, for reasons of inadequate study, a lack of researchers, or difficulties in taxonomy, have thus far been little used by paleolimnologists. Major opportunities await new students in the field who are willing to take up the challenges of studying these clades.

Cyanobacteria, Dinoflagellates, and Cryptophytes

Despite their importance in lacustrine communities, cyanobacteria remain a relatively unexploited source of information for paleolimnology. Isolated cells have poor preservation potential, and fossil cyanobacterial cells are preserved in Late Quaternary lake muds primarily by their more resistant reproductive spores (*akinetes*), or occasionally by filaments. Planktonic cyanobacteria are only rarely recorded in older sediments. In contrast, benthic cyanobacterial communities are well represented in ancient lake beds by their constructional deposits, lithified algal mats, stromatolites, and thrombolites.

Planktonic Cyanobacteria

Although their body fossils have been used only rarely to solve paleolimnological problems, planktonic cyanobacteria have great potential for this purpose, given their obvious importance in many lacustrine communities. Relatively resistant akinetes might be very useful for understanding changes in plankton communities, especially in cases where better-studied siliceous microfossils (diatoms and chrysophytes) are not well preserved, for example, in very alkaline lakes. However, almost nothing is known of the taphonomic biases that control the planktonic cyanobacterial fossil record. Stratigraphical sequences of fossil cyanobacteria record a combination of responses to nutrient, temperature, and stratification conditions of the water column (Cronberg, 1986). Very promising results have also been obtained in the study of N-limitations on lake productivity from planktonic cyanobacteria (Van Geel et al., 1994; Kling, 1998). The ability of cyanobacteria to fix nitrogen allows them to outcompete other phytoplankton in N-limited surface waters, for example in strongly stratified lakes. Talbot and Johannessen (1992) used the presence of abundant *Anabaena* filaments, coupled with comparatively low ^{15}N ratios, to argue for water column stability and uniform climate at Lake Bosumtwi (Ghana) during the Early Holocene.

Dinoflagellates and Cryptophytes

Dinoflagellates and cryptophyte algae are occasionally recovered as fossils in lake sediments.

Dinoflagellate resting cysts are common in palynological preparations (Norris and McAndrews, 1970; Evitt, 1985). Although important components of planktonic ecosystems in modern lakes, little is known about their paleoecological interpretation or taphonomy. This results from both the small number of specialists in the field, and because their remains are quite delicate, and therefore easily destroyed in many standard microfossil cleaning procedures. Sample processing techniques are available that can greatly increase the potential for obtaining and interpreting meaningful records from these and other easily destroyed microfossils. When this is done it may be possible to reconstruct valuable information from these algal groups, particularly with reference to changing lake trophic state (Pollingher et al., 1992; Findlay et al., 1998).

Benthic Stromatolites and Thrombolites

Layered stromatolites and unlayered thrombolites, hereafter collectively referred to as stromatolites, are calcareous (occasionally siliceous) constructions, formed primarily by cyanobacteria. The durability of these structures allows them to be used as paleolimnological indicators in much older sediments than typically preserve isolated akinetes and filaments; cyanobacterial stromatolites provide evidence for the very earliest known lacustrine ecosystems, extending back to at least 2.7 ga, in the Archaean Era (Buick, 1992). Because most stromatolites are attached and cemented features, they are not easily transported, reducing the taphonomic complications associated with their paleolimnological interpretation. Furthermore, when stromatolites have been transported, as small unattached structures or broken features, this is usually evident sedimentologically.

Stromatolites are characteristic features of the littoral zone in hardwater lakes and streams. The precise role of autotrophic microbial communities in forming stromatolites, whether as trappers, binders, or accreters of calcium carbonate, is controversial (Bertrand-Sarfati and Monty, 1994). Although stromatolites are formed in deep-water environments by nonphotosynthesizing bacteria and other heterotrophs, these features are relatively rare (Dromart et al., 1994) and practically unknown in lakes. Therefore, the presence of stromatolites in lake deposits provides a first-order indication of the extent of the nearshore photic zone, particularly when accompanied by preserved algal filaments.

Stromatolites display a wide variety of external forms and internal structures (Grey, 1989). This complexity is characterized by variation in large-scale configuration, shape and size of individual heads, domes and mats, and types or shapes of laminae and fabric. Most of the rather bewildering variation in head shape and size seen in these structures appears to be controlled by environmental factors during growth, rather than by the composition of the microbial community itself (Monty and Mas, 1981; Casanova, 1994; Lindqvist, 1994; Moore and Burne, 1994). This is fortunate from the perspective of the paleolimnologist, because these features are most readily preserved, and are also evident to the nonspecialist. The role of specific microbes or grazers on stromatolite morphology increases at the smaller scale of internal fabric and laminae shape (Bertrand-Sarfati et al., 1994; Winsborough et al., 1994). For example, the internal layering of many stromatolites consists of alternating light (sparry, coarse-crystalline calcite) and dark (finer, micritic calcite) bands. In their study of East African stromatolites, Casanova and Hillaire-Marcel (1993) found that the thicker, light-colored laminae comprise numerous erect cyanobacterial filaments, woven into an interlaced meshwork. These appear to form during periods of runoff. The dark laminae are similar in thickness or slightly thinner, but comprise heterotrophic bacterial films.

Stromatolite head shape and size varies systematically with water depth in a number of lakes and paleolakes (figure 11.1) (Monty and Mas, 1981; Cohen and Thouin, 1987; Bertrand-Sarfati et al., 1994; Casanova, 1994). The reasons for this probably involve a combination of declining light and water agitation with depth, with small, mobile growths (oncoids) dominating in very shallow water, and larger encrustations, heads and domes, in progressively deeper water. Available nutrient levels and underlying lake-floor morphology may also be important factors in some lakes.

It is impossible to generalize between lakes about the precise depth ranges occupied by specific stromatolite zones. In small or turbid lakes, the vertical range of stromatolites is a few meters or less. Here the use of stromatolites as depth indicators is essentially limited to identifying paleoshorelines. In contrast, stromatolite growth in large lakes, especially on rocky and nonturbid coastlines, may extend to depths of 20 m or more. Under these conditions the existence of vertical zonation in stromatolite growth form allows the reconstruction of semiquantitative lake-level curves (Cohen et al., 1997a).

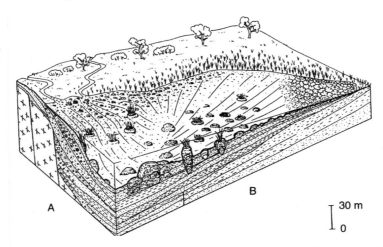

Figure 11.1. Block diagram reconstruction of a lacustrine paleoshoreline in the Tertiary Limagne graben, France, illustrating the shapes and positions of different types of stromatolites observed in outcrop. Systematic differences in stromatolite morphology occur with varying paleowater depth. Profile A is along a steeply sloping margin. Oncolites and decimeter-scale stromatolite heads occur in fluvial channels and along the lake shoreline, interbedded with oolites. Profile B represents a more gently sloping lake margin, with encrusted mudflats, and both club-shaped and columnar stromatolites. In slightly deeper-water areas, globular or columnar stromatolites formed as pseudocolumnar laminae around accumulations of caddisfly cases, and around submerged macrophytes. More complex mounds formed in the deepest areas, where stromatolite buildups became superimposed with fewer caddisfly casing layers. From Bertrand-Sarfati et al. (1994).

Stromatolites also provide clues about paleosalinity. For example, bacterial stromatolites are characteristic of hypersaline lakes (Casanova, 1987). These structures form under conditions of relatively low degrees of biological competition, and display a limited range of morphology, with smooth, continuous laminae, and very fine-grained textures. Stromatolites growing in freshwater conditions have more mixed microbial assemblages. They are characterized by more complex, clotted, or variable microstructure, in part caused by interspecific competition of growth forms and variable grazing intensity.

them well suited to address many types of paleolimnological questions (Battarbee et al., 2001) (figure 11.2).

The lacustrine fossil record of diatoms extends back to the Late Cretaceous. Evolutionary conservatism in morphology that characterizes many diatom clades allows quite detailed paleoenvironmental reconstructions to be derived from these algae, even in Tertiary sediments. However, it is in Late Quaternary lake deposits where the greatest advances have been made in applying diatom fossil data to the quantitative determination of past environments.

Diatoms

Diatoms are probably the single most valuable group of fossils for paleolimnological reconstruction. Because of the excellent preservation potential of their frustules, and the wide range of lacustrine habitats in which they occur, diatoms are among the most common lacustrine fossils. In addition to being abundant, diatom species assemblages are well known for their sensitivity to a variety of limnological and hydroclimate parameters, making

Taphonomy

Diatom assemblages are subject to a number of significant taphonomic biases, including fragmentation, dissolution, and diagenetic overprinting. These processes have systematic effects on preserved diatoms, which differ based on both the morphology of the fossils and the type of lake in which they are preserved. Because of their relatively delicate structure, diatoms are readily broken by waves and collisions with coarse sediment grains in turbulent water, especially in large lakes.

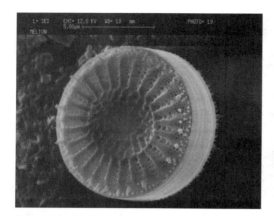

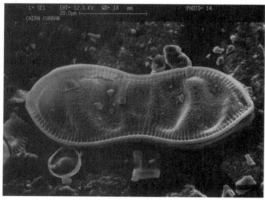

Figure 11.2. Scanning electron microscope photographs of two common species of freshwater diatoms. (A) The centric diatom *Cyclostephanos tholiformis* (12 µm in diameter). (B) The pennate diatom *Cymatopleura solea* (75 µm in length). Both are strong indicators of nutrient-rich, fresh to brackish waters.

Flower (1993) found that diatoms preserved in macrophyte stands were better preserved than those at similar depths in unvegetated areas, suggesting the importance of wave baffling effects. Bioturbation, grazing, and, unfortunately, sample cleaning also result in fragmentation.

Fossil diatoms are often chemically corroded, signaling partial dissociation of the valves. The physical manifestation of this is usually an enlargement of pore openings on the valve surface and corrosion along exposed edges, which eventually cause the valve to fall apart. Diatom dissolution is regulated by several factors (Shemesh et al., 1989; Barker et al., 1994a). At high pH and salinity (> 9.0) dissolution increases dramatically. Elevated temperatures and pressures also increase dissolution rates.

Highly corroded diatom assemblages are characteristic of lake deposits that have accumulated under high alkalinity and salinity conditions (Barker et al., 1990). Pressure effects have also been demonstrated to cause significant dissolution in deep lakes (Flower, 1993).

Experiments with living diatoms and both modern and fossil diatom frustules in saline and alkaline solutions have shown that the greatest amount of dissolution occurs under high pH conditions, in $NaCO_3$-rich fluids, and more generally, in most fluids rich in monovalent cations (Na^+, Li^+, and K^+) (Barker, 1992; Barker et al., 1994a). In contrast, saline solutions dominated by divalent cations (Ca^{2+} and Mg^{2+}), with their lower dissociation constants, produce less dissolution. Weak acids are generally good preservational environments and produce less dissolution than distilled water. Other saline fluids can also cause increased rates of dissociation, even in the absence of high alkalinity. Dissolution is highly correlated with the surface area:volume ratio (SA : V) of individual diatoms. Diatom species with high SA : V are rapidly eliminated from assemblages in corrosive fluids, whereas low SA : V species become overrepresented (figure 11.3). Strong dissolution can lead to the elimination of certain species, the creation of "no-analog assemblages," and systematic biases in quantitative paleochemical inferences, especially in alkaline lakes (Reed, 1998).

The most dramatic dissolution effects on diatoms occur either in the water column or within the first few decades after deposition (Flower, 1993). Early diagenetic dissolution results in supersaturation of pore waters by silicic acid; much of this silica is eventually redeposited on or around the remaining sedimented diatoms. Diagenetic alteration that coats or cements diatoms shortly after deposition confers resistance to subsequent dissolution, and there seems to be little difference in probability of preservation with respect to age once this early diagenetic interval passes.

There is usually a better correspondence between the living diatom community and the surface sedimentary assemblage of diatoms in small lakes versus large ones (Battarbee, 1981; Haberyan, 1990). During settling in large, deep lakes, diatoms are exposed to greater turbulence, time for dissolution, grazing during settling, and greater probability of pressure dissolution. Diatom assemblages in deep lakes display systematic biases toward species with higher sinking rates and those preferentially incorporated into rapidly settling fecal pellets. Time averaging of sedimentary assemblages may also be greater in larger or deeper lakes. To some extent

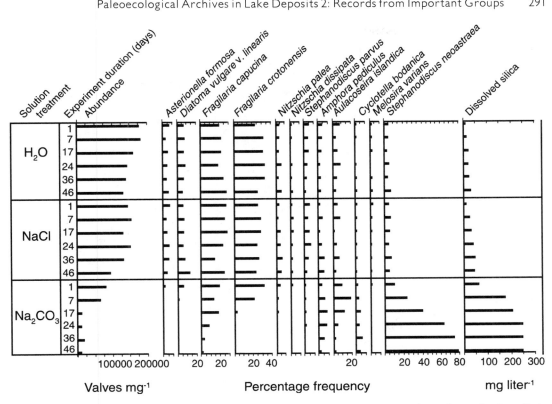

Figure 11.3. Summary of changes in the composition of diatom assemblages in cleaned samples from Lake Geneva (southeast France) for various chemical treatments (distilled H_2O, 3 M NaCl, and 3 M Na_2CO_3) and experiment durations. The plot illustrates changes in percentages of the 12 most abundant species, valve abundances, and measured dissolved silica in the solutions. Modified from Barker et al. (1994a).

these factors are offset in very shallow lakes by increased and continuous turbulence and physical fragmentation. This suggests that diatom preservation is probably best under conditions of intermediate water depths, below the storm wave base, but above depths at which pressure dissolution and settling time become significant.

pH

The determination of changes in pH is one of the most common paleoecological applications of fossil diatom assemblages. The recognition by Hustedt (1939) that diatom assemblages are extremely sensitive to acidity sparked great interest in their use to both monitor and reconstruct pH changes in lakes and ponds. This effort intensified greatly after the 1960s, with the discovery of widespread acid precipitation caused by fossil fuel emissions, and the subsequent anthropogenic acid-

ification of lakes, prompting the development of numerous regional diatom training sets for pH reconstruction. By the 1980s diatom-based pH reconstruction in northeastern North America and northern Europe had become something of a cottage industry, with prodigious effort directed toward both quantification of acidification trends and assuring the quality of these inferences (Charles and Whitehead, 1986; Smol et al., 1986; Battarbee and Renberg, 1990; Charles, 1990; Battarbee et al., 1999). Changes in pH-sensitive diatom assemblages in numerous lakes in North America and Europe over the past 100 years provide strong evidence for historical period acidification (figure 11.4). Under conditions of decreasing pH, fossil diatom floras often shift toward an increasing proportion of benthic, acid-tolerant species, consistent with the relationship between increased acidity, decreased dissolved organic carbon (DOC), and increased water clarity, as discussed in chapter 5 (N.J. Anderson

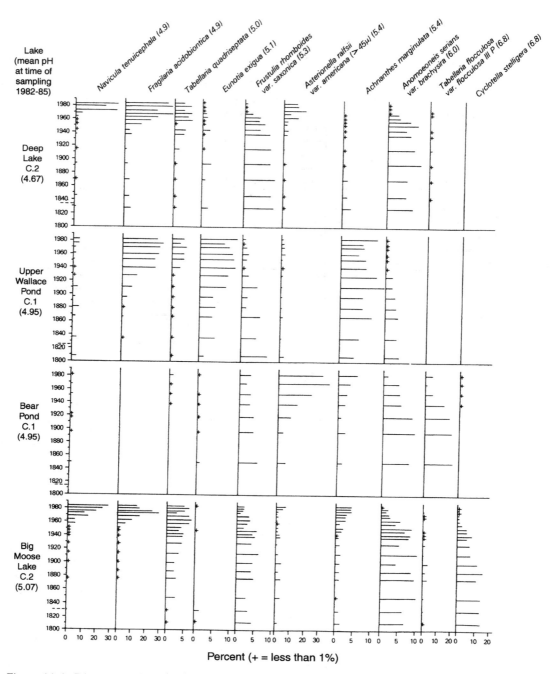

Figure 11.4. Diatom stratigraphy for pH-sensitive species (weighted average pH values given after species names), for several lakes in the Adirondack Mountains (New York), showing the increase in low-pH species and decline in high-pH species over the past 200 years. From Charles et al. (1990).

and Renberg, 1992). Many longer-term records of diatom-inferred pH have been obtained from Late Glacial–Holocene lake records, recording such processes as changes in watershed weathering under warmer Holocene climatic conditions, and the introduction of agriculture and soil disturbance within watersheds (Renberg, 1990) (figure 11.5).

Salinity and Alkalinity

Diatom communities respond strongly to changes in salinity and alkalinity (chapter 5). Since the 1960s it has been evident that fossil diatom assemblages could be used to differentiate marine from lacustrine deposits, and to infer lacustrine paleosalinity or paleoalkalinity (e.g., Hustedt, 1957; Richardson, 1968; Ehrlich, 1975). Since the 1980s paleosalinity estimates using diatoms have become quantitative, with the collection of regional training set data and the introduction of the transfer function approach (e.g., Gasse et al., 1987; Fritz, 1990; Fritz et al., 1991; Cumming and Smol, 1993a; Sylvestre et al., 2001). Today, diatom-based salinity

inference models are being used successfully in many parts of the world.

Because of the linkage between changing salinity or alkalinity and aridity, especially as lake basins become hydrologically closed, salinity-sensitive diatom assemblages can be used indirectly to infer past precipitation/evaporation ratios and regional climate (Fritz et al., 1999). Where instrumental records are available to compare with diatom-inferred climate change from short cores, the records are often strikingly coherent (figure 11.6). Diatoms have proven useful as quantitative recorders of aridity on various time scales throughout the Quaternary. Diatom-based reconstructions of salinity change have also been made on pre-Quaternary lake deposits, although here the inferences are necessarily qualitative or semiquantitative (e.g., Bao et al., 1999).

Nutrient Availability, Stratification, and Productivity

As we saw in chapter 5, nutrient availability plays a major role in structuring algal communities.

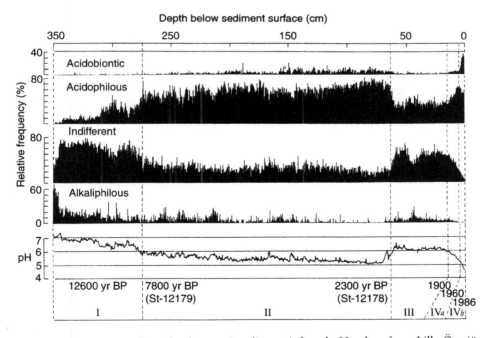

Figure 11.5. pH categories and weighted averaging diatom-inferred pH values from Lilla Öresjön, southwest Sweden, showing the long-term acidification of the lake at the end of Pleistocene, an abrupt increase in pH about 2000 years ago (advent of agriculture?) and the recent acidification of the lake related to industrialization. Acidobiontic species are those with optimal pH < 5.5. Acidophilous taxa prefer pH slightly under neutral, and alkaliphilous prefer pH slightly greater than neutral. Age dates are calibrated radiocarbon values and, for the most recent portion of the record, ^{210}Pb dates. Modified from Renberg (1990).

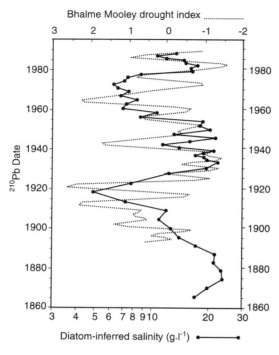

Figure 11.6. Diatom-inferred salinity estimates ($g\,l^{-1}$, solid line) for Moon Lake, North Dakota, derived from a transfer function developed by Fritz et al. (1993). Chronology is based on ^{210}Pb. The Bhalme Mooley Drought Index (dotted line) is based on data for the Great Plains region and monthly summer precipitation records for nearby climate stations, and has been smoothed with a four-point smoothing filter. From Laird et al. (1998).

Individual diatom species have distinct optima and tolerance ranges for particular levels of nutrients (especially P and Si), and also for particular nutrient ratios (Hall and Smol, 1999). The physiological reasons for these different optima and ranges are rarely known. However, the correlations between specific diatom taxa or assemblages and either nutrient-poor, oligotrophic conditions, or nutrient-rich, eutrophic conditions have been known for many years. Much of the early work applying these observations to fossil diatom assemblages was qualitative, but after the 1970s, diatomists began to systematically determine the nutrient conditions in which various diatom species occurred (e.g., Brugam, 1979). This was followed by studies of the vertical profiles of fossil species abundances and ratios, for those taxa shown in field studies to be diagnostic of particular nutrient levels (e.g., Bradbury and Dieterich-Rurup, 1993).

Over the past 10 years diatomists interested in nutrient histories have been increasing their efforts to generate quantitative reconstructions of actual nutrient concentrations, through the training set transfer-function approach. In large part this has come about to satisfy the needs of the lake-management community to better understand historical patterns of nutrient loading (particularly for P) into currently eutrophic lakes. Questions as to whether these lakes had high inputs of phosphorus prior to the advent of extensive agricultural fertilization and sewage runoff are crucial for decision-makers trying to determine courses of treatment in nutrient runoff remediation. As we saw in chapter 9, simple measurements of changes in sediment phosphate levels can be extremely misleading because of the diagenetic complexity and mobility of this element. Paleolimnological records of phosphate-sensitive diatoms offer an alternative route to addressing the question "How much phosphorus was there?" The transfer-function approach to inferring total phosphorus concentrations from fossil diatoms was first pursued in studies by Hall and Smol (1992) in British Columbia, primarily on oligotrophic lakes. About the same time Anderson and his colleagues used the same techniques to study eutrophic lakes in Northern Ireland (N.J. Anderson et al., 1993; N.J. Anderson, 1997b). Subsequent work has attempted to bridge the gap between these two extremes of lake types, by assembling large training sets that span a broad range of phosphorus load conditions.

The Northern Ireland training set studies show that a weighted averaging model can produce total phosphorus inferences that are in good agreement with observed total phosphorus in modern lakes (figure 11.7). N.J. Anderson and Rippey (1994) made a subsequent paleolimnological study of Augher Lough, a small eutrophic lake that was subject to P-rich runoff from a creamery from 1900 to 1974, after which time the effluent was diverted away from the lake (figure 11.8). Reconstructed total phosphorus concentrations in the lake based on diatoms clearly show the rise in nutrient loading and decline after the 1970s. In contrast, the sedimented elemental phosphorus flux shows no simple relationship with the historical pattern of eutrophication.

Temperature

The relationship between temperature and diatom communities is not nearly as well understood as is the case for pH, salinity, and nutrients. Evidence for

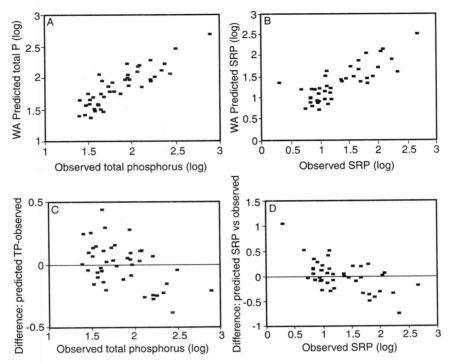

Figure 11.7. Scatter plots for diatom-inferred phosphorus versus observed phosphorus concentrations ($\log_{10} \mu\text{g TP l}^{-1}$) for eutrophic lakes in Northern Ireland, using a weighted averaging inference model. (A) Total phosphorus (TP). (B) Soluble reactive phosphorus (SRP). (C,D) Differences between observed and inferred values of TP and SRP. From N.J. Anderson et al. (1993).

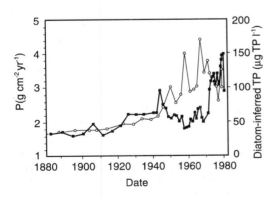

Figure 11.8. Comparison of sedimentary phosphorus fluxes (heavy line with square symbols, $\text{g cm}^{-2} \text{yr}^{-1}$) and diatom-inferred epilimnetic total phosphorus (lighter line with circles) for a single, deep-water core from Augher Loch, Northern Ireland, a lake subject to heavy anthropogenic phosphorus loading from the early twentieth century until the 1970s. From N.J. Anderson and Rippey (1994).

direct temperature controls on diatoms is limited (Battarbee, 1991). Although temperature clearly regulates the rates of many metabolic processes, it is likely that most of the "controls" exerted by temperature on diatom communities are indirect. However, there are well-documented correlations between diatom assemblages and temperature, and these relationships can be exploited by the paleolimnologist interested in climate change. These relationships are strongest in the extremely cold environments of the alpine, subpolar, and polar regions of the world, although even tropical altitudinal gradients show strong changes in diatom floras (Vyverman and Sabbe, 1995). But the polar relationships are of particular interest for paleolimnologists interested in global climate change issues, since global circulation models suggest that these areas will be among the most strongly affected by warming trends in years to come.

Strong altitudinal and latitudinal gradients in diatom floras have been known for many years. Over the past few years, diatom and environmental data have been collected from enough lakes in these regions to develop calibration training sets that relate summer temperatures with diatom floras (Pienitz et al., 1995; Lotter et al., 1997a). These correlations are mediated by such temperature-related variables of cold environments as available light seasonality, determined by duration or summer extent of ice cover, increasing seasonal meltwater runoff, water chemistry, and nutrient availability from forested versus tundra landscapes, turnover processes, and stability of stratification at low temperatures (Lotter et al., 1999).

One particularly intriguing correlate of warmer summer temperatures that may strongly influence fossil diatom assemblages is meltwater runoff. On the assumption that diatom productivity in stream systems increases during years with greater meltwater runoff, Ludlam et al. (1996) developed a *Lotic Index* of stream-derived versus total diatoms for high Canadian Arctic lakes. Warmer summers with greater runoff are associated with increased deposition of stream-derived diatoms in the downstream lake sediments. This index, in association with other variables like varve thickness, can be used in lake deposits in the region to obtain a semiquantitative estimate of past runoff or temperature.

Temperature changes in high latitude and altitude lakes are reflected by diatoms in other ways, besides community composition. Major increases in productivity that accompany the early phases of deglaciation or loss of permanent ice cover are expressed by rapid rises in diatom flux to lake sediments, as has been observed in the high Canadian Arctic during the twentieth century (see figure 1.8) (Douglas et al., 1994; Gajewski et al., 1997; Overpeck et al., 1997). In Canadian Arctic lakes, this increased abundance is accompanied by a shift from dominantly small, araphic diatoms to larger raphe-bearing species. Increasing species diversity, and increasing proportions of large, raphe-bearing diatoms are also often positively correlated with temperature in paleolimnological records (N.J. Anderson et al., 1996; Douglas and Smol, 1999).

Lake Level and Hydroperiod

The recognition that diatoms could be used as paleoindicators of past lake-level fluctuations arose initially from the observation of distinct habitat zonation of diatom communities within lakes. This is most clearly expressed by the distinction between nearshore benthic species, limited to the depth of the photic zone, versus offshore planktonic communities, which occur in the water column at variable depths. This suggests that a simple ratio of benthic to planktonic diatom species might

serve to differentiate shallow from deep-water deposits. Vertical zonation also occurs among benthic diatoms, between shallower and deeper-water species. This indicates that further differentiation might be made based on variations in benthic diatom assemblages.

Unfortunately for the paleolimnologist, both ecological and taphonomic variables complicate such an interpretation. The distribution of epiphytic diatom species is regulated by the distribution of macrophytes, which may or may not extend throughout the potential littoral zone (Wolin and Duthie, 1999). Because the extent of the littoral zone is a function of both water depth and transparency, its extent changes as turbidity increases or decreases. Uncertainty about the assignment of some species to planktonic versus benthic categories is another problem in using such ratios (Barker et al., 1994b). Some diatoms are neither truly planktonic, nor truly benthic, occurring instead in semi-suspended aggregates at intermediate water depths, whereas others are facultatively planktonic, capable of occurring in benthic habitats as well. In some cases the habitat affinities of species are simply unknown.

Taphonomic factors that secondarily affect the distribution of planktonic versus benthic species include downslope transport, bottom currents, and fluvio-deltaic transport. The strength of these variables is in large part determined by bottom morphometry of the lake basin, limiting the usefulness of multilake "training sets" as a basis for interpretation of quantitative lake-level change. Lakes with gently sloping floors and without rapid local transitions in depth are most likely to yield meaningful paleodepth information, since under these conditions, downslope transport is more limited. The magnitude of sediment focusing of diatoms from shallow to deep water is also likely to vary, depending on lake level and lake-level change, creating a "no-analog taphonomic situation," a difficulty for training set-based inferences that are derived from a lake at a single, modern, elevation.

Notwithstanding these complexities, significant headway has been made in the interpretation of lake-level histories (e.g., Barker et al., 1994b). On a qualitative or semiquantitative level this is done by comparing proportions of diatoms from various habitats. Long-term records, prior to the Late Pleistocene, generally rely on interpretation of general trends in benthic versus planktonic species (Bradbury, 1991). Semiquantitative indices of relative lake-level change have also been developed based on proportions of multiple diatom habitat categories (Metcalfe et al., 1991).

Quantitative lake-level estimates using weighted averaging regression have been made in a number of recent studies (e.g., Duthie et al., 1996). These have the advantage of using all species and depth-related effects on assemblages, rather than just the ratio of categorized groups. For example, aside from affecting the benthic/planktonic ratio, water depth changes also impact planktonic diatom communities by changing water turbulence, by increasing or decreasing nutrient releases from sediments, and by causing correlated changes in water chemistry (Wolin and Duthie, 1999). WA regression models are tailored for individual lakes, to avoid the morphometric problems of diatom transport discussed above, but difficulties still remain in making these quantitative reconstructions. Several studies have found that WA models of water depth tend to overestimate true depth in shallow water and underestimate it in deep water (Yang and Duthie, 1995; Brugam et al., 1998). This probably comes about as a result of differential diatom fossil focusing during lake-level fluctuations.

In temporary water bodies and seasonal wetlands, diatom assemblages may be useful indicators of the hydroperiod, the annual duration of inundation. Gaiser et al. (1998) developed a weighted averaging model for predicting hydroperiod for wetlands of the southeastern U.S. coastal plain. Some wetland diatoms are capable of living in essential "dry" conditions for extended periods, in soils, moss or bark, whereas others require more or less permanent inundation. This type of model could be used for determining past drought conditions in seasonally flooded areas.

Chrysophytes

The scales and resting spores of the dominantly planktonic golden-brown algae (Chrysophyceae) are commonly preserved as siliceous microfossils in lake sediments. Most research on this group to date has focused on Late Quaternary fossils, although chrysophytes are known from pre-Quaternary sediments (Van Landingham, 1964b; Srivastava and Binda, 1984). Like the other major group of siliceous algae, the diatoms, they are very useful paleoenvironmental indicators (Smol, 1995a; Zeeb and Smol, 2001). However, paleoecological research on chrysophytes has a shorter history and has been more intermittent than that of diatoms, dating to the studies of Nygaard (1956). Only since the 1980s has the potential of these fossils for paleoenvironmental reconstruction been

fully recognized, and there is still relatively little known about the taphonomy of chrysophytes.

Only a minority of chrysophyte species produce readily-preserved siliceous scales and bristles, and all of these are planktonic members of two major clades, the Synurophyceae and some Chrysophyceae. Usually these fossils can be identified to the species level. Individual species produce variable numbers of scales, complicating the interpretation of fossil flux rates, and overweighting the importance of species that produce more scales (Siver, 1991; Cumming and Smol, 1993b). All chrysophytes produce siliceous resting spores (also called statospores or stomatocysts). However, only a minority of spores have been linked to the chrysophyte species that produced them, and the application of statospores in paleoecological reconstruction has correspondingly lagged behind that of scales. Chrysophyte specialists have developed informal classification schemes to address this problem, allowing the spores to be used for paleoenvironmental inference purposes despite uncertainty as to their origins (Cronberg and Sandgren, 1986).

The principal paleoenvironmental applications of chrysophytes to date have been to infer paleo-pH, monomeric aluminum, salinity/conductivity, and trophic history, much of this work spurred by lake acidification studies in the 1980s (Charles et al., 1989; Battarbee et al., 1990). As with diatoms these investigations have yielded chrysophyte-inferred pH inference models of recently acidified lakes. Chrysophyte-inferred records of pH changes are often larger than those recorded by diatoms, and chrysophytes may actually be more sensitive to the early stages of lake acidification than diatoms (figure 11.9) (Smol, 1990; Uutala et al., 1994). This comes about for two reasons. First, acid-sensitive benthic diatoms are partly buffered from open-water pH changes by sediment chemistry, and therefore respond more slowly to lakewide changes than the exclusively planktonic chrysophyte scale records. Second, in cool–cold temperate lakes, most chrysophytes develop in the spring, when relatively acidic snowmelt dominates the water chemistry signal.

Most species of chrysophytes are intolerant of eutrophic lake conditions (Smol, 1995a). As a result, trends toward more eutrophic conditions are frequently marked by declines in total chrysophyte flux, or in the ratio of chrysophyte cysts to diatoms (Sandgren, 1988; Engstrom et al., 1991; Yang et al., 1993). Studies relating chrysophyte abundance or assemblage composition have been largely qualitative to date. Attempts to develop

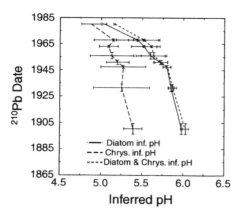

Figure 11.9. Diatom-inferred, chrysophyte-inferred, and combined diatom/chrysophyte-inferred pH for eight matched samples from three cores from Big Moose Lake, New York (also see figure 11.4). Each point shown is the average of three values with corresponding ^{210}Pb dates. Horizontal bars represent the average standard deviations calculated from sets of three points for each profile. From Charles et al. (1991).

quantitative models of chrysophyte-inferred total phosphorus are still in their early stages, limited to date by the small sizes of existing training sets, as well as interference from the much larger effects of pH on species assemblages (Wilkinson et al., 1999). Scaled chrysophyte assemblages have also been used successfully to analyze changes in conductivity (Lott et al., 1994).

Green Algae

Green algae (chlorophytes) are known as fossils from the Late Precambrian onward. However, only a few groups are well represented in the fossil record, dominantly the benthic charophytes, and several genera of acid-resistant algae (*Pediastrum* and *Botryococcus*). Both charophytes and acid-resistant algae are widely documented in lake deposits, and are potentially very valuable fossils for reconstructing past environments (Crisman, 1978; Cronberg, 1986). However, with the exception of a few key variables discussed below, their precise ecological requirements are still largely anecdotal, and relatively little is known about their taphonomy. Both of these factors have limited their paleoecological applications to date and most

inferences based on these organisms remain qualitative or semiquantitative.

Charophytes

Charophytes are known from Silurian to Recent sediments. They are preserved primarily as calcareous reproductive structures (gyrogonites), and to a lesser extent by calcified molds of stems (figure 5.3). Charophytes are common today along the margins of mildly alkaline to saline ponds and lakes. Various species are documented in conditions ranging from freshwater up to quite saline conditions, and are well known from brackish or estuarine conditions. In many cases the salinity optima and ranges of individual species are known quite accurately. Some charophytes such as *Nitella* are generally restricted to less saline and alkaline waters, whereas others occur in highly saline conditions (Burne et al., 1980). However, charophytes do not appear to occur in continuous marine conditions, probably because they do not tolerate high sulfate levels (Hutchinson, 1975). Also, most species that tolerate high salinities appear to require periodic supplies of freshwater to maintain growth. In addition to species assemblage changes, the morphology of gyrogonites from individual species also changes along salinity gradients.

Most applications of charophyte assemblages for salinity reconstruction have, not surprisingly, emphasized arid regions and closed basins. In semihumid to semiarid regions, rising alkalinities are often associated with the rapid expansion of *Chara* meadows in shallow ponds and lakes (Vance et al., 1993, 1997; Yansa, 1998). Qualitative and semiquantitative inferences of salinity have been made using charophytes from both Quaternary and much older lake beds (Soulié-Märsche, 1991; Schudack, 1995; García, 1999).

Interpretation of charophyte records from coastal lakes presents special problems, since these environments are subject to a mixture of saline water sources, from marine surface incursions, from marine or terrestrial groundwaters, or from evaporative concentration. Not surprisingly, these assemblages often contain mixed assemblages of charophyte fossils, brackish water mollusks, foraminiferans, and ostracodes (De Deckker, 1988a; Anadón, 1992; Dini et al., 1998).

Charophytes are useful indicators of past lake-level changes, water body permanence, and, indirectly, aridity, particularly when coupled with the study of other vascular plant macrophytes (Petit-Maire and Riser, 1981). Charophyte populations require at least three months of submersion to produce gyrogonites; the presence of the latter therefore provides some minimum indication of water body permanence (Soulié-Märsche, 1991). Individual species of charophytes are sometimes sufficiently zoned by depth to be useful lake-level indicators. Numeric abundance of *Chara* gyrogonites and stem casts has sometimes been used as an indicator of shallow water conditions in calcareous lakes (Lamb and van der Kaars, 1995). This must be done in conjunction with other lines of evidence, as gyrogonites are relatively durable, and can be readily transported downslope.

The use of charophytes to determine paleotemperature has been largely restricted to broad inferences based on biogeographical range extensions (e.g., Kröpelin and Soulié-Märsche, 1991; Soulié-Märsche, 1991). Willemse and Törnqvist (1999) obtained an intriguing high-resolution record from a West Greenland lake suggesting a correlation between *Chara* stem abundances and other temperature indicators.

Other Green Algae

Acid-resistant green algae such as the chlorophytes *Botryococcus* and *Pediastrum* are commonly found as Quaternary fossils during routine fossil pollen preparation and are well known from pre-Quaternary lake deposits as old as the Late Proterozoic (Guy-Ohlson, 1992). Less resistant green algae, such as *Cladophora* and the fossils (coenobia) of the common genus *Scenedesmus* are also preserved in Holocene sediments, particularly in cold-water lakes, where decay rates are relatively slow (e.g., Eisner et al., 1995). Under exceptional circumstances these fossils have also been documented in pre-Quaternary deposits (Fleming, 1989). Green algae are also represented as fossil spores.

Despite their importance in modern lake ecosystems and their abundance in Quaternary lake sediments, our understanding of the paleoecological significance and taphonomy of the commonly fossilized chlorophytes remains anecdotal (Nielsen and Sørensen, 1992). Species have been broadly categorized as indicative of eutrophic versus oligotrophic conditions, or warm versus cold conditions, but these terms are relative to local conditions, and a systematic and quantitative interpretation of the significance of green algal remains for paleoecology is still elusive.

Botryococcus is a common green alga, which forms large, floating clusters of botryoidal-shaped colonies. Sometimes these aggregations form sedimented, gelatinous horizons that can be preserved as organic-rich layers in lake sediments. *Botryococcus* occurs in a wide spectrum of lakes from the tropics to subpolar regions, but is most common in shallow-water environments of semiarid to arid regions. As a result, its abundant occurrence has sometimes been used as evidence for arid–semiarid climates in pre-Pleistocene lake sediments (Mángano et al., 1994). *Botryococcus* colonies also take on a variety of growth forms that may be related to growth seasonality and stress on the colony during growth. Guy-Ohlson (1998) used these growth form distinctions to interpret changes in paleolimnological conditions in Jurassic–Cretaceous lake deposits from southern Sweden.

Pediastrum can occur as a planktonic, benthic, or periphytic alga, although it is most commonly associated with macrophytes in the littoral zone. *Pediastrum* and *Botryococcus* abundances often vary inversely in cores, a situation that has been generally ascribed to differences in nutrient requirements (Lamb et al., 1999). *Pediastrum* peaks are frequently taken as indications of nutrient-rich, relatively freshwater conditions (Fredskild, 1983; Warner et al., 1984). However, the optimal nutrient conditions for the individual species of the two genera are not well known, and cases are well documented where an interpretation of low nutrient loads based on declines in the *Pediastrum/ Botryococcus* ratio are at odds with other indications of trophic state (Lamb et al., 1999). Taphonomic explanations for peaks in these algal species must also be taken into consideration. Algal fossil abundances can be inflated by their flocculation and adherence to clay particles during sedimentation (Avnimelech et al., 1982).

Less resistant chlorophyte fossils are occasionally useful indicators of environmental change in Holocene lake records (e.g., O.K. Davis and Shafer, 1992; Einarsson et al., 1993). In modern lakes, hypereutrophic conditions are often associated with sustained periods of dominance by certain species of green algae. This suggests the use of these assemblages as indicators of past hypereutrophy. Huber (1996) made a detailed study of the green algal stratigraphy in Holocene deposits of Lake Gegoka, Minnesota (figure 11.10). Periodic spikes in the abundance of several chlorophyte

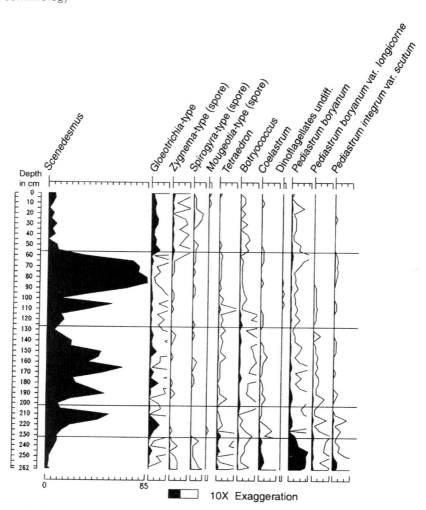

Figure 11.10. Algal (chlorophyte and dinoflagellate) stratigraphy for selected common taxa in a Holocene core from Gegoka Lake, Minnesota. From Huber (1996).

species, especially *Scendesmus*, suggest periods of hypereutrophic conditions.

Bryophytes and Vascular Plants

Bryophytes and vascular plant remains, of both endogenic and exogenic origin, are commonly preserved in lake sediments. Fossil pollen, spores, and macrofossils are among the most informative and extensively studied fossil remains found in lake sediments. Here we can do no more than review a few highlights of this research, concentrating on the less well-studied, endogenic fossil remains of aquatic macrophytes. Pollen and spores are generally trea-

ted separately from macrofossils by paleobotanists, owing to the different sampling and analytical techniques required for their study. Pollen grains are the microscopic (< 10 to > 200 μm), male gametophytes of seed-bearing plants, whereas spores are the immature gametophytes of mosses, ferns, and related plants. A highly resistant complex of polymers, known as sporopollenin, surrounds both pollen and spores, and comprises the preservable fossil material. Plant macrofossils include the entire array of macroscopic plant reproductive organs and vegetative structures, although the more abrasion and decay-resistant plant parts, such as stems, rhizomes, seeds, and thick types of leaves, usually dominate the fossil record.

Taphonomy of Pollen Preserved in Lake Deposits

It is tempting to interpret quantitative shifts in pollen abundances simply in terms of the relative abundance of local vegetation. However, many years of experience by palynologists studying both pollen distribution data and collections from surficial pollen traps show that this would be a mistake. The pollen grains and spores deposited in lake muds are a complex mixture of fossils, derived from aquatic plants within the lake, from lake margin vegetation, from upland areas within the lake's watershed, and from sources outside the watershed. Aside from plant abundance or cover in proximity to the depositional site, at least four major factors regulate the frequency of pollen and spore types found in a lake mud sample; pollen production, size and shape of grains, local pollen transport mechanisms, and the nature of the receiving lake basin.

The amount of pollen produced by an individual plant is related to its pollination strategy and variations in growth conditions (MacDonald, 1990). Wind-pollinated plants, such as pine trees, can produce billions of pollen grains each season. This results in overrepresentation of such plants in the pollen record. Animal-pollinated plants produce orders of magnitude fewer grains, and may be unrepresented in a pollen record even if the plant is locally abundant.

Both short-term seasonality and regional or long-term climate variation also affect pollen production. A single plant will also produce more pollen during some years than others, generally as a function of interannual variation in rainfall and temperature. Pollen trap data show that flowering seasonality is evident in initial pollen deposition, but this can be quickly homogenized by sediment resuspension in a well-mixed lake (Bonny and Allen, 1984; M.B. Davis et al., 1984). Climate regulates the overall terrestrial productivity of plants, reducing pollen concentrations in Arctic lake sediments by several orders of magnitude relative to the temperate zone and the tropics (Gajewski et al., 1995).

Because of their small size, pollen grains are subject to transport by both wind and water. However, the wide range of size and shape variation among pollen grains results in variable transport efficiencies between species. Species represented by small, aerodynamic pollen grains can be transported great distances, and wind transport of such pollen allows it to move across watershed divides. Relatively large or dense pollen grains, for example from spruce and fir trees, are carried short distances, tending to settle out in more proximal parts of a lake. In water, pollen grains behave much like clastic silt particles. However, their unusual shapes (for example, wing-like projections in pine pollen) cause some of them to remain in suspension for long periods, causing an increasing overrepresentation of aerodynamic shapes and light grains as the distance from the pollen source increases. In an open-basin lake context it may also determine whether pollen grains will settle through the thermocline and be deposited, or be carried out of the lake in suspension (M.B. Davis and Brubaker, 1973).

Dispersal of pollen to lakes is also dependent on the interaction between spatial scale of the lake and watershed and local/regional climate (Jacobsen and Bradshaw, 1981). In both small lakes in moderately humid forested areas and more generally in arid climates, wind tends to be the dominant transport mechanism, and stream-borne pollen is dominated by deteriorated grains from soils (Tauber, 1977; Hedges et al., 1982; M.B. Davis et al., 1984). In forests, pollen enters a lake from rapidly moving air above the forest canopy, from slowly moving air through the trees, or by directly falling from trees overhanging the lake. The winds blowing over a forest canopy are carrying pollen from a broad areal range, and are usually moving fast enough that pollen probably does not settle in less than a few hundred meters radius. This results in underrepresentation of this source in nearshore environments or very small ponds. Pollen settling nearshore is transported from the latter two sources by slow winds or no wind, and thus tends to reflect more local sources. However, these latter transport mechanisms are much less effective given their low velocities. These observations suggest that, for forested areas at least, the pollen source area in a lake record is correlated with the size of the lake, and that spatial scaling also interacts with pollen size (Jacobsen and Bradshaw, 1981; Prentice, 1985). In small ponds, pollen sources differ greatly depending on pollen dispersal properties (Jackson, 1990). Larger pollen grains mostly come from within 500 m of shore, whereas smaller pollen grains may be derived from more than a kilometer away. In contrast, a relatively small proportion of pollen comes from vegetation that directly overhangs the lakeshore, or from plants within 100 m of shore. Pollen produced at the lakeshore is mostly deposited within a short distance of shore, with the more delicate local pollen grains subject to greater destruction by nearshore wave activity (Bonny and Allen, 1984). As a whole, these results suggest that both spatial averaging and time averaging increase with lake area, especially in the central or deepest parts of the lake, where pollen grains are most distant from any potential source.

In very wet climates, or in large lakes fed by large river systems, fluvial transport becomes increasingly important, and the overall pattern of pollen accumulation becomes more complex (Bonny, 1978; Pennington, 1979; M.B. Davis et al., 1984; De Busk, 1997). Differences in transport and sorting mechanisms, geomorphology, and the proximity of large river systems can outweigh vegetational differences in determining lacustrine pollen assemblages in these types of lakes.

Pollen transport and accumulation in arid areas tends to be dominated by wind-borne sources, with overland runoff being relatively unimportant (Luly, 1997). The absence of canopy forest around playa or desert lakeshores, coupled with wind reworking of lake muds during dry intervals, allows pollen to be more readily and continuously homogenized over the lake surface. This reduces the likelihood of playa lakes yielding very high-resolution records, in contrast to small, deep lakes, but at the same time may increase their likelihood of producing a spatially accurate representation of the time-averaged vegetation.

Transport mechanisms of the pollen, spores, and macrofossils of aquatic and semiaquatic plants have received much less study than those of terrestrial plants. Pollen of aquatic and semiaquatic plants are normally recorded by palynologists, but in most analyses they are subject to more limited analysis than upland pollen species. Some groups are very difficult to identify to lower taxonomic levels, for example among lake margin grasses and sedges. Also, some macrophytes reproduce with rhizomes, only producing pollen under unusual circumstances. On a more positive side, the low stature of emergent macrophytes causes their pollen to have a limited dispersal range, and most probably accumulates very close to its place of origin (Kratinger, 1975; H.H. Birks, 1980). This has allowed palynologists to use declining proportions of nearshore aquatic pollen in a core record to indicate a progressive deepening of the lake, as the source increasingly becomes just long-distance, wind-blown pollen (e.g., Kershaw, 1979). Declining littoral pollen ratios must be interpreted carefully, however, because the thin outer coating of pollen from many aquatic plant species limits their preservability (H.H. Birks, 1980; Collinson, 1988).

Following their initial surface accumulation in lakes, pollen assemblages undergo further taphonomic modification, the most important of which involves sediment focusing (M.B. Davis et al., 1984). Pollen proportions in cores taken at various water depths may show similar profiles, indicating a homogenization of pollen types. However, sediment focusing can artificially inflate the apparent rates of pollen flux and local accumulation for some sites, and result in very different rates between sites within a lake.

Plant Macrofossil Taphonomy in Lakes

Plant macrofossils comprise a wide array of vegetative parts and reproductive organs, whose preservability varies greatly between clades. Bryophyte macrofossils are normally preserved as leafy stems, or in the case of *Sphagnum*, as detached stem and branch leaves (Janssens, 1990). Vascular plants are represented by leaves, stems, roots, seeds, or fruits (Collinson, 1988). All of these are subject to selective decomposition between species, and transport of older remains, resulting in both biased and mixed assemblages. Fortunately, mixed assemblages are often recognizable based on the preservation quality of the plant fragments. Under exceptional preservation of anoxia and rapid burial, whole organ systems and the more delicate tissues of aquatic plants remain intact. These types of assemblages provide very detailed pictures of plant macrofossil communities, even in lake sediments that are quite old (e.g., Cevallos-Ferriz et al., 1991).

Local plants dominate the plant macrofossil record in lakes, and aquatic species are usually well represented. Terrestrial leaves are generally only transported short distances by wind, with smaller ones deposited further offshore (Roth and Dilcher, 1978; Spicer, 1981). Once in a lake they can be carried for distances of up to a few kilometers, depending on their size and density, and by the strength of local currents (Rich, 1989). This results in a systematic reduction of the proportion of preserved larger or denser plant parts in an offshore direction. Water-borne terrestrial plant debris deposited in lakes is also usually deposited within a few kilometers of its origin. In this instance however longer-distance transport does occur in areas of high topographical gradients, or with more durable plant parts (MacGinitie, 1969; Rich, 1989).

Most aquatic macrofossils are only transported short distances from their origin (H.H. Birks, 1973; Gastaldo et al., 1989). Plant macrofossil studies therefore have been traditionally used to compliment the more regional vegetation picture provided by palynology, particularly for lakeshore or aquatic plants that produce little pollen (H.H. Birks, 1984).

The only important exceptions to this generalization of local provenance seem to be seeds from shoreline plants, which can float indeterminate distances (Hannon and Gaillard, 1997), and floating vegetation mats, particularly in the tropics. Although a primary transport bias is not prominent in the aquatic macrophyte record, there are other important factors modifying their accumulation. Submerged and floating macrophyte tissues are delicate, and many types do not preserve readily. Also, the leaves of many aquatic macrophytes decay on the plant, rather than being shed like terrestrial plants, thereby reducing the likelihood of their preservation. Finally, both pollen and more resistant macrophyte components are subject to reworking and secondary burial, although this is usually evident based on pollen deterioration and abraded macrofossil surfaces (N.G. Miller and Calkin, 1992).

Lake Level and Water Availability

The most common paleoecological application of aquatic plants is the inference of water depth. Some plants, like the sphenopsid *Equisetum*, grow at or very close to the water–air interface. Where plants like this are preserved in growth position, they provide compelling evidence for a lake margin environment, even in deposits as old as the Late Paleozoic (Collinson, 1988). However, aside from those species adapted for growth in the zone of interannual exposure and inundation (0–1 m), the depth ranges of most aquatic plants are not highly specific. Variations in light intensity as a result of water depth, turbidity, and shading cause depth ranges of most species to vary between lakes (Hannon and Gaillard, 1997). As a result, paleoecological interpretations based on entire assemblages rather than indicator species are most useful in the inference of water depth changes.

Paleoecological transitions between extremely shallow-water assemblages (emergent macrophytes such as *Carex*, *Typha*, *Equisetum*, and *Phragmites*), intermediate depth, submerged, and floating macrophytes (*Nymphaea*, *Potamogeton*, *Myriophyllum*, and *Ceratophyllum*), and low-light, sublittoral assemblages (*Chara*, *Nitella*, and some bryophytes) can be used as indications of rising or lowering relative lake levels (figure 11.11). In larger lakes, with more vigorous nearshore circulation, paleoecological interpretations of water depth are complicated by the fact that fruits and seeds are readily transported outside of the littoral zone, and deposited near wave base.

In making an interpretation of relative water depth change it is important to bear in mind that factors other than absolute lake level may be involved; littoral macrophyte zones will also shift in response to local processes like delta progradation or seral succession (Van Dijk et al., 1978; Rich, 1989). Response to relative lake level will also be a function of the morphometry of the lake floor, since the flooding of a broad bench of relatively low relief can also cause a proportionate increase in littoral habitat. Changes in total lake area will also cause the total area of marsh to fluctuate, and this can be reflected in the proportion of pollen from marsh plants such as sedges (Lamb and van der Kaars, 1995).

Infilling by terrigenous or chemical sediments, or by plant debris, often results in lake shallowing over time, especially in smaller lakes (Vanhoorne and Ferguson, 1997). The infilling of bogs or abandoned oxbow lakes is reflected by vertical sequences comprising open-water aquatics, giving way to littoral fringe aquatics, and eventually to swamp and mire species (e.g., Jasinski et al., 1998). Successional patterns in infilling oxbow lakes of this type, that are comparable to modern analogs, have been documented in lakes as old as the Eocene (Hickey, 1977). However, neither the rates nor the direction of evolution in bogs and oxbow lakes are uniform. Multiple pathways of vegetation change are seen as these depressions infill, a result of variations in local edaphic features, groundwater hydrology, and climate (Nicholson and Vitt, 1994; Green-Winkler and Sanford, 1995; Willemsen et al., 1996).

Only very subtle changes in lake or groundwater level are required to initiate vegetational successions in wetlands that are evident in paleolimnological records (Vardy et al., 1997). In swamps, mires, and peat bogs, standing or subsurface water is normally a limiting factor for vegetation. Bryophytes are particularly sensitive to these changes (Jannsens, 1990). Changes in both species composition and growth morphology are associated with variations in water stress.

When local processes can be ruled out as the cause of shifts in the littoral belt, macrophytes can be a powerful tool for inferring eustasy. The initial flooding during rising lake stands is normally signaled by an abrupt appearance of aquatic macrofossils and a decline of terrestrial plant macrofossils. Declining proportions of nearshore macrophytes, peaty sediments, aquatic/lake margin pollen, and increasing planktonic algae or deep water mosses all indicate continued lake-level rise and passage of a site through the littoral depth zone (e.g.,

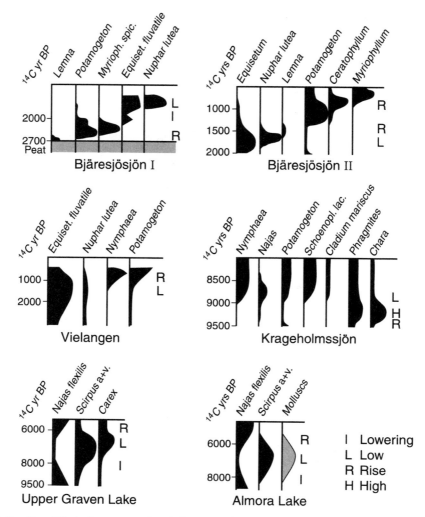

Figure 11.11. Simplified plant–macrofossil diagrams from Holocene lake deposits investigated for lake-level changes in Sweden and the United States. Schoenopl. lac., *Schoenoplectus lacustris*; a + v, *acutus + validus*. From Hannon and Gaillard (1997).

Wright et al., 1985; O.K. Davis and Shafer, 1992; R.S. Anderson, 1993; Björck et al., 1993; Yu and McAndrews, 1994). In long records the repeated occurrence of lake margin aquatic species can provide evidence for repeated fluctuations in lake levels, and indirectly, for climate cyclicity (Partridge et al., 1993; O.K. Davis and Moutoux, 1998).

Other Paleoecological Inferences from Aquatic Macrophytes

Variations in salinity tolerance among macrophytes have been used to determine changes in lake chem-istry. The most commonly cited indicators are fossils of the lacustrine halophytes *Ruppia* and *Zanichellia palustris*. Shifts to drier climates, or for coastal lakes, marine incursions, are often marked by increases in the representation of these species in closed basin lakes (Barnosky, 1989; Stevenson and Battarbee, 1991; Vance et al., 1997).

The distributions of many moss species are highly correlated with changes in pH, especially in the acidic to circum-neutral range. As we saw in chapter 5, this partly arises from the ability of some mosses to exchange base cations and in the process reduce pH (Clymo, 1967; Glime et al., 1982). This allowed Janssens (1990) to develop a weighted-averaging inference model for acidity based on modern North American distributions, which has been applied to Holocene deposits.

Most aquatic macrophytes have broad ranges of temperature tolerance. As a result, the use of macrophytes to directly infer temperature has been limited to extremes of temperature ranges in lakes, particularly to colonization of polar or subpolar regions following deglaciation. Morphological variation in some aquatic plants also seems to be expressed as a result of temperature differences. In a number of independent clades, the seeds from plants living in temperate lakes are smooth, whereas those from tropical–subtropical climates are spiny, or covered by small papillae and tubercles (Collinson, 1988).

In some circumstances an indirect link between temperature and macrophyte distribution comes about through the intermediate step of correlated shifts in lake nutrient status. Aquatic macrophytes are sensitive to overall nutrient availability and this is reflected paleolimnologically by shifts in species composition. Even stronger inferences of trophic conditions and climate can be derived when macrofossil and pollen data are coupled with other paleolimnological indicators, such as fossil algae (e.g., Hoek et al., 1999). Strong productivity and climate signals from macrophytes are evidenced in high-latitude and alpine lakes following deglaciation, when nutrient availability rises dramatically as a result of rapid silt influx, from extremely low

prior levels. Eisner et al. (1995) observed this pattern in a combined study of algae, pollen, spores, and macrofossils from a Middle to Late Holocene core collected from a small lake in West Greenland (figure 11.12). High abundances of *Pediastrum* followed by *Myriophyllum* (a submerged aquatic) during the early phase of the lake's history are indicative of relatively high, but declining nutrient inputs shortly after deglaciation. As the input of minerogenic nutrients slowed, there was a transition to a more mesotrophic to oligotrophic macrophyte community, dominated by the submerged angiosperms *Potamogeton* and *Hippuris*, and the bryophytes *Menyanthes* and *Sphagnum*.

H.H. Birks (2000) documented a remarkable record of changing aquatic macrophyte communities for the Late Glacial and Early Holocene of Kråkenes Lake in western Norway (figure 11.13). From 12.15–10.9 ^{14}C yr BP, early postglacial pioneers occur, the charophyte *Nitella* and the angiosperm *Ranunculus* sect. *Batrachium*. These taxa are known from modern studies to be rapid colonizers in Arctic aquatic environments. Following a period of relatively rare aquatic plants during the Younger Dryas (10.9–10 k ^{14}C yr BP), an expansion of a variety of submerged and emergent aquatic species occurred. This includes a succession of isoetid pteridophytes (quillworts), indicative of disturbed, but low sedimentation environments, as well as

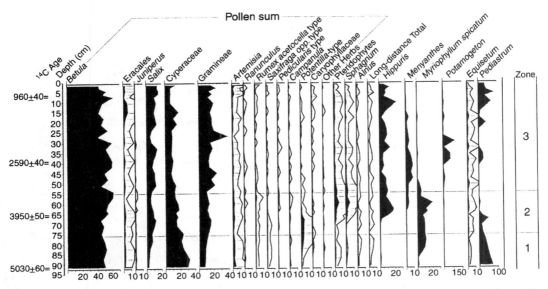

Figure 11.12. Terrestrial and aquatic pollen, spore, and algal stratigraphy in a Holocene record from West Greenland. From Eisner et al. (1995).

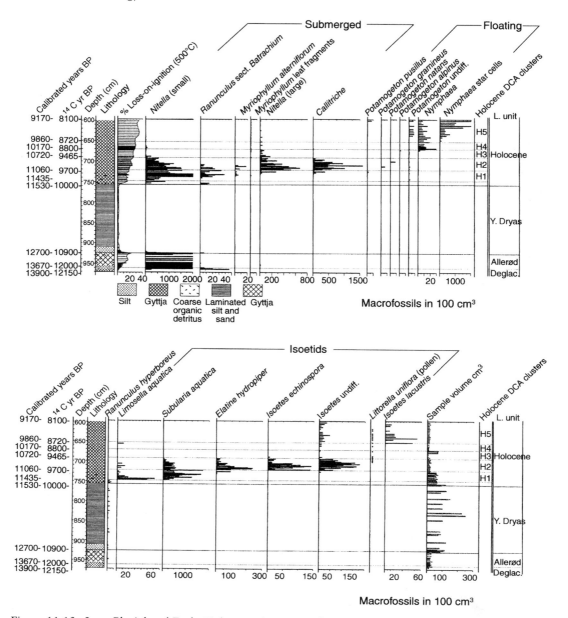

Figure 11.13. Late Glacial and Early Holocene plant macrofossil stratigraphy for Kråkenes Lake, western Norway. This record illustrates a striking succession of aquatic plant species following deglaciation after the Younger Dryas. From H.H. Birks (2000).

changes in conductivity, trophic state, and seasonality in climate. A period of general macrophyte decline followed, from 9.5–8.8 k ^{14}C yr BP. After this time a new, more productive community appeared, dominated by the rooted, floating water lily *Nymphaea*, a strong competitor that shades out submerged aquatics.

Constraints on Combining Terrestrial Pollen and Aquatic Plant Macrofossil Records

Interpretation of paleolimnological change at both local and regional scales normally involves a careful consideration of interactions between the lake and its surrounding terrestrial environment. Fossil pollen, terrestrial plant macrofossils, and terrestrial insects preserved in lake deposits provide us with some of our most important and most continuous records of terrestrial ecosystems, especially for watersheds surrounding lakes. Because paleolimnologists often work closely with palynologists, it is important to understand how these two types of records, often collected from the same cores, differ, and how they can be integrated. Significant differences in the time and space averaging of terrestrial and aquatic fossils are particularly important for paleolimnologists to understand as we attempt to interpret both high and low-resolution records of environmental and climate change from lake deposits.

Systematic discrepancies often exist between upland pollen and lake records of climate. Apart from simple misinterpretation, there are several reasons why this occurs (Cohen et al., 2000a; Wigand and Rhode, in press):

1. *Differing sensitivity and/or response rates to the same climatic factor.* Upland plants, especially large trees, have long life spans and their response to climate change is probably more time-averaged than the rapid (subannual) responses observed in algae or invertebrates. This may result in climate responses evident in the paleolimnological record that are not apparent from the pollen record (e.g., Forester et al., 1989; Cwynar and Levesque, 1995). On the other hand, the modulation of climate by water also causes temperature ranges experienced by aquatic organisms to be reduced, in comparison with air temperature fluctuations.

2. *Variable response of terrestrial versus aquatic communities to the same change in climate.* In an alpine environment, a glacial advance caused by slightly cooler temperatures and higher precipitation might significantly reduce the habitable area for trees within a watershed, while at the same time causing the downstream lake to expand but not change significantly in temperature.

3. *Drainage capture.* If a stream system is diverted away from a lake, a major change in hydroclimate can be registered in the aquatic fossil biota, as a result of declining lake levels or increasing salinity. Simultaneously, the air-borne pollen record would be little affected by this change.

Fungi

Fungal spores and vegetative hyphae are common in lacustrine sediments (Sherwood-Pike, 1988). Most of these are exogenic fossils, derived from soil or animal dung, and introduced through runoff from the lake's watershed. Some smaller fungal spores are also transported by wind. Exogenic fungal remains in lake sediments have occasionally been used to interpret terrestrial ecosystem processes (O.K. Davis, 1987), and to record the seral succession of open-water ponds (few fungal remains) into semiaquatic fens or bogs (common fungal fossil spores and hyphae) (Kuhry, 1997). Some exogenic species preserved in lake sediments are also useful indicators of paleotemperature (Prasad and Ramesh, 1984; Van Geel, 1986; Pirozynski, 1990). The potential of using aquatic fungi for paleoecological interpretation, however, is effectively unknown. Aquatic fungi are relatively common fossils, and transitional sequences from dominantly terrestrial to dominantly aquatic species in strata can be used to infer the development of marsh or open-lake conditions (Pirozynski, 1990). However, little data is available regarding the specific ecological tolerances or distribution of fossilizable aquatic species, and there are very few specialists in the field capable of identifying specimens. Semiaquatic fungi, especially those in bog deposits, have received slightly more attention. Van Geel (1986) has suggested that the morphology of asexual spores (conidia) may be a useful indicator of peat moisture conditions.

Testate Amoebans and Foraminiferans

Testate amoebans and foraminiferans have considerable potential as paleoecological indicators in lake deposits, although neither of these groups of amoeboid protists has received much attention by paleolimnologists to date. Although micropaleontologists have studied marine foraminiferans extensively, nonmarine species are not well known, either in terms of their ecology or taphonomy, and the same is true of the testate amoebans.

Based on their morphology and phylogenetic relationships, the testate amoebans are clearly an ancient clade, dating back at least to the Late Precambrian, although early members may have been dominantly marine rather than lacustrine organisms (Porter and Knoll, 2000). However, their pre-Late Quaternary fossil record is extremely sparse (Medioli and Scott, 1988; Porter and Knoll, 2000); almost all paleolimnological studies to date have focused on Late Quaternary fossils. Most applications of fossil testate amoebans have been made in temperate lakes and mires of North America and Europe. These studies have looked at the relationship of testate amoebans with sediment input and indirectly, climate, with moisture content in bogs and wetlands, and with pH (Tolonen, 1986; Warner, 1990b; Kuhry, 1997).

Testate amoebans are benthic, particulate detritus feeders, and therefore sensitive to the quality and moisture content of organic matter in surrounding sediments. Some paleolimnological investigations have interpreted changes in amoeban assemblages to be driven by variable food resources, and indirectly, lake trophic state (Burbridge and Schröder-Adams, 1998; Beyens and Meisterfeld, 2001). The correlation between quantity/quality of lacustrine organic matter and climate has also allowed investigators to use testate amoebans as paleotemperature indicators, particularly across the Pleistocene/Holocene boundary (McCarthy et al., 1995). In temperate bog environments, testate amoebans can be used to infer moisture availability, since some are capable of living on bog surfaces that are desiccated during the summer months (Beyens, 1985; Woodland et al., 1998).

The distribution of modern testate amoebans is frequently correlated with pH (Ellison and Ogden, 1987). In a study of Ullswater, a tarn basin in the English Lake District, Ellison (1995) used the declining proportions of high pH taxa through the Early Holocene to argue for a long-term, postglacial acidification trend.

In contrast with the testate amoebans, the foraminifera have a long and rich fossil record, a result of their much more durable tests, which are calcareous or more strongly agglutinated than testate amoebans. Foraminiferans are generally restricted to saline waters, and their utility in paleolimnology comes about primarily as indicators of saline lake conditions. They are relatively common fossils in the deposits of saline lakes, especially those of sulfate-dominated, low alkalinity systems. Fossil lacustrine foraminiferans have been documented from saline lake deposits throughout the arid parts of the world, even in lakes far removed from the oceans, with no prior seaway connections (Cann and De Deckker, 1981; Anadón, 1989, 1992; Stevenson and Battarbee, 1991; Boomer, 1993; Plaziat, 1993; J.E. Spencer and Patchett, 1997). However, most lacustrine species are closely related to, or in many cases identical to estuarine and coastal species, capable of withstanding variable salinity waters. Colonizing species most likely enter saline lakes along established bird flyways (Patterson et al., 1997).

Fossil successions of foraminiferans and testate amoebans can provide evidence for the history of progressive isolation of coastal lakes from the marine environment. For example, as a basin undergoes isostatic uplift, or is cut off from the sea by the formation of coastal barriers, its paleoecological record will show an increasing proportion of testate amoeabans to foraminiferans, whereas a basin undergoing marine flooding during relative sea-level rise will display the reverse trend (Laidler and Scott, 1996).

Sponges

Freshwater sponges of the family Spongillidae have a fossil record extending back to the Cretaceous (Harrison, 1990), although the vast majority of recorded freshwater sponge fossils are of Quaternary age. Normally, sponges are preserved as isolated siliceous spicules, and occasionally as reproductive gemules, which themselves bear spicules (Turner, 1985). Three spicule types are normally preserved (Harrison, 1990). Megascleres are larger skeletal spicules that make up the supporting meshwork of the sponge. These are the most common fossil spicules, but are of limited used in identification. Gemmoscleres are smaller spicules that protect the assexual reproductive gemule, and are the most valuable spicules for species identification. Microscleres are small dermal spicules, not found in all species, which are also sometimes useful for identification purposes. Rare occurrences of intact freshwater sponge skeletons are also known as fossils (Racek and Harrison, 1974).

Although sponges are sensitive indicators of water quality, very little is known about the taphonomy of sponge spicule preservation. This, along with the small number of specialists in the field, has limited the use of these fossils. This is unfortunate, because sponge fossils are very commonly reported, and those studies that have been done have shown them to be potentially informative. Sponges are known to rapidly colonize freshwater habitats, providing potentially valuable indicators for differentiating marine versus freshwater or lentic from lotic habitats, and for understanding the ontogeny of individual lakes (Solem et al., 1997).

Sponge distribution is affected by such factors as temperature, turbidity, light availability, pH, and alkalinity (Harrison, 1974; Frost, 2001). Some attempts have been made to apply these findings in semiquantitative paleochemical reconstructions (Racek, 1974; Hall and Herman, 1980; Turner, 1985; Harrison and Warner, 1986; Harrison, 1988, 1990). In the Holocene sediments of Lake Baikal (Siberia, Russia), the abundance of spicules is correlated with other indicators of warmer temperature (Karabanov et al., 2000). In modern habitats some, although not all, sponge species are known to be quite intolerant of turbidity, because of its negative effects on sponge filter feeding. The occurrences of silt-intolerant species might therefore be used as an indicator of reduced clastic sedimentation loads (Harrison, 1990). This must be done carefully, since some species respond to turbidity by inhabiting cryptic habitats (overhangs, caves, etc.) where they may still produce spicules that can later be sedimented.

Light intensity also affects sponge distribution, in part because some species possess photosynthetic symbionts. This does not automatically translate into depth gradients; some sponges that thrive under low light conditions do so by growing on the undersurfaces of rocks or plants. In heavily eutrophied lakes, declining light penetration, coupled with excessive suspended particulate matter, is sometimes correlated with declining upcore abundance of sponges (Findlay et al., 1998).

Rotifers

Rotifer fossils are known exclusively from Quaternary sediments (Warner, 1990c). Their preserved body parts include flask-shaped, proteinaceous shells and resting eggs, though the latter are probably not diagnostic for identification at the species level. Although rotifer fossils are commonly noted in Late Quaternary sediments, they have been little used in paleoecology, and almost nothing is known of their taphonomy. This is unfortunate as rotifers inhabit a wide range of aquatic and semiaquatic environments, and are important components of modern lacustrine ecosystems. What little work has been done on their paleoecology suggests they could be useful for understanding paleofood webs, as they are important dietary components for planktivorous fish (e.g., Jeppesen et al., 1996). Some studies demonstrate that rotifer fossils can be used as indices of paleomoisture in bogs (Warner and Chengalath, 1988), as eutrophication indicators (Findlay et al., 1998), or to track the evolution of wetland systems (Kuhry, 1997).

Bryozoans

The status of freshwater bryozoan paleoecology is similar to that for rotifers; very little work has been done in the field, despite the fact that bryozoan fossils are not uncommon in Quaternary lake sediments. Our knowledge of freshwater bryozoan fossils is almost entirely restricted to the Late Quaternary, although a few exceptional specimens are known from lake beds as old as the Early Cretaceous (Jell and Duncan, 1986). Bryozoans are preserved as asexual reproductive statoblasts, characteristic of freshwater taxa. These microscopic structures are formed in large numbers by most freshwater bryozoan species (Warner, 1990c). Their occurrence therefore provides evidence of freshwater conditions (Solem et al., 1997). Because of their small size, they are readily distributed by floatation, and are thought to be dispersed between lakes on birds' feet. Unfortunately, relatively little is currently known of the hydrodynamics of statoblast dispersal as it relates to sedimentation.

Studies of fossil statoblasts have focused on their potential role in delimiting littoral zone area, and for inferring siltation and climate change (Francis, 2001). In Florida lakes, the accumulation of statoblasts is correlated with the extent of the littoral zone, although this has yet to be demonstrated with fossil specimens (Crisman et al., 1986). Because they are benthic or epiphytic filter feeders, most bryozoans are negatively affected by high concentrations of suspended sediment; some paleolimnological records demonstrate this correlation (Francis, 1997). Finally, there is some prospect for using bryozoans to infer climate conditions based on biogeographical range changes over time (Kuc, 1973).

Cladocerans and Other Branchiopods

Since the pioneering work by Frey (1955, 1958), cladocerans have been recognized as one of the most valuable Quaternary paleoecological indicators in lakes. They are abundantly preserved by disarticulated skeletal parts (*chitin*), including head shields, shells, claws, ephippia, and other chitinous skeletal parts. These are shed during molting or after death, and many are easily identifiable. Cladocerans are widespread in planktonic, epibenthic, and epiphytic/littoral settings, making them useful as paleolimnological tools in numerous contexts.

A large number of families of cladocerans exist. However, only a small number of these are regularly preserved and therefore of interest to paleolimnologists. These more strongly sclerotized or abundant clades include the following groups:

1. *The Chydoridae.* These are predominantly benthic taxa, including littoral macrophyte crawlers, plus some benthic species that occur on sand or mud–water interface. Chydorids are the largest family of cladocerans and the best-known group as fossils.
2. *The Bosminidae.* These are mostly planktonic cladocerans, which are common as fossils but represented by only a few species.
3. *The Daphniidae.* These are common zooplankters in modern lakes but are less commonly fossilized.

Cladoceran Taphonomy

Taphonomic processes affecting cladoceran fossils are fairly well understood. Bosminid and chydorid skeletal parts are well represented as fossils, with a broad representation of body parts and molt instars (Frey, 1988b). Daphniidae are preserved erratically, and often are poorly represented by smaller instars (Culver et al., 1981). Body parts of littoral taxa are often redistributed downslope, resulting in partial but incomplete mixing with offshore, planktonic

bosminid remains in small lakes (Mueller, 1964; Amoros and Jacquet, 1987; Frey, 1988b) (figure 11.14). This transport results in the production of time-averaged samples on time scales of a year to several years, and spatial averaging that broadly integrates the fauna of the littoral zone in small lakes. It does not result in total mixing, however, and offshore planktonic assemblages remain relatively distinct from littoral ones or areas strongly influenced by nearshore transport.

Amalgamation of nearshore chydorids and planktonic bosminids occurs in larger lakes as well (Hann and Karrow, 1993), although the extent of such large lake mixing is much less well understood. Littoral remains are likely to produce proportionately smaller fossil contributions to the offshore fossil record in large lakes, despite the greater potential for offshore transport by stronger currents. Kerfoot (1995) showed that considerable population and species-level heterogeneity exists in the bay and littoral fringes of larger lakes, and that this heterogeneity is reflected in local differences in fossil assemblages, but that very little of this nearshore skeletal material finds its way to offshore core locations.

Paleoecological Interpretations from Cladocera

Cladoceran assemblages respond directly and indirectly to a wide array of environmental variables

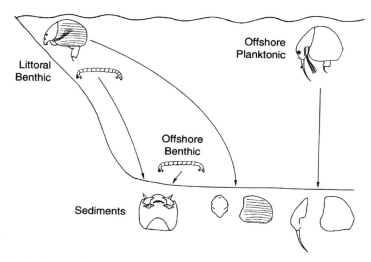

Figure 11.14. Major habitats of Cladocera and Chironomidae (midges) and the accumulation of their remains in deep-water sediments. Chydorid cladocera and most midges are benthic in the littoral zone, whereas *Bosmina* is planktonic in the open water. Some midge larvae also can occur in deep-water sediments, the species present being controlled by ambient temperature and the oxygen concentration at the end of stratification. From Frey (1988).

(Frey, 1988b; Whiteside and Swindoll, 1988; Korhola and Rautio, 2001). Those offering the greatest promise to paleolimnology include differences in planktonic versus littoral habitat, productivity, temperature, salinity, pH, and predation.

Planktonic/Littoral Ratios and Lake-Level Fluctuations

Differences in the abundance of planktonic (dominantly bosminid) versus littoral (dominantly chydorid) cladoceran fossils provides valuable information on the distribution of littoral habitat at the time of deposition. A common way of expressing this information has been as the ratio of planktonic/littoral species or individuals (Goulden, 1964; Alhonen, 1970, 1971; Whiteside, 1970). It would be useful if such a ratio simply reflected the distance onshore or offshore, or paleowater depth where the fossils were collected, but unfortunately nature is not quite so simple. Planktonic/littoral (P/L) ratios of cladoceran remains have been found to correlate strongly with the relative area of the littoral zone in a lake (Hoffman, 1998), and have occasionally been used as a semiquantitative index of water depth (Alhonen, 1970, 1971). Littoral taxa are more strongly affected by lake-level fluctuations than planktonic ones, suggesting that littoral cores are most likely to provide a P/L ratio record interpretable in terms of lake level (Hoffman, 1998). However, even the early proponents of the P/L ratio noted that it is only indirectly related to water depth. Because this index is widely cited in the literature, and because of the natural interest in using this data as an indicator of lake level, it is worthwhile considering the factors that cause it to vary aside from water depth.

Chydorid cladocerans are most abundant in the littoral zone, especially on or around macrophytes. When these animals die or molt, their skeletal fragments can be partially transported offshore, but horizontal transect studies consistently show a pattern of declining abundance of these remains in an offshore direction. Planktonic *Bosmina* remains are not, however, found most abundantly in offshore regions, but rather in the water column in relatively shallow water, or where the lake bottom intersects the upper hypolimnion. They are less common in deep water simply because this area usually provides less food. Thus a summation of all planktonic species (nearshore planktonic species plus truly pelagic ones) confounds the significance of the P/L ratio as a simple index of proximity to the littoral

zone. The P/L ratio is also affected by several factors that are not predicted simply by the areal extent of the littoral zone (Frey, 1976, 1986). These include:

1. *The topography of the littoral zone.* A lake surrounded by flat topography will have a proportionately increasing littoral area. This will cause the P/L ratio to decline as lake level rises (Binford, 1986).
2. *Top-down effects of new predators.* The selective grazing of planktonic cladoceran species can reduce the P/L ratio without lake-level change.
3. *Nutrient loading.* Increased nutrient loads and phytoplankton blooms may cause planktonic cladocerans to increase, and declining water transparency will cause macrophytes and littoral cladocerans to decrease without any lake-level change. Some modifications of the P/L indices attempt to compensate for these trophic effects by separately analyzing those zooplanktonic species known to respond strongly to nutrient loading (Szeroczynska, 1998).
4. *Deep-water mosses.* The significance of P/L ratios can be clouded by the occurrence of aquatic mosses. Bryophytes adapted to low light levels effectively extend the habitat of "littoral" cladocera into much deeper water than would be expected from the range of larger macrophyte species distribution (Sarmaja-Korjonen and Alhonen, 1999).

These limitations suggest that a more profitable approach to lake-level inference is likely to come out of quantitative techniques that move away from the use of a simple P/L ratio, to methods such as weighted averaging, which treat the data available from each species independently (figure 11.15). Microhabitat differences between individual chydorid species can also be exploited to determine water depth histories (Jurasz and Amoros, 1991).

Productivity and Temperature

Many of the influential studies on cladoceran paleoecology have centered on their use as indicators of changing trophic state. Changes in planktonic *Bosmina* species have been most commonly cited as indicative of changing nutrient loads,

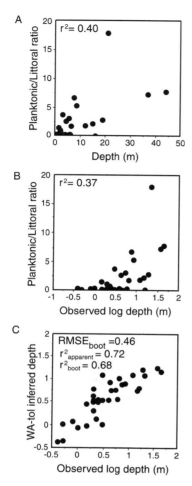

Figure 11.15. Comparison of goodness-of-fit of observed water depth with water depths inferred from cladoceran remains, based on planktonic/littoral ratios (A,B) and weighted averaging inferred depths (C), for modern lakes from British Columbia (Canada). The WA-tol model downweights taxa with wide tolerances when predicting salinity. Because the planktonic/littoral ratio cannot yield negative values, both untransformed (A) and log (B) depth models using the P/L ratio are shown. In either case the WA model (C) provides a better estimator of water depth. From Bos et al. (1999).

conditions on macrophyte communities. In a study of Florida lakes, Binford (1986) found that fossil accumulation rates of most cladoceran species (both chydorids and bosminids) were positively correlated with lake productivity. At very high productivities, however, chydorid diversity tends to decline. This is probably a threshold phenomenon, which sets in when macrophyte habitat is eliminated on a large scale under very eutrophic and turbid conditions (Frey, 1988b).

Hann et al. (1994), in their study of an experimentally manipulated Canadian lake, found distinctive changes in fossil cladoceran community composition associated with various phases of fertilization and types of fertilization treatment (specific N : P ratio). For littoral chydorid species, changes seem to have been in response to such indirect factors as changes in microhabitat, for example available algal mats, or reduction in species dependent on macrophytes, which declined with decreasing water transparency. Planktonic *Bosmina* species also responded to nutrient enrichment through the intermediate effects of elevated productivity on pH or oxygen availability.

Because of the strong interactions between temperature and productivity, especially on long time scales, it has not always been a simple matter to differentiate the effects of these two variables on changes in cladoceran communities during major climate changes (e.g., Hann and Karrow, 1984, 1993; Hoffman, 1986b; Guilizzoni et al., 2000). Some fairly unambiguous evidence does exist for cladoceran community change in response to non-anthropogenic eutrophication. Boucherle et al. (1986) studied the response of cladoceran communities to the widespread disappearance of hemlock from watersheds in eastern North America about 4800 yr BP. This event resulted in widespread soil erosion and presumably rapid nutrient discharge in many watersheds prior to the re-establishment of deciduous tree cover (Likens and Davis, 1975; M.B. Davis, 1981). Changes in cladoceran fossils following this event in smaller lakes were more profound and occurred more quickly than in larger lakes, as would be predicted from a nutrient loading model.

However, much research on the linkage between cladocerans and nutrient loads has been ambiguous. Quantitative studies by Lotter et al. (1997a, 1998) found only a relatively weak relationship between cladoceran community structure and nutrient loading in Swiss Alpine lake faunas. They found that changes in total community structure could only secondarily be linked to total phosphorus for benthic chydorid remains, and that for planktonic

although the specifics of these associations are often controversial, especially when applied to paleolimnological reconstructions (Boucherle and Züllig, 1983; Hoffman, 1986b; Frey, 1988b; Lotter et al., 1998). Littoral cladocerans also respond to nutrient loading, although here the impact is probably even more indirect, mediated by the effect of eutrophic

cladocera the relationship was insignificant. In the same lakes, temperature was a much stronger predictor of cladoceran distribution patterns, sufficient to allow the researchers to develop WA transfer functions for cladoceran-inferred temperatures, especially for benthic species.

Salinity and pH

Several recent studies have shown that cladocerans can sometimes be good indicators of paleochemistry (Steinberg et al., 1988; Bos et al., 1999). In the deposits of highly saline lakes cladoceran fossils are dominated by halophilous species, or species capable of surviving highly variable conditions. Bos et al. (1999) found a good relationship between cladoceran species turnover and salinity in British Columbian lakes, adequate for the development of a WA training set (figure 5.8). Verschuren (1994) documented the disappearance or decline of fossils for a number of swamp-dwelling chydorid taxa in Oloidien Lake, Kenya associated with declining lake levels and rising salinity. Simultaneous with rising salinity he found a rise in the abundance of the planktonic species *Moina micrura*, which is widespread in African lakes of high but fluctuating salinity.

Chydorids have also been used as indicators of recent acidification. Early studies of this kind were based on the differentiation of pH categories, and application of multiple regression models to these categories (Krause-Dellin and Steinberg, 1986). The early attempts to use chydorids in this fashion met with some skepticism (Hoffman, 1986b). As with the P/L ratio, categorization approaches have the potential to underestimate the importance of other pH-correlated factors on the abundance of apparently pH-sensitive species. In particular these include the effects of metal concentrations and fish predation (Nilssen and Sandøy, 1990). However, further work in the field has tended to solidify the view that at least some chydorids are sensitive to pH changes, and that pH can be inferred quantitatively from cladocerans (Steinberg et al., 1988). This view is supported by studies that have simultaneously reconstructed similar pH changes from cladocera and other groups of organisms (Van Dam et al., 1988).

Despite the correlation that exists between pH and chydorid assemblage change, attempts to attribute the loss of cladoceran species diversity to long-term declines in pH in acidified lakes are not always straightforward (Korhola, 1992). Where acidification has been associated with a loss of species there

are usually other correlated factors, particularly higher metal concentrations, which may be responsible for the loss (e.g., Manca and Comoli, 1995). In rock basins undergoing organic sedimentation and natural acidification, species diversity has sometimes been seen to increase with acidification.

Ecological Interactions

Planktonic cladoceran fossils are useful indicators of the intensity of predation, and many studies have used them to understand the history of ecological interactions in the water column of lakes (Frey, 1986; Nilssen and Sandøy, 1990; Kitchell and Sanford, 1992; Räsänen et al., 1992; Sanford, 1993; Kerfoot, 1995; Szeroczynska, 1998). As large-bodied zooplankters, they are susceptible to predation by fish and some larger invertebrates. Heavy predation results in a decline in the abundance of the fossils of large-bodied zooplankton species (e.g., Leavitt et al., 1994a, see figure 1.7). In contrast, lakes that lack fish often have large populations of slow-moving zooplankters; this relationship is quantitatively recorded in cladoceran fossil assemblages (Jeppesen et al., 1996).

Other Branchiopods

Aside from cladocerans, the paleoecological interpretation of fossils from the other major groups of branchiopods presents something of a puzzle to paleolimnologists. None of these groups have received anywhere near the attention given by ecologists to cladocerans, and their distributions and habitat requirements, with the exception of the brine shrimp, *Artemia salina* (figure 5.11), remain poorly known. Most conchostracans, anostracans, and notostracans are occupants of rather extreme aquatic habitats, such as shallow, temporary pools in the case of most conchostracans and notostracans, saline lakes for most anostracans, and some notostracans and conchostracans, or both.

Conchostracans are epibenthic branchiopods. Some species are slow swimmers, whereas others crawl on or burrow in sediment. Modern species distributions are largely limited to freshwater or slightly saline and alkaline temporary pools, playas, spring seeps, and floodplains. Conchostracan species often have very patchy distributions, with large aggregations in certain sites and no individuals in seemingly similar ones. This probably relates to their reproductive strategy, involving the production of large numbers of resting eggs that for most species can withstand long intervals of desiccation. Local diversity tends to be quite low; usually a water body contains a single species or a small number of species.

The large body sizes and generally slow-moving behavior of all noncladoceran branchiopods make them easy prey for active predators, and all are extremely uncommon in permanent water bodies that contain fish. The modern distribution pattern of conchostracans (i.e., restriction to temporary water bodies) has been used by some authors to make similar habitat inferences for conchostracan-bearing sediments (Tasch and Gafford, 1968; Tasch, 1969, 1979). In Quaternary and probably Tertiary deposits, these types of interpretations for the noncladoceran branchiopods are quite reasonable and supported by independent sedimentological and paleontological evidence. Certainly the fossil occurrence of the distinctive eggs and fecal pellets of the brine shrimp *Artemia salina* provides excellent evidence for high paleosalinity (Bradbury et al., 1989; Gell et al., 1994; Kowalewska and Cohen, 1998; Bos et al., 1999). However, it is unclear that modern distribution patterns are an accurate guide for the interpretation of pre-Tertiary fossil conchostracans, anostracans, and notostracans (Webb, 1979; Frank, 1988; Knox and Gordon, 1999). Fish fossils commonly co-occur with both conchostracans and anostracans in these older sediments (Waldman, 1971; Gore, 1986, 1988; Gray, 1988b).

Ostracodes

Ostracodes are an extremely important group of fossils for the paleolimnologist. This comes about because of several factors:

1. Excellent preservability. Ostracodes possess calcite rather than just chitin in their carapace.
2. Abundance. Ostracodes are often the dominant, mesoscopic, benthic invertebrates in lakes and a single individual can produce numerous fossils through its lifetime, as it molts progressively larger exoskeletons (called *instars*).
3. A long fossil record, extending back to the Paleozoic in lakes.
4. Relative ease of identification.

Lacustrine ostracodes are, with a few exceptions, benthic (epifaunal and shallow infaunal) or epibenthic crawlers and weak swimmers, found in both the littoral and profundal regions of lakes. Because most species are detritus feeders, their life distribution and fossils are highly correlated with available food sources. In addition, ostracodes are rarely found in very high-energy littoral environments. Limited food availability also causes most ostracode populations to decline offshore in large lakes (Cohen, 1984). As a result, ostracode fossil abundances in these settings tend to be very low (e.g., Forester et al., 1994; Kowalewska and Cohen, 1998).

Ostracode Taphonomy

Because their carapaces can disaggregate, and are variable in size, ostracode fossils are subject to differential transport. As with other benthic fossils, an amalgamation of shallow and deep-water species also occurs in ostracodes. Adult ostracode fossils behave sedimentologically like sand grains, and because of their relatively large size and mass, downslope transport can cause considerable littoral/profundal species mixing in lakes wherever there are relatively steeply sloping lake floors.

Ostracodes are preserved as complete carapaces, isolated valves, and valve fragments. Valve/carapace ratios are both species-specific, and a function of transport and burial conditions (Whatley, 1988). The weak hinges of some taxa allow them to fall apart readily after death. Extensive transport or slow burial also serves to disarticulate ostracodes. As a result, within-species comparisons of carapace/valve ratios provide a useful taphonomic indicator of environmental conditions, comparable to other indications such as degree of breakage or physical abrasion. Only under exceptional circumstances of mass mortalities will juvenile carapaces be preserved in large proportions relative to disarticulated valves (Danielopol et al., 1986). Disintegration of ostracode valves is driven by a combination of microboring (predominantly by fungi), chemical alteration and dissolution, and physical abrasion, all of which are inhibited under situations of rapid burial and high sedimentation rate (Danielopol et al., 1986).

Because they are time-averaged over the course of a year to several years, surface sediment populations of ostracodes that have not undergone differential sorting and transport contain a broad assortment of instar sizes (Whatley, 1988). Sorting of ostracode fossils by wave and currents separates different sized individuals between species, within species (adults versus juveniles), or within individuals for organisms that disarticulate into multiple preservable fragments (Brouwers, 1988).

Ostracode fossils are poorly preserved in even mildly acidic lakes, because their exoskeletons are dominantly calcitic. Valve dissolution also occurs in organic-rich sediments, as a result of reduced pore water pH during organic matter oxidation (e.g., G.L. Smith, 1997; Holmes et al., 1998).

Water Chemistry

It has been known for many years that ostracode assemblages and individual valve morphology are sensitive to variation in water chemistry, both solute concentration and composition (Delorme, 1969, 1971; Carbonel and Peypouquet, 1979; Cohen et al., 1983). As a result, ostracode assemblage variation can be used to infer changing water chemistry and, at a more derivative level, climate change (Forester, 1986; De Deckker and Forester, 1988). Many studies have shown that ostracode species occupy distinct portions of water solute "space," specifically for the major anions, along the brine evolution scheme of Eugster and Hardie (1978) (figures 4.12 and 11.16). As a result, ostracode assemblages can be used to discriminate, for example, between paleolakes that were alkaline or hardwater. These changes in water chemistry result in both species composition and abundance shifts (De Deckker, 1982; De Deckker and Forester, 1988; A.J. Smith, 1993b; Holmes et al., 1998; Curry, 1999). Both diversity and abundance tend to be low at salinities below the calcite branch

point, probably as a result of Ca deficiency. As ionic concentration rises toward the branch point, both species diversity and abundance rise. At very high salinities (and particularly at elevated alkalinities) diversity falls again, but the abundance of remaining species may be very high. These findings are significant for paleoclimate studies, since the evolution of a particular water body along a brine pathway toward higher salinity and more chemically fractionated composition is strongly tied to aridity.

Indicator species approaches have been commonly used to semiquantitatively bracket a water chemistry inference based on overlapping tolerance ranges of two or more ostracode species. This technique has been most commonly used when the environmental tolerances and optima of only some of the species present are known (e.g., Anadón et al., 1986, 1994). The greatest value of this method today lies in extending water chemistry inferences into pre-Quaternary lake sediments, where no-analog assemblages and extinct taxa are commonly encountered. In older sediments, the water chemistry inferences from ostracode taxa tend to become more generalized, and are often based on the qualitative characteristics (e.g., brackish, freshwater, etc.) of more inclusive clades (Neale, 1988; Forester, 1991b; Gliozzi, 1999).

For Late Quaternary ostracodes, transfer functions and a quantitative approach to environmental inference of water chemistry and other variables are also possible. The most extensive ostracode training set was developed using literally thousands of Canadian lakes by Delorme and his colleagues. This inference model, later extended to other parts of North America, uses a regression technique similar to weighted averaging (Delorme, 1969, 1971, 1990, 1996, Forester et al., 1987; A.J. Smith et al., 1992; Curry, 1999).

For older, no-analog ostracode assemblages, alternative methods of inferring paleosalinity are available, notably shell chemistry (discussed in chapter 9) and carapace ornamentation. Carapace ornamentation appears to be affected by salinity in a number of groups of brackish or inland water/saline lake ostracode species (Sandberg, 1964; Kilenyi, 1972; Carbonel, 1980; Carbonel et al., 1988; Van Harten, 1996). The causes of this, whether genetic or ecophenotypic, are debatable. Regardless of the reason, the occurrence of variable ornamentation is of great importance for ostracode-based paleochemistry inference, since it provides a measure of water chemistry that is independent from species assemblages. In the pre-Quaternary, this is often one of the best, and sometimes only,

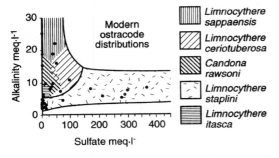

Figure 11.16. Ostracode assemblages from 38 midwestern U.S. lakes, superimposed on major anion (sulfate vs. alkalinity) composition, showing the common relationship between ostracode occurrence and lake-water chemistry. Only a few of the more diagnostic or abundant species are plotted. From A.J. Smith (1993a).

available indicators of paleochemistry (e.g., do Carmo et al., 1999).

Spring Discharge

A variety of ostracodes live both in groundwater and around groundwater discharge points into lakes, and are therefore useful indicators of groundwater influence. This is of interest to paleolimnologists because shallow groundwater discharge can introduce solutes from anthropogenic sources into lakes. Groundwater species that are sensitive to such inputs can provide a record of the timing or magnitude of such disturbances (Taylor and Howard, 1993). Changes in groundwater flow and spring discharge over time are sometimes marked by increased abundance of groundwater species (Scharf, 1998), or a transition between species assemblages typical of entirely different groundwater compositions and brine evolution pathways (Curry, 1997). Frequently, groundwater discharge increases during intervals of higher precipitation, and this may be evident from the ostracode assemblage change (Forester, 1991a). Finally, the temperature of a shallow spring approximates that of local mean annual air temperature, providing a potential temperature indicator for temperature-sensitive groundwater taxa (Mezquita and Roca, 1999).

Temperature, Oxygen, Water Depth, and Nutrients

Different ostracode assemblages characterize the littoral versus profundal zones, although macrophyte cover per se is probably less important overall in structuring the shallow/deep-water distinctions between ostracodes than cladocerans. Some ostracodes are epiphytic, and therefore likely to respond directly to aquatic plant distribution (e.g., Mourguiart and Carbonel, 1994; Bridgewater et al., 1999), but the majority of commonly preserved species are sediment-dwelling detritivores. In the temperate zone, shallow/deep differences between assemblages are better correlated with temperature and oxygenation levels, both of which probably exert direct physiological controls on ostracodes. Temperature differences are closely correlated with the distinction between taxa of the littoral zone adapted to warm summer temperature and broad temperature ranges, and cold water ones that are abundant below the summer thermocline, where temperatures hover around the 4°C range year round. In turn, these differences can be used

to record the thermal changes in a lake's deep-water environment (Colman et al., 1990).

Some species are sensitive to low oxygen levels, and most drop off in abundance with depth (e.g., Danielopol, 1990; Danielopol et al., 1993). For example, *Cytherissa lacustris*, a cosmopolitan, deep-water ostracode of the North Temperate Zone, requires both cold and well-oxygenated conditions to thrive. These conditions are typical of the profundal zone in smaller lakes, but occur over a wider range of depths in large, temperate, and boreal lakes. In eutrophic lakes, with high concentrations of low-density organic oozes in deep-water sediments, crawling species with heavy carapaces, such as *Cytherissa lacustris*, are eliminated, even when oxygen levels just above the sediment–water interface are adequate for the survival of swimming species. The occurrence of this species as fossils has therefore been considered diagnostic of relatively oxygen-rich, oligotrophic conditions (e.g., Löffler, 1975; Karrow et al., 1995). These environmental interactions highlight the difficulty that ostracodologists have had in separating the interactive effects of oxygen, organic matter accumulation, food availability, and temperature in this regard, since strong gradients exist for all of these in most temperate lakes. Fossil ostracode assemblages respond to productivity changes because of their sensitivity to oxygen levels at or near the sediment–water interface. As with cladocerans, a loss of macrophytes under highly eutrophic conditions also results in the elimination of periphytic species (Scharf, 1998). However, the proportional impact of this process is smaller for ostracode species assemblages than for chydorids because of the weaker linkage of most ostracode species to vegetation.

Ostracode faunas vary with regional climate because of their sensitivity to the hydroclimate parameters of water temperature and solute concentration. The training set of Delorme and his colleagues (Delorme, 1969, 1971, Delorme et al., 1977) produced transfer functions that directly relate air temperature and precipitation to ostracode assemblages. These studies have documented not only variation in mean annual temperature and precipitation trends, but also seasonal variability in climate variables. This is possible because winter and summer precipitation have different impacts on lake water chemistry. Using the Delorme model, researchers have been able to quantitatively reconstruct climate variables for northern North America during the Quaternary (figure 11.17). Qualitative reconstructions of paleoclimate may also be possible for older, no-analog ostracode assemblages using shell morphology criteria (Forester, 1991b).

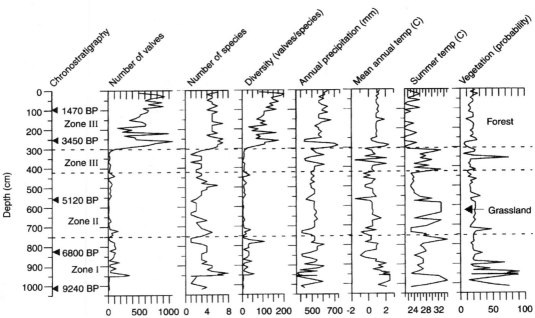

Figure 11.17. Summary ostracode stratigraphy and ostracode-inferred climatic variables for a Holocene record from Harris Lake, Saskatchewan (Canada). From Porter et al. (1999).

Littoral /Open-Water Species and Lake Level

Numerous authors have constructed lake-level curves based on the proportion of littoral, sublittoral, and profundal ostracode species, and in some cases it has been possible to extend this reasoning to pre-Pleistocene lake beds (Mourguiart and Carbonel, 1994; Kowalewska and Cohen, 1998; Mourguiart et al., 1998; Belis et al., 1999). Semiquantitative depth indices, based on proportions of dominantly littoral to sublittoral fossil taxa, have been used in some studies of fossil ostracode assemblages (e.g., Forester et al., 1994; Bridgewater et al., 1999). However, the adoption of such an approach has never been systematized among ostracode workers to the extent used for fossil cladocerans.

Siltation

As benthic detritus feeders, ostracode communities can be impacted by siltation. At moderate levels of increased sedimentation, ostracode diversity can actually rise, presumably as a result of increased food resources. However, under conditions of very rapid sedimentation or a rapid change in sediment texture toward coarse-grained (food-poor) particles, ostracode diversity quickly declines. Rapid soil erosion is commonly associated with human activities,

such as road construction or deforestation. Such impacts have been recorded in the paleolimnological records of several lakes in both the temperate zone and the tropics (Danielopol et al., 1985; Wells et al., 1999; Cohen, 2000). At Lake Tanganyika, extremely high sedimentation rates in areas adjacent to deforested watersheds have resulted in wholesale elimination of many taxa. A limited pool of sedimentation-tolerant species repeatedly colonizes core localities offshore from high sedimentation watersheds, although the core ages demonstrate that the impact did not begin during the twentieth century rise in human population (figure 11.18). This is in strong contrast to the much higher diversity and variability in the available species pool providing colonists to core sites undisturbed by sedimentation.

Other Crustaceans

Several other groups of crustaceans are occasionally preserved as fossils in lake sediments, and when found provide valuable paleoenvironmental information. Copepod spermatophores or egg sacs are sometimes found in lake beds, and may be useful indicators of cultural eutrophication (Frey, 1964; Warner, 1990c; Bennike, 1998; Findlay et al.,

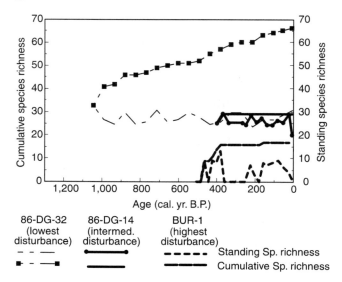

Figure 11.18. Changes in ostracode standing species richness (species per sample–time increment) and cumulative species richness (total number of species observed from the base of the core to the plotted age) for three sites at Lake Tanganyika, Africa. The cores show distinct differences between sites affected by high rates of siltation disturbance, associated with deforestation, versus low disturbance sites. The difference also clearly predates the twentieth century increase in human population in the region, indicating that the high disturbance region, because of its watershed characteristics, was already susceptible to sediment loading. Modified from Cohen (2000).

1998). Barnacles are known as isolated plates or cemented fragments in some lake beds, where they provide an indication of saline conditions (e.g., Spencer and Patchett, 1997). Freshwater decapods, primarily crayfish and crabs, are documented from numerous Mesozoic and Cenozoic lake deposits (Gray, 1988b; Babcock et al., 1998). Body fossils and burrows of these animals are sometimes found in association with floodplain lake and lake margin deposits, providing indications of the degree of seasonal fluctuation in water table in semiarid and arid environments (Bown, 1982; Hassiotis and Honey, 2000).

Insects and Aquatic Mites

Fossil insects have a long and rich fossil history in lake deposits, extending back to the Middle Paleozoic. Because of their extraordinary diversity and range of habitat and feeding specializations, the interpretation of fossil insect faunas has the potential to provide very valuable information to the paleolimnologist.

Fossil insect assemblages are commonly preserved in lake deposits, as three-dimensional frag-

ments (especially in Quaternary sediments, or in concretions), or as bedding plane compressions in older lake beds. Insects are normally preserved only by their skeletal cuticle, which is relatively resistant to chemical degradation. Fresh cuticle comprises a mixture of chitin and proteins, although the latter normally degrades rapidly in fossils. Low pH conditions accelerate the rate of chitin degradation, as does oxidation, making organic-rich deposits in the laminated muds of meromictic (neutral to alkaline) lakes favorable preservational environments (Briggs et al., 1998). Insect cuticle is also quite brittle, does not preserve well under conditions of repeated wetting and drying, and is destroyed by compression.

As with plant fossils, the insects preserved in lake beds include a mixture of endogenic and exogenic species. Somewhat surprisingly, it is the exogenic, terrestrial insects that are often better represented in lake beds, especially in pre-Quaternary deposits (Wilson, 1988; Bajc et al., 1997; D.M. Smith, 2000). The reasons for this are not entirely understood, given the limited research on insect fossil taphonomy to date, but probably relate to the heavier sclerotization of terrestrial insects, coupled with the destruction of three-dimensional chitinous structures such as chironomid head capsules under compression. Delicate, lightweight, or flying species are proportionately more abundant in offshore assemblages (Wilson, 1988).

Midges (Lake Flies)

Several families of aquatic fly larvae are commonly represented as fossils in lake beds, most notably the Chironomidae, Ceratopogonidae, and Chaoboridae. Lake flies are known as fossils from the Late Triassic/Early Jurassic onward, although they are really only abundant in Quaternary lake deposits, for reasons alluded to above. Of these, the chironomids are usually both most numerous and most diverse. Chironomid larvae in particular inhabit a wide range of predominantly benthic habitats, with many species restricted to either nearshore or profundal environments. To date, however, the greatest attention by paleolimnologists has been given to profundal species.

Taphonomy

Chironomids are preserved primarily by their robust, chitinous larval head capsules, and to a lesser extent by their larval feeding structures. Head capsules are sand-sized structures, which can be differentially transported offshore depending on lake wave and current conditions (Frey, 1988b). In smaller or protected lakes, littoral remains closely reflect the original life distribution patterns, whereas in larger, more turbulent lakes there is considerable offshore movement, and littoral chironomid species are commonly found as fossils in deep water. Surface sediment transect studies of head capsule distribution in the offshore environments of Lake Constance (Germany), a relatively large and eutrophic lake, illustrate the importance of offshore transport (Schmäh, 1993). Some profundal regions of this lake are subject to persistent anoxia, and therefore lack living chironomids, whereas other regions have oxygenated profundal zones. Because of their high relative production rates and significant downslope transport, littoral chironomid fossils make up the majority of fossils at all depths, with relatively little variation between samples. However, there is a trend toward an increasing proportion of profundal taxa in the oxygenated regions down to the local maximum depth, whereas profundal taxa drop off as a proportion in the deepest water in the anoxic basins.

Chironomid downslope transport is also sensitive to lake mixing processes (Mees et al.,

1991). Littoral species are reworked into deepwater assemblages during periods of whole lake mixing. When lakes are meromictic, however, the rate of downslope transport of littoral remains into the monimolimnion tends to decline toward deep water. This suggests that paleoecologists should be careful in making ecological interpretations of fossil abundance data when changes in chironomid abundance are associated with facies transitions between laminated and bioturbated sediments.

Controversies in Chironomidland

The underlying environmental factors that have controlled midge distributions in the past have been more strongly debated than for almost any other group of lacustrine fossils (I.R. Walker and Mathewes, 1989a,b; Warwick, 1989; Hann et al., 1992; I.R. Walker et al., 1992; Olander et al., 1999). Most disagreement centers around the extent to which temperature, as opposed to other environmental variables (especially productivity, oxygen availability, and sediment type), has shaped midge communities. This hearkens back to the point raised in chapter 10 about direct versus indirect causality. Since productivity and sediment discharge are partly regulated by climate, the question of "which is responsible for changes in chironomid assemblages?" may be time scale-dependent, without a simple, one-or-the-other answer. However, these other factors are not entirely under climate control, since they also depend on bedrock geology, lake morphology, and in the modern world, cultural eutrophication. Our best hope for addressing this problem lies with further development of quantitatively analyzed training sets that contain long environmental gradients across all of the variables in question (e.g., Olander et al., 1999).

Much of the controversy about chironomid distribution has focused on the critical Late Pleistocene/Holocene transition. Many of the early paleoecological studies of chironomid fossils in Europe and North America noted the linkages observed by ecologists between midge associations and lake trophic state, and tried to extend these observations to the past. Deep benthic midges, like *Heterotrissocladius*, that require high oxygen levels and cold water, were thought to characterize the Late Glacial interval, when lakes were assumed to be uniformly oligotrophic (figure 11.19). In contrast, those species tolerant of lower oxygen levels within the profundal environment typified more eutrophic Holocene lakes. As a generalization for

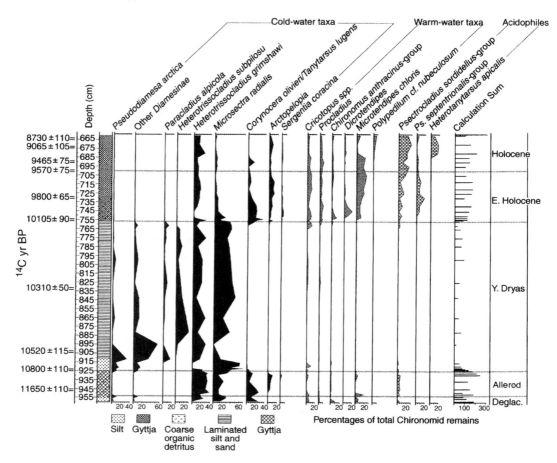

Figure 11.19. Changes in the percentage abundances of the major fossil chironomid taxa during the Late Glacial and Early Holocene at Kråkenes Lake, western Norway. Deglac., deglaciation phase; Y. Dryas, Younger Dryas; E. Holocene, earliest Holocene. From Brooks and Birks (2000).

understanding the Pleistocene–Holocene transition this model has proven problematical for several reasons. First, the last glacial/interglacial transition was not a continuous warming event, but rather an oscillatory change that involved several episodes of cooling and warming. Chironomids appear to have responded to these oscillations quickly, although the question of what drove this response (nutrients, temperature, etc.) remains controversial (I.R. Walker, 1995). There has been an increasing recognition by chironomid workers over the past few decades that much of what was originally interpreted as a response to oxygen stress may in fact be a combined response to O_2, temperature, and other variables. Lower temperatures both increase dissolved oxygen availability and decrease biological oxygen demand, and not surprisingly, midges with high oxygen demands are likely to also be obligate

cold stenotherms. Higher temperatures have the opposite effects. As a result, in many lakes, and especially in fossil interpretations, the effects of these two variables have been confounded. Warming trends of the Early Holocene would have had many effects on temperate zone lake ecosystems. These include changing sediment yield, releasing more nutrients from vegetated watersheds, making higher water columns more productive, expanding opportunities for warm-water littoral species, reducing or eliminating deep-water, high-oxygen habitats, and reducing the extent of cold-water habitat (Warwick, 1989). The nearly simultaneous shifts in faunal composition across the Pleistocene–Holocene boundary in lakes with very different watershed characteristics, soils, vegetative cover, and bedrock geology argue for temperature as the underlying key variable in driving these changes (I.R. Walker and Mathewes, 1989a). However, it is also clear that lake productivity and sedimented organic matter are not uniformly tied to temperature change, and in some cases are even stronger predictors of variation in chironomid assemblages (Olander et al., 1999). In low latitude and arid regions, where temperature variation is small, the importance of temperature for chironomid assemblages is almost certain to be overshadowed by factors like salinity, To date, however, the amount of information available on chironomids (modern or fossil) from these regions is extremely limited.

Temperature

The correlation between chironomid distribution and temperature has been recognized for many years (e.g., Brundin, 1949). Temperature may determine the ability of dipteran larvae to complete their life cycle, both for cold and warm-water species (Oliver, 1971). Temperature also directly affects such life history variables as timing of pupation and emergence, rate of growth, feeding, and hatching (I.R. Walker and Mathewes, 1989c). Changes in species diversity and assemblage turnover are associated with temperature. In both warm and cold-

temperate climates, low midge diversity is normally correlated with both high elevation and cold lakes (Thorp and Chesser, 1983; I.R. Walker and Mathewes, 1989c; Levesque et al., 1996).

Studies on Canadian and European lakes supporting the primary role of temperature in regulating chironomid community structure have stimulated considerable research into their potential as paleoclimate indicators. Training sets and weighted-averaging regression/calibration models using chironomids have now been developed for a number of regions in Europe and North America (I.R. Walker et al., 1991b, 1997; Olander et al., 1997; Brooks and Birks, 2000). The small error statistics associated with these models suggest that chironomids are probably the best quantitative indicators of temperature currently available to the Quaternary paleolimnologist working in cool–cold climate lakes (e.g., Levesque et al., 1993; Cwynar and Levesque, 1995; Brooks et al., 1997) (figure 11.20). Interestingly, this relationship is often stronger with air rather than water temperatures. Possibly this is because long-lived aquatic larvae must be able to accommodate a wider range of temperature than the short-lived (few weeks) adults. Successful colonization may be dependent on appropriate temperature conditions at the precise time the adults

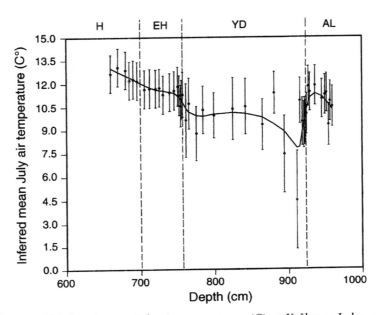

Figure 11.20. Chironomid-inferred mean July air temperatures (C) at Kråkenes Lake, western Norway during the Late Glacial and Early Holocene. The sample-specific prediction errors are shown, along with a smoothed curve of the data. The main time intervals indicated are: AL, Allerød; YD, Younger Dryas; EH, earliest Holocene; H, Early Holocene. From Brooks and Birks (2000).

emerge from the lakes. It bears noting, however, that the lakes chosen for training sets in most of these studies were selected to emphasize long environmental gradients of temperature, for example across latitudinal or altitudinal transects. A lack of equivalent gradient coverage along nutrient, sediment quality, or other gradients keeps the question of "which factor was most important for chironomid distribution in lakes of the past?" unresolved.

Oxygen, Productivity, Macrophytes, and Sediments

Changes in nutrient loading to lakes, whether through natural variability or cultural eutrophication, affect chironomids. This is mediated by the intermediate relationships of productivity with oxygen saturation in the hypolimnion, development of macrophyte cover, and accumulation of organic-rich sediment food sources. Although these processes affect chironomids in different ways, they are not independent variables, and it is often difficult to separate their effects on midge populations (Lotter et al., 1998). The most direct relationships between nutrient loading and chironomid communities seems to be expressed in profundal faunas; littoral species appear to be buffered from many short-term variations in water quality by the uptake capacities of macrophytes (e.g., Little and Smol, 2000). Chironomids vary greatly in their ability to tolerate low oxygen levels, suggesting their potential use in the reconstruction of water column oxidation state in the hypolimnion, and secondarily, to reconstruct trophic conditions (Saether, 1979, 1980; Lang and Lodz-Crozet, 1997). Some taxa possess hemoglobin (*Chironomus*) and can thrive under low oxygen conditions, whereas members of the *Tanytarsus lugens* community have high oxygen requirements.

Differences between modern midge communities from oligotrophic versus eutrophic lakes have been widely applied to interpret paleolimnological trends in terms of cultural eutrophication in both North American and European lakes (Wiederholm and Eriksson, 1979; Warwick, 1980; Kansanen, 1986). Most of these differences seem to relate to oxygenation state, rather than nutrient loads per se (e.g., Kansanen, 1986; Meriläinen and Hamina, 1993). Some attempts have been made to develop quantitative inference models of anoxia using chironomids (Quinlan et al., 1998). Other studies have found a close link between midge faunas and sediment quality, specifically TOC and sediment texture, which affect feeding patterns (M.J. Smith et al., 1998; Olander et al., 1999). However, the direct role of nutrient loads on chironomids remains uncertain. In their study of Swiss alpine lakes, Lotter et al. (1998) found that the relationship between surface sediment chironomid fossil assemblages and total phosphorus, while statistically significant, was nevertheless weaker than that for diatoms or cladocerans.

Salinity, Lake Level, and Other Factors

In arid or tropical climates dipteran successions are almost certainly controlled by environmental gradients other than temperature. These include the persistence of species tolerant of high salinities, and correlations with macrophyte cover, which also declines at the high salinities associated with low lake stands. The investigation of tropical and subtropical dipteran fossils is still in its infancy and only a handful of studies have been completed to date, but these few studies show the potential for detailed lake-level and salinity reconstructions (e.g., Mees et al., 1991). Verschuren et al. (1999, 2000) have documented changes in midge and crustacean assemblages in Lake Oloidien (Kenya) over the last few hundred years (figure 11.21). This lake has no surface outlet, and is fed by subsurface inflow from a larger freshwater lake. The lake has experienced considerable fluctuations in level, known from the historical record, with high lake stands marked by expansion of freshwater macrophyte cover and associated littoral invertebrates.

Changes in chironomid fossil assemblages are associated with lake acidification, recorded in fossils by declining diversity and stability in midge communities (e.g., Brodin, 1990). However, the causal link, whether direct or indirect, remains uncertain. Dipteran larval populations are also affected indirectly by low pH because of the elimination of fish from acidified lakes, and changes in their fossil assemblages have been used as a record of the timing of sufficient acidification to eliminate predatory fish (M.G. Johnson et al., 1990; Uutala, 1990).

Other Aquatic Insects and Arthropods

Aside from chironomids, the most commonly preserved and studied aquatic insect fossils are trichopteran (caddisfly) larvae and coleopterans (beetles). Caddisflies are preserved by their larval exoskeletal elements (sclerites), normally as disarticulated fragments (Williams, 1988a). One individual undergoes five instars, and therefore potentially produces many fossil fragments. Occasionally, agglutinated larval cases are also preserved.

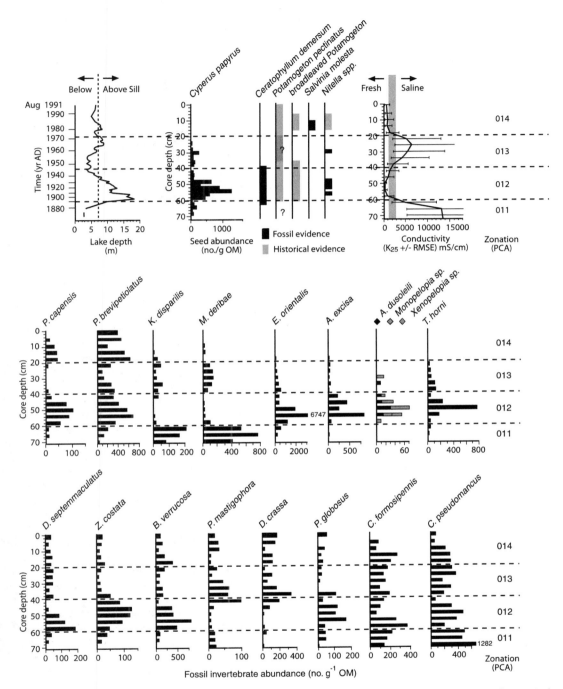

Figure 11.21. Fossil assemblages of aquatic invertebrates (chironomids, cladocerans, ostracodes) in Lake Oloidien, Kenya; sediments, in relation to lake depth at the time of deposition. Abbreviations: Ostracodes *Physocypria* (*P.*) *capensis*, *Zonocypris* (*Z.*) *costata*, *Potamocypris* (*P.*) *mastigophora*. Chironomids *Procladius* (*P.*) *brevipetiolatus*, *Kiefferulus* (*K.*) *disparilis*, *Microchironomus* (*M.*) *deribae*, *Ablabesmyia* (*A.*) *dusoleili*, *Monopelopia* (*M.*) sp., *Xenopelopia* sp., *Tanytarsus* (*T.*) *horni*, *Dicrotendipes* (*D.*) *septema-culatus*, *Chironomus* (*C.*) *formosipennis*, *Cladotanytarsus* (*C.*) *pseudomancus*. Chydorid cladocerans *Euryalona* (*E.*) *orientalis*, *Alonella* (*A.*) *excisa*, *Biapertura* (*B.*) *vericosa*, *Dunhevedia* (*D.*) *crassa*, *Pseudochydorus* (*P.*) *globosus*. From Verschuren et al. (2000).

The application of caddisfly fossils to solving paleoecological problems is still in its early stages of development. Caddisflies are entirely aquatic and their fossils therefore provide automatic evidence of an aquatic habitat. Many families are largely or completely restricted to running water, and some of these lotic taxa are also specific to flow rates. Genera from a smaller number of families are restricted to lakes. Some are quite specific in substrate type (mostly nearshore) or feeding behavior (Williams and Eyles, 1995). Modern caddisfly biogeography shows that they may be useful indicators of paleotemperature, through interpretation of range shifts (Elias and Wilkinson, 1983; Williams and Eyles, 1995; Solem and Birks, 2000, Elias, 2001).

Beetles are an extremely diverse group of organisms, which inhabit a remarkable range of habitats, making them excellent paleoenvironmental indicators, especially for paleoclimate reconstruction. Most beetle fossils preserved in lake deposits are actually terrestrial species, particularly those that inhabit lake margin environments or are strong fliers over lakes. These taxa can provide detailed information on subtle changes in local lakeshore habitats, especially because beetles are known to respond quickly to climate change (Coope, 1979, 1986; Morgan and Morgan, 1990). Aquatic species may be informative about changes in vegetation pattern, or in the case of invertebrate predators, suitable food items. Some temperature sensitivity is evident in aquatic beetle assemblages in colder climates, insofar as temperature controls the duration of ice cover, which in turn affects reproduction (Lemdahl, 2000).

Both aquatic and terrestrial mites are common as Quaternary fossils in lakes and bogs, preserved by their resistant adult cuticle (Erickson, 1988; Drouk, 1997; Solhøy and Solhøy, 2000; Solhøy, 2001). Aquatic species are commonly associated with aquatic macrophytes, and less commonly, found on exposed coastlines or on unvegetated muddy bottoms. A wide variety of species also inhabit wetland mires, and fossils of these species may be useful indicators of moisture availability (Markkula, 1986).

Mollusks

Gastropods and bivalves are ubiquitous fossils in lacustrine sediments, with a long and rich fossil record (lacustrine prosobranchs, Early Devonian–Recent; lacustrine pulmonates, Early Carbonife-rous–Recent; lacustrine bivalves, Early Devonian–Recent), in large part owing to their excellent preservation potential (Lozek, 1986; Taylor, 1988; B.B. Miller and Bajc, 1990). Mollusks are potentially very valuable indicators for pre-Quaternary lake deposits, where they are often the most conspicuous and readily identified fossil remains. Lacustrine mollusks are also found in a wide range of habitats and rapidly colonize new, suitable habitats, further enhancing their representation in the fossil record, although they are most frequently and abundantly preserved in nearshore deposits. Yet despite these mostly positive attributes, freshwater mollusks have received less attention by paleoecologists than other common groups of freshwater fossils.

Taphonomy

Mollusks are preserved in lake beds as external shells, molds, and casts. Occasionally, prosobranch opercula and eggs also preserved (B.B. Miller and Bajc, 1990; M.J.C. Walker et al., 1993). Preservation of mollusks is water chemistry-dependent. Mollusk fossils are rare in acidic or poorly buffered lakes, or in lakes that are undersaturated with respect to $CaCO_3$, although this can be counteracted by rapid burial (e.g., Colman et al., 1990). Conversely, mollusk abundance and fossil abundance is normally higher in association with marl deposition (e.g., B.B. Miller et al., 1998). Lacustrine fossil mollusk assemblages commonly comprise a mixture of endogenic and exogenic species, the latter derived from inflowing streams or, in the case of pulmonate gastropods, terrestrial and amphibious species. Smaller gastropods from nearshore or amphibious habitats are also known to float (sometimes in large rafts) eventually being deposited in offshore locations (Schwalb et al., 1998). Because of their durability, mixed assemblages of varying aged mollusk fossils are common, particularly in nearshore or floodplain deposits, where erosion and reworking of older fossils is most common (B.B. Miller et al., 1985; Cohen, 1989b; Dubiel et al., 1991). However, these types of mixing quickly decline in importance in deeper water or in small, isolated lakes with less vigorous wave reworking. When opercula can be identified to species, their abundance relative to the shells of the same species has been used as a reworking index, since the two skeletal parts have very different hydrodynamic properties (Marcusson, 1967; Griffiths et al., 1994; Hoek et al., 1999). In one of the most detailed taphonomic studies of freshwater mollusks conducted to date, Cummins (1994) found that sediment surface unionid bivalve shell assemblages provide a relatively accurate, albeit time-averaged, picture of living communities in small streams and reservoirs.

Water Body Characteristics and Temperature

One of the primary uses of freshwater mollusks in paleolimnological reconstruction has been to differentiate moving versus standing water environments, and shallow and stagnant ephemeral ponds or wetlands versus permanent lakes (Lozek, 1986; Miller and Tevesz, 2001). Typically, investigators assess the proportions of species or individuals in each habitat guild, although increasingly this is being supplemented by shell isotopic analysis (Dettman and Lohmann, 1993; Karrow et al., 1995; Sharpe et al., 1998).

Terrestrial mollusk species are more sensitive to climate change than aquatic species, the latter having broad distributions (Lozek, 1986). However, aquatic species can occasionally provide useful paleotemperature information (e.g., Alexandrowicz, 1981; Chaix, 1983). The presence of what are presently high-latitude or deep-water, glacial relict taxa in shallow-water deposits provides qualitative or semiquantitative information on colder-water temperatures in the Late Quaternary (B.B. Miller and Bajc, 1990; Karrow et al., 1995; Schwalb et al., 1998). Conversely, relatively warm-water conditions during interglacial episodes have also been inferred from fossil mollusk assemblages in the Laurentian Great Lakes region (Kerr-Lawson et al., 1992).

Macrophytes and Lake Level

Many species of gastropods graze on or around macrophytes, and therefore are restricted in distribution to the littoral zone. Some aquatic gastropods require fairly dense macrophyte cover; their presence as fossils has been used as an indication of vegetation patterns (Magny et al., 1995; Schwalb et al., 1998). At sublittoral and profundal depths some mollusks are more sensitive than others to low oxygen conditions, and this too provides relative paleodepth information. The infilling of lake basins is often well marked by changes in mollusk faunas, with transitions from faunas dominated by sediment-dwelling bivalves and gastropods, to increasing proportions of littoral species, and a rise in terrestrial and amphibious pulmonates

(Alexandrowicz, 1987; Keen et al., 1988; B.B. Milller and Thompson, 1990; M.J.C. Walker et al., 1993; Griffiths et al., 1994; B.B. Miller et al., 1998, 2000) (figure 11.22). Although these approaches are best developed for Quaternary lake deposits, they are also applicable in Tertiary lake beds, because of the evolutionary conservatism of major freshwater molluskan clades (e.g., Hanley, 1976).

Salinity, Dispersal, and Associations with Fish

Freshwater mollusks can be used as an indicator of salinity in arid or semiarid environments. The mere indication that water bodies existed, from the presence of eurytopic freshwater mollusk species, can be very informative in establishing the climatic histories of regions that are hyperarid today (e.g., Petit-Maire and Riser, 1981). Given their size, mollusks are surprisingly transportable between water bodies in arid and semiarid environments, probably mostly by birds. Estuarine and marine–coastal gastropods and bivalves (especially the ubiquitous cardiid bivalve *Cerastoderma glaucum*) are sometimes found in saline lakes far from any marine influence or modern water body interconnections (Plaziat, 1993; Patterson et al., 1997). The same taxa are often found in Tertiary and Quaternary saline lake deposits of northern Africa and southern Europe, where they can be used as semiquantitative paleosalinity indicators (Anadón et al., 1994).

Bivalves of the superfamilies Unionacea and Mutelacea are dispersed in their larval stage as ectoparasites on fish. Thus, their presence provides an indicator of the presence of host fish at the time of their recruitment (e.g., Firby et al., 1997).

Fish

The use of fish fossils in lacustrine paleoecology has been very limited, despite the obvious importance of fish in lake ecosystems. Fish are potentially very useful paleoecological indicators, and a great deal is known about their modern ecology, but paleoecological research on freshwater fish has been limited, primarily because of preservational problems. Although fish are occasionally preserved as complete skeletons, more commonly they are represented as disarticulated bones, scales, and otoliths, and occasionally, coprolites, all of which are difficult to identify. Furthermore, in core studies, whole

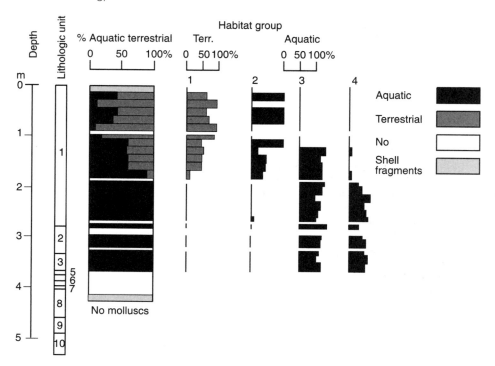

Figure 11.22. Relative frequencies of mollusks from various habitat guilds in a core from the Cowles Bog area, Lake Michigan Basin. Group 1 species inhabit terrestrial habitats and marshes. Group 2 species inhabit temporary water bodies. Group 3 species are aquatics associated with eutrophic water bodies. Group 4 species inhabit permanent, unpolluted water bodies. From B.B. Miller and Thompson (1990).

fish skeletons are unlikely to be retrieved as a result of their size. These problems result in small sample population sizes, and often place limitations on the degree to which fish fossil data can be analyzed statistically in paleolake deposits. In contrast, much more attention has been paid to well-preserved lacustrine fish fossils in the context of evolutionary studies, discussed in chapter 14.

Taphonomy

Taphonomic processes in fish preservation serve as both an information filter and a source of paleoenvironmental information for the paleolimnologist. When fish die suddenly their corpses often undergo a form of rigor mortis known as *tetany* (Elder, 1985; Ferber and Wells, 1995). Tetany results in corpses with gaping mouths and gills, fanned and stiffened fins, and occasionally arched bodies. Common causes of tetany include respiratory stress from anoxia, heat shock, or salinity shock, although only anoxia is likely to preserve the fish

in this configuration. Under conditions of bottom current or wave activity, bones from decaying fish can be reworked in down-current configurations, or concentrated into lags. In nearshore settings, or in large lakes with vigorous bottom currents, fish skeletons normally disarticulate (Wilson, 1988).

Water temperature and water chemistry also have major effects on decay rate and disarticulation processes. Decay rates rise with increasing temperature. Above about 16°C, fish carcasses will float, causing vertebral displacement, abdominal separation, and partial disarticulation in deep water deposits (Elder and Smith, 1988). Below 16°C, fragmentation primarily depends on bottom currents, or disturbance by scavengers on the lake floor, the latter related to oxygen availability. Bone degradation in fish fossils is accelerated under acidic lake sediment conditions, especially in peats (Casteel et al., 1977).

Whole preserved fish skeletons are most commonly preserved under anoxic conditions in meromictic lakes, where they can both accumulate and remain undisturbed by secondary scavenging and current activity (e.g., Elder and Smith, 1988; Wilson, 1988; Trewin and Davidson, 1999). Occasionally, spectacular mass mortality accumulations of very well-preserved fish fossils are found associated with open-lake, varved sediments (Waldman, 1971; Elder and Smith, 1988), or less commonly in saline mudflat deposits (Ferber and Wells, 1995).

Temperature and Seasonality

A number of fish clades have well-defined optimal temperature ranges in modern lakes, notably trout and whitefish in North American and Eurasian lakes, which make their fossils useful paleotemperature indicators (G.R. Smith and Patterson, 1994). Modern distributional limits for both warm and cold-water species have been used occasionally as paleoclimate indicators (e.g., G.R. Smith, 1978; Firby et al., 1997). Based on the previously discussed taphonomic reasoning, preservational condition can also be used as an indication of temperature for whole fish fossils in deep water (Elder, 1985). Another potential paleotemperature record from fish fossils comes from the analysis of scales. Scale growth is a function of fish growth rate, and the fish size at the time of growth. By standardizing comparisons to a specific age class, paleoichthyologists can determine variations in optimal growth conditions. Depending on the species involved, this optimum may or may not be highly correlated with temperature (Hopkirk, 1988).

The information value of fossil fish for paleolimnology can be greatly enhanced when faunal analysis is coupled with geochemistry (Patterson and Smith, 2001). G.R. Smith and Patterson (1994) combined the study of fish community structure and oxygen isotope analysis of otoliths to infer climatic conditions for Tertiary deposits of Paleolake Idaho of the Snake River Plain (northwestern United States). In Miocene deposits they found a surprising combination of warm-water sunfish and catfish, along with cold-water salmon and trout, suggesting a much more equitable climate (warmer winters and cooler summers) than exists in the region today. The hypothesized more equitable climates from these "incongruent" species distributions are consistent with oxygen isotopic variability in the otoliths, suggesting more limited temperature ranges for the Late Miocene than today.

Salinity, Lake Levels, and Hydrological Interconnections

Some salinity inferences are possible from fish fossils, based on the occurrence of fossils of halotolerant species (De Deckker, 1982, Firby et al., 1997). Most of these taxa, however, are usually capable of tolerating a broad range of salinities, rather than being true halophiles. Elevated fluorinity in saline/alkaline lakes is also associated with growth abnormalities in fish, which may be evident in fossil bones (Schlüter et al., 1992).

Fish communities in large lakes are commonly differentiated into nearshore, littoral communities and offshore, open-water or pelagic communities. Additional differences between nearshore and offshore environments include a preference for sheltered nearshore habitats by juveniles of many species, and better fossil preservation in offshore settings. These differences are recorded in fossil assemblages, and occasionally can be used as a basis for paleoenvironmental inference (Buchheim and Surdam, 1981).

Fish fossils are extremely valuable indicators of past surface water paleohydrology, and especially hydrological interconnections between paleolakes and rivers (Patterson and Smith, 2001). Because fish are large organisms, and because most require permanent water, they are not readily transported between disjunct aquatic habitats, as is possible for many smaller invertebrates (G.R. Smith, 1978, 1981, 1987). The distribution of distinctive fish clades within circumscribed regions can provide direct evidence of past watershed interconnections.

Summary

Our conclusions from this chapter can be summarized as follows:

1. Cyanobacteria are primarily represented in lake beds by benthic stromatolites, and other encrustations, and are informative about lake level, and occasionally salinity. Planktonic cyanobacteria are occasionally preserved, mostly as resistant spores or filaments.

2. Diatoms are probably the most important fossil group for paleolimnological interpretation. Their siliceous frustules are easily preserved in a wide range of lake environments. They can be used to quantitatively

reconstruct pH, salinity/alkalinity (and indirectly climate), nutrient loading, temperature, and lake level. Siliceous chrysophyte scales and spores are also useful for many of the same purposes, although they have been less extensively studied than diatoms.

3. Charophytes are useful indicators of salinity, especially in hardwater lakes. Some planktonic green algae are also commonly preserved in lake muds, where they may provide information about lake trophic conditions.

4. Mosses and vascular plants are represented in lake beds by both *exogenic* and *endogenic* fossils. Fossil pollen from lake deposits provides a record of local and regional vegetation, and indirectly, climate. However, its preservation in lake beds is determined by such factors as grain size, transport mechanisms, and internal lake processes, in addition to vegetation communities. Plant macrofossils typically provide a much more local representation of vegetation than pollen. Aquatic macrophytes are useful for determining lake-level history, and occasionally salinity.

5. Fossil testate amoebans are useful indicators of water availability in wetlands. Lacustrine foraminiferans are commonly associated with saline or coastal lakes.

6. Sponges, rotifers, and bryozoans are occasionally preserved in lake beds, but are poorly known in terms of either taphonomy or paleoecology.

7. Cladocerans are extremely valuable indicators for reconstructing the extent of littoral and open-water habitats in Quaternary paleolakes. Cladoceran communities may also be useful for reconstructing productivity, water temperature, and nutrient loading, although the interacting and indirect effects of these variables on cladocerans has made it difficult to distinguish them in paleorecords.

8. Ostracodes are also very important paleolimnological indicators and, along with diatoms and mollusks, are one of the most abundant fossil groups in pre-Quaternary lake deposits. They are especially sensitive to variation in major ion water chemistry, and this relationship can be used to infer paleoclimates.

9. Fossil insects are abundant in lake beds, with a surprisingly large representation by terrestrial rather than aquatic species, reflecting differential preservation properties. The most important group of insects for Quaternary paleolimnologists are the chironomids, midge larvae whose fossils can be used to quantitatively infer paleotemperature with considerable precision in boreal and temperate climate belts. In arid and tropical areas, chironomids are also useful, indirect indicators of aridity.

10. Because they occupy terrestrial, amphibious, and various aquatic habitats, mollusk fossils are useful for interpreting lake habitat zonation, lake level, and infilling history. Fish fossils have proven most useful for interpreting paleotemperatures and watershed interconnections.

12

Paleolimnology at the Local to Regional Scale: Records of Changing Watersheds and Industrialization

Paleolimnologists have developed an impressive track record documenting the history of human influence on lakes and their surroundings, and using these historical inferences to help policy makers establish lake and ecosystem management goals. Our ability to do this depends on both a comparative analysis of multiple lake records, and a firmly established chronology. The comparative approach to paleolimnology allows us to differentiate local phenomena resulting from peculiarities of study watersheds from regional phenomena. Comparison of records also allows the timing of events to be placed in a regional context, where explanations of processes that affect large areas, like lake acidification, regional patterns of air pollution, or landscape disturbance may be more broadly interpretable. Comparative paleolimnology allows the researcher to study the multiple effects of local to regional-scale phenomena and differentiate them from global phenomena. Closely coupled with our requirement for a comparative approach to paleolimnology is the need to place events in a highly resolved chronology, especially over the past 200 years, the period of greatest interest to understanding major human alternations of the environment.

In many parts of the world, including the highly industrialized and relatively well-"monitored" environments of North America and Europe, instrumental records of water quality are either spotty or unavailable. Until the 1960s, the number of lakes with regular monitoring programs for even basic limnological parameters was extremely small. And in regions with numerous water bodies, selection criteria for the investigation of lakes often has had more to do with proximity to major research facilities or peculiarities of road access than with the needs of society. Paleolimnological records integrate ecological signals at scales that are relevant to the interests of lake managers, who need to understand the timing and magnitude of human

activities. Even when limnological monitoring is available, paleolimnological approaches can answer questions at temporal and spatial scales that are unattainable by the monitoring regime in place.

The difficulty of understanding the history of human impacts on ecosystems is particularly acute in underdeveloped regions of the world, where access to monitoring equipment is limited. For lakes in these regions, paleolimnology may provide the only practical and relatively inexpensive means of reconstructing impact histories. In some industrialized regions where lake monitoring has been actively discouraged in the past, paleolimnological records may be the only source of reliable historical data on pollutant histories and baseline conditions available to lake managers (Vile et al., 2000).

Relating changes in lake archives to particular climatic events or human activities is a complex task. Lake ecosystems are subject to multiple external forces simultaneously, any or all of which may affect our archives of interest. It is probably the norm rather than the exception for lakes to be subject to numerous anthropogenic impacts, whose paleolimnological "effects" are interactive. Even seemingly simple applications of paleolimnological records to infer human impacts have been controversial. Ascribing particular lake archives to particular human activities is also complicated because lake ecosystems undergo profound and rapid changes even in the absence of direct human intervention. In fact, a major contribution of high-resolution paleolimnology has been the demonstration of this fact (Renberg, 1990; Smol, 1995b). Also, the effects of local to regional-scale human activities on lake systems are increasingly being overlain during the twentieth to twenty-first centuries by the effects of global climate change. In practice this has meant that the response of lakes to remediation efforts, for example, reductions in local nutrient loads, may be counteracted by factors such as simultaneously rising water temperatures, or internal nutrient loading caused by a background of changing climate. All of these caveats should not discourage us. Rather, they should prod us to be both realistic about the limitations of the paleolimnological approach to historical reconstruction and to search for ever better methods of disentangling historical records.

Land-Use Change and Erosion

Someone from another planet who understood nothing about the purpose of our activities on the earth's surface would surely come to the conclusion that we like to shovel up and move dirt. The number of human activities that mobilize sediment is truly astounding; tilling soil, cutting forests, and constructing roads and structures are just a few of the most obvious examples. All of these activities also make loose sediment available to the whims of water, air, and gravity, for transport downslope to receiving basins such as lakes. Soil scientists have produced an overwhelming body of evidence showing that the initiation of landscape modification in previously undisturbed areas leads to significant increases in the short-term rates of soil erosion. Thus, lake deposits are a natural place to look for an historical record of how, when, and at what rate we are transforming our landscape through our propensity for digging.

The relationship between watershed erosion rates and lacustrine sediment accumulation rates is complicated by the fact that sediments have long and variable periods of storage within alluvial systems. Therefore, human impacts on watersheds such as deforestation and urbanization do not produce simple responses in terms of sediment or nutrient yield; they reflect both the geomorphology and short-term climate variability within the basin, as well as its degree of soil degradation. Increased land clearing within a watershed can yield a lake response of either increasing or decreasing nutrient loading and productivity, depending on the nutrient composition and concentration of eroding soils. Furthermore, the effects of anthropogenic and climatic forcing on sedimentation rates and sediment composition strongly overlap, making it difficult to differentiate local from regional processes. For example, increased allochthonous inputs of sediments can cause spikes in magnetic susceptibility records (e.g., Lami et al., 1994). This might result from anthropogenic or climate processes, or both, in a single lake. This type of complication implies that a regional and comparative approach is required to discriminate the two types of processes.

Lake-level fluctuations caused by climate change can also produce many of the same signals of varying productivity, sediment accumulation rate, and changing biota that are caused by watershed disturbance by humans. Only through very detailed examination of the exact spatial patterns and timing of paleolimnological change can these be separated.

Processes that affect lakes at the watershed scale may also operate over broader regions, as a result of parallel land-use histories, and similar patterns of settlement, expansion of agriculture, and urban development, leading to similar lake histories. Even under these circumstances, however, the unique features of each watershed and lake are likely to cause subtle differences in the timing or magnitude of its responses to sediment or nutrient loading. Variations in factors such as the size of inflowing rivers, slope, watershed area, and water residence time can cause one lake to respond more quickly or dramatically than another to the change in sediment or nutrient flux (Alefs and Müller, 1999).

These concerns notwithstanding, there are many circumstances where valuable records of the history of soil and watershed erosion can be deduced from paleolimnological archives. Under uniform climatic and geomorphic conditions, a variety of human activities accelerate lacustrine sediment accumulation rates. These rate changes are often the clearest signal of human disturbance on landscapes, particularly within smaller watersheds. Furthermore, this pattern has been documented over the entire spectrum of climate regimes. In Finland, ditching of peatlands, intended to drain vast areas for forestry and agriculture, has strongly accelerated soil erosion, recorded in lake sedimentation rates (Sandman et al., 1990). In the southeastern United States, rapid gully erosion in deforested watersheds has led to increased rates of floodplain aggradation; occasionally this is sufficient to dam alluvial valleys and create new lakes, or increase the rate of infilling of pre-existing ones (Hyatt and Gilbert, 2000). In rural areas of the English Lake District, increased traffic by livestock has driven an approximately eightfold increase in mass sedimentation rate in Blenham Tarn during the late twentieth century (Van der Post et al., 1997). Intense grazing pressure has also been shown to greatly accelerate accumulation rates in lakes in the semiarid tropics (e.g., El-Daoushy and Eriksson, 1998). And conversion of dense, tropical woodland to intensive and unterraced agriculture on hillslopes has resulted in greatly accelerated sediment accumulation rates in front of disturbed watersheds in the northern part of Lake Tanganyika (figure 12.1). Here, the degree

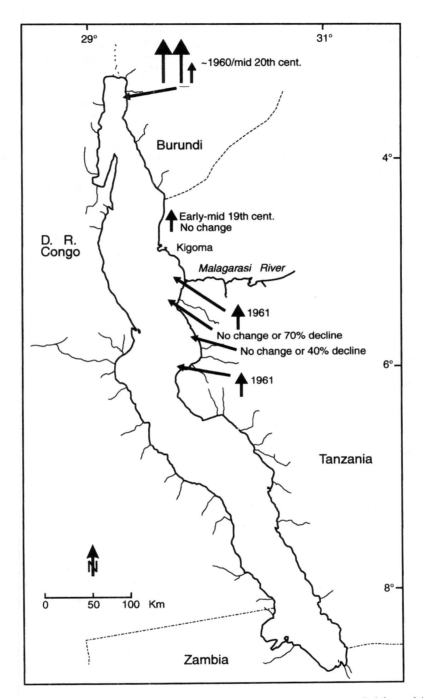

Figure 12.1. Sediment accumulation rate changes during the past 150 years recorded from deltas of Lake Tanganyika. Arrows indicate direction (up = increase) and magnitude of change. Times of the start of increasing rates are also given. Note the coincidence of early 1960s dates, corresponding to the extremely wet year 1961. Sites registering no change or rate declines are all forested, whereas all deforested sites show increased rates.

of lake ecosystem response to deforestation varies because of watershed area, watershed slope, and lake floor slope. The simultaneous increase in sedimentation rates around 1961 in deforested areas corresponds with extraordinary rainfall and high lake levels of that year. These processes triggered shoreline mass wasting and gully incision of previously accumulated soil, which continues to the present. This emphasizes the need for paleolimnologists to examine the interrelationship between human activities and climate processes.

Anthropogenic increases in erosion rates are most commonly associated with the rapid growth of human population since the early twentieth century. However, clear examples of the effects of deforestation have also been documented in preindustrial landscapes. Often the earliest phases of land clearing are marked by extraordinary rises in sedimentation rates, particularly in watersheds where the ratio of land:lake surface area is large. This initial spike of sediment accumulation is frequently followed by a "settling down" of accumulation rates to some rate that is intermediate between the preland-clearing phase and the deforestation spike (M.B. Davis, 1976; Brown and Barber, 1985).

A variety of archives are associated with changes in watershed erosion rates aside from changing sedimentation rates. Variations in the depth and intensity of erosion produce different suites of dominant iron minerals depending on whether soil or deeper bedrock is being eroded. This in turn affects the concentrations of minerals with different magnetic susceptibility properties (e.g., Lott et al., 1994). Magnetic profiles are particularly useful in identifying pulses of sedimentation that may be linked to episodic human activities such as drainage ditching (Flower et al., 1989). However, the interpretation of magnetic profiles as a record of erosion and deposition is complicated by the fact that certain bacteria precipitate magnetic minerals, especially magnetite, and these bacterial communities are redox-sensitive (Oldfield and Wu, 2000).

Pollen records are frequently coupled with lake sedimentation rate histories to infer patterns of watershed disturbance. This approach can yield clues about the nature or cause of changing land use, for example from the appearance of agricultural or weed/disturbance species, or the disappearance of forest species. However, taphonomic complications sometimes preclude simple interpretation of palynological profiles during episodes of deforestation; changes in the proportions of forest to agricultural or weed species pollen during defor-

estation is dependent on relative pollen production and transport rates, as well as vegetative cover.

The interplay between climatic and anthropogenic effects on land cover, erosion, and lake sedimentation is complex, and evidence exists for them both being the dominant forcing factor on short-term changes in accumulation rates. Changing land-use patterns are often as much a consequence of climate change as they are of cultural development. Differences in cultivation practices in Europe were probably strongly influenced by decadal to century-scale changes in climate, particularly prior to the industrial era, and this in turn would have affected soil erosion and nutrient runoff rates to lakes. In the Mayan region of Central America, intervals of deforestation and aforestation, linked to human population growth and decline, affected sediment yields to lakes (Brenner et al., 1990). Yet some of these human-induced changes may themselves have resulted from varying precipitation and runoff over the same period, altering the numbers of people that could inhabit any watershed.

A study of the small Lake Bjärsejösjön watershed in southern Sweden illustrates the potential for interpreting various human impacts at the watershed scale, and at least partially separating climate effects, using a multi-indicator approach (Gaillard et al., 1991). The watershed of Lake Bjärsejösjön has been occupied by human settlements since the Neolithic, about 6000 years, although settlements have only been continuous since the Iron Age, about 2500 BP. A record of likely anthropogenic impacts on the lake ecosystem and sedimentation can be inferred from an independent archeological record from the region (figure 12.2). Several archives provide a lake record of land-use changes that can be compared against this archeological history. Magnetic susceptibility records provide information on the changing proportions of magnetic carriers, which vary in this watershed as a function of the extent of topsoil disturbance from plowing. Episodes of increased soil erosion can be inferred from sediment yields, which are based on accumulation rates of minerogenic sediments. Diatoms and sediment geochemistry record pH and trophic state changes, and a combination of pollen and plant macrofossils record lake-level changes. The record at Bjärsejösjön suggests that lake-level fluctuations, driven by changing climate, were responsible for the most profound changes in the lake prior to the Viking Age, about 1100–1000 yr BP. Since that time, however, land-use practices have strongly affected both trophic status and sedimentation rates in the lake.

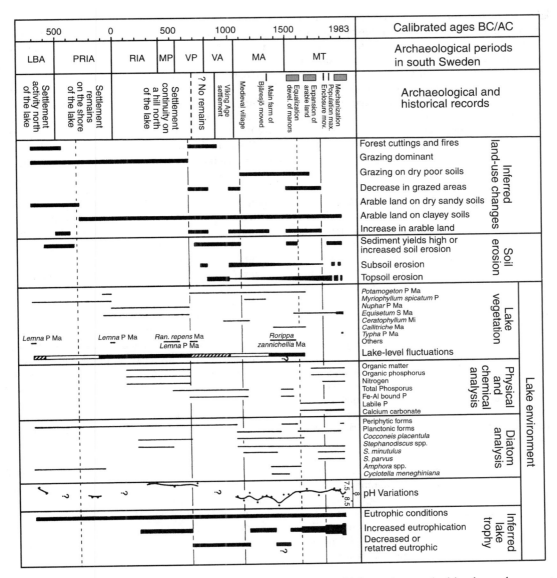

Figure 12.2. Summary of paleolimnological records and inferred lake and watershed land use changes at Bjärsejösjön (southern Sweden). Black segments indicate the occurrence of an inferred process, such as soil erosion, lake-level fluctuations, eutrophication. A rise in lake level is shown with oblique lines, a lowering and low levels in black, and high levels in white. Black lines indicate periods with high values of particular taxa or chemical parameters. From Gaillard et al. (1991).

Not all watershed-scale changes that affect lakes are the result of human intervention; for the vast majority of earth history human effects on erosion processes were nonexistent. It is important for managers to have some frame of reference from places and times with little or no human influence on erosion rates. Extreme events from such periods are particularly instructive in establishing the bounds of natural variability. A major decline of hemlock (*Tsuga canadensis*) populations throughout eastern North America in the middle Holocene, probably driven by a forest pathogen, resulted in the deforestation of vast areas, with accompanying rises in evapotranspiration, water and nutrient discharge, and soil erosion (M.B. Davis, 1981). Paleolimnological analyses of southeastern Ontario lakes during this time show that lakes with relatively large catchment areas were more susceptible to the effects of this deforestation, manifested through increased nutrient input and eutrophication (Hall and Smol, 1993). Catchment slope also appears to have been important in determining lake response; steeply sloping watersheds with exposed bedrock slopes generated accelerated rates of minerogenic sedimentation, and lesser organic matter sedimentation, in downslope lakes.

Nutrient Loading and Eutrophication

The exponential growth in human population density, the use of agricultural fertilizers, and the runoff of urban sewage into lakes throughout the world have resulted in the eutrophication of enormous numbers of lakes over the past century. Even though nutrient loading problems are often inferred to have developed "after the fact," from twentieth century instrumental data and observations, their development is rarely documented from the onset of eutrophication. Also, background variability in nutrient loading has rarely been documented in lakes prior to historical intervals of accelerated anthropogenic loading. These limitations present a serious problem to water quality managers, who are mandated by law in many regions of the world to establish targets for nutrient load reductions to some predetermined level. What this level of nutrient load should be is in turn influenced by perceptions of what background or preimpact level may have been, as well as their natural variability, resulting from changes in local climate or lake level. These values are unlikely to be available from instrumental data. For these reasons, paleolimnology can play an important role in defining the timing and magnitude of human-induced changes in lake trophic conditions (e.g., Wolfe et al., 2001), and in determining realistic goals for lake restoration (e.g., Bennion et al., 1996). Paleolimnological data can also define differing response pathways of various types to nutrient loading or other anthropogenic disturbances (Little and Smol, 2000). Once paleolimnological reconstructions indicate that water quality deterioration has occurred, restoration efforts can be contemplated, by integrating historical data with mathematical models of watershed nutrient sources. Paleolimnological data may indicate that a lake was naturally eutrophic prior to human activities. In this case, the water quality of a hypereutrophic system, polluted by sewage or other artificial nutrient loads, might only be im-

provable to a limited extent (Brenner et al., 1996). Where limited financial resources are available for lake restoration, this provides an important means of prioritizing cleanup efforts.

Diatoms have proven particularly useful paleolimnological tools in reconstructing past nutrient inputs, and determining the timing of both eutrophication and improvement in lake water quality (N.J. Anderson, 1997a,b; Lotter, 1998) (figures 11.7 and 11.8). Diatom analyses can be supplemented by sedimentary phosphorus analyses, although, as discussed in chapter 9, the interpretation of $P_{sediment}$ records as indicators of eutrophication is complicated by the variable mobility of phosphorus under different limnological or sedimentological conditions. However, lakes in regions of naturally low phosphorus loading from their watersheds frequently show significant increases in phosphorus retention following the development of intensive agriculture or cattle ranching (Brezonik and Engstrom, 1998).

Several other indicators have proven useful in tracking changes in trophic status associated with nutrient loading, including declining C : N ratios, increasing N : P ratios, changes in pigment concentrations, and chironomid assemblages. In some lakes, the onset of eutrophication is also accompanied by changes in varve thicknesses and the mean grain size of authigenic calcite crystals, the latter related to the effects of water column PO_4 on calcite crystal nucleation and growth (Lotter et al., 1997b).

Land-use changes are often manifested through a combination of watershed and lake effects, which are archived independently in lake sediments. Organic geochemical indicators can provide a valuable means of inferring these complex, and frequently interactive, changes. For instance, land clearing in previously forested areas often results in increased nutrient delivery to lakes from degraded watersheds, and accompanying increases in lake productivity along with organic matter signals from the changing watershed vegetation itself. All of these effects are potentially recorded by C : N ratio, C and N isotopes and biomarker, especially geolipid, archives (Bourbonniere and Meyers, 1996a).

Tracing the History of Multiple Land-Use Changes from Lake Sediments

The use of multiple indicators of land-use change is particularly important when documenting the history of multiple impacts. Many areas of the world have experienced rapid population growth, urbanization, and other forms of both intentional and unintentional ecosystem manipulation over the past few hundred years. This has commonly led to a pattern of sequential changes in pollutant, nutrient, and sediment runoff and lake system change, all occurring against a backdrop of varying climate, the impacts of which are anything but stable. An historical analysis of key paleolimnological indicators that differentiate between these effects can clarify when and how such changes took place. Two case studies serve to illustrate the power of this approach.

Lake Victoria (East Africa)

Paleolimnological evidence can be used to constrain alternative hypotheses of the causes of eutrophication in lakes, by recording the timing of ecological change. Lake Victoria (East Africa) has undergone extremely rapid and profound ecological changes since the early twentieth century. Separating the effects of the numerous impacts to this ecosystem (increased nutrient loading, introduction of predatory fish and water hyacinth, long-term climate change) requires historical data available only from paleolimnology, which can be incorporated into realistic models of ecosystem feedbacks.

Stratigraphical studies of sediment geochemistry, biogenic silica, and diatom assemblages from short cores provide a record of this history. Evidence for early increases in productivity are evident even in the late nineteenth century from organic geochemical (geolipid) records. Since 1900 there has been almost a doubling of algal *n*-alkane (*n*-C15 and *n*-C17) accumulation rates, accompanying other indications of increased primary productivity, especially by cyanobacteria (Lipiatou et al., 1996) (figure 12.3). By the 1920s the clearing and burning of land around Lake Victoria had begun to affect nitrogen inputs to the lake (Hecky, 1993). The twentieth century record also shows a significant increase in the concentration of low molecular weight PAH compounds, which probably formed in low temperature combustion such as cooking fires or charcoal production.

By the 1930s, diatom production had begun to increase in certain parts of the lake (Verschuren et al., 1998). Phosphorus inputs from soil erosion appear to have accelerated during the 1950s. Enormous increases in diatom production eventually led to the depletion of silica in the water column, starting in the 1960s. By the late 1980s,

the effects of this silica depletion in the face of increasing nutrient loads had become evident in both the paleolimnology record and modern observations. Diatom fossils show a transition to dominance by species that can take advantage of high N and P supplies, and that can grow smaller and more thinly silicified frustules. Simultaneously, limnologists began to observe the explosive growth of cyanobacterial blooms in the lake. Although some observers initially attributed these changes exclusively to top-down effects from the introduction of the predatory Nile Perch, the historical record shows that they are actually part of a longer-term process.

Longer-term studies of Lake Victoria's history provide a record of background variability against which the twentieth century records can be assessed. For example, T.C. Johnson et al. (1998) showed that any short-term trends in biogenic silica in Lake Victoria must be interpreted against a background of Quaternary change related to lake-level fluctuations and lake outflow. Holocene diatom records suggest that the lake has experienced considerable long-term trophic instability, presumably as a result of climatic forcing effects (Stager, 1998). Furthermore, these ancient changes seem to have occurred on time scales comparable to the rates of change observed during the twentieth century. These records cautions us about interpreting twentieth century signals strictly in terms of watershed impacts, and emphasizes the need for impact-specific indicators, such as unusual organic geochemicals that are only produced by human activities.

Separating Local Anthropogenic and Regional Climate Effects: The Qu'Appelle River Valley of Southern Saskatchewan, Canada

In some studies, multivariate statistical analyses have been used to partition the relative impacts of local and regional-scale processes on lake ecosystems. For instance, Hall et al. (1999) made a comprehensive study of the history of land use, climate, and water quality in the Qu'Appelle River Valley of southern Saskatchewan, Canada, in the northern Great Plains region of North America. Prairie lakes of this watershed have been influenced by a major expansion of agriculture, fish stocking, and commercial fish harvesting, urban population (the cities of Regina and Moosehead), and sewage discharge over the past 100 years (figure 12.4). All of these changes may influence lake water quality and they may also interact with the long-term warming

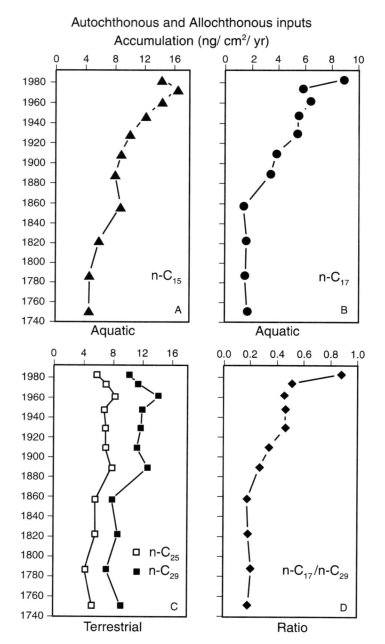

Figure 12.3. Sedimentary profiles for selected geolipids from a short core from Eastern Lake Victoria, Kenya. All profiles are in $ng \, cm^{-2} \, yr^{-1}$. Increasing accumulation rates of aquatic biomarkers (short chain length C_{15} and C_{17} n-alkanes) relative to long chain length (C_{25} and C_{29}) n-alkanes typical of terrestrial organic matter tracks the increasing algal productivity since the late nineteenth or early twentieth centuries. From Lipiatou et al. (1996).

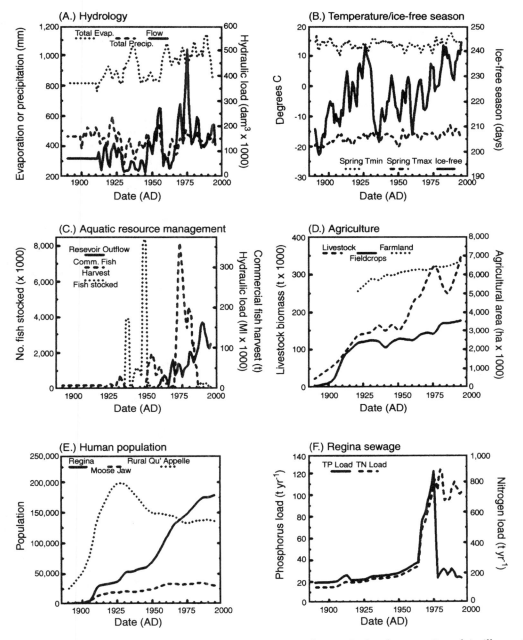

Figure 12.4. Historical data for the Qu'Appelle drainage basin (Saskatchewan, Canada), illustrating increasing intensity of human activity since the early twentieth century and background climate variability. Regina (F) represents nutrient loading from the urban areas within the watershed. From Hall et al. (1999).

trend and lengthening of the ice-free season of the region's lakes. Hall and his colleagues have used the fossil history of pigments, diatoms, and chironomids to understand changes in one of these lakes (Pasqua Lake) from the pre-European settlement period to the present. Their analyses showed that this lake, currently highly eutrophic, was nevertheless quite productive in the period prior to European-style agriculture. Pigment analyses also showed that the lake has been subject to regular cyanobacterial blooms over the past 200 years. However, fossil indicators suggest a serious decline in water quality during the mid-twentieth century, consistent with the timing of rapid expansion in many of the anthropogenic impactors within the watershed. Significantly, this decline in water quality was not substantially improved by sewage treatment installations, even with tertiary treatment schemes that lowered phosphorus levels to those of the 1930s. Using a multivariate statistical technique known as *variance partitioning analysis* (Borcard et al., 1992), Hall and his colleagues were able to explain the patterns of changing lake ecosystem structure, as the combined result of several key factors (climate, urban factors, and resource use) or their interactions (figure 12.5). Interestingly, but perhaps not surprisingly, different components of the ecosystem have responded more or less strongly to each of these factors over the past 100 years.

This study shows the direct management potential paleolimnological studies can have at the local to regional scale. Pasqua Lake's pretwentieth-century history as a naturally eutrophic lake suggests that the lake should not be managed with low productivity as a water quality target. Similarly, the fossil record of algal blooms and water quality both prior to and following the installation of tertiary sewage treatment provides evidence that nutrient abatement schemes for the region need to focus on nitrogenous waste control, the apparent limiting nutrient in the Qu'Appelle River valley lakes, rather than phosphorus.

Heavy Metal and Pesticide Accumulation

Lakes are excellent sites for obtaining historical records of the atmospheric and watershed inputs of metal or organic pollutants. However, to correctly interpret these records, it is necessary to take into consideration the "taphonomy" of pollutant archives in lakes. There is normally a high degree of spatial heterogeneity in deposition of atmospheric, nonpoint-source, pollutants, even at the watershed scale. Even after deposition, pollutants are subject to sediment focusing. The interpretation of lake sediment profiles of metals can be complicated by variable input and erosion rates from the surrounding watershed. In some cases, this problem can be circumvented by obtaining pollutant flux records from raised peat mires, which receive all of their heavy metals from the atmosphere, as opposed to runoff sources (e.g., Weiss et al., 1999).

Not all lakes provide equally useful records of pollutant fluxes to the sediments, primarily because of the extremely large range of diagenetic environments that exists at the sediment–water interface of lakes. Given the potential mobility of metals between the sediments and the water column, the most accurate and highly resolved records of metal pollutants generally come from very high sedimentation rate environments, such as in reservoirs, where metals are effectively isolated from remobilization into the water column (Callender, 2000). Although diagenesis certainly continues in these rapid burial situations, the remobilized metals are effectively trapped in the sediments, making geochemical profiles much more interpretable.

Lake sediments archive a pattern of increasing metal and organic combustion byproduct deposition in numerous lakes throughout the world. Much of this increase is attributable to the rapid expansion of heavy industry, internal combustion engines, and power plants since the late nineteenth century. Increasing concentrations of heavy metals, magnetic minerals, and combustion-derived particulates have been found in lake sediments throughout the world, even in relatively remote areas at high altitudes or latitudes (Oldfield and Richardson, 1990; Rognerud and Fjeld, 1993; D.R. Smith and Flegal, 1995; Wik and Renberg, 1996). In addition to these global patterns, more local or regional effects are evident within specific watersheds, reflecting local patterns of industrialization. Often, this is expressed by rising concentrations of specific metals, which are correlated with local patterns of mining or manufacturing. In regions with intensive programs of comparative lake sediment sampling, it is possible to use paleolimnological records to infer the spatial pattern of regional contamination by air-borne pollutants, and to produce regional models of pollutant metal "enrichment."

The interpretation of metal profiles is complicated by the fact that two quasi-independent processes related to industrialization have both caused

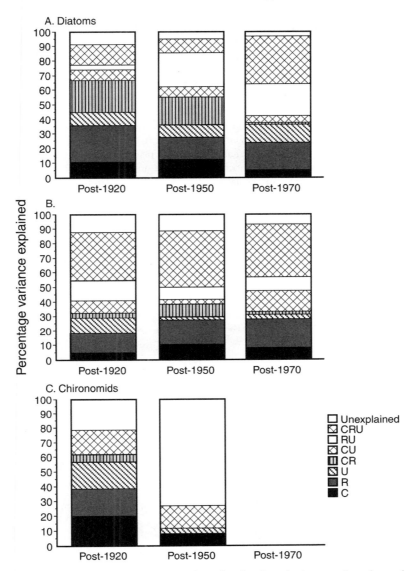

Figure 12.5. Results of variance partitioning analysis for fossil and pigment data from the Qu'Appelle Valley, Saskatchewan (Canada). Graphs display the proportion of variance in diatom (A), pigment (B), and chironomid (C) records explained by the effects of climate (C), resource use (R), urban factors (U), and various combinations of these three factors (CRU, RU, etc.) for the periods 1920–1993 (left columns), 1950–1993 (middle columns), and 1970–1993 (right columns). No variance could be explained for post-1970s chironomids. From Hall et al. (1999).

a widespread rise in the concentration of many metals in lake sediments. Part of the increase is clearly attributable to actual increases in the atmospheric and point source fluxes of metals to lakes. But in acid-sensitive watersheds subject to high levels of acid precipitation, part of the increase is also probably a result of increased leaching rates (Rognerud and Fjeld, 1993).

Mercury

Because of its importance as a health risk, a great deal of effort has been put into documenting the accumulation of mercury in lake sediments. Mercury, highly toxic in some of its salt compounds, is derived from manufacturing, coal combustion, waste incineration, and nonferrous metal smelting. Mercury can be transported long distances in the atmosphere, and even lakes in regions that are remote from industrial activity show a common pattern of rising concentrations of mercury in the sediments starting in the mid-nineteenth century (Lorey and Driscoll, 1999) (figure 12.6). In Minnesota lakes, mercury deposition peaked in the 1960s or 1970s, reflecting the progressive control of mercury pollution from industrial and municipal point sources (Engstrom et al., 1994; Engstrom and Swain, 1997; Balogh et al., 1999). In stark contrast, mercury deposition continued to rise in the lakes of southeastern Alaska into the 1990s. These obser-

vations suggest two things. First, a decline has occurred in the regional production and mercury emissions in the eastern United States. Manufacturing records indicate that mercury production and consumption in the United States has declined since the 1960s, in large part because of the use of alternative materials, and because emission standards for industrial sources of mercury have become more stringent. Second, the ongoing rise in mercury levels in pristine Alaskan lakes implies that the total global atmospheric flux of mercury has not followed the trend in the United States. Similar findings of rapid increases in sedimented mercury from rela-

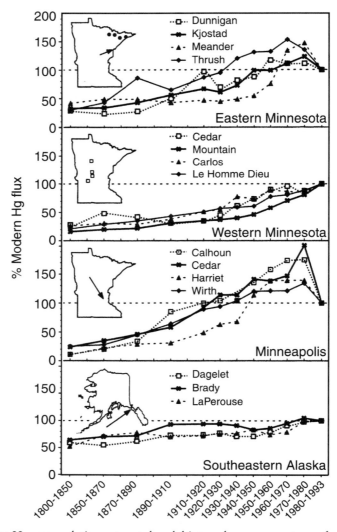

Figure 12.6. Average Hg accumulation rates at decadal intervals as a percentage of modern (1980–1993) Hg accumulation rates in sediment cores from lakes in rural and urban Minnesota and southeastern Alaska (United States). From Engstrom and Swain (1997).

tively remote lakes of southern Florida provide further evidence supporting a background rise in atmospheric deposition rates of mercury from regional to global sources (Rood et al., 1995, 1998).

Lead

As with mercury, interest in tracking lead accumulation in the biosphere results from its toxicity, coupled with its tendency to accumulate in the earth's surface environments (D.R. Smith and Flegal, 1995). Lead readily substitutes for calcium in many biochemical pathways, often with devastating results for growth and development. Lead production by humans began about 4500 yr BP and remained at significant but relatively stable levels until about 500 years ago, when production increased dramatically. In the United States, lead production rose drastically in the twentieth century with the introduction of leaded gasoline. Lead use in fuels declined sharply after 1970 in the United States and Canada as leaded fuels were progressively eliminated from production. European conversion to nonleaded fuels occurred about 10 years after the United States. However, total consumption actually continued to rise during the 1980s and 1990s, as lead use in batteries supplanted lead in fuel.

Compelling evidence exists that the initial rise in anthropogenic lead concentrations from atmospheric deposition occurred during the preindustrial period (figure 12.7) (Veselý et al., 1993; Renberg et al., 2000). This occurred as a byproduct of mining precious metals, and also as an outcome of the leading of glass. In lake sediments this is recognizable both through the analysis of total lead concentrations and the $^{206}Pb/^{207}Pb$ ratio, which differs between natural and anthropogenic sources. A remarkable record of the history of European lead production and consumption is captured in varved Swedish lake sediments (Brännvall et al., 1999; Renberg et al., 2000). There, natural sources have $^{206}Pb/^{207}Pb$ ratios of about 1.5, whereas anthropogenic sources are in the range 1.15–1.19. High-resolution studies of varved sediments show that prior to 4000 yr BP there was no anthropogenic lead entering Swedish lakes. With the onset of silver smelting about that time, total lead concentrations began to rise and the $^{206}Pb/^{207}Pb$ ratio began to fall. Mining of silver from various sulfide minerals often produces lead as a byproduct, which can enter the atmosphere and be transported up to several thousand kilometers. Lead levels and isotopes continued to change into the period of extensive silver

production during the Roman period, and dropped dramatically at the fall of the Roman Empire, when production of coinage declined. A new period of increased lead concentrations began during the Medieval period, again punctuated by the decline of human population during the Black Death plague of the 1300s. Industrial-era emissions, and the use of leaded fuels resulted in further increases in lead pollution, although this rise was clearly underway prior to the nineteenth century. Since the introduction of lead-free gasoline in the 1970s there has been a remarkable decline in lead concentrations, although alternative and rising uses of Pb in manufacturing have prevented these values from declining to preindustrial levels (Bergbäck et al., 1992).

Recent lake sediments have also provided a high-resolution record of the late twentieth century decline in atmospheric lead emissions (Moor et al., 1996). In numerous lakes, total lead emissions during the industrial era were closely traced by sedimentary lead concentrations. In the mid-twentieth century this can be linked to fossil fuel consumption by the $^{206}Pb/^{207}Pb$ isotopic ratio in lake sediments, which continued to decline until the introduction of unleaded gasoline. Since that time sedimentary lead concentrations have declined and the $^{206}Pb/^{207}Pb$ has risen in lakes where gasoline was the primary source of lead. However, some paleolimnological records from urban areas provide evidence that sources other than leaded gasoline have locally dominated the anthropogenic release of lead. Core records of a peak in lead concentrations in the 1930s from Central Park Lake in New York City are temporally consistent with an historical peak in municipal solid waste incineration in New York, and are inconsistent with the history of leaded fuel consumption (figure 12.8, Chillrud et al., 1999).

Fossil Fuels

Both gaseous and particulate products of fossil fuel combustion are a ubiquitous feature of our modern atmosphere. Many of the emission byproducts from fossil fuel combustion, such as SO_2, NO_x, organic and metallic compounds, and complex particulates, are ultimately redeposited on the earth's surface and become incorporated in lakes, altering water chemistry and ecology, and leaving a sedimentary record of their flux. Industrial gases and particulates are removed from the atmosphere either by precipitation, also called *wet deposition*, or by

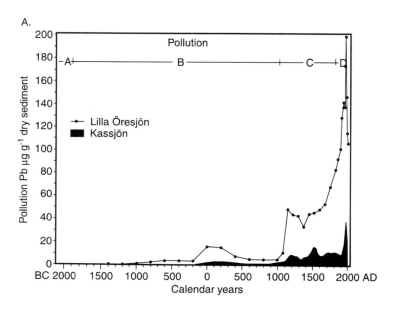

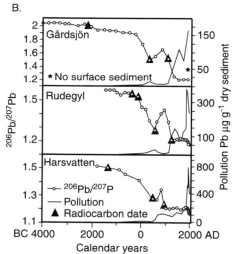

Figure 12.7. Temporal trends in lead pollution in Sweden determined from lake records. (A) Trends in total Pb accumulation from two lakes, Lilla Öresjon (southern Sweden) and Kassjon (northern Sweden), during four periods: A, the unpolluted period; B, the ancient period of pollution byproducts of silver mining; C, the Medieval period to the Industrial Period increased silver mining, decline during the Black Death, and renewed economic expansion; D, the Industrial Period. (B) Lead isotope ratios and pollutant lead concentrations in lake sediments from three Swedish lakes which, because of low background lead concentrations and high $^{206}Pb/^{207}Pb$ ratios, were sensitive to influx of air-borne lead pollution. From Renberg et al. (2000).

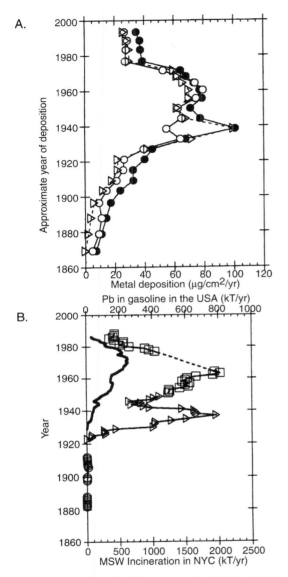

direct settling without water being involved (*dry deposition*). Transport-related biases exist in the sedimentary records of atmospherically borne pollutants, with implications for paleolimnology (Larsen, 2000). Wet deposition, particularly important for combustion byproduct heavy metals, tends to act regionally, producing spatially averaged records. Dry deposition tends to occur more locally, better reflecting local pollutant sources.

Spheroidal Carbonaceous Particles

The organic and metallic residues of coal and oil combustion produce a variety of microscopic fly ash particles, the vast majority of which are round and porous. The deposition of these *spheroidal carbonaceous particles* (SCPs) shows a strong correlation with the twentieth century increase in the use of fossil fuels during the Industrial Era (Wik and Renberg, 1996). SCP accumulation rate in lake sediments is strongly influenced by the proximity of particulate sources, especially power plants, but may also be influenced by soil erosion rates and sediment focusing. Marked increases in the abundance of SCPs in lake sediments appear throughout much of North America and Northern Europe after the 1950s–1960s. Increasing rates of SCP deposition in lakes from the atmosphere are often accompanied by other indicators of industrial burning, for example rising magnetic intensities in postindustrial sediments caused by the ferromagnetic minerals component of fly ash (Kodama et al., 1997). Records of SCPs therefore provide an independent estimator of atmospheric deposition of particulates, including industrially produced acids, against which paleoecological archives of the effects of pollution on lake ecosystems can be compared. The fact that SCP records show strong correlations with fossil fuel consumption, regional SO_2 emission patterns, and regional energy production, even in very remote lakes, shows that SCPs are primarily deposited from the atmosphere (Odgaard, 1993; Doubleday et al., 1995).

A detailed study of SCP concentrations in surface sediments and ^{210}Pb-dated core profiles from Lake Baikal illustrates their potential for recording atmospheric inputs of fossil fuel particulates (Rose et al., 1998). Coal and power production in the Irkutsk area and other nearby regions of Siberia has increased dramatically since the early twentieth century. A major oil-fired power plant at Angarsk, as well as other emissions sources at the southern end of Lake Baikal, contribute substantial loads of atmospheric particulates and SO_2 to the atmo-

Figure 12.8. Metal pollution trends in New York City. (A) Pb (●), Zn (○), and Sn (△) deposition rates, normalized to unsupported ^{210}Pb delivery from a core in Central Park Lake, New York City. (B) History of incineration of municipal solid waste (MSW) in NYC at municipal facilities (symbols) and total production of Pb for leaded gasoline in the United States (heavy line). The 1930s peak in MSW incineration accounted for almost half of the total MSW disposed by NYC at that time. During the mid-1960s MSW incineration comprised 30–40% of the total. The progressive decline in incineration rates for 1964–1975 is consistent with the history of plant closings. From Chillrud et al. (1999).

sphere in this region (figure 12.9). Cores from various parts of the lake show the potential of SCP concentrations to record this history of industrial inputs (figure 12.10). After the 1930s the rise in SCP concentrations closely paralleled the rise in coal production and regional power station heating capacity. At the northern end of Lake Baikal the development of industrial emissions is a much more recent phenomenon, and this is reflected in both a shorter history of SCP accumulation and much lower SCP concentrations in lake sediments.

Polycyclic Aromatic Hydrocarbon (PAH) and Magnetic Intensity Records of Fossil Fuel Combustion

Polycyclic aromatic hydrocarbons, discussed briefly in chapter 9, are of particular concern to environmental toxicologists because of their known carcinogenic properties. Like SCPs, PAHs form from

incomplete combustion of fossil fuels. They have been accumulating in lakes in significant quantities since the nineteenth century. As with SCPs, their records can be compared between lakes at a regional scale to understand both the timing and magnitude of increases in fossil fuel emissions. In Europe, for example, an increase in PAHs is registered simultaneously throughout the region between 1870 and 1910. However, lakes in Eastern Europe, a region with a long history of heavy reliance on coal burned under conditions of relatively uncontrolled emissions, register orders of magnitude higher levels of sedimented PAH compounds than do areas that are remote from coal burning, such as northern Scandinavia (Fernández et al., 2000). PAH records can also be used to identify dominant sources of industrial emissions, to differentiate local sources between watersheds, or to determine changes in emission sources over time (Simcik et al., 1996; Karls and Christensen, 1998).

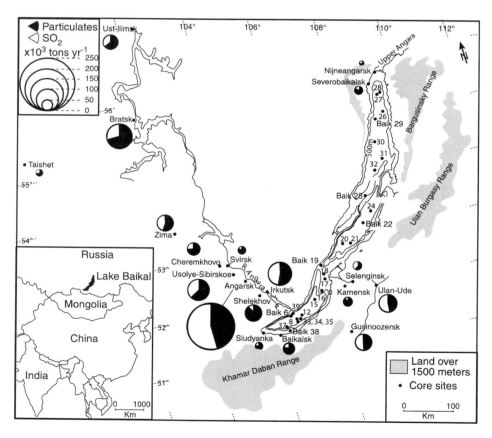

Figure 12.9. Sediment coring sites in Lake Baikal for study of spheroidal carbonaceous particles discussed in text, together with the locations of the main industries in the region. Size of circle corresponds to annual emissions, subdivided to show the fraction relating to particulates and SO₂. From Rose et al. (1998).

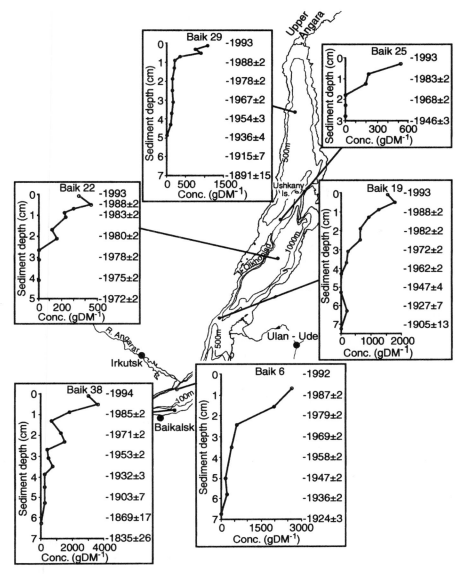

Figure 12.10. SCP profiles for six ^{210}Pb-dated cores from Lake Baikal (Siberia, Russia). From Rose et al. (1998).

The flux of magnetic particles (magnetite and hematite) from power plant and heavy industry emissions often parallels that of SCPs and PAHs (Oldfield, 1990). In lakes where watersheds can be shown to provide very little magnetic mineral flux, the profiles of magnetic properties of cores can provide a parallel record of atmospheric inputs from industrial emissions. Because these magnetic properties can be measured more quickly and easily than SCP or PAH separations, they provide an important alternative means of establishing atmo-spheric pollutant histories in appropriate water-sheds.

Polychlorinated Dioxins and Furans

Polychlorinated dioxins and furans are classes of highly toxic compounds released from a variety of industrial sources. During the 1980s paper mills were discovered to be a significant source of these compounds, produced during the bleaching of pulp with chlorine. Paleolimnological records have been used to establish the history of production of these compounds, their relationship to pulp processing or other sources, and the efficacy of local emission controls (Kahkonen et al., 1998; Macdonald et al., 1998).

Lake Acidification

Given the ease with which pH is measured in rainwater, rivers, and lakes today, it is somewhat surprising to learn that pH measurements were not made directly in lakes until the 1920s, nearly 60 years after large-scale industrial emissions of acid-producing compounds had begun (R.B. Davis, 1987). And it was not until the early 1960s that pH data was collected in North American and European lakes with sufficient regularity and accuracy to provide an ongoing record of changes in lake acidity. By that time many poorly buffered lakes in heavily industrialized parts of the world were already experiencing serious problems with acidification.

The study of long-term lake acidification is undoubtedly one of the great "success stories" of paleolimnology, in that it convinced both the scientific community and the public at large of the usefulness of lake sediments as repositories of high-resolution environmental records. Paleolimnological data also proved decisive in testing the many hypotheses about ultimate causes of recent lake acidification in acid-sensitive lakes, placing the lion's share of blame directly on an increase in acidic precipitation from industrial sources (Battarbee, 1990).

Early paleolimnological investigations of lake acidification during the Industrial Era began in Scandinavia in the 1970s (U. Miller, 1973; Berge, 1975). This work quickly spread to North America and other parts of Europe, and by the mid-1980s numerous studies had been undertaken or were underway to quantify the history of lake acidification, primarily through the use of fossil diatom assemblages. Two extremely successful studies of lake acidification merit special attention, the Paleoecological Investigation of Recent Lake Acidification (PIRLA) Project in North America, and the similar paleolimnological component of the Surface Water Acidification Programme (SWAP) of Northern Europe (Battarbee and Renberg, 1990; Whitehead et al., 1990). Both programs took a regional, multidisciplinary approach

to the problem of understanding the history of lake acidification and its probable causes in those parts of the world, Eastern North America and Northern Europe, with the longest and most intensive histories of industrialization. The strength of these studies rested in three factors:

1. Their broad scope, taking advantage of numerous lines of geochemical, sedimentological, and paleoecological evidence. These studies not only documented the changes in indicators known to be sensitive to pH, but also tracked the response of other aspects of the ecosystem, such as invertebrates, pigments, sediments, and metals, whose response to acidification was uncertain.
2. Their extremely careful site selection, to investigate lakes representative of regional and watershed variation in such factors as bedrock geology (acid sensitivity), climate, and depositional patterns of acid precipitation in each study region.
3. Their strong emphasis on quantitative reconstruction and quality assurance of the inferences of pH change made by the studies. This factor was mandated by the contentious political climate surrounding the issue of lake acidification and the possible role of coal-fire power plants or other industrial sources of emissions. The statistical rigor used in inferring paleo-pH has served as a model for quantitative paleolimnological reconstruction efforts throughout the discipline.

The general paleolimnological approach in these studies has been to document the history of lake acidity and correlated limnological variables in lakes of varying susceptibilities to acidification and varying inputs of twentieth century atmospheric acid loads (figures 12.11 and 12.12). Using well-dated cores, these studies demonstrated statistically significant declines in pH in numerous lakes that had experienced little or no other types of land use/watershed impacts other than acidic precipitation (figures 11.4 and 1.9) (Birks et al., 1990a; Charles et al., 1990; Smol and Dixit, 1990). Records of these changes include both paleoecological and geochemical indicators. Reconstructions of lake acidity have also shown that significant changes in acidity have not occurred over the same time interval in lakes that are either well-buffered lakes, or that lie in regions far from important sources of strong acids from industrial sources. As

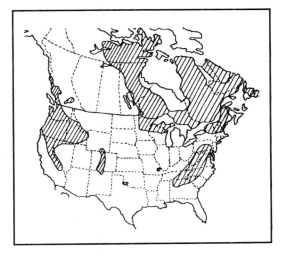

Figure 12.11. Acid-sensitive areas in North America, based on bedrock geology. From Whitehead et al. (1990).

we saw in chapter 5, acidification also entails major community level changes in lakes, resulting from such changes as decreasing dissolved organic carbon levels, and increased light levels in both benthic and deep-water planktonic environments (Leavitt et al., 1999). Diatom records have proven useful in understanding lake acidification in ways that go well beyond simply tracking the history of pH change. They can also be used to define the point in time when acid-sensitive lakes exceeded a "critical load" of atmospheric acidity from sulfur deposi-

tion (Battarbee et al., 1996). Diatom records have been combined with acid (S) and base (Ca) flux information to determine thresholds of critical loading for individual lakes. Simultaneous increases in the sedimentation of both acidification indicators and the deposition of diagnostic air pollutants such as Pb and PAHs in remote lakes provided further evidence that acidification was a consequence of anthropogenic burning of fossil fuels (R.B. Davis et al., 1990).

Not All Lakes Show Twentieth Century Acidification

In evaluating the importance of nineteenth to twenty-first century acid deposition from industrial sources for lake water chemistry, and the value of lake records for tracking the production of atmospheric acids, it is important to bear in mind that many lakes worldwide do not show a simple trend of increasing postindustrial acidity. On a global basis, some, and perhaps most lakes show either no clear trends in pH, or show much longer-term trends toward acidification. A major challenge for applied paleolimnology is to both consider alternative explanations of acidification trends, and to explain why some lakes undergo such changes while others do not.

The PIRLA Project study lakes covered other regions of the United States besides the Adirondack Mountains and New England, including lakes in the Upper Midwest and Florida. Although some of these lakes showed strong acidification trends in the late nineteenth and twentieth centuries, not all do so. In some regions the effects of increased, anthropogenic erosion of soils have been sufficient to counteract the effects of strong atmospheric acid inputs. Lakes in these areas have not registered any directional change in pH over the past two centuries (Kingston et al., 1990). Acidification is frequently expressed in clusters of lakes within circumscribed regions, characterized by similar bedrock lithology, hydrology, and watershed soil types (Sweets, 1992). In some regions, artificial drawdown of the regional water table may be partly responsible for pH declines, by varying the proportion of acid-neutralizing cations derived from seepage (Sweets et al., 1990), although more recent studies in Florida suggest that this is unlikely to be a primary cause of decadal-scale acidification trends. Paleolimnological data obtained by the PIRLA Project also showed that some study lakes had long histories of being acidic, extending back several thousand years.

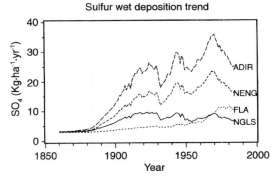

Figure 12.12. Sulfur wet deposition trends for various parts of the eastern United States. ADIR (Adirondack Mountains), NENG (Northern New England), NGLS (Northern Great Lakes States), and FLA (Florida). From Whitehead et al. (1990).

Similar indications of long and complex pH histories come from the lakes of northernmost Sweden (Korsman, 1999). Here, paleolimnological records show a mixture of decreasing, increasing, or stable pH values, all in a region that is acid-susceptible and subject to substantial local industrial emissions. Furthermore, in lakes that do record acidification, generally among the most acidic of the lakes in the region, the trend toward acidification has occurred over a several thousand year period, long before any industrial activity.

Numerous other hypotheses have been suggested to explain both short and long-term trends in lake acidity. Land use, fire, and deforestation have been documented to counter the trend toward acidification in some European and North American lakes, and conversely the cessation of lumbering and renewed soil stabilization has accelerated the acidifying effects of atmospheric precipitation (R.B. Davis et al., 1994). This causes complex signals, sometimes preventing unidirectional declines in pH in otherwise acid-susceptible lakes subject to strong atmospheric acid deposition.

Aforestation, particularly by tree species that produce strong humic acids, has been suggested as a possible explanation for long-term acidification trends in both boreal and temperate lakes (Hallbäcken and Tamm, 1986; Huvane and Whitehead, 1996). However, not all studies of forest lake histories find support for this relationship (e.g., Korsman et al., 1994). Also, a long-term dominance of humic acid-producing trees within a watershed may establish organic acid buffering systems and base cation depleted soils, which are not conducive to the re-establishment of circum-neutral pH conditions once the vegetative cover changes. Thus, the ability of lakes to respond to forest cover change may be contingent on the nature of local soil conditions prior to a disturbance.

Many North American lakes in regions sensitive to acid precipitation show surprising signs of elevated pH levels during early phases of European settlement, with precipitous declines in pH occurring much later, during the industrial era (Rhodes, 1991). The most common explanation put forward for this phenomenon is that it results from the effects of intensive logging, significantly reducing the loading of organic acids from the watershed to the lake. Paleolimnological signals of logging effects (increased pH) appear to be time lagged after the actual land-use impacts. This occurs both as an artifact of the time required for the diatom signals to be registered in lake sediments (as a result of sediment focusing) and perhaps as a real effect of the time required for soil-derived sources of alkali-

nity to be delivered to a lake (R.B. Davis et al., 1994).

Acidic Debates Over Recently Acidified Lakes

The notion that changes in climate, forest cover, and soil erosion can have profound effects on lake water pH has also stimulated debate among researchers studying nineteenth and twentieth century acidification trends. During the nineteenth and twentieth centuries watersheds in many acid-sensitive regions were subject to many land-use practices that, in principle, could have caused lake acidification. Is this an important component of the modern trends toward declining pH of poorly buffered lakes that we observe in so many industrialized parts of the world? The question, of course, is of far more than academic interest; it gets at the heart of paleolimnology's utility, as not only a recorder of past events but also a management tool for identifying probable *causes* of environmental change.

In some cases, the effects of watershed organic acid loading versus atmospheric (SO_4/NO_3) acid loading are separable, both in timing and magnitude. Many small and poorly buffered lakes in northern New England experienced acidic conditions prior to the period of European settlement, the primary source of acidity being organic, as opposed to modern sulfate sources from the atmosphere. Acidification during periods of forest recovery however has been a relatively small contributor to the present status of remote, poorly buffered lakes, where rates of acidification since about the 1920s have been unprecedented.

The most heated debates about the relative importance of land use or soil changes versus acid rain in driving pH declines have taken place in Northern Europe. Two hypotheses have been put forward as alternatives to acid precipitation for promoting lake acidity. First, a conversion of pastoral grasslands to forests or heathlands might have prompted the growth of plant species (especially the heath *Calluna vulgaris*) that cause the accumulation of humus and drainage acidification, the so-called "land-use hypothesis" (Rosenqvist et al., 1980). Second, postglacial soil pedogenesis during the late stages of interglacial periods, including the present one, promote gradual soil acidification, which might be responsible for such trends in lakes (Pennington, 1984).

Fortunately, these hypotheses make specific predictions about the timing of changes in acidification and its association with vegetational change, hypotheses that are testable with paleolimnological data. In many parts of Northern Europe, acidification has either increased simultaneously with a decline in *Calluna vulgaris*, or else lake acidity has been uncorrelated with changes in heathland cover in the watershed (Battarbee et al., 1988; Patrick et al., 1990). In either case these results are the opposite of what would be predicted by the land-use hypothesis. Furthermore, the timing of acidification in many Northern European lakes is an essentially modern (late nineteenth to twentieth century) phenomenon, occurring at the wrong temporal scale to be primarily a result of long-term interglacial soil acidification. In an ingenious test of the land-use hypothesis in Scandinavia, Birks et al. (1990a) examined lake histories from hilltop lake basins, where land-use effects would be expected to be minimal, the only likely source of acidity being atmospheric (figure 12.13). This study, and similar ones elsewhere, indicates that even hilltop lakes have undergone strong twentieth century acidification, and runs counter to the land-use hypothesis (figure 12.14).

Records of Reductions in Acid Precipitation Since the 1980s

The recognition that acidic precipitation was causing widespread watershed and lake acidification prompted legislation in many industrialized nations to mitigate this problem, either through controls on the sources of industrial acidity, or by local interventions in individual lakes, such as the addition of lime to buffer lake water pH (Lynch et al., 2000). Perhaps surprisingly, even for this period of remediation, paleolimnology has an important role to play in providing historical records of "What happened?" Between the 1970s and late 1980s the establishment of sulfate emissions controls in the United Kingdom resulted in a 40% decline in industrial sulfate emissions. This in turn significantly reduced the flux of nonmarine sulfate to lakes in the region. High-resolution, paleo-pH records from diatom fossils in lakes in southwestern Scotland show that over a mere 10-year time period pH rose and sulfate concentration fell significantly

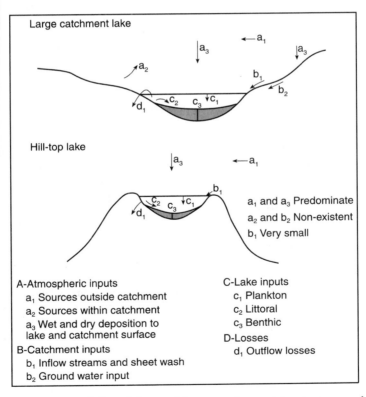

Figure 12.13. Contrast between hilltop lakes and large catchment lakes in terms of major sources of material preserved in their sediments. Hilltop lakes provide ideal sites for testing models of atmospheric inputs because of their very limited watershed areas. From H.J.B. Birks et al. (1990).

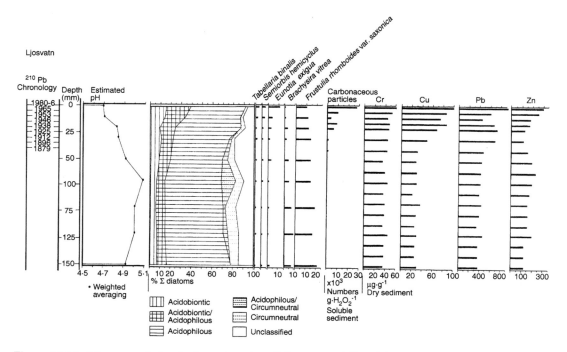

Figure 12.14. Diatom and chemical stratigraphy of Ljosvatn, a hilltop lake in southern Norway, illustrating pronounced increase in acidophilous and acidobiontic diatoms, decline in estimated pH, and increase in metal concentrations since the late nineteenth century, with the most pronounced change occurring over the last 50 years. See figure 11.5 for definitions of diatom acidity classes. From H.J.B. Birks et al. (1990).

(Battarbee et al., 1988). Rapid rises in inferred "paleo" pH are similarly evident in midwestern Canadian lakes that have experienced reductions in acid inputs (Dixit et al., 1993). In southern Sweden, the problem of lake acidification has been in part attacked since the 1970s through the use of liming (Renberg et al., 1993). An almost instantaneous effect of this type of treatment is evident in fossil diatom assemblages from cores taken from these lakes.

In the United States, the 1990 Clean Air Act Amendments mandated a reduction in acid-producing emissions, which the U.S. Environmental Protection Agency has monitored by measuring precipitation for the principal ions (SO_4^{-2} and NO_3^{-}) that contribute to acidity. For various political, economic, and scientific reasons, compliance with the act and the legal evaluation of its effectiveness is not monitored by changes in acidity in lakes and rivers themselves, but only through analysis of atmospheric levels of acidity, which closely track actual emissions. Because of this legislation, major declines in SO_4^{-2} and H^{+} concentrations in precipitation had occurred in the eastern United States by the late

1990s, relative to the 1980s and early 1990s, although NO_3^{-} levels in precipitation did not decline significantly over the same period (Lynch et al., 2000). Because there is no legal requirement to directly monitor lakes for continuing trends in pH following implementation of the Clean Air Act, funding for such efforts has been limited. Paleolimnology therefore provides a relatively cheap, retrospective way of evaluating the efficacy of Clean Air legislation where its impacts are chiefly hoped to occur, in lakes and ponds. There has, unfortunately, been little recovery in pH in some strongly acidified lakes, such as those of the Adirondack Mountains, since the reduction in sulfur emissions. In such regions, long time periods may be required to counteract the extreme depletion of base cations from soil resulting from decades of acidic precipitation (Majewski and Cumming, 1999).

Summary

Our conclusions from this chapter can be summarized as follows:

1. Paleolimnology provides an effective means of determining the history of human impacts to ecosystems and addressing questions about the timing and magnitude of those impacts. In some circumstances, where instrumental records are of poor quality or are nonexistent, paleolimnology may provide the only means of addressing these questions. Human impact studies in paleolimnology require both a comparative, interlake or interwatershed approach, and excellent chronology to be effective.

2. Paleolimnologists need to be realistic about the limitations of their approach to historical reconstruction, because different anthropogenic or climatic forcing processes can result in the same paleolimnological archive, and almost all lakes are subject to several types of human impacts.

3. The history of land-use change is recorded by such paleolimnological variables as changes in sedimentation rate, pollen profiles, magnetic susceptibility records, and changes in lake fossil assemblages. Paleolimnological data has documented significant watershed alteration by human activities well before the burst of industrialization and population growth of the last few centuries. The impacts of human-induced sediment erosion are modulated by variations in climate and drainage basin geomorphology.

4. Paleolimnological data can be used to determine the history of nutrient loading from sewage, fertilizer application, or other sources, and to establish realistic targets for remediation. Most lakes that have been subject to anthropogenic nutrient loading have also experienced other impacts, such as soil erosion or fish introductions. Paleolimnological data can be used to partition the impacts of these various environmental changes.

5. Discharges of heavy metals (especially Hg and Pb), fossil fuel byproducts, and industrially produced toxins are all archived in lake sediments. Interpreting these archives requires an understanding of the transport and sediment focusing mechanisms of their accumulation. Comparative studies between pollutant profiles in cores from different lakes may allow us to distinguish between local versus regional/global sources of readily dispersed pollutants.

6. Determining the timing and probable causes of lake acidification has been one of the great "success stories" of paleolimnology. The political need to accurately document the history of lake acidification, primarily using fossil phytoplankton and geochemical indicators, has set a standard of quantitative reconstruction and quality assurance throughout paleolimnology. Paleolimnological data has shown that lake acidification can occur for reasons unrelated to fossil fuel consumption, such as vegetative cover change and soil compositional change related to climate. However, the rate of lake acidification since the industrial revolution began is unprecedented, especially in remote, poorly buffered lakes. This indicates that atmospheric precipitation of sulfates and nitrates, derived from fossil fuel combustion in power plants, is the most important modern source of excess acidity in lakes.

13

Paleolimnology at the Regional to Global Scale: Records of Climate Change

Reconstructing climatic change is perhaps the single most common application of paleolimnology. Paleoclimatology is a vast subject, and several entire books have been written on this subject alone (e.g., Crowley and North, 1991; Parrish, 1998; Bradley, 1999). Here we can only touch on some of the more important, interesting, and controversial aspects of climate history that are potentially recorded in lake sediments. As with human impact histories, archives of paleoclimate from individual lakes record responses from both local and regional events (e.g., Giraudi, 1998); teasing the two apart from a single basin often poses a difficult problem. In order to differentiate regional from global-scale changes in climate from lake deposits, it is also necessary that local influences on hydrology, such as drainage diversions, or changes in groundwater flow fields unrelated to climate, be understood. The problem of identifying regionally significant events becomes even more acute when the goals are to assess the *rate* at which climate changed from lake records or to assess the synchroneity of events between locations. All of these issues accentuate the importance of excellent geochronometry for paleoclimatic interpretation. Also, biological or physical mixing of sediments in any individual core record may mislead us into thinking a change was gradual when in fact it was rapid, whereas unrecognized small-scale unconformities in a single core could mislead us in the opposite direction (Dominik et al., 1992). Conversely, some lakes act to amplify climatic signals, particularly when they cross a threshold of limnological response to some climate variable (for example the transition from closed to open-lake conditions that might accompany an increasing precipitation:evaporation ratio). In this case a "gradual" climatic process might appear rapid from its depositional record.

As with human impact studies, a common solution to these problems is to use a comparative-lake and/or comparative-indicator approach, identifying coherent patterns of change in indicators of precipitation, temperature, windiness, or other climate variables of interest throughout a region. This can be done using many of the types of biotic, geochemical, geophysical, or geomorphic indicators we have discussed in chapters 7–11. In Late Quaternary deposits, one of the most widely used comparative methods involves looking at regional patterns of lake-level fluctuations, especially in closed basins, as they are recorded in terraces and sedimentary facies changes. In some areas, however, multiple lake records may simply be unavailable, or may not cover the same intervals of time. In such cases paleoclimate indicators may still be highly informative, but also need to be interpreted more cautiously.

Regional and intercontinental comparisons have great power to help us understand the chronology of climate change. However, they must be interpreted in the context of an understanding of global climate processes, and with consideration of the quality of our geochronology. A discordance of climatic indications between regions, for example evidence for wet conditions in one lake basin and dry lakes 1000 km away, may be perfectly understandable when viewed in the context of a realistic model for how the regional climate system works in that area. Similarly, as the temporal spacing of events we wish to compare gets shorter, we need ever better geochronological tools to make a strong case that two depositional "events" from widely spaced localities are truly synchronous.

Robust interpretation of paleoclimate signals from lake deposits requires that the lake basin under study be situated in a setting that is conducive to capturing climatic signals. Choosing a lake system that will amplify a climate signal in its paleorecords, versus one that will dampen the signal, may make the difference between a successful and an unsuccessful paleoclimatic study. Lakes located on climatic boundaries, or in regions of known, large magnitude climate shifts are good candidates for this requirement. As we have seen, lake and watershed size are also strongly correlated with the geographical scale of climate records; a smaller lake has a higher probability of providing a more precise record of environmental change, but is also highly sensitive to local events and the particularities of individual drainage basins. As a result, both large and small lake deposits can provide useful paleoclimate records, but their archives may be better suited for different types of questions.

There are numerous instances where, at first glance, indicator records from different sources appear to be providing contradictory indications of climatic conditions. Although this situation might cause a paleolimnologist to despair, a careful reading of such "contradictions" may actually provide a more detailed picture of climate change than any one signal alone. For example, in lake beds from the Middle Atlas Mountains (Morocco), pollen data suggests a Mid-Holocene increase in effective moisture and decreasing summer temperatures (Lamb and van der Kaars, 1995). However, this same time interval is marked by a series of pronounced lake-level falls, of 200–400 years duration each. Although seemingly contradictory, these two signals are probably recording changes in the seasonality of effective moisture and its effects, at different time scales, on providing moisture to groundwater aquifers feeding lakes, as opposed to watering plants. Systematic differences in climate signals can occur for many reasons, some reflecting subtle complexities in the effects of climate on different parts of the ecosystem, and some reflecting the inherent lag times required for certain signals to be expressed in paleolimnological records (e.g., Mayle et al., 1999).

In this chapter we will first examine aspects of modern climate dynamics of importance to paleoclimatic interpretation of lake deposits, and the sources of climate variability. Then we will consider the use of time series analysis for detecting periodicities in past climate change that may relate to specific forcing mechanisms, and the role of paleoclimate modeling in driving hypothesis-based inquiry in paleolimnological studies of past climates. Finally, we will examine lake sediment records of past climate variability from several regions of the world, moving from short to progressively longer time scales.

General Features of the Earth's Present-Day Climate System

The earth's climate is driven by the general circulation system of the atmosphere and its interactions with heat sources and sinks on the earth's surface, the oceanic water masses, land masses, and polar ice caps. This circulation is driven by both vertical and horizontal (especially latitudinal) imbalances in the distribution of heat over the surface of the earth (S.E. Nicholson and Flohn, 1980; Martens and Rotmans, 1999). Heat is transferred vertically by convection in the atmosphere up to an altitude of about 12 km, and laterally through circulation of the atmosphere and the oceans, the latter being particularly important in the subtropics (figure 13.1). A set of continuously moving wind systems and both high and low pressure cells blanket the earth at relatively low elevations, driven by the combination of unequal heat distribution on the earth's surface and the Coriolis effects of the earth's rotation. These systems distribute heat and moisture both *zonally* (within defined latitudinal belts) and *meridionally* (between latitudes). The much colder con-

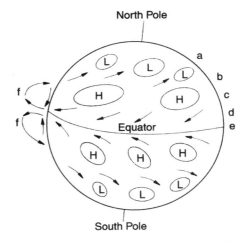

Figure 13.1. Major features of the general atmospheric circulation system: winds (arrows), high-pressure cells (H), low-pressure cells (L), subpolar lows (a), westerlies and mid-latitude cyclones (b), subtropical highs (c), surface easterlies (trade winds) (d), intertropical convergence zone (ITCZ) (e), vertical Hadley cells with rising motion near equator and subsidence in the subtropics (f). From Nicholson and Flohn (1980).

ditions of the Antarctic versus the Arctic polar regions, as well as the much greater proportion of water south of the equator, strengthens circulation systems in the Southern Hemisphere, and causes air masses to flow more zonally than in the Northern Hemisphere.

Near the equator a warm, low-pressure belt of rising air, variable winds, and abundant rainfall, known as the *Intertropical Convergence Zone* (ITCZ) occurs. Although the diagram shows the ITCZ as centered on the geographical equator, it actually moves seasonally. Because of the unequal distribution of land masses and heat in the Southern versus Northern Hemispheres, this ITCZ migration is not symmetrical about the equator, but rather migrates from about 20–30°N in the Northern Hemisphere summer, to near the equator during the Southern Hemisphere summer. This motion drives the migration of pressure cells, winds, and warm-season, continental precipitation associated with large-scale *Tropical Monsoon* systems, wind systems that move air masses meridionally. This seasonal variability is most pronounced where summer/winter thermal contrasts are most strongly developed, in Asia and Africa, but is clearly evident in the subtropics of other continents as well.

Rising equatorial air masses are replaced by air from higher latitudes. Coriolis effects deflect this air toward the west as it moves equatorward, forming a zone of easterly equatorial surface winds in both hemispheres (the northeast and southeast *trade winds*). This generates a screw-like motion toward the west. Ascending air generates precipitation, which is normally concentrated in 100–300-km scale disturbances in the tropics, or in much larger disturbances (500–3000 km wide) at mid and high latitudes. Subsiding air on the descending limb of these helical cells creates subtropical high-pressure belts, which are accompanied by relatively stationary, rotating wind systems (*anticyclones*). As with the ITCZ, these subtropical highs migrate seasonally, in the polar direction of the prevailing summer hemisphere, although their displacement is typically smaller than that of the ITCZ. The descending air masses of the subtropical highs are generally depleted of moisture, and the position of these highs therefore marks zones of seasonal to permanent aridity, for example the desert regions of North Africa and Arabia.

The northern sides of the subtropical highs are flanked by zones of mid-latitude surface westerlies, which separate warm subtropical air from cold polar air. This *polar front* is relatively unstable, however, and is accompanied by transient and migratory cyclones and anticyclones. These cyclonic

systems generate winter moisture in the Northern Hemisphere, for example over North Africa, when they are cooler and displaced equatorward. The westerlies broaden in the upper troposphere (9–14 km), blanketing both the subtropical high-pressure belt and the surface easterlies, and interacting with the latter.

Poleward of the surface westerlies lies a subpolar low-pressure belt, which is also marked by migratory cyclones and generally abundant precipitation. The precise position of this belt is unstable, varying both seasonally and on longer time spans, as a result of both equator to pole temperature gradients and the position of circumpolar wind circulation. Precipitation and wind patterns in the subpolar and polar regions are complicated by the combined effects of interacting atmospheric circulation systems, permanent glacial icecaps, and seasonal winter sea ice. Precipitation in the polar regions is generally low, and decreases poleward. Moisture is concentrated in the regions of convergence between circulation systems and migrates both seasonally and on longer time intervals as these circulation systems strengthen or weaken. Ice reflects solar radiation from the earth's surface and, in the case of sea ice, limits potential heat exchange between the oceans and atmosphere. In the North Atlantic, sea ice affects the positions of major Arctic circulation systems, as well as the migration paths of major circulatory systems. Melting sea ice is also responsible for the persistent cloudiness of the polar oceans.

Heat is also redistributed on the earth's surface by oceanic circulation. To a first approximation, surface winds control the direction of surface ocean currents. However, there are notable exceptions to this generalization, especially at the equator. Horizontal circulation of the oceans is also constrained to circular *gyres* by the shape of the ocean basins and by Coriolis effects. Oceanic redistribution of heat also occurs through vertical motion, driven primarily by slight differences in density that result from variations in water temperature and salinity. This so-called *thermohaline circulation* affects buoyancy of the deep ocean waters on a global scale, driving a series of oceanic *conveyor belts* that circulate throughout the ocean basins and in the process, global climate (e.g., Boyle, 2000; Bjornsson and Mysak, 2001; Fieg and Gerdes, 2001). Cold and relatively dense North Atlantic Deep Water (NADW) normally forms east of Central Greenland, and then flows down and southward, driving a conveyor belt of oceanic circulation that extends far to the south, and is replaced by returning surface warm waters from south of the equator. This thermohaline circulation pumps the global redistribution of heat, warming coastal northern Europe in the process. Conversely, any episodic factor that interferes with this NADW thermohaline pump will also slow down this heat exchange, with profound and sometimes surprising results. Disruptive factors of this type include the position or extent of North Atlantic sea ice, the discharge of massive volumes of freshwater into the North Atlantic, or the precipitation/evaporation ratio in the North Atlantic. Freshening of surface seawater, or more extensive sea ice formation in the North Atlantic stabilizes the water column, thereby reducing cross-equatorial heat transfer of heat. Not surprisingly, this results in colder temperatures in Northern Europe, but it also appears to cause warmer than normal sea surface temperatures at low latitudes, and drought in the North African and North American tropics (e.g., Street-Perrott and Perrott, 1990). Other distant influences, such as Pacific Ocean sea surface temperatures, may also be affecting the African climate (Hunt, 2000). These types of long-distance interrelationships, or *teleconnections*, between climate forcing events and their impacts on climate are probably important factors in explaining the apparent synchroneity of many climate events that occur on a global scale (e.g., Zhou et al., 1999, 2001).

Regional features, such as mountain belts and proximity to oceans or moisture sources also profoundly affect climate. Rainshadowing effects of coastal mountain belts are responsible for major intrazonal differences in precipitation, even over relatively small distances. Relatively minor differences in moisture transport across such *orographic* barriers can push interior regions over or under significant thresholds of precipitation, causing lake levels or salinity to change rapidly (e.g., Bindlish and Barros, 2000; Pienitz et al., 2000). Proximity to ocean basins also reduces the temperature contrast between winter and summer (maritime climate) and reduces summertime evaporation.

Sources of Climate Variability Over Time

The climate systems described above change on varying time scales because of a large number of forcing factors (Overpeck et al., 2002 in press). At annual to centennial time scales, internal variations in the climate system, driven by factors like solar variability and volcanic eruptions, can cause changes in the migration paths of major atmospheric circulation cells, oceanic circulation, sea surface temperatures, the extent of sea ice, and changes in anthropogenic inputs of atmospheric greenhouse gases. One of the most interesting and important of these short-term climatic phenomena is the El Niño-Southern Oscillation (ENSO). ENSO events affect much of the world's climate system, involving major changes in oceanic and atmospheric circulation linked to the position of the subtropical highs residing over the oceans. Normally, the eastern sides of these highs, especially along west coastal South America, are marked by a combination of upwelling, cold seawater and descending air masses, producing pronounced aridity along the western margins of the continents at these latitudes. Conversely, the convergence of warm surface waters and unstable, rising air masses on the western sides of the subtropical highs, especially the western Pacific, produces humid climates along the eastern margins of continents in the subtropics. This pattern of rising air masses and humidity on the western margin of low-latitude ocean basins and descending air masses and aridity on the eastern margins is referred to as *Walker Circulation*. For reasons still not entirely understood, Walker Circulation periodically breaks down, resulting in reversals in the pressure relationships across the ocean basins. This is accompanied by a weakening of oceanic upwelling along the eastern sides of the subtropical ocean basins and associated warming sea surface temperatures, all signals of El Niño events. Although ENSO is primarily a tropical and subtropical phenomenon of the Pacific Ocean, its effects reverberate into the Atlantic and Indian Oceans, and also impact the mid-latitude climate system. Depending on location, El Niño years are times of anomalous rainfall or drought, storminess, and changes in marine productivity, all cause for serious societal and economic concern throughout the affected regions of the world. Over the past 50-year period of instrumental records, ENSO events have been quasiperiodic, occurring approximately every 3–8 years. To obtain longer archives of ENSO variability, paleorecords, including those from lakes, are required.

Historical records from Europe and Eastern North America provide evidence for the existence of decadal to century-scale temperature fluctuations during the Late Holocene. These fluctuations have been linked with a variety of possible causes, including variations in solar output, volcanic eruptions, trace gas concentrations in the atmosphere, and instabilities in the coupling of the oceans and atmosphere, and many leave archived paleolimnological records (Rind and Overpeck, 1993; Kirby et al., 2001). Records of solar activity variability are particularly amenable to time series studies by paleolimnologists, because they appear to be cyclical. The best supported of these from a statistical standpoint are the 11-year sunspot (Schwabe) cycle, the 22-year solar magnetic (Hale) cycle, a 44-year multiple of the Hale cycle, and an 80–90-year (Gleisburg) cycle also related to sunspot activity, and a ~ 200–year (Suess) cycle. These cycles appear to be correlated with both temperature and precipitation changes on earth, although the reasons for these correlations are only partially understood (Lean et al., 1995; Lean, 1996). For example, intervals of higher sunspot numbers are associated with generally warmer climatic conditions. Over longer time intervals of the Late Holocene, sunspots have gone through prolonged, century-scale episodes of increased or decreased abundance, that seem to correspond with generally warmer and cooler periods, although these fluctuations do not appear to be of uniform length. Changes in the production rate of cosmogenic radionuclides in the earth's upper atmosphere are linked to solar activity and records of change in these fluxes, especially for ^{14}C and ^{10}Be, can be compared to lacustrine paleoclimate indicator records to determine if a temporal relationship exists between solar activity and short-term climate change (e.g., Björck et al., 2001).

Millennial-scale climatic fluctuations during glacial periods are evidenced in the ice core and marine ice-rafted sediment records, referred to as Dansgaard-Oeschger (DO) cycles (Broecker and Denton, 1989). These events reflect warm (interstadial) events superimposed on a cooler background. Based on ice core records, the onset and termination of these DO events is remarkably rapid, perhaps lasting only years to decades (Alley et al., 1993; Hammer et al., 1997). Similar duration events have been recorded from numerous marine, lacustrine, and ice cores, and speleothems. If such events are truly global in nature they may also have resulted from rapid changes in the global conveyor belt of thermohaline circulations, with warming and cooling in the North Atlantic or eastern equatorial Pacific being the most likely triggers (Van Kreveld et al., 2000). However, whether individual events of this duration can be correlated globally prior to the terminal Pleistocene is controversial. Some very high-resolution lake records in regions that would be expected to be sensitive to major North Atlantic cooling and warming events do not show any signal associated with these DO cycles (Guiot et al., 1992).

Over longer time scales (10^3 to 10^6 years), changes in the extent of continental ice sheets, variation in the seasonal cycle of insolation, and long-term storage and release of carbon play major roles in determining atmospheric circulation, global climate change, and lake responses. For example, 7–10-ka duration oscillations in climate, known as Bond cycles, were superimposed on the glacial buildup phase of the most recent glacial cycle, and similar events may have occurred during earlier glacial cycles (Bond et al., 1993; Broecker, 1994). The Bond cycle involved episodes of climatic cooling, culminating in enormous pulses of iceberg discharge into the North Atlantic from both the Laurentide and Fennoscandian Ice Sheets. Times of debris discharge from these icebergs, known as Heinrich Events, were followed by renewed warming of the North Atlantic sea surface (Heinrich, 1988; Broecker et al., 1992). One possible explanation for the rapid onset of interglacial warming following the maximal ice sheet advances of glacial stages is that the Heinrich-type iceberg discharges at such times become so enormous as to destabilize and deplete their sources of continental shelf grounded ice. This would have terminated the phase of freshwater discharge into the North Atlantic and reconstituted the thermohaline circulation pump typical of interglacial conditions (Denton et al., 1999). Similarly, the duration of the glacial stages themselves may have been a function not only of direct orbital forcing (discussed below), but also the length of time required to build up the ice sheets to the point at which they could undergo these large collapses.

Over long time scales the total incoming solar radiation received by the earth varies as a result of cyclical variations in the distance from the earth to the sun and the orientation of the earth with respect to the sun, the Milankovitch cycles (Milankovitch, 1920, 1930; Hays et al., 1976; Berger, 1978; Imbrie and Imbrie, 1980) (figure 13.2). The orbit of the earth is elliptical and the degree to which this ellipse varies from a circle (eccentricity) varies on regular cycles of about 400,000 and 100,000 years. When the earth's orbit is more eccentric, the total amount of short-wavelength radiation received by the earth increases during the period of the year when it is closer to the sun. This increases seasonal variation in heating in one hemisphere and decreases variation in the other. The earth–sun distance at any given time of the year is also controlled by the orientation of the earth's axis of rotation, which varies cyclically in space every 19,000 or 23,000 years, referred to as precession. The precessional cycle determines the

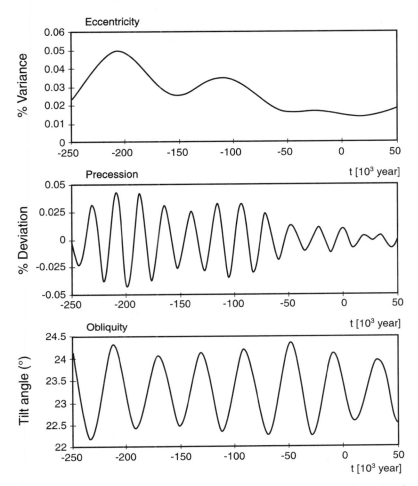

Figure 13.2. Long-term variations of the three Milankovitch orbital parameters from 250 ka BP to 50 ka in the future. (Top) The eccentricity cycle, given as percent variance from a circular orbit. (Middle) The precession cycle, given as percent deviation from the mean distance earth–sun on June 21. (Bottom) The obliquity cycle, given as changes in the tilt angle of the earth. From Jansen (1999).

season when the distance between the sun and earth is at minimum or maximum, thereby increasing or decreasing the difference between winter and summer temperatures in each hemisphere. Changes in the tilt angle of the earth's axis of rotation relative to the ecliptic plane (*obliquity*) also vary in a regular fashion, between 22 and 24.5° every 41,000 years. This obliquity cycle causes more extreme seasonality during periods of greater tilt, increasing the duration of summer daylight at the pole oriented toward the sun during maximum tilt.

All of the Milankovitch cycles contribute some component of change to the varying balance of solar insolation over the earth's surface, especially at high latitudes. However, a simple theory derived from the relative contributions of each of the above

cycles to variability in insolation does not always accord with the actual record of long-term climate variation. The largest variable in determining summer radiation at the poles is obliquity. Therefore we might expect that the 41,000-year cycle would dominate the pattern of long-term climate change at high latitudes. In fact, for the Pleistocene at least this does not appear to be the case. Deep-sea isotopic records, ice cores, and increasingly lake cores all support the notion that the dominant beat of Quaternary glacial/interglacial climate variation is about 100,000 years (e.g., figures 1.12, 9.1, and 9.7), despite the fact that the 100,000-year eccentricity cycle is both small in amplitude, and out of phase to explain the actual pattern of paleoclimate variation observed. Furthermore, there is considerable variation from the 100,000-year average in these cycles, ranging from 84–120 ka (Raymo, 1997). Several ideas have been proposed to explain this apparent discrepancy. Some researchers have argued that orbital eccentricity is ultimately responsible for the 100-ka cycles, but that internal feedbacks between the oceans, atmosphere and ice sheets cause time lags in how and when these cycles are expressed (Imbrie et al., 1993). Others have noted that a 100-ka cycle also exists in the earth's orbital inclination, caused by the earth's gravitational interaction with other planets (Muller and MacDonald, 1997a,b). A third intriguing possibility is that the glacial/interglacial cycles are tied to the combined pattern of frequency and amplitude pulses in the pattern of obliquity change, which also varies at a 100-ka beat (Liu, 1999).

Another difficulty in understanding the causes behind the 100-ka glacial cycles arises in their temporal asymmetry. Although marine, ice core, and lake records all support the notion that the termination of full glacial conditions occurred abruptly, involving the collapse of the ice sheets over time periods of a few thousand years, the entrance into full glacial conditions appears to be much more gradual, and the intensification of glaciation far from a linear trend. This asymmetry has characterized the transitions between glacial and interglacial conditions for at least the past 650 ka and perhaps for the last 960 ka (Imbrie et al., 1993).

At the longest ($> 10^6$ −year) time scales, tectonic processes, such as the formation of major mountain belts and the movement, collision, or separation of continents become important variables, possibly coupled with some very long period celestial cycles driven by planetary interactions (Laskar et al., 1993; Olsen, 2001). Regardless of time scale, these changes may be unidirectional or random over time, resulting from nonperiodic forcing events, or they may be cyclical in nature, the result of some underlying periodic forcing mechanism. And the interaction of periodic and nonperiodic events may induce even more complex and often abrupt changes in the earth's climate system.

Periodic Climate Events and Spectral Analysis of Paleolimnological Data

Some of the most powerful tools available for detecting cyclical or periodic behavior in a time series of data fall into a class of statistical techniques known as *spectral analysis*. Spectral analyses make use of the fact that any time series of data, such as varve thicknesses in lake beds, ice core band thicknesses, and so on, can be expressed as an approximation of a sum of cosine and sine functions, called a *Fourier series*. Various types of *Fourier transforms* can be used to identify whether a range of frequencies are either in or out of phase with the variability in the time series data set. If the data are in phase with a specific frequency, the Fourier transform results in a large Fourier amplitude, expressed as the amplitude squared, or *spectral power*, since the amplitude can be either positive or negative. A detailed discussion of the various methods of spectral analysis and time series data transformation is beyond the scope of this book; readers are referred to Muller and MacDonald (2000) for a good introduction to the application of these various techniques to paleoclimatology, and Paillard et al. (1996) and Mann and Lees (1996) for some specific methods.

Spectral analyses have been used to study many types of climate-sensitive paleorecords such as marine and lake sediments, ice cores, and corals. A spectral analysis can be used to examine any type of quantitative variation in facies, geochemistry, or biotic indicators in lake sediments. The pattern of interest is ideally investigated directly against an absolute frame, producing information on the most common recurrence intervals or frequencies of the variable of interest, as well as the coherence or phasing in time between multiple time series. For example, in a varved sequence of lake beds, the thickness of annual deposits could be analyzed to determine the frequencies that dominate the formation of thicker beds. Such a record might then be used to provide evidence for or against the importance of a cyclical forcing phenomenon, such as extreme precipitation during El Niño years, which could cause more runoff and sediment transport to the lake to occur. In reality, most lake records do not have a continuous time record of events against which sediment thickness or other variables can be compared. In these cases the time series must be compared against an age model derived from some assumptions about the relationship of stratigraphical thickness and age, for example, linear or interpolated sediment accumulation rates. The value of any spectral analysis is strongly controlled by both the quality of the age model behind the analysis, and both the number of data points and their distribution over time, upon which the analysis is based. Time series analyses based on fewer than 50–100 data points generally yield unsatisfactory results, and power spectra cannot detect frequencies below the frequency of sampling.

Most types of spectral analysis in common use by paleolimnologists calculate a power spectrum of the variable of interest across a range of frequencies (either events/unit time or events/unit stratigraphical thickness) over the entire time (thickness) interval of interest (figure 13.3). *Evolutionary* (evolutive) spectral methods chart the changes in spectral power over the entire time (or thickness) series of the data set. This is accomplished by calculating a series of individual spectra over moving time windows (e.g., C. Cohen, 1992; Percival and Walden, 1993). The display of an evolutionary spectral analysis permits the investigator to determine changes in the dominant event frequencies over the interval of interest (figure 13.4). The analysis of infrequent events in evolutionary spectral analyses is limited by the duration of the time window used to calculate the individual spectra. The duration of the moving time window of the analysis therefore is a tradeoff between the potential to resolve less frequent events, as permitted with longer windows, versus the potential to detect changes in the dominant frequency of events, permitted with shorter windows.

Climate Models

Models explaining climate change have existed since the mid-nineteenth century, almost as long as our recognition that the earth's climate system has changed over time. For most of that time, however, these hypotheses were generally framed in such loose ways as to defy testing. *Conceptual models*, based on consideration of general principles of climate dynamics, provided the paleoclimate community with some ability to test hypotheses about ancient climates, albeit in a qualitative fashion. Relatively simple numerical models have also been available for many years, all of which rely on a great deal of parameterization of the internal dynamics of the earth's climate to a small number of variables (Shukla et al., 1999). *Zero-dimensional energy balance models* can be used to predict changes in the earth's climate at a single location or as a global average, given some change in solar radiation at the top of the atmosphere. More complex, *one or two-dimensional radiative–convective models* take into consideration the vertical structure of the atmosphere, especially the motion of air and

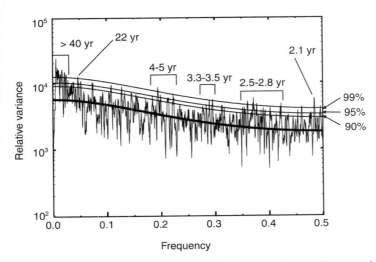

Figure 13.3. Spectral analysis (multitaper method) on thicknesses of annual varves from glacial Lake Hitchcock, New England, covering the interval 17,500 to 13,500 cal yr BP. Relative variance (strength) of the spectral signal is shown as a function of the frequency (year^{-1}). The heavy line represents the median red noise (a null model of climate spectral variability) associated with the spectrum, and the 90, 95, and 99% confidence intervals are indicated for the various parts of the spectrum. Statistically significant variability occurs at several time bands, shown with arrows or brackets. Notable among these are the 22-year scale (comparable to the Hale solar cycle), and several intervals in the ENSO band (2.5–7 years). Modified from Rittenour et al. (2000).

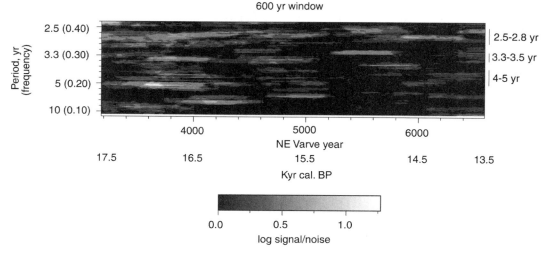

New England Varve chronology
ENSO-Band (2.5 - 10 Yr) evolutive spectrum

Figure 13.4. Evolutionary spectral analysis of the Lake Hitchcock varve record, covering the same interval as figure 13.3. Contours indicate level of statistical significance above the 50% confidence interval. Whereas the 2.5–2.8-year component is relatively persistent over the entire 4000-year varve record, the other, lower-frequency components (3.3–3.5 and 4–5 years) of variation weaken over the Late Pleistocene, consistent with some model predictions of fewer long-duration ENSO events as the Laurentide Ice waned.

the redistribution of heat to and from the earth's surface.

More sophisticated *Atmospheric General Circulation Models* (GCMs) consider climate interactions in three dimensions. These types of models explicitly consider spatial interactions of the atmosphere at scales larger than some resolved level, the horizontal and vertical grid size of the model. Interactions at scales below that of the grid cell are parameterized into a set of mathematical rules, derived from a combination of theory and climatological observation. At the larger, resolved scale, atmospheric interactions, and ultimately the distribution of heat and winds, are governed by laws of momentum, fluid motion, and gravitation. These motions are described by a series of partial differential equations, whose derivatives can be approximated as discrete algebraic equations and then solved by a computer.

Resolved grid cells are affected by variable inputs of external heat from the sun, and both affect and are affected by adjacent cells through heat, air, and moisture transfer. The output of a GCM model is often a gridded map that specifies some set of conditions (for example, mean annual temperatures, or changes in regional precipitation patterns)

over the model area (figure 13.5). Using GCMs, climate modelers are able to describe the behavior of factors such as atmospheric circulation, temperature structure, precipitation and evaporation, as they are influenced by variables such as changing solar radiation, sea surface temperatures, atmospheric composition, especially the proportion of greenhouse gases such as CO_2 or water vapor, ice volume or reflectivity of the earth's surface (albedo). This is done over a series of time steps, starting with some specified initial conditions.

The quality of a particular GCM can be tested by setting its inputs (e.g., atmospheric trace gas concentrations, solar input, sea surface temperatures, etc.) to simulate the current climate and comparing the simulation with actual climate conditions. Follow-up experiments, known as sensitivity tests, typically consist of varying one or a few variables of interest, such as changing solar insolation or atmospheric CO_2 levels, to observe their consequences. Modelers then proceed on to more complex simulations altering numerous variables. However, solving the large number of equations involved in the model simultaneously over a large grid is computationally very complex. The development of realistic GCMs therefore had to

A.

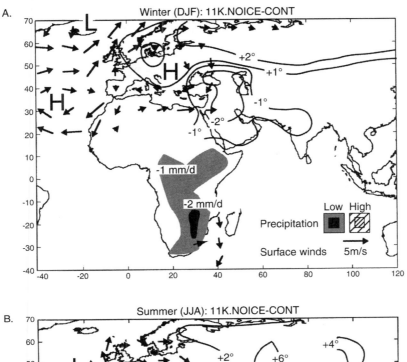

B.

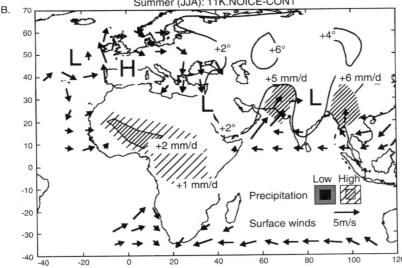

Figure 13.5. GCM simulation outputs for Eurasian and African paleoclimate at 11 ka, illustrating surface wind patterns, temperature, and precipitation differences. One model was run with 11 ka BP insolation conditions, modern sea surface temperatures, and modern ice conditions (11K.NOICE), and then compared with a model control (CONT.), simulating exclusively modern conditions. The difference (11K.NOICE – CONT.) reflects temperature, wind, and precipitation anomalies predicted to result from 11 ka–modern insolation differences alone, for both Northern Hemisphere winter (A) and summer (B). From deMenocal and Rind (1996).

await the advent of supercomputers, and even today there are severe computational limits on what can be put into a model. Limitations on the "realism" of the model simulations fundamentally rest on two major constraints. First, do the parame-terized (subresolution) processes affecting the model accurately describe actual physical processes of the transfer of heat and momentum? The trend in recent years in GCM development has been to con-sider the coupled interactions of the atmosphere, the oceans, and the earth's surface vegetation and hydrology in ever more sophisticated ways. Second, is the resolution of the model sufficiently fine scale, in a spatial sense, to fully describe the atmosphere–earth surface interactions? The highest resolution GCMs today have horizontal grid scales on the order of a few degrees of latitude and longitude. This puts the orographic effects of many mountain ranges and the moisture effects of most lakes at below-resolution scale, and limits our ability to realistically predict the local effects of climate on a lake and compare those simulations with paleorecords of temperature and precipitation. As a result, GCMs are often coupled with higher-resolution, regional-scale models, that parameterize the external influences on a region using GCM predictions, and concentrate computational power on local climate interactions.

GCMs have proven to be of tremendous importance to paleolimnology, principally because they serve as benchmarks of comparison for our interpretation of climatic events from lake archives (Kutzbach and Street-Perrott, 1985; Barron, 1990; Kutzbach et al., 1993; deMenocal and Rind, 1996; Overpeck et al., 1996; Wright, 1996; Felzer et al., 2000; Morrill et al., 2001). By definition, models are simplifications of reality that abstract the essential elements of an entity or process sufficient to understand its organization or operation. From the paleolimnologist's perspective, a useful GCM is one that makes a series of predictions about such things as changing P/E ratios, which might be reflected in paleolake archives. For example, evidence for rising lake levels in a closed basin would be consistent with the predictions of a GCM indicating increased P/E for that region at a given time under some set of boundary conditions, whereas a fall in lake level for the same region would be inconsistent with the same model outcome. The complexity of the earth's climate system forces modelers to greatly simplify the inputs and interactions used in modeling; even with the speed of modern supercomputers no GCM can capture all the processes and variables of significance to the earth's climate. Lake records are of importance for climate modeling therefore not only because they lend credence to, or falsify, a particular model's predictions about climate history for some region, but also by helping identify what aspect(s) of a GCM need improvement or modification. A synergy has therefore developed between paleoclimate modelers and the paleoclimate archivist scientific communities, with data-model comparisons being used to both better understand the significance of historical patterns and to refine or

reject particular models (e.g., COHMAP, 1988; MacDonald et al., 2000).

Lake deposits and climate model simulations can be studied and compared at varying time scales. For the remainder of this chapter we will examine what can be learned about the earth's climate system from this synergy, starting at the shortest time periods resolvable in lake sediments and working our way to progressively longer time scales.

Records of Annual to Decadal Time Scale Climate Events

Given the profound societal implications of ENSO events, it is natural that climatologists would want to know how persistent this periodicity has been in the earth's climate. Did the frequency of ENSO events change in association with the transition from glacial to interglacial climate regimes, and has global warming over the past century impacted the intensity or frequency of ENSO events? And, to what extent was the strength of ENSO variability affected by the very different global climatic conditions present during glacial intervals? Paleoprecipitation and paleoflood indications from highly resolved lake records can address these questions. Spectral analysis of appropriate paleolimnological records can be used to determine the occurrence or periodicity of ENSO and other quasiperiodic variations in atmospheric conditions that have historically affected precipitation indicators.

The most promising lakes for studying quasiperiodic climate cycles lie in areas that are sensitive to fluctuations in the relevant atmospheric phenomena. In the case of ENSO, this might mean a concerted effort to look for its signal in southeast Asia and northeast Australia, where a "warm pool" of oceanic waters is strongly influenced by ENSO-driven circulation, or in western South America, where the thermal and precipitation inversions associated with El Niño events are felt most strongly (e.g., McGlone et al., 1992; Rodbell et al., 1999).

In the high Andes, stronger-than-normal precipitation during El Niño years can trigger an increase in the frequency of debris flow events, which in turn generate turbidity flows in alpine lakes. These flows leave behind a record of episodic turbidite beds. Twentieth-century core records from Laguna Pallacocha (Ecuador) show a near-perfect match between severe El Niño years known from historical data and the occurrence of turbidites, suggesting that this lake can serve as a high-resolution barometer for El Niño activity in western South America

(Rodbell et al., 1999). Spectral analyses of the Late Pleistocene–Holocene record from Laguna Pallacocha show that the modern El Niño variability has only been in place since about 5000 yr BP. Prior to that time the periodicity of clastic turbidite deposition was ~15 years. Prior to ~ 5000 yr BP the east-central Pacific Ocean was probably cooler than today, leading to the infrequency of ENSO events recorded in Andean lakes (Cole, 2001). The increasing frequency of warm sea surface temperature anomalies in the western Pacific and the establishment of modern El Niño may have been driven by enhanced trade wind circulation and a steeper zonal sea surface temperature gradient after 5000 yr BP.

ENSO is known to influence climate well beyond the Pacific Ocean Basin, and even in other regions its signals are recorded in lake deposits. ENSO signals may also be present in lake deposits in regions where the ENSO effect today is quite weak, suggesting substantial changes in the global ENSO teleconnections over time. For example, during the Late Pleistocene, the Laurentide Ice Sheet appears to have localized the effects of ENSO-induced storminess in eastern North America to a much greater extent than is observed today, producing strong ENSO band periodicities in the runoff indicators of lakes in the region (Godsey et al., 1999; Rittenour et al., 2000).

Records of Decadal to Century-Scale Events

There is compelling evidence that global climate has warmed substantially in recent decades. Some of this trend can be attributed to natural climatic variability. However, it now seems firmly established that a large component of this warming, particularly since the 1920s, is attributable to increasing concentrations of greenhouse gases in the atmosphere, driven in large part by industrial activity and the burning of fossil fuels (Houghton et al., 2001).

Paleolimnological data has contributed to our understanding of this process in two ways. First, paleolimnological data can be used to directly estimate climate change during the past few centuries. The importance of this approach is greatest for areas of the world that have been highly responsive to the effects of twentieth century global climate change, and where good instrumental records are lacking. Polar and alpine regions best fit this prescription. Paleolake records in the Arctic and Antarctic have the greatest potential to track global warming because the absolute temperature increases in these regions have been large, and because they exceed thresholds for qualitatively different behavior of lakes (see figure 1.9) (Douglas et al., 1994; Overpeck et al., 1997; Hughen et al., 2000; Sorvari, 2002 in press). In some remote polar and alpine regions, significant increases in weathering rates resulting from warmer conditions may be controlling pH as much as, or even more than atmospheric acid deposition. Under these conditions diatom-inferred pH becomes highly correlated with air temperature, and may become a useful indicator of climate warming since the Industrial Revolution (Sommaruga-Wögrath et al., 1997). Second, paleolimnological data can be used to estimate the degree of natural variability in climate systems, and their rates of change prior to the twentieth century. Such information is critical for assessing the resilience of climate systems to global perturbations and has direct implications for interpreting our present-day climate change.

Prior to the Industrial Revolution deposition in many lacustrine sequences appears to have been cyclical in the solar frequency bands discussed earlier, especially in regions where sediment discharge is sensitive to slight changes in runoff, or productivity is sensitive to slight insolation changes (Glenn and Kelts, 1991; Verschuren et al., 2000; Hodell et al., 2001; Livingstone and Hajdas, 2001). For example, paleoshoreline and delta elevations, from Mono Lake, California show a striking series of lake-level fluctuations, cycling in phase with the 200-year Suess cycle of upper atmospheric production of ^{14}C (figure 13.6).

Two century-scale events of the Late Holocene that have sometimes been attributed to fluctuations in solar activity are the Medieval Warm Period (MWP, ~ 1000–1300 AD) and the Little Ice Age (LIA, ~ 1300–1850). In Europe, the LIA was marked by a period of alpine glacial advance and generally cold winter conditions; this is reflected in high-resolution lake and bog records by a variety of geochemical, magnetic, and paleoecological indicators, especially for the more restricted time interval of ~ 1700–1850 (e.g., Barber et al., 1999). Some data suggests that the MWP and LIA were global in scale, and involved relatively continuous intervals of comparatively warm or cold conditions (Lamb, 1982; Luckman, 1993). However, other studies suggest more complex patterns of temperature variation over the past 1000 years (Bradley and Jones, 1993; Hughes and Diaz, 1994).

What can we learn about these key Late Holocene climate interludes from paleolimnological records?

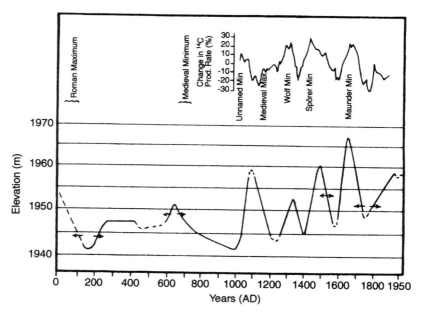

Figure 13.6. Comparison between the fluctuations of Mono Lake level and the known sunspot maxima and minima for the past ~ 2000 years. From Stine (1990).

A comparison of records from lakes in midwestern North America is instructive here. Accretionary $\delta^{18}O$ records from fish otoliths suggest that between 985–1200 AD the Lake Erie area of central North America experienced warmer summers but colder winters than during the twentieth century (Patterson, 1998). Core records from the Northern U.S. Great Plains show that lake levels fluctuated greatly and droughts were common prior to 1200 AD, although the MWP was not unique in this regard (Vance et al., 1992; Laird et al., 1998). During the fifteenth century, extreme intra- and interannual temperature variation occurred in the Lake Erie area, simultaneous with an interval of large seasonal and interannual variation in Europe. The LIA record at Moon Lake, North Dakota, while containing evidence for some extreme freshening events, was evidently not a period of continuous low salinity conditions. This record, when compared with other regional lake and tree ring records, supports the notion that the "Little Ice Age" was not a regionally synchronous or uniform event. At nearby Elk Lake, Minnesota, eolian dust influx, indicative of aridity, shows cyclical variation throughout the Late Holocene, with an entire wet–dry cycle encompassed within the LIA period (Dean, 1997).

Perhaps this asynchroneity in LIA moisture fluctuations should not surprise us (Fritz et al., 1994).

The Upper Midwest and Northern Great Plains regions of the United States are areas of complex interactions between warm, moist air from the Gulf of Mexico, warm dry air from the Pacific, and cold dry Arctic air masses. Minor readjustments in the positions where these air masses interact could have caused lakes in some areas to receive much greater amounts of precipitation, while nearby lakes were relatively dry.

Global Teleconnections of Abrupt Events in the Late Pleistocene and Holocene

Detailed investigations of ice cores provided dramatic evidence that the transition from the Last Glacial Maximum to the Holocene was accompanied by a series of profound and very rapid changes in climate (Dansgaard et al., 1989). Although evidence for profound climate switching had been slowly accumulating from Late Quaternary stratigraphical studies for many years (e.g., Coope and Brophy, 1972), evidence that such events were extremely rapid had been limited until recently by the absence of sufficient numbers of high-resolution Late Glacial records. This situation began to change with the collection in the 1980s and 1990s of continuously varved high-resolution lake records covering this time interval for many parts of the world. Chironomid records from North American lakes provide not only control on the timing of millennial-scale climate oscillations of the Late Quaternary, but also give us a quantitative picture of their magnitude (Cwynar and Levesque, 1995; Brooks and Birks, 2000) (figure 11.20).

Late Quaternary paleolimnological records in many parts of the North American and North African tropics show remarkably abrupt lake-level fluctuations, which appear to be synchronous, even between the two continents (Street-Perrott and Perrott, 1990; Street-Perrott, 1994). Abrupt lake-level declines, over time intervals of 10–100 years and persisting for hundreds of years, occurred between 10,800–10,000 yr BP and again at about 8000 yr BP. These events correspond with the periods of significant meltwater discharge from the receding Laurentide Ice Sheets. Some researchers hypothesize that teleconnections between these events were driven by the effect that ice meltwater pulses would have had in suppressing the formation of North Atlantic Deep Water, thereby reducing Atlantic thermohaline circulation, increasing the thermal gradient from the equator northward, and causing tropical drought.

Even longer-distance teleconnections are suggested by the synchroneity of abrupt climate events between the Southern Hemisphere tropics and mid-latitudes and the North Atlantic. Some pollen records from lakes and bogs in the Southern Andes and New Zealand indicate a remarkable coincidence in timing of millennial–multimillennial scale events, such as the individual Heinrich Events and the Younger Dryas (discussed below), with the chronology of glaciation and deglaciation at high latitudes in the Northern Hemisphere (Denton et al., 1999; Baker et al., 2001). Some evidence suggests that the linkage between climates over long distances might ultimately be driven by variability in thermohaline circulation. However, these southern lakes and bogs lie in regions far removed from the influences of thermohaline circulation, and an intermediate step would have to exist that could transmit climate signals globally. The driving forces behind these apparent interhemispherical teleconnections remain a source of controversy, especially given that solar insolation models based on Milankovitch cyclicity would suggest that precipitation in these southerly regions should be out of phase with their northern counterparts. Some paleolimnological records from the Southern Hemisphere tropics show just such an out-of-phase pattern of precipitation (Gasse and Van Campo, 1998; Seltzer et al., 2000) (figure 13.7). However, other lake records from the southern continents are inconsistent with the hemispherical predictions of solar forcing, providing either ambiguous records, or as in the cases discussed earlier, cycle with Northern Hemisphere warming trends. At present, no simple explanation exists to explain these seemingly contradictory conclusions about interhemispherical climate variation during the Pleistocene. Perhaps there are thresholds of sensitivity to the global "teleconnectors" that make some geographical settings (maritime vs. continental; presence or absence of orographic effects, latitudinal belts) more or less susceptible to these ice sheet-induced events. Ultimately the key to testing models of interhemispherical climate teleconnections or solar forcing must come from comparison of many more high-resolution lake records across the geographical, latitudinal, and continental spectrum than are presently available, to determine just how "synchronous" supposedly global climate events actually were at a variety of time scales (Markgraf, 2000). Many abrupt "events" may have explanations that lie outside of the predictable effects of global general circulation modeling. For example, some researchers have proposed other causes such as regional fire occurrence to explain the existence of high frequency changes in "climatic" indicators in the southern parts of South America at the end of the Pleistocene (Markgraf, 1993). Such events however would not be expected to occur simultaneously in different continents unless their history was underlain by some type of global forcing mechanism.

The Bølling-Allerød Warming Event and Younger Dryas Cooling Event

The end of the Pleistocene and the onset of deglaciation was marked by a series of abrupt, decadal to century-scale, warming and cooling events, and wet/dry events, which were first recognized in the North Atlantic region. Although the first strong evidence for the timing of these events came from marine sediments and ice cores, it is also important to ask to what extent such events were felt on the ice-free parts of continents. Over the past 20 years lake cores have provided increasingly voluminous records for near synchronous changes in climates between at least some of the continents. Between ~ 14.7–12.8 cal kyr BP a first strong pulse of glacial melting, the *Bølling-Allerød interstade*, occurred, marked by changes in lake status and indications of warming in many regions (e.g., Van Huissteden et al., 2001). Although *Bølling-Allerød* warming is most evident in the circum-North Atlantic region (figure 11.20), it is also evident from other regions of the globe (e.g., J.W.C. White et al., 1994).

Immediately following the Allerød interstadial, an interval of extreme cooling occurred through a series of steps, lasting from about 12.8–11.5 kcal yr BP, referred to as the Younger Dryas stadia (e.g., Alley, 2000; Hughen et al., 2000). This event affected large parts of Europe, North America, Africa, and possibly other parts of the world. It is marked in lake deposits by pronounced lake-level changes, paleoecological and isotopic indications of cooling, and changes in depositional characteristics. Some varved lake records indicate that the transitions out of the Younger Dryas stade into the warm Early Holocene may have occurred over intervals as short as a few years (fig-

ure 6.12). Numerous lake records in Europe and eastern North America indicate summer cooling of as much as 12°C during the Younger Dryas, and strong evidence exists that cooling was synchronous around the North Atlantic region (Von Grafenstein et al., 1994; Levesque et al., 1994, 1996, 1997; Mayle et al., 1999). At tropical and subtropical latitudes in North America and Africa, the Younger Dryas is marked by pulses of extreme drying in lake basins (Gasse et al., 1990; Street-Perrott and Perott, 1990; Metcalfe et al., 1997). However the termination of the Younger Dryas aridity does not appear to progress consistently from the equator northward, as would be predicted by a strict solar insolation model driving the intensification of the African monsoon.

Broecker et al. (1988, 1989) proposed that the Younger Dryas cooling resulted from an increased

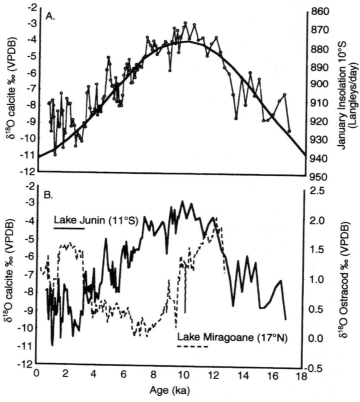

Figure 13.7. (A) $\delta^{18}O_{calcite}$ and January insolation for Lake Junin (Peru, 10°S) versus calendar age, showing strong relationship with modeled solar insolation. Note that scale for January insolation has been inverted. (B) $\delta^{18}O_{calcite}$ of Lake Junin and $\delta^{18}O_{ostracode\ calcite}$ for Lake Miragoane, Haiti, in the Northern Hemisphere tropics. These trends are interpreted to reflect asymmetry in moisture change in northern and southern tropics during the Late Glacial and Holocene. VDB is Vienna Peedee belemnite standard. From Seltzer et al. (2000).

discharge of meltwater from the dwindling Laurentide Ice Sheet into the North Atlantic. They argued that this freshwater pulse would have caused a shutdown in the formation of North Atlantic Deep Water (NADW), effectively refrigerating much of the Northern Hemisphere. Some evidence points to repeated catastrophic outbursts of Lake Agassiz as the source of this water (Duplessy et al., 1992; Clark et al., 2001). This hypothesis has proven controversial among some paleoclimatologists, in part stemming from uncertainty about the correlation in timing between meltwater pulses and regional cooling, and because of the need to call on complex teleconnections to explain Younger Dryas-aged cooling and drying events in regions of the world far removed from the North Atlantic.

Evidence from lake deposits can be drawn on to address both of these sources of uncertainty. Stable isotope records from fossil ostracodes from the Great Lakes during the ice melt period confirm the notion of generally increased discharge during the Younger Dryas period, and are consistent with the idea of freshwater discharge contributing to the suppression of the formation of NADW (Moore et al., 2000). However, these records show that the timing of maximal outflow does not precisely correlate with the timing of the Younger Dryas, as inferred from Greenland ice core records. This suggests that other factors besides the rate of melting of the Laurentide Ice Sheet must have also played contributing roles in the initiation and termination of the Younger Dryas episode. Core records from Lake Erie, while also supporting the notion of Younger Dryas-aged cooling, illustrate the importance that local events may have had in creating positive feedbacks on the regional climatic expression of global events (Lewis and Anderson, 1992). The paleolimnological record of the Great Lakes region for this time shows that the general climatic warming associated with the end of the glacial era was accompanied by increased meltwater discharge from proglacial Lake Agassiz to the northwest. This discharge was episodic, reflecting the discontinuous nature of the melting of the ice sheet, and was routed to different exit points on the North American continent as the ice sheet waned, with variable effects on NADW formation and global climate (Clark et al., 2001; Teller et al., 2001). While small flood events are evidenced in lake cores primarily from the Lake Superior Basin, the largest discharges left deposits even in remote "downstream" regions (Colman et al., 1994). In the Late Wisconsin period this water was increasingly diverted from flowing down the Mississippi River, and instead toward the Great Lakes and the St.

Lawrence or Hudson Rivers. This massive increase in the volume of very cold water in the Great Lakes would have greatly increased the surface area of Great Lakes region water bodies, in turn increasing both their capacity to generate regional cooling and the duration of winter ice cover. These local effects would have been superimposed on the more global effects of Younger Dryas climatic cooling.

Abrupt Events of the Holocene

One of the most striking facts to emerge from the study of Holocene paleolakes is the discovery that rapid climate shifts continued on the continents even after the terminal Pleistocene deglaciation (e.g., Street-Perrott and Perrott, 1990). This finding has been somewhat surprising, since the Holocene had long been pictured as a period of relatively gradual climate change in comparison with the Late Pleistocene. For example, both lake and ice core records support the occurrence of an abrupt and short-lived (several decades) interval of global climatic instability at 8.2 ka (Dansgaard et al., 1993; Alley et al., 1997; Von Grafenstein et al., 1998; Tinner and Lotter, 2001). High-resolution lake records are now showing that many Middle-Holocene climate changes were also quite abrupt, with strong fluctuations in temperatures on decadal to century time scales (MacDonald et al., 1993; Mullins, 1998; Willemse and Törnqvist, 1999). The Middle Holocene had long been thought to be an interval of globally warm temperatures, in accord with both low-resolution paleorecords and GCM simulations. However, high-resolution lake records have shown that, whereas Middle Holocene warming was widespread, it was probably not synchronous on a global scale, and proceeded at different rates in different areas.

A dense network of paleolimnological records from a region may help us sort out the apparent complexities of regional climate change, and may provide explanations for asynchronous changes in temperature or precipitation between adjacent regions. This will be particularly important for areas with complex orographic controls on climate, like the Andean region of South America, whose paleoclimate records are often "contradictory" even over short distances. Interpreting paleoclimate records from lakes, or any other source for that matter, is therefore more than a matter of "wiggle matching" abrupt events; it also requires an understanding of the underlying climate controls that might simultaneously make one site dry and another just a few hundred kilometers away wet.

Complexities and apparent "contradictions" in climate records are probably most pronounced in regions that lie near the intersections of air masses and moisture sources, or that have seasonally variable climate controls. For example, in the north-central United States, paleolake evidence for asynchronous episodes of aridity during the Holocene most likely record the complex interaction between (a) zonal, dry westerlies, (b) meridional, Arctic air masses, and (c) meridional, moist Gulf of Mexico air where they converge (Fritz et al., 1994; Haskell et al., 1996). The precise zones of convergence of these three air masses in the central United States and their degree of seasonal dominance in particular regions have apparently varied throughout the Holocene. This has resulted in relatively high-frequency switches between zonal and meridional circulation during the Holocene at any given locality (Haskell et al., 1996; Schwalb and Dean, 1998).

Global Records of Late Quaternary Climate Change from Lakes

Before the advent of high-precision dating techniques, many Quaternary geologists believed that high lake stands of apparent Late Pleistocene age could be correlated globally. This idea became untenable as our understanding of both the global climate system and geochronology improved. It is true that for small, glacially formed lake basins in alpine settings, the beginning of lacustrine sediment accumulation frequently marks the local retreat of glaciers to elevations higher than the lake. In such settings, a series of well-dated records from lakes at various elevations provides us with a chronology of glacial retreat. However, interpreting the climatic significance of lake expansion and contraction in the low-altitude tropics, in semiarid regions, in temperate lowlands, or even in proglacial environments adjacent to continental ice sheets is considerably more complicated. GCM simulations suggest that both the presence of high-latitude ice sheets, and the differences in solar insolation that existed during the Late Pleistocene would have produced circulation and precipitation conditions quite different from those of today, and that different parts of the globe would have reacted in different ways to these changes. It is therefore worthwhile considering how lakes in several regions of the world recorded the profound climate changes that accompanied the end of the Pleistocene.

My review here is intentionally noncomprehensive. Lake records from some parts of the world, most notably South America, have yielded strongly contrasting pictures of climate change, even from records in closely adjacent basins (Betancourt et al., 2001). Regions like the South American Altiplano, which are both topographically complex and areas where contrasting climate systems intersect, are probably the most difficult areas from which to unravel regionally coherent paleoclimate histories. Undoubtedly our understanding of the paleoclimate history of such regions will evolve rapidly in years to come, making a synthesis here premature. For the purpose of this book I have chosen to concentrate on Africa and Eurasia, where more coherent regional climate patterns have emerged from lake studies.

African and Southern Asian Lakes

The African climate today is controlled by a complex interplay of regional and global forcing mechanisms. At a continental scale, climate is controlled by the seasonal migration of the Intertropical Convergence Zone, and its interaction with westerly and southwesterly winds from the Atlantic (Nicholson, 1994, 1996). Atlantic moisture is drawn across central Africa by the heating of North Africa, providing precipitation to a wide equatorial swath of the continent (Hastenrath, 1985). This moist air is partially blocked from incursion into East Africa by orographic rainshadows. Weaker moisture systems enter East Africa through the seasonal movement of northeastern and southeastern monsoons. The northern and southern limits of abundant moisture are controlled by the positions of the subtropical highs, currently centered over the Northern African Sahara Desert and Southern African Kalahari Desert. Deviations from this general pattern can occur over both short and long time scales, as a result of factors such as ENSO events, changes in solar insolation driven by Milankovitch forcing, or changes in sea surface temperatures, that serve to strengthen, weaken, or move the position of moisture-laden air masses entering the continent.

The paleolakes of North and East Africa have proven to be a particularly rich testing ground for the evaluation of competing hypotheses about global climate controls during the Late Pleistocene and Early Holocene. A major question that has vexed paleoclimatologists has been the relative importance of solar forcing of African climates versus the long-term effects of the ice sheets, their meltwater discharges, and especially their role in regulating the global thermohaline conveyor belt and heat pump and sea surface temperatures (e.g., Kutzbach and Street-Perrott; deMenocal and Rind, 1996; Overpeck et al., 1996). About 9–10 ka BP the Northern Hemisphere summer solar radiation on the African and Asian land masses was at its highest point of the past 18,000 years, as a result of a combination of perihelion occurring during summer, relatively high earth axial tilt, and high orbital eccentricity. GCM simulations predict that such heating would intensify the strength of the summer monsoon, increasing precipitation in the Afro/Asian tropics and subtropics at that time. Most likely this effect of increased monsoonal precipitation would have been further enhanced by the reduction in albedo that would have accompanied increasing vegetation cover during this period in the arid parts of northern Africa (Street-Perrott et al., 1990).

To a first-order approximation the pattern of Late Quaternary lake-level fluctuations in internally drained lakes from North and North Equatorial Africa is consistent with the hypothesis that solar insolation is a primary forcing factor driving major changes in precipitation in this region (Street and Grove, 1979; Nicholson and Flohn, 1980; Street-Perrott and Roberts, 1983; Street-Perrott and Perrott, 1993; T.C. Johnson, 1996) (figure 13.8). From 18,000 yr BP to the end of the Pleistocene, most African lakes near or north of the equator were generally quite low or even completely dry (figure 9.17). Interestingly, this period of tropical and subtropical aridity corresponds with the time of maximum lake stands in the glacial/pluvial lakes of North America. The picture is somewhat complicated by contradictory evidence from lakes south of the equator, as we will see later in this chapter. Simple solar forcing models would suggest that these lakes should respond to precipitation levels that would be out of phase with those north of the equator, as appears to be the case in the Western Hemisphere.

The dominance of low lake stands and dry lakes in Northern and Tropical Africa began to reverse itself after about 12.5 ka BP, perhaps a bit earlier in the equatorial region, culminating in a period of nearly continent-wide high lake stands between 10–9 ka BP (figure 13.8). This rise in lake levels occurred earliest in more southerly regions and migrated northward until about the mid-Holocene, presumably in response to a northward migration of the ITCZ (Street-Perrott and Roberts, 1983). Lake levels declined again on a regional scale

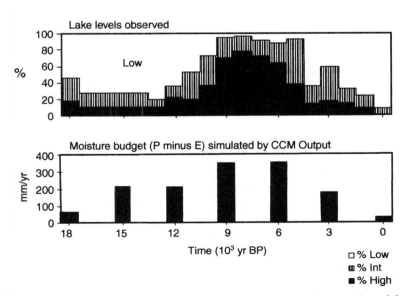

Figure 13.8. African paleoclimate records from lake-level histories in Africa and model results. (Top) Frequency of high lake levels in North and East Africa since 18 ka, as compared with model simulations of moisture budgets for the same time range for the latitude belt 8.9–26.6°N. (Bottom) Distribution of lakes in Africa showing high lake levels 9000 yr BP. Modified from Wright (1996), after Street-Perrott and Perrott (1993) and COHMAP (1988).

during the mid-Holocene, with broad indications of aridification starting about 5500 yr BP in the northern tropics (Gasse, 2001).

Disagreement exists about relative roles of direct solar insolation forcing, which would have gradually driven the intensification of the monsoon, versus other factors in changing precipitation patterns over Africa and Asia, especially the abrupt effects of changes in thermohaline circulation on the Afro/Asian monsoon (Kutzbach and Street-Perrott, 1985; Overpeck et al., 1996; Broecker et al., 1998). High-resolution chronologies from numerous lakes in Africa and Asia should provide a means of testing the relative importance various processes may have had in ending Late Pleistocene aridity in the region. A number of paleolake records from eastern China and Tibet support a strengthening of the Indian monsoon from about 12,500 yr BP until the Middle Holocene, simultaneous with its intensification in Africa. At this time however Central Asia appears to have become drier (e.g., Lister et al., 1991; Van Campo and Gasse, 1993). The southwest Asian story may have been even more complicated because of the added influence of westerly moisture sources. For example, highly resolved playa lake records from western India also show an increase in precipitation from the Early to Middle Holocene, in opposition to what might be expected from a solar insolation forcing model alone (Enzel et al., 1999). These data suggest that other factors such as sources of winter precipitation may have also played a major role in controlling the precipitation/evaporation balance within the monsoon belt during the Holocene.

Eurasian Lakes

Modern Eurasian climates are largely controlled by the interactions of the mid-latitude, seasonally migratory westerlies and the Asian monsoon (Harrison et al., 1996). Today, the westerlies penetrate as far as Eastern Siberia, the Northern European mountain ranges being inadequate to cause complete rainshadowing. In winter this brings precipitation across much of Southern Europe. In Siberia, strong pressure cells develop because of seasonally variable heating and cooling, resulting in the formation of monsoonal circulation patterns. Wintertime high pressure keeps much of that region quite dry. In summer, the lower equator–pole thermal gradient causes the westerlies to weaken and move northward, to the latitude of southern Scandinavia, bringing moisture to the Baltic region

and Eastern Europe. Simultaneously, high-pressure cells further south generate dry summer conditions across Southern Europe, while in central Asia and Siberia, monsoonal air flow from the Pacific draws moisture into the continent, causing abundant summer precipitation in all but the most interior portions of Asia. The combined effects of these summer and winter climate regimes provides only limited moisture to Southern Europe today, keeping lake levels in closed basins of the Mediterranean region relatively low, while providing a much more positive moisture balance to northwestern Europe, where lake stands are high.

Lake level and other paleolimnological records from throughout Eurasia suggest that very different climate regimes dominated Eurasia during most of the Late Quaternary, and provide a means of evaluating the quality of existing GCM simulations (Harrison et al., 1996; Tarasov and Harrison, 1998; Qin and Yu, 1998; Yu, 1998). During the Eurasian LGM (~ 18,000 yr BP) and continuing into the earliest Holocene, lakes around the southern periphery of the Fennoscandian Ice Sheet, Central Russia, South and East Asia were generally low whereas lake levels were quite high in the Mediterranean region, in West Asia and at least parts of western China (figure 13.9). This resulted from the development of a major anticyclonic and dry circulation system over the ice sheet, which forced the Atlantic westerly jetstream and associated moisture-laden westerlies to penetrate Eurasia further south than today. Interestingly, lake-level records from France suggest that similar north–south shifts in the incursion point of the westerly jet and cyclonic activity over Western Europe has occurred on shorter time scales through the Holocene (Magny, 1998).

Rising paleolake levels from Northern Europe between 9500–7500 yr BP, and simultaneous declines in lake levels in the circum-Mediterranean region, show that after the retreat and disappearance of the ice sheet, the westerlies began to migrate northward. Lakes in southern Scandinavia appear to have undergone rather significant fluctuations in level during the Early Holocene, although hydrological modeling suggests that at least some of the lake were "amplifier" systems, and these changes may have been the result of quite modest changes in precipitation (Vassiljev and Harrison, 1998). Simultaneously, East, Central, and South Asia were all becoming wetter than at present, as indicated by regionally higher lake levels, the result of much stronger Early Holocene insolation and monsoonal circulation than during either the Pleistocene or today.

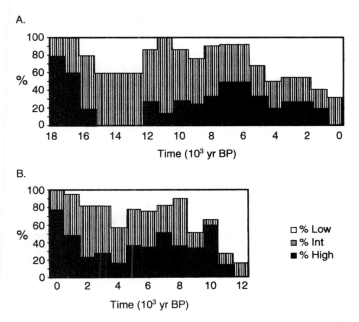

Figure 13.9. Temporal pattern of changes in European lake levels. (A) From 18,000 yr BP to present in the Mediterranean region. (B) From 12,000 yr BP to present in northern Europe. Modified from Harrison et al. (1996).

Longer-Term Quaternary Lake Records of Millennial-Scale Events

Evidence for millennial-scale climatic events recorded on a near-global scale seems fairly compelling for the Pleistocene/Holocene transition. But what can lake records tell us about earlier global climate variation on these time scales? Interpreting such oscillations from pre-Late Quaternary lake deposits has proven difficult for several reasons. First, our ability to discriminate events at the millennial scale, and to correlate them between different regions of the globe, diminishes greatly for deposits that are too old to be dated using ^{14}C methods, making proposed correlations between climate indicator curves much less secure. Even within the radiocarbon time-scale interval, the uncertainty of individual dates may be as long as the events we are trying to infer. Finally, the number of investigated lake records that date to more than 30 ka BP is simply much smaller than the number of younger records, limiting the opportunities for comparing records on a regional or global basis.

These problems notwithstanding, there are a few places where long and detailed paleolimnological records are now available that cover the last

glacial cycle or even multiple cycles, where we can begin to test hypothesized global teleconnections at a variety of time scales (Hooghiemstra and Cleef, 1995; Partridge et al., 1997; Reille et al., 1998; Benson, 1999; Watts et al., 2000; Rioual et al., 2001; Trauth et al., 2001). Although we are still at the early stages of being able to make precise correlations between events at all these lake basins, the best-dated sequences show two noteworthy patterns, which may link them to the high-latitude marine and ice core records. First, many of these lake records show patterns of very abrupt switches in climate regimes. Second, in each case where a high-resolution record exists, only some of the high-frequency switches can be correlated with specific marine-based paleoclimate markers, such as the DO events, and conversely, many DO or similar events are unrecorded in the lake sequences. This should make us pause; it is possible that many high-latitude events are simply too weak for their influence to be felt globally, but it is also possible, given our present state of knowledge, that some of these global "correlations" are spurious, driven by the likelihood that random variation in regional climate systems will occasionally result in global matches. This latter point argues for the need for more synchronous and long records from various parts of the world, so that we can begin to compare intercontinental patterns of lake level, lake chemistry, and pollen records, and tease out which correlations are meaningful and which are random.

An Example from the Great Basin, United States

GCM simulations suggest that the Basin and Range region of the United States would have been much wetter during the LGM than it is today (Hostetler and Benson, 1990). Simulations predict that permanent high-pressure systems associated with the Laurentide Ice Sheet would have forced a southerly displacement of the westerly storm tracks and the polar jetstream over the Basin and Range region, increasing precipitation/evaporation ratios and in the process raising lake levels. At the maximum ice extent (~ 18 ka BP) simulations suggest that the westerly jet would have been displaced as far south as California from where it presently enters North America, in British Columbia. As the ice sheet began to recede the zone of jet penetration would have gradually moved northward across the North American Great Basin.

This model interpretation is in broad agreement with the large number of records indicating high lake stands and fresher water conditions in the Basin and Range during Late Pleistocene, with wet events in the southwest United States around 18 ka BP, generally preceding those of the Northern Great Basin (14–15 ka BP). Most likely, the average latitude of jetstream passage would have acted as a precipitation firehose, focusing more intense precipitation, and at higher elevations, promoting alpine glaciation, over narrow latitudinal belts for a few centuries–millennia as it moved north or south (Benson, 1993). For this reason, over long time periods (10^4 to 10^5 years) precipitation records are not precisely in phase between latitudinal belts within the Great Basin, as inferred from lake-level records or other precipitation archives (Bradbury, 1997).

What remains to be determined is the sensitivity of Great Basin climate to smaller pulses in the extent or height of the Laurentide Ice Sheet. Would the polar jetstream have responded in western North America latitudinally to the scales of perturbations responsible for Bond or even OD cycles? Higher-resolution records are required to address these questions than what can be gleaned from a simple analysis of the maximum high and low lake stand histories, and some lake records from the region are beginning to shed light on

these questions. Decadal resolution records from a number of lake basins support the notion that numerous millennial-scale oscillations in climate occurred within this region between about 53–9 ka BP. These indicators give us a record of precipitation in the Great Basin watersheds, and of glacial advance/retreat headwaters regions of the Sierra Nevada (Benson, 1999).

At Owens Lake, fluctuations in lacustrine indicators of glacial advance and retreat show an intriguing asymmetry in the history of each cycle (figure 13.10). Each glacial advance was apparently marked by an increasing flux of glacial rock flour, which both diluted available organic carbon and reduced productivity, thereby decreasing TOC. Conversely, glacial retreat is indicated by increasing TOC. The Owens Lake record suggests that the glacial advances were progressive phenomena, taking about 1000 years to reach their maximum, whereas the retreats were much more abrupt, occurring over only a few hundred years.

To what extent can these millennial-scale lake records of glacial advance and retreat in the Sierra Nevada be correlated with the North Atlantic marine and Greenland ice core records of millennial-scale events? Some researchers have attempted to correlate the multimillennial Bond cycles, which culminated in Heinrich events, with records of climate oscillation in the Great Basin (Oviatt, 1997; Benson et al., 1998). The Great Basin lakes appear to have been low and Great Basin climate both cold and dry at the times of several of the cold-indicator Heinrich events. Furthermore, the number of millennial-scale glacial advances and retreats recorded in the Owens Lake sediments are about the same as the number of DO cycles over the same interval (Benson et al., 1996). However, there is not a one-to-one match between all the peaks and troughs at the submillennial scale, and the timing of these records is still insufficiently resolved to determine what proportion of these shorter-duration climatic oscillations are synchronous.

Can this pattern of Great Basin/North Atlantic climatic consistency at the multimillennial scale, but not necessarily at the shorter time scale, be reconciled with what we know about potential teleconnections between the two regions? Assuming that the absence of precise wiggle matches is not simply the result of imprecise geochronology, then the lack of a direct match could result from two factors. Either these centurial–millennial scale events are not truly global in scale, or there may be lag times associated with the transmittal of

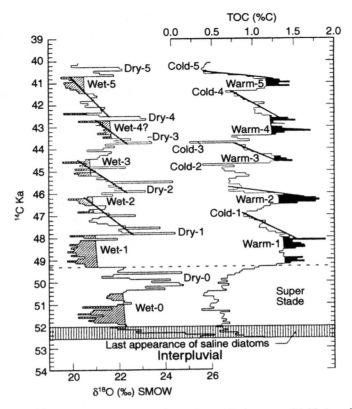

Figure 13.10. TOC and $\delta^{18}O$ values of sediments from Owens Lake cores OL90-1 and OL90-2 between 52.6 and 40.0 ^{14}C ka, indicating the transition to pluvial conditions. Relatively warm glacial interstadials indicated by high TOC concentrations are marked WARM. Each of the five interstadials occurred when $\delta^{18}O$ values were low, indicating WET, overflowing conditions. Between 52.3 and 49.3 ^{14}C ka a wet–dry oscillation occurred during an exceptionally long stade. From Benson (1999).

high-frequency teleconnection signals. Recall that periodic increases in the size of the Laurentide Ice Sheet would have had the effect of forcing the polar jet and the locus of precipitation toward the south. So the difference between multimillennial record consistency and centurial–millennial record inconsistency may reflect a threshold in the ability of this polar jet teleconnection to be transmitted to lower latitudes. The shorter, DO events may be records of climatic oscillations that were simply too weak to be felt in the Great Basin. At present we are far from the point of having sufficient numbers of high-resolution lake records across the latitudinal extent of the Great Basin to establish the true sensitivity of the region to shorter-term variations in the ice sheet, but given the large number of lake basins throughout the region this is clearly an achievable goal for paleolimnologists of the future.

Lake Archives as Long-Term Records of Astronomical Forcing and Glacial Cycles

One of the most exciting results to come out of intercontinental comparisons of long Pleistocene lake records is the growing body evidence for 100-ka cyclicity in moisture balance, salinity from faunal and geochemical signals, and vegetation, synchronous with the history of global ice volume, as inferred from the marine oxygen isotope record (Hooghiemstra and Sarmiento, 1991; Narcisi et al., 1992; Meyers and Horie, 1993; Peck et al., 1994; Colman et al., 1995; Lezzar et al., 1996; Bischoff et al., 1997a; Cohen et al., 1997b, 2000b; Kashiwaya et al., 1997; O.K. Davis, 1998; Chen et al., 1999; Williams et al., 1999). These records hold the promise of showing us the unique continental responses to Pleistocene climate change. They suggest that the expression of Milankovitch cycles in continental interiors was more complex than what is recorded in the marine and ice cap isotopic records, probably as a result of the complex climatic feedbacks created by the topography and vegetation on the continents (Bischoff et al., 1997a; Menking et al., 1997; Lowenstein et al., 1999).

However, long-term changes in Pleistocene climates cannot be understood solely in reference to simple variation in insolation. Additional factors that are partly or wholly decoupled from the direct effects of solar forcing have almost certainly been important as well, including variations in the size of the ice sheets, North Atlantic sea surface temperatures, and the uplift of the Himalayas, which undoubtedly affected Afro/Asian monsoonal circulation (deMenocal and Rind, 1996).

African and Asian climates and lake records provide instructive examples of how these longer-term forcing processes may have operated. Although 21,000-year cycles are recorded in some African lake-level histories (Trauth et al., 2001), for much of the Pleistocene, the climatic effects of these precessional cycles appear to have been overshadowed by longer-term rhythms (~ 100 ka), associated with the expansions and contractions of the ice sheets (deMenocal, 1995). Several long lake records from Central Asia support the dominance of a $\sim 100 - $ ka cyclicity in climate change, more or less synchronous with global ice volume records (Chen et al., 1999; Williams et al., 1999). Some data suggest that the establishment of the ice sheets about 2.8 Ma fundamentally altered the response of the African climate to precessional variations in insolation. After the ice sheets formed, the African climate appears to have become much more dependent on the long-distance effects that ice sheets and cooler sea surface temperatures in the North Atlantic had on global circulation. Model simulations suggest that the effect of both ice sheets and cooler sea surface temperatures on Africa would have been to weaken the flow of moisture-laden monsoonal air masses into the continent, thereby causing aridification and declining lake levels. Significantly, GCMs make differing predictions as to precisely which regions of the African continent would be most affected by aridifying influences of ice sheets versus cooling sea surface temperatures—allowing the models to be further tested by comparing paleoclimate records from different regions of the continent.

Variations in pollen, freshwater diatoms, and desert dust that have been transported into marine sediments off the coasts of Africa provide some support for the importance of the 100-ka climate beat in the African tropics, as do low-resolution terrestrial records from paleosols and vertebrate fossils. However long and highly resolved lake records hold out the greatest promise for determining the relative importance of varying climate rhythms over this time scale (Cohen et al., 2000b). Long cores from the oldest rift lakes almost certainly hold the key for unlocking the long-term climate puzzle of the Afrotropics. Seismic stratigraphical evidence has already provided some tantalizing evidence for the importance of 100-ka climate cycles in East Africa. The repeated packages of unconformity-bound sedimentary sequences, discussed in chapter 8, reflect alternations between low and high lake stand conditions in Lakes Tanganyika and Malawi. These packages appear to have formed over 100-ka intervals, based on reasonable estimates of sediment accumulation rates (figures 8.2, 13.11, and 13.12) (Cohen et al., 1997b).

Lake Records of Pre-Pleistocene Climate Forcing by Astronomical Cycles and Plate Tectonics

Significant advances have been made in recent years in interpreting pre-Quaternary climate change using lake records. However, two limitations exist on our ability to pursue paleoclimate questions from pre-Quaternary lake beds. The first of these comes from the greater imprecision in radiometric age dates that typically accompanies older sediments, making precise interlacustrine correlations difficult. The second problem is that there is an increasing level of uncertainty about continental configurations, sea floor bathymetry, and continental topography prior to the Quaternary. All of these are required to accurately parameterize GCMs, especially the more sophisticated coupled ocean/atmosphere models. As a result, the paleoclimate models against which we compare our paleolake data must necessarily be more generalized, and the range of testable hypotheses must be more limited. Despite these limitations, there are notable cases where important questions in pre-Quaternary paleoclimatology have been addressed using lake deposits, some of which are highlighted here.

One of the most intriguing paleoclimate problems we can address with paleolimnological data relates to the importance of solar forcing mechanisms in preglacial times. Does evidence exist for Milankovitch time-scale cycles before the development of the Quaternary ice sheets,

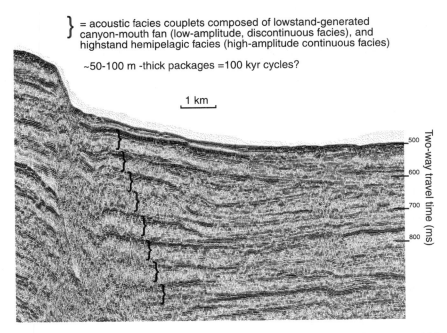

= acoustic facies couplets composed of lowstand-generated canyon-mouth fan (low-amplitude, discontinuous facies), and highstand hemipelagic facies (high-amplitude continuous facies)

~50-100 m -thick packages =100 kyr cycles?

1 km

Figure 13.11. Acoustic facies cycles evident in a seismic profile from Lake Malawi, East Africa. Figure courtesy of Chris Scholz (see Cohen et al., 2000).

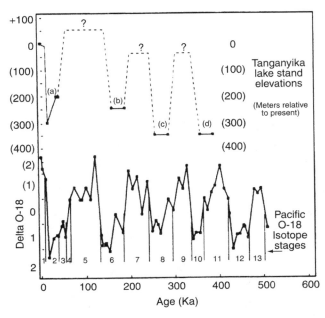

Figure 13.12. Estimated lake-level chronology for Lake Tanganyika superimposed over the marine oxygen isotope chronology for the northwestern Pacific. Age estimates for L. Tanganyika are based on intersite comparisons of realistic sediment accumulation rates, coupled with reflection seismic lines illustrating sequential high and low lake stands. Modified from Cohen et al. (1997).

and if so, are the dominant cycles similar to the present? Many ancient lacustrine sequences world-wide display evidence for cyclical lake-level fluc-tuations of varying durations, with high-frequency cycles clustered into longer-duration cycles. In some cases the precise periodicity of these higher and lower-frequency cycles is difficult to establish, or only mean cycle durations can be estimated (e.g., Dam and Surlyk, 1993; Stollhofen et al., 1999). In other instances, time series analy-sis of the thickness patterns of the high and low-frequency cycles provides evidence for Milankovitch forcing (Juhász et al., 1997, 1999). In some cases, however, independent age control is available that provides a geochronology of cycli-city independent of any a priori assumptions of sedimentation rates. One of the most intriguing examples of this kind of record comes from the Lower Pliocene Ptolemais Fm., which occupies an elongate fault-bounded basin in northern Greece. The upper part of this sequence consists of a spec-tacular lacustrine sequence of highly repetitive cycles of interbedded carbonate marls and organic-rich lignites (Van Vugt et al., 1998; Steenbrink et al., 1999) (figure 13.13). A radiome-trically derived age model for these deposits is con-strained by a series of 22 high-precision Ar/Ar dates, and gives an average cycle duration of 21.8 ±0.8 ka, in excellent agreement with a model of

precession-driven cyclicity (figures 6.13 and 13.14).

To what extent have glacial ice caps amplified or distorted the global climatic effects of the Milankovitch cycles? One way to get at this pro-blem is to examine the history of lacustrine cyclicity for parts of earth history when glacial ice was lim-ited in extent or nonexistent. Triassic lake beds from many parts of the world provide an excellent opportunity to do just that, since the Triassic was a period when ice sheets were absent even at the poles (Van Houten, 1962; Olsen, 1986; Olsen and Kent, 1996; Clemmensen et al., 1998; Hoffman et al., 2000; Reinhardt and Ricken, 2000). Cyclical lacus-trine sediments, recording alternately deep and shal-low or desiccated lake conditions, were deposited in rift valley paleolakes that formed in eastern North America, Greenland, western Europe, and northern Africa during the breakup of Pangaea, at paleolati-tudes of 2.5 to 9.5°N (figure 1.11). The extreme variation seen in lake levels over this time period is a testimony to the sensitivity of the Pangaean supercontinent to variability in monsoonal preci-pitation. As in tropical Africa during the Late Quaternary, this pattern of precipitation would have been closely linked to the seasonal variation in solar insolation. Using the Quaternary as a model, the strongest rainy seasons and deepest lakes would be predicted to have formed whenever

Figure 13.13. Repetitive cycles of interbedded carbonate marls and organic-rich lignites from the Ptolemais Fm., northern Greece, reflecting strong Milankovitch forcing on lacustrine facies.

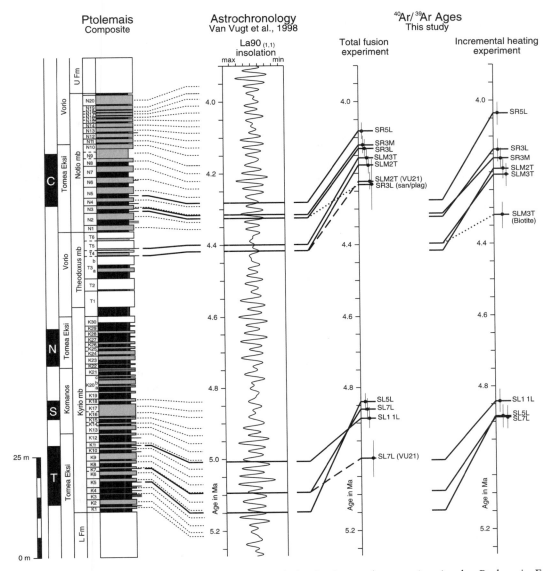

Figure 13.14. Composite stratigraphical section of the lignite–marl succession in the Ptolemais Fm. Magnetic polarity measurements and formation names are given to the left of the sedimentary cycle numbers. Black (white) bands indicate normal (reversed) polarity. T, S, N, and C are Thvera, Sidufjall, Nunivak and Cochiti subchrons. Also shown is the comparison of step-heating and total fusion $^{40}Ar/^{39}Ar$ ages of nine volcanic ashes (see figure 6.9) with astronomical-tuned insolation curve and age model for the same beds. From Steenbrink et al. (1999).

perihelion occurred in the Northern Hemisphere simultaneous with the maximum eccentricity of the earth's orbit. In fact, spectral analysis of these facies alternations shows that they are cyclical and vary at some of the frequencies predicted by Milankovitch theory, with two at about a 20,000-year frequency, two at ~ 100,000 years, one at ~ 410,000 years, and one at about 2 million years.

Interestingly, the ~ 41 – ka obliquity cycle is largely unrecorded in this record. Newark Basin history cannot be completely understood as a result of celestial forcing mechanisms. At the longest time scales (~ 10^7 years) tectonic processes related to the evolution of Newark Basin rifting must also play a role in controlling the expression of lake-level cyclicity.

At extremely long time scales (10^7 to 10^8 years) the positions and orientations of continents clearly play a major role in defining global climate systems. Unlike the shorter-term, astronomically forced cycles, changes in climate related to continental positions are effectively aperiodic. Although sea floor spreading and continental collisions do undergo cyclical behavior, the time scales over which these events occur are highly variable. Most research into tectonically driven climate change therefore has focused on understanding directional changes, such has those that might be predicted to occur as a land mass becomes larger or more elevated through continental collision, or the migration of a land mass across a latitudinal gradient. Long-lived lake systems that accompany these plate movements provide valuable records for interpreting the climate response to continental reorganization. For example, during the Triassic the Pangaean supercontinent reached its maximum size, with an enormous land mass that symmetrically straddled the equator (Dubiel et al., 1991). Atmospheric circulation models for Pangaea suggest it would have been a favorable setting for the development of extremely strong cross-equatorial or seasonal monsoonal circulation (Parrish et al., 1986; Dubiel et al., 1991; Kutzbach, 1994). This *Pangaean megamonsoon* hypothesis makes four predictions that can potentially be tested through the analysis of qualitative paleoclimate indicators, including lake deposits:

1. There should be evidence for strong and widespread seasonality in precipitation across much of equatorial and subequatorial Pangaea.
2. The eastern equatorial region of Pangaea should have been dry compared with both higher latitudes at the same time, or equatorial regions at nonsupercontinent times. When the megamonsoon reached its maximum intensity it may have reversed the normal equatorial air flow and generated relatively humid conditions in the western part of the supercontinent.
3. Climatic belts would not have paralleled latitude.
4. Patterns 1–3 above would have reached their maximum expression when the Pangaean supercontinent was both at its largest extent and most equally distributed across the equator. This occurred during the Triassic.

Many paleoclimate indicators of Triassic age conform to these predictions. In the Colorado Plateau of the western United States (what would have been western north-equatorial Pangaea) there is abundant evidence for perennial Late Triassic lakes and marshes associated with semiarid climate soils. Fish, plant, and invertebrate body fossils and trace fossils are all consistent with a warm climate and highly seasonal moisture, and a local water table that underwent strong seasonal fluctuations. Bivalve growth banding also suggests a strong seasonality in growth, which in the tropics is most likely associated with moisture or salinity stress. Similar aged Triassic deposits from the margins of the equatorial Tethys Seaway also provide evidence for the megamonsoon hypothesis (Reinhardt and Ricken, 2000).

Summary

Our conclusions from this chapter can be summarized as follows:

1. Comparative studies of lake archives between regions and continents allow paleolimnologists to accurately reconstruct climate history, and potentially address important questions about climate forcing mechanisms over various time scales. Lake basins must be chosen carefully for these types of studies, because some lakes amplify climate signals, whereas others mute them.
2. Understanding modern climate processes, and how atmospheric circulation is affected by global and regional heat sources and sinks, is critical for interpreting paleorecords of climate. Lakes respond to changes in atmospheric circulation and accompanying precipitation changes through variations in lake level, salinity, and biota. Some of these responses reflect relatively localized changes in atmospheric conditions, whereas others involve *teleconnections*, reflecting global, and often complex, interconnections in climate systems.
3. Changes in climate can be either aperiodic (e.g., regional mountain building) or cyclical (e.g., astronomical forcing). Spectral analysis methods can be used to detect periodic changes in paleolake archives, driven by factors such as the various *Milankovitch cycles*, and to determine the relative impor-

tance of cyclical events in driving regional climate history.

4. Paleoclimate models provide explicit hypotheses of historical change, which can be compared with paleolimnological records, to test model predictions, and to identify aspects of the model that require modification.

5. High-resolution lake records are ideal archives for addressing questions about annual–decadal scale climate processes, such as the historical record of ENSO. Paleolimnological records are valuable for tracking the directional climate changes that have occurred since the advent of the Industrial Revolution, and provide essential records of natural climate variability for comparable intervals prior to the nineteenth century. Many lakes have responded strongly to variations in solar output at decadal to centurial time scales. However, these responses are frequently asynchronous on regional scales, especially in regions where air mass interactions are complex.

6. Lake deposits provide some of the most detailed information available about the complex climatic changes that occurred during the Pleistocene–Holocene transition. Much of our understanding of these changes come from interregional comparisons of paleolake level trends, but detailed quantitative paleoclimate reconstructions from fossils and geochemistry are becoming increasingly important for this purpose. Paleolimnological records have also demonstrated that the Holocene, far from being a period of relatively constant climate conditions as once thought, has actually been a time of significant climatic variability on the continents.

7. Paleolimnological archives are well suited to study the global synchroneity of century–millennial scale climate events inferred from high-latitude marine and ice core records. Lake records lend support for the existence of intercontinental, and even interhemispherical, teleconnections in the coupled ocean–atmosphere circulation system, and, in principle, can be used to determine lead and lag times of climate response between regions. An increasing number of long lake records, obtained from drilling lakes in continuous existence since the Early–Mid Pleistocene or longer, are providing a means of verifying the cyclical nature of millennial scale, and even orbital forcing events, and testing specific paleoclimate models that cover these longer time frames.

8. Paleolake records from pre-Pleistocene deposits can also be used to investigate astronomically forced climate cycles, as well as directional changes in climate caused by mountain building or changing continental configuration. Despite limitations imposed by uncertainties about age dating or paleocontinental configuration, our ability to reconstruction climate from pre-Pleistocene lakes is rapidly improving. Such records are exceptionally important as they provide the most resolved archives of climate change over the long haul of earth history.

14

Paleolimnology in Deep Time: The Evolution of Lacustrine Ecosystems

Most lakes are geologically ephemeral; even the longest-lived individual lakes persist only for tens of millions of years. However there is a continuity to lake systems that transcends the geologically short history of individual lake basins. This continuity comes from the long-term biological evolution of life in freshwater, and fittingly, forms the final subject of this treatment of paleolimnology.

Like the oceans, lakes have provided habitats for living organisms for most of the earth's history. Yet the patterns of aquatic ecosystem evolution in rivers and lakes have differed dramatically from those of the oceans. In large part this can be traced to the fundamentally ephemeral nature of most continental aquatic habitats and the "disconnectedness" in both time and space that exists between individual lakes and rivers compared with the world ocean. This pattern of temporal and spatial patchiness in water body distribution on the continents has shaped the evolution of lacustrine species and communities. Some understanding of this history can be gleaned from the study of modern ecology and molecular genetics of living freshwater organisms. But to understand long-term trends in lacustrine biodiversity and their relationship to the history of the lacustrine environment we must turn to the pre-Quaternary fossil record. Understanding this history, the timing and tempo of major species diversification and extinction events, and the evolution of key ecological innovations is critical for correctly interpreting ancient lake deposits.

The fossil record of pre-Quaternary lakes is more difficult to interpret than that of more recent lake basins. Robust phylogenies are largely unavailable for clades of ancient lacustrine fossils, hindering our ability to test hypotheses of evolutionary ecology, although that situation hopefully will improve in coming years. Many major clades of fossil lacustrine organisms are extinct, and ecologies must be inferred from their depositional context. Even for organisms that have close-living relatives,

our certainty in making inferences about habitat and relationship with other species weakens as we go back in time. Also the record we have to work with deteriorates with age, the result of (a) a declining volume of lake beds available for study with increasing age, (b) difficulties associated with processing lithified lake beds for their fossil content, and (c) an increasing likelihood of destruction by diagenesis with increasing age.

For the average pre-Quaternary lake deposit, we must contend with poor preservation of important members of the lacustrine community that are not heavily skeletonized. This biases our record toward clades like mollusks with robust skeletons, and limits our view of the structure of a paleolake's ecosystem. Under special circumstances, however, a wider variety of more delicate fossils are preserved, presenting us with key windows into ecosystems of the past. Exceptionally preserved fossil assemblages that retain both delicate fossils and/or the soft tissues of many organisms are referred to as Lagerstätten (*lit.* "mother lode"). A well-known marine example would be the Cambrian Burgess Shale fauna, in which exceptionally preserved arthropods and various types of "worms" are preserved in organic-rich black shales. Lagerstätten are perhaps even more important for interpreting the history of freshwater ecosystems because of the relatively smaller proportion of strongly skeletonized freshwater organisms compared with marine communities. Fortunately, Lagerstätten are quite common in lake deposits, because of the abundance of reducing and rapid burial environments in lakes compared with oceans. Most meromictic lakes or lakes with high rates of organic matter sedimentation meet these criteria. Anoxia, the formation of bacterial mats, and rapid burial are all common in lakes, and all favor the preservation and/or mineralization of soft tissues and delicate fossils (Maisey, 1991; Martill, 1993; Meléndez, 1995; Ortega et al., 1998; Martínez and Meléndez, 2000). Lacustrine

Lagerstätten of this type are typically preserved in microlaminated oil shales or calcareous mudstones as two-dimensionally compressed fossils. An alternation of redox states or episodic ventilation in benthic lake environments allows lacustrine Lagerstätten to occasionally incorporate benthic body and trace fossils (De Gibert et al., 2000). Lagerstätten can also form in shallow, saline lakes through the rapid formation of calcareous or phosphatic concretions around fossils, allowing for three-dimensional preservation (Park and Downing, 2001). Lacustrine Lagerstätten are known from deposits throughout the Phanerozoic (table 14.1). Collectively, they improve our estimates of the ages of particular lacustrine clades, or the structure of ancient lake ecosystems.

Constraints on Evolution in Lakes

Most lakes and small rivers are temporary, biogeographical "islands" in a sea of land. Evolution in the lacustrine habitat selects for organisms to be able to cope with demands of this ephemeral world. Lacustrine organisms must be able to disperse between lakes when they dry up or become filled by sediments. Alternatively, they must be able to either leave the lake habitat at some phase of their life cycle or, in some cases where water bodies are ephemeral on seasonal to decadal time scales, be able to persist as desiccation resistant eggs, seeds, spores, or aestivating adults. In comparison with the marine environment, lacustrine organisms must cope with demands of a chemically and thermally variable environment, and the relative unpredictability imposed by the unique limnological, bathymetric, and nutrient loading conditions of each lake, its watershed vegetation and its geological surroundings. This combination of habitat unpredictability and requirement for dispersal ability between water bodies has suppressed many long-term evolutionary trends in lake ecosystems, particularly in comparison with oceans. Major changes in lake ecosystems appear to have lagged behind similar events in the oceans during the Phanerozoic, in terms of the evolution of intricate species interactions, the expansion of biodiversity, and the functional exploitation of resources by organisms. This conservative nature of lake ecosystems has also tended to buffer them against the most severe effects of profound mass extinction events that have fundamentally restructured marine and terrestrial environments during the past 550 Ma. Mass extinctions have left their mark on the biota of lakes but in a more muted and generally protracted way than in terrestrial or marine ecosystems.

Given the observations above, it is somewhat paradoxical that repeatedly, wherever lakes and rivers have managed to persist for geologically lengthy periods, their physical isolation from other continental water bodies has served as an engine for exuberant speciation. Large, old lake and river systems, like Lake Baikal and Lake Tanganyika, or the Mississippi or Mekong River Basins, are effectively self-contained biogeographical provinces, with extraordinary levels of diversity that have evolved in situ, geographically isolated from their neighboring regions and biotas. The isolating effects of oceanic barriers and, on continents, both arid regions and mountain range watershed divides have produced a characteristic pattern of local endemism in some modern lakes and rivers that is truly exceptional, even in comparison with better-known oceanic islands. And the fossil record of paleolakes demonstrates that this type of rampant and often rapid speciation has occurred repeatedly throughout the Phanerozoic.

"Deep-time" paleolimnological research addresses questions concerning the phylogenetic history of lacustrine clades, and the assembly of and long-term changes in lake ecosystems (e.g., Gray, 1988b). Here our interest lies in documenting the historical pattern of evolution and extinction of major clades of freshwater organisms and, to the extent possible with a fragmentary fossil record, interpreting their probable associations and interactions throughout earth history. However, before we can begin to make sense of 3+ billion years of lacustrine ecosystem history we need to consider some more general questions about controls on dispersal and diversification in the continental aquatic environment. How does dispersal of continental aquatic organisms affect speciation and diversification, and what roles do long-term changes in climate and tectonic history play in regulating dispersal? Under what circumstances do freshwater habitats become isolated from each other? And, under what circumstances and at what rates do clades diversify in lakes?

Dispersal of Continental Aquatic Organisms

Dispersal of organisms between lake systems is strongly regulated by life history characteristics of aquatic organisms. Many lacustrine clades cope

Table 14.1. Selected examples of lacustrine Lagerstätten

Age	Formation Name	Location	Depositional Setting	Major Clades Represented	References
Late Devonian (Frasnian)	Escuminac	Gaspé Peninsula, Quebec, Canada	Narrow, elongate lake, up to 200 m deep	Various jawless fish groups (Anaspida, Osteostraci), Placoderms, Chondrichthyes, Acanthodi, Actinopterygii (ray-finned bony fishes), Sarcopterygii (lobe-finned bony fishes), Chelicerates, terrestrial vascular plants	Schultze and Cloutier (1996)
Pennsylvanian	Braidwood	Mazon Creek, midwestern United States	Brackish coastal lakes and lagoons	Crustaceans, terrestrial vascular plants	Schram (1981, 1984)
Late Triassic (Carnian)	Cow Branch Fm.	Virginia, North Carolina	Microlaminated organic-rich shales from a meromictic rift lake	Insects, Actinopterygii, terrestrial vascular plants	Krzeminski (1992)
Early Cretaceous (Late Barremian)	Rambla de Las Cruces	Las Hoyas, E. Spain	Laminated mudstones from anoxic lake, rapid turbidite burial in shallow lake	Bryophytes, Pteridophytes, Cheirolepidiacean conifers, crustaceans, insects, Sarcopterygian and Actinopterygian fishes, amphibians, turtles, crocodiles, dinosaurs/birds	Meléndez (1995); Ortega (1998); Martínez and Meléndez (2000)
Early Cretaceous (Late Barremian)	Santana and Crato Fms.	Northeastern Brazil	Laminated calcareous mudstone. Offshore setting in a shallow lake	Ostracodes, decapods, conchostracans, insects, bivalves, Actinopterygian and Sarcopterygian fish, Pelomedusid turtles, crocodylians, aquatic sauropods	Maisey (1991); Martill (1993); Mabesoone et al. (2000)
Eocene	Allenby Fm. and other unnamed units	British Columbia, Canada	Laminated mudstones interpreted as varves	Fish, macrophytes, insects	Wilson (1988)
Oligocene	Florrisant lake beds	Colorado, United States	Laminated mudstones from anoxic lake. Preservation possibly enhanced by diatom mat coatings	Macrophytes, terrestrial plants, insects, fish	Harding and Chant (2000)
Miocene	Barstow Fm.	Southeastern California, United States	Concretions in saline/alkaline lake microlaminated mudstones	Crustaceans, insects	Park and Downing (2001)
Miocene	Rubielos de Mora	Eastern Spain	Laminated mudstones from anoxic lake	Macrophyte angiosperms, salamanders	Anadón et al. (1988)
Miocene	Shangwang Fm.	Shandong Province, E. China	Laminated diatomites from shallow, stratified, eutrophic lake	Macrophyte angiosperms, insects, teleost fish, salamanders, frogs, water snakes, alligators	Yang (2000)

with environmental catastrophes in individual lakes or lake districts through the possession of features such as desiccation resistant eggs, parthenogenesis, long dormant phases, or a tendency to aggregate in shallow water areas visited by birds. These clades often show little biogeographical provinciality, with species widely distributed, even between continents (e.g., Wesselingh et al., 1999). Migratory ability at various life stages, for example the adults of many aquatic insects, or the strong dispersive capabilities of some fish clades, also serves to limit genetic fragmentation of populations and formation of new species. In contrast, patterns of regional endemism and intralacustrine speciation are more evident among clades that either have more limited powers of dispersal, or that for ecological and behavioral reasons do not migrate. These distinctions are evident in the modern biogeography of lakes, and in the lacustrine fossil record. For example, the breakup and separation of continents during the Phanerozoic has repeatedly resulted in more profound dislocations in lake biotas and establishment of distinct biogeographical provinces among poor dispersers (Feist and Schudack, 1991; Schudack, 1996). Differences in the capacity for dispersal of various "freshwater" fish clades across the widening marine barriers of the post-Pangaean world resulted in very different diversification patterns (Lundberg et al., 2000). At least some apparently freshwater fishes were capable of crossing broad and fully marine barriers, and these groups show patterns of diversification that do not neatly follow the pattern of continental breakup and establishment of independent freshwater systems. Even within continents dispersal pathways of freshwater organisms are often complex. Some pathways are regulated by geographically predictable phenomena, such as the pattern of glacial ice retreat in boreal regions during the Pleistocene (Wilson and Hebert, 1996). However, many others have probably resulted from the occurrence of essentially chance events.

Freshwater Habitat Isolation

Populations of freshwater organisms are easily fragmented by tectonic and climatic processes, increasing the probability of speciation. At the largest scale, continental fragmentation has led to the redistribution of descendant lineages of numerous freshwater clades scattered around the globe. This is most evident for relatively ancient clades whose distributions were affected by the Pangaean supercontinent breakup. Unique lineages of freshwater fishes and invertebrates have evolved on fragments of Pangaea that formed over the last 250 Ma. At a regional scale, formation of mountain belts and tectonic lake basins within continents has regulated the biogeography of freshwater diversification. Major, persistent river and lake systems form as a result of continental-scale orogenic activity and play major roles in intracontinental diversification in freshwater. Rivers such as the Mississippi or the Mekong and their associated systems of tributaries, floodplain lakes, and ponds have acted both as centers of diversification, and as stable refuges for the repopulation of less permanent water bodies during periods of drought or glacial advance (G.R. Smith, 1981; G.M. Davis, 1982). In the intermontane western United States, even small drainage basins have become sufficiently isolated by localized uplifts to cause population fragmentation and speciation.

Episodes of tectonic lake formation, from rifting, collision, or persistent transtension along major strike–slip fault systems, have the same effect, creating lacustrine habitats that are geologically persistent and at the same time isolated from other continental water bodies. Intervals of earth history like the Early Cretaceous, marked by the formation of numerous tectonic lakes, were also periods of exuberant lacustrine speciation (Neustrueva, 1993). However, the extent to which species-level diversification in such tectonically formed lakes has served as an incubator for higher-level diversification, such as the formation of widely distributed freshwater families, is unknown, largely because of the inherently greater probability of preservation of both lake deposits and their enclosed fossils in tectonic basins. For most time intervals we lack sufficient geographical coverage of the global distribution of major freshwater clades to pinpoint their place of origin to a specific lake basin.

Climatic events have both fragmented populations of freshwater organisms and generated corridors for dispersal throughout the Phanerozoic. Amplifier lakes, whose surface areas or salinities respond strongly to changes in precipitation/evaporation ratios, show this most clearly. For example, wet and dry climate cycles recorded in the Triassic basins of eastern North America are accompanied not only by alternating lake ecosystem states, but also by repeated pulses of evolution and extinction (McCune, 1996). Similar patterns of diversification during lake filling and extinction during lake drying are evident in the rift lakes of East Africa during the Neogene. But climatic effects on diversification in lakes are by no means restricted to arid or tropical regions. In boreal regions of North America and Eurasia, glacial advances and retreats have both created and broken lake interconnections, providing conditions for both dispersal and diversification (Colbourne et al., 1998; Witt and Hebert, 2000).

Lacustrine Species Flocks and Speciation in Lakes

One of the most interesting evolutionary features observed in lakes is the occurrence of *species flocks*, groups of species limited to a particular lake or watershed, referred to as *endemics*, and united by a common ancestor that also lived in that lake basin (Martens et al., 1994; Rossiter and Kawanabe, 2000). Species flocks are known from numerous modern lakes, such as the amphipods of Lake Baikal, the gastropods of Lake Tanganyika, and cichlid fishes of Lake Malawi. Some lakes house multiple species flocks within closely related lineages. For example in Lake Tanganyika, both morphological and genetic evidence clearly show that endemic cichlid fish are derived from several independent ancestral lineages that entered the lake at different times. Cichlids therefore form not one but numerous species flocks in that lake.

Species flocks occur only in a subset of lakes and a subset of major groups of lacustrine organisms, prompting researchers to speculate about environmental and biotic factors that may be responsible for their formation. Most, but not all lakes with extensive species flocks are geologically ancient (> 1 Ma duration) and relatively deep, suggesting that long-term habitat stability may be important for species flock evolution. High levels of endemic biodiversity are particularly common in tectonic lakes that have persisted for long periods of time, buffered from climatic change by great depth. A subset of tectonic lakes, including rift basins, trapped oceanic basins, and back arc basins, particularly in humid or semihumid climates, is probably the most favorable setting for diversification. Under such deep lake conditions, where subsidence rates frequently outpace sediment accumulation, lakes can develop a wide array of deep-water habitats, leading to considerable opportunities for habitat segregation and deep water speciation (Bate, 1999; Sideleva, 2000). Exceptions to this ancient + deep model certainly exist. For example, Lake Victoria, despite its hundreds of species of modern cichlids, is thought to have dried out in the latest Pleistocene and is a relatively shallow, flat basin, subject to rapid sediment infilling.

A preponderance of species flocks in lakes occurs among benthic invertebrates and fish groups, which by virtue of feeding or breeding behavior have a strong relationship to specific substrate habitats. Certain groups of organisms, such as cichlid and cottoid fish, benthic ostracode and amphipod crustaceans, and various gastropods seem particularly prone to evolving into species flocks. Others, such as aquatic insects, bryozoans, hydrozoans, and diatoms rarely undergo endemic diversification. In contrast to the benthic habitat, most pelagic lacustrine communities contain relatively few endemics, and many lakes with nearshore, benthic species flocks are characterized by a pelagic zone that is notably species-poor (Dumont, 1994). These observations suggest that intrinsic characteristics of organisms, such as poor dispersal capabilities, may make them more prone to genetic isolation and the formation of species flocks (Cohen and Johnston, 1987).

Modern species flocks often show strong differentiation in terms of feeding preferences, substrate distribution or water depth, suggesting that they represent *adaptive radiations*, with selective pressure driving the exploration of new ecological niches, leading to speciation. However, many members of modern lacustrine species flocks have widespread distribution within lakes and many are ecological generalists. There is also considerable evidence for intralacustrine geographical divergence in modern species flocks (Michel et al., 1992; Verheyen et al., 1996). Speciation driven by the establishment of barriers to dispersal or gene flow (in lakes caused by lake-level fluctuations or habitat disruption) is referred to as *allopatric speciation*. Lakes that undergo substantial drying events may experience periods when the water body becomes subdivided into two or more basins. Subdivided populations may undergo genetic divergence sufficient for distinct species lineages to form before renewed lake-level rise rejoins the basins. Alternatively, in tectonically formed lakes, habitat disruptions might develop because of the control exerted by faults on habitat distribution. Faulted lake margins typically result in rocky shorelines; species that require sandy or muddy lake floors may be unable to disperse across such areas, thereby limiting gene flow and promoting speciation. Unfaulted lake margins with sandy or muddy substrates would in turn present barriers to rocky habitat specialists. In this way intralacustrine barriers can be established even without lake-level fluctuations. However, strong evidence exists in some modern species flocks that genetic divergence and *sympatric speciation* can also occur in populations not geographically separated by habitat or other barriers (Turner, 2000).

Although species flock formation is a hallmark feature of deep, tectonically formed lake basins, rampant speciation is by no means limited to these types of habitats. Rapid diversification has also been recognized in other more ephemeral lacustrine environments. In Canada and northern Eurasia lacustrine speciation has been driven by the isolation of numerous proglacial and postglacial lake systems following glacial retreat. Most likely this has occurred multiple times following each glacial/interglacial cycle of the middle and late Pleistocene (e.g., Witt and Hebert, 2000).

Fossil Species Flocks

Fossil examples of mollusk, ostracode, bony fish, and even amphibian species flocks are known from paleolakes throughout the world and through much of the Phanerozoic (G.R. Smith, 1975, 1981, 1987; Moura, 1988; Bate, 1999; Stollhofen et al., 2000). In some cases it is evident that multiple bursts of diversification have occurred within a single lake or in closely linked lake basins (e.g., Nishino and Watanabe, 2000). One of the best-studied fossil examples of the evolution of endemic lacustrine species flocks comes from the Neogene Paratethyan lakes north of the Mediterranean (Krstic, 1990; Müller et al., 1999; Geary et al., 2000). As we saw in chapter 2, a series of large lake basins extending from Switzerland to the Caspian Basin developed in foreland basins and trapped back-arc basins, in response to the collision of Africa with Eurasia. These basins became isolated from marine interconnections and were progressively infilled with sediments from west to east, such that the easternmost ones, the Black, Caspian, and Aral Seas, are still extant. The Pannonian Basin of Central Europe persisted as a gradually freshening body of water from its initial restricted marine condition, about 12 Ma, to its final infilling as a freshwater lake, about 4 Ma. During this time a spectacular radiation of hundreds of species of gastropods, bivalves, and ostracode crustaceans evolved.

The molluskan species flocks of the Pannonian Basin evolved from both freshwater and marine ancestors. Marine-derived clades, especially the cardiid bivalves (cockles), radiated extensively between 12–4.5 Ma, when the lake was relatively saline, large, and deep. These conditions were similar to the modern Caspian Sea, where endemic invertebrate species flocks are dominated by clades with a high degree of tolerance to salinity variation (Dumont, 2000). Pannonian marine-derived species, along with some freshwater-derived endemic species flocks, radiated to occupy a wide variety of habitats, from shallow lagoonal settings, to wave swept rocky coasts, to deep profundal environments (figure 14.1). Some of these clades underwent very rapid bursts of evolution, whereas other contemporary, and often closely related clades evolved gradually, over periods of 1–2 Ma.

During the latest Miocene, the Pannonian basin was filled by sediments from the northwest toward the southeast, reflected in a progressive migration of the locus of speciation for many lineages toward the southeast (Pogácsás et al., 1993). Through this interval Lake Pannon also became progressively less saline, driving the extinction of most marine-derived endemic mollusks. However, the effects of Lake Pannon radiations were felt in Eurasian faunas long after the lake itself disappeared. Lake Pannon served as an incubator for subsequent dispersal events and radiations in the eastern Paratethyan basins and dispersal into the circum-Mediterranean region (Nuttall, 1990). The origins of many of the peculiar endemic mollusks and crustaceans of the Caspian Sea today can be traced back to this origin, along with some more widespread elements of the modern European benthic invertebrate fauna. By the early Pliocene (5–4 Ma) the Pannonian Basin was transformed into a freshwater lake, and other, exclusively freshwater, mollusk families diversified during the final phases of basin infilling.

Tempo and Mode of Speciation in Lakes

The evolution of species flocks in lakes and paleolakes affords evolutionary biologists an outstanding window for observing the tempo and mode of speciation. Sequences of fossils from paleolake deposits contribute valuable information to debates over the tempo and mode of speciation because of the potential of lake beds to yield such highly resolved stratigraphical records. This is particularly true in lake settings where sedimentation rates have been high and relatively continuous. Correctly interpreting such fossil sequences however requires not only a continuous record of fossils through the lake bed sequence but also an understanding of how such fossils are transported and sorted by lake hydrodynamics after death. For example, in the Lake Turkana Basin of the East African Rift Valley, numerous endemic mollusks evolved during the lake's Plio-Pleistocene history, and are well pre-

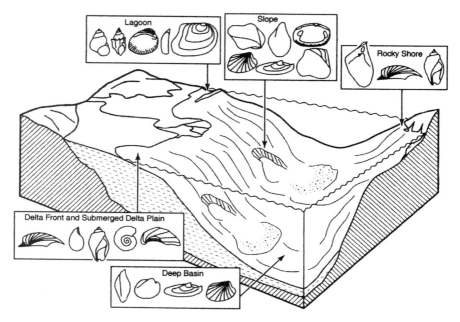

Figure 14.1. Inferred paleoenvironmental distributions of endemic mollusks of the Pannonian Basin. From Geary et al. (2000).

served in prominent shell beds. Williamson (1981) argued that these evolutionary events occurred extremely rapidly during low lake stands, based on apparently abrupt morphological transitions in shell characteristics through a brief stratigraphical sequence, without obvious intermediate forms. These events punctuated otherwise long intervals of little or no morphological change (termed *stasis*), consistent with the macroevolutionary concept of punctuated equilibrium (Eldredge and Gould, 1972). However, the extent to which shells in the shell beds may have been hydrodynamically sorted is unknown, complicating the evolutionary interpretation of these fossils (Cohen and Schwartz, 1983; Cohen, 1989b). For this reason, and because the mollusks of the Turkana basin occur in episodic and unconformity-bound stratigraphical sequences, they do not lend themselves to temporally continuous sampling required to resolve questions about the dynamics of speciation.

An ideal setting to overcome this problem lies in annually laminated lake deposits accumulating in deeper water environments of meromictic lakes. In a detailed study of morphological variation through a lacustrine stratigraphical sequence, Bell et al. (1985) examined stickleback fish fossils from a varved sequence of Miocene diatomites. Sampling at approximately 5-ka intervals over an estimated 110-ka time period they found both intervals of gradual and rapid change in fish morphology, but little evidence for prolonged periods of stasis.

Some modern evidence does support Williamson's original contention that speciation can occur extraordinarily quickly in certain lakes and among certain clades, although not his idea that diversification was linked with environmental/salinity stresses of low lake stands. There is good evidence from cores and seismic stratigraphy that Lake Victoria in East Africa was completely dry prior to about 14 ka BP (Johnson et al., 1996). Prior to the 1970s, hundreds of species of cichlid fish inhabited this lake. A simplistic reading of this pattern would suggest that all of these species evolved from a very small number of ancestors that reinvaded the lake as it began to refill in the latest Pleistocene and Holocene. More likely a significant number of ancestral lineages persisted in small aquatic refuges within the lake's watershed. Nevertheless, extremely high levels of endemism found in the Lake Victoria cichlid fish fauna today argues that many, if not most of its cichlid species have evolved in the past 14,000 years.

It is important to note that although the Lake Victoria species flocks includes an astounding number of species, genetic divergence underpinning this diversification is small, an observation consistent with the idea that these speciation events have occurred very rapidly. However it would be a mistake to conclude that all species flocks are so closely interrelated. In the extremely long-lived lakes like Baikal and Tanganyika, the levels of both genetic and morphological divergence between sister species are much greater, reflecting longer intervals since speciation.

Both stratigraphical and genetic evidence suggest that most lacustrine fish speciate over time ranges of 1500–300,000 years (McCune, 1997; McCune and Lovejoy, 1998). In modern lakes, where speciation mechanisms are more easily inferred, the shortest time-for-speciation intervals are associated with apparent sympatric speciation. Highly variable rates of speciation are evident in most paleontological examples as well. Fish, particularly cichlid fish, may be exceptional in their rapid rates of speciation in lakes, whereas most invertebrate species flocks have probably evolved much more slowly (e.g., Park and Downing, 2000).

A remarkable radiation of semionotid fishes in the Late Triassic and Early Jurassic rift lakes of the Newark Basin in eastern North America demonstrates the controls that climate and lake-level change can exert on this process (McCune, 1996). Semionotid fish fossils are distinguished by their large variety of body forms and variation in their dorsal ridge scale morphology, both preservable in fossils. As we saw in chapter 1, the Newark lakes underwent cyclical lake-level fluctuations, which may have been forced by earth's orbital cyclicity. Within well-studied deposits, diversity of endemic semionotid species has varied inversely with that of other fishes. During periods when semionotids were most diverse in the Newark Basin lake cycles, other fish species are either rare or absent, suggesting some form of ecological displacement. Semionotid speciation occurred rapidly during the early infilling phases of the Newark Basin lakes, similar to the pattern of transgressive lake-phase diversification observed in other well-studied fossil species flocks (e.g., G.R. Smith, 1987; Bate, 1999). Among Triassic semionotid fishes of the Newark Basin lakes, time for speciation intervals as short as 5000–8000 years occurred during some periods of the basin's history. These bursts of speciation however followed a very long interval of ~ 25 Ma during which only a single species of semionotid fish occupied the basin. Similar patterns of long intervals of little or no change followed by periods of diversification have been noted in other lacustrine organisms, such as the gastropods of Paleolake Pannon (Geary, 1990). In the latter case, however, the subsequent diversification proceeded at much

slower rates, with gradual morphological change in numerous lineages protracted over a 1–2-Ma period.

Pannonian gastropod lineages, along with other lacustrine mollusks, crustaceans, and fish, display repeated evolution of particular suites of body characteristics from a generalized ancestral morphology (Michel, 1994). Many cases of this type of *iterative evolution* in lakes are probably driven by the coevolution of predators and their prey. For example, among gastropods, most river or pond species have relatively thin and nonornate shells, evolved under conditions where calcium carbonate availability is low and unpredictable, and predation pressure is extremely variable. Under such conditions the metabolic investment involved in producing a heavy, predator-resistant or wave-resistant shell is not a favorable evolutionary strategy. Widespread and readily dispersed species such as these probably serve as original colonizers of larger and longer-lived lakes. Under conditions where long-term environmental stability of water chemistry and/or increasing predation pressure from fish or larger invertebrates occurs, ancestral morphological strategies may no longer be favorable. Among gastropods this can lead toward the evolution of heavier shells, resistant to predation or wave damage (e.g., West and Cohen, 1996). Alternative strategies to cope with predation however may occur in clades with smaller body sizes; among ostracodes, which are directly ingested by fish rather than crushed, it is common to observe trends toward smaller body sizes and greater mobility among large lake species flocks.

Iterative evolutionary patterns are strongly correlated with expansion and contraction histories of paleolakes. In Lake Pannon, strong shell sculpturing and thickening evolved during periods of maximal lake stability, and the disappearance of the stable, brackish lake phase spelled the death knell for many of these exotic taxa. Relatively small numbers of generalized ancestral taxa may give rise to numerous radiations, either within a single lake basin or in closely situated lakes within a biogeographical province (Van Damme, 1984; Michel, 1994; West and Michel, 2000). In Asian Mesozoic lakes, trigonoidoid bivalves underwent a series of iterative evolutionary events from smooth to heavily ribbed and ornamented shells (Barker et al., 1997; Gu, 1998; Guo, 1998a,b). Repeated trends toward more ornamented conditions seem to have accompanied the colonization of this group of bivalves into a succession of long-lived lakes, as they spread from their center of origin in China throughout Eurasia.

The History of Lacustrine Ecosystems

Lacustrine ecosystems including both eubacterial oxygenic photosynthesizers and archebacterial methanogenic carbon recyclers, have existed since at least the Late Archaean, the age of the earliest definitive lacustrine stromatolites (2700 Ma; Buick, 1992). However, for most of the Precambrian and Early Paleozoic, our knowledge of these ecosystems is extremely limited. Even identification of sedimentary deposits as definitively lacustrine prior to the advent of clearly identifiable metazoan and charophyte body fossils is often controversial, based as it must be on somewhat ambiguous sedimentary facies and geochemical clues.

This uncertainty extends well into the Phanerozoic, over 100 Ma after the first undoubted skeletons of multicellular animals are known from marine sediments. Even the identification of Cambrian through Early Silurian deposits as lacustrine is often controversial, precisely because these sediments are normally barren of fossils. Some paleontologists have argued for a substantial phase of metazoan evolution in lakes prior to the Silurian (Gray, 1988b). However, almost all records of freshwater, multicellular animals and plants from deposits prior to the Early Silurian ($\sim$ 440 Ma) have proven controversial. Trace fossils, burrows, and trails left by benthic or epibenthic organisms are also exceedingly rare in demonstrably lacustrine or fluvial sediments prior to the Silurian (Maples and Archer, 1989; Buatois and Mángano, 1993).

Early–Mid-Paleozoic Lacustrine Ecosystems: Invasion of Freshwater by Multicellular Organisms

Most likely, species diversification in lakes in the Early Paleozoic, prior to the Devonian ($\sim$ 410 Ma), was limited by a peculiar combination of low nutrient loads and high sediment loads. Nutrient delivery to lakes prior to the advent of abundant upland terrestrial plant cover must have been extremely inefficient; in all likelihood the vast majority of early Paleozoic lakes, even in warm climates and at low elevations, were ultraoligotrophic, extremely phosphorus-limited systems. Simultaneously, an absence of terrestrial plant cover would have reduced temporary storage of sediments in river valleys and probably increased rates of both erosion and siltation in lakes (Ponomarenko, 1996). It is significant in this regard that the vast majority of lakes of pre-Devonian age

are interpreted to have been very shallow and easily overfilled by sediments. The closest modern analogs would probably be some high-altitude or polar proglacial lakes, with extremely slow biochemical weathering rates, low nutrient inputs, and high terrigenous sediment loads, although the pre-Devonian lakes of this type would have occurred at all altitudes and latitudes. Nitrogen fixation was probably less of a problem, given the abundance of nitrogen fixing cyanobacteria present in early lacustrine stromatolites and phytoplankton. Under such conditions the supportable biomass at higher trophic levels in a typical lake would have been exceedingly small, accounting for the rarity of metazoans in these ecosystems. The contrast of these lakes with the contemporary, rich marine communities of the Cambrian and Ordovician is striking; the world ocean, as the ultimate sink for terrestrially weathered phosphates, would have slowly accumulated an internal load of nutrients that could support higher trophic level production, but no such phosphate reservoir would have been available in a short-lived lake. This may in part explain why most early records of freshwater animals are from coastal lakes and estuaries, and increases the likelihood that partially entrapped ocean basins similar to the Black Sea would have served as the original colonization arenas for the adaptation to freshwater. Not only would these habitats have been in close proximity to potential colonizers, but they also would have been the only continental waters to both retain the potential for significant internal nutrient loading and at least the occasional potential to receive a flux of nutrients from the spillover or mixing of marine waters.

Starting in the Silurian, body fossils of early freshwater vertebrates, arthropods, and charophytes, as well as crawling and swimming trace fossils are evident from numerous localities worldwide (Pollard, 1985). The congruence of both trace and body fossil records of early Paleozoic freshwater ecosystems indicates that even if larger, multicellular organisms did inhabit lakes prior to the Silurian, they were exceedingly rare, and most likely accidental entrants from marine habitats. After the early Silurian it becomes possible to make global tabulations of freshwater families of animals and plants, as a measure of standing biodiversity and biotic turnover (figures 14.2 and 14.3). A tally of fossil occurrences of freshwater families must be interpreted cautiously. Earliest occurrences and measured diversity of fossil taxa for a given time interval are known to be sensitive to a variety of factors aside from changes in actual biodiversity, such as area or volume of rock exposure, number

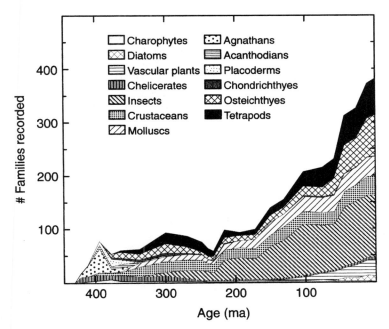

Figure 14.2. Standing family-level diversity for freshwater fossil organisms through the Phanerozoic. Compiled from data in Aguirre-Urreta (1992); Babcock et al. (1998); Berman and Reisz (1992); Briggs et al. (1993); Carbonel et al. (1988); Carpetta et al. (1993); Chacón-Baca et al. (2002); Cevallos-Ferriz et al. (1991); Cleal (1993); Collinson (1988); Collinson et al. (1993); G.M. Davis (1982); Dunagan (1999); Evans et al. (1996); Fraser et al. (1996); Friedman and Lundin (1998); Friis et al. (2001); Gardiner (1993a,b); Gray (1988b); Gu (1998); Halstead (1993); Hart and Williams (1993); Kohring and Hornig (1998); LaBandiera (1998, 1999, 2002); Maxwell (1992); Morris (1985); Patterson (1993); Retallack (1997); Ross and Jarzembowski (1993); Schultze (1993); Silantiev (1998); Skelton and Benton (1993); Sohn and Rocha-Campos (1990); Tracey et al. (1993); Whatley et al. (1993); Wilson et al. (1992); Wooton (1988); Zidek (1993).

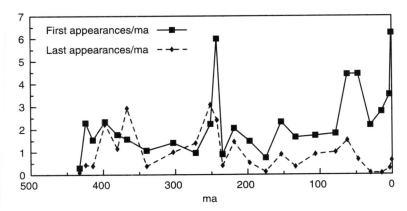

Figure 14.3. First and last appearances of nominal families of freshwater organisms per million years during the Phanerozoic. Based on the same sources as figure 14.2.

of fossil Lagerstätten of a given age, and intensity of collection effort by paleontologists. Similarly, the record of first and last appearances in the fossil record of individual families may differ from precise timing of evolution and extinction because of incomplete sampling of the group's actual longevity. Phylogenetic evidence may indicate that particular clades may have existed during a given time interval, despite a lack of fossils. And, counts of "families" may be artificially high or low estimates of clade diversification, because of classification systems that are not based on phylogenetic relationships. Nevertheless, such compilations provide a useful starting point for understanding long-term diversity changes. After an initial rise during the Late Silurian–Early Devonian, familial level diversity plateaued, with first and last appearances of families closely tracking each other. Throughout the Phanerozoic, periods of increasing numbers of first appearances are mirrored by increases in last appearances, and periods of declining first and last appearances are also correlated. However, there is a notable, proportional dropoff in last appearances over time, particularly over the last 200 Ma. This decline in apparent extinction rates has caused familial diversity to increase rapidly since the Triassic. Thus the overall record of freshwater organisms is marked not so much by major turnover events in biodiversity, but rather by the retention of older, morphologically conservative clades, along with the diversification of new ones. A remarkable proportion of extant families of well-skeletonized freshwater organisms can be identified from Early Mesozoic or even Late Paleozoic lake deposits.

Freshwater ecosystems of the Devonian (408–360 Ma) were marked by a rapid increase in both abundance and diversity of freshwater animals and charophytes. A remarkable feature of this diversification is its close parallel in time with the terrestrial diversification of land plants and terrestrial arthropods. This is almost certainly not a coincidence (Maples and Archer, 1989; Buatois and Mángano, 1993; Friedman and Chamberlain, 1995). As rates of phosphate weathering and terrestrial humic matter influx to lakes began to rise in the Siluro-Devonian, nutrient loading would have risen dramatically for the average lake, with immediate implications for the types of trophic structures that were supportable in short-lived continental water bodies. Upper trophic order predators, especially those with strong phosphate demands, would have been particularly limited by pre-Silurian phosphate limitations in lakes. The earliest evidence for strong seasonality in

phosphate availability comes from Middle Devonian lakes, where annual cycles of scale and bone phosphate resorbtion are evident from various lobe-finned fish (Donovan, 1980).

Mid–Late Devonian (~ 370 Ma) Lagerstätten show that shallow-water freshwater fish and arthropod communities were quite diverse, encompassing a wide array of jawless fish, cartilaginous elasmobranchs (shark and ray relatives), early jawed fishes (acanthodians, placoderms, and both ray-finned and lobe-finned bony fishes), and large chelicerates (eurypterids, early aquatic scorpions, and limuloids—relatives of modern horseshoe crabs) (Schultze and Cloutier, 1996). In addition, a low diversity of conchostracans, leperditocopid crustaceans (close relatives of ostracodes), bivalves, and gastropods was present in lakes of that time (Morris, 1985; Gray, 1988b; Friedman and Lundin, 1998). The earliest amphibians make their first appearance in the Late Devonian.

Devonian lacustrine communities seem to have been tightly linked to benthic habitats, with little evidence for complex open-water or pelagic food webs (Ponomarenko, 1996). Analysis of fossil fecal material and gut morphologies indicates that Devonian fishes comprised a mixture of near-bottom predators and bottom-feeding detritivores, and contemporaneous arthropods probably had a similar array of diets. No direct evidence exists for herbivory in these assemblages. Similarly, the earliest freshwater gastropods of the Late Devonian and Early Carboniferous, a mixture of viviparids, physids, and possibly thiarids, all extant families, were probably mixed detritus/bacterial feeders, with algal scraping herbivory apparently unimportant. However by the Late Devonian particulate organic matter and phytoplankton was abundant enough in lakes to support substantial and large populations of suspension-feeding bivalves (Friedman and Chamberlain, 1995). Fish, arthropods, and mollusks from Devonian freshwater deposits all show little differentiation from their marine relatives, and many, if not most of these "freshwater" species were capable of tolerating a wide range of salinities, from freshwater to fully marine (Morris, 1985; Schultze and Cloutier, 1996; Knox and Gordon, 1999; Vannier et al., 2001). Fossil burrows in Devonian lake beds are overwhelmingly restricted to lake margin, deltaic, and ephemeral lake deposits (Pollard et al., 1982; Buatois and Mángano, 1993). This observation, coupled with a paucity of organic-rich deep lake sediments suggests that offshore or deep lake environments remained largely uninhabited prior to the Middle Devonian.

Carboniferous–Early Triassic Lacustrine Ecosystems (360–240 Ma): Invasion of Deep-Water Benthic Habitats, Climate Change, and Freshwater Community Turnover

The onset of the Carboniferous Period was accompanied by significant changes in the composition of freshwater communities, following a major episode of extinction, especially among vertebrate and chelicerate clades. Following a period of rebound, typical lacustrine fossil assemblages of the Late Paleozoic (Carboniferous–Permian Periods) have far higher species diversity and evidence for endemism than their Middle Paleozoic counterparts (Stollhofen et al., 2000; Wartes et al., 2000; Zhao and Tang, 2000). By the Permian there was also a decline in the number of clades whose preference for fresh or marine waters is ambiguous. The Late Paleozoic was marked by the near-total disappearance of numerous freshwater vertebrates (the placoderms and several groups of jawless fish), and major increases in the diversity of freshwater crustaceans and mollusks (Gray, 1988b; Briggs and Clarkson, 1990). The invasion of freshwater by various shrimp-like crustaceans (Hoplocarids and Eumalacostracans) started in the early Carboniferous (Visean, ~ 350 Ma), although no genera of shrimps appear restricted to freshwater before the Namurian (~ 325 Ma). By the Late Carboniferous (Westphalian, ~ 310 Ma) shrimps had diversified into a wide variety of trophic niches, including rapacious carnivory, scavenging, and perhaps some detritus feeding (Schram, 1981, 1984; Briggs and Clarkson, 1990). During the Late Carboniferous numerous vertebrates also make their first appearances and subsequently diversify in freshwater, including chondrichthyans (shark-like fishes), paleonisciforms (early ray-finned fishes), and early amniotes. By the Late Permian (~ 280 Ma) there is abundant evidence for complex food webs among lacustrine vertebrates including higher-level vertebrate predators that probably filled some of the roles of modern active piscivorous fish and crocodiles (Sennikov, 1996). However, the intensity of predation in Paleozoic freshwater ecosystems was probably much lower than today, as evidenced by the abundance of large, epibenthic crustacean fossils, such as conchostracans and notostracans in freshwater lake deposits, in association with fish fossils. As we saw in chapter 5, such slow-moving animals today are almost entirely restricted to temporary, fishless water bodies, where predation pressure is minimized.

Terrestrial plant communities were evolving rapidly during the Carboniferous but little evidence exists that vascular plants formed anything like the littoral vegetation belts that we associate with aquatic angiosperms today. By the Late Carboniferous equisetophytes (horsetails) and tree-like lycophytes (giant club mosses) occupied lake margin habitats, often rooted in mud or swampy substrates (Scott, 1978, 1979; Stollhofen et al., 2000). Some of these plants (especially members of the widespread genus *Calamites*) may have also grown as emergent plants some distance from shore. Many fern spores from the Late Paleozoic may also represent aquatic plants. However, apart from algal charophytes, evidence for a fully submerged littoral macrophyte community is extremely poor prior to the Middle Mesozoic.

One of the most profound, Late Paleozoic, evolutionary events affecting freshwater ecosystems was the evolution of aquatic insects. Ancestors of modern aquatic insects apparently invaded freshwater secondarily from terrestrial ancestors. Early insects with aquatic larvae, including Odonata (dragonflies), Ephemeroptera (mayflies), and a variety of extinct Late Paleozoic clades, are known starting in the Late Carboniferous, about 300 Ma (Bashkirian–Moscovian) (LaBandeira, 1999). However, aquatic insects were probably rare until the Early Permian. The earliest aquatic insect fossils are found in river and floodplain deposits, suggesting that they primarily inhabited running water; the invasion of lakes seems to have occurred later (Wiggins and Wichard, 1989). This body fossil evidence is paralleled by a sequential increase in the trace fossil evidence for insect burrowing, spreading from river and floodplain environments progressively into lake margin and offshore habitats during the Triassic (Buatois et al., 1998).

Most Permian aquatic insect clades were either predators (Odonata, Megaloptera, and Coleoptera) or detritivores (many Ephemeroptera and Plecoptera) (Wooton, 1990; LaBandeira, 2002). It is unknown whether direct algal herbivory existed among early aquatic insects. Evidence for aquatic herbivory among extant orders of insects is almost nonexistent, but several major extinct clades of aquatic insects are also known from Permian lake deposits, whose larval feeding habits are unknown.

A significant extinction event occurred among lacustrine organisms at the end of the Paleozoic, coinciding with the massive marine and terrestrial Permo-Triassic extinctions. However in freshwater, this extinction, like the earlier Devonian event, appears to have been a protracted event, with successive waves of extinctions, especially among aquatic vertebrates extending from the Early

Permian (Artinskian Age, ~ 265 Ma) to the Early Triassic (Scythian Age, ~ 245 Ma) (Maxwell, 1992). Unfortunately the Permo-Triassic boundary is extremely poorly preserved in lacustrine stratigraphical sequences, making it difficult to determine the tempo of extinction. Major extinctions of both crustaceans and aquatic insects also occurred in the Late Permian (LaBandeira, 1999). As with Devonian extinctions, the Permo-Triassic events were also accompanied by bursts of diversification at the familial level, especially in the earliest Triassic. This resulted in a turnover of the fauna and a gradual loss of total freshwater diversity from the Late Carboniferous/Early Permian until the Middle/Late Triassic, an interval of about 70 Ma, rather than simply a catastrophic decline in freshwater biodiversity at the end of the Permian.

The protracted and global nature of this series of events suggests it was related to planetary climate changes occurring over this time interval, including glacial episodes in the early phases, and later, pronounced warming and drying events associated with the development of extreme continental climates during the assembly of Pangaea. Globally, a large proportion of Late Permian lake deposits record saline or hypersaline conditions and the complete disappearance of shallow freshwater lakes (Molostovskaya, 2000). Late Paleozoic freshwater ecosystems simply may have been victims of prolonged global aridification.

Middle Triassic–Cenozoic Lacustrine Ecosystems (240–0 Ma): Continental Fragmentation, Diversification, and the Evolution of the "Modern" Lacustrine Biota

The decline in familial level lacustrine biodiversity of the Late Paleozoic–Middle Triassic was abruptly reversed in the Late Triassic. At that time diversity began a rise that apparently has continued to the Quaternary. Almost every major freshwater clade appears to have increased in diversity since the Late Triassic, and familial level extinction rates have generally been lower than during the Paleozoic. Some of this apparent increase in diversity is probably an artifact of better preservation and more outcrops of younger lacustrine sediments, but this probably cannot explain the entire trend of increasing numbers of families, since fossil assemblages from individual Lagerstätten of younger ages are typically more diverse than their older, Paleozoic counterparts.

Some diversification is probably attributable to the fragmentation of the Pangaean supercontinent,

starting in the Triassic. Prior to the fragmentation of Pangaea, most fossil lacustrine genera, and even species, had quite broad geographical distributions (K.G. McKenzie, 1981). The breakup of Pangaea led to greatly increased *provinciality*, the separation of lake-rich regions onto discrete continental blocks separated by widening ocean basins. Unique, modern, freshwater biotas of isolated continental blocks like Australia are in part a testimony to this process of provincial isolation (G.M. Davis, 1982). The rifting process also produced numerous lake belts along the zones of continental fragmentation, many of which became centers of diversification, especially during the early, intense phases of continental breakup, from the Late Triassic to the Early Cretaceous (Bate, 1999; Carvalho et al., 2000). Also, the fragmentation of Pangaea led to more extensive humid, maritime climates, further increasing the spread of freshwater habitats.

Three key evolutionary events in lakes accompanied the Middle Mesozoic continental breakup, the diversification of aquatic insects, aquatic macrophytes (especially angiosperms), and teleosts (advanced bony fishes). Each of these developments had profound ramifications for the establishment of modern lacustrine ecosystems and food-web linkages by the Late Cretaceous. Among insects, this included the radiation of groups that today frequently function as keystone species in stream, pond, and shallow lake communities, including numerous extant families of the Trichoptera (caddisflies) in the Middle Triassic, and aquatic Coleoptera (water beetles), Hemiptera (water bugs), and Diptera (true flies) in the Late Triassic to Early Jurassic (Wooton, 1988; Krzeminski, 1992; Ponomarenko, 1996; Fraser et al., 1996; J.M. Anderson et al., 1998). Steady increases in the diversity of aquatic insect feeding strategies occurred during the Middle–Late Triassic, including the first good evidence for benthic filter feeding, pelagic predation, and herbivory (LaBandeira, 2002). In the Jurassic, insects feeding strategies extended into other modes, such as water surface film feeding among various Hemipteran families, mid-water predation, benthic deposit feeding, and infaunal deposit feeding (LaBandeira, 1998). Early herbivorous insects in lakes were scrapers and shredders of plant matter, mostly feeding on fallen leaf litter derived from terrestrial sources in running waters. A major expansion of lacustrine insect herbivory in lakes began in the Jurassic and by the Cretaceous there is good evidence for insects occupying the wide range of feeding preferences observed today, including macrophytes, algal scraping, and phytoplankton feeding. Probably one of the most important impacts of the aquatic insect radiation was the evolution of mobile, dipteran larvae, especially chironomids and chaoborids, which acted to both consume settling organic matter or plankton and, upon surfacing, redistributing nutrients and food resources to the upper water column. The upward transfer of food resources was probably limited prior to the evolution of this type of behavior, and may have acted in turn to limit the exploitation of the open-water or pelagic environment of lakes, particularly large ones.

An increase in structural diversity of the littoral zone also began in the Triassic. Initially this involved the evolution of several small groups of nonflowering vascular plants, such as relatives of *Isoetes* in the Early Triassic, *Azolla* in the Jurassic, and hydropterid ferns in the Early Cretaceous. The expansion of flowering plants in freshwater followed quickly on their rapid terrestrial diversification in the Early Cretaceous ($\sim$ 120 Ma). By the Late Cretaceous (Coniacian, $\sim$ 88 Ma), flowering aquatic macrophytes occupied the entire range of submerged, emergent, floating attached, and floating unattached habitats observed today. Eocene Lagerstätten reveal macrophyte communities that are almost indistinguishable from modern ones (Cevallos-Ferriz et al., 1991). Aside from providing new potential food resources, this diversification greatly increased habitat complexity of shallow littoral environments, providing refuges for invertebrates from predators and new types of water column attachment surfaces for algae and filter feeders. For example, the earliest freshwater bryozoans occur in Late Triassic lake deposits (Kohring and Hörnig, 1998). In addition to insects, macroherbivory expanded in the Triassic among gastropods (ampullarids). Most major gastropod clades of periphyton herbivores, such as the ancylids, lymnaeids, neritids, thiarids, and pleurocerids, also either evolved during the Mesozoic, or first entered freshwater at that time.

Intensification of freshwater predation also seems to have occurred in the Middle Mesozoic, with the evolution of teleost fishes, frogs, crocodilians, and aquatic birds. Evolutionary effects of intensified predation are also evident in the disappearance of many slow-moving, branchiopod crustaceans, especially conchostracans, from most permanent water bodies starting in the Jurassic (Webb, 1979; Kohring, 1998). The types of top-down trophic cascade effects that characterized so many fish-bearing lakes today, such as the exclusion of numerous large-bodied cladocerans from the pelagic zone and large ostracodes from the benthos of many large lakes probably dates from this time (Kerfoot and Lynch, 1987).

The combined effect of macrophyte and aquatic insect diversification, and increased predation during the Middle–Late Mesozoic, greatly strengthen linkages between benthic, nearshore, and openwater food webs in lakes. This must have increased the efficiency of nutrient recycling in lakes, and probably opened the door for much greater utilization of open-water habitats by fish and macroinvertebrates. This Lacustrine Mesozoic Revolution in many ways parallels simultaneous changes in the marine environment (Vermeij, 1987). However the intensity of ecological linkages in lakes has generally not proceeded to the levels observed in the oceans. Lacustrine ecosystem evolution in the Mesozoic, like today, was retarded by the ephemeral nature of most lakes, and lacustrine biotas have remained relatively "grubby" in comparison with marine ecosystems, particularly those of the tropics. In exceptionally long-lived lakes, like modern Lake Tanganyika, or Pliocene Paleolake Idaho, predation episodically drives the evolution of exceptional predator–prey arms races, with strongly armored prey (generally mollusks and crustaceans) evolving simultaneous with their more heavily armed predators (G.R. Smith, 1987; West and Cohen, 1996). These relationships can develop over a few million years into intricate species interactions comparable to those observed in marine ecosystems, but are almost inevitably doomed by long-term infilling or drying in even the longest-lived lakes. Heavily armed species and species that are highly dependent on specific interactions made possible by the relatively stable habitats of long-lived lakes become endangered species when those habitats disappear and the more normal world of temporary lakes and ponds is their only available refuge.

Ironically, it was probably the lack of strong-species interactions and broad habitat tolerances of many lacustrine species, coupled with the physical disconnectedness of individual lakes, which allowed the basic structure of lake ecosystems to overcome the worst effects of the Cretaceous/ Tertiary (K/T) boundary (65 Ma) mass extinction. Compelling evidence exists that at the K/T boundary the earth experienced a major impact by a large meteorite or comet. Many paleontologists have argued that this event caused mass extinctions in both marine and terrestrial environments, through a cascading series of ecological impacts on the world's major biogeochemical cycles. One predicted consequence of such an impact was the planetary-scale atmospheric release of enormous quantities of nitric and sulfuric acid, producing the mother of all acid rains! In lakes, acid-sensitive taxa, such as bony fish, or calcium carbonate secreting mollusks and ostracodes, would have been particularly vulnerable to such acid loading. Impacts would have been most severe in poorly buffered stream and lake systems, and proportionately less severe in lakes lying on limestone or other carbonate bedrock. Because most lacustrine organisms are quite widely distributed at the family level with respect to details of bedrock geology the response of family level biodiversity in lakes to the K/T impact event may have been relatively muted, even if local endemics were severely impacted. Unfortunately, research into the tempo of ecosystem change in freshwater at the K/T boundary has been hampered by the extremely small number of continuous paleolake records that span the mass extinction boundary (Hartman, 1998; Yo, 2000). In the Williston Basin of Montana, a diverse and morphologically complex assemblage of mollusks persisted in latest Cretaceous rivers and lakes of the region right up to the K/T boundary (Hartman, 1998). Almost none of these species occur in Paleocene sediments above the boundary. However essentially all mollusk families do survive the K/T event in the general region to repopulate Paleocene rivers and ponds of the region. Intriguingly, these survivors are almost exclusively morphologically simple taxa, lacking the shell sculpture typical of Late Cretaceous taxa.

Tertiary lacustrine ecosystem evolution has been strongly impacted by three major events, biogeographical and paleolimnological effects of Tertiary continental collisions in Eurasia, India, and Africa, global cooling and increasing aridity in many parts of the world after the Eocene, and the diversification of freshwater diatoms. Major continental collisions of the Tertiary created vast new lake and river systems in Eurasia. Some, like the Pannonian Basin, served as entry points for invasions of continental waters by marine organisms and radiation of their descendants, whereas others, like the Mekong River and its floodplain lakes, were incubators for species flock formation from freshwater ancestors. At the end of the Eocene, global cooling resulted in the regional elimination of many tropical freshwater clades from mid-latitude lakes (e.g., Nury, 2000). At this time modern distinctions between tropical and temperate lacustrine biotas become evident. Further cooling and aridification in the Late Tertiary isolated freshwater ecosystems in many parts of the world, particularly in the arid, intermontane interiors of continents.

Probably the most profound event in Tertiary lacustrine evolutionary history has been the invasion of freshwater by diatoms. Fossil marine diatoms are known from Early Cretaceous, and possibly Jurassic strata and some authors have argued, based on phylogenetic evidence, that the clade may be much older than this (Round and Crawford, 1981). The oldest known lacustrine diatoms are from Late Cretaceous sediments (Chacón-Baca et al., 2002), but given their rarity in pre-Tertiary lakes they were probably unimportant components of lacustrine ecosystems prior to ~ 50 Ma. It was not until the Miocene that freshwater diatom assemblages became diverse and diatomite deposits globally abundant (Kociolek and Stoermer, 1989, 1993; Khursevich, 1994; Krebs and Bradbury, 1995).

Even in the earliest lacustrine diatom floras modern genera are recognizable; the extent of morphological evolution among groups, and apparently the habitat differentiation recognized today between species were well established early in the history of freshwater diatoms (Lohman and Andrews, 1968; Krebs, 1994; Bradbury and Krebs, 1995). Intervals of rapid speciation have occurred for the diatoms, particularly in the Late Miocene. But the morphological, and probably the ecological differences that accompanied these speciation events were minor.

The ecological impact of the invasion of diatoms into freshwater must have been substantial, given their importance in lacustrine communities today. As prominent members of the phytoplankton, diatom invasion of lakes would have had a profound effect on the biogeochemical cycling of silica, given that no common siliceous phytoplankton are known from older lake sediments. In lakes where diatoms outcompeted other phytoplankton groups they would have also caused diversion of other nutrients into new trophic pathways, particularly where diatoms became preferred food resources for specific algal consumers. At present, we do not know if overall productivity in lakes was influenced by freshwater diatom evolution. However, a connection between the expansion of diatoms as keystone organisms and overall increasing productivity in the freshwater environment is suggested by the fact that during the Miocene, when freshwater planktonic diatoms were undergoing a major diversification, there was a simultaneous, worldwide increase in the abundance of thick lacustrine diatomites, indicative of high productivity conditions.

Long-Term Functional Change in Lake Ecosystems

Evolutionary paleoecologists disagree as to whether ecological interactions in lakes have changed fundamentally over the course of the Phanerozoic. Some scientists argue that modern modes of species interactions, ecological functions, and energy cycling were established early in the history of lakes, perhaps as early as the Middle Paleozoic, and certainly by the Late Paleozoic, and there has been no long-term trend in lake ecosystem complexity over the last 300 Ma (Gray, 1988b). This view suggests that the number of functional or feeding guilds, especially at upper trophic levels or among larger organisms, has been relatively stable since at least the Late Paleozoic, and that these organisms provide an overwhelming, top-down structural control on lake ecosystems. Proponents of this viewpoint note that our ability to infer functional roles of major groups of lacustrine organisms with no living representatives is quite limited, and that extinct organisms may have occupied niches similar to those of completely unrelated organisms living today. The apparent increase in familial level diversity through the Phanerozoic may be simply a consequence of better preservation of more recent sediments and fossils.

I take a fundamentally different view of long-term history of lake ecosystems; that there have been substantial functional changes in lakes over time. These changes are marked by an increasing connectivity of lake ecosystem components through the Phanerozoic. The case for this argument comes not only from the body fossil record of lacustrine organisms, but also from trace fossils and sedimentology, both of which are much less subject to vagaries of preservation. Increased abundance of burrowing and progressive changes in sediment types observed in lakes through the Phanerozoic support the idea that nutrients and food have become more widely distributed in lakes over time, as a result of both watershed inputs and more efficient internal cycling. The rising abundance of nearshore, and eventually offshore lacustrine burrowing after the Middle Paleozoic is a strong signal that, as terrestrial plant communities evolved, they had an increasing effect on nutrient availability in lakes. In this sense the average lake has probably become more "eutrophic" over time. It is almost certainly not a coincidence that the carbonate/clastic varve couplets formed in many productive, stratified lakes today are relatively rare in pre-Mesozoic lake beds, when offshore nutrient input and productivity of most lake systems was probably below the threshold for accumulation of open-water micrite produced by productivity blooms. This would have inhibited the formation of seasonal biogeochemical varves. Before the Mesozoic most varved lake beds were silty and glaciolacustrine in origin. Similarly the delivery of fine clays to offshore environments was probably more limited prior to widespread terrestrial plant-induced weathering in the Middle Paleozoic.

Increasing food availability in the form of terrestrially derived organic matter was mirrored by a progressive invasion of the open and upper-water column by phytoplankton, zooplankton, and ultimately, by higher-level consumers in the Mesozoic. The exploration of sheltered, cryptic, or deeply buried habitats, and the elimination of many large-bodied, slow-moving, epibenthic organisms from open-bottom habitats and fish-bearing lakes are all probable consequences of improved predation efficiency in the Mesozoic. This process must have accelerated with the great Cretaceous increase in littoral vegetation. Perhaps the most important outcome of increased nutrient release to lakes and increased predation by the Late Mesozoic was the establishment of strong connections between benthic and planktonic food webs, epitomized by the evolution of organisms that migrate between the two realms, such as dipteran larvae or meroplanktonic diatoms. Nutrient exchange between the water column and the lake floor must have increased because of the evolution of aquatic dipterans, macrophytes, and the deeper burrowing benthos trying to avoid predation. Modern lacustrine ecosystems, with their wide range of species interactions and energy transfer systems between habitats, are a legacy of this evolutionary process.

Summary

1. Although individual lakes are ephemeral ecosystems, continuity exists in the biological evolution of life in standing inland waters. The paleolimnological record of this continuity largely comes from a limited number of well-skeletonized clades, but is "fleshed out" by more complete whole ecosystem "snapshots," provided by exceptionally preserved *Lagerstätten*.

2. Lacustrine ecosystem evolution is constrained by the relative unpredictability of the lake environment, the disconnectedness of inland water habitats, and the require-

ment that organisms be able to disperse between these ephemeral water bodies.

3. Where tectonic lake systems have persisted for long periods, they sometimes give rise to the evolution of *species flocks*, groups of closely related species that have evolved from a common ancestor, wholly within the lake system. Rates of intralacustrine evolution in these circumstances can be extraordinarily rapid, especially during transgressive intervals. Similar evolutionary trends are sometimes repeated through multiple lake phases in response to similar environmental conditions and organism interactions.

4. Lacustrine organisms are recorded from rocks as old as the Archaean, although little is known in detail about their ecology prior to the Silurian. Prior to the evolution of land plants, lakes were probably characterized by ultraoligotrophy and high sediment loading.

5. A rapid increase in freshwater diversity parallels the evolution of land plants in the Siluro-Devonian, signaling a dramatic rise in productivity and complexity of lacustrine ecosystems. Paleozoic lake communities were tightly linked to benthic or epibenthic habitats; open-water habitats were largely unexploited because of inadequate nutrient recycling.

6. Lacustrine diversity and ecosystem complexity increased greatly in the Late Paleozoic, especially among crustaceans, and with the evolution of aquatic insects. A protracted extinction between the Early Permian–Early Triassic caused freshwater diversity to decline sharply.

7. Freshwater diversity rebounded in the mid-Mesozoic with the fragmentation of the Pangaean supercontinent. Major diversification events among insects, macrophytes, and advanced bony fish profoundly reorganized freshwater communities at this time, promoting a *Lacustrine Mesozoic Revolution* that involved greatly expanded herbivory, intensified predation, and much stronger ecological coupling between the nearshore and offshore regions of lakes.

8. Tertiary lacustrine ecosystems were profoundly altered by tectono-climatic events associated with major continental collisions, which both created vast new freshwater systems and isolated pre-existing ones. The early Tertiary evolution of freshwater diatoms had a major effect on biogeochemical cycling in lakes.

9. Body fossil, trace fossil, and sedimentological evidence all suggest that there have been substantial functional changes in lake ecosystems since the Paleozoic, especially a progressive exploration of lake environments away from the littoral zone and toward the open water-column regions of lakes, coinciding with greater offshore nutrient availability.

15

Paleolimnology: The Past Meets the Future

Exciting days lie ahead for paleolimnology. As we embark on a new millennium, the opportunities and challenges in this field are extremely bright. As an epilogue to this book, it seems appropriate to conclude with a few of the developments that seem to me particularly promising for the near future.

1. *Increasing application of paleolimnological data to address problems in global climate change.* Paleolimnologists need to make governments and societies aware of the importance of high-resolution paleorecords from lakes for providing information about baseline variability of the biosphere, consequences and histories of past climate change events, and past responses of our precious aquatic resources to such changes. Paleolimnology should and will increasingly play a role in providing decision-makers with critical information about earth system history as they formulate policies to cope with these changes. Few, if any, paleoenvironmental records provide earth history records in environments as intimately associated with human activity as lake deposits. Lakes and wetlands are increasingly recognized as potentially important components of the global carbon cycle, especially as environments for sequestering large volumes of carbon, and future research will undoubtedly quantify the magnitude and dynamics of this role. Paleolimnologists will need to work even more closely with climate modelers, hydrologists, and atmospheric scientists in years to come, to insure that the paleorecords we study will help resolve important questions about the earth's climate system.
2. *Advances in geobiology.* The rapid developments of new and automated tools in molecular biology and organic geochemistry for analyzing small sample volumes and extracting compound-specific isotopic information from organic compounds have important implications for paleolimnology. In years to come we will increasingly rely on organic geochemistry and microbial geobiology to help decipher the organic record of algal primary producers, decomposers, and other elements of the microbial food web. These are components of a lake's ecosystem that ecologists recognize as immensely important in biogeochemical cycles and as being on the front line of lake responses to changes in climate and watershed processes, but which have heretofore been largely intractable to any detailed interpretation by paleolimnologists. Vast improvements in identifying the specific organisms responsible for producing fossilized organic matter will revolutionize our understanding of past lake ecosystems.
3. *Long records of past climate and environmental change.* The recent development of dedicated, modular, and portable lake drilling systems, especially the Global Lakes Drilling (GLAD-800) system (figure 1.2g), promises to revolutionize the nature and duration of lake history records accessible to the paleolimnological community (Dean et al., 2002). This system is capable of taking high-quality, continuous cores hundreds of meters in length, and can now do so in water depths of hundreds of meters. Several lakes (Great Salt Lake, Utah; Bear Lake, Utah; Lake Titicaca, and Lake Malawi) either have already or will shortly be explored using this extraordinary community scientific tool, and many others are slated for study in years to come. Long cores from deep, ancient lakes promise to provide high-resolution records covering hundreds of thou-

sands to millions of years, comparable to the ice cores. Even a cursory inspection of the impact that the long ice core records from Greenland and Antarctica have had on our understanding of global climate history, or the Ocean Drilling Program cores records have had for paleoceanography, make it abundantly clear just how much more we stand to gain from having similar records at low and mid-latitudes on the continents, the zones of human habitation. As long lake records come on line, paleolimnologists should be proactive in insuring that cross-disciplinary studies link the long continental and marine records with appropriate model-driven questions, to make the most of these exceptional archives. In the case of the African Great Lakes, long core records may unlock for us a detailed record of the environmental context of human evolution. Long core records from ancient lake basins may also help resolve long-standing debates concerning the biotic and tectonic evolution of lake systems (Cohen et al., 2000b). And finally, long core records from pre-Quaternary lake beds may hold the key to understanding the rhythms of continental climate change over millions to tens of millions of years (Olsen, 2001).

4. *New developments in geochronometry.* Rapid advances in our ability to accurately date lake sediments will play a critical role in advancing paleolimnology, by allowing us to increasingly refine the interpretation of time series data, and to use those data to test quantitative model (especially GCM) predictions. Improvements in the calibration of ^{14}C records and extension of good calibration corrections over longer intervals will be of particular importance to paleolimnology. As more long paleolimnological records come on line it will become increasingly important to refine and advance the toolkit of techniques amenable to the preradiocarbon time scale.

5. *Extension of "neo"-paleolimnological approaches to quantitative paleoenvironmental reconstruction to much longer time scales.* The extraordinary successes that paleolimnology has enjoyed in quantitatively reconstructing climate, water chemistry, and human impact histories for the Holocene, and especially for the last few hundred years, beg to be duplicated over longer time scales. Clearly hurdles exist in

this challenge; many of the quantitative transfer functions based on microfossils in use today may not be applicable to longer time series data. But the potential also exists for similar multivariate transfer function approaches to be applied to geochemical or sedimentological data, unaffected by the whims of biological evolution (e.g., Noon et al., 2001). Sedimentologists and geochemists should take note of this potential, so that in the future the indicator records that we often erroneously refer to today as "proxies" for specific past environmental conditions will eventually become just that.

6. *Paleolimnology meets planetary geology.* The recognition that lakes probably once existed on Mars, and perhaps elsewhere in the solar system, beckons the next generation of paleolimnologists with extraordinary opportunities to apply terrestrially gained knowledge of how "small" water bodies record their histories through their sedimentary records. Undoubtedly planetary geologists will be reaching into the paleolimnological literature to help them unravel these records. But our community should be proactive, reaching to the forefront in suggesting how such deposits should be explored, and devising analog studies of lake histories from extreme terrestrial environments that mimic those of other planets.

7. *Acquisition of regional, high-resolution paleolimnological data sets.* To take advantage of the full *power of paleolimnology,* an increasingly regional approach to data collection will continue to grow in importance. Targeting clusters of lakes for regional comparison has already gained considerable popularity in the paleoclimate community (e.g., Markgraf et al., 2000). Paleolimnologists need to maintain an active voice in proposing future studies that specifically target lakes for study that amplify the signals of interest and have the potential to provide the most highly resolved temporal records possible.

Clearly, this short shopping list reflects my own biases and interests; others sitting in my seat undoubtedly would produce a different list of where we may be headed. But almost all scientists engaged in the rapidly developing field of paleolimnology would agree that the most exciting discoveries and innovations are yet to come.

Glossary

Abandoned Channel Lake (Oxbow or Cutoff Lake) A lake formed by the abandonment of a river channel, either gradually or suddenly, to a new position, leaving a standing water body.

Accommodation Space The volume of air + water in a basin available for infilling by sediment.

Accommodation Zone The structurally complex region between two adjacent *half-grabens* in a rift valley, often forming a topographical ridge, or less commonly a depression.

Aggradation The vertical accumulation of sediments in a basin.

Airshed The geographical region of the atmosphere that can contribute material (dust, aerosols, pollen, etc.) to a lake.

Alkaline Lake A lake with a high relative proportion of anions contributing to alkalinity, typically carbonate and bicarbonate.

Allochthonous Derived from outside of the general site of burial.

Alluvial Fan A fan-shaped body of sediment that builds outward where a river discharges from a mountainous region into a flat-lying lowland.

Amictic Lake A lake that never undergoes vertical mixing.

Aragonite A mineral form of $CaCO_3$.

Authigenic A mineral that formed as a chemical precipitate or reaction product in situ, as opposed to having been transported to its burial site.

Autochthonous Derived from the general area of burial.

Autotroph Organisms capable of fixing carbon, such as plants or photosynthesizing bacteria, using an external energy source, such as sunlight or geothermal heat.

Back-Arc Basin Lake A tectonic lake formed in an area of crustal extension, generally on the continental side of a volcanic/magmatic arc.

Bathymetry The measurement of water depth.

Benthic Zone The lake floor habitat.

Bioclastic Sedimentary particles composed of biological materials, most commonly shell or other skeletal fragments.

Biogenic Silica (Bsi) Amorphous silica in sediments derived from biological sources. In lakes most BSi is derived from diatoms, chrysophytes, and sponges.

Biomarkers Organic geochemicals in sediments that are useful in determining the precursor organisms, environment of deposition or particle origin, or diagenetic history of the sediment.

Biostratigraphy The study of relative age determination in sediments using fossils.

Bioturbation Postdepositional mixing of sediment layers by the movement of organisms, particularly from burrowing.

Box Corer A type of coring device used to collect a large volume of undisturbed sediment from the sediment–water interface.

Branchiopods A group of crustaceans including the cladocera, brine shrimp, etc. united by their flattened locomotory appendages.

Brine Evolution The change in water chemistry that occurs as a result of evaporative concentration and mineral precipitation in a saline lake.

Calcite A common mineral form of $CaCO_3$, and the principal component of limestone.

Caldera A large and often complex volcanic crater, formed from both eruption and the posteruptive collapse of the top of a volcano.

Canonical Correspondence Analysis (CCA) A type of multivariate statistical analysis used commonly by biological paleolimnologists to explore the possible relationships between species abundances and the environmental factors that may be regulating them.

Carbonate (Geol.) Any of a large number of (primarily) sedimentary minerals that incorporates carbonate or bicarbonate ions, for example calcite, dolomite, or trona. Also used to describe sediments composed of such minerals.

Charophyte A macroscopic, benthic green algae, found commonly in the littoral zone of lakes.

Chironomid(ae) A family of lake flies (midges) whose larvae are aquatic, commonly preserved as fossils in Quaternary lake deposits.

Chrysophyte Small, flagellated eukaryotic algae, which siliceous scales or cysts are commonly preserved as fossils in Quaternary lake deposits.

Clade A group of organisms that includes all and only descendants of an ancestral species. A monophyletic group.

Cladocera A group of small branchiopod crustaceans, commonly found in both littoral and planktonic habitats of lakes, whose skeletal parts are often preserved in Quaternary lake deposits.

Closed-Basin Lake A lake with no surficial outlet.

Coated Grain (also see Ooid) A sedimentary particle consisting of cemented coatings of a precipitated mineral, most commonly calcite or aragonite.

Coleoptera The insect order of beetles. Many adult and larval coleopterans inhabit the littoral zones of lakes and their exoskeletons are commonly preserved as fossils.

Copepods An important group of zooplankton crustaceans in lakes (some also live in benthic habitats). Rarely preserved as fossils.

Coriolis Effect The apparent force, driven by the earth's rotation, that causes moving objects or fluids (e.g., water, air) on the earth's surface to be deflected clockwise in the Northern Hemisphere and counterclockwise in the Southern Hemisphere.

Correlation (of stratigraphical horizons) The determination of the equivalence of two stratigraphical bodies (beds, formations, etc.) in space or time.

Cratonic Basin Lake A lake formed in the tectonically stable central portion of a continent (the craton), generally from mild downwarping of the continental surface.

Cross-Stratification A common type of layering in sediments in which individual layers dip downward at high angles onto underlying layers. Formed as a consequence of the development of ripples or dunes.

Cryptic Habitat Habitats that are hidden from direct exposure to sunlight or waves, for example in caves, crevices, or overhangs.

Cyanobacteria The blue–green "algae." A group of prokaryotic bacteria that possess photosynthetic pigments and that are common in lakes.

Debris Flow An extremely viscous gravity flow, comprising a mixture of variable-sized sediment particles and water.

Deflation Erosion of a surface (often a playa) by the action of wind.

Denitrification The reduction of nitrate to various N gases, primarily N_2. A major pathway for the loss of fixed nitrogen from lakes.

Depocenter The zone of maximal vertical accumulation of sediment in a lake. Often but not always the deepest part of a lake.

Detritivore An organism that feeds on particulate organic matter.

Diagenesis The low-temperature and low-pressure formation of new minerals or textures in sediments after burial.

Diapir A mass of salt, mud, or other materials that rises upward through a sequence of sediments.

Diatom (Bacillariophyceae) Common unicellular or colonial algae in lakes, possessing distinctive siliceous cell walls that are readily fossilized in Tertiary and Quaternary sediments.

Diatomite A sediment composed almost entirely of diatom fossils.

Diptera The insect order of true flies. Dipteran larvae are common lacustrine organisms and also commonly preserved as fossils in lake deposits.

Distal (Geol.) Far from the source, as in the distal part of a delta being the portion furthest offshore from the river mouth.

Dolomite The carbonate mineral $CaMg(CO_3)_2$.

Dysaerobic An aquatic environment with very low (but not zero) free oxygen levels, typically defined as $< 1\,\mathrm{mg}\,\mathrm{l}^{-1}\,O_2$.

Endemic Restricted to a limited geographical region, as in a species that is endemic to one lake.

Endogenic (Fossils) Fossils that are derived from within the lake environment where they were buried.

ENSO (El Niño Southern Oscillation) A quasicyclical (3–8 year) change in sea surface temperatures in the Pacific that drives major global climatic fluctuations in precipitation over a large portion of the earth.

Eolian Pertaining to the action of wind.

Ephemeroptera The insect order of mayflies.

Epibenthic Referring to organisms that live just above the lake floor.

Epilimnion The upper, well-mixed portion of a lake's water column.

Epilithic Attached to the surfaces of stones or rocks.

Epiphytic Attached to the surfaces of aquatic plants (macrophytes).

Eukaryote Unicellular or multicellular organisms whose cells possess a nucleus.

Eurytopic An organism able to tolerate a wide variety of environmental conditions.

Eutrophic Lake A lake that is generally characterized by high planktonic biomass, high nutrient loading, and high rates of primary production.

Eutrophication The process of conversion of a lake to a eutrophic state, for either natural or anthropogenic reasons.

Evaporite A mineral precipitate formed by the evaporative concentration of a water body.

Exogenic (Fossils) A fossil derived from outside of the lake in which it was buried, for example, terrestrial pollen grains.

Facies The mineralogical, textural, geochemical, or fossil characteristics of a sedimentary deposit.

Fetch The maximal distance wind can move across the surface of a water body, thereby driving mixing and the formation of surface waves.

Filter (of paleolimnological records) Any factor that distorts our interpretation of original environmental conditions based on a sedimentary, geochemical, or paleontological indicator.

Flocculation The accumulation of clay-sized particles in a water body because of electrostatic attraction or mechanical aggradation.

Fluvial Pertaining to the action of running water.

Flux (of particles or fossils) The rate of accumulation of some material per unit time, per unit area.

Foraminifera A group of pseudopod-bearing, unicellular protists. Some benthic species occur in saline and/or coastal lakes and their skeletons are readily preserved as fossils.

Foreland Basin Lake A lake occupying a depression formed by the loading of mass from mountain building on the crust.

Foreset Inclined layers of sediment sloping downward away from a source, as on the front of a delta.

Freeze Corer A sediment coring device used to obtain undisturbed stratigraphies at the sediment–water interface by freezing the sediment to an extremely cold plate or rod.

Geochronology The study of both relative and absolute age determination for rocks and sediments.

Geochronometry The study of absolute age determination (in years) for rocks and sediments.

Geometry (Stratigraphical) The three-dimensional shape of a body of rock or sediment.

Geostrophic Circulation Circulation in a water body driven by the Coriolis Effect.

General Circulation Model (GCM) Any of a number of sophisticated mathematical models that describe the circulation and behavior of the earth's atmosphere in three dimensions, and the effect of that circulation on winds, precipitation, and temperature.

Graded Bedding A sedimentary deposit characterized by a gradual change in average particle size, fining upward through the deposit.

Gravity Corer A hollow coring device that operates by penetrating the sediments from its weight alone, allowing the sediment to displace water at the top of the coring tube.

Gravity Flow A class of sediment/fluid mixtures that move down slopes as a result of gravity, eventually becoming sedimented.

Groundwater Discharge The flow of groundwater from a subterranean source to the surface, via springs and seeps.

Groundwater Recharge The flow of surface water into the subsurface.

Gypsum The mineral form of $CaSO_4 \cdot 2H_2O$. A common evaporite mineral formed in saline lakes.

Gyttja A gray or dark brown lake sediment composed of a mix of humic material, plant fragments, terrigenous sediments, diatoms, and other algal and zooplankton remains.

Half-Graben A tectonic basin formed by crustal extension, in which one side of the basin is bounded by a major fault and the other side by minor faults and more gentle flexure of the crust.

Halite The mineral form of $NaCl$, a common evaporite mineral formed in saline lakes.

Halophyte A plant adapted to living in saline to hypersaline water.

Hardwater Effects (on ^{14}C) Offsets between a measured ^{14}C age of some lake-derived carbon source and the calendar age of the material, caused by the assimilation of old HCO_3^- in the dated material from the lake water.

Hardwater Lake A lake whose cation chemistry is dominated by calcium and magnesium.

Hemipelagic Extremely fine-grained, terrestrially derived sediments that accumulate from suspension settling in the offshore portions of a lake.

Heterotroph Organisms such as animals, which are incapable of fixing carbon directly using an external energy source.

Hydroclimate The physical environmental conditions such as temperature, salinity, etc. that occur in a water body.

Hydrothermal Groundwater that has been heated by surrounding rocks (geothermally).

Hypolimnion The deeper part of the water column in a lake that is not undergoing mixing with surface waters.

Illite A three-layered clay mineral with the approximate formula $K_2(Si_6Al_2)Al_4O_{20}(OH)_4$.

Indicator Data, such as fossil presence/absence, isotopic data, etc. that provides information about some paleoenvironmental variable, but which may also be influenced by other, indeterminate factors. A weaker source of inference than a *proxy*.

Infauna Organisms that live below the sediment–water interface.

Interbedding Interlayering of two distinct sediment types, for example sand and mud beds.

Interflow (Homopycnal Flow) A density flow that upon entering a lake, reaches a depth of equivalent density at some intermediate depth above the lake floor.

Internal Waves Waves that move within a water body.

Interstitial The habitat between sedimentary particles (generally sands) on the floor of a water body.

Intertropical Convergence Zone (ITCZ) The zone of convergence between the Northerly and Southerly Trade Winds. A warm and seasonally migratory belt of rising air (low pressure), variable winds, and abundant rainfall, located near the Equator.

Isostasy The balance of portions of the earth's crust based on their "floating" on a dense underlying layer (the earth's aesthenosphere, in the upper mantle), causing zones of thicker crust to have higher surface elevations.

Isotopic Fractionation Differential uptake of light versus heavy isotopes of an element during physicochemical processes such as evaporation, causing the ratio of isotopes to differ in each of the end-products relative to the initial component.

Karst Irregular topography, including formation of caves and sinkholes in carbonate bedrock terrain, caused by dissolution.

Kettle Hole A topographical basin caused by the collapse of large melting fragments of ice within a moraine, following the retreat of a glacier.

Keystone Species A species whose presence or absence strongly affects the structure of an ecological community.

Lagerstätte (Konservat-Lagerstätte) An exceptionally preserved assemblage of fossils, including delicate or poorly skeletonized organisms, not normally found as fossils.

Laminated (sediments) Sedimentary deposits consisting of extremely thin layers of (generally very fine-grained) sediment.

Limiting Nutrient (Factor) A nutrient, such as phosphorus, which is in such limited supply in an ecosystem as to limit the growth rate of a particular species or assemblage of species.

Limnogeology The study of the geology of lakes, their origins, and deposits.

Littoral Zone The shallow portion of a lake within the photic zone, where macrophytes and/or benthic algae occur.

Loading (of nutrients or pollutants) The supply of nutrients (or pollutants) to a lake from both internal (upwelling, recycling from sediments) and external (watershed, airshed) sources.

Longshore Current The flow of water parallel to a lake's shoreline, caused by the lateral momentum of waves breaking on beaches.

Maar Lake A lake that occupies a maar, a simple, generally small volcanic crater formed by an explosive eruption.

Macrophyte A macroscopic plant or alga growing in a lake, either attached or floating.

Magnetic Susceptibility The measure of the degree to which a substance is attracted to a magnet.

Marl A fine-grained carbonate mud typical of hardwater lakes.

Massive (sediments) Describing a sediment with extremely thick beds or lacking obvious layering.

Meromictic Lake A lake that is permanently stratified, with an upper water column that is well mixed (the *mixolimnion*) and a lower water column (the *monimolimnion*) that either does not mix, or mixes extremely slowly with the mixolimnion.

Metalimnion A transitional zone in a meromictic (or seasonally stratified lake) between the mixolimnion (epilimnion) and the monimolimnion (hypolimnion).

Milankovitch Cycle One of several types of cycles in the earth's orbital characteristics around the sun and with respect to its rotation (eccentricity, obliquity, precession). A major determinant of the earth's long-term climate history.

Mixed Layer (of lake). See *epilimnion*.

Mixolimnion See *Meromictic Lake*.

Monimolimnion See *Meromictic Lake*.

Monophyletic A group of organisms that includes all descendants of a single ancestor and only descendants of that ancestor.

Moraine A pile of unstratified or poorly stratified sediment deposited at the terminus or on the flanks of a glacier.

Morphometric Describing the shape of a thing, for example, a lake or an organism.

Nitrification (Nitrogen Fixation) Any of several reactions that fix nitrogen into a more oxidized state.

Normal Fault A fault produced by extension of the earth's crust, in which one block of rocks lying above the fault plane (the hanging wall) slips downward, relative to the other block, lying below the fault plane (the foot wall).

Nutrient A compound or element that is necessary for growth of an organism.

Nutrient Loading See *Loading (of nutrients)*.

Odonata The insect order including the dragonflies.

Oil Shale Any very fine-grained sedimentary rock with high concentrations of liquid hydrocarbons.

Oligomictic A lake that rarely or irregularly mixes.

Oligotrophic Lake A lake that is generally characterized by low planktonic biomass, low nutrient loading, and low rates of primary production.

Ooid A type of sand-sized *coated grain*.

Orogenic Mountain building. A single episode of orogenic activity is referred to as an orogeny.

Orographic Elevation, typically in reference to the effects of mountain heights on atmospheric processes.

Osmoregulation The ability of organisms to control the salt concentration (osmotic pressure) of their bodily fluids.

Ostracode A small, crawling, or swimming crustacean, commonly found in lakes, whose bivalved shells are often preserved as fossils.

Overdeepened Lake Lake basins that formed by repeated episodes of glacial scour in valleys that are particularly prone to erosion, often as a result of prior tectonic history.

Overflow (Hypopycnal Flow) An inflow to a lake that is lighter than the lake water, and flows outward at the surface of the lake.

Overturn The process of vertical mixing of the entire water column of a lake at such time as the lake reaches a uniform density profile from top to bottom.

Oxbow Lake See *Abandoned Channel Lake*.

Paleohydrology The study or inference of ancient conditions of water discharge, water chemistry, or surface/groundwater interactions from sedimentological or geomorphic evidence.

Paleomagnetic Dating The dating of sediments or volcanic rocks based on their recorded sequences of magnetic polarity reversals or shorter-term excursions in the magnetic field.

Paleoshoreline An ancient shoreline, inferred from sedimentological, geomorphic, or geophysical evidence.

Pangaea A supercontinent that formed from the coalescence of numerous continents as a result of plate tectonic motion toward the end of the Paleozoic.

Pelagic Zone The open-water, offshore region of a large lake or an ocean.

Periglacial Lake Lakes formed in regions of permafrost from ice melting and collapse, or as a result of isostatic rebound of irregular land surfaces following deglaciation.

Periphyton Microfloral (algal, bacterial) growth on any substrate.

Phanerozoic A time-stratigraphical term encompassing the Paleozoic, Mesozoic, and Cenozoic Eras, approximately the last 544 Ma.

Photic Zone The region of the water column penetrated by sufficient light to permit photosynthesis and net primary production.

Phylogenetics The study of the evolutionary history of organisms and generation of evolutionary trees to depict their relationships.

Phytolith Siliceous, secondary cell walls of plants, especially grasses, which are commonly preserved as fossils.

Phytoplankton Autotrophic, unicellular, or colonial algae and bacteria that float in the water column.

Pigments (in plants and algae) Compounds used by algae, bacteria, and higher plants in photosynthesis.

Piggyback Basin Lake A type of foreland basin lake that forms in an area of subsidence on top of an advancing fold–thrust belt.

Piscivore An animal that feeds primarily on fish.

Piston Corer A sediment coring device that uses a piston inside the core barrel to maintain equivalent inside/outside hydrostatic pressure, thereby avoiding the deformation effects of inequal pressure.

Playa A lake that fills only irregularly. Generally found in arid environments.

Polymictic Lake A lake that undergoes frequent or continuous vertical mixing.

Productivity The rate of formation of organic matter per unit area per unit time.

Profundal Zone The deep-water, benthic, or near-bottom habitat of a lake, below the photic zone and the littoral zone.

Proglacial Lake A lake that lies in front of a glacier.

Progradation The building outward of sediments, as in a delta.

Prokaryote Single celled or colonial organisms, including the bacteria, whose cells lack a nucleus.

Protist Any single-celled organism.

Proximal Close to the source, as in sediment inputs.

Proxy A direct substitute for another thing. In paleoenvironmental interpretation, often used (incorrectly) to describe an *indicator* record.

Pull-Apart Basin (Transtensional Basin) Lake A lake basin formed at a kink along a strike slip (transform) fault, where tensional fractures result in subsidence.

Radiometric Dating Any absolute dating method relying on the measurement of the concentration of radiogenic isotopes or their daughter products, whose radioactive decay rates are known.

Redox (Potential) The reduction/oxidation state. In water, defined as the electrical (voltage) potential to transfer electrons, with the capacity of the water to oxidize being defined as having a positive potential, or to reduce defined as having a negative potential relative to a standard.

Reservoir Effects (on ^{14}C) Offsets between a measured ^{14}C age of some lake-derived carbon source and the calendar age of the material, caused by the accumulation of radiogenically "old" carbon in a stratified, or poorly mixed water body, that is not in equilibrium with the atmosphere.

Residence Time The time an average atom or particle of a substance remains in a water body before being deposited, discharged, or otherwise lost.

Rhythmite A rhythmically bedded sediment, which may or may not be annually laminated.

Rift Lake A lake that formed in a rift valley, an extensional tectonic environment.

Sapropel Black, fine-grained organic-rich lake sediments, containing abundant CH_4 and H_2S and other products of microbial reduction.

Sediment Focusing The preferential accumulation of sediment particles in a particular part of a lake.

Seiche A standing wave in a lake, often caused by the gravitational relaxation of the water body following prolonged, directionally persistent winds, which can occur at the surface (surface seiche) or below the surface (internal seiche).

Seismic Reflection Profile A cross-sectional or 3D visualization of the subsurface of the earth, produced by the geophysical collection and interpretation of reflected seismic (sound) waves.

Shoreface The submerged zone of a beach that is affected by continuous wave activity.

Siliciclastic Sediment composed of terrigenous, silicate minerals, such as quartz or feldspar.

Smectite A family of clay minerals with expandable, three-layer lattice structures and a generalized formula $R_{0.33}Al_2Si_4O_{10}(OH)_2 \cdot nH_2O$, with R typically being K^+, Na^+, Ca^{2+}, or Mg^{2+}.

Species Flock A group of species limited to a particular region (for example, a lake or watershed) and united by a common ancestor, which also lived in that region.

Spheroidal Carbonaceous Particles (SCPs) Round, microscopic fly ash particles, produced by combustion, and often used in paleolimnological studies as an indicator of fossil fuel burning.

Stenotopic An organism that is unable to tolerate a wide variety of environmental conditions.

Stoke's Law A mathematical model describing the settling rate of spherical particles through a fluid.

Stratification (of lake water) Layering of a water body, caused by density differences, that inhibit vertical mixing.

Stratigraphical Completeness The degree to which strata provide a record of deposition and earth history for defined intervals over some time period.

Stratigraphical (Sedimentary) Sequence A body of strata bounded by unconformities (depositional breaks), or their correlative conformable surfaces.

Strike–Slip Fault A fault in which the direction of motion (slip) is parallel to the map-view orientation of the fault.

Stromatolite A layered organic buildup (usually of calcite or aragonite) formed by the growth and accretion of microbial organisms (primarily cyanobacteria).

Sublacustrine Fan A fan-shaped body of sediment formed at the base of an underwater slope, usually from the accumulation of gravity flow deposits.

Taphonomy The study of the range of processes that affect fossils between death and recovery by a paleontologist.

Tarn A small lake, formed in an ice-gouged basin.

Teleconnection A connection in climatic processes that is manifest over a long distance, often through intermediate effects on the oceans.

Temporal (Time) Resolution The degree to which discrete events can be inferred from a stratigraphical record.

Tephra Volcanic ash.

Terrace A horizontal or subhorizontal geomorphic surface, formed by either erosion or deposition.

Terrigenous Derived from terrestrial sources, through weathering.

Thermocline The depth zone in a water body of most rapid vertical temperature change.

Thermohaline Circulation Water circulation driven by density differences (from temperature or salinity) and resulting gravitational instability.

Thrust Fault A reverse fault, in which a mass of rocks is pushed (thrust) over other rocks, generally at a low angle with respect to the horizontal.

Till Unsorted and unstratified debris, carried and deposited by a glacier.

Time Averaging The mixing of particles or fossils of different age into a single sample, either by natural processes after deposition, or analytically by a scientist during an investigation.

Trace Fossil Burrows, borings, trails, etc. produced by the activity or growth of organisms.

Training Set A data set consisting of modern environmental and species abundance data, used to develop a *transfer function* for paleoecological inferences.

Transfer Function A mathematical model that is used to quantitatively infer some past environmental conditions based on archived variables (generally fossil species abundances).

Transform Fault A type of strike–slip fault that connects two other major fault systems, generally along tectonic plate boundaries.

Transtensional Basin Lake See *Pull-Apart Basin*.

Trichoptera The order of insects including the caddisflies.

Trophic Referring to feeding or energy transfer state in organisms.

Tufa Mound-like buildup of calcium carbonate, generally around a spring discharge.

Turbidity Flow A gravity flow, comprising a relatively low-density, sediment/water mixture.

Underflow (Hyperpycnal Flow) An inflow to a water body that is denser than the surrounding water.

Ultramodern (nuclear testing) ^{14}C The excess loading of ^{14}C from atmospheric nuclear testing, evident in ^{14}C analyses of post-1950s sediments.

Varve An annual set of layers (couplets or triplets) in sediments.

Watershed The area which provides surface runoff to a stream or lake.

Weighted Averaging A unimodal regression technique used as a transfer function to quantitatively infer past environmental conditions from fossil assemblages.

Younger Dryas A climate oscillation of very cold climate conditions during the Pleistocene–Holocene transition ($\sim$ 10.8–10.1 ka ^{14}C yr BP), which followed and preceded much warmer conditions.

Zooplankton Animals that are suspended in the water column, and are incapable of swimming against a current.

Bibliography

Aaby, B. and Digerfeldt, B., 1986, Sampling techniques for lakes and bogs. In Berglund, B.E. (ed.), *Handbook of Holocene Paleoecology and Paleohydrology*. J. Wiley and Sons, New York, pp. 181–194.

Abbott, M.R. and Stafford, T.W., 1996, Radiocarbon geochemistry of modern and ancient Arctic lake systems, Baffin Island, Canada. *Quat. Res.* 45:300–311.

Abelson, P.H., 1954, Organic constituents of fossils. *Carnegie Inst. Yearbook* 53:97–101.

Abrahão, D. and Warme, J., 1990, Lacustrine and associated deposits in a rifted continental margin–lower Cretaceous Lagoa Feia Formation, Campos Basin, Offshore Brazil. In Katz, B.J. (ed.), *Lacustrine Basin Exploration: Case Studies and Modern Analogues*. Amer. Assoc. Petrol. Geol. Mem. 50:287–305.

Adams, K.D. and Wesnousky, S.G., 1998, Shoreline processes and the age of the Lake Lahontan highstand in the Jessup embayment, Nevada. *Geol. Soc. Amer. Bull.* 110:1318–1332.

Agassiz, L., 1850, *Lake Superior: Its Physical Character, Vegetation, and Animals Compared to Those of Other and Similar Regions*. Gould, Kendall, and Lincoln, Boston.

Ager, D.V., 1973, *The Nature of the Stratigraphic Record*. J. Wiley and Sons, New York.

Aguirre-Urreta, M.B., 1992, Tertiary freshwater Decapoda (Crustacea: Parastacidae) from the Ñirihuau Basin, Patagonia, Argentina. *Jour. Paleo.* 66:817–825.

Aitken, M.J., 1994, Optical dating: a nonspecialist review. *Quat. Sci. Rev.* 13:503–508.

Aitken, M.J., 1998, *Introduction to Optical Dating*. Oxford University Press, Oxford.

Alefs, J. and Müller, J., 1999, Differences in the eutrophication dynamics of Ammersee and Starnberger See (Southern Germany) reflected by the diatom succession in varve-dated sediments. *Jour. Paleolim.* 21:395–407.

Alexandrowicz, S.W., 1981, Malacofauna of the Late Quaternary lacustrine deposits in the Krosno Depression. *Bull. Acad. Pol. Sci.* 28:243–255.

Alexandrowicz, S.W., 1987, Malacofauna of the Late Vistuian and Early Holocene lacustrine chalk from Roztoki near Jaslo (Jaslo–Sanok Depression). *Acta Palaeobot.* 27:67–74.

Alhonen, P., 1970, On the significance of the planktonic/littoral ratio in the cladoceran stratigraphy of lake sediments. *Soc. Sci. Fenn. Comm. Biol.* 35:1–9.

Alhonen, P., 1971, The Flandrian development of the pond Hyrynlampi, southern Finland, with special reference to the pollen and cladoceran stratigraphy. *Acta Bot. Fenn.* 95:1–19.

Alin, S.R., Cohen, A.S., Bills, R., Gashagaza, M.M., Michel, E., Tiercelin, J.J., Martens, K., Coveliers, P., Mboko, S.K., West, K., Soreghan, M., Kimbadi, S., and Ntakimazi, G., 1999a, Effects of landscape disturbance on animal communities in Lake Tanganyika, East Africa. *Conserv. Biol.* 13:1017–1033.

Alin, S.R., Palacios-Fest, M.R., and Cohen, A.S., 1999b, The live, the dead, and the very dead; translating live ostracod assemblages into the fossil record in Lake Tanganyika, East Africa. *Geol. Soc. Amer. Ann. Mtg. Abstr. w/prog.* 31:438.

Alin, S.R., Cohen, A.S., Palacios-Fest, M., Msaky, E., Dettman, D., and McKee, B., 2000, A multi-indicator study of watershed disturbance and paleoecological change in Lake Tanganyika, East Africa, through c. 500 years. *AGU 2000 Ann. Fall Mtg. Abstr. w/prog.* 81(48):F263.

Alley, N.F., 1998, Cainozoic stratigraphy, palaeoenvironments and geological evolution of the Lake Eyre Basin. *Palaeogeog., Palaeoclim., Palaeoecol.* 144:239–263.

Alley, R., 2000, The Younger Dryas cold interval as viewed from central Greenland. *Quat. Sci. Rev.* 19:213–226.

Alley, R.B., Meese, D.A., Shuman, C.A., Gow, A.J., Taylor, K.C., Grootes, P.M., White, J.W.C., Ram, M., Waddington, E.D., Mayewski, P.A., and Zielinski, G.A., 1993, Abrupt increases in Greenland snow accumulation at the end of the Younger Dryas event. *Nature* 362:527–529.

Alley, R.B., Mayewski, P.A., Sowers, T., Stuiver, M., Taylor, K.C., and Clark, P.U., 1997, Holocene climatic instability: a prominent, widespread event 8200 yr ago. *Geology* 25:483–486.

Amoros, C. and Jacquet, C., 1987, The dead arm evolution of river systems: a comparison between the information provided by living Copepoda and Cladocera populations and by Bosminidae and Chydoridae remains. *Hydrobiologia* 145:333–341.

Anadón, P., 1989, Los lagos salinos interiores con faunas de afinidad marina del Cenozoico de la Península Ibérica. *Acta Geol. Hisp.* 24:83–102.

Anadón, P., 1992, Composition of inland waters with marine-like fauna and inferences for a Miocene lake in Spain. *Palaeogeog., Palaeoclim., Palaeoecol.* 99:1–8.

Anadón, P., 1994, The Miocene lacustrine evaporite system of La Bureba (western Ebro Basin, Spain). In Gierlowski-Kordesch, E. and Kelts, K. (eds.), *Global Geological Record of Lake Basins* v. 1, Cambridge University Press, Cambridge, pp. 311–314.

Anadón, P., De Deckker, P., and Julià, R., 1986, The Pleistocene lake deposits of the NE Baza Basin (Spain): salinity variations and ostracod succession. *Hydrobiologia* 143:199–208.

Anadón, P., Cabrera, L., Julià, R., 1988a, Anoxic–oxic cyclical lacustrine sedimentation in the Miocene Rubielos de Mora Basin, Spain. In Fleet, A.J., Kelts, K., and Talbot, M.R. (eds.), *Lacustrine Petroleum Source Rocks.* Geol. Soc. Lond. Spec. Publ. 40:353–367.

Anadón, P., Cabrera, L., Julià, R., Roca, E., and Rosell, L., 1988b, Lacustrine oil–shale basins in Tertiary grabens from NE Spain (Western European Rift System). *Palaeogeog., Palaeoclim., Palaeoecol.* 70:7–28.

Anadón, P., Cabrera, L., Julià, R., and Marzo, M., 1991, Sequential arrangement and asymmetrical fill in the Miocene Rubielos de Mora Basin (northeast Spain). In Anadón, P., Cabrera, L., and Kelts, K. (eds.), *Lacustrine Facies Analysis.* Intl. Assoc. Sedimentol. (IAS) Spec. Publ. 13. Blackwell Science, London, pp. 257–275.

Anadón, P., Utrilla, R., and Julià, R., 1994, Palaeoenvironmental reconstruction of a Pleistocene lacustrine sequence from faunal assemblages and ostracode shell geochemistry. *Palaeogeog., Palaeoclim., Palaeoecol.* 111:191–205.

Anati, D.A., Stiller, M., Shasha, S., and Gat, J.R., 1987, The thermohaline structure of the Dead Sea: 1979–1984. *Earth and Planet. Sci. Lett.* 84:109–121.

Anderson, F.W., 1973, The Jurassic–Cretaceous transition: the non-marine ostracod faunas. *Geol. Jour., Spec. Issue* 5:101–110.

Anderson, J.M., Anderson, H.M., and Cruickshank, A.R.I., 1998, Late Triassic ecosystems of the Molteno/Lower Elliot Biome of Southern Africa. *Palaeontology* 41:387–421.

Anderson, N.J., 1995a, Temporal scale, phytoplankton ecology and palaeolimnology. *Freshwater Biol.* 34:367–378.

Anderson, N.J., 1995b, Using the past to predict the future: lake sediments and the modeling of limnological disturbance. *Ecol. Modeling* 78:149–172.

Anderson, N.J., 1997a, Reconstructing historical phosphorus concentrations in rural lakes using diatom models. In Tunney, H., Carton, O.T., Brookes, P.C., and Johnston, A.E. (eds.), *Phosphorus Loss from Soil to Water.* CAB International, Wallingford, UK, pp. 95–118.

Anderson, N.J., 1997b, Historical changes in epilimnetic phosphorus concentrations in six rural lakes in Northern Ireland. *Freshwater Biol.* 38:427–440.

Anderson, N.J. and Renberg, I., 1992, A palaeolimnological assessment of diatom production responses to lake acidification. *Envir. Pollut.* 78:113–119.

Anderson, N.J. and Rippey, B., 1994, Monitoring lake recovery from point-source eutrophication: the use of diatom-inferred epilimnetic total phosphorus and sediment chemistry. *Freshwater Biol.* 32:625–639.

Anderson, N.J., Rippey, B., and Gibson, C.E., 1993, A comparison of sedimentary and diatom-inferred phosphorus profiles: implications for defining pre-disturbance nutrient conditions. *Hydrobiologia* 253:357–366.

Anderson, N.J., Odgaard, B.V., Segerstrom, U., and Renberg, I., 1996, Climate–lake interactions recorded in varved sediments from a Swedish boreal forest lake. *Global Change Biol.* 2:399–405.

Anderson, R.S., 1993, A 35,000 year vegetation and climate history from Potato Lake, Mogollon Rim, Arizona. *Quat. Res.* 40:351–359.

Anderson, R.Y., 1986, The varve microcosm: propagator of cyclic bedding. *Paleoceanography* 1:373–382.

Anderson, R.Y. and Dean, W.E., 1988, Lacustrine varve formation through time. *Palaeogeog., Palaeoclim., Palaeoecol.* 62:215–235.

Andrejko, M.J. and Upchurch, B.S., 1977, Silicate structures and mineralogy of sawgrass (*Cladium jamlcensis* Orgnatz) and its associated peats from the Florida Everglades. *Fla. Sci.* 40:24.

Andren, A.W. and Strand, J.W., 1981, Atmospheric deposition of particulate organic matter and polyaromatic hydrocarbons in Lake Michigan.

In Eisenreich, S.J. (ed.), *Atmospheric Pollutants in Natural Waters*. Ann Arbor Science Publishers, Ann Arbor, MI, pp. 459–479.

Anstey, N.A., 1982, *Simple Seismics*. International Human Resources Development Corporation, Boston.

Antevs, E., 1922, The recession of the last ice sheet in New England. *Amer. Geogr. Soc. Res. Ser.* 11.

Anthony, R.S., 1977, Iron-rich rhythmically laminated sediments in Lake of the Clouds, Northeastern Minnesota. *Limnol. Oceanogr.* 22:45–54.

Appleby, P.G. and Oldfield, F., 1977, The calculation of ^{210}Pb dates assuming a constant rate of supply of unsupported ^{210}Pb to the sediment. *Catena* 5:1–8.

Appleby, P.G. and Oldfield, F., 1983, The assessment of ^{210}Pb dates from sites with varying sedimentation rates. *Hydrobiologia* 103:29–35.

Appleby, P.G., Oldfield, F., Thompson, R., Huttunen, P., and Tolonen, K., 1979, ^{210}Pb dating of annually laminated lake sediments from Finland. *Nature* 280:53–55.

Appleby, P.G., Richardson, N., and Smith, J.T., 1993, The use of radionuclides from Chernobyl and weapons test fallout for assessing the reliability of ^{210}Pb in dating very recent sediments. *Verh. Int. Ver. Limnol.* 25:266–269.

Appleby, P.G., Jones, V.J., and Ellis-Evans, J.C., 1995, Radiometric dating of lake sediments from Signy Island (maritime Antarctic): evidence of recent climatic change. *Jour. Paleolimnol.* 13:179–191.

Appleby, P.G., Flower, R.J., Mackay, A.W., and Rose, N.L., 1998, Paleolimnological assessment of recent environmental change in Lake Baikal: sediment chronology. *Jour. Paleolim.* 20:119–133.

Applegate, D., 1995, Transform-normal extension on the Northern Death Valley fault system, California–Nevada. *Basin Res.* 7:269–280.

Aravena, R., Warner, B.G., MacDonald, G., and Hanf, K., 1992, Carbon isotope composition of lake sediments in relation to lake productivity and radiocarbon dating. *Quat. Res.* 37:333–345.

Arenas, C., Casanova, J., and Pardo, G., 1997, Stable-isotope characterization of the Miocene lacustrine systems of Los Montenegros (Ebro Basin, Spain): palaeogeographic and palaeoclimatic implications. *Palaeogeog., Palaeoclim., Palaeoecol.* 128:133–155.

Arnold, L.D., 1995, Conventional radiocarbon dating. In Rutter, N.W. and Catto, N.R. (eds.), *Dating Methods for Quaternary Deposits*.

GEOTEXT 2, Geological Society of Canada, pp. 107–115.

Ashley, G.M., 1975, Rhythmic sedimentation in glacial Lake Hitchcock, Massachusetts–Connecticut. In Jopling, A.V. and McDonald, B.C. (eds.), *Glaciofluvial and Glaciolacustrine Sedimentation*. Soc. Econ. Paleo. Mineral. Spec. Publ. 23:304–320.

Ashley, G.M., 1995, Glaciolacustrine environments. In Menzies, J. (ed.), *Modern Glacial Environments: Processes, Dynamics and Sediments*. Butterworth-Heinemann, Burlington, MA, pp. 417–444.

Ashley, G.M., Shaw, J., and Smith, N.D., 1985, *Glacial Sedimentary Environments*. SEPM Short Course Notes 16.

Assel, R. and Robertson, D.M., 1995, Changes in winter air temperatures near Lake Michigan, 1851–1993, as determined from regional lake-ice records. *Limnol. Oceanogr.* 40:165–176.

Atwater, B.F., 1986, Pleistocene glacial-lake deposits of the Sanpoil River Valley, Northeastern Washington. *U.S. Geol. Surv. Bull.* 1661.

Atwater, B.F., Adam, D.P., Bradbury, J.P., Forester, R.M., Mark, R.K., Lettis, W.R., Fisher, G.R., Gobalet, K.W., and Robinson, S.W., 1986, A fan dam for Tulare Lake, California, and implications for the Wisconsin glacial history of the Sierra Nevada. *Geol. Soc. Amer. Bull.* 97:97–109.

Atwater, T. and Molnar, P., 1973, Relative motion of the Pacific and North American plates deduced from sea-floor spreading in the Atlantic, Indian and South Pacific Oceans. In Kovach, R.L. and Nur, A. (eds.), *Proc. Conf. on Tectonic Problems of the San Andreas Fault System*. Stanford Univ. Publ. Geol. 13:136–148.

Avnimelech, Y., Troeger, B.W., and Reed, L.W., 1982, Mutual flocculation of algae and clay: evidence and implications. *Science* 216:63–65.

Avouac, J.P., Dobremez, J.F., and Bourjot, L., 1996, Paleoclimatic interpretation of a topographic profile across middle Holocene regressive shorelines of Longmu Co (Western Tibet). *Palaeogeog., Palaeoclim., Palaeoecol.* 120:93–104.

Babcock, L.E., Miller, M.F., Isbell, J.L., Collinson, J.W., and Hasiotis, S.T., 1998, Paleozoic-Mesozoic crayfish from Antarctica: earliest evidence of freshwater decapod crustaceans. *Geology* 26:539–542.

Baccini, P., 1985, Phosphate interactions at the sediment–water interface. In Stumm, W. (ed.), *Chemical Processes in Lakes*. Wiley Interscience, New York, pp. 189–205.

Back, S., De Batist, M., Kililov, P., Strecker, M.R., and Vanhauert, P., 1998, The Frolikha Fan: a large Pleistocene glaciolacustrine outwash fan in northern Lake Baikal, Siberia. *Jour. Sed. Res.* 68:841–849.

Bada, J.L., 1984, In vivo racemization in mammalian proteins. *Methods Enzymol.* 106:98–115.

Bada, J.L. and Schroeder, R.A., 1975, Amino acid racemization reactions and their geochemical implications. *Naturwissenschaften* 62:71–79.

Bajc, A.F., Morgan, A.V., and Warner, B.G., 1997, Age and paleoecological significance of an early postglacial fossil assemblage near Marathon, Ontario, Canada. *Can. Jour. Earth Sci.* 34:687–698.

Baker, L.A., Brezonik, P.L., and Kratzere, C.R., 1981, Nutrient loading–trophic state relationships in Florida lakes. *Florida Wat. Res. Ctr. Publ.* 56.

Baker, P.A., Seltzer, G.O., Fritz, S.C., Dunbar, R.B., Grove, M.J., Tapia, P.M., Cross, S.L., Rowe, H.D., and Broda, J.P., 2001, The History of South American tropical precipitation for the past 25,000 years. *Science* 291:640–643.

Baker, V.R., 1973, Paleohydrology and sedimentology of Lake Missoula flooding in eastern Washington. *Geol. Soc. Amer Spec. Pap.* 144.

Baker, V.R. and Bunker, R.C., 1985, Cataclysmic Late Pleistocene flooding from Glacial Lake Missoula: a review. *Quat. Sci. Rev.* 4:1–41.

Baker, V.R. and Numedal, D. (eds.), 1978, *The Channeled Scabland.* Comparative Planetary Geology Field Conference, NASA.

Balescu, S. and Lamothe, M., 1994, Comparison of TL and IRSL age estimates of feldspar coarse grains from waterlain sediments. *Quat. Sci. Rev.* 13:437–444.

Balogh, S.J., Engstrom, D.R., Almendinger, J.E., Meyer, M.L., and Johnson, D.K., 1999, History of mercury loading in the upper Mississippi River reconstructed from the sediments of Lake Pepin. *Envir. Sci. Tech.* 33:3297–3302.

Baltanás, A., Montes, C., and Martino, P., 1990, Distribution patterns of ostracodes in Iberian saline lakes. Influence of ecological factors. *Hydrobiologia* 197:207–220.

Baltzer, F., 1991, Late Pleistocene and Recent detrital sedimentation in the deep parts of northern Lake Tanganyika (East African Rift). *Spec. Publ. Intl. Assoc. Sediment.* 13:147–173.

Bao, R., Sáez, A., Servant-Vildary, S., and Cabrera, L., 1999, Lake level and salinity reconstruction from diatom analyses in Quillagua Formation (late Neogene, Central Andean forearc, northern Chile). *Palaeogeog., Palaeoclim., Palaeoecol.* 153:309–335.

Barber, K.E., Battarbee, R.W., Brooks, S.J., Eglinton, G., Haworth, E.Y., Oldfield, F., Stevenson, A.C., Thompson, R., Appleby, P.G., Austin, W.E.N., Cameron, N.G., Ficken, K.J., Golding, P., Harkness, D.D., Holmes, J.A., Hutchinson, R., Lishman, J.P., Maddy, D., Pinder, L.C.V., Rose, N.L., and Stoneman, R.E., 1999, Proxy records of climate change in the UK over the last two millenia: documented change and sedimentary records from lakes and bogs. *Jour. Geol. Soc. London* 156:369–380.

Barbetti, M.F. and McElhinny, M.W., 1976, The Lake Mungo geomagnetic excursion. *Phil Trans. R. Soc. London* 281:515.

Barbour, C.D. and Brown, J.H., 1974, Fish diversity in lakes. *Amer. Nat.* 108:473–489.

Barendregt, R.W., 1995, Paleomagnetic dating methods. In Rutter, N.W. and Catto, N.R. (eds.), *Dating Methods for Quaternary Deposits.* GEOTEXT 2, Geological Society of Canada, pp. 29–49.

Barker, M.J., Munt, M.C., and Radley, J.D., 1997, The first recorded Trigonioidoidean bivalve from Europe. *Palaeontology* 40:955–963.

Barker, P., 1992, Differential diatom dissolution in Late Quaternary sediments from Lake Manyara, Tanzania: an experimental approach. *Jour. Paleolim.* 7:235–251.

Barker, P., Gasse, F., Roberts, N., and Taieb, M., 1990, Taphonomy and diagenesis in diatom assemblages; a Late Pleistocene palaeoecological study from Lake Magadi. *Hydrobiologia* 214:267–272.

Barker, P., Fontes, J.C., Gasse, F., and Druart, J.C., 1994a, Experimental dissolution of diatom silica in concentrated salt solutions and implications for paleoenvironmental reconstruction. *Limnol. Oceanogr.* 39:99–110.

Barker, P.A., Roberts, N., Lamb, H.F., van der Kaars, S., and Benkaddour, A., 1994b, Interpretation of Holocene lake-level change from diatom assemblages in Lake Sidi Ali, Middle Atlas, Morocco. *Jour. Paleolim.* 12:223–234.

Barker, P.A., Street-Perrott, F.A., Leng, M.J., Greenwood, P.B., Swain, D.L., Perrott, R.A., Telford, R.J., and Ficken, K.J., 2001, A 14,000 year oxygen isotope record from diatom silica in two alpine lakes on Mt. Kenya. *Science* 292:2307–2310.

Barnes, R.S. and Schell, W.R., 1973, Physical transport of trace metals in the Lake Washington watershed. In Murray, M.G. and Giggliotti, G.M. (eds.), *Cycling and Control of Metals.* National Environmental Research Center, Cincinnati, OH, pp. 45–53.

Barnosky, C.W., 1989, Postglacial vegetation and climate in the northwestern Great Plains of Montana. *Quat. Res.* 31:57–73.

Barrell, J., 1917, Rhythms and the measurement of geologic time. *Geol. Soc. Amer. Bull.* 28:745–904.

Barron, E., 1990, Climate and lacustrine petroleum source prediction. In Katz, B.J. (ed.), *Lacustrine Basin Exploration: Case Studies and Modern Analogues*. Amer. Assoc. Petrol. Geol. Mem. 50:1–18.

Bartell, S.M., Brenkert, A.L., O'Neill, R.V., and Gardner, R.H., 1988, Temporal variation in regulation of production in a pelagic food web model. In Carpenter, S. (ed.), *Complex Interactions in Lake Communities*. Springer-Verlag, New York, pp. 101–118.

Bate, R.H., 1999, Non-marine ostracode assemblages of the Pre-Salt rift basins of West Africa and their role in sequence stratigraphy. In Cameron, N.R., Bate, R.H., and Clure, V.S. (eds.), *The Oil and Gas Habitats of the South Atlantic*. Geol. Soc. Lond. Spec. Publ. 153:283–292.

Bates, A.L., Spiker, E.C., Hatcher, P.G., Stout, S.A., and Weintraub, V.C., 1995, Sulfur geochemistry of organic-rich sediments from Mud Lake, Florida, USA. *Chem. Geol.* 121:245–262.

Bates, B.H., 1953, Rational theory of delta formation. *Amer. Assoc. Petrol. Geol. Bull.* 37:2119–2162.

Battarbee, R.W., 1981, Changes in the diatom microflora of a eutrophic lake since 1900 from a comparison of old algal samples and the sedimentary record. *Holarc. Ecol.* 4:73–81.

Battarbee, R.W., 1984, Diatom analysis and the acidification of lakes. *Phil. Trans. R. Soc. London, Ser. B* 305:451–477.

Battarbee, R.W., 1990, The causes of lake acidification, with special reference to the role of acid deposition. *Phil. Trans. R. Soc. London, Ser. B* 327:339–347.

Battarbee, R.W., 1991, Paleolimnology and climate change. In Frenzel, B., Pons, A., and Gläser, B. (eds.), *Evaluation of Climate Proxy Data in Relation to the European Holocene*. Paläoklimaforschung Akad. Wiss. Lit. Mainz 6:149–157.

Battarbee, R.W. and Renberg, I., 1990, The Surface Water Acidification Project (SWAP) Palaeolimnology Programme. *Phil. Trans. R. Soc. London, Ser. B* 327:227–232.

Battarbee, R.W., Flower, R.J., Stevenson, A.C., Jones, V.J., Harriman, R., and Appleby, P.G., 1988, Diatom and chemical evidence for reversibility of acidification of Scottish lochs. *Nature* 332:530–532.

Battarbee, R.W., Renberg, I., Talling, J.F., and Mason, J. (eds.), 1990, *Paleolimnology and Lake Acidification*. Phil. Trans. R. Soc. London, Ser. B, London, v. 327.

Battarbee, R.W., Allott, T.E.H., Juggins, S., Kreiser, A.M., Curtis, C., and Harriman, R., 1996, Critical loads of acidity to surface waters: an empirical diatom-based paleolimnological model. *Ambio* 25:366–369.

Battarbee, R.W., Charles, D.F., Dixit, S.S., and Renberg, I., 1999, Diatoms as indicators of surface water acidity. In Stoermer, E. and Smol, J. (eds.), *The Diatoms: Applications for the Environmental and Earth Sciences*. Cambridge University Press, Cambridge, pp. 85–127.

Battarbee, R.W., Jones, V.J., Flower, R.J., Cameron, N.G., Bennion, H., Carvalho, L., and Juggins, S., 2001, Diatoms. In Smol, J.P., Birks, H.J.B., and Last, W.M. (eds.), *Tracking Environmental Change Using Lake Sediments. Vol. 3. Terrestrial, Algal and Siliceous Indicators*. Kluwer Academic Publishers, Dordrecht, pp. 155–202.

Beadle, L.C., 1981, *The Inland Waters of Tropical Africa*, 2nd ed. Longman, New York.

Bechara, J., 1996, The relative importance of water quality, sediment composition and floating vegetation in explaining the macrobenthic community structure of floodplain lakes (Parana River, Argentina). *Hydrobiologia* 333:95–109.

Beck, C., Van Rensbergen, P., De Batist, M., Berthier, F., Lallier, S., and Manalt, F., 2001, The Late Quaternary sedimentary infill of Lake Annecy (northwestern Alps): an overview from two seismic reflection surveys. *Jour. Paleolim.* 25:149–161.

Behbehani, A.R., Muller, J., Schmidt, R., Schneider, J., Schroeder, H.G., Strackenbrock, I., and Sturm, M., 1986, Sediments and sedimentary history of Lake Attersee (Salzkammergut, Austria). *Hydrobiologia* 143:233–246.

Behrensmeyer, A.K., Damuth, J. D., DiMichele, W.A., Potts, R., Sues, H.D., and Wing, S. (eds.), 1992, *Terrestrial Ecosystems Through Time*. University of Chicago Press, Chicago.

Beierle, B. and Smith, D.G., 1998, Severe drought in the early Holocene (10,000–6800 BP) interpreted from lake sediment cores, southwestern Alberta, Canada. *Palaeogeog. Palaeoclim., Palaeoecol.* 140:75–83.

Belis, C.A., Lami, A., Guilizzoni, P., Ariztegui, D., and Geiger, W., 1999, The late Pleistocene ostracod record of the crater lake sediments from Lago di Albano (Central Italy): changes in trophic status, water level and climate. *Jour. Paleolim.* 21:151–169.

Bell, M.A., Baumgartner, J.V., and Olson, E.C., 1985, Patterns of temporal change in single morphological characters of a Miocene stickleback fish. *Paleobiology* 11:258–271.

Bellanca, A., Calvo, J.P., Censi, P., Neri, R., and Pozo, M., 1992, Recognition of lake-level changes in Miocene lacustrine units, Madrid Basin, Spain. Evidence from facies analysis, isotope geochemistry and clay mineralogy. *Sed. Geol.* 76:135–153.

Ben-Avraham, Z., 1997, Geophysical framework of the Dead Sea: structure and tectonics. In Niemi, T.M., Ben-Avraham, Z., and Gat, J.R. (eds.), *The Dead Sea: The Lake and its Setting.* Oxford University Press, Oxford, pp. 22–35.

Bengtsson, L. and Malm, J., 1997, Using rainfall-runoff modeling to interpret lake level data. *Jour. Paleolim.* 18:235–248.

Bennett, C.L., Beukens, R.P., Clover, M.R., Gove, H.E., Liebert, R.P., Litherland, A.E., Purser, K.H., and Sondheim, W.E., 1977, Radiocarbon dating using electrostatic accelerators: negative ions provide the key. *Science* 198:508–510.

Bennett, M.R., Hambrey, M.J., Huddart, D., Glasser, N.F., and Crawford, K., 1998, The ice dammed lakes of Ossian Sarsfjellet (Svalbard): their geomorphology and significance. *Boreas* 27:25–43.

Bennike, O., 1998, Fossil egg sacs of *Diaptomus* (Crustacea: Copepoda) in Late Quaternary lake sediments. *Jour. Paleolim.* 19:77–79.

Bennion, H., Duigan, C.A., Haworth, E.Y., Allott, T.E.H., Anderson, N.J., Juggins, S., and Monteith, D.T., 1996, The Anglesey lakes, Wales, UK—changes in trophic status of three standing waters as inferred from diatom transfer functions and their implications for conservation. *Aquat. Conserv.: Mar. Freshw. Ecosyst.* 6:81–92.

Benson, B.J. and Magnuson, J.J., 1992, Spatial heterogeneity of littoral fish assemblages in lakes: relation to species diversity and habitat structure. *Can. Jour. Fish. Aquat. Sci.* 49:1493–1500.

Benson, L., 1993, Factors affecting ^{14}C ages of lacustrine carbonates: timing and duration of the last highstand lake in the Lahontan Basin. *Quat. Res.* 39:163–174.

Benson, L., 1999, Records of millennial-scale climate change from the Great Basin of the Western United States. In Clark, P., Webb, R., and Keigwin, L. (eds.), *Mechanisms of Global Climate Change at Millennial Time Scales.* Amer. Geophys. Union Monogr. 112:203–225.

Benson, L.V., Burdett, J.W., Kashgarian, M., Lund, S.P., Phillips, F.M., and Rye, R.O., 1996, Climatic and hydrologic oscillations in the Owens Lake Basin and adjacent Sierra Nevada, California. *Science* 274:746–751.

Benson, L.V., May, H.M., Antweiler, R.C., Brinton, T.I., Kashgarian, M., Smoot, J.P., and Lund, S., 1998, Continuous lake-sediment records of glaciation in the Sierra Nevada between 52,600 and 12,500 ^{14}C yr B.P. *Quat. Res.* 50:113–127.

Bergbäck, B., Anderberg, S., and Lohm, U., 1992, Lead load: historical pattern of lead use in Sweden. *Ambio* 21:159–165.

Berge, F., 1975, pH-Forandringer og sedimentasjon av diatomeer I Langtjern. SNSF Project IR 11/75. Aas, Norway.

Berger, A., 1978, Long-term variation of daily insolation and Quaternary climate changes. *Jour. Atmos. Sci.* 35:2362–2367.

Berger, G.W., 1990, Effectiveness of natural zeroing of the thermoluminescence in sediments. *Jour. Geophys. Res.* 95:12,375–12,397.

Berger, G.W., 1994, Thermoluminescence dating of sediments older than ~100ka. *Quat. Sci. Rev.* 13:445–455.

Berger, G.W., 1995, Progress in luminescence dating methods for Quaternary sediments. In Rutter, N.W. and Catto, N.R. (eds.), *Dating Methods for Quaternary Deposits.* GEOTEXT 2, Geological Society of Canada, pp. 81–106.

Berger, G.W. and Anderson, P.M., 2000, Extending the geochronometry of Artic lake cores beyond the radiocarbon limit by using thermoluminescence. *Jour. Geophys Res.* 105(D12):15,439–15,455.

Berger, G.W. and Doran, P., 2001, Luminescence dating zeroing tests in Taylor Valley (McMurdo Dry Valleys), Antarctica. *Jour. Paleolim.* 25:519–529.

Berger, G.W. and Easterbrook, D.J., 1993, Thermoluminescence dating tests for lacustrine, glaciomarine and floodplain sediments from western Washington and British Columbia. *Can. Jour. Earth Sci.* 30:1815–1828.

Berger, G.W. and Eyles, N., 1994, Thermoluminescence chronology of Toronto area Quaternary sediments and implications for extent of the midcontinent ice sheet(s). *Geology* 22:31–34.

Berglund, B.E. (ed.), 1986, *Handbook of Holocene Paleoecology and Paleohydrology.* J. Wiley and Sons, New York.

Bergquist, P.R., 1998, The Porifera. In Anderson, D.T. (ed.), *Invertebrate Zoology.* Oxford University Press, Melbourne, pp. 10–27.

Berman, D.S. and Reisz, R.R., 1992, *Dolabrosaurus aquitilis*, a small lepidosauromorph reptile from the Upper Triassic Chinle Formation of north-central New Mexico. *Jour. Paleo.* 66:1001–1009.

Bernardi, R., Giussani, G., and Grimaldi, E., 1984, Lago Maggiore. In Taub, F.B. (ed.), *Lakes and Reservoirs.* Ecosystems of the World 23. Elsevier, New York, pp. 247–266.

Bernat, M. and Allègre, C.J., 1974, Systematics in uranium–thorium dating of sediments. *Earth Planet. Sci. Lett.* 21:310–314.

Berner, R.A., 1984, Sedimentary pyrite formation: an update. *Geochim. Cosmochim. Acta* 48:605–615.

Berner, R.A. and Raiswell, R., 1984, C/S method for distinguishing freshwater from marine sedimentary rocks. *Geology* 12:365–368.

Bertrand-Sarfati, J. and Monty, C. (eds.), 1994, *Phanerozoic Stromatolites II*. Kluwer, Dordrecht.

Bertrand-Sarfati, J., Freytet, P., and Plaziat, J.C., 1994, Microstructures in Tertiary nonmarine stromatolites (France). Comparison with Proterozoic. In Bertrand-Sarfati, J. and Monty, C. (eds.), *Phanerozoic Stromatolites II*. Kluwer, Dordrecht, pp. 155–191.

Betancourt, J., Quade, J., and Seltzer, G. (eds.), 2001, Central Andean Paleoclimate Workshop, Tucson, AZ, abstr. w/prog.

Beuning, K.R.M., Talbot, M.R., and Kelts, K., 1997, A revised 30,000-year paleoclimatic and paleohydrologic history of Lake Albert, East Africa. *Palaeogeog., Palaeoclim., Palaeoecol.* 136:259–279.

Beyens, L., 1985, On the Sub-boreal climate of the Belgian Campine as deduced from diatom and testate amoebae analyses. *Rev. Palaeobot. Palynol.* 46:9–31.

Beyens, L. and Meisterfeld, R., 2001, Protozoa; Testate Amoebae. In Smol, J.P., Birks, H.J.B., and Last, W.M. (eds.), *Tracking Environmental Change Using Lake Sediments. Vol. 3. Terrestrial, Algal and Siliceous Indicators*. Kluwer Academic Publishers, Dordrecht, pp. 121–153.

Bindlish, R. and Barros, A., 2000, Disaggregation of rainfall for one-way coupling of atmospheric and hydrological models in regions of complex terrain. *Global Planet. Change* 25:111–132.

Binford, M.W., 1986, Ecological correlates of net accumulation rates of Cladocera remains in lake sediments. *Hydrobiologia* 143:123–128.

Binford, M.W., Deevey, E.S., and Crisman, T.L., 1983, Paleolimnology: an historical perspective on lacustrine ecosystems. *Ann. Rev. Ecol. Syst.* 14:255–286.

Binford, M.W., Kahl, J.S., and Norton, S.A., 1993, Interpretation of ^{210}Pb profiles and verification of the CRS dating model in PIRLA project lake sediment cores. *Jour. Paleolim.* 9:275–296.

Birchfield, G.E. and Grumbine, R.W., 1985, "Slow" physics of large continental ice sheets and underlying bedrock and its relation to the Pleistocene ice ages. *Jour. Geophys. Res.* 90:11,294–11,302.

Bird, P., 1998, Kinematic history of the Laramide orogeny in latitudes 35–49N, western United States. *Tectonics* 17:780–801.

Birks, H.H., 1973, Modern macrofossil assemblages in lake sediments in Minnesota. In Birks, H.J.B. and West, R.G. (eds.), *Quaternary Plant Ecology*. Blackwell Science, Oxford, pp. 173–190.

Birks, H.H., 1980, Plant macrofossils in Quaternary lake sediments. *Arch. Hydrobiol. Beih. Ergeb. Limnol.* 15:1–60.

Birks, H.H., 1984, Late-Quaternary pollen and plant macrofossil stratigraphy at Lochan an Druim, north-west Scotland. In Haworth, E.Y. and Lund, J.W.G. (eds.), *Lake Sediments and Environmental History*. University of Minnesota Press, Minneapolis, MN, pp. 377–405.

Birks, H.H., 2000, Aquatic macrophyte vegetation development in Kråkenes Lake, western Norway, during the late-glacial and early Holocene. *Jour. Paleolim.* 23:7–19.

Birks, H.J.B., 1998, Numerical tools in palaeolimnology—progress, potentialities, and problems. *Jour. Paleolim.* 20:307–332.

Birks, H.J.B. and Gordon, A.D., 1985, *Numerical Methods in Quaternary Pollen Analysis*. Academic Press, London.

Birks, H.J.B., Webb, T. III and Berti, A.A., 1975, Numerical analysis of surface pollen samples from central Canada: a comparison of methods. *Rev. Palaeobot. Palynol.* 20:133–169.

Birks, H.J.B., Berge, F., Boyle, J.F., and Cumming, B.F., 1990a, A palaeoecological test of the land-use hypothesis for recent lake acidification in South-West Norway using hill-top lakes. *Jour. Paleolim.* 4:69–85.

Birks, H.J.B., Line, J.M., Juggins, S., Stevenson, A.C., and Ter Braak, C.J.F., 1990b, Diatoms and pH reconstruction. *Phil. Trans. R. Soc. London, Ser. B* 327:263–278.

Bischoff, J.L. and Fizpatrick, J.A., 1991, U-series dating of impure carbonates: an isochron technique using total sample dissolution. *Geochim. Cosmochim. Acta* 55:543–554.

Bischoff, J.L., Rosenbauer, R.J., and Smith, G.I., 1985, Uranium series dating of sediments from Searles Lake: differences between continental and marine climate records. *Science* 227:1222–1224.

Bischoff, J.L., Juliá, R., Shanks, W.C., and Rosenbauer, R.J., 1994, Karstification without carbonic acid: bedrock dissolution by gypsum-driven dedolomitization. *Geology* 22:995–998.

Bischoff, J.L., Fitts, J.P., and Fitzpatrick, J.A., 1997a, Responses of sediment geochemistry to climate change in Owens Lake sediment: an 800-k.y. record of saline/fresh cycles in core OL-2. In Smith, G.I. and Bischoff, J.L. (eds.), *An 800,000 Year Paleoclimate Record from Core OL-2, Owens Lake, Southeast California*. Geol. Soc. Amer. Spec. Pap. 317:37–47.

Bischoff, J.L., Stafford, T.W., and Rubin, M., 1997b, A time–depth scale for Owens Lake sediments of core OL-92: radiocarbon dates and constant mass-accumulation rates. In Smith, G.I. and Bischoff, J.L. (eds.), *An 800,000 Year Paleoclimate Record from Core OL-2, Owens Lake, Southeast California.* Geol. Soc. Amer. Spec. Pap. 317:91–98.

Björck, S., 1995, A review of the history of the Baltic Sea, 13.0–8.0 ka B.P. *Quat. Intl.* 27:19–40.

Björck, S., Cato, I., Brunnberg, L., and Strömberg, S., 1992, The clay-varve based Swedish time scale and its relation to the Late Weichselian radiocarbon chronology. In Bard, E. and Broecker, W.S. (eds.), *The Last Deglaciation: Absolute and Radiocarbon Chronologies.* NATO ASI Ser., Springer-Verlag, New York, pp. 25–44.

Björck, S., Håkansson, H., Olsson, S., Barekow, L., and Janssens, J., 1993, Palaeoclimatic studies in South Shetland Islands, Antarctica, based on numerous stratigraphic variables in lake sediments. *Jour. Paleolim.* 8:233–272.

Björck, S., Muscheler, R., Kromer, B., Andresen, C.S., Heinemeier, J., Johnsen, S.J., Conley, D., Koc, N., Spurk, M., and Veski, S., 2001, High-resolution analyses of an early Holocene climate event may imply decreased solar forcing as an important climate trigger. *Geology* 29:1107–1110.

Bjornsson, H. and Mysak, L., 2001, Present day and last glacial maximum ocean thermohaline circulation in a zonally averaged coupled ocean–sea–ice–atmosphere model. *Jour. Climate* 14:1422–1439.

Blackman, R.B. and Tukey, J.W., 1958, *The Measurement of Power Spectra From The Point of View of Communication Engineering.* Dover Publications, New York.

Blackwell, B. and Schwarcz, H.P., 1995, The uranium series disequilibrium dating methods. In Rutter, N.W. and Catto, N.R. (eds.), *Dating Methods for Quaternary Deposits.* GEOTEXT 2, Geological Society of Canada, pp. 167–208.

Blair, T.C., 1999, Sedimentology of gravelly Lake Lahontan highstand shoreline deposits, Churchill Butte, Nevada, USA. *Sed. Geol.* 123:199–218.

Blair, T.C. and Bilodeau, W.L., 1988, Development of tectonic cyclotherms in rift, pull-apart and foreland basins: sedimentary response to episodic tectonism. *Geology* 16:517–520.

Blair, T.C. and McPherson, J.G., 1994, Historical adjustments by Walker River to lake level fall over a tectonically tilted half-graben floor, Walker Lake Basin, Nevada. *Sed. Geol.* 92:7–16.

Blais, J.M., Kalff, J., Cornett, R.J., and Evans, R.D., 1995, Evaluation of ^{210}Pb dating in lake sediments using stable Pb, *Ambrosia* pollen, and [137]Cs. *Jour. Paleolim.* 13:169–178.

Blake, W.H., Plater, A.J., and Boyle, J.F., 1998, Seasonal trends in the uranium-series isotopic signatures of lake water and sediments: Hawes Water, northwest England. *Jour. Paleolim.* 20:1–14.

Bloesch, J. and Uehlinger, U., 1986, Horizontal sedimentation differences in a eutrophic Swiss lake. *Limnol. Oceanogr.* 31:1094–1109.

Blunt, D.J. and Kvenvolden, K.A., 1988. Amino-acid diagenesis and its implications for late Pleistocene lacustrine sediment, Clear Lake, California. In Sims, J.D. (ed.), *Late Quaternary Climate, Tectonism and Sedimentation in Clear Lake.* Geol. Soc. Amer. Spec. Pap. 214:161–170.

Blunt, D.J., Kvenvolden, K.A., and Sims, J.D., 1981, Geochemistry of amino acids in sediments from Clear Lake, California. *Geology* 9:378–382.

Bodergat, A.M., Carbonnel, G., Rio, M., and Keyser, D., 1993, Chemical composition of *Leptocythere* (Crustacea: Ostracoda) as influenced by winter metabolism and summer supplies. *Mar. Biol.* 117:53–62.

Bohacs, K.M., Carroll, A.R., Neal, J.E., and Mankiewicz, P.J., 2000, Lake-basin type, source potential, and hydrocarbon character: an integrated sequence-stratigraphic–geochemical framework. In Gierlowski-Kordesch, E.H. and Kelts, K.R. (eds.), *Lake Basins Through Space and Time.* AAPG Stud. Geol. 46:3–33.

Bohncke, S., Vandenberghe, J., and Wijmstra, T.A., 1988, Lake level changes and fluvial activity in the Late Glacial lowland valleys. In Lang, G. and Schlüchter, C., *Lake, Mire and River Environments.* Balkema, Rotterdam, pp. 115–121.

Bonacina, C., Bonomi, G., and Monti, C., 1986, Oligochaete cocoon remains as evidence of past lake pollution. *Hydrobiologia* 143:395–400.

Bond, G., Broecker, W., Johnsen, S., McManus, J., Labeyrie, L., Jouzel, J., and Bonani, G., 1993, Correlations between climate records from North Atlantic sediments and Greenland ice. *Nature* 365:143–147.

Bondevik, S., Svendsen, J.I., and Mangerud, J., 1997, Tsunami sedimentary facies deposited by the Storegga tsunami in shallow marine basins and coastal lakes, western Norway. *Sedimentology* 44:1115–1131.

Bonifay, E., 1991, Stratigraphie des dépôts du dernier cycle climatique et vue d'ensemble sur la séquence sédimentaire du Lac Du Bouchet: implications paléogéographiques. In Bonifay, E., Vanbesien, C., and Decobert, M. (eds.), *Le Lac du Bouchet (I): Environnement Naturel et*

Étude des Sediments du Dernier Cycle Climatique. Mem. 2, doc. CERLAT, pp. 127–158.

Bonifay, E. and Truze, E., 1991, Histoire gèologique du lac Bouchet. In Bonifay, E., Vanbesien, C., and Decobert, M. (eds.), *Le Lac du Bouchet (I): Environnement Naturel et Étude des Sediments du Dernier Cycle Climatique.* Mem. 2, doc. CERLAT, pp. 35–62.

Bonny, A.P., 1978, The effect of pollen recruitment processes on pollen distribution over the sediment surface of a small lake in Cumbria. *Jour. Ecol.* 66:385–416.

Bonny, A.P. and Allen, P.V., 1984, Pollen recruitment to the sediments of an enclosing lake in Shropshire, England. In Haworth, E.Y. and Lund, J.W.G. (eds.), *Lake Sediments and Environmental History.* University of Minnesota Press, Minneapolis, MN, pp. 231–259.

Boomer, I., 1993, Palaeoenvironmental indicators from Late Holocene and contemporary ostracoda of the Aral Sea. *Palaeogeog., Palaeoclim., Palaeoecol.* 103:141–153.

Borcard, D., Legendre, P., and Drapeau, P., 1992, Partialling out the spatial component of ecological variation. *Ecology* 73:1045–1055.

Borchardt, M.A., 1996, Nutrients. In Stevenson, R.J., Bothwell, M.L., and Lowe, R.L. (eds.), *Algal Ecology: Freshwater Benthic Ecosystems.* Academic Press, New York, pp. 183–227.

Borkent, A., 1981, The distribution and habitat preferences of the Chaoboridae (Culcomorpha: Diptera) of the Holarctic region. *Can. Jour. Zool.* 59:122–133.

Bos, D.G., Cumming, B.F., and Smol, J.P., 1999, Cladocera and Anostraca from the Interior Plateau of British Columbia, Canada, as paleolimnological indicators of salinity and lake level. *Hydrobiologia* 392:129–141.

Bouchard, D.P., Kaufman, D.S., Hochberg, A., and Quade, J., 1998, Quaternary history of the Thatcher Basin, Idaho, reconstructed from the $^{87}Sr/^{86}Sr$ and amino acid composition of lacustrine fossils: implications for the diversion of the Bear River into the Bonneville Basin. *Palaeogeog., Palaeoclim., Palaeoecol.* 141:95–114.

Boucherle, M.M., Smol, J.P., Oliver, T.C., Brown, S.R., and McNeely, R., 1986, Limnologic consequences of the decline in hemlock 4800 years ago in three Southern Ontario lakes. *Hydrobiologia* 143:217–225.

Boucherle, M.M. and Züllig, H., 1983, Cladoceran remains as evidence of change in trophic state in three Swiss lakes. *Hydrobiologia* 103:141–146.

Bourbonniere, R.A. and Meyers, P.A., 1996a, Sedimentary geolipid records of historical changes in the watersheds and productivities of Lakes Ontario and Erie. *Limnol. Oceanogr.* 41:352–359.

Bourbonniere, R.A. and Meyers, P.A., 1996b, Anthropogenic influences on hydrocarbon contents of sediments deposited in eastern Lake Ontario since 1800. *Envir. Geol.* 28:22–28.

Bouroullec, J.L., Rehault, J.P., Rolet, J., Tiercelin, J.J., and Mondeguer, A., 1991, Quaternary sedimentary processes and dynamics in the northern part of the Lake Tanganyika Trough, East African Rift System. Evidence of lacustrine eustatism. *Bull. Centr. Rech. Explor.-Prod. Elf Aquitaine* 15:343–368.

Bouroullec, J.L., Thouin, C., Tierceline, J.J., Rolet, J., Rehault, J.P., and Mondeguer, A., 1992, Séquences sismiques haute résolution du fossé nord-Tanganyika, Rift Est-africain. Implications climatiques, tectoniques et hydrothermales. *C. R. Acad. Sci. Paris* 315:601–608.

Bowler, J.M., 1983, Lunettes as indices of hydrologic change: a review of Australia evidence. *Proc. R. Soc. Victoria* 95:147–168.

Bowler, J.M. and Teller, J.T., 1986, Quaternary evaporites and hydrological changes, Lake Tyrell, northwest Victoria. *Austral. Jour. Earth Sci.* 33:43–63.

Bown, T.M., 1982, Ichnofossils and rhizoliths of the nearshore fluvial Jebel Qatrani Formation (Oligocene) Fayum Province, Egypt. *Palaeogeog., Palaeoclim., Palaeoecol.* 40:255–309.

Boyer, B.W., 1981, Tertiary lacustrine sediments from Sentinel Butte, North Dakota and the sedimentary record of ectogenic meromixis. *Jour. Sed. Petrol.* 51:429–440.

Boygle, J., 1993, The Swedish varve chronology: a review. *Prog. Phys. Geogr.* 17:1–9.

Boyle, E.A., 2000, Is ocean thermohaline circulation linked to abrupt stadial/interstadial transitions? *Quat. Sci. Rev.* 19:255–272.

Boyle, J.F., 1994, Acidification and sediment aluminium: palaeolimnological interpretation. *Jour. Paleolim.* 12:181–187.

Boyle, J.F., 2001a, Redox remobilization and the heavy metal record in lake sediments: a modeling approach. *Jour. Paleolim.* 26:423–431.

Boyle, J.F., 2001b, Inorganic geochemical methods in paleolimnology. In Last, W.M. and Smol, J.P. (eds.), *Tracking Environmental Change Using Lake Sediments. Vol. 2. Physical and Geochemical Methods.* Kluwer Academic Publishers, Dordrecht, pp. 83–141.

Boyle, J.F., Mackay, A.W., Rose, N.L., Flower, R.J., and Appleby, P.G., 1998, Sediment heavy metal record in Lake Baikal: natural and anthropogenic sources. *Jour. Paleolim.* 20:135–150.

Bradbury, J.P., 1991, The late Cenozoic diatom stratigraphy and paleolimnology of Tule Lake, Siskiyou Co. California. *Jour. Paleolim.* 6:205–255.

Bradbury, J.P., 1997, A diatom record of climate and hydrology for the past 200KA from Owens Lake, California with comparison to other Great Basin records. *Quat. Sci. Rev.* 16:203–216.

Bradbury, J.P. and Dieterich-Rurup, K.V., 1993, Holocene diatom paleolimnology of Elk Lake, Minnesota. In Bradbury, J.P. and Dean, W.E. (eds.), *Elk Lake, Minnesota: Evidence for Rapid Climate Change in the North-Central United States.* Geol. Soc. Amer. Spec. Pap. 276:215–237.

Bradbury, J.P. and Krebs, W.N., 1982, Neogene and Quaternary lacustrine diatoms of the western Snake River Basin, Idaho–Oregon, USA. *Acta Geol. Acad. Sci. Hungaricae* 25:97–122.

Bradbury, J.P. and Krebs, W.N., 1995, Fossil continental diatoms: paleolimnology, evolution and biochronology. In Blome, C.D. et al. (eds.), *Siliceous Microfossils.* Paleo. Soc. Short Course Paleontol. 8:119–138.

Bradbury, J.P., Forester, R.M., and Thompson, R.S., 1989 Late Quaternary paleolimnology of Walker Lake, Nevada. *Jour. Paleolim.* 1:249–267.

Bradley, R.S., 1999, *Quaternary Paleoclimatology.* Academic Press, San Diego.

Bradley, R.S. and Jones, P.D., 1993, "Little Ice Age" summer temperature variations: their nature and relevance to recent global warming trends. *The Holocene* 3:367–376.

Bradley, W.H., 1929, The varves and climate of the Green River epoch. *U.S. Geol. Surv. Prof. Pap.* 158:87–110.

Bradley, W.H. and Eugster, H.P., 1969, Geochemistry and paleolimnology of the trona deposits and associated authigenic minerals of the Green River Formation of Wyoming. *U.S. Geol. Surv. Prof. Pap.* 496-B.

Braithwaite, C.J.R. and Zedef, V., 1994, Living hydromagnesite stromatolites from Turkey. *Sed. Geol.* 92:1–5.

Brännvall, M.L., Bindler, R., Renberg, I., Emteryd, O., Bartnicki, J., and Billström, K., 1999, The Medieval metal industry was the cradle of modern large-scale atmospheric lead pollution in northern Europe. *Envir. Sci. Tech.* 33:4391–4395.

Brehmer, A.C., 1988, Smeltevandsaflerjringer I NV Sjælland. Thesis, University of Copenhagen.

Brenner, M., 1994, Lakes Salpeten and Quexil, Peten, Guatemala, Central America. In Gierlowski-Kordesch, E. and Kelts, K. (eds.), *Global Geological Record of Lake Basins* v. 1. Cambridge University Press, Cambridge, pp. 377–380.

Brenner, M., Leyden, B., and Binford, M.W., 1990, Recent sedimentary histories of shallow lakes in the Guatemalan savannas. *Jour. Paleolim.* 4:239–252.

Brenner, M., Whitmore, T.J., Flannery, M.S., and Binford, M.W., 1993, Paleolimnological methods for defining target conditions in lake restoration: Florida case studies. *Lake Res. Mgmnt.* 7:209–217.

Brenner, M., Whitmore, T.J., and Schelske, C.L., 1996, Paleolimnological evaluation of historical trophic state conditions in hypereutrophic Lake Thonotosassa, Florida, USA. *Hydrobiologia* 331:143–152.

Brenner, M., Whitmore, T.J., Curtis, J.H., Hodell, D.A., and Schelske, C.L., 1999a, Stable isotope ($\delta^{13}C$ and $\delta^{15}N$) signatures of sedimented organic matter as indicators of historic lake trophic state. *Jour. Paleolim.* 22:205–221

Brenner, M., Whitmore, T.J., Lasi, M.A., Cable, J.E., and Cable, P.H., 1999b, A multi-proxy trophic state reconstruction for shallow Orange Lake, Florida, USA: possible influence of macrophytes on limnetic nutrient concentrations. *Jour. Paleolim.* 21:215–233.

Bretz, J.H., 1923, The Channeled Scabland of the Columbia Plateau. *Jour. Geol.* 31:617–649.

Bretz, J.H., 1930, Lake Missoula and the Spokane Flood. *Geol. Soc. Amer. Bull.* 41:92–93.

Brezonik, P.L. and Engstrom, D.R., 1998, Modern and historical accumulation rates of phosphorus in Lake Okechobee, Florida. *Jour. Paleolim.* 20:31–46.

Bridge, J.S. and Leeder, M.R., 1979, A simulation model of alluvial stratigraphy. *Sedimentology* 26:617–644.

Bridgewater, N.D., Heaton, T.H.E., and O'Hara, S.L., 1999, A late Holocene palaeolimnological record from central Mexico, based on faunal and stable isotope analysis of ostracod shells. *Jour. Paleolim.* 22:383–397.

Briggs, D.E.G. and Clarkson, E.N.K., 1990, The late Palaeozoic radiation of malacostracan crustaceans. In Taylor, P.D. and Larwood, G.P. (eds.), *Major Evolutionary Radiations.* Syst. Assoc. Spec. Vol. 42:165–186.

Briggs, D.E.G., Weedon, M.J., and Whyte, M.A., 1993, Arthropoda (Crustacea excluding Ostracoda). In Benton, M.J. (ed.), *The Fossil Record 2.* Chapman and Hall, London, pp. 321–342.

Briggs, D.E.G., Stankiewicz, B.A., Meischner, D., Bierstedt, A., and Evershed, R.P., 1998, Taphonomy of arthropod cuticles from Pliocene lake sediments, Willershausen, Germany. *Palaios* 13:386–394

Brigham-Grette, J., Cosby, C., Apfelbaum, M.G., and Nolan, M., 2001, The Lake El'gygytgyn sediment core—a 300ka climate record of the terrestrial Arctic. *Eos Trans. AGU* 82:F752.

Brinkhurst, R.O. and Cook, D.G., 1974. In Hart, C.W. and Fuller, S.L.H., *Pollution Ecology of Freshwater Invertebrates*. Academic Press, New York, pp. 143–156.

Brinkhurst, R.O. and Gelder, S.R., 1991, Annelida: Oligochaeta and Branchiobdellida. In Thorp, J.H. and Covitch, A.P. (eds.), *Ecology and Classification of North American Freshwater Invertebrates*. Academic Press, New York, pp. 401–435.

Brock, T.D., 1978, *Thermophilic microorganisms and life at high temperature*. Springer-Verlag, New York.

Brodin, Y.W., 1990, Midge fauna development in acidified lakes in northern Europe. *Phil. Trans. R. Soc. London* 327:295–298.

Brodzikowski, K. and van Loon, A.J., 1991, *Glacigenic Sediments*. Developments in Sedimentology Series 49. Elsevier, Amsterdam.

Broecker, W.S., 1994, Massive iceberg discharge as triggers for global climatic change. *Nature* 372:421–424.

Broecker, W.S. and Denton, G.H., 1989, The role of ocean–atmosphere reorganizations in glacial cycles. *Geochim. Cosmochim Acta* 53:2465–2501.

Broecker, W.S. and Olson, E.A., 1959, Lamont radiocarbon measurements VI. *Amer. Jour. Sci.* 1:111–132.

Broecker, W.S. and Walton, A.F., 1959, The geochemistry of ^{14}C in freshwater systems. *Geochim. Cosmochim. Acta* 16:15–38.

Broecker, W.S., Andree, M., Wölfi, W., Oeschger, H., Bonani, G., Jennett, J., and Peteet, D., 1988, The chronology of the last deglaciation: implications to the cause of the Younger Dryas event. *Paleoceanography* 3:1–19.

Broecker, W.S., Kennett, J.P., Flower, B.P., Teller, J.T., Trumbore, S., Bonani, G., and Wölfi, W., 1989, Routing of meltwater from the Laurentide ice sheet during the Younger Dryas cold episode. *Nature* 341:318–321.

Broecker, W.S., Bond, G., Klas, M., Clark, E., and McManus, J., 1992, Origin of the northern Atlantic Heinrich events. *Climate Dyn.* 6:265–273.

Broecker, W.S., Peteet, D., Hajdas, I., Lin, J., and Clark, E., 1998, Antiphasing between rainfall in Africa's rift valley and North America's Great Basin. *Quat. Res.* 50:12–20.

Brönmark, C., 1985, Freshwater snail diversity: effects of pond area, habitat heterogeneity and isolation. *Oecologia* 67:127–131.

Brooks, A., Kokis, J.E., Hare, P.E., Miller, G.H., Ernst, R.E., and Wendorf, F., 1990, Dating Pleistocene archaeological sites by protein diagenesis in ostrich eggshell. *Science* 248:60–64.

Brooks, S.J. and Birks, H.J.B., 2000, Chironomid-inferred late-glacial and early Holocene mean July air temperatures for Kråkenes Lake, western Norway. *Jour. Paleolim.* 23:77–89.

Brooks, S.J., Mayle, F.E., and Lowe, J.J., 1997, Chironomid-based Lateglacial climatic reconstruction for southeast Scotland. *Jour. Quat. Sci.* 12:161–167.

Brown, A.G. and Barber, K.E., 1985, Late Holocene paleohydrology and sedimentary history of a small lowland catchment in Central England. *Quat. Res.* 24:87–102.

Brown, F.H., McDougall, I., Davies, I., and Maier, R., 1985, An integrated Plio-Pleistocene chronology for the Turkana Basin. In Delson, E. (ed.), *Ancestors, The Hard Evidence*. Alan Liss Publications, New York, pp. 82–90.

Brown, K.M., 1991, Mollusca:Gastropoda. In Thorp, J. and Covich, A. (eds.), *The Ecology and Classification of North American Freshwater Invertebrates*. Academic Press, New York, pp. 291–320.

Brown, S.R., McIntosh, H.J., and Smol, J.P., 1984, Recent paleolimnology of a meromictic lake: fossil pigments of photosynthetic bacteria. *Verh. Internat. Verein. Limnol.* 22:1357–1360.

Brouwers, E.M., 1988, Sediment transport detected from the analysis of ostracod population structure: an example from the Alaskan continental shelf. In De Deckker, P., Colin, J.P., and Peypouquet, J.P. (eds.), *Ostracoda in the Earth Sciences*. Elsevier, Amsterdam, pp. 231–244.

Brugam, R.B., 1979, A re-evaluation of the Araphidinae/Centrales index as an indicator of lake trophic status. *Freshwater Biol.* 9:451–460.

Brugam, R.B., McKeever, K., and Kolesa, L., 1998, A diatom-inferred water depth reconstruction for an Upper Peninsula, Michigan lake. *Jour. Paleolim.* 20:267–276.

Brundin, L., 1949, Chironomiden und andere Bodentiere der Südschweidischen Urgebirgsseen. *Inst. Freshw. Res. Drott. Rep.* 32:32–42.

Brunskill, G.J., 1969, Fayetteville Green Lake, New York. II. Precipitation and sedimentation of calcite in a meromictic lake with laminated sediments. *Limnol. Oceanogr.* 14:831–847.

Brusca, R.C. and Brusca, G.J., 1990, *Invertebrates*. Sinauer, Sunderland, MA.

Buatois, L.A. and Mángano, M.G., 1993, Ecospace utilization, paleoenvironmental trends and the evolution of early nonmarine biotas. *Geology* 21:595–598.

Buatois, L.A. and Mángano, M.G., 1994, Lithofacies and depositional processes from a

Carboniferous lake, Sierra de Narváez, northwest Argentina. *Sed. Geol.* 93:25–49.

Buatois, L.A., Mángano, M.G., Genise, J.F., and Taylor, T.N., 1998, The ichnologic record of the continental invertebrate invasion: evolutionary trends in environmental expansion, ecospace utilization, and behavioral complexity. *Palaios* 13:217–240.

Bucheim, P. and Surdam, R., 1981, Palaeoenvironments and fossil fishes of the Laney Meber, Green River Formation, Wyoming. In Gray, J. et al. (eds.), *Communities of the Past.* Hutchinson Ross, Stroudsburg, PA, pp. 415–452.

Büchel, G., 1993, Maars of the Westeifel, Germany. In Negendank, J.F.W. and Zolitschka, B. (eds.), *Paleolimnology of European Maar Lakes.* Lecture Notes in Earth Science v. 49. Springer-Verlag, New York, pp. 1–13.

Büchel, G. and Pirrung, M., 1993, Tertiary maars of the Hocheifel Volcanic Field, Germany. In Negendanck, J. and Zolitschka, B. (eds.), *Paleolimnology of European Maar Lakes.* Lecture Notes in Earth Science v. 49. Springer-Verlag, New York, pp. 447–465.

Buhay, W.M. and Betcher, R.N., 1998, Paleohydrologic implications of ^{18}O enriched Lake Agassiz water. *Jour Paleolim.* 19:285–296.

Buick, R., 1992, The antiquity of oxygenic photosynthesis: evidence from stromatolites in sulfate-deficient Archaean lakes. *Science* 255:74–77.

Bujalesky, G.G., Heusser, C.J., Coronato, A.M., Roig, C.E., and Rabassa, J.O., 1997, Pleistocene glaciolacustrine sedimentation at Lago Fagnano, Andes of Tierra del Fuego, Southernmost South America. *Quat. Sci. Rev.* 16:767–778.

Burbridge, S.M. and Schröder-Adams, C.J., 1998. Thecamoebians in Lake Winnipeg: a tool for Holocene paleolimnology. *Jour. Paleolim.* 19:309–328.

Burchart, J., Dakowski, M., and Galazka, J., 1975, A technique to determine extremely high fission track densities. *Bull. Acad. Pol. Sci. Ser. Sci. de la Terre* 23:1–7.

Burgis, M.J. and Morris, P., 1987, *The Natural History of Lakes.* Cambridge University Press, Cambridge.

Burkholder, J.M., 1996, Interactions of benthic algae with their substrata. In Stevenson, R.J., Bothwell, M.L., and Lowe, R.L. (eds.), *Algal Ecology: Freshwater Benthic Ecosystems.* Academic Press, San Diego, pp. 253–297.

Burky, A.J., 1983, Physiological ecology of freshwater bivalves. In Russell-Hunter, W.D. (ed.), *The Mollusca, v. 6: Ecology.* Academic Press, New York, pp. 281–327.

Burky, A.J., Hornbach, D.J., and Way, C.M., 1985, Comparative bioenergetics of permanent and temporary pond populations of the freshwater clam *Musculium partumenium* (Say). *Hydrobiologia* 126:35–48.

Burne, R.V., Bauld, J., and De Deckker, P., 1980, Saline lake charophytes and their geological significance. *Jour. Sed. Pet.* 50:281–293.

Bushnell, J.H., 1966, Environmental relations of Michigan Ectoprocta, and the dynamics of natural populations of *Plumatella repens. Ecol. Monogr.* 36:95–123.

Bushnell, J.H., 1968, Aspects of architecture, ecology and zoogeography of freshwater Ectoprocta. *Atti. Soc. Ital. Sci. nat. Museo civ. St. nat. Milano* 108:129–151.

Bushnell, J.H., 1974, Bryozoans. In Hart, C.W. and Fuller, S.L.H. (eds.), *Pollution Ecology of Freshwater Invertebrates.* Academic Press, New York, pp. 157–194.

Butler, R.F., 1992, *Paleomagnetism: Magnetic Domains to Geologic Terrains.* Blackwell, Boston, MA.

Callender, E., 2000, Geochemical effects of rapid sedimentation in aquatic systems: minimal diagenesis and the preservation of historical metal signatures. *Jour. Paleolim.* 23:243–260.

Calvo, J.P., Alonso-Zarza, A.M., and García del Cura, M.A., 1989, Models of Miocene marginal lacustrine sedimentation in response to varied depositional regimes and source areas in the Madrid Basin (Central Spain). *Palaeogeog., Palaeoclim., Palaeoecol.* 70:199–214.

Cambray, R.S., Playford, K., Lewis, G.N.J., Carpenter, R.C., and Gibson, J.A.B., 1987, Observations on radioactivity from Chernobyl accident. *Nucl. Energy* 26:77–101.

Camoin, G., Casanova, J., Rouchy, J.M., Blanc-Valleron, M.M., and Deconinck, J.F., 1997, Environmental controls on perennial and ephemeral carbonate lakes: the central palaeo-Andean Basin of Bolivia during Late Cretaceous to Early Tertiary times. *Sed. Geol.* 113:1–26.

Cande, S.C. and Kent, D.V., 1995, Revised calibration of the geomagnetic polarity timescale for the Late Cretaceous and Cenozoic. *Jour. Geophys. Res.* 100:6093–6095.

Cann, J.H. and De Deckker, P., 1981, Fossil Quaternary and living foraminifera from athalassic (non-marine) saline lakes, southern Australia. *Jour. Paleo.* 55:660–670.

Canter-Lund, H. and Lund, J.W.G., 1995, *Freshwater Algae.* Biopress Ltd., Bristol.

Cantrell, M.A., 1988, Effect of lake level fluctuations on the habitat of benthic invertebrates in a shallow tropical lake. *Hydrobiologia* 58:125–131.

Capesius, I., 1995, A molecular phylogeny of bryophytes based on the nuclear encoded 18S rRNA genes. *Jour. Plant Physiol.* 146:59–63.

Carbonel, J.P. and Peypouquet, J.P., 1979, Les Ostracodes des séries du Bassin de l'Omo. *Bull. Inst. Géol. Bassin d'Aquitaine Bordeaux* 25:167–199.

Carbonel, P., 1978, La zone a *Loxoconcha djaffarovi* Schneider (Ostracoda, Miocène supèrieur) ou le Messinien de la vallée du Rhone. *Rev. Micropaléo.* 21:106–118.

Carbonel, P., 1980, Les ostracodes et leur intérêt dans la définition des écosystèmes estuariens et de plateforme continentale. *Mém. Inst. Géol. Bassin Acquitaine* 11:350.

Carbonel, P., Colin, J.P., Danielopol, D.L., Löeffler, H., and Neustrueva, I., 1988, Paleoecology of limnic ostracodes: a review of some major topics. In Gray, J. (ed.), *Paleolimnology: Aspects of Freshwater Paleoecology and Biogeography.* Elsevier, Amsterdam, pp. 413–461.

Card, V.M., 1997, Varve-counting by the annual pattern of diatoms accumulated in the sediment of Big Watab Lake, Minnesota, AD 1837–1990. *Boreas* 26:103–112.

Carmack, E.C., Gray, C.B.J., Pharo, C.H., and Daley, R.J., 1979, Importance of lake–river interaction on the physical limnology of the Kamloops Lake/Thompson River system. *Limnol. Oceanogr.* 24:634–644.

Carmack, E.C., Wiegand, R.C., Daley, R.J., Gray, C.B.J., Jasper, S., and Pharo, C.H., 1986, Mechanisms influencing the circulation and distribution of water mass in a medium residence-time lake. *Limnol. Oceanogr.* 31:249–265.

Carnignan, R. and Flett, R.J., 1981, Postdepositional mobility of phosphorus in lake sediments. *Limnol. Oceanogr.* 26:361–366.

Carpenter, S.R., 1988, *Complex Interactions in Lake Communities.* Springer-Verlag, New York.

Carpenter, S.R. and Kitchell, J.F., 1993, *The Trophic Cascade in Lakes.* Cambridge University Press, Cambridge.

Carpenter, S.R. and Lodge, D.M., 1986, Effects of submersed macrophytes on ecosystem processes. *Aquat. Bot.* 26:341–370.

Carpenter, S.R., Elser, M.M., and Elser, J.J., 1986, Chlorophyll production, degradation, and sedimentation. Implications for paleolimnology. *Limnol. Oceanogr.* 31:112–124.

Carpetta, H., Duffin, C., and Zidek, J., 1993, Chondrichthyes. In Benton, M.J. (ed.), *The Fossil Record 2.* Chapman and Hall, London, pp. 593–610.

Carroll, A.R. and Bohacs, K.M., 1999, Stratigraphic classification of ancient lakes: balancing tectonic and climatic controls. *Geology* 27:99–102.

Carroll, A.R., Liang, Y., Graham, S.A., Xiao, X., Hendrix, M.S., Chu, J., and McKnight, C.L., 1990, Junggar Basin, northwest China: trapped Late Paleozoic Ocean. *Tectonophysics* 181:1–14.

Carter, L.D., Heginbottom, J.A., and Woo, M., 1987, Artic lowlands. In Graf, W.L. (ed.), *Geomorphic Systems of North America.* Geol. Soc. Amer. Centenn. Spec. Vol. 2:583–628.

Carvalho, M.D., Praça, U.M., Da Silva-Telles, A.C., Jahnert, R.J., and Dias, J.L., 2000, Bioclastic carbonate lacustrine facies models in the Campos Basin (Lower Cretaceous), Brazil. In Gierlowski, E.H. and Kelts, K.R. (eds.), *Lake Basins Through Space and Time.* AAPG Stud. Geol. 46:245–256.

Carvalho-Cunha, M.C. and Alves-Moura, J., 1979, Especies novas de ostracodes nao-marinhos da serie do Reconcavo: Paleontologia e bioestratigrafia. *Bol. Tec. Petrobras. Rio de Janeiro* 22:87–100.

Casanova, J., 1987, Limnologie des stromatolites en milieu continental. *Trav. CERLAT* 1:145–164.

Casanova, J., 1994, Stromatolites from the East African Rift: a synopsis. In Bertrand-Sarfati, J. and Monty, C. (eds.), *Phanerozoic Stromatolites II.* Kluwer, Dordrecht, pp. 193–226.

Casanova, J. and Hillaire-Marcel, C., 1993, Carbon and oxygen isotopes in African lacustrine stromatolites: palaeohydrological interpretation. In Swart, P.K., Lohmann, K.C., McKenzie, J., and Savin, S. (eds.), *Climate Change in Continental Isotopic Records.* Amer. Geophys. Union Geophys. Monogr. 78:123–133.

Casanova, J. and Nury, D., 1989, Biosédimentologie des stromatolites fluvio-lacustres du fosse oligocene de Marseille. *Bull. Soc. Géol. France* 8:1173–1184.

Casteel, R.W., Adam, D.P., and Sims, J.D., 1977, Late Pleistocene and Holocene remains of *Hysterocarpus traski* (tule perch) from Clear Lake, California, and inferred Holocene temperature fluctuations. *Quat. Res.* 7:133–143.

Castle, J.W., 1990, Sedimentation in Eocene Lake Uinta (Lower Green River Formation), Northeastern Uinta Basin, Utah. In Katz, B. (ed.), *Lacustrine Basin Exploration: Case Studies and Modern Analogues.* Amer. Assoc. Petrol. Geol. Mem. 50:243–263.

Cato, I., 1987, On the definitive connection of the Swedish geochronological time scale with the present. *Sveriges Geol. Undersökning* 68:1–55.

Cerling, T.E. and Quade, J., 1993, Stable carbon and oxygen isotopes in soil carbonates. In Swart, P., Lohmann, K.C., McKenzie, J., and Savin, S. (eds.), *Climate Change in Continental Isotopic Records*. AGU Monogr. 78:217–231.

Cevallos-Ferriz, S.R.S., Stockey, R.A., and Pigg, K.B., 1991, The Princeton chert: evidence for *in situ* aquatic plants. *Rev. Palaeobot. Palynol.* 70:173–185.

Chacón-Baca, E., Beraldi-Campesi, H., Cevallos-Ferriz, S.R.S., Knoll, .H., and Golubic, S., 2002, 70 Ma nonmarine diatoms from northern Mexico. *Geology* 30:279–281.

Chafetz, H.S., Utech, N.M., and Fitzmaurice, S.P., 1991, Differences in the $\delta^{18}O$ and $\delta^{13}C$ signatures of seasonal laminae comprising travertine stromatolites. *Jour. Sed. Petrol.* 61:1015–1028.

Chaix, L., 1983, Malacofauna from the Late Glacial deposits of Lobsigensee (Swiss Plateau). *Rev. Paleobiol.* 2:211–216.

Chapman, E.J., 1861, Some notes on drift deposits of Western Canada and on the ancient extension of the lake area of that region. *Can. J. (new series)* 6:221–229.

Charles, D.F., 1990, Effects of acidic deposition on North American lakes: palaeolimnological evidence from diatoms and chrysophytes. *Proc. Trans. R. Soc. London, Ser. B* 327:403–412.

Charles, D.F. and Smol, J.P., 1994, Long-term chemical changes in lakes: quantitative inferences using biotic remains in the sediment record. In Baker, L. (ed.), *Environmental Chemistry of Lakes and Reservoirs*. Amer. Chem. Soc., Adv. Chem. Ser. 237:3–31.

Charles, D.F. and Whitehead, D.R., 1986, The PIRLA project: paleoecological investigation of recent lake acidification. *Hydrobiologia* 143:13–20.

Charles, D.F., Battarbee, R.W., Renberg, I., van Dam, H., and Smol, J.P., 1989, Paleoecological analysis of diatoms and chrysophytes for reconstructing lake acidification trends in North America and Europe. In Norton, S.A., Lindberg, S.E., and Page, A.L. (eds.), *Acid Precipitation v. 4 of Soils, Aquatic Processes and Lake Acidification*. Springer-Verlag, New York, pp. 207–276.

Charles, D.F., Binford, M.W., Furlong, E.T., Hites, R.A., Mitchell, M.J., Norton, S.A., Oldfield, F., Patterson, M.J., Smol, J.P., Uutala, A.J., White, J.R., Whitehead, D.R., and Wise, R.J., 1990, Paleoecological investigation of recent acidification in the Adirondack Mountains, NY. *Jour. Paleolim.* 3:195–241.

Charles, D.F., Dixit, S.S., Cumming, B.F., and Smol, J.P., 1991, Variability in diatom and chrysophyte assemblages and inferred pH: paleolimnological studies of Big Moose Lake, New York, USA. *Jour. Paleolim.* 5:267–284.

Chekunov, A.V., Pustovichenko, B.G., and Kul'chitskiy, V.E., 1994, Seismicity and deep tectonics of the Black Sea depression and its margins. *Geotectonics* 28:221–225.

Chen, F.H., Bloemendal, J., Zhang, P.Z., and Liu, G.X., 1999, An 800 ky proxy record of climate from lake sediments of the Zoige Basin, eastern Tibetan Plateau. *Palaeogeog., Palaeoclim., Palaeoecol.*, 151:307–320.

Chen, X.Y., 1995, Geomorphology, stratigraphy and thermoluminescence dating of the lunette dune at Lake Victoria, western New South Wales. *Palaeogeog., Palaeoclim., Palaeoecol.* 113:69–86.

Chenggao, G. and Renaut, R.W., 1994, The effect of Tibetan uplift on the formation and preservation of Tertiary lacustrine source rocks in eastern China. *Jour. Paleolim.* 11:31–40.

Cherdyntsev, V.V., 1971, *Uranium 234*. Israel Program for Scientific Translations, Jerusalem.

Chesner, C.A. and Rose, W.I., 1991, Stratigraphy of the Toba Tuffs and the evolution of the Toba Caldera Complex, Sumatra, Indonesia. *Bull. Volcanol.* 53:343–356.

Chillrud, S.N., Bopp, R.F., Simpson, H.J., Ross, J.M., Wsuster, E.L., Chaky, D.A., Walsh, D.C., Choy, C.C., Tolley, L.R., and Yarme, A., 1999, Twentieth century atmospheric metal fluxes into Central Park Lake, New York City. *Envir. Sci. Tech.* 33:657–662.

Chivas, A., De Deckker, P., and Shelley, J.M.G., 1983, Magnesium, strontium and barium partitioning in nonmarine ostracode shells and their use in paleoenvironmental reconstructions. A preliminary study. In Maddocks, R.F. (ed.), *Applications of Ostracoda*. 8th Intl. Symp. Ostracoda, University of Houston, Houston, TX, pp. 238–249.

Chivas, A., De Deckker, P., and Shelley, J.M.G., 1986a, Magnesium content of nonmarine ostracode shells: a new paleosalinometer and paleothermometer. *Palaeogeogr., Palaeoclim., Palaeoecol.* 54:43–61.

Chivas, A., De Deckker, P., and Shelley, J.M.G., 1986b, Magnesium and strontium in nonmarine ostracode shells as indicators of paleosalinity and paleotemperature. *Hydrobiologia* 143:135–142.

Chivas, A.R., De Deckker, P., Cali, J.A., Chapman, A., Kiss, E., and Shelley, J.M., 1993, coupled stable-isotope and trace element measurements of lacustrine carbonates as paleoclimatic indicators. In Swart, P.K., Lohmann, K.C., McKenzie, J., and Savin, S. (eds.), *Climate Change in Continental Isotopic Records*. Amer. Geophys. Union Geophys. Monogr. 78:113–121.

Cipollari, P., Cosentino, D., Esu, D., Girotti, O., Gliozzi, E., and Praturlon, A., 1999, Thrust top lacustrine–lagoonal basin development in accretionary wedges: late Messinian (Lago Mare) episode in the central Apennines. *Palaeogeogr., Palaeoclim., Palaeoecol.* 151:149–166.

Clark, J.A., Hendriks, M., Timmerman, T.J., Struck, C., and Hilverda, K.J., 1994, Glacial isostatic deformation of the Great Lakes region. *Geol. Soc. Amer. Bull.* 106:19–31.

Clark, P.U. and Rudolf, G.A., 1990, Sedimentology and stratigraphy of late Wisconsin deposits, Lake Michigan bluffs, northern Illinois. In Schneider, A.F. and Fraser, G.S. (eds.), *Late Quaternary History of the Lake Michigan Basin.* Geol. Soc. Amer. Spec. Pap. 251:29–41.

Clark, P.U., Marshall, S.J., Clarke, G.K.C., Hostetler, S.W., Licciardi, J.M., and Teller, J.T., 2001, Freshwater forcing of abrupt climate change during the last glaciation. *Science* 293:283–287.

Clarke, R.G., Matthews, W.H., and Pack, R.T., 1984, Outburst floods from glacial Lake Missoula. *Quat. Res.* 22:289–299.

Clayton, L. and Attig, J.W., 1989, *Glacial Lake Wisconsin.* Geol. Soc. Amer. Mem. 173.

Cleal, C.S., 1993, Pteridophytes. In Benton, M.J. (ed.), *The Fossil Record 2.* Chapman and Hall, London, pp. 779–794.

Clemmensen, L.B., Kent, D.V., and Jenkins, F.A., 1998, A Late Triassic lake system in East Greenland: facies, depositional cycles and palaeoclimate. *Palaeogeog., Palaeoclim., Palaeoecol.* 140:135–159.

Clymo, R.S., 1967, Control of cation concentrations, and in particular pH, in *Sphagnum* dominated communities. In Golterman, H.L. and Clymo, R.S. (eds.), *Chemical Environment in the Aquatic Habitat.* Noord-Hollandsche Uitgevers Maatschappij, Amsterdam, pp. 273–284.

Coakley, J.P. and Karrow, P.F., 1994, Reconstruction of post-Iroquois shoreline evolution in western Lake Ontario. *Can. Jour. Earth Sci.* 31:1618–1629.

Cogbill, D.W. and Likens, G.E., 1974, Acid precipitation in the northeastern United States. *Wat. Res.* 10:1133–1137.

Cohen, A.S., 1984, Effect of zoobenthic standing crop on laminae preservation in tropical lake sediment, Lake Turkana, East Africa. *Jour. Paleontology* 58:499-510.

Cohen, A.S., 1989a, Facies relationships and sedimentation in large rift lakes and implications for hydrocarbon exploration: examples from Lakes Turkana and Tanganyika. *Palaeogeog., Palaeoclim., Palaeoecol.* 70:65–80.

Cohen, A.S., 1989b, The taphonomy of gastropod shell accumulations in large lakes: an example

from Lake Tanganyika, Africa. *Paleobiology* 15:26–45.

Cohen, A.S., 1990, Tectonostratigraphic model for sedimentation in Lake Tanganyika, Africa. In Katz, B.J. (ed.), *Lacustrine Basin Exploration: Case Studies and Modern Analogues.* Amer. Assoc. Petrol. Geol. Mem. 50:137–150.

Cohen, A.S., 2000, Linking spatial and temporal changes in the diversity structure of ancient lakes: examples from the ostracod ecology and paleoecology of Lake Tanganyika. In Rossiter, A. and Kawanabe, H. (eds.), *Ancient Lakes: Biodiversity, Ecology and Evolution.* Advances in Ecological Research 31. Academic Press, New York, pp. 521–537.

Cohen, A.S. and Johnston, M.R., 1987, Speciation in brooding and poorly dispersing lacustrine organisms. *Palaios* 2:426–435.

Cohen, A.S. and Schwartz, H., 1983, Speciation in mollusks from Turkana Basin. *Nature* 304:659–660.

Cohen, A.S. and Thouin, C., 1987, Nearshore carbonate deposits in Lake Tanganyika. *Geology* 15:414–418.

Cohen, A.S., Dussinger, R., and Richardson, J., 1983, Lacustrine paleochemical interpretations based on East and South African ostracodes. *Palaeogeog., Palaeoclim., Palaeoecol.* 43:129–151.

Cohen, A.S., Chase, C.G., and Anadón, P., 1993a, Three dimensional sequence stratigraphic modeling of the Miocene Rubielos de Mora Basin, Spain. *Geol. Soc. Amer. Ann. Mtg. Abstr. w/prog.* 25:113–114.

Cohen, A.S., Soreghan, M.J., and Scholz, C.A., 1993b, Estimating the age of formation of lakes: an example from Lake Tanganyika, East African Rift system. *Geology* 21:511–514.

Cohen, A.S., Talbot, M.R., Awramik, S.M., Dettman, D.L., and Abell, P., 1997a, Lake level and paleoenvironmental history of Lake Tanganyika, Africa, as inferred from late Holocene and modern stromatolites. *Geol. Soc. Amer. Bull.* 109:444–460.

Cohen, A.S., Lezzar, K.E., Tiercelin, J.J., and Soreghan, M., 1997b, New palaeogeographic and lake-level reconstructions of Lake Tanganyika: implications for tectonic, climatic and biological evolution in a rift lake. *Basin Res.* 9:107–132.

Cohen, A.S., Palacios, M., Negrini, R.M., Wigand, P.E., and Erbes, D.B., 2000a, A paleoclimate record for the past 250,000 years from Summer Lake, Oregon, USA: II. Sedimentology, paleontology and geochemistry. *Jour. Paleolim.* 24:151–182.

Cohen, A.S., Scholz, C.A., and Johnson, T.C., 2000b, The International Decade of East African Lakes (IDEAL) drilling initiative for the African Great Lakes. *Jour. Paleolim.* 24:231–235.

Cohen, C., 1992, Introduction. A primer on time frequency analysis. In Boashash, B. (ed.), *Time-Frequency Signal Analysis Methods, Applications.* Longman Cheshire, Melbourne, pp. 3–42.

COHMAP, 1988, Climate changes in the last 18,000 years: observations and simulations. *Science* 241:1043–1052.

Colbourne, J.K., Crease, T.J., Weider, L.J., Hebert, P.D.N., Dufresne, F., and Hobaek, A., 1998, Phylogenetics and evolution of a circumarctic species complex (Cladocera: *Daphnia pulex*) *Biol. Jour. Linn. Soc.* 65:317–365.

Cole, G.A., 1979, *Textbook of Limnology*, 2nd ed. C.V. Mosby, St. Louis, MO.

Cole, J., 2001, A slow dance for El Niño. *Science* 291:1495–1497.

Cole, J.J., Caraco, N.F., Kling, G.W., and Kratz, T.K., 1994, Carbon dioxide supersaturation in the surface waters of lakes. *Science* 265:1568–1570.

Colin, J.P. and Lethiers, F., 1988, The importance of ostracodes in biostratigraphic analysis. In De Deckker, P., Colin, J.P., and Peypouquet, J.P. (eds.), *Ostracoda in the Earth Sciences.* Elsevier, Amsterdam, pp. 27–45.

Collinson, M.E., 1988, Freshwater macrophytes in palaeolimnology. In Gray, J. (ed.), *Paleolimnology: Aspects of Freshwater Paleoecology and Biogeography.* Elsevier, Amsterdam, pp. 317–342.

Collinson, M.E., Boulter, M.C., and Holmes, P.L., 1993, Magnoliophyta ('Angiospermae'). In Benton, M.J. (ed.), *The Fossil Record 2.* Chapman and Hall, London, pp. 809–842.

Collister, J.W. and Hayes, J.M., 1991, A preliminary study of the carbon and nitrogen isotopic biogeochemistry of lacustrine sedimentary rocks from the Green River Formation, Wyoming, Utah and Colorado. *U.S. Geol. Surv. Bull.* 1973-A-G:C1–16.

Colman, S.M., 1998, Water-level changes in Lake Baikal, Siberia: tectonism versus climate. *Geology* 26:531–534.

Colman, S.M., Jones, G.A., Forester, R.M., and Foster, D.S., 1990, Holocene paleoclimatic evidence and sedimentation rates from a core in southwestern Lake Michigan. *Jour. Paleolim.* 4:269-284.

Colman, S.M., Keigwin, L.D., and Forester, R.M., 1994, Two episodes of meltwater influx from glacial Lake Agassiz into the Lake Michigan basin and their climatic contrasts. *Geology* 22:547–550.

Colman, S.M., Peck, J.A., Karabanov, E.B., Carter, S.J., Bradbury, J.P., King, J.W., and Williams, D.F., 1995, Continental climate response to orbital forcing from biogenic silica records in Lake Baikal. *Nature* 378:769–771.

Colman, S.M., Jones, G.A., Rubin, M., King, J.W., Peck, J.A., and Orem, W.H., 1996, AMS radiocarbon analyses from Lake Baikal, Siberia: challenges of dating sediments from a large, oligotrophic lake. *Quat. Sci. Rev.* 15:669–684.

Colman, S.M., Peck, J.A., Hatton, J., Karabanov, E.B., and King, J.W., 1999, Biogenic silica records from the BDP93 drill site and adjacent areas of the Selenga Delta, Lake Baikal, Siberia. *Jour. Paleolim.* 21:9–17.

Cook, R.B. and Schindler, D.W., 1983, The biogeochemistry of sulfur in an experimentally acidified lake. *Ecol. Bull.* 35:115–127.

Coope, G.R., 1979, Late Cenozoic fossil Coleoptera: evolution, biogeography and ecology. *Ann. Rev. Ecol. Syst.* 10:247–267.

Coope, G.R., 1986, Coleoptera analysis. In Berglund, B.E. (ed.), *Handbook of Holocene Paleoecology and Paleohydrology.* J. Wiley and Sons, New York, pp. 703–713.

Coope, G.R. and Brophy, J.A., 1972, Late Glacial environmental changes indicated by a coleopteran succession from North Wales. *Boreas* 1:94–142.

Cooper, C.M., 1984, The freshwater bivalves of Lake Chicot, an oxbow of the Mississippi in Arkansas. *Nautilus* 98:142–145.

Coplen, T.B., Kendall, C., and Hipple, J., 1983, Comparison of stable isotope reference samples. *Nature* 302:236–238.

Cottingham, K.L. and Carpenter, S.R., 1998, Population, community, and ecosystem variates as ecological indicators: phytoplankton responses to whole-lake enrichment. *Ecol. Appl.* 8:508–530.

Cowell, B.C. and Dawes, C.J., 1991, Nutrient enrichment experiments in three central Florida lakes of different trophic status. *Hydrobiologia* 220:217–231.

Cox, E.J., 1993, Freshwater diatom ecology: developing an experimental approach as an aid to interpreting field data. *Hydrobiologia* 269/270:447–452.

Craig, H., 1961, Isotopic variations in meteoric waters. *Science* 133:1702–1703.

Cranwell, P.A., 1984, Organic geochemistry of lacustrine sediments: triterpenoids of higher plant origin reflecting post-glacial vegetational succession. In Haworth, E.Y. and Lund, J.W.G. (eds.), *Lake Sediments and Environmental History.* University of Minnesota Press, Minneapolis, MN, pp. 69–92.

Crisman, T.L., 1978, Reconstruction of past lacustrine environments based on the remains of aquatic invertebrates. In Walker, D. and Guppy, J.C. (eds.), *Biology and Quaternary*

Environments. Australian Academy of Science, Canberra, pp. 69–101.

Crisman, T.L., Crisman, U.A.M., and Binford, M.W., 1986, Interpretation of bryozoan microfossils in lacustrine sediment cores. *Hydrobiologia.* 143:113–118.

Cronberg, G., 1986, Blue–green algae, green algae and chrysophyceae in sediments. In Berglund, B.E. (ed.), *Handbook of Holocene Palaeoecology and Palaeohydrology.* J. Wiley and Sons, New York, pp. 507–526.

Cronberg, G. and Sandgren, C.D., 1986, A proposal for the development of standardized nomenclature and terminology for chrysophycean statospores. In Kristiansen, J. and Andersen, R.A. (eds.), *Chyrsophytes: Aspects and Problems.* Cambridge University Press, Cambridge, pp. 317–328.

Crowell, J.C. and Link, M.H (eds.), 1982, *Geologic History of Ridge Basin, Southern California.* Pacific Section Soc. Econ. Paleo. Mineral.

Crowley, T.J. and North, G.R., 1991, *Paleoclimatology.* OUP Monograph on Geology and Geophysics 18. Oxford University Press, Oxford.

Crusius, J. and Anderson, R.F., 1995, Evaluating the mobility of ^{137}Cs, $^{239+240}$Pu and ^{210}Pb from their distributions in laminated lake sediments. *Jour. Paleolim.* 13:119–141.

Culver, D.A., Vaga, R.M., Munch, C.S., and Harris, S.M., 1981, Paleoecology of Hall Lake, Washington: a history of meromixis and disturbance. *Ecology* 62:848–863.

Cumming, B.F. and Smol, J.P., 1993a, Development of diatom-based salinity models for paleoclimate research from lakes in British Columbia, Canada. *Hydrobiologia* 269/270:179–196.

Cumming, B.F. and Smol, J.P., 1993b, Scaled chrysophytes and pH inference models: the effect of converting scale counts to cell counts and other species transformations. *Jour. Paleolim.* 9:147–153.

Cumming, B.F., Wilson, S.E., and Smol, J.P., 1993, Paleolimnological potential of chrysophyte cysts and scales, and sponge spicules as indicators of lake salinity. *Intl. Jour. Salt Lake Res.* 2:87–92.

Cummins, R.H., 1994, Taphonomic processes in modern freshwater molluscan death assemblages: implications for the freshwater fossil record. *Palaeogeog., Palaeoclim., Palaeoecol.* 108:55–73.

Currey, D.R., 1990. Quaternary paleolakes in the evolution of semidesert basins, with special emphasis on Lake Bonneville and the Great Basin, USA. *Palaeogeog., Palaeoclim., Palaeoecol.* 76:189–214.

Currey, D.R. and Oviatt, C.G., 1985. Durations, average rates and probable cause of Lake

Bonneville expansions during the last deep-lake cycle, 32,000 to 10,000 years ago. In Kay, P.A. and Diaz, H.F. (eds.), *Problems of and Prospects for Predicting Great Salt Lake Levels.* Center for Public Affairs and Administration, Utah University, pp. 1–9.

Curry, B., 1997, Paleochemistry of Lake Agassiz and Manitoba based on ostracodes. *Can. Jour. Earth Sci.* 34:699–708.

Curry, B., 1999, An environmental tolerance index for ostracodes as indicators of physical and chemical factors in aquatic habitats. *Palaeogeog., Palaeoclim., Palaeoecol.* 148:51–63.

Curry, B., Anderson, T.F., and Lohmann, K.C., 1997, Unusual carbon and oxygen isotopic ratios of ostracodal calcite from last interglacial (Sangamon episode) lacustrine sediment in Raymond Basin, Illinois, USA. *Jour. Paleolim.* 17:421–435.

Curtis, J.H. and Hodell, D.A., 1993, An isotopic and trace element study of ostracodes from Lake Miragoane, Haiti: a 10,500 year record of paleosalinity and paleotemperature changes in the Caribbean. In Swart, P.K., Lohmann, K.C., McKenzie, J., and Savin, S. (eds.), *Climate Change in Continental Isotopic Records.* Amer. Geophys. Union Geophys. Monogr. 78:113–121.

Curtis, J.H., Brenner, M., Hodell, D.A., Balser, R.A., Islebe, G.A., and Hooghiemstra, H., 1998, A multi-proxy study of Holocene environmental change in the Maya lowlands of Peten, Guatemala. *Jour. Paleolim.* 19:139–159.

Cwynar, L.C. and Levesque, A.J., 1995, Chironomid evidence for Late-Glacial climatic reversals in Maine. *Quat. Res.* 43:405–413.

D'Agostino, A., 1980, The vital requirements of *Artemia*: physiology and nutrition. In Persoone, G., Sorgeloos, P., Roels, O., and Jaspers, E. (eds.), *The Brine Shrimp Artemia v. 2. Physiology, Biochemistry, Molecular Biology.* Universa Press, Wetteren, Belgium, pp. 56–82.

Daley, R.J., 1973, Experimental characterization of lacustrine chlorophyll diagenesis. 2. Bacterial, viral and herbivore grazing effects. *Arch. Hydrobiol.* 72:409–439.

Dam, G. and Surlyk, F., 1993, Cyclic sedimentation in a large wave- and storm-dominated anoxic lake: Kap Stewart Formation (Rhaetian-Sinemurian), Jameson Land, East Greenland. In Posmentier, H., Summerhayes, C.P., Haq, B.U., and Allen, G.P. (eds.), *Sequence Stratigraphy and Facies Associations.* Int. Assoc. Sedimentol. Spec. Publ. 18:419–448.

Damon, P.E., Lerman, J.C., and Long, A., 1978, Temporal fluctuations of atmospheric ^{14}C:

causal factors and implications. *Ann. Rev. Earth Planet Sci.* 6:457–494.

Damon, P.E., Cheng, S., and Linick, T.W., 1989, Fine and hyperfine structure in the spectrum of secular variations of atmospheric ^{14}C. In Long, A., Kra, R., and Srdoc, D. (eds.), Proc. 13th Intl. Radiocarbon Conf. *Radiocarbon* 31:704–718.

Danielopol, D.L., 1990, On the interest of the *Cytherissa* Project, and on the present state of research. *Bull. Inst. Géol. Bassin Acquitaine* 47/48:15–26.

Danielopol, D.L., Geiger, W., Tölderer-Farmer, M., Orellana, C.P., and Terrat, M.N., 1985, The ostracoda of Mondsee: spatial and temporal changes during the last fifty years. In Danielopol, D.L., Schmidt, R., and Schultze, E. (eds.), *Contribution to the Paleolimnology of the Trumer Lakes and the Lakes Mondsee, Attersee and Traunsee (Upper Austria).* Limnological Institute, Monsee, pp. 99–121.

Danielopol, D.L., Casale, L.M., and Olteanu, R., 1986, On the preservation of carapaces of some limnic ostracods: an exercise in actuoplaeontology. *Hydrobiologia* 143:143–157.

Danielopol, D.L., Handl, M., and Yu, Y., 1993, Benthic ostracods in the pre-alpine deep lake Mondsee. Notes on their origin and distribution. In McKenzie, K.G. and Jones, P.J. (eds.), *Ostracoda in the Earth and Life Sciences.* Balkema, Rotterdam, pp. 465–480.

Dansgaard, W., 1964, Stable isotopes in precipitation. *Tellus* 16:436–468.

Dansgaard, W., White, J.W.C., and Johnson, S.J., 1989, The abrupt termination of the Younger Dryas climate event. *Nature* 339:532–534.

Dansgaard, W., Johnsen, S.J., Clausen, H.B., Dahl-Jensen, D., Gundestrup, N.S., Hammer, C.U., Hvidberg, C.S., Steffensen, J.P., Sveinbjornsdottir, A.E., Jouzel, J., and Bond, G., 1993, Evidence for general instability of past climate from a 250-kyr ice core record. *Nature* 364:218–220.

Darling, W.G., Allen, D.J., and Armannsson, H., 1990, Indirect detection of subsurface outflow from a rift valley lake. *Jour. Hydrol.* 113:297–305.

Darragi, F. and Tardy, Y., 1987, Authigenic trioctahedral smectites controlling pH, alkalinity, silica and magnesium concentrations in alkaline lakes. *Chem. Geol.* 63:59–72.

Davies, R.W., 1991, Annelida: leeches, polychaetes and acanthobdellids. In Thorp, J.H. and Covitch, A.P. (eds.), *Ecology and Classification of North American Freshwater Invertebrates.* Academic Press, New York, pp. 437–479.

Davis, G.M., 1982, Historical and ecological factors in the evolution, adaptive radiation, and biogeography of freshwater molluscs. *Amer. Zool.* 22:375–395.

Davis, M.B., 1976, Erosion rates and land-use history in southern Michigan. *Envir. Conserv.* 3:139–148.

Davis, M.B., 1981, Outbreaks of forest pathogens in Quaternary history. *Proc. IVth Intl. Palynol. Congr., Lucknow* 3:216–227.

Davis, M.B., 1986, Climatic instability, time lags, and community disequilibrium. In Diamond, J. and Case, T. (eds.), *Community Ecology.* Harper and Row, New York, pp. 269–294.

Davis, M.B. and Brubaker, L.B., 1973, Differential sedimentation of pollen grains in lakes. *Limnol. Oceanogr.* 18:635–646.

Davis, M.B. and Ford, M.S., 1982, Sediment focusing in Mirror Lake, New Hampshire. *Limnol. Oceanogr.* 27:137–150.

Davis, M.B., Moeller, R.E., and Ford, J., 1984, Sediment focusing and pollen influx. In Haworth, E. and Lund, J.W.G. (eds.), *Lake Sediments and Environmental History.* University of Minnesota Press, Minneapolis, MN, pp. 260–293.

Davis, O.K., 1987, Spores of the dung fungus *Sporormiella*: increased abundance in historic sediments and before Pleistocene megafaunal extinction. *Quat. Res.* 28:290–294.

Davis, O.K., 1998. Palynological evidence for vegetation cycles in a 1.5 million year pollen record from the Great Salt Lake, Utah, USA. *Palaeogeog., Palaeoclim., Palaeoecol.* 138:175–185.

Davis, O.K. and Moutoux, T.E., 1998, Tertiary and Quaternary vegetation history of the Great Salt Lake. *Jour. Paleolim.* 19:417–427.

Davis, O.K. and Shafer, D.S., 1992, A Holocene climatic record for the Sonoran Desert from pollen analysis of Montezuma Well, Arizona. *Palaeogeog., Palaeoclim., Palaeoecol.* 92:107–119.

Davis, R.B., 1987, Paleolimnological diatom studies of acidification of lakes by acid rain: an application of Quaternary science. *Quat. Sci. Rev.* 6:147–163.

Davis, R.B., 1989, The scope of Quaternary paleolimnology. *J. Paleolim.* 2:263–283.

Davis, R.B., Hess, C.T., Norton, S.A., Hanson, D.W., Hoagland, K.D., and Anderson, D.S., 1984, ^{137}Cs and ^{210}Pb dating of sediments from soft water lakes in New England (USA) and Scandinavia, a failure of ^{137}Cs dating. *Chem. Geol.* 44:151–185.

Davis, R.B., Anderson, D.S., Whiting, M.C., Smol, J.P., and Dixit, S.S., 1990, Alkalinity and pH of three lakes in northern New England, USA over the past 200 years. *Phil. Trans. R. Soc. London, Ser. B* 327:413–421.

Davis, R.B., Anderson, D.S., Norton, S.A., and Whiting, M.C., 1994, Acidity of twelve northern New England (USA) lakes in recent centuries. *J. Paleolim.* 12:103–154.

Davis, W.M., 1882, On the classification of lake basins. *Proc. Boston Soc. Nat. Hist.* 21:315–381.

Davis, W.M., 1887, On the classification of lake basins. *Science* 10:142.

Davison, W., 1988, Interactions of iron, carbon and sulfur in marine and lacustrine sediments. In Fleet, A.J., Kelts, K., and Talbot, M.R. (eds.), *Lacustrine Petroleum Source Rocks.* Geol. Soc. Lond. Spec. Publ. 40:131–137.

Dean, J.M., Kemp, A.E.S., Bull, D., Pike, J., Patterson, G., and Zolitschka, B., 1999, Taking varves to bits; scanning electron microscopy in the study of laminated sediments and varves. *Jour Paleolim.* 22:121–136.

Dean, W.E., 1997, Rates, timing and cyclicity of Holocene eolian activity in north-central United States: evidence from varved lake sediments. *Geology* 25:331–334.

Dean, W.E., 1999, The carbon cycle and biogeochemical dynamics in lake sediments. *Jour. Paleolim.* 21:375–393.

Dean, W.E. and Gorham, E., 1976, Major chemical and mineral components of profundal surface sediments in Minnesota lakes. *Limnol. Oceanogr.* 21:259–284.

Dean, W.E. and Gorham, E., 1998, Magnitude and significance of carbon burial in lakes, reservoirs, and peatlands. *Geology* 26:535–538.

Dean, W.E., Gorham, E., and Swaine, D.J., 1993, Geochemistry of surface sediments in Minnesota lakes. In Bradbury, J.P. and Dean, W.E. (eds.), *Elk Lake, Minnesota: Evidence for Rapid Climate Change in the North-central United States.* Geol. Soc. Amer. Spec. Pap. 276:115–134.

Dean, W., Rosenbaum, J., Haskell, B., Kelts, K., Schnurrenberger, D., Valero-Garcés, B., Cohen, A., Davis, O., Dinter, D., and Nielsen, D., 2002, Progress in global lake drilling holds potential for global change research. *Eos. Trans. Amer. Geophys. Union* 83:85, 90–91.

Dearing, J.A., 1991, Lake sediment records of erosional processes. *Hydrobiologia* 214:99–106.

Dearing, J.A. and Foster, I.D.L., 1993, Lake sediments and geomorphological processes: some thoughts. In McManus, J. and Duck, R.W. (eds.), *Geomorphology and Sedimentology of Lakes and Reservoirs.* J. Wiley and Sons, New York, pp. 5–14.

Deaton, L.E. and Greenberg, M.J., 1991, The adaptation of bivalve molluscs to oligohaline and fresh waters: phylogenetic and physiological aspects. *Malacol. Rev.* 24:1–18.

De Busk, G.H., 1997, The distribution of pollen in the surface sediments of Lake Malawi, Africa, and the transport of pollen in large lakes. *Rev. Palaeobot. Palynol.* 97:123–153.

DeCelles, P.G. and Giles, K.A., 1996, Foreland basin systems. *Basin Res.* 8:105–123.

De Deckker, P., 1982, Holocene ostracods, other invertebrates and fish remains from cores of four maar lakes in southeastern Australia. *Proc. R. Soc. Victoria* 94:183–220.

De Deckker, P., 1983, The limnological and climatic environment of modern halobiont ostracodes in Australia, a basis for paleoenvironmental reconstruction. In Maddocks, R.F. (ed.), *Applications of Ostracoda.* 8th Intl. Symp. Ostracoda, University of Houston, Houston, TX, pp. 250–254.

De Deckker, P., 1988a, Biological and sedimentary facies of Australian salt lakes. In Gray, J. (ed.), *Paleolimnology: Aspects of Freshwater Paleoecology and Biogeography.* Elsevier, Amsterdam, pp. 237–270.

De Deckker, P., 1988b, Large Australian lakes during the last 20 million years: sites for petroleum source rock or metal ore deposition, or both? In Fleet, A.J., Kelts, K., and Talbot, M.R. (eds.), *Lacustrine Petroleum Source Rocks.* Geol. Soc. Lond. Spec. Publ. 40:45–58.

De Deckker, P. and Forester, R.M., 1988, The use of ostracodes to reconstruct continental paleoenvironmental records. In De Deckker, P., Colin, J.P., and Peypouquet, J.P. (eds.), *Ostracoda in the Earth Sciences.* Elsevier, Amsterdam, pp. 176–199.

De Deckker, P. and Last, W.M., 1988, A newly discovered region of modern dolomite deposition in western Victoria, Australia. *Geology* 16:29–32.

Deevey, E.S., 1942, Studies on Connecticut lake sediments III. The biostratonomy of Linsley Pond. *Amer. Jour. Sci.* 240:233–264.

Defu, S., Qibin, L., Weiming, W., and Chenggao, G., 1988, Tertiary lacustrine deposits, Shandong Province. *Int. Assoc. Sedimentol. Symp. on Sedimentary Mineral Deposits,* Beijing, Excursion Guidebook B3.

De Geer, G., 1882, Om en postglacial landsänkning i södra och mellersta Sverige. *Geol. Fören. Förh. Stockholm* 4:149–162.

De Geer, G., 1912, A geochronology of the last 12,000 years. *XI Int. Geol. Congr. Stockholm (1910), Compte Rendu* 1:241–258.

De Geer, G., 1940, Geochronologica Suecia Principles. *Kungliga Svenska Vetenskapsakadmiens Handlingar* 3:1–367.

De Gibert, J.M., Fregenal-Martinez, M.A., Buatois, L.A., and Mángano, M.G., 2000, Trace fossils and their palaeoecological significance in Lower Cretaceous lacustrine conservation deposits, El Montsec, Spain. *Palaeogeog., Palaeoclim., Palaeoecol.* 156:89–101.

Delorme, L.D., 1969, Ostracodes as Quaternary paleoecological indicators. *Can. Jour. Earth Sci.* 6:1471–1476.

Delorme, L.D., 1971, Palaeoecological determinations using Pleistocene freshwater ostracodes. *Centr. Rech. Pau SNPA Bull.* 5:341–347.

Delorme, L.D., 1982, Lake Erie oxygen: the prehistoric record. *Can. Jour. Fish. Aquat. Sci.* 39:1021–1029.

Delorme, L.D., 1990, Freshwater ostracodes. In Warner, B.G. (ed.), *Methods in Quaternary Ecology.* Geoscience Canada Reprint Ser. 5, Geological Association of Canada, pp. 93–100.

Delorme, L.D., 1991, Ostracoda. In Thorp, J.H. and Covitch, A.P. (eds.), *Ecology and Classification of North American Freshwater Invertebrates.* Academic Press, New York, pp. 691–722.

Delorme, L.D., 1996, Burlington Bay, Lake Ontario, its paleolimnology based on fossil ostracodes. *Water Qual. Res. Jour. Canada* 31:643–671.

Delorme, L.D., Zoltai, S.C., and Kalas, L.L., 1977, Freshwater shelled invertebrate indicators of paleoclimate in Northwestern Canada during late glacial times. *Can. Jour. Earth Sci.* 14:2029–2046.

DeMaster, D.J., 1981, The supply and accumulation of silica in the marine environment. *Geochim. Cosmochim. Acta* 45:1715–1732.

deMenocal, P.B., 1995, Plio-Pleistocene African climate. *Science* 270:53–59.

deMenocal, P.B. and Rind, D., 1996, Sensitivity of subtropical African and Asian climate to prescribed boundary condition changes: model implications for the Plio-Pleistocene evolution of low-latitude climate. In Johnson, T.C. and Odada, E. (eds.), *The Limnology, Climatology and Paleoclimatology of the East African Lakes.* Gordon and Breach, Amsterdam, pp. 57–77.

Dendy, F.E., Champion, W.A., and Wilson, R.B., 1973, Reservoir sedimentation surveys in the United States. In Ackerman, W.C., White, G.F., and Worthington, E.B. (eds.), *Man-Made Lakes: Their Problems and Environmental Effects.* Amer. Geophys. Union Monogr. 17:349–357.

Den Hartog, C. and Van Der Velde, G., 1988, Structural aspects of aquatic plant communities. In Symoens, J.J. (ed.), *Vegetation of Inland Waters.* Kluwer Academic, Dordrecht, pp. 113–153.

DeNicola, D.M., 1996, Periphyton response to temperature at different ecological levels. In Stevenson, R.J., Bothwell, M.L., and Lowe, R.L. (eds.), *Algal Ecology: Freshwater Benthic Ecosystems.* Academic Press, New York, pp. 149–181.

Denton, G.H., Heusser, C.J., Lowell, T.V., Moreno, P.I., Anderson, B.G., Heusser, L.E., Schlüchter, C., and Marchant, D.R., 1999, Interhemisphere linkage of paleoclimate during the last glaciation. *Geograf. Annal.* 81:107–153.

Dettman, D. and Lohmann, K.C., 1993, Seasonal change in Paleogene surface water $\delta^{18}O$: Fresh-water bivalves of western North America. In Swart, P.K., Lohmann, K.C., McKenzie, J., and Savin, S. (eds.), *Climate Change in Continental Isotopic Records.* Amer. Geophys. Union Geophys. Monogr. 78:153–163.

Dettman, D. and Lohman, K.C., 2000, Oxygen isotope evidence for high-altitude snow in the Laramide Rocky Mountains of North American during the Late Cretaceous and Paleogene. *Geology* 28:243–246.

Dettman, D.L., Smith, A.J., Rea, D.K., Moore, T.C., and Lohmann, K.C., 1995, Glacial meltwater in Lake Huron during Early Postglacial time as inferred from single-valve analysis of oxygen isotopes in ostracodes. *Quat. Res.* 43:297–310.

de Vries, H., 1958, Variation in concentration of radiocarbon with time and location on earth. *Kon. Ned. Akad. Wet. Proc. Ser. B* 61:94–102.

Dickinson, W.R., 1974, Plate tectonics and sedimentation. In Dickinson, W.R. (ed.), *Tectonics and Sedimentation.* Soc. Econ. Paleo. Mineral. Spec. Publ. 22:1–27.

Dickinson, W.R., 1981, Plate tectonics and the continental margin of California. In Ernst, W.G. (ed.), *The Geotectonic Development of California.* Prentice-Hall, Englewood Cliffs, NJ, pp. 1–28.

Dickinson, W.R., 1993, Basin geodynamics. *Basin Res.* 5:195–196.

Dickinson, W.R. and Snyder, W.S., 1978, Plate tectonics of the Laramide Orogeny. In Matthews, V. (ed.), *Laramide Folding Associated with Basement Block Faulting in the Western United States.* Geol. Soc. Amer. Mem. 151:355–365.

Dickinson, W.R., Klute, M.A., Hayes, M.J., Janecke, S.U., Lundin, E., McKittrick, M.A., and Olivares, M.D., 1988, Paleogeographic and paleotectonic setting of Laramide sedimentary basins in the central Rocky Mountain region. *Geol. Soc. Amer. Bull.* 100:1023–1039.

Dickson, J.H., 1986, Bryophyte analysis. In Berglund, B.E. (ed.), *Handbook of Holocene*

Paleoecology and Paleohydrology. J. Wiley and Sons, New York, pp. 627–643.

Didyk, B.M., Simoneit, B.R., Brassell, S.C., and Eglinton, G., 1978, Organic geochemical indicators of palaeoenvironmental conditions of sedimentation. *Nature* 272:216–222.

Dillon, P.J. and Rigler, F.H., 1974, The phosphorus–chlorophyll relationship in lakes. *Limnol. Oceanogr.* 19:767–773.

Dillon, R.T., 2000, *The Ecology of Freshwater Molluscs.* Cambridge University Press, Cambridge.

Dini, M., Tunis, G., and Venturini, S., 1998, Continental, brackish and marine carbonates from the Lower Cretaceous of Kolone-Barbariga (Istria–Croatia): stratigraphy, sedimentology and geochemistry. *Palaeogeog., Palaeoclim., Palaeoecol.* 140:245–269.

Dinsmore, W.P. and Prepas, E.E., 1997, Impact of hypolimnetic oxygenation on profundal macroinvertebrates in a eutrophic lake in central Alberta. I. Changes in macroinvertebrate abundance and diversity. *Can. Jour. Fish. Aquatic Sci.* 54:2157–2169.

Dinsmore, W.P., Scrimgeour, G.J., and Prepas, E.E., 1999, Empirical relationships between profundal macroinvertebrate biomass and environmental variables in boreal lakes of Alberta, Canada. *Freshwater Biol.* 41:91–100.

Dixit, S.S., Cumming, B.F., Birks, H.J.B., Smol, J.P., Kingston, J.C., Uutala, A.J., Charles, D.F., and Camburn, K.E., 1993, Diatom assemblages from Adirondack lakes (New York, USA) and the development of inference models for retrospective environmental assessment. *Jour. Paleolimnol.* 8:27–47.

Dobson, D.M., Moore, T.C., and Rea, D.K., 1995, The sedimentation history of Lake Huron and Georgian Bay: results from analysis of seismic reflection profiles. *Jour. Paleolimnol.* 13:231–249.

do Carmo, D.A., Whatley, R.C., and Timberlake, S., 1999, Variable noding and palaeoecology of a Middle Jurassic limnocytherid ostracod: implications for modern brackish water taxa. *Palaeogeog., Palaeoclim., Palaeoecol.* 148:23–35.

Dodson, S. and Frey, D., 1991, Cladocera and other Branchiopoda. In Thorp, J.H. and Covitch, A.P. (eds.), *Ecology and Classification of North American Freshwater Invertebrates.* Academic Press, New York, pp. 723–786.

Dominik, J., Loizeau, J.L., and Span, D., 1992, Radioisotopic evidence of perturbations of recent sedimentary record in lakes: a word of caution in climate studies. *Clim. Dynam.* 6:145–152.

Donnelly, R. and Harris, C., 1989, Sedimentology and origin of deposits from a small ice-dammed lake, Leirbreen, Norway. *Sedimentology* 36:581–600.

Donovan, R.N., 1975, Devonian lacustrine limestones at the margin of the Orcadian Basin, Scotland. *Quat. Jour. Geol. Soc. London* 131:489–510.

Donovan, R.N., 1980, Lacustrine cycles, fish ecology and stratigraphic zonation in the Middle Devonian of Caithness. *Scott. Jour. Geol.* 16:35–50.

Doran, P.T., Berger, G.W., Lyons, W.B., Wharton, R.A., Davisson, M.L., Southon, J., and Dibb, J.E., 1999, Dating Quaternary lacustrine sediments in the McMurdo dry valleys, Antarctica. *Palaeogeog., Palaeoclim., Palaeoecol.* 147:223–239.

Dott, R.H., 1983, Episodic sedimentation; how normal is average? How rare is rare? Does it matter? *Jour. Sed. Pet.* 53:5–23.

Doubleday, N.C., Douglas, M.S.V., and Smol, J.P., 1995, Paleoenvironmental studies of black carbon deposition in the High Arctic: a case study from Northern Ellesmere Island. *Sci. Tot. Env.* 161:661–668.

Douglas, M.S.V. and Smol, J.P., 1995, Paleolimnological significance of observed distribution patterns of chrysophyte cysts in arctic pond environments. *Jour. Paleolim.* 13:79–83.

Douglas, M.S.V. and Smol, J.P., 1999, Freshwater diatoms as indicators of environmental change in the High Arctic. In Stoermer, E.F. and Smol, J.P. (eds.), *The Diatoms: Applications for the Environmental and Earth Sciences.* Cambridge University Press, Cambridge, pp. 227–244.

Douglas, M.S.V., Smol, J.P., and Blake, W., 1994, Marked post-18th century environmental change in high-arctic ecosystems. *Science* 266:416–419.

Dowd, J.F., Werner, D.B., and Clark, S.H., 1992, Factors controlling the $^{18}O/^{16}O$ and D/H composition of water within a forested hillslope watershed. *EOS-AGU Abstr.* 93:139.

Dowdeswell, J.A. and Siegert, M.J., 1999, The dimensions and topographic setting of Antarctic subglacial lakes and implications for large-scale water storage beneath continental ice sheets. *Geol. Soc. Amer. Bull.* 111:254–263.

Drabløs, D. and Tolan, A. (eds.), 1980, *Ecological Impact of Acid Precipitation.* Proc. Int. Conf, SNSF Project, Oslo, Norway.

Dreimanis, A., 1992, Early Winsconsinian in the north-central part of the Lake Erie basin: a new interpretation. In Clark, P.U. and Lea, P.D. (eds.), *The Last Interglacial–Glacial Transition in North America.* Geol. Soc. Amer. Spec. Pap. 270:109–118.

Drever, J.I., 1997, *The Geochemistry of Natural Waters*, 3rd ed. Prentice-Hall, Upper Saddle River, NJ.

Dromart, G., Gaillard, C., and Jansa, L.F., 1994, Deep-marine microbial structures in the Upper Jurassic of Western Tethys. In Bertrand-Sarfati, J. and Monty, C. (eds.), *Phanerozoic Stromatolites II.* Kluwer, Dordrecht, pp. 295–318.

Drouk, A.Y., 1997, Acarological analysis: problems of palaeoecological reconstructions. In Edwards, M.E., Sher, A.V., and Guthrie, R.D. (eds.), *Terrestrial Paleoenvironmental Studies in Beringia.* Alaska Quaternary Center, Fairbanks, AK, pp. 91–97.

Drummond, C.N., Patterson, W.P., and Walker, J.C.G., 1995, Climatic forcing of carbon-oxygen isotopic covariance in temperate-region marl lakes. *Geology* 23:1031–1034.

Drummond, C.N., Wilkinson, B.H., and Lohmann, K.C., 1996, Climatic control of fluvial-lacustrine cyclicity in the Cretaceous Cordilleran Foreland Basin, western United States. *Sedimentology* 43:677–689.

Dubiel, R.F., Parrish, J.T., Parrish, J.M., and Good, S.C., 1991, The Pangaean Megamonsoon—Evidence from the Upper Triassic Chinle Formation, Colorado Plateau. *Palaios* 6:347–370.

Duff, K.E. and Smol, J.P., 1991, Morphological description and stratigraphic distributions of the chrysophysean stomatocysts from a recently acidified lake (Adirondack Park, NY). *Jour. Paleolim.* 5:73–113.

Dumont, H.J., 1994, Ancient lakes have simplified pelagic food webs. *Arch. Hydrobiol. Ergeb. Limnol.* 44:223–234.

Dumont, H.J., 2000, Endemism in the Ponto-Caspian fauna, with special emphasis on the Onychopoda (Crustacea). In Rossiter, A. and Kawanabe, H. (eds.), *Ancient Lakes: Biodiversity, Ecology and Evolution.* Advances in Ecological Research 31. Academic Press, New York, pp. 181–196.

Dunagan, S.P., 1999, A North American freshwater sponge (*Eospongilla morrisonensis* new genus and species) from the Morrison Formation (Upper Jurassic), Colorado. *Jour. Paleo.* 73:389–393.

Duncan, A. and Hamilton, R.F.M., 1988, Palaeolimnology and organic geochemistry of the Middle Devonian in the Orcadian Basin. In Fleet, A.J., Kelts, K., and Talbot, M.R. (eds.), *Lacustrine Petroleum Source Rocks.* Geol. Soc. Lond. Spec. Publ. 40:173–201.

Dungworth, G. (and reply by Blunt, D.J., Kvenvolden, K.A., and Sims, J.D.), 1982, Comment and reply on "Geochemistry of amino acids in sediments from Clear Lake, California. *Geology* 10:124–125.

Duplessy, J.C., Labeyrie, L., Arnold, M., Paterne, M., Duprat, J., and van Weering, T.C.E., 1992, Changes in surface salinity of the North Atlantic during the last deglaciation. *Nature* 358:485–487.

Duthie, H.C., Yang, J.R., Edwards, T.W.D., Wolfe, B.B., and Warner, B.G., 1996, Hamilton Harbour, Ontario: 8300 years of environmental change inferred from microfossil and isotopic analyses. *Jour. Paleolim.* 15:79–97.

Dutkiewicz, A. and Prescott, J.R., 1997, Thermoluminescence ages and palaeoclimate from the Lake Malata–Lake Greenly complex, Eyre Peninsula, South Australia. *Quat. Sci. Rev.* 16:367–385.

Eadie, B.J. and Meyers, P.A., 1992, Carbon flux and remineralization in Lake Michigan. *EOS* 73:197.

Eadie, J.M. and Keast, A., 1984, Resource heterogeneity and fish species diversity in lakes. *Can. Jour. Zool.* 62:1689–1695.

Eardley, A.J., 1938, Sediments of the Great Salt Lake, Utah. *Amer. Assoc. Petrol. Geol. Bull.* 22:1305–1411.

Eardley, A.J., Shuey, R., Gvodetsky, V., Nash, W.P., Picard, M.D., Grey, D.C., and Kukla, G.J., 1973. Lake cycles in the Bonneville Basin, Utah. *Geol. Soc. Amer. Bull.* 84:211–215.

Eckert, W., Nishiri, A., and Paprova, R., 1997, Factors regulating the flux of phosphate at the sediment water interface of a subtopical calcareous lake: a simulation study with intact sediment cores. *Water Air Soil Pollut.* 99:401–409.

Edgington, D.N., Robbins, J.A., Colman, S.M., Orlandini, K.A., and Gustin, M.P., 1996, Uranium-series disequilibrium, sedimentation, diatom frustules and paleoclimate change in Lake Baikal. *Earth Planet. Sci. Lett.* 142:29–42.

Edwards, R.L., Chen, J.H., and Wasserburg, G.J., 1986/87, ^{238}U–^{234}U–^{230}Th–^{232}Th systematics and the precise measurement of time over the past 500,000 years. *Earth Planet. Sci. Lett.* 81:175–192.

Edwards, T.W.D., 1993, Interpreting past climate from stable isotopes in continental organic matter. In Swart, P.K., Lohmann, K.C., McKenzie, J., and Savin, S. (eds.), *Climate Change in Continental Isotopic Records.* Amer. Geophys. Union Geophys. Monogr. 78:333–341.

Ehrlich, A., 1975, The diatoms from the surface sediments of the Bardawil Lagoon (Northern Sinai). Paleoecological significance. *Nova Hedwigia* 53:253–277.

Einarsson, T., 1986, Tephrochronology. In Berglund, B.E. (ed.), *Handbook of Holocene Paleoecology and Paleohydrology*. J. Wiley and Sons, New York, pp. 329–342.

Einarsson, Á., Óskarsson, H., and Haflidason, H., 1993, Stratigraphy of fossil pigments and *Cladophora* and its relationship with deposition of tephra in Lake Mvatn, Iceland. *Jour. Paleolim.* 8:15–26.

Einsele, G. and Hinderer, M., 1997, Terrestrial sediment yield and the lifetimes of reservoirs, lakes and larger basins. *Geol. Rundsch.* 86:288–310.

Einsele, G. and Hinderer, M., 1998, Quantifying denudation and sediment accumulation systems (open and closed lakes): basic concepts and first results. *Palaeogeog., Palaeoclim., Palaeoecol.* 140:7–21.

Eisner, W.R., Törnqvist, T., Koster, E.A., Bennike, O., and van Leeuwen, J.F.N., 1995, Paleoecological studies of a Holocene record from the Kangerlussaq (Søndre Strømfjord) region of West Greenland. *Quat. Res.* 43:55–66.

Eklöv, P., 1997, Effects of habitat complexity and prey abundance on the spatial and temporal distributions of perch and pike. *Can. Jour. Fish. Aquatic Sci.* 54:1520–1531.

El-Daoushy, F. and Eriksson, M.G., 1998, Radiometric dating of recent lake sediments from a highly eroded area in semiarid Tanzania. *Jour. Paleolim.* 19:377–384.

Elder, R.L., 1985. *Principles of aquatic taphonomy with examples from the fossil record*. Unpublished Ph.D. dissertation, University of Michigan.

Elder, R.L. and Smith, G.R., 1988, Fish taphonomy and environmental inference in paleolimnology. *Palaeogeog., Palaeoclim., Palaeoecol.* 62:577–592.

Eldredge, N. and Gould, S.J., 1972, Punctuated equilibria: an alternative to phyletic gradualism. In Schopf, T.J. (ed.), *Models in Paleobiology*. Freeman, Cooper, San Francisco, pp. 82–115.

Elias, S.A., 2001, Coleoptera and Trichoptera. In Smol, J.P., Birks, H.J.B., and Last, W.M. (eds.), *Tracking Environmental Change Using Lake Sediments. Vol. 4. Zoological Indicators*. Kluwer Academic Publishers, Dordrecht, pp. 67–80.

Elias, S.A. and Wilkinson, B., 1983, Lateglacial insect fossil assemblages from Lobsigensee (Swiss Plateau). *Studies in the Late Quaternary of Lobsigensee, 3. Rev. Paleobiol.* 2:184–204.

Elliot, T., 1974, Interdistributary bay sequences and their genesis. *Sedimentology* 21:611–622.

Ellison, R.L., 1995, Paleolimnological analysis of Ullswater using testate amoebae. *Jour. Paleolim.* 13:51–63.

Ellison, R.L. and Ogden, C.G., 1987, A guide to the study and identification of fossil testate amoebae in Quaternary lake sediments. *Int. Rev. ges. Hydrobiol.* 72:639–652.

Endoh, S., Okumura, Y., and Okamoto, I., 1995, Field observations in the North Basin. In Okuda, S., Imberger, J., and Kumagai, M. (eds.), *Physical Processes in a Large Lake: Lake Biwa, Japan*. Amer. Geophys. Union Coastal Est. Stud. Ser. 48:15–29.

Engstrom, D.R., 1983, *Chemical stratigraphy of lake sediments as a record of environmental change*. Ph.D. dissertation, University of Minnesota.

Engstrom, D.R. and Nelson, S.R., 1991, Paleosalinity from trace metals in fossil ostracodes compared with observational records at Devils Lake, North Dakota, USA. *Palaeogeog., Palaeoclim., Palaeoecol.* 83:295–312.

Engstrom, D.R. and Swain, E.B., 1986, The chemistry of lake sediments in time and space. *Hydrobiologia* 143:37–44.

Engstrom, D.R. and Swain, E.B., 1997, Recent declines in atmospheric mercury deposition in the Upper Midwest. *Envir. Sci. Tech.* 31:960–967.

Engstrom, D.R. and Wright, H.E. Jr., 1984, Chemical stratigraphy of lake sediments as a record of environmental change. In Haworth, E.Y. and Lund, J.W.G. (eds.), *Lake Sediments and Environmental History*. University of Minnesota Press, Minneapolis, MN, pp. 11–68.

Engstrom, D.R., Whitlock, C., Fritz, S.C., and Wright, H.E., 1991, Recent environmental changes inferred from the sediments of small lakes in Yellowstone's northern range. *Jour. Paleolim.* 5:139–174.

Engstrom, D.R., Swain, E.B., Henning, T.A., Brigham, M.E., and Brezonik, P.L., 1994, Atmospheric mercury deposition to lakes and watersheds. In Baker, L.A. (ed.), *Environmental Chemistry of Lakes and Reservoirs*. Amer. Chem. Soc. Adv. Chem. Ser. 237:33–66.

Enzel, Y., Ely, L., Mishra, S., Ramesh, R., Amit, R., Lazar, B., Rajaguru, S.N., Baker, V.R., and Sandler, A., 1999, High-resolution Holocene environmental changes in the Thar Desert, Northwestern India. *Science* 284:125–128.

Epstein, S., Thompson, P., and Yapp, C.J., 1977, Oxygen and hydrogen isotopic ratios in plant cellulose. *Science* 198:1209–1215.

Erdtman, G., 1943, *Pollen Analysis*. Chronica Botanica, Waltham, MA.

Erickson, J.M., 1988, Fossil oribatid mites as tools for Quaternary paleoecologists: preservation quality, quantities and taphonomy. In Laub, R.S., Miller, N.G., and Steadman, D.W. (eds.),

Late Pleistocene and Early Holocene Paleoecology and Archaeology of the Eastern Great Lakes Region. Bull. Buffalo Soc. Nat. Sci. 33:207–226.

Erten, H.N., von Gunten, H.R., Rossler, E., and Sturm, M., 1985, Dating of sediments from Lake Zurich (Switzerland) with ^{210}Pb and ^{137}Cs. Schweisz. Z. Hydrol. 47:5–11.

Espitalié, J., Laporte, J.L., Madec, M., Marquis, F., Leplat, P., Paulet, J., and Boutefeu, A., 1977, Méthode rapide de caractérisation des roches mères, de leur potential pétrolier et de leur degré d'évolution. Rev. Inst. Franc. Pétrole 32:23–42.

Eugster, H.P., 1967, Hydrous sodium silicate from Lake Magadi. Precursors of bedded chert. Science 157:1177–1180.

Eugster, H.P., 1980, Lake Magadi and its precursors. In Nissenbaum, A. (ed.), Hypersaline Brines and Evaporitic Environments. Elsevier, Amsterdam, pp. 195–232.

Eugster, H.P. and Hardie, L.A., 1975, Sedimentation in an ancient playa-lake complex—The Wilkins Peak Member of the Green River Formation of Wyoming. Geol. Soc. Amer. Bull. 86:319–334.

Eugster, H.P. and Hardie, L.A., 1978, Saline lakes. In Lerman, A. (ed.), Lakes: Chemistry, Geology, Physics. Springer-Verlag, New York, pp. 237–293.

Eugster, H.P. and Jones, B.F., 1979, Behavior of major solutes during closed-basin brine evolution. Amer. Jour. Sci. 279:609–631.

Evans, S.E., Milner, A.W.R., and Werner, C., 1996, Sirenid salamanders and a gymnophionan amphibian from the Cretaceous of the Sudan. Palaeontology 39:77–95.

Évin, J., 1987, Carbone 14. In Miskovsky, J.-C. (ed.), 1987, Géologie de la préhistoire. Géopré, Paris, pp. 1041–1060.

Evitt, W.R., 1985, Sporopollenin dinoflagellate cysts: their morphology and interpretation. American Association of Stratigraphic Palynology.

Eyles, N., 1987, Late Pleistocene debris-flow deposits in large glacial lakes in British Columbia and Alaska. Sed. Geol. 53:33–71.

Eyles, N. and Williams, N.E., 1992. The sedimentary and biological record of the last interglacial–glacial transition at Toronto, Canada. In Clark, P.U. and Lea, P.D. (eds.), The Last Interglacial–Glacial Transition in North America. Geol. Soc. Amer. Spec. Pap. 270:119–137.

Eyles, N., Eyles, C.H., and Miall, A.D., 1983, Lithofacies types and vertical profile models; an alternative to the description and environmental interpretation of glacial diamict and diamictite sequences. Sedimentology 30:393–410.

Eyles, N., Mullins, H.T., and Hine, A.C., 1991, The seismic stratigraphy of Okanagan Lake, British Columbia; a record of rapid deglaciation in a deep "fjord-lake" basin. Sed. Geol. 73:13–41.

Fabryka-Martin, J., Bentley, H., Elmore, D., and Airey, P.L., 1985, Natural iodine-129 as an environmental tracer. Geochim. Cosmochim. Acta 49:337–347.

Fabryka-Martin, J., Davis, S.N., and Elmore, D., 1987, Applications of ^{129}I and ^{36}Cl in hydrology. Nucl. Instr. Meth. Phys. Res. B 29:361–371.

Falkner, A.J. and Fielding, C.R., 1993, Geometrical facies analysis of a mixed influence deltaic system: the Late Permian German Creek Formation, Bowen Basin, Australia. In Marzo, M. and Puigdefàbregas, C. (eds.), Alluvial Sedimentation. Int. Assoc. Sediment. Spec. Publ. 17:195–209.

Farquharson, G.W., 1982, Lacustrine deltas in a Mesozoic alluvial sequence from Camp Hill, Antarctica. Sedimentology 29:717–725.

Faure, G., 1986, Principles of Isotope Geology, 2nd ed. J. Wiley and Sons, New York.

Fee, E.J., Hecky, R.E., Kasian, S.E.M., and Cruikshank, D.R., 1996, Effects of lake size, water clarity and climatic variability on mixing depths in Canadian Shield lakes. Limnol. Oceanogr. 41:912–920.

Fehn, U., Tullai, S., Teng, R.T.D., Elmore, D., and Kubik, P.W., 1987 Determination of ^{129}I in heavy residues of two crude oils. Nucl. Instrum. Meth. Phys. Res. B 29:380–382.

Feist, M. and Schudack, M.E., 1991, Correlation of charophyte assemblages from the nonmarine Jurassic–Cretaceous transition of NW Germany. Cret. Res. 12:495–510.

Feist-Castel, M., 1975, Répartition des Charophytes dans le Paléocène et l'Eocène du bassin d'Aix-en-Provence. Bull. Soc. Géol. France 17:88–97.

Feist-Castel, M. and Columbo, F., 1983, La limite Crétacé–Tertiare dans le nord-est de l'Espagne du point de vue des charophytes. Géol. Médit. 10:303–325.

Fels, E. and Keller, R., 1973, World register on man-made lakes. In Ackerman, W.C., White, G.F., and Worthington, E.B. (eds.), Man-Made Lakes: Their Problems and Environmental Effects. Amer. Geophys. Union Monogr. 17:43–49.

Felzer, B., Thompson, S.L., Pollard, D., and Bergengren, J.C., 2000, GCM-simulated hydrology in the Arctic during the past 21,000 years. Jour. Paleolim. 24:15–28.

Ferber, C.T. and Wells, N.A., 1995, Palaeolimnology and taphonomy of some fish deposits in "Fossil" and "Uinta" Lakes of the

Eocene Green River Formation, Utah and Wyoming. *Palaeogeog., Palaeoclim., Palaeoecol.* 117:185–210.

Fernández, P., Vilanova, R.M., Martínez, C., Appleby, P., and Grimalt, J.O., 2000, The historical record of atmospheric pyrolitic pollution over Europe registered in the sedimentary PAH from remote mountain lakes. *Envir. Sci. Tech.* 34:1906–1913.

Fieg, K. and Gerdes, R., 2001, Sensitivity of the thermohaline circulation to modern and glacial surface boundary. *Jour. Geophys. Res. C. Oceans* 106:6853–6867.

Fielding, C.R., 1984, Upper delta plain lacustrine and fluviolacustrine facies from the Westphalian of the Durham coalfield, NE England. *Sedimentology* 31:547–567.

Fillippi, M.L., Lambert, P., Hunziker, J.C., and Kübler, B., 1998, Monitoring detrital input and resuspension effects on sediment trap material using mineralogy and stable isotopes: the case of Lake Neuchâtel. *Palaeogeog., Palaeoclim., Palaeoecol.* 140:33–50.

Fillipi, M.L., Lambert, P., Hunziker, J., Kübler, B., and Bernasconi, S., 1999, Climatic and anthropogenic influence on the stable isotope record from bulk carbonates and ostracodes in Lake Neuchâtel, Switzerland, during the last two millennia. *Jour. Paleolim.* 21:19–34.

Findlay, D.L., Kling, H.J., Rönicke, H., and Findlay, W.J., 1998, A paleolimnological study of eutrophied Lake Arendsee (Germany). *Jour. Paleolim.* 19:41–54.

Firby, J.R., Sharpe, S.E., Whelan, J.F., Smith ,G.R., and Spaulding, W.G., 1997, Paleobiotic and isotopic analysis of mollusks, fish and plants from core OL-92: indicators for an open or closed lake system. In Smith, G.I. and Bischoff, J.L. (eds.), *An 800,000-Year Paleoclimate Record from Core OL-92, Owens Lake, Southeast California.* Geol. Soc. Amer. Spec. Pap. 317:121–125.

Fischer, P. and Eckmann, R., 1997, Spatial distribution of littoral fish species in a large European lake, Lake Constance, Germany. *Arch. Hydrobiol.* 140:91–116.

Fisher, T.G. and Smith, D.G., 1994, Glacial Lake Agassiz: its northwest maximum extent and outlet in Saskatchewan. *Quat. Sci. Rev.* 13:845–858.

Fleet, A., Kelts, K., and Talbot, M.R. (eds.), 1988, *Lacustrine Petroleum Source Rocks.* Geol. Soc. Lond. Spec. Publ. 40.

Fleming, R.F., 1989, Fossil *Scenedesmus* (Chlorococcales) from the Raton Formation, Colorado and New Mexico, USA. *Rev. Palaeobot. Palynol.* 59:1–6.

Flower, R.J., 1993, Diatom preservation: experiments and observations on dissolution and breakage in modern and fossil material. *Hydrobiologia* 269/270:473–484.

Flower, R.J., Stevenson, A.C., Dearing, J.A., Foster, I.D.L., Airey, A., Rippey, B., Wilson, J.P.F., and Appleby, P.G., 1989, Catchment disturbance inferred from paleolimnological studies of three contrasted sub-humid environments in Morocco. *Jour. Paleolim.* 1:293–322.

Fontes, J.C. and Gasse, F., 1991, Chronology of the major palaeohydrological events in NW Africa during the late Quaternary: PALHYDAF results. In Smith, J.P., Appleby, P.G., Battarbee, R.W., Dearing, J.A., Flower, R., Haworth, E.Y., Oldfield, F., and O'Sullivan, P.E.O. (eds.), *Environmental History and Paleolimnology.* Kluwer Academic, Norwell, MA, pp. 367–372.

Fontes, J.C. and Gonfiantini, R., 1967, Comportement isotopique au cours de l'evaporation du deux bassins sahariens. *Earth Planet. Sci. Lett.* 3:258–266.

Forel, F.A., 1885, Les ravins sous-lacustre des fleuves glaciaires. *C. R. hebd. Séanc. Acad. Sci. Paris* 101:725–728.

Forester, R.M., 1986, Determination of the dissolved anion composition of ancient lakes from fossil ostracodes. *Geology* 14:796–798.

Forester, R.M., 1991a, Ostracode assemblages from springs in the western United States: implications for paleohydrology. *Mem. Entomol. Soc. Canada* 155:181–201.

Forester, R.M., 1991b, Pliocene-climate history of the Western United States derived from lacustrine ostracodes. *Quat. Sci. Rev.* 10:133–146.

Forester, R.M., Delorme, L.D., and Bradbury, J.P., 1987, Mid-Holocene climate in northern Minnesota. *Quat. Res.* 28:263–273.

Forester, R.M., Delorme, L.D., and Ager, T.A., 1989, A lacustrine record of Late Holocene climate change from South-Central Alaska. *Amer. Geophys. Union Geophys. Monogr.* 55:33–40.

Forester, R.M., Colman, S.M., Reynolds, R.L., and Keigwin, L.D., 1994, Lake Michigan's Late Quaternary limnological and climate history from ostracode, oxygen isotope, and magnetic susceptibility. *Jour. Great Lakes Res.* 20:93–107.

Fouch, T.D., Hanley, J.H., and Forester, R.M., 1979, Preliminary correlation of Cretaceous and Paleogene lacustrine and related nonmarine sedimentary and volcanic rocks in parts of the Eastern Great Basin of Nevada and Utah. In Newman, G.W. and Goode, H.D. (eds.), *1979 Basin and Range Symp.* Rocky Mountain Association of Geology/Utah Geological Association, pp. 305–312.

Francis, D.R., 1997, Bryozoan statoblasts in the recent sediments of Douglas Lake, Michigan. *Jour. Paleolim.* 17:255–261.

Francis, D.R., 2001, Bryozoan Statoblasts. In Smol, J.P., Birks, H.J.B., and Last, W.M. (eds.), *Tracking Environmental Change Using Lake Sediments. Vol. 4. Zoological Indicators.* Kluwer Academic Publishers, Dordrecht, pp. 105–123.

Frank, P.W., 1988, Conchostraca. In Gray, J. (ed.), *Paleolimnology: Aspects of Freshwater Paleoecology and Biogeography.* Elsevier, Amsterdam, pp. 399–403.

Fraser, N.C., Grimaldi, D.A., Olsen, P.E., and Axsmith, B., 1996. A Triassic Lagerstätte from eastern North America. *Nature* 380:615–620.

Fredskild, B., 1983, The Holocene development of some low and high arctic Greenland lakes. *Hydrobiologia* 103:217–224.

Freeman, N.G., Murthy, T.S., and Haras, W.S., 1972, A study of a storm surge on Lake Huron. *Collected Abstr. 3rd Canadian Oceanogr. Symp.*, Burlington, Ontario.

Frey, D.G., 1955, Längsee: a history of meromixis. *Mem. Ist. Ital. Idriobiol.* Suppl. 8:141–164.

Frey, D.G., 1958, The late-glacial cladoceran fauna of a small lake. *Arch. Hydrobiol.* 54:209–275.

Frey, D.G., 1964, Remains of animals in Quaternary lake and bog sediments and their interpretation. *Arch. Hydrobiol. Beih. Ergebn. Limnol.* 2:1–114.

Frey, D.G., 1974, Paleolimnology. *Mitt. Int. Ver. Limnol.* 20:95–123.

Frey, D.G., 1976, Interpretation of Quaternary paleoecology from Cladocera and midges, and prognosis regarding usability of other organisms. *Can. Jour. Zool.* 54:2208–2226.

Frey, D.G., 1986, Cladocera analysis. In Berglund, B.E. (ed.), *Handbook of Holocene Paleoecology and Paleohydrology.* J. Wiley and Sons, New York, pp. 667–692.

Frey, D.G., 1988a, What is paleolimnology? *Jour. Paleolim.* 1:5–8.

Frey, D.G., 1988b, Littoral and offshore communities of diatoms, cladocerans and dipterous larvae, and their interpretation in paleolimnology. *Jour. Paleolim.* 1:179–191.

Frey, D.G., 1993, The penetration of cladocerans into saline waters. *Hydrobiologia* 267:233–248.

Freytet, P. and Plet, A., 1996, Modern freshwater microbial carbonates: the *Phormidium* stromatolites (Tufa-Travertines) of southeastern Burgundy (Paris Basin, France). *Facies* 34:219–238.

Friedman, G.M. and Chamberlain J.A. Jr., 1995, *Archanodon catskillensis* (Vanuxem): freshwater clams from one of the oldest backswamp fluvial facies (Upper Middle Devonian), Catskill Mountains, New York. *Northeast. Geol. Environ. Sci.* 17(4): 431–443.

Friedman, G.M. and Lundin, R.F., 1998, Freshwater ostracodes from Upper Middle Devonian fluvial facies, Catskill Mountains, New York. *Jour. Paleo.* 72:485–490.

Friedman, I. and O'Neil, J.R., 1977, Compilation of stable isotope fractionation factors of geochemical interest. In Fleischer, M. (ed.), *Data on Geochemistry.* U.S. Geol. Surv. Prof. Pap. 440KK.

Friedman, S.J. and Burbank, D.W., 1995, Rift basins and supradetachment basins: intracontinental extensional end-members. *Basin Res.* 7:109–127.

Friis, E.M., Pedersen, K.R., and Crane, P.R., 2001, Fossil evidence of water lilies (Nymphaeales) in the Early Cretaceous. *Nature* 410:357–360.

Fritz, P., Anderson, T.W., and Lewis, C.F.M., 1975, Late Quaternary climatic trends and history of Lake Erie from stable isotope studies. *Science* 190:267–269.

Fritz, S.C., 1990, Twentieth century salinity and water level fluctuations in Devils Lake, North Dakota: test of a diatom transfer function. *Limnol. Oceanogr.* 35:1771–1781.

Fritz, S.C. and Carlson, R.E., 1982, Stratigraphic diatom and chemical evidence for acid-strip mine lake recovery. *Water Air Soil Pollut.* 17:151–163.

Fritz, S.C., Juggins, S., Battarbee, R.W., and Engstrom, D.R., 1991, Reconstruction of past changes in salinity and climate using a diatom-based transfer function. *Nature* 352:706–708.

Fritz, S.C., Engstrom, D.R., and Haskell, B.J., 1994, Little Ice Age aridity in the North American Great Plains: a high resolution reconstruction of salinity fluctuations from Devils Lake, North Dakota, USA. *The Holocene* 4:69–73.

Fritz, S.C., Cumming, B.F., Gasse, F., and Laird, K.R., 1999, Diatoms as indicators of hydrologic and climatic change in saline lakes. In Stoermer, E.F. and Smol, J.P. (eds.), *The Diatoms: Applications for the Environmental and Earth Sciences.* Cambridge University Press, Cambridge, pp. 41–72.

Frost, T.M., 1991, Porifera. In Thorp, J.H. and Covich, A.P. (eds.), *Ecology and Classification of North American Freshwater Invertebrates.* Academic Press, New York, pp. 95–124.

Frost, T.M., 2001, Freshwater Sponges. In Smol, J.P., Birks, H.J.B., and Last, W.M. (eds.), *Tracking Environmental Change Using Lake Sediments. Vol. 4. Zoological Indicators.* Kluwer Academic Publishers, Dordrecht, pp. 253–263.

Frostick, L.E. and Reid, I., 1987, Tectonic control of desert sediment in rift basins ancient and modern. In Frostick, L.E. and Reid, I. (eds.),

Desert Sediments, Ancient and Modern. Geol. Soc. Lond. Spec. Publ. 35:53–68.

Fuller, S.L.H., 1974, Clams and mussels. In Hart, C.W. and Fuller, S.L.H. (eds.), *Pollution Ecology of Freshwater Invertebrates.* Academic Press, New York, pp. 215–273.

Fürsich, F. and Aberhan, M., 1990, Significance of time-averaging for palaeocommunity analysis. *Lethaia* 23:143–152.

Gaillard, M.J., Dearing, J.A., El-Daoushy, F., Enell, M., and Håkanson, H., 1991, A late Holocene record of land-use history, soil erosion, lake trophy and lake-level fluctuations at Bjärsejösjön (South Sweden). *Jour. Paleolim.* 6:51–81.

Gaiser, E.E., Philippi, T.E., and Taylor, B.E., 1998, Distribution of diatoms among intermittent ponds on the Atlantic Coastal Plain: development of a model to predict drought periodicity from surface-sediment assemblages. *Jour. Paleolim.* 20:71–90.

Gajewski, K., Garneau, M., and Bourgeois, J.C., 1995, Paleoenvironments of the Canadian high arctic derived from pollen and plant macrofossils: problems and potentials. *Quat. Sci. Rev.* 14:609–629.

Gajewski, K., Hamilton, P.B., and McNeely, R., 1997, A high-resolution proxy climate record from an arctic lake with annually-laminated sediments on Devon Island, Nunavut, Canada. *Jour. Paleolim.* 17:215–225.

Galbrun, B., Feist, M., Colombo, F., Rocchia, R., and Tambareau, Y., 1993, Magnetostratigraphy and biostratigraphy of Cretaceous–Tertiary continental deposits, Ager Basin, Province of Lerida, Spain. *Palaeogeog., Palaeoclim., Palaeoecol.* 102:41–52.

Gall, Q. and Hyde, R., 1989, Analcime in lake and lake-margin sediments of the Carboniferous Rocky Brook Formation, Western Newfoundland, Canada. *Sedimentology* 36:875–887.

Gams, H., 1927, Die Geschichte der Lunzer Seen, Moore, und Wälder. *Int. Rev. ges. Hydrobiol. Hydrogr.* 18:304–387.

Ganf, G.G., Heaney, S.I., and Corry, J., 1991, Light absorbtion and pigment content in natural populations and cultures of a non-gas vacuolate cyanobacterium *Oscillatoria bourrellyi* (=*Tychonema bourrellyi*). *Jour. Plankton Res.* 13:1101–1121.

Garbary, D.J., Rezaglia, K.S., and Duckett, J.G., 1993, The phylogeny of land plants: a cladistic analysis based on male gametogenesis. *Plant Syst. Evol.* 188:237–269.

García, A., 1999, Quaternary charophytes from Salina del Bebedero, Argentina: their relation with extant taxa and palaeolimnological significance. *Jour. Paleolim.* 21:307–323.

Gardiner, B.G., 1993a, Placodermi. In Benton, M.J. (ed.), *The Fossil Record 2.* Chapman and Hall, London, pp. 583–588.

Gardiner, B.G., 1993b, Osteichthyes: basal actinopterygians. In Benton, M.J. (ed.), *The Fossil Record 2.* Chapman and Hall, London, pp. 611–620.

Gardosh, M., Kashai, E., Salhov, S., Shulman, H., and Tannenbaum, E., 1997, Hydrocarbon exploration in the southern Dead Sea area. In Niemi, T.M., Ben-Avraham, Z., and Gat, J.R (eds.), *The Dead Sea: The Lake and Its Setting.* Oxford University Press, Oxford, pp. 57–72.

Garfunkel, Z., 1997, The history and formation of the Dead Sea Basin. In Niemi, T.M., Ben-Avraham, Z., and Gat, J.R (eds.), *The Dead Sea: The Lake and Its Setting.* Oxford University Press, Oxford, pp. 36–56.

Garfunkel, Z. and Ben-Avraham, Z., 1996, The structure of the Dead Sea Basin. *Tectonophysics* 266:155–176.

Garrels, R.M. and Mackenzie, F.T., 1967, Origin of the chemical composition of some springs and lakes. In Gould, R.F. (ed.), *Equilibrium Concepts in Natural Water Systems.* Advances in Chemistry Series 67:222–242.

Gasse, F., 1990, Tectonic and climatic controls on lake distribution and environments in Afar from Miocene to present. In Katz, B.J. (ed.), *Lacustrine Basin Exploration—Case Studies and Modern Analogues.* Amer. Assoc. Petrol. Geol. Mem. 50:19–41.

Gasse, F., 2001, Hydrological changes in Africa. *Science* 292:2259–2260.

Gasse, F., Fontes, J.C., Plaziat, J.C., Carbonel, P., Kaczmarska, I., De Deckker, P., Soulié-Marsche, I., Callot, Y., and Dupeuble, P.A., 1987, Biological remains, geochemistry and stable isotopes for the reconstruction of environmental and hydrological changes in the Holocene lakes from North Sahara. *Palaeogeog., Palaeoclim., Palaeoecol.* 60:1–46.

Gasse, F. and Tekaia, F., 1983, Transfer functions for estimating paleoecological conditions (pH) from East African diatoms. *Hydrobiologia* 103:85–90.

Gasse, F. and Van Campo, E., 1998, A 40,000-yr pollen and diatom record from Lake Tririvakely, Madagascar, in the southern tropics. *Quat. Res.* 49:299–311.

Gasse, F., Téhet, A., Durand, E., Gibert, E., and Fontes, J.Ch., 1990, The arid–humid transition in the Sahara and the Sahel during the last deglaciation. *Nature* 346:141–146.

Gastaldo, R.A., Bearce, S.C., Degges, C.W., Hunt, R.J., Peebles, M.W., and Violette, D.L., 1989, Biostratinomy of a Holocene oxbow lake: a backswamp to mid-channel transect. *Rev. Palaeobot. Palynol.* 58:47–58.

Gat, J.R., 1981, Lakes. In Gat, J.R. and Gonfiantini, R. (eds.), *Stable Isotope Hydrology: Deuterium and Oxygen-18 in the Water Cycle*. Intl. Atomic Energy Agency Tech. Rep. 210:203–221.

Gat, J.R., Bowser, C.J., and Kendall, C., 1994, The contribution of evaporation from the Great Lakes to the continental atmosphere: estimate based on stable isotope data. *Geophys. Res. Lett.* 21:557–560.

Gavrieli, I., 1997, Halite deposition from the Dead Sea: 1960–1993. In Niemi, T., Ben-Avraham, Z., and Gat, J.R. (eds.), *The Dead Sea: The Lake and Its Setting*. OUP Monograph on Geology and Geophysics 36. Oxford University Press, New York, pp. 161–170.

Gawthorpe, R.L., Fraser, A.J., and Collier, R.E., 1994, Sequence stratigraphy in active extensional basins: implications for the interpretation of ancient basin-fills. *Mar. Petrol. Geol.* 11:642–658.

Geary, D., Magyar, I., and Müller, P., 2000, Ancient Lake Pannon and its endemic molluscan fauna (Central Europe; Mio-Pliocene). In Rossiter, A. and Kawanabe, H. (eds.), *Ancient Lakes: Biodiversity, Ecology and Evolution*. Advances in Ecological Research 31. Academic Press, New York, pp. 463–482.

Gell, P.A., Barker, P.A., De Deckker, P., Last, W.M., and Jelicic, L., 1994, The Holocene history of West Basin Lake, Victoria, Australia; chemical changes based on fossil biota and sediment mineralogy. *Jour. Paleolim.* 12:235–258.

Geyh, M.A. and Schleicher, H., 1990, *Absolute Age Determination*. Springer-Verlag, New York.

Gilbert, G.K., 1890, Lake Bonneville. U.S. Geol. Surv. Monogr. 1:1–275.

Giovanoli, F., 1990, Horizontal transport and sedimentation by interflows and turbidity currents in Lake Geneva. In Tilzer, M.M. and Serruya, C. (eds.), *Large Lakes: Ecological Structure and Function*. Springer-Verlag, New York, pp. 175–195.

Giraudi, C., 1998, Late Pleistocene and Holocene lake-level variations in Fucino Lake (Abruzzo, Central Italy) inferred from geological, archaeological and historical data. In Harrison, S.P., Frenzel, B., Huckriede, U., and Weiß, M.M. (eds.), *Palaeohydrology as Reflected in Lake Level Changes as Climatic Evidence for Holocene Times*. Gustav Fischer-Verlag, Stuttgart, pp. 1–17.

Given, K. and Wilkinson, B., 1985, Kinetic control of morphology, composition, and mineralogy of abiotic sedimentary carbonates. *Jour. Sed. Petrol.* 55:109–119.

Gleadow, A.J.W., 1980, Fission track age of the KBS Tuff and associated hominid remains in northern Kenya. *Nature* 284:225–230.

Glen, J.M. and Coe, R.S., 1997, Paleomagnetism and magnetic susceptibility of Pleistocene sediments from drill hole OL-92, Owens Lake, California. In Smith, G.I. and Bischoff, J.L. (eds.), *An 800,000 Year Paleoclimate Record from Core OL-2, Owens Lake, Southeast California*. Geol. Soc. Amer. Spec. Pap. 317:67–78.

Glenn, C. and Kelts, K., 1991, Sedimentary rhythms in lake deposits. In Einsele, G., Ricken, W., and Seilacher, A. (eds.), *Cycles and Events in Stratigraphy*. Springer-Verlag, Berlin, pp. 188–221.

Glime, J.M., Wetzel, R.G., and Kennedy, B.J., 1982, The effects of bryophytes on succession from alkaline marsh to *Sphagnum* bog. *Amer. Midland Nat.* 108:209–223.

Gliozzi, E., 1999, A late Messinian brackish water ostracod fauna of Paratethyan aspect from Le Vicenne Basin (Abruzzi, central Apennines, Italy). *Palaeogeogr., Palaeoclim., Palaeoecol.* 151:191–208.

Godsey, H.S., Moore, T.C., Rea, D.K., and Shane, L.C.K., 1999 Post-Younger Dryas seasonality in the North American mid-continent region as recorded in Lake Huron varved sediments. *Can. Jour. Earth Sci.* 36:533–547.

Goldberg, E.D., 1963, Geochronology with ^{210}Pb. In *Radioactive Dating*. International Atomic Energy Agency, Vienna, pp. 121–131.

Goodfriend, G., 1992, Rapid racemization of aspartic acid in mollusc shells and potential for dating over recent centuries. *Nature* 357:399–401.

Goodfriend, G., Collins, M., Fogel, M., Macko, S., and Wehmilller, J. (eds.), 2000, *Perspectives in Amino Acid and Protein Geochemistry*. Plenum Press, New York.

Gore, P.J.W., 1986, Triassic notostracans in the Newark Supergroup, Culpeper Basin, Northern Virginia. *Jour. Paleo.* 60:1086–1096.

Gore, P.J.W., 1988, Paleoecology and sedimentology of a Late Triassic lake, Culpeper Basin, Virginia, USA. *Palaeogeog., Palaeoclim., Palaeoecol.* 62:593–608.

Gorham, E., 1960, Chlorophyll derivatives in surface muds from the English lakes. *Limnol. Oceanogr.* 5:29–33.

Gorham, E. and Sanger, J.E., 1967, Plant pigments in woodland soils. *Ecology* 48:306–308.

Gorham, E., Lund, J.W.G., Sanger, J.E., and Dean, W.E., 1974, Some relationships between algal standing crop, water chemistry, and sediment chemistry in the English Lakes. *Limnol. Oceanogr.* 19:601–617.

Görür, N., Tüysüz, O., Aykol, A., Sakinç, M., Yigtbas, E., and Akkök, R., 1993, Cretaceous red pelagic carbonates of northern Turkey: their place in the opening history of the Black Sea. *Eclog. geol. Helv.* 86:819–838.

Gou, Y. and Cao, M., 1983, Stratigraphic and biogeographic distribution of the *Cypridea*-bearing faunas in China. In Maddocks, R.F. (ed.), *Applications of Ostracoda.* Proc. 8th Intl. Symp. Ostracoda, University of Houston, Houston, TX, pp. 381–393.

Goulden, C.E., 1964, The history of the cladoceran fauna of Esthwaite Water (England) and its limnological significance. *Arch. Hydrobiol.* 60:1–52.

Graham, S., Hendrix, M.S., Wang, L.B., and Carroll, A.R., 1993, Collisional successor basins of western China: impact of tectonic inheritance on sand composition. *Geol. Soc. Amer. Bull.* 105:323–344.

Grahn, O., Hultberg, H., and Landner, L., 1974, Oligotrophication—a self-accelerating process in lakes subjected to excessive supply of acid substances. *Ambio* 3:431–433.

Gray, J. (ed.), 1988a, *Paleolimnology: Aspects of Freshwater Paleoecology and Biogeography.* Elsevier, Amsterdam.

Gray, J., 1988b, Evolution of the freshwater ecosystem: the fossil record. In Gray, J. (ed.), *Paleolimnology: Aspects of Freshwater Paleoecology and Biogeography.* Elsevier, Amsterdam, pp. 1–214.

Green-Winkler, M. and Sanford, P.R., 1995, Coastal Massachusetts pond development; edaphic, climatic and sea level impacts since deglaciation. *Jour. Paleolim.* 14:311–336.

Greenwood, P.H., 1961, A revision of the genus *Dinopterus* Blqr. (Pisces, Clariidae) with notes on the comparative anatomy of the superbranchial organs in the Clariidae. *Bull. Brit. Mus. Nat. Hist. Zool.* 7:217–241.

Grekoff, N. and Krömmelbein, K., 1967, Etude comparée des ostracodes mésozoiques continentaux des bassins atlantiques: série de Cocobeach, Gabon, et série de Bahia, Brésil. *Rev. Inst. Fran. Pétrole* 22:1307–1353.

Grey, K., 1989, Handbook for the study of stromatolites and associated structures. In Kennard, J.M. and Burne, R.V. (eds.), *Stromatolite Newsletter.* Australian Bureau of Mines Research, Geology and Geophysics, pp. 82–171.

Griffiths, H.I., Ringwood, V., and Evans, J.G., 1994, Weichselian Late-glacial and early Holocene molluscan and ostracod sequences from lake sediments at Stellmoor, north Germany. *Arch. Hydrobiol./Suppl.* 99:357–380.

Griffiths, M., Perrott, P.S., and Edmundson, W.T., 1969, Oscillaxanthin in the sediment of Lake Washington. *Limnol. Oceanogr.* 14:317–326.

Grimm, E.C., 1987, CONISS: a Fortran 77 program for stratigraphically constrained cluster analysis by the method of incremental sum of squares. *Comput. Geosci.* 13:13–35.

Grootes, P.M., 1993, Interpreting continental oxygen isotope records. In Swart, P.K., Lohmann, K.C., McKenzie, J., and Savin, S. (eds.), *Climate Change in Continental Isotopic Records.* Amer. Geophys. Union Geophys. Monogr. 78:37–46.

Grosdidier, E., Braccini, E., Dupont, G., and Moron, J.M., 1996, Non-marine Lower Cretaceous biozonation of the Gabon and Congo Basins. In Jardine, S., de Klasz, I., and Debeney, J.P. (eds.), Géologie de l'afrique et de l'Atlantique Sud. *Bull. Cent. Rech. Elf-Aquitaine Mém.* 16:67–82.

Groshopf, P., 1936, Die postglaziale Entwicklung des Grossen Ploner See in Ostholstein auf Grund pollenanalytischer Sedimentuntersuchungen. *Arch. Hydrobiol.* 30:1–84.

Grosse-Brauckmann, G.G., 1986, Analysis of vegetative plant macrofossils. In Berglund, B.E. (ed.), *Handbook of Holocene Paleoecology and Paleohydrology.* J. Wiley and Sons, New York, pp. 591–618.

Grossman, E.L. and Ku, T.L., 1986, Oxygen and carbon isotope fractionation in biogenic aragonite: temperature effects. *Chem. Geol. (Isotope Sect.)* 59:59–74.

Grover, N.C. and Howard, C.S., 1938, The passage of turbid water through Lake Mead. *Trans. Amer. Soc. Civil Eng.* 103:720–790.

Gu, Z., 1998, Evolutionary trends in nonmarine Cretaceous bivalves of northeast China. In Johnston, P.A. and Haggart, J.W. (eds.), *Bivalves: An Eon of Evolution.* University of Calgary Press, Calgary, pp. 267–276.

Guilizzoni, P. and Lami, A., 1988, Sub-fossil pigments as a guide to the phytoplankton history of the acidified Lake Orta (N. Italy). *Verh. Int. Verein. Limnol.* 23:874–879.

Guilizzoni, P., Lami, A., and Marchetto, A., 1992, Plant pigment ratios from lake sediments as indicators of recent acidification in alpine lakes. *Limnol. Oceanogr.* 37:1565–1569.

Guilizzoni, P., Marchetto, A., Lami, A., Oldfield, F., Manca, M., Belis, C.A., Nocentini, A.M., Comoli, P., Jones, V.J., Juggins, S., Chondrogianni, C., Aristegui, D., Lowe, J.J., Ryves, D.B., Battarbee, R.W., Rolph, T.C., and Massaferro, J., 2000, Evidence for short-lived oscillations in the biological records from the sediments of Lago Albano (Central Italy) spanning the period ca. 28–17k yr BP. *Jour. Paleolim.* 23:117–127.

Guiot, J., Reille, M., Beaulieu, J.L., and Pons, A., 1992, Calibration of the climate signal in a new pollen sequence from La Grande Pile. *Clim. Dynam.* 6:259–264.

Guo. F., 1998a, Origin and phylogeny of the Trigonioidoidea (nonmarine Cretaceous bivalves). In Johnston, P.A. and Haggart, J.W. (eds.), *Bivalves: An Eon of Evolution.* University of Calgary Press, Calgary, pp. 277–289.

Guo. F., 1998b, Sinonaiinae, a new subfamily of Asian nonmarine Cretaceous bivalves. In Johnston, P.A. and Haggart, J.W. (eds.), *Bivalves: An Eon of Evolution.* University of Calgary Press, Calgary, pp. 291–294.

Gustavson, T.C., 1975, Sedimentation and physical limnology in proglacial Malaspina Lake, southeastern Alaska. In Jopling, A. and McDonald, B.C. (ed.), *Glaciofluvial and Glaciolacustrine Sedimentation.* Soc. Econ. Paleo. Mineral. Spec. Publ. 23:249–263.

Guy-Ohlson, D., 1992, *Botryococcus* as an aid in the interpretation of palaeoenvironment and depositional process. *Jour. Palaeobot. Palynol.* 71:1–15.

Guy-Ohlson, D., 1998, The use of the microalga *Botryococcus* in the interpretation of lacustrine environments at the Jurassic–Cretaceous transition in Sweden. *Palaeogeog., Palaeoclim., Palaeoecol.* 140:347–356.

Haberyan, K.A., 1985, The role of copepod fecal pellets in the deposition of diatoms in Lake Tanganyika. *Limnol. Oceanogr.* 30:1010–1023.

Haberyan, K.A., 1990, The misrepresentation of the planktonic diatom assemblage in traps and sediments: southern Lake Malawi, Africa. *Jour. Paleolim.* 3:35–44.

Haberyan, K.A. and Hecky, R.E., 1987, The late Pleistocene and Holocene stratigraphy and paleolimnology of Lakes Kivu and Tanganyika. *Palaeogeog., Palaeoclim., Palaeoecol.* 61:169–197.

Hajdas, I., Zolitschka, B., Ivy-Ochs, S.D., Beer, J., Bonanai, G., Leroy, S.A.G., Negendanck, J.W., Ramrath, M., and Suter, M., 1995, AMS radiocarbon dating of annually laminated sediments from Lake Holzmaar, Germany. *Quat. Sci. Rev.* 14:137–143.

Håkanson, L., 1982, Lake bottom dynamics and morphometry: the dynamic ratio. *Wat. Res. Res.* 18:1444–1450.

Håkanson, L. and Jansson, M., 1983, *Principles of Lake Sedimentology.* Springer-Verlag, Heidelberg.

Hall, B.V. and Hermann, S.J., 1980, Paleolimnology of three species of fresh-water sponges (Porifera: Spongillidae) from a sediment core of a Colorado semidrainage mountain lake. *Trans. Amer. Micros. Soc.* 99:93–100.

Hall, R.I. and Smol, J.P., 1992, A weighted-averaging regression and calibration model for inferring total phosphorus concentration from diatoms in British Columbia (Canada) lakes. *Freshwater Biol.* 27:417–434.

Hall, R.I. and Smol, J.P., 1993, The influence of catchment size on lake trophic status during the hemlock decline and recovery (4800 to 3500 BP) in southern Ontario. *Hydrobiologia* 269/270:371–390.

Hall, R.I. and Smol, J.P., 1999, Diatoms as indicators of lake eutrophication. In Stoermer, E.F. and Smol, J.P. (eds.), *The Diatoms: Applications for the Environmental and Earth Sciences.* Cambridge University Press, Cambridge, pp. 128–168.

Hall, R.I., Leavitt, P.R., Smol, J.P., and Zirnhelts, N., 1997, Comparison of diatoms, fossil pigments and historical records as measures of lake eutrophication. *Freshwater Biol.* 38:401–417.

Hall, R.I., Leavitt, P.R., Quinlan, R., Dixit, A.S., and Smol, J.P., 1999, Effects of agriculture, urbanization and climate on water quality in the Northern Great Plains. *Limnol. Oceanogr.* 44:739–756.

Hallbäcken, L. and Tamm, C.O., 1986, Changes in soil acidity from 1927 to 1982–1984 in a forest area of south-west Sweden. *Scand. Jour. Forest Res.* 1:219–232.

Halstead, L.B., 1993, Agnatha. In Benton, M.J. (ed.), *The Fossil Record 2.* Chapman and Hall, London, pp. 573–582.

Hamilton-Taylor, J. and Davison, W., 1995, Redox-driven cycling of trace elements in lakes. In Lerman, A., Imboden, D.M., and Gat, J. (eds.), *Physics and Chemistry of Lakes*, 2nd ed. Springer-Verlag, New York, pp. 217–263.

Hammarlund, D., Aravena, R., Barnekow, L., Buchardt, B., and Possnert, G., 1997, Multi-component carbon isotope evidence of early Holocene environmental change and carbon-flow pathways from a hard-water lake in northern Sweden. *Jour. Paleolim.* 18:219–233.

Hammer, C., Mayewski, P.A., Peel, D., and Stuiver, M., 1997, Greenland summit ice cores. *Jour. Geophys. Res.* 102:C12.

Hanley, J., 1976, Paleosynecology of nonmarine molluscs from the Green River and Wasatch Fms., SW Wyoming and NW Colorado. In Scott, R. and West, R. (eds.), *Structure and Classification of Paleocommunities.* Dowden, Hutchinson and Ross, Stroudsburg, PA, pp. 235–261.

Hann, B.J., 1990, Cladocera. In Warner, B.G. (ed.), *Methods in Quaternary Ecology.* Geoscience

Canada Reprint Ser. 5. Geological Association of Canada, pp. 81–91.

Hann, B.J. and Karrow, P.F., 1984, Pleistocene paleoecology of the Don and Scarborough Formations, Toronto, Canada, based on cladoceran microfossils at the Don Valley Brickyard. *Boreas* 13:377–391.

Hann, B.J. and Karrow, P.F., 1993, Comparative analysis of cladoceran microfossils in the Don and Scarborough Formations, Toronto, Canada. *Jour. Paleolim.* 9:223–241.

Hann, B.J., Warner, B.G., and Warwick, W.F., 1992, Aquatic invertebrates and climate change: a comment on Walker *et al.* (1991). *Can. Jour. Fish. Aquat. Sci.* 49:1274–1276.

Hann, B.J., Leavitt, P.R., and Chang, P.S.S., 1994, Cladocera community response to experimental eutrophication in Lake 227 as recorded in laminated sediments. *Can. Jour. Fish. Aquatic. Sci.* 51:2312–2321.

Hannon, G. and Gaillard, M.J., 1997, The plant macrofossil record of past lake-level changes. *Jour. Paleolim.* 18:15–28.

Hansel, A.K., Mickelson, D.M., Schneider, A.F., and Larsen, C.E., 1985, Late Wisconsinan and Holocene history of the Lake Michigan Basin. In Karrow, P.F. and Calkin, P.E. (eds.), *Quaternary Evolution of the Great Lakes.* Geol. Assoc. Can. Spec. Pap. 30:39–53.

Hansen, K., Mouridsen, S., and Kristensen, E., 1998, The impact of *Chironomus plumosus* larvae on organic matter decay and nutrient (N,P) exchange in a shallow eutrophic lake sediment following a phytoplankton sedimentation. *Hydrobiologia* 364:65–74

Hanson, P.H., Hanson, C.S., and Yoo, B.H., 1992, Recent Great Lakes ice trends. *Bull. Amer. Meteor. Soc.* 73:577–584.

Hao, Y., Su, D., and Li, Y., 1983, Late Mesozoic nonmarine ostracodes in China. In Maddocks, R.F. (ed.), *Applications of Ostracoda.* Proc. 8th Intl. Symp. Ostracoda, University of Houston, Houston, TX, pp. 372–380.

Hardie, L.A. and Eugster, H.P., 1970, The evolution of closed-basin brines. *Mineral. Soc. Amer. Spec. Publ.* 3:273–290.

Hardie, L.A., Smoot, J.P., and Eugster, H.P., 1978, Saline lakes and their deposits: a sedimentological approach. In Matter, A. and Tucker, M.E. (eds.), *Modern and Ancient Lake Sediments.* Intl. Assoc. Sedimentol. Spec. Publ. 2:7–41.

Harding, I.C. and Chant, L.S., 2000, Self-sedimented diatom mats as agents of exceptional fossil preservation in the Oligocene Florissant lake beds, Colorado, United States. *Geology* 28:195–198.

Hardy, D.R., Bradley, R.S., and Zolitschka, B., 1996, The climatic signal in varved sediments from Lake C2, northern Ellesmere Island, Canada. *Jour. Paleolimnol.* 16:227–238.

Hardy, E.P., 1977, Final tabulation of monthly ^{90}Sr fallout data, 1954–1976. *Envir. Quat.* HASL-329, U.S. Dept. Energy, New York.

Hare, P.E. and Abelson, P.H., 1966, Racemization of amino acids in fossil shells. *Carnegie Inst. Yearbook* 66:526–528.

Hare, P.E. and Mitterer, R.M., 1967, Laboratory simulation of amino acid diagenesis in fossils. *Carnegie Inst. Yearbook* 67:205–207.

Harman, W.N., 1974, Snails (Mollusca: Gastropoda) In Hart, C.W. and Fuller, S.L.H. (eds.), *Pollution Ecology of Freshwater Invertebrates.* Academic Press, New York, pp. 275–312.

Harris, N.B., Sorriaux, P., and Toomey, D.F., 1994, Geology of the Lower Cretaceous Viodo Carbonate, Congo Basin: a lacustrine carbonate in the South Atlantic rift. In Lomando, A.J., Schreiber, B.C., and Harris, P.M. (eds.), *Lacustrine Reservoirs and Depositional Systems.* Soc. Econ. Paleo. Mineral. Core Workshop 19:143–172.

Harrison, F.W., 1974, Sponges (Porifera: Spongillidae) In Hart, C.W. and Fuller, S.L.H., *Pollution Ecology of Freshwater Invertebrates.* Academic Press, New York, pp. 29–66.

Harrison, F.W., 1988, Utilization of freshwater sponges in paleolimnological studies. In Gray, J. (ed.), *Paleolimnology: Aspects of Freshwater Paleoecology and Biogeography.* Elsevier, Amsterdam, pp. 387–397.

Harrison, F.W., 1990, Freshwater sponges. In Warner, B.G. (ed.), *Methods in Quaternary Ecology.* Geoscience Canada Reprint Ser. 5. Geological Association of Canada, pp. 75–80.

Harrison, F.W. and Warner, B.G., 1986, Fossil freshwater sponges (Porifera: Spongillidae) from Western Canada: an overlooked group of Quaternary paleoecological indicators. *Trans. Amer. Micros. Soc.* 105:110–120.

Harrison, S.P., Yu, G., and Tarasov, P.E., 1996, Late Quaternary lake-level record from Northern Eurasia. *Quat. Res.* 45:138–159.

Harrsch, E.C. and Rea, D.K., 1982, Composition and distribution of suspended sediments in Lake Michigan during summer stratification. *Envir. Geol. Water Sci.* 4:87–98.

Hart, M.B. and Williams, C.L., 1993, Protozoa. In Benton, M.J. (ed.), *The Fossil Record 2.* Chapman and Hall, London, pp. 43–70.

Hartley, R.W. and Allen, P.A., 1994, Interior cratonic basins of Africa: relation to continental break-up and role of mantle convection. *Basin Res.* 6:95–113.

Hartman, J.H., 1998, The biostratigraphy and paleontology of Latest Cretaceous and Paleocene freshwater bivalves from the Western Williston Basin, Montana, USA. In

Johnston, P.A. and Haggart, J.W. (eds.), *Bivalves: An Eon of Evolution*. University of Calgary Press, Calgary, pp. 317–346.

Hartman, J.H. and Roth, B., 1998, Late Paleocene and Early Eocene nonmarine molluscan faunal change in the Bighorn Basin, Northwestern Wyoming and South-Central Montana. In Aubry, M.P., Lucas, S., and Berggren, W.A. (eds.), *Late Paleocene–Early Eocene Climatic and Biotic Events in the Marine and Terrestrial Record*. Columbia University Press, New York, pp. 323–379.

Hartung, J. and Koeberl, C., 1994, In search of the Australasian tektite source crater: the Tonle Sap hypothesis. *Meteoritics* 29:411–416.

Haskell, B.J., Engstrom, D.R., and Fritz, S.C., 1996, Late Quaternary paleohydrology in the North American Great Plains inferred from the geochemistry of endogenic carbonate and fossil ostracodes from Devils Lake, North Dakota, USA. *Palaeogeogr., Palaeoclim., Palaeoecol.* 124:179–193.

Hasler, A.D. and Einsele, W.G., 1948, Fertilization for increasing productivity of natural inland waters. *Trans. 13th N. Amer. Wildlife Conf.*, pp. 527–554.

Hassiotis, S.T. and Honey, J.G., 2000, Paleohydrologic and stratigraphic significance of crayfish burrows in continental deposits: examples from several Paleocene Laramide basins in the Rocky Mountains. *Jour. Sed. Res.* 70:127–139.

Hastenrath, S., 1985, *Climate and Circulation of the Tropics*. D. Reidel, Boston.

Haworth, E.Y. and Lund, J.W.G. (eds.), 1984, *Lake Sediments and Environmental History*. University of Minnesota Press, Minneapolis, MN.

Hay, R.L., 1966, Zeolites and zeolitic reactions in sedimentary rocks. *Geol. Soc. Amer. Spec. Pap.* 85.

Hay, R.L., 1968, Chert and its sodium-silicate precursors in sodium carbonate lakes of East Africa. *Contr. Mineral. Petrol.* 17:225–274.

Hayes, J.M., Takigiku, R., Ocampo, R., Callot, H.J., and Albrecht, P., 1987, Isotopic compositions and probable origins of organic molecules in the Eocene Messel shale. *Nature* 329:48–51.

Hays, J., Imbrie, J., and Shackleton, N., 1976, Variations in the Earth's orbit: pacemaker of the ice ages. *Science* 194:1121–1132.

Hearn, B.C., McLaughlin, R.J., and Donnelly-Nolan, J.M., 1988, Tectonic framework of the Clear Lake basin, California. In Sims, J.D. (ed.), *Late Quaternary Climate, Tectonism and Sedimentation in Clear Lake*. Geol. Soc. Amer. Spec. Pap. 214:9–20.

Hecky, R.E., 1993, The eutrophication of Lake Victoria. *Verh. Int. Verein. Limnol.* 25:39–48.

Hecky, R.E., Campbell, P., and Hendzel, L.L., 1993, The stoichiometry of carbon, nitrogen and phosphorus in particulate matter of lakes and oceans. *Limnol. Oceanogr.* 38:709–724.

Hecky, R.E., Bootsma, H.A., Mugidde, R.M., and Bugenyi, F.W.B., 1996, Phosphorus pumps, nitrogen sinks, and silicon drains: plumbing nutrients in the African Great Lakes. In Johnson, T.C. and Odada, E.O. (eds.), *The Limnology, Climatology and Paleoclimatology of the East African Lakes*. Gordon and Breach, Amsterdam, pp. 205–224.

Hedberg, H., 1976, *International Stratigraphic Guide. A Guide to Stratigraphic Classification, Terminology and Procedure*. John Wiley & Sons, New York.

Hedderson, T.A., Chapman, R.L., and Rootes, W.L., 1996, Phylogenetic relationships of bryophytes inferred from nuclear-encoded rRNA gene sequences. *Plant Syst. Evol.* 200:213–224.

Hedges, J.L., Ertel, J.R., and Leopold, E.B., 1982, Lignin geochemistry of a Late Quaternary core from Lake Washington. *Geochim. Cosmochim. Acta* 46:1869–1877.

Heer, O., 1865, *Die Urwelt der Schweiz*. F. Schulthess, Zürich.

Heinis, F., Sweerts, J.P., and Loopik, E., 1994, Micro-environment of chironomid larvae in the littoral and profundal zones of Lake Maarsseveen I, The Netherlands. *Arch. Hydrobiol.* 130:53–67.

Heinrich, H., 1988, Origin and consequences of cyclic ice rafting in the northeast Atlantic Ocean during the past 130,000 years. *Quat. Res.* 29:142–152.

Heiri, O., Lotter, A., and Lemke, G., 2001, Loss on ignition as a method for estimating organic and carbonate content in sediments: reproducibility and comparability of results. *Jour. Paleolim.* 25:101–110.

Heirtzler, J.R., Dickson, G.O., Herron, E.M., Pitman, W.C., and LePichon, X., 1968, Marine magnetic anomalies, geomagnetic field reversals and motions of the ocean floor and continents. *Jour. Geophys. Res.* 73:2119.

Heller, P.L., Angevine, C.L., Winslow, N.S., and Paola, C., 1988, Two phase stratigraphic model of foreland basin sequence. *Geology* 16:501–504.

Henderson, P.J. and Last, W.M., 1999, Holocene sedimentation in Lake Winnipeg, Manitoba, Canada; implications of compositional and textural variations. *Jour. Paleolim.* 19:265–284.

Herczeg, A.L. and Fairbanks, R.G., 1987, Anomalous carbon isotope fractionation between atmospheric CO_2 and dissolved inorganic carbon induced by intense photosynthesis. *Geochim. Cosmochim. Acta* 51:895–899.

Herczeg, A.L. and Lyons, W.B., 1991, A chemical model for the evolution of Australian sodium chloride lake brines. *Palaeogeog., Palaeoclim., Palaeoecol.* 84:43–53.

Herdendorf, C.E., 1990, Distribution of the world's large lakes. In Tilzer, M.M. and Serruya, C. (eds.), *Large Lakes*. Springer-Verlag, New York, pp. 3–38.

Hershey, A.E., 1985, Effects of predatory sculpins on the chironomid communities in an arctic lake. *Ecology* 66:1131.

Hickey, L.J., 1977, *Stratigraphy and Paleobotany of the Golden Valley Formation (Early Tertiary) of Western North Dakota*. Geol. Soc. Amer. Mem. 150.

Hicks, R.E., Owen, C.J., and Aas, P., 1994, Deposition, resuspension, and decomposition of particulate organic matter in the sediments of Lake Itasca, Minnesota, USA. *Hydrobiologia* 284:79–91.

Hill, J., 1975, The origin of southern African coastal lakes. *Trans. R. Soc. S. Afr.* 41:225–240.

Hill, W.R., 1996, Effects of light. In Stevenson, R.J., Bothwell, M.L., and Lowe, R.L. (eds.), *Algal Ecology: Freshwater Benthic Ecosystems*. Academic Press, New York, pp. 121–148.

Hillaire-Marcel, C. and Casanova, J., 1987, Isotopic hydrology and paleohydrology of the Magadi (Kenya)–Natron (Tanzania) Basin during the Late Quaternary. *Palaeogeogr., Palaeoclim., Palaeoecol.* 58:155–181.

Hillaire-Marcel, C., Carro, O., and Casanova, J., 1986, ^{14}C and Th/U dating of Pleistocene and Holocene stromatolites from East African paleolakes. *Quat. Res.* 25:312–329.

Hillaire-Marcel, C., Aucour, A.M., Bonnefille, R., Riollet, G., Vincens, A., and Williamson, D., 1989, ^{13}C/Palynological evidence of differential residence times of organic carbon prior to its sedimentation in East African rift lakes and peat bogs. *Quat. Sci. Rev.* 8:207–212.

Hillsenhoff, W.L., 1991, Diversity and classification of insects and collembola. In Thorp, J.H. and Covitch, A.P. (eds.), *Ecology and Classification of North American Freshwater Invertebrates*. Academic Press, New York, pp. 593–663.

Hilton, J., 1985, A conceptual framework for predicting the occurrence of sediment focusing and sediment redistribution in small lakes. *Limnol. Oceanogr.* 30:1131–1143.

Hinga, K.R., Arthur, M.A., Pilson, M.E.Q., and Whitaker, D., 1994, Carbon isotope fractionation by marine phytoplankton in culture: the effects of CO_2 concentration, pH, temperature, and species. *Global Biogeochem. Cycles* 8:91–102.

Hobbie, J.E., 1984, Polar limnology. In Taub, F.B. (ed.), *Lakes and Reservoirs*. Ecosystems of the World 23. Elsevier, New York, pp. 63–106.

Hobday, D.K., 1979, Geological evolution and geomorphology of the Zulu-land coastal plain. In Allanson, B.R. (ed.), *Lake Sibaya*. Monographiae Biologicae 36. Junk Publishers, Boston, pp. 1–21.

Hodell, D.A. and Schelske, C.L., 1998, Production, sedimentation and isotopic composition of organic matter in Lake Ontario. *Limnol. Oceanogr.* 43:200–214.

Hodell, D.A., Curtis, J.J., and Brenner, M., 1995, Possible role of climate in the collapse of Classic Maya civilization. *Nature* 375:391–394.

Hodell, D.A., Schelske, C.L., Fahnenstiel, G.L., and Robbins, L.L., 1998, Biologically induced calcite and its isotopic composition in Lake Ontario. *Limnol. Oceanogr.* 43:187–199.

Hodell, D.A., Brenner, M., Curtis, J.H., and Guilderson, T., 2001, Solar forcing of drought frequency in the Maya lowlands. *Science* 292:1367–1370.

Hodgkin, E.P. and Hesp, P., 1998, Estuaries to salt lakes: Holocene transformation of the estuarine ecosystems of south-western Australia. *Mar. Freshw. Res.* 49:183–201.

Hodgson, D.A., 1999, The formation of flocculated clay laminae in the sediments of a meromictic lake. *Jour. Paleolim.* 21:263–269.

Hodgson, D.A., Wright, S.W., and Davies, N., 1997, Mass spectrometry and reverse phase HPLC techniques for the identification of degraded fossil pigments in lake sediments and their application in paleolimnology. *Jour. Paleolimnol.* 18:335–350.

Hodgson, D.A., Wright, S.W., Tyler, P.A., and Davies, N., 1998, Analysis of fossil pigments from algae and bacteria in meromictic Lake Fidler, Tasmania, and its application to lake management. *Jour. Paleolimnol.* 19:1–22.

Hoek, W.Z., Bohnke, S.J.P., Ganssen, G.M., and Meijer, T., 1999, Late glacial environmental changes recorded in calcareous gyttja deposits at Gulickshof, southern Netherlands. *Boreas* 28:416–432.

Hoffman, A., Tourani, A., and Gaupp, R., 2000, Cyclicity of Triassic to Lower Jurassic continental red beds of the Argana Valley, Morocco: implications for palaeoclimate and basin evolution. *Palaeogeog., Palaeoclim., Palaeoecol.* 161:229–266.

Hoffman, W., 1986a, Chironomid analysis. In Berglund, B.E. (ed.), *Handbook of Holocene Palaeoecology and Palaeohydrology*. J. Wiley and Sons, New York, pp. 715–727.

Hoffman, W., 1986b, Developmental history of the Grosser Plönen See and the Schönsee (north Germany): cladoceran analysis, with special reference to eutrophication. *Arch. Hydrobiol. Suppl. Bd.* 74:259–287.

Hoffman, W., 1988, The significance of chironomid analysis (Insecta: Diptera) for palaeolimnological research. In Gray, J. (ed.), *Paleolimnology: Aspects of Freshwater Paleoecology and Biogeography*. Elsevier, Amsterdam, pp. 501–509.

Hoffman, W., 1996, Empirical relationships between cladoceran fauna and trophic state in thirteen northern German lakes: analysis of surficial sediments. *Hydrobiologia* 318:195–201.

Hoffman, W., 1998, Cladocerans and chironomids as indicators of lake level changes in north temperate lakes. *Jour. Paleolim.* 19:55–62.

Hollander, D.J., 1989, *Carbon and nitrogen isotopic cycling and organic geochemistry of eutrophic Lake Greifen: implications for preservation and accumulation of ancient organic carbon-rich sediments*. Ph.D. Thesis, ETH, Zürich.

Holliday, V.T., Hovorka, S.D., and Gustavson, T.C., 1996, Lithostratigraphy and geochronology of fills in small playa basins on the Southern High Plains, United States. *Geol. Soc. Amer. Bull.* 108:953–965.

Holloway, P., 1980, A criterion for thermal stratification in a wind mixed system. *Jour. Phys. Oceanogr.* 10:861–869.

Holmes, J.A., 1992, Nonmarine ostracodes as Quaternary palaeoenvironmental indicators. *Prog. Phys. Geogr.* 16:425–431.

Holmes, J.A., 1998, A late Quaternary ostracod record from Wallywash Great Pond, a Jamaican marl lake. *Jour. Paleolim.* 19:115–128.

Holmes, J.A., Fothergill, P.A., Street-Perrott, F.A., and Perrott, R.A., 1998, A high-resolution Holocene ostracod record from the Sahel zone of Northeastern Nigeria. *Jour. Paleolim.* 20:369–380.

Holmquist, B. and Wohlfarth, B., 1998, An evaluation of the Late Weichselian Swedish varve chronology based on cross-correlation analysis. *GFF (Geol. Survey of Sweden)* 120:35–46.

Hooghiemstra, H. and Cleef, A., 1995, Pleistocene climatic change and environmental and generic dynamics in the north Andean montane forest and paramo. In Churchill, S.P. et al (eds.), *Biodiversity and Conservation of Neotropical Montane Forests*. New York Botanical Garden, New York, pp. 35–49.

Hooghiemstra, H. and Sarmiento, G., 1991, Long continental pollen record from a tropical intermontane basin: Late Pliocene and Pleistocene history from a 540-meter core. *Episodes* 14:107–115.

Hopkirk, J.D., 1988, Fish evolution and the late Pleistocene and Holocene history of Clear Lake. In Sims, J.D. (ed.), *Late Quaternary Climate, Tectonism and Sedimentation in Clear Lake*. Geol. Soc. Amer. Spec. Pap. 214:183–194.

Horton, B.K. and DeCelles, P.G., 1997, The modern foreland basin system adjacent to the Central Andes. *Geology* 25:895–898.

Hosn, W.A. and Downing, J.A., 1994, Influence of cover on the spatial distribution of littoral-zone fishes. *Can. Jour. Fish. Aquat. Sci.* 51:1832–1838.

Hostetler, S.W. and Benson, L.V., 1990, Paleoclimatic implications of the high stand of Lake Lahontan derived from models of evaporation and lake level. *Clim. Dynam.* 4:207–217.

Houghton, J.T., Ding, Y., Griggs, D.J., Noguer, M., van der Linden, P.J., and Xiaosu, D., (eds.), 2001, *Climate Change 2001: The Scientific Basis (Contribution of Working Group I to the Third Assessment Report of the Intergovernmental Panel on Climate Change (IPCC)*. Cambridge University Press, Cambridge.

Hovorka, S.D., 1997, Quaternary evolution of ephemeral playa lakes on the Southern High Plains of Texas, USA: cyclic variation in lake level recorded in sediments. *Jour. Paleolimnol.* 17:131–146.

Howarth, R.W., Marino, R., and Cole, J.J., 1988, Nitrogen fixation in freshwater, estuarine and marine ecosystems. I. Rates and importance. *Limnol. Oceanogr.* 33:669–687.

Hsü, K., 1978, Stratigraphy of the lacustrine sedimentation in the Black Sea. In Ross, D.A. et al. (eds.), *Initial Reports of the Deep Sea Drilling Project Leg 42*. U.S. Government Printing Office, pp. 509–524.

Hsü, K., 1988, Relict back-arc basins: principles of recognition and possible new examples form China. In Kleinspehn, K. and Paola, C. (eds.), *New Perspectives in Basin Analysis*. Springer-Verlag, New York, pp. 245–264.

Hsü, K., Kelts, K., and Giovanoli, F., 1984, Quaternary geology of the Lake Zürich region. *Contrib. Sedimentol.* 13:187–203.

Huang, Y., Street-Perrott, F.A., Metcalfe, S.E., Brenner, M., Moreland, M., and Freeman, K.H., 2001, Climate change as the dominant control on glacial–interglacial variations in C3 and C4 plant abundance. *Science* 293:1647–1651.

Huber, J.K., 1996, A post-glacial pollen and nonsiliceous algae record from Gegoka Lake, Lake County, Minnesota. *Jour. Paleolim.* 16:23–35.

Huddart, D., 1983, Flow tills and ice-walled lacustrine sediments, the Petteril Valley, Cumbria, England. In Evenson, E.B., Schlüchter, C., and Rabassa, J. (eds.), *Tills and Related Deposits*. Balkema, Rotterdam, pp. 81–94.

Hughen, K.A., Overpeck, J.T., and Anderson, R.F., 2000a, Recent warming in a 500 year palaeotemparature record from varved sediments, Upper Soper Lake, Baffin Island, Canada. *The Holocene* 10:9–19.

Hughen, K.A., Southon, J.R., Lehman, S.J., and Overpeck, J.T., 2000b, Synchronous radiocarbon and climate shifts during the last deglaciation. *Science* 290:1951–1954.

Hughes, M.K. and Diaz, H.F., 1994, Was there a 'Medieval Warm Period', and if so, where and when? *Clim. Change* 26:109–142.

Hunt, B.G., 2000, Natural climate variability and Sahelian rainfall trends. *Global Planet. Change* 24:107–131.

Hunter, R.E., Reiss, T.E., Chin, J.L., and Anima, R.J., 1990, Coastal depositional and erosional effects of 1985–1987 high lake levels in Lake Michigan. *U.S. Geol. Surv. Open File Rep.* 90:27–29.

Hustedt, F., 1939, Systematische und okologische Untersuchungen uber den Diatomeen-Flora von Java, Bali, Sumatra. *Arch. Hydrobiol. Suppl.* 16:274–394.

Hustedt, F., 1957, Die Diatomeenflora des Flussystems des Weser im Gebiet des Hansestadt Bremen. *Abh. Naturwiss. Ver. Bremen* 34:18–140.

Hutchinson, G.E., 1957, *A Treatise on Limnology, v. 1, Geography, Physics and Chemistry*. J. Wiley and Sons, New York.

Hutchinson, G.E., 1967, *A Treatise on Limnology, v. 2, Introduction to Lake Biology and the Limnoplankton*. J. Wiley and Sons, New York.

Hutchinson, G.E., 1975, *A Treatise on Limnology, v. 3, Limnological Botany*. J. Wiley and Sons, New York.

Hutchinson, G.E. and Loeffler, H., 1956, The thermal classification of lakes. *Proc. Nat. Acad. Sci.* 42:84–86.

Huvane, J.K. and Whitehead, D.R., 1996, The paleolimnology of North Pond: watershed–lake interactions. *Jour. Paleolim.* 16:323–354.

Hyatt, J.A. and Gilbert, R., 2000, Lacustrine sedimentary record of human-induced gully erosion and land-use changes at Providence Canyon, southwest Georgia, USA. *Jour. Paleolim.* 23:421–438.

Hyne, N.J., Laidig, L.W., and Cooper, W.A., 1979, Prodelta sedimentation on a lacustrine delta by clay mineral flocculation. *Jour. Sed. Petrol.* 49:1209–1216.

Imboden, D.M. and Wüest, A., 1995, Mixing mechanisms in lakes. In Lerman, A., Imboden, D.M., and Gat, J. (eds.), *Physics and Chemistry of Lakes*, 2nd ed. Springer-Verlag, New York, pp. 83–138.

Imbrie, J. and Imbrie, J.Z., 1980, Modeling the climate response to orbital variations. *Science* 207:943–952.

Imbrie, J. and Kipp, N.G., 1971, A new micropaleontological method for quantitative paleoclimatology: application to a late Pleistocene Caribbean core. In Turekian, K. (ed.), *The Late Cenozoic Glacial Ages*. Yale University Press, New Haven, CT, pp. 71–181.

Imbrie, J., Berger, A., Boyle, E.A., Clemens, S.C., Duffy, A., Howard, W.R., Kukla, G., Martinson, D.G., McIntyre, A., Mix, A.C., Molfino, B., Morley, J.J., Peterson, L.C., Pisias, N. G., Prell, W.L., Raymo, M.E., Shackleton, N.J., and Toggweiller, J.R., 1993, On the structure and origin of major glaciation cycles. 2. The 100,000 year cycle. *Paleoceanography* 5:699–735.

Ishiwatari, R. and Uzaki, M., 1987, Diagenetic changes of lignin compounds in a more than 0.6 million-year old lacustrine sediment (Lake Biwa, Japan). *Geochim. Cosmochim Acta* 51:321–328.

Israelson, C., Björck, S., Hawkesworth, C.J., and Possnert, G., 1997, Direct U–Th dating of organic and carbonate rich lake sediments from southern Scandinavia. *Earth Planet. Sci. Lett.* 153:251–263.

Itkonen, A. and Salonen, V.P., 1994, The response of sedimentation in three varved lacustrine sequences to air temperature, precipitation and human impact. *Jour. Paleolim.* 11:323–332.

Izett, G.A., 1981, Volcanic ash beds: recorders of upper Cenozoic silicic pyroclastic volcanism in the western United States. *Jour. Geophys. Res.* 86:10,200–10,222.

Jackson, S.T., 1990, Pollen source area and representation in small lakes of the northeastern United States. *Rev. Paleobot. Palynol.* 63:53–76.

Jacobs, J.A., 1994, *Reversals of the Earth's Magnetic Field*. Cambridge University Press, Cambridge.

Jacobsen, G.L. and Bradshaw, R.H.W., 1981, The selection of sites for paleovegetational studies. *Quat. Res.* 16:80–96.

James, M.R., Weatherland, M., Stanger, C., and Graynoth, E., 1998, Macroinvertebrate distribution in the littoral zone of Lake Coleridge, South Island, New Zealand—effects of habitat stability, wind exposure, and macrophytes. *New Zealand Jour. Mar. Freshw. Res.* 32:287–305.

Janaway, T.M. and Parnell, J., 1989, Carbonate production within the Orcadian Basin, northern Scotland: a petrographic and geochemical study. *Palaeogeog., Palaeoclim., Palaeoecol.* 70:89–105.

Jansen, D., 1999, The climate system. In Martens, P. and Rotmans, J. (eds.), *Climate Change: An Integrated Perspective.* Kluwer Academic, Dordrecht.

Janssens, J.A., 1990, Bryophytes. In Warner, B.G. (ed.), *Methods in Quaternary Ecology.* Geoscience Canada Reprint Ser. 5. Geological Association of Canada, pp. 23–36.

Järnefelt, H., 1956, Zur Limnologie einiger Gewässer Finnlands. *Ann. Zool. Soc. Vancimo* 17:1–201.

Jasinski, J.P.P., Warner, B.G., Andreev, A.A., Aravena, R., Gilbert, R., Zeeb, B.A., Smol, J.P., and Velichko, A.A., 1998, Holocene environmental history of a peatland in the Lena River valley, Siberia. *Can. Jour. Earth Sci.* 35:637–648.

Jell, P.A. and Duncan, P.M., 1986, Invertebrates, mainly insects, from the freshwater Lower Cretaceous, Koonwarra Fossil Bed (Korumburra Group), South Gippsland, Victoria. In Jell, P.A. and Roberts, J. (eds.), *Plants and Invertebrates from the Lower Cretaceous Koonwarra Fossil Bed, South Gippsland, Victoria.* Assoc. Australas. Palaeo. Mem. 3:111–205.

Jensen, H.S., Kristensen, P., Jeppesen, E., and Skytthe, A., 1992, Iron:phosphorus ratio in surface sediment as an indicator of phosphate release from aerobic sediments in shallow lakes. *Hydrobiologia* 235/236:731–743.

Jeppesen, E., Madsen, E.A., Jensen, J.P., and Anderson, N.J., 1996, Reconstructing the past density of planktivorous fish and trophic structure from sedimentary zooplankton fossils: a surface sediment calibration data set from shallow lakes. *Freshwater Biol.* 36:115–127.

Jobling, M., 1995, *Environmental Biology of Fishes.* Chapman and Hall, New York.

Johansen, K.A. and Robbins, J.A., 1977, Fallout ^{137}Cs in sediments of southern Lake Huron and Saginaw Bay. *Proc. w/abstr. 20th Conf. Great Lakes Res.*

Johnson, M.D., Addis, K.L., Ferber, L.R., Hemstad, C.B., Meyer, G.N., and Komai, L.T., 1999, Glacial Lake Lind, Wisconsin and Minnesota. *Geol. Soc. Amer. Bull.* 111:1371–1386.

Johnson, M.G., Kelso, J.R.M., McNeil, O.C., and Morton, W.B., 1990, Fossil midge associations and the historical status of fish in acidified lakes. *Jour. Paleolim.* 3:113–127.

Johnson, P.G., 1997, Spatial and temporal variability of ice-dammed lake sediments in alpine environments. *Quat. Sci. Rev.* 16:635–647.

Johnson, R.K. and Wiederholm, T., 1989, Classification and ordination of profundal macroinvertebrate communities in nutrient poor, oligo-mesohumic lakes in relation to environmental data. *Freshwater Biol.* 21:375–386.

Johnson, T.C., 1984, Sedimentation in large lakes. *Ann. Rev. Earth Planet. Sci.* 12:179–204.

Johnson, T.C., 1996, Sedimentary processes and signals of past climatic change in the large lakes of the East African Rift Valley. In Johnson, T.C. and Odada, E.O. (eds.), *The Limnology, Climatology and Paleoclimatology of the East African Lakes.* Gordon and Breach, Amsterdam, pp. 367–412.

Johnson, T.C. and Eisenreich, S.J., 1979, Silica in Lake Superior: mass balance considerations and a model for dynamic response to eutrophication. *Geochim. Cosmochim. Acta* 43:77–91.

Johnson, T.C. and Hecky, R.E., 1988, A silica budget for Lake Malawi: net fluvial input, biogenic opal preservation and burial. *EOS* 69:1144.

Johnson, T.C. and Ng'ang'a, P., 1990, Reflections on a rift lake. In Katz, B.J. (ed.), *Lacustrine Basin Exploration: Case Studies and Modern Analogues.* Amer. Assoc. Petrol. Geol. Mem. 50:113–135.

Johnson, T.C., Wells, J.T., and Scholz, C.A., 1995, Deltaic sedimentation in a modern rift lake. *Geol. Soc. Amer. Bull.* 107:812–829.

Johnson, T.C., Scholz, C.A., Talbot, M.R., Kelts, K., Ricketts, R.D., Ngobi, G., Beuning, K., Ssemanda, I., and McGill, J.W., 1996, Late Pleistocene desiccation of Lake Victoria and rapid evolution of cichlid fishes. *Science* 273:1091–1093.

Johnson, T.C., Chan, Y., Beuning, K., Kelts, K., Ngobi, G., and Verschuren, D., 1998, Biogenic silica profiles in Holocene cores from Lake Victoria: implications for lake level history and initiation of the Victoria Nile. In Lehman, J.T. (ed.), *Environmental Change and Response in East African Lakes.* Kluwer, Dordrecht, pp. 75–88.

Jónasson, P.M., 1978, Zoobenthos in lakes. *Verh. Int. Ver. Limnol.* 20:13–37.

Jones, B.F., 1965, The hydrology and mineralogy of Deep Springs Lake, Inyo County, California. *U.S. Geol. Surv. Prof. Pap.* 502-A.

Jones, B.F., Rettig, S.L., and Eugster, H.P., 1967, Silica in alkaline brines. *Science* 158:1310–1314.

Jones, B.F., Hanor, J.S., and Evans, W.R., 1994, Sources of dissolved salts in the central Murray Basin, Australia. *Chem. Geol.* 111:135–154.

Jones, F.G. and Wilkinson, B.H., 1978, Structure and growth of lacustrine pisoliths from recent Michigan marl lakes. *Jour. Sed. Pet.* 48:1103–1110.

Jones, R., 1984, Heavy metals in the sediments of Llangorse Lake, Wales, since Celtic-Roman times. *Verh. Int. Ver. Limnol.* 22:1377–1382.

Jones, W.B., Bacon, M., and Hastings, D.A., 1981, The Lake Bosumtwi impact crater, Ghana. *Geol. Soc. Amer. Bull.* 8:5–46.

Jopling, A.V., 1975, Early studies on stratified drift. In Jopling, A.V. and McDonald, B.C. (eds.), *Glaciofluvial and Glaciolacustrine Sedimentation*. Soc. Econ. Paleo. Mineral. Spec. Publ. 23:4–21.

Jordan, T.E. and Allmendinger, R.W., 1986, The Sierras Pampeanas of Argentina: a modern analogue of Rocky Mountain foreland deformation. *Amer. Jour. Sci.* 286:737–764.

Jørgensen, N.B., 1982, Turbidites and associated resedimented deposits from a tilted glaciodeltaic sequence, Denmark. *Danm. geol Unders. Årbog.* 1981:47–72.

Joshi, S.R., 1987, Nondestructive determination of lead-210 and radium-226 in sediments by direct photon analysis. *Jour. Radioanal. Nucl. Chem.* 116:169–182.

Juhász, E., Kovács, L.O., Müller, P., Tóth-Makk, A., Phillips, L., and Lantos, M., 1997, Climatically driven sedimentary cycles in the Late Miocene sediments of the Pannonian Basin. *Tectonophysics* 282:257–276.

Juhász, E., Phillips, L., Müller, P., Ricketts, B., Tóth-Makk, A., Lantos, M., and Kovács, L.O., 1999, Late Neogene sedimentary facies and sequences in the Pannonian Basin, Hungary. In Durand, B., Jolivet, L., Horvath, F., and Séranne, M. (eds.), *The Mediterranean Basins: Tertiary Extension within the Alpine Orogen.* Geol. Soc. Lond. Spec. Publ. 156:335–356.

Juliá, R., 1980, *La conca lacustre de Banyoles-Besalú: Banyolas, Spain.* Monografies del Centre d'Estudis Comarcals de Banyoles.

Juliá, R. and Bischoff, J.L., 1991, Radiometric dating of Quaternary deposits and the hominid mandible of Lake Banyoles, Spain. *Jour. Archaeol. Sci.* 18:707–722.

Jurasz, W. and Amoros, C., 1991, Ecological succession in a former meander of the Rhône River, France, reconstructed by Cladocera remains. *Jour. Paleolim.* 6:113–122.

Kahkonen, M.A., Sominen, K.P., Manninen, P.K.G., and Salkonoja-Salonen, N.S., 1998, 100 years of sediment accumulation history of organic halogens and heavy metals in recipient and nonrecipient lakes of pulping industry in Finland. *Envir. Sci. Tech.* 32:1741–1746.

Kalindekafe, L.S.N., Dolozi, M., and Yuretich, R., 1996, Distribution and origin of clay minerals in the sediments of Lake Malawi. In Johnson, T.C. and Odada, E.O. (eds.), *The Limnology, Climatology and Paleoclimatology of the East African Lakes.* Gordon and Breach, Amsterdam, pp. 443–460.

Kansanen, P.H., 1986, Information value of chironomid remains in the uppermost sediment layers of a complex lake basin. *Hydrobiologia* 14:159–165.

Karabanov, E.B., Prokopenko, A.A., Williams, D.F., and Colman, S.M., 1998, Evidence from Lake Baikal for Siberian Glaciation during oxygen-isotope substage 5d. *Quat. Res.* 50:46–55.

Karabanov, E.B., Prokopenko, A.A., Williams, D.F., and Khursevich, G.K., 2000, A new record of Holocene climate change from the bottom sediments of Lake Baikal. *Palaeogeog., Palaeoclim., Palaeoecol.* 156:211–224.

Karlin, R., 1990, Magnetite diagenesis in marine sediments from the Oregon continental margin. *Jour. Geophys. Res.* 95:4405–4419.

Karlin, R. and Levi, S., 1983, Diagenesis of magnetic minerals in Recent hemipelagic sediments. *Nature* 303:327–330.

Karls, J.F. and Christensen, E.R., 1998, Carbon particles in dated sediments from Lake Michigan, Green Bay and tributaries. *Envir. Sci. Tech.* 32:225–231.

Karrow, P.F. and Calkin, M. (eds.), 1985, *Quaternary Evolution of the Great Lakes.* Geol. Assoc. Canada Spec. Pap. 30.

Karrow, P.F., Anderson, T.W., Delorme, L.D., Miller, B.B., and Chapman, L.J., 1995, Late-glacial paleoenvironments of lake Algonquin sediments near Clarksburg, Ontario. *Jour. Paleolim.* 14:297–309.

Kashiwaya, K., Nakamura, T., Takamatsu, N., Sakai, H., Nakamura, M., and Kawai, T., 1997, Orbital signals found in physical and chemical properties of bottom sediments from Lake Baikal. *Jour. Paleolim.* 18:293–297.

Kaszycki, C., 1985, History of Glacial Lake Algonquin in the Haliburton region, south central Ontario. In Karrow, P.F. and Calkin, P.E. (eds.), *Quaternary Evolution of the Great Lakes.* Geol. Assoc. Canada Spec. Pap. 30:109–123.

Katz, B.J., 1988, Clastic and carbonate lacustrine systems: an organic geochemical comparison (Green River Formation and East African lake sediments). In Fleet, A.J., Kelts, K., and Talbot, M.R. (eds.), *Lacustrine Petroleum Source Rocks.* Geol. Soc. Lond. Spec. Publ. 40:81–90.

Katz, B.J., 1990, Controls on distribution of lacustrine source rocks through time and space. In Katz, B.J. (ed.), *Lacustrine Basin Exploration: Case Studies and Modern Analogues.* Amer. Assoc. Petrol. Geol. Mem. 50:61–76.

Katz, B.J., 2001, Lacustrine basin hydrocarbon exploration—current thoughts. *Jour. Paleolim.* 26:161–179.

Kaufman, D.S., 2000, Amino acid racemization in ostracodes. In Goodfriend, G., Collins, M.,

Fogel, M., Macko, S., and Wehmiller, J. (eds.), *Perspectives in Amino Acid and Protein Geochemistry*. Oxford University Press, New York, pp. 145–160.

Kaufman, D.S. and Manley, W.F., 1998, A new procedure for determining DL amino acid ratios in fossils using reverse phase liquid chromatography. *Quat. Sci. Rev. (Quat. Geochron.)* 17:987–1000.

Kaufman, D.S. and Miller, G.H., 1992, Overview of amino acid geochronology. *Comp. Biochem. Physiol.* 102:199–204.

Kazanci, N., Gevrek, A.I., and Varol, B., 1995, Facies changes and high calorific peat formation in a Quaternary maar lake, central Anatolia, Turkey: the possible role of geothermal processes in a closed lacustrine basin. *Sed. Geol.* 94:255–266.

Keen, D.H., Jones, R.L., Evans, R.A., and Robinson, J.E., 1988, Faunal and floral assemblages from Bingley Bog, West Yorkshire and their significance for Late Devensian and early Flandrian environmental change. *Proc. Yorkshire Geol. Soc.* 47:125–138.

Kehew, A.E., 1993, Glacial-lake outburst erosion of the Grand Valley, Michigan, and impacts on glacial lakes in the Lake Michigan Basin. *Quat. Res.* 39:36–44.

Kelts, K., 1988, Environments of deposition of lacustrine petroleum source rocks: an introduction. In Fleet, A.J., Kelts, K., and Talbot, M.R. (eds.), *Lacustrine Petroleum Source Rocks*. Geol. Soc. Lond. Spec. Publ. 40:3–26.

Kelts, K. and Hsü, K., 1978, Freshwater carbonate sedimentation. In Lerman, A. (ed.), *Lakes: Chemistry, Geology, Physics*. Springer-Verlag, New York, pp. 295–323.

Kerfoot, W.C., 1995, *Bosmina* remains in Lake Washington sediments: qualitative heterogeneity of bay environments and quantitative correspondence to production. *Limnol. Oceanogr.* 40:211–225.

Kerfoot, W.C. and Lynch, M., 1987, Branchiopod communities: associations with planktivorous fish in time and space. In Kerfoot, W.C. and Sih, A. (eds.), *Predation: Direct and Indirect Impacts on Aquatic Communities*. University Press of New England, Hanover, NH, pp. 367–378.

Kerr-Lawson, L.J., Karrow, P.F., Edwards, T.W.D., and Mackie, G.L., 1992, A paleoenvironmental study of the molluscs from the Don Formation (Sangamonian?) Don Valley Brickyard, Toronto, Ontario. *Can. Jour. Earth Sci.* 29:2406–2417.

Kershaw, A.P., 1979, Local pollen deposition in aquatic sediments on the Atherton Tableland, Northeastern Australia. *Austral. Jour. Ecol.* 4:253–263.

Kessels, H.J. and Dungworth, G., 1980, Necessity of reporting amino acid compositions of fossil bones where racemization rates are used for geochronological applications. In Hare, P.E., Hoering, T.C., and King, K. (eds.), *Biogeochemistry of Amino Acids*. J. Wiley and Sons, New York, pp. 527–541.

Khursevich, G., 1994, Evolution of freshwater centric diatoms within the Euroasian continent. *13th Intl. Diatom. Symp.*, pp. 507–520.

Kietzke, K.K., 1989, Microfossil zonation and correlation of nonmarine Triassic of southwestern United States. *Amer. Assoc. Petrol. Geol. Bull.* 73:1163.

Kilenyi, T.I., 1972, Transient and balanced genetic polymorphism as an explanation of variable noding in the ostracode *Cyprideis torosa*. *Micropaleo.* 18:47–64.

Kilgour, B.W. and Mackie, G.L., 1988, Factors affecting the distribution of sphaeriid bivalves in Brittania Bay of the Ottawa River. *Nautilus* 102:73–77.

Kilham, P., 1971, A hypothesis concerning silica and the freshwater planktonic diatoms. *Limnol. Oceanogr.* 16:10–18.

Kimble, B.J., Maxwell, J.R., Philip, R.P., Eglinton, G., Albrecht, P., Ensminger, A., Arpino, P., and Ourisson, G., 1974, Tri- and tetraterpenoid hydrocarbons in the Messel oil shale. *Geochim. Cosmochim. Acta* 58:879–893.

King, J. and Peck, J., 2001, Use of paleomagnetism in studies of lake sediments. In Last, W.M. and Smol, J.P. (eds.), *Tracking Environmental Change Using Lake Sediments. Volume 1: Basin Analysis, Coring, and Chronological Techniques*. Kluwer Academic, Dordrecht, pp. 371–389.

King, J.W., Banerjee, S.K., Marvin, J., and Lund, S., 1983, Use of small-amplitude paleomagnetic fluctuations for correlation and dating of continental climate change. *Palaeogeog., Palaeoclim., Palaeoecol.* 42:167–183.

King, R.H., 1991, Paleolimnology of a polar oasis, Truelove Lowland, Devon Island, NWT, Canada. *Hydrobiologia* 214:317–325.

Kingston, J.C., Cook, R.B., Kreis, R.G., Camburn, K.E., Norton, S.A., Sweets, P.R., Binford, M.W., Mitchell, M.J., Schindler, S.C., Shane, L.C.K., and King, G.A., 1990, Paleoecological investigation of recent lake acidification in the northern Great Lake states. *Jour. Paleolim.* 4:153–201.

Kirby, M.E., Mullins, H.T., Patterson, W.P., and Burnett, A.W., 2001, Lacustrine isotopic evidence for multidecadal natural climate variability related to the circumpolar vortex over the northeast United States during the past millennium. *Geology* 29:807–810.

Kitchell, J.F. and Carpenter, S.R., 1993, Cascading trophic interactions. In Carpenter, S.R. and Kitchell, J.F. (eds.), *The Trophic Cascade in Lakes.* Cambridge University Press, Cambridge, pp. 1–14.

Kitchell, J.F. and Sanford, P.R., 1992 Paleolimnology and evidence of food web dynamics in Lake Mendota. In Kitchell, J.F. (ed.), *Food Web Management—A Case Study of Lake Mendota.* Springer-Verlag, New York, pp. 31–48.

Kitagawa, H. and van der Plicht, J., 1998, Atmospheric radiocarbon calibration to 45,000 yr. BP: Late Glacial fluctuations and cosmogenic isotope production. *Science* 279:1187–1190.

Kjensmo, J., 1968, Late and post-glacial sediments in the small meromictic Lake Svinsjøen. *Arch. Hydrobiol.* 65:125–141.

Klerkx, J., Abdrachmatov, K., Buslov, M., De Batist, M., Vermeesch, P., Imbo, Y., Hus, R., and Delvaux, D., 1999, Active deformation in the Issyk-Kul Basin (Kyrghyz Tien Shan): from pull-apart to transpressional basin. Lennou, 2nd Intl. Congress of Limnogeology, Brest, France, Abstr. w/prog. p. 36.

Kling, G.W., Kipphut, G.W., and Miller, M.C., 1991, Arctic lakes and streams as gas conduits to the atmosphere: implications for tundra carbon budgets. *Science* 251:298–301.

Kling, H.J., 1998, A summary of past and recent plankton of Lake Winnipeg, Canada using algal fossil remains. *Jour. Paleolim.* 19:297–307.

Knighton, D., 1984, *Fluvial Form and Process.* Edward Arnold, Baltimore, MD.

Knox, L.W. and Gordon, E.A., 1999, Ostracodes as indicators of brackish water environments in the Catskill Magnafacies (Devonian) of New York State. *Palaeogeog., Palaeoclim., Palaeoecol.* 148:9–22.

Kociolek, J. and Stoermer, E., 1989, Phylogenetic relationships and evolutionary history of the diatom genus *Gomphoneis. Phycologia* 28:438–454.

Kociolek, J. and Stoermer, E., 1993, Freshwater gomphonemoid diatom phylogeny: preliminary results. *Hydrobiologia* 269/270:31–38.

Kodama, K.P., Lyons, J.C., Siver, P., and Lott, A.M., 1997, A mineral magnetic and scaled chrysophyte paleolimnological study of two northeastern Pennsylvania lakes: records of fly ash deposition, land use change, and paleorainfall variation. *Jour. Paleolim.* 17:173–189.

Koeberl, C., Reimwold, W.U., Blum, J.D., and Chamberlain, C.P., 1998, Petrology and geochemistry of target rocks from the Bosumtwi impact structure, Ghana, and comparison with Ivory Coast tektites. *Geochim. Cosmochim Acta* 62:2179–2196.

Kohn, B.P., Pillans, B., and McGlone, M.S., 1992, Zircon fission track age for middle Pleistocene Rangitawa Tephra, New Zealand: stratigraphic and paleoclimatic significance. *Palaeogeog., Palaeoclim., Palaeoecol.* 95:73–94.

Kohn, B.P., Farley, K.A., and Pillans, B., 2000, (U–Th)/He and fission track dating of the Pleistocene Rangitawa Tephra, North Island, New Zealand: a comparative study. 9th Intl. Conf. on Fission Track Dating and Thermochronology, Lorne, Australia, abstr. w/ prog. pp. 207–208.

Kohring, R.R., 1998, Conchostraca (Arthropoda: Crustacea) from the Molteno Formation (upper Triassic, South Africa) palaeoecological suggestions. *Jour. Afr. Earth Sci.* 27:124–125.

Kohring, R.R. and Hòrnig, A.C.F., 1998, The earliest freshwater Bryozoa: evidence from the Upper Triassic Molteno Formation (South Africa). *Jour. Afr. Earth Sci.* 27:125.

Koide, M., Bruland, K., and Goldberg, E.D., 1973, ^{228}Th/^{232}Th and ^{210}Pb geochronology in marine and lake sediments. *Geochim. Cosmochim. Acta* 37:1171–1187.

Kolbe, R.W., 1927, Zur Okologie, Morphologie und Systematik der Brackwasser-Diatomeen. *Pflanzenforschung* 7:1–146.

Komar, P.D. and Miller, M.C., 1975, On the comparison between the threshold of sediment motion under waves and unidirectional currents with a discussion of the practical evaluation of the threshold. *Jour. Sed. Petrol.* 51:362–367.

Komor, S.C., 1994, Bottom sediment chemistry in Devils Lake, northeastern North Dakota. In Renaut, R.W. and Last, W.M. (eds.), *Sedimentology and Geochemistry of Modern and Ancient Saline Lakes.* Soc. Econ. Paleo. Mineral. Spec. Publ. 50:21–32.

Korhola, A., 1992, The Early Holocene hydrosere in a small acid hill-top basin studied using crustacean sedimentary remains. *Jour. Paleolim.* 7:1–22.

Korhola, A. and Rautio, M., 2001, Cladocera and other branchiopod remains. In Smol, J.P., Birks, H.J.B., and Last, W.M. (eds.), *Tracking Environmental Change Using Lake Sediments. Vol. 4. Zoological Indicators.* Kluwer Academic Publishers, Dordrecht, pp. 5–41.

Korsman, T., 1999, Temporal and spatial trends of lake acidity in northern Sweden. *Jour. Paleolim.* 22:1–15.

Korsman, T., Renberg, I., and Anderson, N.J., 1994, A palaeolimnological test of the influence of Norway spruce (*Picea abies*) immigration on lake-water acidity. *The Holocene* 4:132–140.

Koskenniemi, E., 1994, Colonization, succession and environmental conditions of the macrozoobenthos in a regulated, polyhumic reservoir, Western Finland. *Int. Rev. ges. Hydrobiol.* 79:521–555.

Kowalewska, A. and Cohen, A.S., 1998, Reconstruction of paleoenvironments of the Great Salt Lake Basin during the late Cenozoic. *Jour. Paleolimnol.* 20:381–407.

Kozarski, S., Gonera, P., and Antczak, B., 1988, Valley floor development and paleohydrological changes: the Late Vistulian and Holocene history of the Warta River (Poland). In Lang, G. and Schlüchter, C., *Lake, Mire and River Environments.* Balkema, Rotterdam, pp. 185–203.

Krabbenhoft, D.P., Bowser, C.J., Anderson, M.P., and Valley, J.W., 1990, Estimating groundwater exchange with lakes 1. The stable isotope mass balance method. *Water Res. Res.* 26:2445–2453.

Kratinger, K., 1975, Genetic mobility in *Typha. Aquat. Bot.* 1:57–70.

Krause, W.E., Krbetschek, M.R., and Stolz, W., 1997, Dating of Quaternary lake sediments from the Schirmacher Oasis (East Antarctica) by infra-red stimulated luminescence (IRSL) detected at the wavelength of 560nm. *Quat. Sci. Rev.* 16:387–392.

Krause-Dellin, D. and Steinberg, C., 1986, Cladocera remains as indicators of lake acidification. *Hydrobiologia* 143:129–134.

Krebs, W.N., 1994, The biochronology of freshwater planktonic diatom communities in western North America. *Proc. 11th Int. Diatom Symp. 1990.* California Academy of Science, pp. 458–499.

Krebs, W.N. and Bradbury, J.P., 1995, Geological ranges of lacustrine *Actinocyclus* species, western United States. *U.S. Geol. Surv. Prof. Pap.* 1543-B:53–61.

Krebs, W.N., Bradbury, J.P., and Theriot, E., 1987, Neogene and Quaternary lacustrine diatom biochronology, Western USA. *Palaios* 2:505–513.

Krishnamurthy, R.V., Syrup, K.A., Baskaran, M., and Long, A., 1995, Late glacial climate record of midwestern United States from the hydrogen isotope ratio of lake organic matter. *Science* 269:1565–1567.

Krishnaswami, S. and Lal, D., 1978, Radionuclide limnochronology. In Lerman, A. (ed.), *Lakes Chemistry, Geology, Physics.* Springer-Verlag, New York, pp. 153–178.

Krishnaswami, S., Lal, D., Martin, J.M., and Meybeck, M., 1971, Geochronology of lake sediments. *Earth Planet. Sci. Lett.* 11:407–414.

Krömmelbein, K., 1966, On "Gondwana Wealden" ostracoda from NE Brazil and West Africa. *Proc. 2nd W. Afr. Micropaleo. Coll., Ibadan,* pp. 112–119.

Kroonenberg, S.B., Rusakov, G.V., and Svitoch, A.A., 1997, The wandering of the Volga delta: a response to rapid Caspian sea-level change. *Sed. Geol.* 107:189–209.

Kröpelin, S. and Souliè-Märsche, I., 1991, Charophyte remains from Wadi Howar as evidence for deep Mid-Holocene freshwater lakes in the Eastern Sahara of Northwest Sudan. *Quat. Res.* 36:210–223.

Krstic, N., 1990, Contribution by ostracods to the definition of the boundaries of the Pontian in the Pannonian Basin. In Stevanovic, P.M., Nevesskaja, L.A., Marinescu, F.L., Sokac, H., and Iambor, A. (eds.), *Chronostratigraphie und Neostratotypen Neogen der Westlichen (Zentrale) Paratethys VIII.* IAZU and SANU, Zagreb-Belgrade, pp. 45–47.

Krzeminksi, W., 1992, Triassic and Lower Jurassic stage of Diptera evolution. *Mitt. Schweiz. Entom. Gessel.* 65:39–59.

Ku, T.L., Luo, S., Lowenstein, T.K., Li, J., and Spencer, R.J., 1998, U-series chronology of lacustrine deposits in Death Valley, California. *Quat. Res.* 50:261–275.

Kuc, M., 1973, Fossil statoblasts of *Cristatella mucedo* Cuvier in the Beaufort Formation and in interglacial and post-glacial deposits of the Canadian Arctic. *Geol. Surv. Canada Pap.* 72-28:1–12.

Kuenzi, W.D., Horst, O.H., and McGehee, R.V., 1979, Effect of volcanic activity on fluvial–deltaic sedimentation in a modern arc-trench gap, southwestern Guatemala. *Geol. Soc. Amer. Bull.* 90:827–838.

Kuhry, P., 1997, The palaeoecology of a treed bog in western boreal Canada: a study based on microfossils, macrofossils and physicochemical properties. *Rev. Palaeobot. Palynol.* 96:183–224.

Kumar, S. and Rzhetsky, A., 1996, Evolutionary relationships of eukaryotic kingdoms. *Jour. Mol. Evol.* 42:183–193.

Kutas, R.I., Kobolev, V.P., and Tsvyashchenko, V.A., 1998, Heat flow and geothermal models of the Black Sea depression. *Tectonophysics* 291:91–100.

Kutzbach, J.E., 1994, Idealized Pangaean climates: sensitivity to orbital change. In Klein, G.D. (ed.), *Pangaea Paleoclimate, Tectonics and Sedimentation During Accretion, Zenith and Breakup of a Supercontinent.* Geol. Soc. Amer. Spec. Pap. 288:41–55.

Kutzbach, J.E. and Street-Perrott, F.A., 1985, Milankovitch forcing of fluctuations in the level of tropical lakes from 18 to 0 kyr BP. *Nature* 317:130–134.

Kutzbach, J.E., Guetter, P.J., Behling, P.J., and Selin, R., 1993, Simulated climatic changes:

results of the COHMAP climate-model experiments. In Wright, H.E., Kutzbach, J.E., Webb III, T., Ruddiman, W.F., Street-Perrott, F.A., and Bartlein, P.J. (eds.), *Global Climates Since the Last Glacial Maximum*. University of Minnesota Press, Minneapolis, MN, pp. 24–93.

LaBandeira, C., 1998, The role of insects in Late Jurassic to Middle Cretaceous ecosystems. In Lucas, S., Kirkland, J.I., and Estep, J.W. (eds.), *Lower and Middle Cretaceous Terrestrial Ecosystems*. New Mexico Mus. Nat. Hist. Sci. Bull. 14:105–124.

LaBandeira, C.C., 1999, Insects and other hexapods. In Singer, R. (ed.), *Encyclopedia of Paleontology*. Fitzroy Dearborn, London, v. 1, pp. 603–624.

LaBandeira, C.C., 2002, The history of associations between plants and animals. In Herrera, C. and Pellmyr, O. (eds.), *History of Plant–Animal Interactions*. Blackwell Science, Oxford, pp. 26–74.

LaBerge, G.L., 1994, *Geology of the Lake Superior Region*. Geoscience Press, Phoenix, AZ.

Laidler, R.B. and Scott, D.B., 1996, Foraminifera and Arcellacea from Porters Lake, Nova Scotia: modern distribution and paleodistribution. *Can. Jour. Earth Sci.* 33:1410–1427.

Laird, K.R., Fritz, S.C., Grimm, E.C., and Mueller, P.G., 1996, Century-scale paleoclimatic reconstruction from Moon Lake, a closed basin in the northern Great Plains. *Limnol. Oceanogr.* 41:890–902.

Laird, K.R., Fritz, S.C., and Cumming, B.F., 1998, A diatom-based reconstruction of drought intensity, duration, and frequency from Moon Lake, North Dakota: a sub-decadal record of the last 2300 years. *Jour. Paleolim.* 19:161–179.

Lalou, C., 1987, Déséquilibres radioactifs dans la famille de l'uranium. In Miskovsky, J.-C. (ed.), 1987, *Géologie de la préhistoire*. Géopré, Paris, pp. 1073–1085.

Lamb, H., 1982, *Climate History and the Modern World*. Methuen, London.

Lamb, H.F. and van der Kaars, S., 1995, Vegetational response to Holocene climatic change: pollen and palaeolimnological data from the Middle Atlas, Morocco. *The Holocene* 5:400–408.

Lamb, S., Hoke, L., Kennan, L., and Dewey, J., 1997, Cenozoic evolution of the Central Andes in Bolivia and northern Chile. In Burg, J.P. and Ford, M. (eds.), *Orogeny Through Time*. Geol. Soc. Lond. Spec. Publ. 121:237–264.

Lamb, H., Roberts, N., Leng, M., Barker, P., Benkaddour, A., and van der Kaars, S., 1999, Lake evolution in a semi-arid montane environment: responses to catchment change and hydroclimate variation. *Jour. Paleolim.* 21:325–343.

Lambert, A., 1982, Trübeströme des Rheins am Grund des Bodensees. *Wasserwirtschaft* 72:1–4.

Lambert, A.M. and Hsü, K.J., 1979, Varve-like sediments of the Wallensee, Switzerland. In Schlüchter, C. (ed.), *Moraines and Varves*. Balkema, Rotterdam, pp. 287–294.

Lambiase, J., 1990, A model for tectonic control of lacustrine stratigraphic sequences in continental rift basins. In Katz, B.J. (ed.), *Lacustrine Basin Exploration: Case Studies and Modern Analogues*. Amer. Assoc. Petrol. Geol. Mem. 50:265–276.

Lami, A., Niessen, F., Guilizzoni, P., Masaferro, J., and Belis, C., 1994, Palaeolimnological studies of the eutrophication of volcanic Lake Albano (Central Italy). *Jour. Paleolim.* 10:181–197.

Lamoureux, S., 1999, Spatial and interannual variations in sedimentation patterns recorded in nonglacial varved sediments from the Canadian High Arctic. *Jour. Paleolim.* 21:73–84.

Lampert, W. and Sommer, U., 1997, *Limnoecology*. Oxford University Press, New York.

Landmann, G., Reimer, A., Lemcke, G., and Kempe, S., 1996, Dating Late Glacial abrupt climate changes in the 14,750 yr long continuous varve record of Lake Van, Turkey. *Palaeogeog., Palaeoclim., Palaeoecol.* 122:107–118.

Lang, C. and Lodz-Crozet, B., 1997, Oligochaetes versus chironomids as indicators of trophic state in two Swiss lakes recovering from eutrophication. *Arch. Hydrobiol.* 139:187–195.

Langbein, W.B. and Schumm, S.A., 1958, Yield of sediment in relation to mean annual precipitation. *Trans. Amer. Geophys. Union* 39:1076–1084.

Lao, Y. and Benson, L., 1988, Uranium series age estimates and paleoclimatic significance of Pleistocene tufas from the Lahontan Basin, California and Nevada. *Quat. Res.* 30:165–176.

Larsen, D. and Crossey, L.J., 1996, Depositional environments and paleolimnology of an ancient caldera lake: Oligocene Creede Formation, Colorado. *Geol. Soc. Amer. Bull.* 108:526–544.

Larsen, J., 2000, Recent changes in diatom-inferred pH, heavy metals and spheroidal carbonaceous particles in lake sediments near an oil refinery at Mongstad, western Norway. *Jour. Paleolim.* 23:343–363.

Laskar, J., Joutel, F., and Boutin, F., 1993, Orbital, precessional, and insolational quantities for the

Earth from 20 Myr to +10 Myr. *Astron. Astrophys.* 270:522–533.

Last, W.M., 1982, Holocene carbonate sedimentation in Lake Manitoba, Canada. *Sedimentology* 29:691–704.

Last, W.M., 1992, Petrology of modern carbonate hardgrounds from East Basin Lake, a saline maar lake, southern Australia. *Sed. Geol.* 81:215–229.

Last, W.M., 1994, Deep-water evaporite mineral formation in lakes of Western Canada. In Renaut, R.W. and Last, W.M. (eds.), *Sedimentology and Geochemistry of Modern and Ancient Saline Lakes.* Soc. Econ. Paleo. Mineral. Spec. Publ. 50:51–59.

Last, W.M. and De Deckker, P., 1990, Modern and Holocene carbonate sedimentology of two saline volcanic maar lakes, southern Australia. *Sedimentology* 37:967–981.

Last, W.M. and Smol, J.P. (eds.), 2001, *Tracking Environmental Change Using Lake Sediments. Volumes 1–4.* Kluwer Academic, Dordrecht.

Last, W.M. and Vance, R.E., 1997, Bedding characteristics of Holocene sediments from salt lakes of the northern Great Plains, western Canada. *Jour. Paleolim.* 17:297–318.

Lean, J., 1996, Reconstruction of past solar variability. In Jones, P.D., Bradley, R.S., and Jouzel, J. (eds.), *Global Environmental Change.* NATO ASI Ser. 41:519–532.

Lean, J., Beer, J., and Bradley, R., 1995, Reconstruction of solar irradiance since 1610; implications for climate change. *Geophys. Res. Lett.* 22:3195–3198.

Leavitt, P.R., 1993, A review of factors that regulate carotenoid and chlorophyll deposition and fossil pigment abundance. *Jour. Paleolim.* 9:109–127.

Leavitt, P.R. and Carpenter, S.R., 1989, Effects of sediment mixing and benthic algal production on fossil pigment stratigraphies. *Jour. Paleolim.* 2:147–158.

Leavitt, P.R. and Carpenter, S.R., 1990, Regulation of pigment sedimentation by photo-oxidation and herbivore grazing. *Can. Jour. Fish. Aquatic Sci.* 47:1166–1176.

Leavitt, P.R., Carpenter, S.R., and Kitchell, J.F., 1989, Whole lake experiments: the annual record of fossil pigments and zooplankton. *Limnol. Oceanogr.* 34:700–717.

Leavitt, P.R., Sanford, P.R., Carpenter, S., and Kitchell, J.F., 1994a, An annual fossil record of production, planktivory and piscivory during whole-lake manipulations. *Jour. Paleolim.* 11:133–149.

Leavitt, P.R., Hann, B.J., Smol, J.P., Zeeb, B.A., Christie, C.E., Wolfe, B., and Kling, H.J., 1994b, Paleolimnological analysis of whole-lake experiments: an overview of results from Experimental Lakes Area Lake 227. *Can. Jour. Fish. Aquatic Sci.* 51:2322–2332.

Leavitt, P.R., Schindler, D.E., Paul, A.J., Hardie, A.K., and Schindler, D.W., 1994c, Fossil pigment records of phytoplankton in trout-stocked alpine lakes. *Can. Jour. Fish. Aquat. Sci.* 51:2411–2423.

Leavitt, P.R., Vinebrooke, R.D., Donald, D.B., Smol, J.P., and Schindler, D.W., 1997, Past ultraviolet radiation environments in lakes derived from fossil pigments. *Nature* 388:457–459.

Leavitt, P.R., Findlay, D.L., Hall, R.I., and Smol, J.P., 1999, Algal responses to dissolved organic carbon loss and pH decline during whole-lake acidification: evidence from paleolimnology. *Limnol. Oceanogr.* 44:757–773.

Lebo, M.E., Reuter, J.E., Rhodes, C.L., and Goldman, C.R., 1992, Nutrient cycling and productivity in a desert saline lake: observations from a dry, low-productivity year. *Hydrobiologia* 246:213–229.

Leclerc, A.J. and Labeyrie, L., 1987, Temperature dependence of the oxygen isotopic fractionation between diatom silica and water. *Earth Planet. Sci. Lett.* 84:69–74.

Lee, C.H. and Hawley, N., 1998, The response of suspended particulate material to upwelling and downwelling events in southern Lake Michigan. *Jour. Sed. Res.* 68:819–831.

Leeder, M.R., 1995, Continental rifts and proto-oceanic troughs. In Busby, C.J. and Ingersoll, R. (eds.), *Tectonics of Sedimentary Basins.* Blackwell, Cambridge, MA, pp. 119–148.

Leeder, M.R. and Jackson, J.A., 1993, The interaction between normal faulting and drainage in active extensional basins, with examples from the western United States and central Greece. *Basin Res.* 5:79–102.

Leenheer, M.J. and Meyers, P.A., 1983, Comparison of lipid composition in marine and lacustrine sediments. In Bjoroy, M. (ed.), *Advances in Organic Geochemistry.* J. Wiley, Chichester, pp. 309–316.

Lehman, J.T., Mugidde, R., and Lehman, D.A., 1998, Lake Victoria plankton ecology: mixing depth and climate-driven control of lake condition. In Lehman, J.T. (ed.), *Environmental Change and Response in East African Lakes.* Kluwer, Dordrecht, pp. 99–116.

Leland, H.V. and Berkas, W.R., 1998, Temporal variation in plankton assemblages and physicochemistry of Devils Lake, North Dakota. *Hydrobiologia* 377:57–71.

Lemdahl, G., 2000, Late-glacial and early Holocene Coleoptera assemblages as indicators of local environment and climate at Kråkenes Lake, western Norway. *Jour. Paleolim.* 23:57–66.

Lemoine, R.M. and Teller, J.T., 1994, Late glacial sedimentation and history of the Lake Nipigon Basin, Ontario. *Géogr. Phys. Quat.* 49:239–250.

Lemons, D. and Chan, M., 1999, Facies architecture and sequence stratigraphy of fine-grained lacustrine deltas along the eastern margin of Late Pleistocene Lake Bonneville, Northern Utah and Southern Idaho. *Amer. Assoc. Petrol. Geol.* 83:635–665.

Leng, M., Barker, P., Greenwood, P., Roberts, N., and Reed, J., 2001, Oxygen isotope analysis of diatom silica and authigenic calcite from Lake Pinarbasi, Turkey. *Jour. Paleolim.* 25:343–349.

Leonard, E.M., 1986, Varve studies at Hector Lake, Alberta, Canada, and the relationship between glacial activity and sedimentation. *Quat. Res.* 25:199–214.

Leonard, E.M., 1995, A varve-based calibration of the Bridge River tephra fall. *Can. Jour. Earth Sci.* 32:2098–2102.

Leonard, E.M., 1997, The relationship between glacial activity and sediment production: evidence from a 4450-year varve record of neoglacial sedimentation in Hector Lake, Canada. *Jour. Paleolim.* 17:319–330.

Lerman, A. (ed.), 1978, *Lakes: Chemistry, Geology and Physics*. Springer-Verlag, NewYork.

Lerman, A., Imboden, D.M., and Gat, J. (eds.), 1995, *Physics and Chemistry of Lakes*, 2nd ed. Springer-Verlag, New York.

Leslie, B.W., Lund, S.P., and Hammond, D.E., 1990, Rock magnetic evidence for the dissolution and authigenic growth of magnetic minerals within anoxic marine sediments of the California Continental Borderland. *Jour. Geophys. Res.* 95:4437–4452.

Leverett, F., 1897, The Pleistocene of Indiana and Michigan and the History of the Great lakes. *U.S. Geol. Surv. Monogr.* 53.

Leverett, F., 1902, Glacial formations and drainage features of the Erie and Ohio Basins. *U.S. Geol. Surv. Monogr.* 41.

Levesque, A.J., Mayle, F.E., Walker, I.R., and Cwynar, L.C., 1993, The Amphi-Atlantic Oscillation: a proposed late-glacial climatic event. *Quat. Sci. Rev.* 12:629–643.

Levesque, A.J., Cwynar, L.C., and Walker, I.R., 1994, A multiproxy investigation of late-glacial climate and vegetation change at Pine Ridge Pond, southwest New Brunswick, Canada. *Quat. Res.* 42:316–327.

Levesque, A.J., Cwynar, L.C., and Walker, I.R., 1996, Richness, diversity and succession of late-glacial chironomid assemblages in New Brunswick, Canada. *Jour. Paleolim.* 16:257–274.

Levesque, A.J., Cwynar, L.C., and Walker, I.R., 1997, Exceptionally steep north–south gradients in lake temperatures during the last deglaciation. *Nature* 385:423–426.

Lewis, C.F.M. and Anderson, T.W., 1989, Oscillations of levels and cool phases of the Laurentian Great Lakes caused by inflows from glacial Lakes Agassiz and Barlow-Ojibway. *Jour. Paleolim.* 2:99–146.

Lewis, C.F.M. and Anderson, T.W., 1992, Stable isotope (O and C) and pollen trends in eastern Lake Erie, evidence for a locally-induced climate reversal of Younger Dryas age in the Great Lakes basin. *Clim. Dynam.* 6:241–250.

Lewis, C.F.M., Moore, T.C., Rea, D.K., Dettman, D.L., Smith, A.M., and Mayer, L.A., 1994, Lakes of the Huron Basin: their record of runoff from the Laurentide Ice Sheet. *Quat. Sci. Rev.* 13:891–922.

Lewis, G.W. and Lewin, J., 1983, Alluvial cutoffs in Wales and the Borderlands. In Collinson, J.D. and Lewin, J. (eds.), *Modern and Ancient Fluvial Systems*. Int. Assoc. Sediment. Spec. Publ. 6:145–154.

Lewis, W.M., 1983, Temperature, heat and mixing in Lake Valencia, Venezuela. *Limnol. Oceanogr.* 28:273–286.

Lewis, W.M., 1984, A five-year record of temperature, mixing and stability for a tropical lake (Lake Valencia, Venezuela). *Arch. Hydrobiol.* 99:340–346.

Lewis, W.M., 1987, Tropical limnology. *Ann. Rev. Ecol. Syst.* 18:159–184.

Lezzar, K.E., Tiercelin, J.J., De Batist, M., Cohen, A.S., Bandora, T., Van Rensbergen, P., Le Turdu, C., Mifundu, W., and Klerkx, J., 1996, New seismic stratigraphy and Late Tertiary history of the North Tanganyika Basin, East African Rift system, deduced from multichannel and high-resolution reflection seismic data and piston core evidence. *Basin Res.* 8:1–28.

Lezzar, K.E., Tiercelin, J.J., Le Turdu, C., Cohen, A.S., Reynolds, D.J., Le Gall, B., and Scholz, C., 2002, Control of normal fault interaction on the distribution of major Neogene sedimentary depocenters, Lake Tanganyika, East African Rift. *Amer. Assoc. Petrol. Geol. Bull.* 86:1027–1059.

Li, H.C. and Ku, T.L., 1997, $\delta^{13}C$–$\delta^{18}O$ covariance as a paleohydrological indicator for closed-basin lakes. *Palaeogeog., Palaeoclim., Palaeoecol.* 133:69–80.

Li, H.C., Ku, T.L., and Stott, L.D., 1997, Stable isotope studies on Mono Lake (California). 1. $\delta^{18}O$ in lake sediments as proxy for climate change during the last 150 years. *Limnol. Oceanogr.* 42:230–238.

Li, J., Lowenstein, T.K., Brown, C.B., Ku, T.L., and Luo, S., 1996, A 100ka record of water tables and paleoclimates from salt cores, Death

Valley, California. *Palaeogeog., Palaeoclim., Palaeoecol.* 123:179–203.

Li, J., Lowenstein, T.K., and Blackburn, I.R., 1997, Responses of evaporite mineralogy to inflow water sources and climate during the past 100k.y. in Death Valley, California. *Geol. Soc. Amer. Bull.* 109:1361–1371.

Li, Y., 1984, Some new Late Jurassic to Early Cretaceous nonmarine ostracodes from Sichuan Basin of China. *Jour. Paleo.* 58:217–233.

Libby, W.F., Anderson, E.C., and Arnold, J.R., 1949, Age determination by radiocarbon content. *Science* 109:227–228.

Liddicoat, J.C., Opdyke, N.D., and Smith, G.I., 1980, Paleomagnetic polarity in a 930m core from Searles Valley, California. *Nature* 286:22–25.

Likens, G.E. and Butler, T.J., 1981, Recent acidification of precipitation in North America. *Atmos. Envir.* 15:1103–1109.

Likens, G.E. and Davis, M.B., 1975, Post-glacial history of Mirror Lake and its watershed in New Hampshire, USA. An initial report. *Int. Ver. Theor. Angew Limnol. Verh.* 19:982–993.

Lindqvist, J.K., 1994, Lacustrine stromatolites and oncoids: Manuherikia Group (Miocene), New Zealand. In Bertrand-Sarfati, J. and Monty, C. (eds.), *Phanerozoic Stromatolites II.* Kluwer, Dordrecht, pp. 227–254.

Line, J.M., ter Braak, C.J.F., and Birks, H.J.B., 1994, WACALIB version 3.3—a computer program to reconstruct environmental variables from fossil assemblages by weighted averaging and to derive sample-specific errors of prediction. *Jour. Paleolim.* 10:147–152.

Link, M.H. and Osborne, R.H., 1978, Lacustrine facies in the Pliocene Ridge Basin Group: Ridge Basin, California. In Matter, A. and Tucker, M.E. (eds.), *Modern and Ancient Lake Sediments.* Int. Assoc. Sedimentol. Spec. Publ. 2:169–187.

Lipiatou, E., Hecky, R.E., Eisenreich, S.J., Lockhardt, L., Muir, D., and Wilkinson, P., 1996, Recent ecosystem changes in Lake Victoria reflected in sedimentary natural and anthropogenic organic compounds. In Johnson, T.C., and Odada, E. (eds.), *The Limnology, Climatology and Paleoclimatology of the East African Lakes.* Gordon and Breach, Amsterdam, pp. 523–541.

Lipscomb, D., 1996, A survey of microbial diversity. *Ann. Missouri Bot. Garden* 83:551–561.

Liro, L.M. and Pardus, Y.C., 1990, Seismic facies analysis of fluvial–deltaic lacustrine systems—Upper Fort Union Formation (Paleocene), Wind River Basin, Wyoming. In Katz, B. (ed.), *Lacustrine Basin Exploration—Case Studies*

and Modern Analogs. Amer. Assoc. Petrol. Geol. Mem. 50:225–242.

Lister, G.S., 1988, A 15,000 year isotopic record from Lake Zürich of deglaciation and climatic change in Switzerland. *Quat. Res.* 29:129–141.

Lister, G.S., Kelts, K., Zao, C.K., Yu, J.Q., and Niessen, F., 1993, Lake Qinghai, China: closed basin lake levels and the oxygen isotope record for ostracoda since the latest Pleistocene. *Palaeogeog., Palaeoclim., Palaeoecol.* 84:141–162.

Litherland, A.E. and Beukens, R.P., 1995, Radiocarbon dating by atom counting. In Rutter, N.W. and Catto, N.R. (eds.), *Dating Methods for Quaternary Deposits.* GEOTEXT 2, Geological Society of Canada, pp. 117–123.

Little, J.L. and Smol, J.P., 2000, Changes in fossil midge (Chironomidae) assemblages in response to cultural activities in a shallow, polymictic lake. *Jour. Paleolim.* 23:207–212.

Liu, H.S., 1999, Insolation changes caused by combination of amplitude and frequency modulation of the obliquity. *Jour. Geophys. Res.* 104:25,197–25,206.

Livingstone, D.A., 1991, Edward Smith Deevey 1914–1988. *Hydrobiologia* 214:1–7.

Livingstone, D.M. and Hajdas, I., 2001, Climatically relevant periodicities in the thicknesses of biogenic carbonate varves in Soppensee, Switzerland (9740–6870 calendar yr BP). *Jour. Paleolim.* 25:17–24.

Livingstone, D.M. and Lotter, A.F., 1998, The relationship between air and water temperature in lakes of the Swiss Plateau: a case study with palaeolimnological implications. *Jour. Paleolim.* 19:181–198.

Lodge, D.M., 1985, Macrophyte-gastropod associations: observations and experiments on macrophyte choice by gastropods. *Freshw. Biol.* 15:695–708.

Lodge, D.M., 1986, Selective grazing on periphyton: a determinant of freshwater gastropod microdistribution. *Freshw. Biol.* 16:831–841.

Lodge, D.M., Brown, K.M., Klosiewski, S.P., Stein, R.A., Covich, A.P., Leathers, B.K., and Bronmark, C., 1987, Distribution of freshwater snails: spatial scale and the relative importance of physicochemical and biotic factors. *Amer. Malacol. Bull.* 5:73–84.

Lodz-Crozet, B. and Lachavanne, J.B., 1994, Changes in the chironomid communities in Lake Geneva in relation with eutrophication, over a period of 60 years. *Arch. Hydrobiol.* 130:453–471.

Löffler, H., 1975, The evolution of ostracode fauna in alpine and prealpine lakes and their value as indicators. In Swain, F.M. (ed.), *Biology and*

Paleobiology of Ostracoda. Bull. Amer. Paleo. 65:433–443.

Loftus, G.W.F. and Greensmith, J.T., 1988, The lacustrine Burdiehouse limestone formation; a key to the deposition of the Dinantian oil shales of Scotland. In Fleet, A.J., Kelts, K., and Talbot, M.R. (eds.), *Lacustrine Petroleum Source Rocks.* Geol. Soc. Spec. Publ. 40:219–234.

Lohman, K.E. and Andrews, G.W., 1968, Late Eocene nonmarine diatoms from the Beaver Divide area, Fremont Co. Wyoming. *U.S. Geol. Surv. Prof. Pap.* 593-E.

Loope, D.B., Swinehart, J.B., and Mason, J.P., 1995, Dune dammed paleovalleys of the Nebraska Sand Hills: intrinsic versus climatic controls on the accumulation of lake and marsh sediments. *Geol. Soc. Amer. Bull.* 107:396–406.

Lorey, P. and Driscoll, C.L., 1999, Historical trends of Mercury deposition in Adirondack lakes. *Envir. Sci. Tech.* 33:718–722.

Lott, A.-M., Siver, P.A., Marsicano, L.J., Kodama, K.P., and Moeller, R.E., 1994, The paleolimnology of a small waterbody in the Pocono Mountains of Pennsylvania, USA: reconstructing 19th–20th century specific conductivity trends in relation to changing land use. *Jour. Paleolim.* 12:75–86.

Lotter, A.F., 1998, The recent eutrophication of Baldegersee (Switzerland) as assessed by fossil diatom assemblages. *The Holocene* 8:395–405.

Lotter, A.F., Birks, J.B., Hoffman, W., and Marchetto, A., 1997a, Modern diatom, cladocera, chironomid, and chrysophyte cyst assemblages as quantitative indicators for the reconstruction of past environmental conditions in the Alps. I. Climate. *Jour. Paleolim.* 18:395–420.

Lotter, A.F., Sturm, M., Teranes, J.L., and Wehrli, B., 1997b, Varve formation since 1885 and high resolution varve analyses in hypereutrophic Baldegersee (Switzerland). *Aquat. Sci.* 59:304–325.

Lotter, A.F., Birks, J.B., Hoffman, W., and Marchetto, A., 1998, Modern diatom, cladocera, chironomid, and chrysophyte cyst assemblages as quantitative indicators for the reconstruction of past environmental conditions in the Alps. II. Nutrients. *Jour. Paleolim.* 19:443–463.

Lotter, A.F., Pienitz, R., and Schmidt, R., 1999, Diatoms as indicators of environmental change near arctic and alpine treeline. In Stoermer, E.F., and Smol, J.P. (eds.), *The Diatoms: Applications for the Environmental and Earth Sciences.* Cambridge University Press, Cambridge, pp. 205–226.

Lowenstein, T.K., Li, J., Brown, C., Roberts, S.M., Ku, T.L., Luo, S., and Yang, W., 1999, 200 k.y. paleoclimate record from Death Valley salt core. *Geology* 27:3–6.

Lozek, V., 1986, Mollusca analysis. In Berglund, B.E. (ed.), *Handbook of Holocene Palaeoecology and Palaeohydrology.* J. Wiley and Sons, New York, pp. 729–740.

Luckman, B.H., 1993, Glacier fluctuation and tree-ring records for the last millennium in the Canadian Rockies. *Quat. Sci. Rev.* 12:441–450.

Ludlam, S.D., 1984, Fayetteville Green Lake, N.Y., VII Varve chronology and sediment focusing. *Chem. Geol.* 44:85–100.

Ludlam, S.D., Feeney, S., and Douglas, M.S.V., 1996, Changes in the importance of lotic and littoral diatoms in a high arctic lake over the last 191 years. *Jour. Paleolim.* 16:187–204.

Luly, J.G., 1997, Modern pollen dynamics and surficial sedimentary processes at Lake Tyrell, semi-arid northwestern Victoria, Australia. *Rev. Palaeobot. Palynol.* 97:301–318.

Lund, S.P., 1996, A comparison of Holocene paleomagnetic records from North America. *Jour. Geophys. Res.* 101:8007–8024.

Lundberg, J.G., Kottelat, M., Smith, G.R., Stiassny, M.L.J., and Gill, A., 2000, So many fishes, so little time: an overview of ichthyological discoveries in freshwaters. *Ann. Missouri Bot. Garden* 87:26–62.

Lundqvist, G., 1927, Bodenablagerungen und Entwicklungstypen des Seen. *Die Binnengewässer Bd. II*, Stuttgart.

Luo, S. and Ku, T.L., 1991, U-series isochron dating: a generalized method employing total-sample dissolution. *Geochim. Cosmochim. Acta* 55:555–564.

Lyell, C., 1830, *Principles of Geology*, J. Murray, London.

Lynch, J.A., Bowersox, V.C., and Grimm, J.W., 2000 Acid rain reduced in eastern United States. *Envir. Sci. Tech.* 34:940–949.

Lyons, W.B., Hines, M.E., Last, W.M., and Lent, R.M., 1994, Sulfate reduction in microbial mat sediments of differing chemistries: implications for organic carbon preservation in saline lakes. In Renaut, R.W. and Last, W.M. (eds.), *Sedimentology and Geochemistry of Modern and Ancient Saline Lakes.* Soc. Econ. Paleo. Mineral. Spec. Publ. 50:13–20.

MacArthur, R.H. and Wilson, E.O., 1967, *The Theory of Island Biogeography.* Princeton University Press, Princeton, NJ.

MacDonald, G.M., 1990, Palynology. In Warner, B.G. (ed.), *Methods in Quaternary Ecology.* Geoscience Canada Reprint Ser. 5:37–52.

MacDonald, G.M., Edwards, T.W.D., Moser, K.A., Plenitz, R., and Smol, J.P., 1993, Rapid response of treeline vegetation and lakes to past climate warming. *Nature* 361:243–247.

MacDonald, G.M., Felzer, B., Finney, B.P., and Forman, S.L., 2000, Holocene lake sediment records of Arctic hydrology. *Jour. Paleolim.* 24:1–14.

MacDonald, R.W., Ikonomou, M.G., and Paton, D.W., 1998, Historical inputs of PCDDs, PCDFs and PCBs to a British Columbia interior lake: the effect of environmental controls on pulp mill emissions. *Envir. Sci. Tech.* 32:331–337.

MacGinitie, H.D., 1969, The Eocene Green River Flora of northwestern Colorado and northeastern Utah. *Univ. California Publ. Geo. Sci.* 83.

MacGregor, K.R., Anderson, R.S., Anderson, S.P., and Waddington, E.D., 2000, Numerical simulation of glacial-valley longitudinal profile evolution. *Geology* 28:1031–1034.

Machena, C. and Kautsky, N., 1988, A quantitative diving survey of benthic vegetation and fauna in Lake Kariba, a tropical man-made lake. *Freshwater Biol.* 19:1–14.

Mackereth, F.J.H., 1966, Some chemical observations on post-glacial lake sediments. *Phil. Trans. R. Soc. London* 250:165–213.

Magee, J.W., 1991, Late Quaternary lacustrine, groundwater, aeolian and pedogenic gypsum in the Prungle Lakes, southeastern Australia. *Palaeogeog., Palaeoclim., Palaeoecol.* 84:3–42.

Magee, J.W. and Miller, G.H., 1998, Lake Eyre palaeohydrology from 60 ka to the present: beach ridges and glacial maximum aridity. *Palaeogeog., Palaeoclim., Palaeoecol.* 144:307–329.

Magee, J.W., Bowler, J.M., Miller, G.H., and Williams, D.L.G., 1995, Stratigraphy, sedimentology, chronology and palaeohydrology of Quaternary lacustrine deposits at Madigan Gulf, Lake Eyre, South Australia. *Palaeogeog., Palaeoclim., Palaeoecol.* 113:3–42.

Magnavita, L.P. and Da Silva, H.T.F., 1995, Rift border system: the interplay between tectonics and sedimentation in the Reconcavo Basin, Northeastern Brazil. *Amer. Assoc. Petrol. Geol. Bull.* 79:1590–1607.

Magny, M., 1998, Reconstruction of Holocene lake-level changes in the French Jura: methods and results. In Harrison, S.P., Frenzel, B., Huckriede, U., and Weiß, M.M. (eds.), *Palaeohydrology as Reflected in Lake Level Changes as Climatic Evidence for Holocene Times.* Gustav Fischer-Verlag, Stuttgart, pp. 67–85.

Magny, M., Mouthon, J., and Ruffaldi, P., 1995, Late Holocene level fluctuations of the Lake Ilay in Jura, France: sediment and mollusc

evidence and climatic implications. *Jour. Paleolim.* 13:219–229.

Magyar, I., Geary, D.H., and Müller, P., 1999, Paleogeographic evolution of the Late Miocene Lake Pannon in Central Europe. *Palaeogeog., Palaeoclim., Palaeoecol.* 147:151–167.

Maisey, J.G., 1991, *Santana Fossils: An Illustrated Atlas.* T.F.H. Publications, Neptune City, NJ.

Majewski, S.P. and Cumming, B.F., 1999, Paleolimnological investigation of the effects of post-1970 reductions of acidic deposition on an acidified Adirondack lake. *Jour. Paleolim.* 21:207–213.

Malde, H., 1982, The Yahoo Clay, a lacustrine unit impounded by the McKinney Basalt in the Snake River Canyon Near Bliss, Idaho. In Bonnischen, B. and Breckenridge, R.M. (eds.), *Cenozoic Geology of Idaho.* Idaho Bur. Mines Geol. Bull. 26:617–628.

Maley, J., 1989, Late Quaternary climatic changes in the African rain forest: refugia and the major role of sea surface temperature variation. In Leinen, M. and Sarnthein, M. (eds.), *Modern and Past Patterns of Global Atmospheric Transport.* Kluwer, Dordrecht, pp. 585–616.

Mamedov, P. and Babaev, D., 1995, Seismic stratigraphy of the South Caspian Sea. *Amer. Assoc. Petrol. Geol. Bull.* 79:1233–1234.

Manca, M. and Comoli, P., 1995, Temporal variations of fossil Cladocera in the sediments of Lake Orta (N. Italy) over the last 400 years. *Jour. Paleolim.* 14:113–122.

Mángano, M.G., Buatois, L.A., Wu, X., Sun, J., and Zhang, G., 1994, Sedimentary facies, depositional processes and climatic controls in a Triassic Lake, Tanzhuang Formation, western Henan Province, China. *Jour. Paleolim.* 11:41–65.

Mann, M.E. and Lees, J., 1996, Robust estimation of background noise and signal detection in climatic time series. *Clim. Change* 33:409–445.

Maples, C.G. and Archer, A.W., 1989, The potential of Paleozoic nonmarine trace fossils for paleoecological interpretations. *Palaeogeog., Palaeoclim., Palaeoecol.* 73:185–195.

Marcusson, I., 1967, The freshwater molluscs in the Late-glacial and early Post-glacial deposits in the bog of Barmosen, southern Sjaelland, Denmark. *Medd. Dan. Geol. Foren.* 17:265–283.

Markgraf, V., 1993, Younger Dryas in southernmost South America—an update. *Quat. Sci. Rev.* 12:351–355.

Markgraf, V., 2000, Inter-PEP Workshop Report. *PAGES News* 8:2–5.

Markgraf, V., Baumgartner, T.R., Bradbury, J.P., Diaz, H.F., Dunbar, R.B., Luckman, B.H., Sltzer, G.O., Swetnam, T.W., and Villalba, R., 2000, Paleoclimate reconstruction along the

Pole–Equator–Pole transect of the Americas (PEP1). *Quat. Sci. Rev.* 19:125–140.

Markkula, I., 1986, Comparison of present and subfossil oribatid faunas in the surface peat of a drained pine mire. *Ann. Entom. Fenn.* 52:39–41.

Martens, K., 1988, Seven new species and two new subspecies of *Sclerocypris* Sars, 1924, from Africa, with new records of some other Megalocypridinids. *Hydrobiologia* 162:243–273.

Martens, K., 1994, Ostracode speciation in ancient lakes: a review. In Martens, K., Goddeeris, B., and Coulter, G. (eds.), *Speciation in Ancient Lakes.* Arch. Hydrobiol. Beih. Ergebn. Limnol. 44:203–222.

Martens, K., Goddeeris, B., and Coulter, G. (eds.), 1994, *Speciation in Ancient Lakes.* Arch. Hydrobiol. Beih.Ergebn. Limnol. 44.

Martens, P. and Rotmans, J. (eds.), 1999, *Climate Change: An Integrated Perspective.* Kluwer Academic, Dordrecht.

Martill, D.M., 1993, *Fossils of the Santana and Crato Formations, Brazil.* The Palaeontological Association (London) Field Guides to Fossils, no. 5.

Martin, P., Granina, L., Martens, K., and Goddeeris, B., 1998, Oxygen concentration profiles in sediments of two ancient lakes: Lake Baikal (Siberia, Russia) and Lake Malawi (East Africa). *Hydrobiologia* 367:163–174.

Martin-Closas, C. and Grambast-Fessard, N., 1986, Les Charophytes du Crétacé inférieur de la région du Maestrat. *Paléobio. Cont.* 15:1–66.

Martínez, M.A.F. and Meléndez, N., 2000, The lacustrine fossiliferous deposits of the Las Hoyas Subbasin–Lower Cretaceous, Serranía de Cuenca, Iberian Ranges, Spain. In Gierlowski, E.H. and Kelts, K.R. (eds.), *Lake Basins Through Space and Time.* AAPG Stud. Geol. 46:303–314.

Massari, F. and Colella, A., 1988, Evolution and types of fan-delta systems in some major tectonic settings. In Nemec, W. and Steel, R.J. (eds.), *Fan Deltas: Sedimentology and Tectonic Setting.* Blackie, London, pp. 103–122.

Matisoff, G., 1982, Mathematical models of bioturbation. In McCall, P.L. and Tevesz, M.J. (eds.), *Animal–Sediment Relations: The Biotic Alteration of Sediments.* Plenum, New York, pp. 289–331.

Maxwell, W.D., 1992, Permian and Early Triassic extinction of nonmarine tetrapods. *Palaeontology* 35:571–583.

Mayle, F.E., Bell, M., Birks, H.H., Brooks, S.J., Coope, G.R., Lowe, J.J., Sheldrick, C., Li, S., Turney, C.S.M., and Walker, M.J.C., 1999, Climate variations in Britain during the Last Glacial–Holocene transition (15.0–11.5 cal ka BP): comparisons with the GRIP ice-core record. *Jour. Geol. Soc. London* 156:411–423.

McAndrews, J.H., 1968, Pollen evidence for the prehistoric development of the "Big Woods" in Minnesota, USA. *Rev. Palaeobot. Palynol.* 7:201–211.

McAndrews, J.H., 1988, Human disturbance of North American forests and grasslands: the fossil pollen record. In Huntly, B. and Webb, T. (eds.), *Vegetation History.* Kluwer Academic, Dordrecht, pp. 673–697.

McCabe, M. and Cofaigh, C.Ó., 1994, Sedimentation in a subglacial lake, Enniskerry, eastern Ireland. *Sed. Geol.* 91:57–95.

McCarthy, F.M., Collins, E.S., McAndrews, J.H., Kerr, H.A., Scott, D.B., and Medioli, F.S., 1995, A comparison of postglacial Arcellacean ("Thecamoebian") and pollen succession in Atlantic Canada, illustrating the potential of Arcellaceans for paleoclimatic reconstruction. *Jour. Paleo.* 69:980–993.

McConnaughey, T., 1989a, ^{13}C and ^{18}O isotopic disequilibrium in biological carbonates: I. Patterns. *Geochim. Cosmochim. Acta* 53:151–162.

McConnaughey, T., 1989b, ^{13}C and ^{18}O isotopic disequilibrium in biological carbonates: II. *In vitro* simulation of kinetic isotope effects. *Geochim. Cosmochim. Acta* 53:163–171.

McCourt, R.M., 1995, Green algal phylogeny. *Trends Ecol. Evol.* 10:159–163.

McCoy, W.D., 1987, Quaternary aminostratigraphy of the Bonneville Basin, western United States. *Geol. Soc. Amer. Bull.* 98:99–112.

McCune, A., 1996, Biogeographic and stratigraphic evidence for rapid speciation in semionotid fishes. *Paleobiology* 22:34–48.

McCune, A.R., 1997, How fast do fishes speciate? Molecular, geological and phylogenetic evidence from adaptive radiations of fishes. In Givnish, T.J. and Sytsma, K.J. (eds.), *Molecular Evolution and Adaptive Radiation.* Cambridge University Press, Cambridge, pp. 585–610.

McCune, A.R. and Lovejoy, N.R., 1998, The relative rate of sympatric and allopatric speciation in fishes. In Howard, D. and Berlocher, S. (eds.), *Endless Forms: Species and Speciation.* Oxford University Press, Oxford, pp. 172–185.

McDougall, I., 1995, Potassium–argon dating in the Pleistocene. In Rutter, N.W. and Catto, N.R. (eds.), *Dating Methods for Quaternary Deposits.* GEOTEXT 2, Geological Society of Canada, pp. 1–14.

McDougall, I. and Harrison, T.M., 1988, *Geochronology and Thermochronology by the $^{40}Ar/^{39}Ar$ Method.* Oxford University Press, New York.

McGlone, M.S., Kershaw, A.P., and Markgraf, V., 1992, The El Niño/Southern Oscillation climatic variability in Australasian and South American paleoenvironmental records. In Diaz, H.F. and Markgraf, V. (eds.), *El Niño: Historical and Paleoclimatic Aspects of The Southern Oscillation*. Cambridge University Press, Cambridge, pp. 435–462.

McKenzie, J.A., 1985, Carbon isotopes and productivity in the lacustrine and marine environment. In Stumm, W. (ed.), *Chemical Processes in Lakes*. Wiley Interscience, New York, pp. 99–118.

McKenzie, J.A. and Hollander, D.J., 1993, Oxygen-isotope record in Recent carbonate sediments from Lake Greifen, Switzerland (1750–1986): application of continental isotopic indicator for evaluation of changes in climate and atmospheric circulation patterns. In Swart, P.K., Lohmann, K.C., McKenzie, J., and Savin, S. (eds.), *Climate Change in Continental Isotopic Records*. Amer. Geophys. Union Geophys. Monogr. 78:101–111.

McKenzie, K.G., 1981, Palaeobiogeography of some salt lake faunas. *Hydrobiologia* 82:407–418.

McLeroy, C.A. and Anderson, R.Y., 1966, Laminations of the Oligocene Florissant lake deposits, Colorado. *Geol. Soc. Amer. Bull.* 77:605–618.

McMahon, R.F., 1991, Mollusca: bivalvia. In Thorp, J.H. and Covitch, A.P. (eds.), *Ecology and Classification of North American Freshwater Invertebrates*. Academic Press, New York, pp. 315–399.

McQueen, D.J., Johannes, M.R.S., Post, J.R., Stewart, T.J., and Lean, D.R.S., 1989, Bottom-up and top-down impacts on freshwater pelagic community structure. *Ecol. Monogr.* 59:289–309.

Medioli, F.S. and Scott, D.B., 1988, Lacustrine thecamoebians (mainly arcellaceans) as potential tools for paleolimnological interpretations. In Gray, J. (ed.), 1988, *Paleolimnology: Aspects of Freshwater Paleoecology and Biogeography*. Elsevier, Amsterdam, pp. 361–386.

Mees, F., Verschuren, D., Nijs, R., and Dumont, H., 1991, Holocene evolution of the crater lake at Malha, Northwest Sudan. *Jour. Paleolim.* 5:227–253.

Melack, J., 1988, Aquatic plants in extreme environments. In Symoens, J.J. (ed.), *Vegetation of Inland Waters*. Kluwer Academic, Dordrecht, pp. 341–378.

Meléndez, N., 1995, *Las Hoyas. A Lacustrine Konservat-Lagerstätte, Cuenca, Spain*. Editorial University Complutense, Madrid.

Mello, M.R. and Maxwell, J.R., 1990, Organic geochemical and biological marker characterization of source rocks and oils derived from lacustrine environments in the Brazilian continental margin. In Katz, B.J. (ed.), *Lacustrine Basin Exploration—Case Studies and Modern Analogs*. Amer. Assoc. Petrol. Geol. Mem. 50:77–97.

Melosh, J., 1989, *Impact Cratering: A Geologic Process*. Oxford University Press, New York.

Menking, K.M., Bischoff, J.L., Fitzpatrick, J.A., Burdette, J.W., and Rye, R.O., 1997, Climatic/hydrologic oscillations since 155,000 yr BP at Owens Lake, California, reflected in abundance and stable isotope composition of sediment carbonates. *Quat. Res.* 48:58–68.

Meriläinen, J., 1967, The diatom flora and the hydrogen ion concentration of the water. *Ann. Bot. Fenn.* 4:51–58.

Meriläinen, J. and Hamina, V., 1993, Recent environmental history of a large, originally oligotrophic lake in Finland: a palaeolimnological study of chrionomid remains. *Jour. Paleolim.* 9:129–140.

Merlivat, L. and Jouzel, J., 1979, Global climatic interpretation of the deuterium–oxygen 18 relationship for precipitation. *Jour. Geophys. Res.* 84:5029–5033.

Merrihue, C. and Turner, G., 1966, Potassium–argon dating by activation with fast neutrons. *Jour. Geophys. Res.* 71:2852–2857.

Metcalfe, S.E., Street-Perrott, F.A., Perrott, R., and Harkness, D.D., 1991, Paleolimnology of the Upper Lerma Basin, Central Mexico: a record of climatic change and anthropogenic disturbance since 11600 yr BP. *Jour. Paleolim.* 5:197–218.

Metcalfe, S.E., Bimpson, A., Courtice, A.J., O'Hara, S.L., and Taylor, D.M., 1997, Climate change at the monsoon/Westerly boundary in Northern Mexico. *Jour. Paleolim.* 17:155–171.

Meybeck, M., 1995, Global distribution of lakes. In Lerman, A., Imboden, D.M., and Gat, J. (eds.), 1995, *Physics and Chemistry of Lakes*, 2nd ed. Springer-Verlag, New York, pp. 1–35.

Meyers, P.A., 1997, Organic geochemical proxies of paleoceanographic, paleolimnologic and paleoclimatic processes. *Org. Geochem.* 27:213–250.

Meyers, P.A. and Benson, L.V., 1987, Sedimentary biomarker and isotopic indicators of the paleoclimatic history of the Walker Lake basin, western Nevada. *Adv. Org. Geochem.* 13:807–813.

Meyers, P.A. and Eadie, B.J., 1993, Sources, degradation and recycling of organic matter associated with sinking particles in Lake Michigan. *Org. Geochem.* 20:47–56.

Meyers, P.A. and Horie, S., 1993, An organic carbon isotopic record of glacial–post glacial change in atmospheric $p\mathrm{CO}_2$ in the sediments of Lake Biwa, Japan. *Palaeogeog., Palaeoclim., Palaeoecol.* 105:171–178.

Meyers, P.A. and Ishiwatari, R., 1993, Lacustrine organic geochemistry—an overview of indicators of organic matter sources and diagenesis in lake sediments. *Org. Geochem.* 20:867–900.

Meyers, P.A. and Ishiwatari, R., 1995, Organic matter accumulation records in lake sediments. In Lerman, A., Imboden, D., and Gat, J. (eds.), *Physics and Chemistry of Lakes.* Springer-Verlag, New York, pp. 279–328.

Meyers, P.A. and Lallier-Vergès, E., 1999, Lacustrine sedimentary organic matter records of Late Quaternary paleoclimates. *Jour. Paleolim.* 21:345–372.

Meyers, P.A. and Teranes, J.L., 2001, Sediment organic matter. In Last, W.M. and Smol, J.P. (eds.), *Tracking Environmental Change Using Lake Sediments. Vol. 2. Physical and Geochemical Methods.* Kluwer Academic Publishers, Dordrecht, pp. 239–269.

Meyers, P.A., Leenheer, M.J., and Bourbonniere, R.A., 1980, Changes in spruce composition following burial in lake sediments for 10,000 yr. *Nature* 287:534–536.

Meyers, P.A., Tenzer, G.E., Lebo, M.E., and Reuter, J.E., 1998, Sedimentary record of sources and accumulation of organic matter in Pyramid Lake, Nevada over the past 1,000 years. *Limnol. Oceanogr.* 43:160–169.

Mezquita, F. and Roca, J.R., 1999, Ostracoda from springs on the eastern Iberian Peninsula: ecology, biogeography and palaeolimnological implications. *Palaeogeog., Palaeoclim., Palaeoecol.* 148:65–85.

Miall, A.D., 1990, *Principles of Sedimentary Basin Analysis.* Springer-Verlag, NewYork.

Miall, A.D., 2000, *Principles of Sedimentary Basin Analysis*, 3rd ed. Springer-Verlag, NewYork.

Michard, G., Viollier, E., Jézéquel, D., and Sarazin, G., 1994, Geochemical study of a crater lake: Pavin Lake, France. Identification, location and quantification of the chemical reactions in the lake. *Chem. Geol.* 115:103–115.

Michel, E., 1994, Why snails radiate: a review of gastropod evolution in long-lived lakes, both recent and fossil. In Martens, K., Goddeeris, B., and Coulter, G. (eds.), *Speciation in Ancient Lakes.* Arch. Hydrobiol. Beih.Ergebn. Limnol. 44:285–317.

Michel, E., Cohen, A.S., West, K., Johnston, M.R., and Kat, P.W., 1992, Large African lakes as natural laboratories for evolution: examples from the endemic gastropod fauna of Lake Tanganyika. *Mitt. Int. Ver. Limnol.* 23:85–99.

Milankovitch, M., 1920, *Théorie mathématique des phénomènes thermiques produits par la radiation solaire.* Gauthier-Villars, Paris.

Milankovitch, M., 1930, Mathematische Klimalchre und Astronomische Theorie der Klimaschwankungen 1. *Handbuch des Klimatologie.* Borntrager, Berlin.

Miller, B.B. and Bajc, A.F., 1990, Non-marine molluscs. In Warner, B.G. (ed.), *Methods in Quaternary Ecology.* Geoscience Canada Reprint Ser. 5. Geological Association of Canada, pp. 101–112.

Miller, B.B. and Tevesz, M.J.S., 2001, Freshwater molluscs. In Smol, J.P., Birks, H.J.B., and Last, W.M. (eds.), *Tracking Environmental Change Using Lake Sediments. Vol. 4. Zoological Indicators.* Kluwer Academic Publishers, Dordrecht, pp. 153–171.

Miller, B.B. and Thompson, T.A., 1990, Molluscan faunal changes in the Cowles Bog area, Indiana Dunes National Lakeshore, following the low-water Lake Chippewa phase. In Schneider, A.F. and Fraser, G.S. (eds.), *Late Quaternary History of the Lake Michigan Basin.* Geol. Soc. Amer. Spec. Pap. 251:21–27.

Miller, B.B., Karrow, P.F., and Mackie, G.L., 1985, Late Quaternary molluscan faunal changes in the Huron Basin. In Karrow, P.F. and Calkin, P.E. (eds.), *Quaternary Evolution of the Great Lakes.* Geol. Soc. Canada Spec. Pap. 30:95–107.

Miller, B.B., Tevesz, M.J.S., and Carney, J.S., 1998, Holocene environmental changes in the Whitefish Dunes Area, Door Peninsula, Northern Lake Michigan Basin, USA. *Jour. Paleolim.* 19:473–479.

Miller, B.B., Schneider, A.F., Smith, A.J., and Palmer, D.F., 2000, A 6000 year water level history of Europe Lake, Wisconsin, USA. *Jour. Paleolim.* 23:175–183.

Miller, G.H. and Brigham-Grette, J., 1989, Amino acid geochronology—resolution and precision in carbonate fossils. *Quat. Int.* 1:111–128.

Miller, G.H., Wendorf, F., Ernst, R., Schild, R., Close, A.E., Friedman, I., and Schwarcz, H., 1991, Dating lacustrine episodes in the eastern Sahara by the epimerization of isoleucine in ostrich eggshells. *Palaeogeog., Palaeoclim., Palaeoecol.* 84:175–189.

Miller, J.M.G., 1994, The Neoproterozoic Konnarock Formation, southwestern Virginia, USA: glaciolacustrine facies in a continental rift. In Deynoux, M., Miller, J.M.G., Domack, E.W., Eyles, N., Fairchild, I.J., and Young, G. (eds.), *Earth's Glacial Record.* Cambridge University Press, Cambridge, pp. 47–59.

Miller, L.G. and Aiken, G.R., 1996, Effects of glacial meltwater inflows and moat freezing on mixing in an ice-covered Antarctic lake as interpreted from stable isotope and tritium distributions. *Limnol. Oceanogr.* 41:966–976.

Miller, N.G. and Calkin, P.E., 1992, Paleoecological interpretation and age of an interstadial lake bed in western New York. *Quat. Res.* 37:75–88.

Miller, U., 1973, Diatoméundersökning av bottenproppar från Stora Skarsjön, Ljungskile. SNV Publikationer (Stockholm) 7:42–60.

Mingram, J., 1998, Laminated Eocene maar-lake sediments from Eckfeld (Eifel region, Germany) and their short term periodicities. *Palaeogeog., Palaeoclim., Palaeoecol.* 140:289–305.

Minigawa, M. and Wada, H., 1984, Stepwise enrichment of ^{15}N along food chains: further evidence and the relation between δ^{15}N and animal age. *Geochim. Cosmochim Acta* 48:1135–1140.

Miskovsky, J.-C. (ed.), 1987, *Géologie de la préhistoire*. Géopré, Paris.

Mitchell, M.J., Schindler, S.C., Owen, J.S., and Norton, S.A., 1988, Comparison of sulfur concentrations within lake sediment profiles. *Hydrobiologia* 157:219–229.

Mitchell, M.J., Owen, J.S., and Schindler, S.C., 1990, Factors affecting sulfur incorporation into lake sediments: paleoecological implications. *Jour. Paleolim.* 4:1–22.

Mitchum, R.M. Jr., Vail, P.R., and Sangree, J.B., 1977, Seismic stratigraphy and global changes of sea level. In Payton, C.E. (ed.), *Seismic Stratigraphy—Applications to Hydrocarbon Exploration.* Amer. Assoc. Petrol. Geol. Mem. 26:53–62.

Mittlebach, G.G., 1988, Competition among refuging sunfishes and effects of fish density on littoral zone invertebrates. *Ecology* 69:614–623.

Mohapatra, G.K. and Johnson, R.A., 1998, Localization of listric faults at thrust fault ramps beneath the Great Salt Lake Basin, Utah: evidence from seismic imaging and finite element modeling. *J. Geophys. Res.* 103:10,047–10,063.

Molostovskaya, I.I., 2000, The evolutionary history of Late Permian Darwinulocopina Sohn, 1988 (Ostracoda) from the Russian Plate. *Hydrobiologia* 419:125–130.

Monty, C.L. and Mas, J.R., 1981, Lower Cretaceous (Wealdian) blue–green algal deposits of the Province of Valencia, Eastern Spain. In Monty, C. (ed.), *Phanerozoic Stromatolites.* Springer-Verlag, New York, pp. 85–120.

Moor, H.C., Schaller, T., and Sturm, M., 1996, Recent changes in stable lead isotope ratios in sediments of Lake Zug, Switzerland. *Envir. Sci. Tech.* 30:2928–2933.

Moore, L.S. and Burne, R.V., 1994, The modern thrombolites of Lake Clifton, Western Australia. In Bertrand-Sarfati, J. and Monty, C. (eds.), *Phanerozoic Stromatolites II.* Kluwer, Dordrecht, pp. 3–30.

Moore, T.C., Rea, D.K., and Godsey, H., 1998, Regional variation in modern radiocarbon ages and the hard-water effects in Lake Michigan and Huron. *Jour. Paleolim.* 20:347–351.

Moore, T.C., Walker, J.C.G., Rea, D.K., Lewis, C.F.M., Shane, L.C.K., and Smith, A.J., 2000, Younger Dryas interval and outflow from the Laurentide Ice sheet. *Paleoceanography* 15:4–18.

Morgan, A.V. and Morgan, A., 1990, Beetles. In Warner, B.G. (ed.), *Methods in Quaternary Ecology.* Geoscience Canada Reprint Ser. 5. Geological Association of Canada, pp. 113–126.

Morrill, C., Small, E.E., and Sloan, L.C., 2001, Modeling orbital forcing of lake level change, Lake Gosiute (Eocene), North America. *Global Planet. Change* 29:57–76.

Morrill, C., Overpeck, J.T., and Cole, J.E., 2002, A synthesis of abrupt changes in the Asian summer monsoon since the last deglaciation. *The Holocene* in press.

Morris, N.J., 1985, Other non-marine invertebrates. *Phil. Trans. R. Soc. London, Ser. B* 309:239–240.

Mortimer, C.H., 1974, Lake hydrodynamics. *Mitt. Int. Ver. Theor. Angew. Limnol.* 20:124–197.

Mortlock, M.A. and Froelich, P.N., 1989, A simple method for rapid determination of biogenic opal in pelagic marine sediments. *Deep-Sea Res.* 36:1415–1426.

Moscariello, A., Schneider, A.M., and Fillippi, M.L., 1998, Late glacial and early Holocene palaeoenvironmental changes in Geneva Bay (Lake Geneva, Switzerland). *Palaeogeog., Palaeoclim., Palaeoecol.* 140:51–73.

Mount, J.F. and Cohen, A.S., 1984, Petrology and geochemistry of rhizoliths from Plio-Pleistocene fluvial and marginal lacustrine deposits, East Turkana, Kenya. *Jour. Sed. Pet.* 54:263–275.

Moura, J.A., 1972, Algunas especies e subespecies novas de ostracodes da Bacia Reconcavo/Tucano. *Bol. Tec. Petrobras* 15:245–363.

Moura, J.A., 1988, Ostracodes from non-marine Early Cretaceous sediments of the Campos Basin, Brazil. In Hanai, T., Ikeya, N., and Ishizaki, K. (eds.), *Evolutionary Biology of Ostracoda: Its Fundamentals and Applications.* Kodanska-Elsevier, pp. 1207–1216.

Mourguiart, P. and Carbonel, P., 1994, A quantitative method of palaeolake-level reconstruction using ostracod assemblages: an example from the Bolivian Altiplano. *Hydrobiologia* 288:183–193.

Mourguiart, P., Corrège, T., Wirrman, D., Argollo, J., Montenegro, M.E., Pourchet, M., and Carbonel, P., 1998, Holocene palaeohydrology of Lake Titicaca estimated from an ostracod-based transfer function. *Palaeogeog., Palaeoclim., Palaeoecol.* 143:51–72.

Mouthon, J., 1992, Snail and bivalve populations analysed in relation to physico-chemical quality of lakes in eastern France. *Hydrobiologia* 245:147–156.

Moutoux, T.E., 1995, *Palynological and tephra correlations among deep wells in the modern Great Salt Lake, Utah, USA, implications for Neogene through Pleistocene climatic reconstructions.* Unpublished M.Sc. Thesis, University of Arizona, Tucson, AZ.

Mueller, W.P., 1964, The distribution of cladoceran remains in surficial sediments from three northern Indiana lakes. *Invest. Indiana Lakes and Streams* 1:1–63.

Müller, G., Irion, G., and Förstner, U., 1972, Formation and diagenesis of inorganic Ca–Mg carbonates in the lacustrine environment. *Naturwissenschaften* 59:158–164.

Müller, P., Geary, D., and Magyar, I., 1999, The endemic molluscs of the Late Miocene Lake Pannon: their origin, evolution, and family level taxonomy. *Lethaia* 32:47–60.

Muller, R.A., 1977, Radioisotope dating with a cyclotron. *Science* 196:489–494.

Muller, R.A. and MacDonald, G.J., 1997a, Glacial cycles and orbital inclination. *Science* 277:215–218.

Muller, R.A. and MacDonald, G.J., 1997b, Spectrum of 100kyr glacial cycle. Orbital inclination, not eccentricity. *Proc. Nat. Acad. Sci. U.S.A.* 94:8329–8334.

Muller, R.A. and MacDonald, G.J., 2000, *Ice Ages and Astronomical Causes: Data, Spectral Analysis and Mechanisms.* Springer, Chichester.

Mullins, H.T., 1998, Holocene lake level and climate change inferred from marl stratigraphy of the Cayuga Lake Basin, New York. *Jour. Sed. Res.* 68:569–578.

Mullins, H.T. and Eyles, N. (eds.), 1996, *Subsurface Geologic Investigations of New York Finger Lakes: Implications for Late Quaternary Deglaciation and Environmental Change.* Geol. Soc. Amer. Spec. Pap. 311.

Mullins, H.T., Hinchey, E.J., Wellner, R.W., Stephens, D.B., Anderson, W.T., Dwyer, T.R., and Hine, A., 1996, Seismic stratigraphy of the Finger Lakes: a continental record of Heinrich event H-1 and Laurentide ice sheet instability. In Mullins, H.T. and Eyles, N. (eds.), *Subsurface Geologic Investigations of New York Finger Lakes: Implications for Late Quaternary Deglaciation and Environmental Change.* Geol. Soc. Amer. Spec. Pap. 311:1–35.

Murphy, D. and Wilkinson, B.H., 1980, Carbonate deposition and facies distribution in a central Michigan marl lake. *Sedimentology* 27:123–135.

Naeser, C.W., Briggs, N.D., Obradovich, J.D., and Izett, C.A., 1981, Geochronology of tephra deposits. In Self, S. and Sparks, R.S.J. (eds.), *Tephra Studies.* Reidel, Dordrecht, pp. 13–47.

Naeser, N.D., Westgate, J.A., Hughes, O.L., and Péwé, T., 1982, Fission-track ages of late Cenozoic distal tephra beds in the Yukon Territory and Alaska. *Can. Jour. Earth Sci.* 19:2167–2178.

Narcisi, B. and Anselmi, B., 1998, Sedimentological investigations on a Late Quaternary lacustrine core from the Lagaccione Crater (central Italy): paleoclimatic and paleoenvironmental inferences. *Quat. Int.* 47/48:21–28.

Narcisi, B., Anselmi, F., Catalano, G., Dai Pra, G., and Magri, G., 1992, Lithostratigraphy of the 250,000 year record of lacustrine sediments from the Valle di Castiglione Crater, Roma. *Quat. Sci. Rev.* 11:353–362.

Neale, J.W., 1988, Ostracods and palaeosalinity reconstruction. In De Deckker, P., Colin, J.P., and Peypouquet, J.P. (eds.), *Ostracoda in the Earth Sciences.* Elsevier, Amsterdam, pp. 125–155.

Neev, D. and Emery, K.O., 1967, The Dead Sea: depositional processes and environments of evaporites. *Geol. Surv. Israel Bull.* 41:1–147.

Negendanck, J.F.W. and Zolitschka, B. (eds.), 1993a, *Paleolimnology of European Maar Lakes.* Springer-Verlag, New York.

Negendank, J.F.W. and Zolitschka, B., 1993b, Maars and maar lakes of the Westeifel volcanic field. In Negendank, J.F.W. and Zolitschka, B. (eds.), *Paleolimnology of European Maar Lakes.* Springer-Verlag, New York, pp. 61–80.

Negrini, R.M., Erbes, D.B., Faber, K., Herrera, A., Roberts, A.P., Cohen, A.S., Wigand, P.E., and Foit, F.F., 2000, A paleoclimate record for the past 250,000 years from Summer Lake, Oregon, USA: I. Chronology and magnetic proxies for lake level. *Jour. Paleolim.* 24:125–149.

Neill, W.E., 1994, Spatial and temporal scaling and the organization of limnetic communities. In Giller, P.S., Hildrew, A.G., and Raffaelli, D.G. (eds.), *Aquatic Ecology.* Blackwell Science, Oxford, pp. 189–231.

Nelson, C.H., Bacon, C.R., Robinson, S.W., Adam, D.P., Bradbury, J.P., Barber, J.H., Schwartz, D., and Vagenas, G., 1994, The volcanic, sedimentologic, and paleolimnologic history of the Crater Lake caldera floor, Oregon: evidence for small caldera evolution. *Geol. Soc. Amer. Bull.* 106:684–704.

Nelson, C.H., Karabanov, E.B., Colman, S.M., and Escutia, C., 1999, Tectonic and sediment supply control of deep rift lake turbidite systems: Lake Baikal, Russia. *Geology* 27:163–166.

Nemec, W., 1990, Aspects of sediment movement on steep delta slopes. In Colella, A. and Prior, D. (eds.), *Coarse-Grained Deltas*. Spec. Publ. Int. Assoc. Sediment. 10:29–73.

Nesbitt, W.H., 1974, *The study of some mineral–aqueous solution interactions*. Ph.D. dissertation, Johns Hopkins University, Baltimore.

Neustrueva, I.Y., 1993, Limnic ostracoda communities of the Paleozoic-Mesozoic in connection with the paleogeography of lacustrine basins. In McKenzie, K. and Jones, P. (eds.), *Ostracoda in the Earth and Life Sciences*. Balkema, Rotterdam, p. 663.

Newhall, C.G., Paull, C.K., Bradbury, J.P., Higuera-Gundy, A., Poppe, L.J., Self, S., Bonnar Sharpless, N., and Ziagos, J., 1987, Recent geologic history of Lake Atitlan, a caldera lake in western Guatemala. *Jour. Volcanol. Geotherm. Res.* 33:81–107.

Newrkla, P., 1985, Respiration of *Cytherissa lacustris* (Ostracoda) at different temperatures and its tolerance towards temperature and oxygen concentrations. *Oecologia* 67:250–254.

Nicholas, W.L., 1998, The Ashchelminthes. In Anderson, D.T. (ed.), *Invertebrate Zoology*. Oxford University Press, Melbourne, pp. 86–115.

Nicholson, B.J. and Vitt, D.H., 1994, Wetland development at Elk Island National Park, Alberta, Canada. *Jour. Paleolim.* 12:19–34.

Nicholson, S.E., 1994, Recent rainfall fluctuations in Africa and their relationship to past conditions over the continent. *The Holocene* 4:121–131.

Nicholson, S.E., 1996, A review of climate dynamics and climate variability in eastern Africa. In Johnson, T.C. and Odada, E. (eds.), *The Limnology, Climatology and Paleoclimatology of the East African Lakes*. Gordon and Breach, Amsterdam, pp. 25–56.

Nicholson, S.E. and Flohn, H., 1980, African environmental and climatic changes and the general atmospheric circulation in Late Pleistocene and Holocene. *Clim. Change* 2:313–348.

Nicoud, G. and Manalt, F., 2001, The lacustrine depression at Annecy (France): geological setting and Quaternary evolution. *Jour. Paleolim.* 25:137–147.

Nielsen, H. and Sørensen, I., 1992, Taxonomy and stratigraphy of late-glacial *Pediastrum* taxa from Lymosen, Denmark—a preliminary study. *Rev. Palaeobot. Palynol.* 74:55–75.

Niessen, F., 1987, *Sedimentologische geophysikalische und geochemische Untersuchungen zur entstehung und Ablagerungsgeschichte des Luganoseef*. Dissertation 8354, ETH, Zurich.

Niessen, F., Lami, A., and Guilizzoni, P., 1993, Climatic and tectonic effects on sedimentation in central Italian volcano lakes (Latium). Implications from high resolution seismic profiles. In Negendank, J.F.W. and Zolitschka, B. (eds.), *Paleolimnology of European Maar Lakes*. Springer-Verlag, New York, pp. 129–148.

Nikishin, A.M., Brunet, M.-F., Cloetingh, S., and Ershov, A.V., 1997, Northern Peri-Tethyan Cenozoic intraplate deformations: influence of the Tethyan collisional belt on the Eurasian continent from Paris to Tian-Shan. *C. R. Acad. Sci. Paris II* 324:49–57.

Nilssen, J.P. and Sandøy, S., 1990, Recent lake acidification and cladoceran dynamics: surface sediment and core analysis from lakes in Norway, Scotland and Sweden. *Phil. Trans. R. Soc. London, Ser. B* 327:299–309.

Nipkow, F., 1920, Vorläufige Mitteilungen über Untersuchungen des Schlammabsatzes im Zürichsee. *Schweiz. Z. Hydrol.* 1:100–122.

Nishino, M. and Watanabe, N.C., 2000, Evolution and endemism in Lake Biwa, with special reference to its gastropod mollusc fauna. In Rossiter, A. and Kawanabe, H. (eds.), *Ancient Lakes: Biodiversity, Ecology and Evolution*. Advances in Ecological Research 31. Academic Press, New York, pp. 152–180.

Noon, P.E., Birks, H.J.B., Jones, V.J., and Ellis-Evans, J.C., 2001, Quantiative models for reconstructing catchment ice-extent using physical–chemical characteristics of lake sediments. *Jour. Paleolim.* 25:375–392.

Norris, G. and McAndrews, J.H., 1970, Dinoflagellate cysts from post-glacial lake muds, Minnesota (U.S.A.). *Rev. Palaeobot. Palynol.* 10:131–156.

Norton, S.A., Bienert, R.W., Binford, M.W., and Kahl, J.S., 1992, Stratigraphy of total metals in PIRLA sediment cores. *Jour. Paleolim.* 7:191–214.

Nury, D., 2000, Lacustrine Oligocene basins in Southern Provence, France. In Gierlowski, E.H. and Kelts, K.R. (eds.), *Lake Basins Through Space and Time*. AAPG Stud. Geol. 46:381–388.

Nuttall, C.P., 1990, Review of the Caenozoic heterodont bivalve superfamily Dresissenacea. *Palaeontology* 33:707–737.

Nygaard, G., 1956, Ancient and recent flora of diatoms and chrysophyceae in Lake Gribsø. Studies on the humic acid lake Gribsø. *Fol. Limnol. Scand.* 8:32–94.

Oana, S. and Deevey, E.S., 1960, Carbon-13 in lake waters and its possible bearing on paleolimnology. *Am. Jour. Sci.* 258A:253–272.

O'Brien, N.R. and Pietraszek-Mattner, S., 1998, Origin of the fabric of laminated fine-grained glaciolacustrine deposits. *Jour. Sed. Res.* 68:832–840.

O'Connor, J.E. and Baker, V.R., 1992, Magnitudes and implications of peak discharges from glacial Lake Missoula. *Geol. Soc. Amer. Bull.* 104:267–279.

Odgaard, B.V., 1993, The sedimentary record of spheroidal carbonaceous fly-ash particles in shallow Danish lakes. *Jour. Paleolim.* 8:171–187.

Ogden, J.G., 1966, Forest history of Ohio I. Radiocarbon dates and pollen stratigraphy of Silver Lake, Logan Co., Ohio. *Ohio Jour. Sci.* 66:387.

Okamoto, I., Endoh, S., and Kumagai, M., 1995, Distribution and mixing of the River Seri water. In Okuda, S., Imberger, J., and Kumagai, M. (eds.), *Physical Processes in a Large Lake: Lake Biwa, Japan.* Amer. Geophys. Union Coastal Est. Stud. Ser. 48:101–118.

Okay, A.I., Sengör, A.M.C., and Görür, N., 1994, Kinematic history of the opening of the Black Sea and its effect on the surrounding regions. *Geology* 22:267–270.

Olander, H., Korhola, A., and Blom, T., 1997, Surface sediment Chironomidae (Insecta: Diptera) distributions along an ecotonal transect in subarctic Fennoscandia: developing a tool for palaeotemperature reconstructions. *Jour. Paleolim.* 18:45–59.

Olander, H., Birks, H.J.B., Korhola, A., and Blom, T., 1999, An expanded calibration model for inferring lakewater and air temperatures from fossil chironomid assemblages in northern Fennoscandia. *The Holocene* 9:279–294.

Oldfield, F., 1990, Magnetic measurements of recent sediments from Big Moose Lake, Adirondack Mountains, NY, USA. *Jour. Paleolim.* 4:93–101.

Oldfield, F. and Richardson, N., 1990, Lake sediment magnetism and atmospheric deposition. *Phil. Trans. R. Soc. London, Ser. B* 327:325–330.

Oldfield, F. and Wu, R., 2000, The magnetic properties of the recent sediments of Brothers Water, NW England. *Jour. Paleolim.* 23:165–174.

Olila, O.G. and Reddy, K.R., 1997, Influence of redox potential on phosphate-uptake by sediments in two sub-tropical eutrophic lakes. *Hydrobiologia* 345:45–57.

Oliver, D.R., 1971 Life history of the chironomidae. *Ann. Rev. Entomol.* 16:211–230.

Olsen, P.E., 1986, A 40 million year lake record of early Mesozoic orbital climatic forcing. *Science* 234:842–848.

Olsen, P.E., 1990, Tectonic, climatic and biotic modulation of lacustrine ecosystems. Examples from Newark Supergroup of Eastern North America. In Katz, B.J. (ed.), *Lacustrine Basin Exploration—Case Studies and Modern Analogues.* Amer. Assoc. Petrol. Geol. Mem. 50:209–224.

Olsen, P.E., 2001, Grand cycles of the Milankovitch band. *EOS Trans.* 82 (47): F2.

Olsen, P.E. and Kent, D.V., 1996, Milankovitch climate forcing in the tropics of Pangaea during the Late Triassic. *Palaeogeog., Palaeoclim., Palaeoecol.* 122:1–26.

Olsen, P.E., Kent, D.V., Cornet, B., Witte, W.W., and Schlische, R.W., 1996, High resolution stratigraphy of the Newark Rift Basin (early Mesozoic, eastern North America). *Geol. Soc. Amer. Bull.* 108:40–77.

Olsson, I.U., 1970, *Radiocarbon Variations and Absolute Chronology.* Almqvist Wiksell, Stockholm.

Olsson, I.U., 1986, Radiometric dating. In Berglund, B.E. (ed.), *Handbook of Holocene Palaeoecology and Palaeohydrology.* J. Wiley and Sons, London, pp. 273–312.

Olsson, S., Regnèll, J., Persson, A., and Sandgren, P., 1997, Sediment-chemistry response to land-use change and pollutant loading in a hypertrophic lake, southern Sweden. *Jour. Paleolim.* 17:275–294.

Ortega, F., Sanz, J.L., Barbadillo, L.J., Buscalioni, A.D., de la Fuente, M., Madero, J., Martín Closas, C., Martínez, X., Meléndez, N., Moratalla, B.P., Moreno, P., Pinardo Moya, E., Poyato Ariza, F.J., Rodríguez Lázaro, J., Sanchiz, B., and Wez, S., 1998, El yacimiento de Las Hoyas (La Cierva, Cuenca) Un Konservat-Lagerstätte del Cretácico Inferior. In Aguirre, E. and Rábano, I. (eds.), La Huella del Pasado. Fosiles de Castilla-La Mancha, pp. 195–215.

Osborne, R.H., Licari, G.R., and Link, M.H., 1982, Modern lacustrine stromatolites, Walker Lake, Nevada. *Sed. Geol.* 32:39–61.

Osborne, R.H., Edelman, M.C., Gaynor, J.M., and Waldron, J.M., 1985, Sedimentology of the littoral zone in Lake Tahoe, California–Nevada. Open File Rep. Plan. Env. Coord. California State Lands Commission, Sacramento.

Ostrom, N.E., Long, D.T., Bell, E.M., and Beals, T., 1998, The origin of cycling of particulate and sedimentary organic matter and nitrate in Lake Superior. *Chem. Geol.* 152:13–28.

O'Sullivan, P.E., 1983, Annually laminated sediments and the study of Quaternary environmental changes—a review. *Quat. Sci. Rev.* 1:245–313.

O'Sullivan, P.E., 1991, Paleolimnology, William Morris and *The Magic Flute.* In Smith, J.P.,

Appleby, P.G., Battarbee, R.W., Dearing, J.A., Flower, R., Haworth, E.Y., Oldfield, F., and O'Sullivan, P.E.O. (eds.), *Environmental History and Paleolimnology*. Kluwer Academic, Norwell, MA, pp. 373–382.

Oswald, G.K.A. and Robin, G., 1973, Lakes beneath the Antarctic ice sheet. *Nature* 245:251–254.

Otis, R.M. and Smith, R.B., 1977, Geophysical surveys of Yellowstone Lake, Wyoming. *Jour. Geophys. Res.* 82:3705–3717.

Overpeck, J., Anderson, D., Trumbore, S., and Prell, W., 1996, The southwest Indian Monsoon over the last 18,000 years. *Clim. Dynam.* 12:213–225.

Overpeck, J., Hughen, K., Hardy, D., Bradley, R., Case, R., Douglas, M., Finney, B., Gajewski, K., Jacoby, G., Jennings, A., Lamoureux, S., Lasca, A., MacDonald, G., Moore, J., Retelle, M., Smith, S., Wolfe, A., and Zielinski, G., 1997, Arctic environmental change of the last four centuries. *Science* 278:1251–1256.

Overpeck, J.T., Whitlock, C., and Huntley, B., 2002, Terrestrial biosphere dynamics in the climate system:past and future. In Alverson, K., Bradley, R., and Pedersen, T. (eds.), *Paleoclimate, Global Change and the Future*. IGBP Synthesis Volume, IGBP in press.

Oviatt, C.G., 1987, Lake Bonneville stratigraphy at the Old River Bed, Utah. *Amer. Jour. Sci.* 287:383–398.

Oviatt, C.G., 1988, Paleoclimatic and neotectonic significance of Plio-Pleistocene lake beds in the Sevier Basin, Utah. *Geol. Soc. Amer. 1988 Ann. Mtg. Abstr. w/prog.* A346–347.

Oviatt, C.G., 1997, Lake Bonneville fluctuations and global climate change. *Geology* 25:155–158.

Oviatt, C.G. and Currey, 1987, Pre-Bonneville Quaternary lakes in the Bonneville Basin, Utah. In Kopp, R.S. and Cohenour, R.E. (eds.), *Cenozoic Geology of Western Utah: Sites for Precious Metal and Hydrocarbon Accumulation*. Utah Geol. Assoc. Publ. 16:257–263.

Oviatt, C.G. and Miller, D.M., 1997, New explorations along the northern shores of Lake Bonneville. *Brigham Young Univ. Geol. Stud.* 42:345–371.

Oviatt, C.G., McCoy, W.D., and Reider, R.G., 1987, Evidence for a shallow early or middle Wisconsin-age lake in the Bonneville Basin, Utah. *Quat. Res.* 27:248-262.

Oviatt, C.G., Thompson, R.S., Kaufman, D.S., Bright, J., and Forester, R.M., 1999, Reinterpretation of the Burmester Core, Bonneville Basin, Utah. *Quat. Res.* 52:180–184.

Paillard, D., Labeyrie, L., and Yiou, P., 1996, MacIntosh program performs time series analysis. *Eos* 77:379 (available at http://www.agu.org/eos_elec/96097e.html).

Paine, J.G., 1994, Subsidence beneath a playa basin on the Southern High Plains, U.S.A.: evidence from shallow seismic data. *Geol. Soc. Amer. Bull.* 106:233–242.

Pair, D.L., Muller, E.H., and Plumley, P.W., 1994, Correlation of Late Pleistocene glaciolacustrine and marine deposits by means of geomagnetic secular variation, with examples from northern New York and southern Ontario. *Quat. Res.* 42:277–287.

Palacios-Fest, M., 1996, Geoquímica de la concha de ostrácodos (*Limnocythere staplini*) un método de regressión múltiple como indicador paleoclimático. *GEOS* 16:130–136.

Palacios-Fest, M. and Dettman, D., 2001, Temperature controls monthly variation in ostracode valve Mg/Ca; *Cypridopsis vidua* from a small lake in Sonora, Mexico. *Geochim. Cosmochim. Acta* 65:2499–2507.

Palacios-Fest, M., Cohen, A., Ruiz, J., and Blank, B., 1993, Comparative paleoclimatic interpretations from nonmarine ostracodes using faunal assemblages, trace element shell chemistry and stable isotope data. In Swart, P.K., Lohmann, K.C., McKenzie, J., and Savin, S. (eds.), *Climate Change in Continental Isotopic Records*. Amer. Geophys. Union Geophys. Monogr. 78:179–190.

Palacios-Fest, M., Cohen, A.S., and Anadón, P., 1994, Use of ostracodes as paleoenvironmental tools in the interpretation of ancient lacustrine records. *Rev. Espan. Paleontol.* 9:145–164.

Palmer, D.F., Smith, A., Ito, E., and Forester, R., 1998, Field calibration of oxygen isotope signatures in nonmarine ostracodes. *Geol. Soc. Amer. Ann. Mtg. Abstr. w/prog.* 30:283.

Pang, P.C. and Nriagu, J.O., 1977, Isotopic variations of the nitrogen in Lake Superior. *Geochim. Cosmochim. Acta* 41:811–814.

Park, L.E. and Downing, K.F., 2000, Implications of phylogeny reconstruction for ostracod speciation modes in Lake Tanganyika. In Rossiter, A. and Kawanabe, H. (eds.), *Ancient Lakes: Biodiversity, Ecology and Evolution*. Advances in Ecological Research 31. Academic Press, New York, pp. 303–330.

Park, L.E. and Downing, K.F., 2001, Paleoecology of an exceptionally preserved arthropod fauna from lake deposits of the Miocene Barstow Formation, Southern California, USA. *Palaios* 16:175–184.

Parker, J.I. and Edgington, D.N., 1976, Concentrations of diatom frustules in Lake Michigan sediment cores. *Limnol. Oceanogr.* 21:887–893.

Parrish, J.M., Parrish, J.T., and Ziegler, A.M., 1986, Permian–Triassic paleogeography and paleoclimatology and implications for therapsid distributions. In Hotton, N.H. III, MacLean, P.D., Roth, J.J., and Roth, E.C. (eds.), *The Biology and Ecology of Mammal-like Reptiles*. Smithsonian Press, Washington, DC, pp. 109–132.

Parrish, J.T., 1998, *Interpreting Pre-Quaternary Climate from the Geologic Record*. Columbia University Press, New York.

Partridge, T.C., Kerr, S.J., Metcalfe, S.E., Scott, L., Talma, A.S., and Vogel, J.C., 1993, The Pretoria Saltpan: a 200,000 year Southern African lacustrine sequence. *Palaeogeog., Palaeoclim., Palaeoecol*. 101:317–337.

Partridge, T.C., DeMenocal, P.B., Lorentz, S.A., Paiker, M.J., and Vogel, J.C., 1997, Orbital forcing of climate over South Africa: a 200,000-year rainfall record from the Pretoria Saltpan. *Quat. Sci. Rev*. 16:1125–1133.

Patalas, K., 1984, Mid-summer mixing depths of lakes of different latitudes. *Verh. Int. Verein. Limnol*. 22:97–102.

Paterson, M., 1993, The distribution of microcrustacea in the littoral zone of a freshwater lake. *Hydrobiologia* 263:173–183.

Patrick, S.T., Timberlid, J.A., and Stevenson, A.C., 1990, The significance of land-use and land-management change in the acidification of lakes in Scotland and Norway: an assessment utilizing documentary sources and pollen analysis. *Phil. Trans. R. Soc. London, Ser. B* 327:363–367.

Patterson, C., 1993, Osteichthyes: Teleostei. In Benton, M.J. (ed.), *The Fossil Record 2*. Chapman and Hall, London, pp. 621–656.

Patterson, R.T., McKillop, W.B., Kroker, S., Nielsen, E., and Reinhardt, E.G., 1997, Evidence for rapid avian-mediated foraminiferal colonization of Lake Winnipegosis, Manitoba during the Holocene Hypsithermal. *Jour. Paleolim*. 18:131–143.

Patterson, W.P., 1998, North American continental seasonality during the last millennium: high resolution analysis of sagittal otoliths. *Palaeogeog., Palaeoclim., Palaeoecol*. 138:271–303.

Patterson, W.P. and Smith, G.R., 2001, Fish. In Smol, J.P., Birks, H.J.B., and Last, W.M. (eds.), *Tracking Environmental Change Using Lake Sediments. Vol. 4. Zoological Indicators*. Kluwer Academic Publishers, Dordrecht, pp. 173–187.

Patterson, W.P., Smith, G.R., and Lohmann, K.C., 1993, Continental paleothermometry and seasonality using the isotopic composition of aragonitic otoliths of freshwater fishes. In Swart, P.K., Lohmann, K.C., McKenzie, J., and Savin, S. (eds.), *Climate Change in Continental Isotopic Records*. Amer. Geophys. Union Geophys. Monogr. 78:191–202.

Peck, J.A., King, J.W., Colman, S.M., and Kravchinsky, V.A., 1994, A rock-magnetic record from Lake Baikal, Siberia: evidence for Late Quaternary climate change. *Earth Planet. Sci. Lett*. 122:221–238.

Peck, J.A., King, J.W., Colman, S.M., and Kravchinsky, V.A., 1996, An 84-kyr paleomagnetic record from the sediments of Lake Baikal, Siberia. *Jour. Geophys. Res*. 101:11,365–11,385.

Peck, R.E., 1957, North American Charophyta. *U.S. Geol. Surv. Prof. Pap*. 294A:1–44.

Pedersen, G. and Noe-Nygaard, N., 1995, *Excursion Guide to Late and Postglacial Deposits in NW Sjælland*. Int. Limnogeol. Congress, Copenhagen, Denmark.

Pedley, H.M., 1990, Classification and environmental models of cool freshwater tufas. *Sed. Geol*. 68:143–154.

Penck, A., 1882, *Die Vergletsherung der deutschen Alpen, ihre Uraschen, periodische Wiederkekr und ihr Einfluss auf die Bodengestaltung*. J.A. Barth, Leipzig.

Penck, A., 1894, *Morphologie der Erdoberflache, v. 1*. Engelhorn, Stuttgart.

Pendl, M.P. and Stewart, K.M., 1986, Variations in carbon fractions within a dimictic and a meromictic basin of the Junius Ponds, New York. *Freshwater Biol*. 16:539–555.

Pennak, R.W., 1978, *Freshwater Invertebrates of the United States*, 2nd ed. J. Wiley and Sons, New York.

Pennington, W., 1943, Lake sediments: the bottom deposits of the North Basin of Windemere, with special reference to the diatom succession. *New Phytol*. 42:1–27.

Pennington, W., 1979, The origin of pollen in lake sediments: an enclosed lake compared with one receiving inflow streams. *New Phytol*. 83:189–213.

Pennington, W., 1984, Long-term natural acidification of upland sites in Cumbria: evidence from post-glacial lake sediments. *Rep. Freshwater Biol. Assoc*. 52:28–46.

Pennington, W., Haworth, E.Y., Bonny, A.P., and Lishman, J.P., 1972, Lake sediments in northern Scotland. *Phil. Trans. R. Soc. London, Ser. B* 264:191–294.

Pennington, W., Cambray, R.S., and Fisher, E.M., 1973, Observations on lake sediments using ^{137}Cs as a tracer. *Nature* 242:324–326.

Pentecost, A. and Riding, R., 1986, Calcification in cyanobacteria. In Leadbetter, B.S.C. and Riding, R. (eds.), *Biomineralization in Lower Plants and Animals*. Clarendon Press, Oxford, pp. 73–90.

Percival, D.B. and Walden, A.T., 1993, *Spectral Analysis for Physical Applications: Multitaper and Conventional Univariate Methods*. Cambridge University Press, New York.

Perkins, J.A. and Sims, J.D., 1983, Correlation of Alaskan varve thickness with climatic parameters, and use in paleoclimatic reconstruction. *Quat. Res.* 20:308–321.

Peterson, B.J. and Fry, B., 1987, Stable isotopes in ecosystem studies. *Ann. Rev. Ecol. Syst.* 18:293–320.

Pethick, J., 1984, *An Introduction to Coastal Geomorphology*. Edward Arnold, London.

Petit-Maire, N. and Riser, J., 1981, Holocene lake deposits and palaeoenvironments in Central Sahara, Northeastern Mali. *Palaeogeog., Palaeoclim., Palaeoecol.* 35:45–61.

Petruccione, J.L., Wellner, R.W., and Sheridan, R.E., 1996, Seismic reflection investigation of Montezuma wetlands central New York State: evolution of a Late Quaternary subglacial meltwater channel system. In Mullins, H.T. and Eyles, N. (eds.), *Subsurface Geologic Investigations of New York Finger Lakes: Implications for Late Quaternary Deglaciation and Environmental Change.* Geol. Soc. Amer. Spec. Pap. 311:77–89.

Piboule, M., Bouchez, R., Ma, J.L., Amossé, J., and Vivier, G., 1987, La méthode de datation potassium–argon. In Miskovsky, J.-C. (ed.), *Géologie de la préhistoire.* Géopré, Paris, pp. 1095–1109.

Pickford, M., Senut, B., Vincens, A., Van Neer, W., Ssemmanda, I., Baguma, Z., and Musiime, E., 1991, Nouvelle biostratigraphie du Néogène et du Quaternaire de la région de Nkondo (Bassin du lac Albert, Rift occidental ougandais). Apport à l'évolution des paléomilieux. *C. R. Acad. Paris* 312:1667–1672.

Pickrill, R.A., 1993, Shallow seismic stratigraphy and pockmarks of a hydrothermally influenced lake, Lake Rotoiti, New Zealand. *Sedimentology* 40:813–828.

Pickrill, R.A. and Irwin, J., 1982, Predominant headwater inflow and its control of lake–river interactions in Lake Wakatipu. *New Zealand Jour. Freshw Res.* 16:201–213.

Pickrill, R.A. and Irwin, J., 1983, Sedimentation in a deep glacier-fed lake—Lake Tekapo, New Zealand. *Sedimentology* 30:63–75.

Pienitz, R., Smol, J.P., and Birks, H.J.B., 1995, Assessment of freshwater diatoms as quantitative indicators of past climatic change in the Yukon and Northwest Territories, Canada. *Jour. Paleolim.* 13:21–49.

Pienitz, R., Smol, J.P., Last, W.M., Leavitt, P.R., and Cumming, B.F., 2000, Multi-proxy Holocene palaeoclimatic record from a saline lake in the Canadian Subarctic. *The Holocene* 10:673–686.

Pillmore, C.H., 1976, Deflation origin of Adams and Bartlett Lake Basins, Vermejo Park, New Mexico. *New Mex. Geol. Soc. Guidebook, 27th Field Conf., Vermejo Park*, pp. 121–124.

Pilskaln, C. and Johnson, T.C., 1991, Seasonal signals in Lake Malawi sediments. *Limnol. Oceanogr.* 36:544–557.

Pirozynski, K.A., 1990, Fungi. In Warner, B.G. (ed.), *Methods in Quaternary Ecology.* Geoscience Canada Reprint Ser. 5. Geological Association of Canada, pp. 15–22.

Planas, D., 1996, Acidification effects. In Stevenson, R.J., Bothwell, M.L., and Lowe, R.L. (eds.), *Algal Ecology: Freshwater Benthic Ecosystems.* Academic Press, New York, pp. 497–530.

Platt, N.H., 1989, Climatic and tectonic controls on sedimentation of a Mesozoic lacustrine sequence: the Purbeck of the West Cameros Basin, northern Spain. *Palaeogeog., Palaeoclim., Palaeoecol.* 70:187–197.

Platt, N.H., 1992, Freshwater carbonates from the Lower Freshwater Molasse (Oligocene, western Switzerland): sedimentology and stable isotopes. *Sed. Geol.* 78:81–99.

Platt, N.H., 1994, The western Cameros Basin, northern Spain: Rupelo Formation (Berriasian). In Gierlowski-Kordesch, E. and Kelts, K. (eds.), *Global Geological Record of Lake Basins, v. 1.* Cambridge University Press, Cambridge, pp. 195–202.

Platt, N.H. and Keller, B., 1992, Distal alluvial deposits in a foreland basin setting—the Lower Freshwater Molasse (Lower Miocene), Switzerland: sedimentology, architecture and palaeosols. *Sedimentology* 39:545–565.

Platt, N.H. and Wright, V.P., 1991, Lacustrine carbonates: facies models, facies distribution and hydrocarbon aspects. In Anadón, P., Cabrera, L., and Kelts, K. (eds.), *Lacustrine Facies Analysis.* Int. Assoc. Sediment. Spec. Publ. 13:57–74.

Plaziat, J.C., 1993, Modern and fossil Potamids (Gastropoda) in saline lakes. *Jour. Paleolim.* 8:163–169.

Plaziat, J.C. and Freytet, P., 1978, Le psedomicrokarst pédologique: un aspect particulier des paléo-pédogenèses developpées sur les dépots calcaires lacustres dans le tertiare du Languedoc. *C. R. Acad. Sci. Paris, Ser. D* 286:1661–1664.

Plisnier, P.D., Chitamwebwa, D., Mwape, L., Tshibangu, K., Langenberg, V., and Coenen, E., 1999. Limnological annual cycle inferred from physical–chemical fluctuations at three stations of Lake Tanganyika. *Hydrobiologia* 407:45–58.

Pogácsás, G., Müller, P., and Magyar, I., 1993, The role of seismic stratigraphy in understanding biological evolution in the Pannonian Lake (SE

Europe, Late Miocene). *Geol. Croat.* 46:63–69.

Pollard, J.E., 1985, Evidence from trace fossils. *Phil. Trans. R. Soc. London, Ser. B* 309:241–242.

Pollard, J.E., Steel, R.J., and Undersrud, E., 1982, Facies sequences and trace fossils in lacustrine/fan delta deposits, Hornelen Basin (M. Devonian), Western Norway. *Sed. Geol.* 32:63–87.

Pollingher, U., Ambühl, H., and Bürgi, H.R., 1992, A new method for processing clay-rich unconsolidated sediments for paleoecological investigations. *Jour. Paleolim.* 7:95–101.

Ponomarenko, A.G., 1996, Evolution of continental aquatic ecosystems. *Paleont. Jour.* 30:705–709.

Popp, B.H. and Wilkinson, B.H., 1983, Holocene lacustrine ooids from Pyramid Lake, Nevada. In Peryt, T.M. (ed.), *Coated Grains.* Springer-Verlag, New York, pp. 142–153.

Porter, K.G., Saunders, P.A., Haberyan, K.A., Macubbin, A.E., Jacobsen, T.R., and Hodson, R.E., 1996, Annual cycle of autotrophic and heterotrophic production in a small monomictic Piedmont lake (Lake Oglethrope): analog for the effects of climatic warming on dimictic lakes. *Limnol. Oceanogr.* 41:1041–1051.

Porter, S.C., Sauchyn, D.J., and Delorme, L.D., 1999, The ostracode record from Harris Lake, southwestern Saskatchewan: 9200 years of local environmental change. *Jour. Paleolim.* 21:35–44.

Porter, S.M. and Knoll, A.H., 2000, Testate amoebae in the Neoproterozoic Era: evidence from vase-shaped microfossils in the Chuar Group, Grand Canyon. *Paleobiology* 26:360–385.

Postma, G., 1990, Delta architecture and facies of river and fan deltas: a synthesis. In Colella, A. and Prior, D. (eds.), *Coarse-Grained Deltas.* Spec. Publ. Int. Assoc. Sediment. 10:13–27.

Prasad, M.N.V. and Ramesh, N.R., 1984, First record of *Circiniconis* conidia from the Holocene formation of Tripura, India. *Curr. Sci.* 53:38–39.

Prell, W.L. and Kutzbach, J.E., 1992, Sensitivity of the Indian monsoon to forcing parameters and implications for its evolution. *Nature* 360:647–652.

Premuzic, E.T., Benkovitz, C.M., Gaffney, J.S., and Walsh, J.J., 1982, The nature and distribution of organic matter in the surface sediments of world oceans and seas. *Org. Geochem.* 4:63–77.

Prentice, I.C., 1985, Pollen representation, source area, and basin size: towards a unified theory of pollen analysis. *Quat. Res.* 23:76–86.

Prentice, I.C., 1986, Multivariate methods for data analysis. In Berglund, B.E. (ed.), *Handbook of Holocene Palaeoecology and Palaeohydrology.* J. Wiley and Sons, New York, pp. 775–797.

Pye, K. and Rhodes, E.G., 1985, Holocene development of an episodic transgressive dune barrier, Ramsay Bay, North Queensland, Australia. *Mar. Geol.* 64:189–202.

Qin, B. and Yu, G., 1998, Implications of lake level variations at 6 ka and 18 ka in mainland Asia. *Global Planet. Change* 18:59–72.

Qiping, P. and Whatley, R., 1990, The biostratigraphical sequence of Mesozoic non-marine ostracod assemblages in northern China. In Whatley, R. and Maybury, C. (eds.), *Ostracoda and Global Events.* Chapman and Hall, New York, pp. 239–250.

Quinlan, R., Smol, J.P., and Hall, R.I., 1998, Quantitative inferences of past hypolimnetic anoxia in south-central Ontario lakes using fossil midges (Diptera: Chirononmidae). *Can. Jour. Fish. Aquat. Sci.* 55:587–596.

Rabette, C. and Lair, N., 1998, Influence des facteurs abiotiques sur la sortie des sédiments de *Cyclops vicinus* et *Chaoborus flavicans* dans les zones sub-littorale et profonde d'un lac tempéré eutrophe. *Ann. Limnol.* 34:295–303.

Racek, A.A., 1974, The waters of Merom: a study of Lake Huleh. *Arch. Hydrobiol.* 74:137–158.

Racek, A.A. and Harrison, F.W., 1974, The systematic and phylogenetic position of *Paleospongilla chubutensis* (Porifera: Spongillidae). *Proc. Linn. Soc. N.S.W.* 99:157–165.

Ragotzkie, R., 1978, Heat budgets of lakes. In Lerman, A. (ed.), *Lakes: Chemistry, Geology, Physics.* Springer-Verlag, New York, pp. 1–20.

Rahel, F.J., 1984, Factors structuring fish assemblages along a bog lake successional gradient. *Ecology* 65:1276–1289.

Ramsay, A.C., 1862, *A Physical Geology and Geography of Great Britain*, 3rd ed. Stanford, London.

Räsänen, M., Salonen, V.P., Salo, J., Walls, M., and Sarvala, J., 1992, Recent history of sedimentation and biotic communities in Lake Pyhäjärvi, SW Finland. *Jour. Paleolim.* 7:107–126.

Rasmussen, E.M., 1968, Atmospheric water vapour transport and the water balance of North America, Part I. *Monthly Weather Rev.* 96:720–734.

Rasmussen, J.B., Godbout, L., and Schallenberg, M., 1989, The humic content of lake water and its relationship to watershed and lake morphometry. *Limnol. Oceanogr.* 34:1336–1343.

Raymo, M., 1997, The timing of major climate terminations. *Paleoceanography* 12:577–585.

Raymo, M.E., Ruddiman, W.F., and Clement, B.M., 1986, Pliocene–Pleistocene paleoceanography of the North Atlantic at Deep Sea Drilling Project site 609. *Init. Rep. Deep Sea Drill. Proj.* 94:890–901.

Rea, D.K. and Colman, S.C., 1995, Radiocarbon ages of pre-bomb clams and the hard-water effect in Lakes Michigan and Huron. *Jour. Paleolim.* 14:89–91.

Rea, D.K., Owen, R.M., and Meyers, P.A., 1981, Sedimentary processes in the Great Lakes. *Rev. Geophys. Space Phys.* 19:635–648.

Redfield, A.C., 1958, The biological control of chemical factors in the environment. *Amer. Sci.* 46:205–222.

Reed, J.M., 1998, Diatom preservation in the recent sediment record of Spanish saline lakes: implications for paleoclimate study. *Jour. Paleolim.* 19:129–137.

Reeves, C.C., 1968, *Introduction to Paleolimnology*. Elsevier, New York.

Reid, I. and Frostick, L.E., 1986, Slope processes, sedimentation derivation and landform evolution in a rift valley basin, northern Kenya. In Frostick, L.E., Renaut, R.W., Reid, I., and Tiercelin, J.J. (eds.), *Sedimentation in the African Rifts*. Geol. Soc. Spec. Publ. 25:99–111.

Reille, M., Andrieu, C., de Beaulieu, J.L., Guenet, P., and Goeury, C., 1998, A long pollen record from Lac du Bouchet, Massif Central, France: for the period ca 325 to 100 ka BP (OIS 9c to OIS 5e). *Quat. Sci. Rev.* 17:1107–1123.

Reinhardt, L. and Ricken, W., 2000, The stratigraphic and geochemical record of Playa Cycles: monitoring a Pangaean monsoon-like system (Triassic, Middle Keuper, S. Germany). *Palaeogeog., Palaeoclim., Palaeoecol.* 161:205–227.

Renaut, R.W., 1993, Zeolitic diagenesis of late Quaternary fluviolacustrine sediments and associated calcrete formation in the Lake Bogoria Basin, Kenya Rift Valley. *Sedimentology* 40:271–301.

Renaut, R.W. and Stead, D., 1994, Last Chance Lake, a natric playa-lake in interior British Columbia. In Gierlowski-Kordesch, E. and Kelts, K. (eds.), *Global Geological Record of Lake Basins, v. 1*. Cambridge University Press, Cambridge, pp. 425–431.

Renaut, R.W. and Tiercelin, J.J., 1994, Lake Bogoria, Kenya Rift Valley—a sedimentological overview. In Renaut, R.W. and Last, W.M. (eds.), *Sedimentology and Geochemistry of Modern and Ancient Saline Lakes*. Soc. Econ. Paleo. Mineral. Spec. Publ. 50:101–123.

Renaut, R.W., Stead, D., and Owen, R.B., 1994, The saline lakes of the Fraser Plateau, British Columbia, Canada. In Gierlowski-Kordesch, E. and Kelts, K. (eds.), *Global Geological Record of Lake Basins, v. 1*. Cambridge University Press, Cambridge, pp. 419–423.

Renberg, I., 1990, A 12,600 year perspective of the acidification of Lilla Öresjön, southwest Sweden. *Phil. Trans. R. Soc. London, Ser. B* 327:357–361.

Renberg, I., Korsman, T., and Anderson, N.J., 1993, A temporal perspective of lake acidification in Sweden. *Ambio* 22:264–271.

Renberg, I., Brännvall, N.L., Bindler, R., and Emteryd, O., 2000, Atmospheric lead pollution history during four millenia (2000 BC to 2000 AD) in Sweden. *Ambio* 29:150–156.

Rettalack, G.J., 1997, Earliest Triassic origin of *Isoetes* and Quillwort evolutionary radiation. *Jour. Paleo.* 71:500–521.

Rhodes, T.E., 1991, A paleolimnological record of anthropogenic disturbances at Holmes Lake, Adirondack Mountains, New York. *Jour. Paleolim.* 5:255–262.

Ricciardi, A. and Reiswig, H.M., 1994, Taxonomy, distribution and ecology of the freshwater bryozoans (Ectoprocta) of eastern Canada. *Can. J. Zool.* 72:339–359.

Rich, F.J., 1989, A review of the taphonomy of plant remains in lacustrine sediments. *Rev. Palaeobot. Palynol.* 58:33–46.

Richardson, J.L., 1968, Diatoms and lake typology in East and Central Africa. *Int. Rev. ges. Hydrobiol.* 59:299–338.

Ricketts, R.D. and Anderson, R.F., 1998, A direct comparison between the historical record of lake level and the $\delta^{18}O$ signal in carbonate sediments from Lake Turkana, Kenya. *Limnol. Oceanogr.* 43:811–822.

Ridge, J.C. and Larsen, F.D., 1990, Re-evaluation of Antev's New England varve chronology and new radiocarbon dates of sediments from glacial Lake Hitchcock. *Geol. Soc. Amer. Bull.* 102:889–899.

Riding, R., 1979, Origin and diagenesis of lacustrine algal bioherms at the margin of the Ries Crater, Upper Miocene, southern Germany. *Sedimentology* 26:645–680.

Rieck, H.J., Sarna-Wojcicki, A.M., Meyer, C.E., and Adam, D.P., 1992, Magnetostratigraphy and tephrochronology of an upper Pliocene to Holocene record in lake sediments at Tulelake, northern California. *Geol. Soc. Amer. Bull.* 104:409–428.

Rieley, G., Collier, R.J., Jones, D.M., Eglinton, G., Eakin, P.A., and Fallick, A.E., 1991, Sources of sedimentary lipids deduced from stable carbon-isotope analyses of individual compounds. *Nature* 352:425–427.

Rietti-Shati, M., Shemesh, A., and Karlen, W., 1998, A 3000 year climatic record from biogenic silica oxygen isotopes in an equatorial high-altitude lake. *Science* 281:980–982.

Rind, D. and Overpeck, J., 1993, Hypothesized causes of decade to century scale climate variability: climate model results. *Quat. Sci. Rev.* 12:357–374.

Rioual, P., Andrieu-Ponel, V., Rietti-Shatti, M., Battarbee, R.W., de Beaulieu, J.L., Cheddadi, R., Reille, M., Svobodova, H., and Shemesh, A., 2001, High-resolution record of climate stability in France during the last interglacial period. *Nature* 413:293–296.

Rippey, B. and Jewson, D.H., 1982, The rates of sediment–water exchange of oxygen and sediment bioturbation in Lough Neagh, N. Ireland. *Hydrobiologia* 92:377–382.

Rittenour, T.M., Brigham-Grette, J., and Mann, M., 2000, El Niño-like climate teleconnections in New England during the Late Pleistocene. *Science* 288:1039–1042.

Riveline, J., 1986, Les Charophytes du Paléogene et du Miocéne inférieur d'Europe Occidentale. *Cah. Paléo.*

Roback, S.S., 1974, Insects (Arthropoda: Insecta). In Hart, C.W. and Fuller, S.L.H. (eds.), *Pollution Ecology of Freshwater Invertebrates.* Academic Press, New York, pp. 313–376.

Robbins, J.A., 1978, Geochemical and geophysical applications of radioactive lead. In Nriagu, J.O. (ed.), *Biogeochemistry of Lead in the Environment.* Elsevier, Amsterdam, pp. 285–393.

Robbins, J.A., 1986, A model for particle selective transport of tracers in sediments with conveyor-belt deposit feeders. *Jour. Geophys. Res.* 91:8542–8558.

Robbins, J.A. and Edgington, D.N., 1975, Determination of recent sedimentation rates in Lake Michigan using ^{210}Pb and ^{137}Cs. *Geochim. Cosmochim Acta* 39:285–304.

Robinson, A.G., Rudat, J.H., Banks, C.J., and Wiles, R.L.F., 1996, Petroleum geology of the Black Sea. *Mar. Petrol. Geol.* 13:195–223.

Rodbell, D.T., Seltzer, G.O., Anderson, D.M., Abbott, M.B., Enfield, D.B., and Newman, J.H., 1999, An ~15,000-year record of El Niño-driven alluviation in Southwestern Ecuador. *Science* 283:516–520.

Rogalev, B., Chernov, V., Korjonen, K., and Jungner, H., 1997, Simultaneous thermoluminescence and optically stimulated luminescence dating of late Pleistocene sediments from Lake Baikal. *Radiat. Meas.* 29:441–444.

Rognerud, S. and Fjeld, E., 1993, Regional survey of heavy metals in lake sediments in Norway. *Ambio* 22:206–212.

Romanek, C.S., Grossman, E.L., and Morse, J.W., 1992, Carbon isotopic fractionation in synthetic aragonite and calcite: effects of temperature and precipitation rate. *Geochim. Cosmochim. Acta* 56:419–430.

Romero, J.R. and Melack, J.M., 1996, Sensitivity of vertical mixing in a large saline lake to variations in runoff. *Limnol. Ocean.* 41:955–965.

Rondot, J., 1994, Recognition of eroded astroblemes. *Earth Sci. Rev.* 35:331–365.

Rood, B.E., Gottgens, J.F., Delfino, J.J., Earle, C.D., and Crisman, T.L., 1995, Mercury accumulation trends in Florida Everglades and savannas marsh flooded soils. *Water, Air, Soil Poll.* 80:981–990.

Rood, B.E., Gottgens, J.F., and Delfino, J.J., 1998, Comparison of mercury accumulation in Lake Eden and the Savannas Marsh wetland, Florida, USA. *Verh. Int. Ver. Limnol.* 26:1365–1369.

Rose, N.L., Appleby, P.G., Boyle, J.F., Mackay, A.W., and Flower, R.J., 1998, The spatial and temporal distribution of fossil-fuel derived pollutants in the sediment record of Lake Baikal, eastern Siberia. *Jour. Paleolim.* 20:151–162.

Rosen, M.R. and Warren, J.K., 1990, The origin and significance of groundwater seepage gypsum from Bristol Dry Lake, California. *Sedimentology* 37:983–996.

Rosenbaum, J.G., Reynolds, R.L., Adam, D.P., Drexler, J., Sarna-Wojcicki, A.M., and Whitney, G.C., 1996, Record of middle Pleistocene climate change from Buck Lake, Cascade Range, southern Oregon—Evidence from sediment magnetism, trace element geochemistry, and pollen. *Geol. Soc. Amer. Bull.* 108:1328–1341.

Rosendahl, B.R., 1987, Architecture of continental rifts with special reference to East Africa. *Ann. Rev. Earth Planet. Sci.* 15:445–503.

Rosendahl, B.R., Reynolds, D.J., Lorber, P.M., Burgess, C.F., McGill, J., Scott, D., Lambiase, J.J., and Derksen, S.J., 1986, Structural expressions of rifting: lessons from Lake Tanganyika, Africa. In Frostick, L.E., Renaut, R.W., Reid, I., and Tiercelin, J.J. (eds.), *Sedimentation in the African Rifts.* Geol. Soc. Lond. Spec. Publ. 25:29–43.

Rosenqvist, I.T., Jørgensen, P., and Ruelåtten, H., 1980, The importance of natural H$^+$ production for acidity in soil water. In Drabløs, D. and Tollan, A. (eds.), *Ecological Impact of Acid Precipitation.* SNSF Project, Oslo, pp. 240–241.

Rosenthal, Y. and Katz, A., 1989, The applicability of trace elements in freshwater shells for paleogeochemical studies. *Chem. Geol.* 78:65–76.

Ross, A.J. and Jarsembowski, E.A., 1993, Arthropoda (Hexapoda: Insecta). In Benton, M.J. (ed.), *The Fossil Record 2.* Chapman and Hall, London, pp. 363–426.

Rossaro, B., 1992, Chironomids and water temperature. *Aquat. Insect.* 13:87–98.

Rossier, O., Castella, E., and Lachvanne, J.B., 1996, Influence of submerged vegetation on size class distribution of perch (*Perca fluviatalis*) and roach (*Rutilus rutilus*) in Lake Geneva. *Aquat. Sci.* 58:1–14.

Rossiter, A. and Kawanabe, H. (eds.), 2000, *Ancient Lakes: Biodiversity, Ecology and Evolution.* Advances in Ecology Research 31. Academic Press, New York.

Roth, J.L. and Dilcher, D.L., 1978, Some considerations in leaf size and leaf-margin analysis of fossil leaves. *Cour. Forsch.-Inst. Senckenberg* 30:165–171.

Rouchy, J.M., Servant, M., Fournier, M., and Causse, C., 1996, Extensive carbonate algal bioherms in upper Pleistocene saline lakes of the central Altiplano of Bolivia. *Sedimentology* 43:973–993.

Round, F.E. and Crawford, R.M., 1981, The lines of evolution of the Bacillariophyta. I. Origin. *Proc. R. Soc. London, Ser. B* 211:237–260.

Royackers, R.M.M., 1986, Development and succession of scale-bearing Chrysophyceae in two shallow freshwater bodies near Nijmegen, The Netherlands. In Kristiansen, J. and Andersen, R.A. (eds.), *Chrysophytes, Aspects and Problems.* Cambridge University Press, Cambridge, pp. 241–258.

Royden, L.H., 1988, Late Cenozoic tectonics of the Pannonian Basin system. In Royden, L.H. and Horváth, F. (eds.), *The Pannonian Basin.* Amer. Assoc. Petrol. Geol. Mem. 45:27–48.

Rozanski, K., Araguás-Araguás, L., and Gonfiantini, R., 1993, Isotopic patterns in modern global precipitation. In Swart, P.K., Lohmann, K.C., McKenzie, J., and Savin, S. (eds.), *Climate Change in Continental Isotopic Records.* Amer. Geophys. Union Geophys. Monogr. 78:1–36.

Ruggiu, D., Morabito, G., Panzani, P., and Pugnetti, A., 1998, Trends and relations among basic phytoplankton characteristics in the course of the long-term oligotrophication of Lake Maggiore (Italy). *Hydrobiologia* 369/370:243–257.

Russell, I.C., 1885, *Geological History of Lake Lahontan, a Quaternary Lake of Northwest Nevada.* U.S. Geol. Surv. Monogr. 11.

Russell, I.C., 1895, *Lakes of North America; a Reading Lesson for Students of Geography and Geology.* Ginn, Boston.

Russell, M. and Gurnis, M., 1994, The planform of epeirogeny: vertical motions of Australia during the Cretaceous. *Basin Res.* 6:63–76.

Russell-Hunter, W.D., 1978, Ecology of freshwater pulmonates. In Fretter, V. and Peake, J. (eds.), *The Pulmonates v. 2a: Systematics, Evolution and Ecology.* Academic Press, New York, pp. 335–383.

Rust, B.R. and Romanelli, R., 1975, Late Quaternary subaqueous outwash deposits near Ottawa, Canada. In Jopling, A.V. and McDonald, B.C. (eds.), *Glaciofluvial and Glaciolacustrine Sedimentation.* Soc. Econ. Paleo. Mineral. Spec. Publ. 23:177–192.

Rutter, N. and Blackwell, B., 1995, Amino acid racemization dating. In Rutter, N.W. and Catto, N.R. (eds.), *Dating Methods for Quaternary Deposits.* GEOTEXT 2, Geological Society of Canada, pp. 125–164.

Rutter, N.W. and Catto, N.R. (eds.), 1995, *Dating Methods for Quaternary Deposits.* GEOTEXT 2, Geological Society of Canada.

Ruttner, F., 1952, Plankton studien der Deutschen Limnologisichen Sunda Expedition. *Arch. Hydrobiol. Suppl.* 21:1–274.

Saarnisto, M., 1986, Annually laminated lake sediments. In Berglund, B.E. (ed.), *Handbook of Holocene Paleoecology and Paleohydrology.* J. Wiley and Sons, New York, pp. 343–370.

Sabater, S. and Haworth, E.Y., 1995, An assessment of recent trophic changes in Windemere South Basin (England) based on diatom remains and fossil pigments. *Jour. Paleolim.* 14:151–163.

Sachs, H.M., Webb, T., and Clark, D.R., 1977, Paleoecological transfer functions. *Ann. Rev. Earth Planet. Sci.* 5:159–178.

Sack, D., 1995, The shoreline preservation index as a relative-age dating tool for Late Pleistocene shorelines: an example from the Bonneville Basin, U.S.A. *Earth Surf. Proc. Landforms* 20:363–377.

Sackett, W.M., Eadie, B.J., and Meyers, P.A., 1986, Stable carbon isotope studies of organic matter in Great Lakes sediments. *Trans. AGU* 76:1058.

Sadler, P.M., 1981, Sediment accumulation rates and the completeness of stratigraphic sections. *Jour. Geology* 89:569–584.

Sadolin, M., Pedersen, G.K., and Pedersen, S.A.S., 1997, Lacustrine sedimentation and tectonics: an example from the Weichselian at Lønstrup Klint, Denmark. *Boreas* 26:113–126.

Saether, O.A., 1975, Nearctic chironomids as indicators of lake typology. *Verh. Int. Ver. Limnol.* 19:3127–3133.

Saether, O.A., 1979, Chironomid communities as water quality indicators. *Holarctic Ecol.* 2:65–74.

Saether, O.A., 1980, The influence of eutrophication on deep lake benthic invertebrate communities. *Prog. Water Tech.* 12:161–180.

Saggio, A. and Imberger, J., 1998, Internal wave weather in a stratified lake. *Limnol. Oceanogr.* 43:1780–1795.

Salvany, J.M. and Orti, F., 1994, Miocene glauberite deposits of Alcanadre, Ebro Basin, Spain: sedimentary and diagenetic processes. In Renaut, R.W. and Last, W.M. (eds.), *Sedimentology and Geochemistry of Modern and Ancient Saline Lakes.* SEPM Spec. Publ. 50:203–215.

Salvany, J.M., Muñoz, A., and Pérez, A., 1994, Nonmarine evaporitic sedimentation and associated diagenetic processes of the southwestern margin of the Ebro Basin (Lower Miocene), Spain. *Jour. Sed. Res.* 64:190–203.

Sandberg, P.A., 1964, The ostracode genus *Cyprideis* in the Americas. *Stockholm Contr. Geol.* 12.

Sandberg, P.A., 1978, New interpretations of Great Salt Lake ooids and nonskeletal carbonate mineralogy. *Sedimentology* 22:137–145.

Sandgren, C.D., 1988, The ecology of chrysophyte flagellates: their growth and perennation strategies as freshwater phytoplankton. In Sandgren, C.D. (ed.), *Growth and Reproductive Strategies of Freshwater Phytoplankton.* Cambridge University Press, Cambridge, pp. 9–104.

Sandman, O., Lichu, A., and Simola, H., 1990, Drainage ditch erosion history as recorded in the varved sediment of a small lake in East Finland. *Jour. Paleolim.* 3:161–169.

Sanford, P.R., 1993, *Bosmina longirostris* antennule morphology as an indicator of intensity of planktivory by fish. *Bull. Mar. Sci.* 53:216–227.

Sanger, J.E., 1988, Fossil pigments in paleoecology and paleolimnology. *Palaeogeogr., Palaeoclim., Palaeoecol.* 62:343–359.

Sanger, J.E. and Gorham, E., 1972, Stratigraphy of fossil pigments as a guide to the postglacial history of Kirchner Marsh, Minnesota. *Limnol. Oceanogr.* 17:840–854.

Sarmaja-Korjonen, K. and Alhonen, P., 1999, Cladoceran and diatom evidence of lake-level fluctuations from a Finnish lake and the effect of aquatic moss layers on microfossil assemblages. *Jour. Paleolim.* 22:277–290.

Sarna-Wojcicki, A.M. and 9 others, 1984, Chemical analysis, correlations, and ages of upper Pliocene and Pleistocene ash layers of east-central and southern California. *U.S. Geol. Surv. Prof. Pap.* 1293.

Sarna-Wojcicki, A.M., Meyer, C.E., Adam, D.P., and Sims, J.D., 1988, Correlations and age estimates of ash beds in late Quaternary sediments of Clear Lake, California. In Sims, J.D. (ed.), *Late Quaternary Climate, Tectonism and Sedimentation in Clear Lake.* Geol. Soc. Amer. Spec. Pap. 214:141–150.

Sarna-Wojcicki, A.M., Meyer, C.E., and Wan, E., 1997, Age and correlation of tephra layers, position of the Matuyama–Brunhes chron boundary, and effects of Bishop ash eruption on Owens Lake, as determined from drill hole OL-92, southeast California. In Smith, G.I. and Bischoff, J.L. (eds.), *An 800,000 Year Paleoclimate Record from Core OL-2, Owens Lake, Southeast California.* Geol. Soc. Amer. Spec. Pap. 317.

Sauer, P.E., Miller, G.H., and Overpeck, J.T., 2001, Oxygen isotope ratios of organic matter in arctic lakes as a paleoclimate proxy: field and laboratory investigations. *Jour. Paleolim.* 25:43–64.

Schafran, G.C. and Driscoll, C.T., 1987, Spatial and temporal variations in aluminum chemistry of a dilute acidic lake. *Biogeochemistry* 3:105–120.

Schaller, T., Moor, H.C., and Wehrli, B., 1997, Sedimentary profiles of Fe, Mn, V, Cr, As and Mo as indicators of benthic redox conditions in Baldeggersee. *Aquat. Sci.* 59:345–361.

Scharf, B.W., 1998, Eutrophication of Lake Arendsee (Germany). *Palaeogeog., Palaeoclim., Palaeoecol.* 140:85–96.

Schell, W.R., 1977, Concentrations, physicochemical states and mean residence times of ^{210}Pb and ^{210}Po in marine and estuarine waters. *Geochim. Cosmochim. Acta* 39:285–304.

Schell, W.R., Jokela, T., and Eagle, R., 1973, Natural ^{210}Pb and ^{210}Po in a marine environment. In *Radioactive Contamination of the Marine Environment.* International Atomic Energy Agency, Vienna, pp. 701–724.

Schelske, C.L., 1988, Historic trends in Lake Michigan silica concentrations. *Int. Rev. ges. Hydrobiol.* 73:559–591.

Schelske, C.L., 1991, Historic nutrient enrichment of Lake Ontario: paleolimnological evidence. *Can. Jour. Fish. Aquat. Sci.* 48:1529–1538.

Schelske, C.L. and Hodell, D.A., 1991, Recent changes in productivity and climate of Lake Ontario detected by isotopic analysis of sediments. *Limnol. Oceanogr.* 36:961–975.

Schelske, C.L. and Stoermer, E.F., 1971, Eutrophication, silica and predicted changes in algal quality in Lake Michigan. *Science* 173:423–424.

Schelske, C.L. and Stoermer, E.F., 1972, Phosphorus, silica, and eutrophication of Lake Michigan. In Likens, G.E. (ed.), *Nutrients and Eutrophication, ASLO Special Symp.* Allen Press, Lawrence, KA, pp. 157–171.

Schelske, C.L., Conley, D.J., Stoermer, E.F., Newberry, T.L., and Campbell, C.D., 1986, Biogenic silica and phosphorus accumulation in sediments as indices of eutrophication in the

Laurentian Great Lakes. *Hydrobiologia* 143:79–86.

Schelske, C.L., Peplow, A., Brenner, M., and Spencer, C.N., 1994, Low background gamma counting: applications for ^{210}Pb dating of sediments. *Jour. Paleolim.* 10:115–128.

Schimmelmann, A., Miller, R.F., and Leavitt, S.W., 1993, Hydrogen isotopic exchange and stable isotope ratios in cellulose, wood, chitin, and amino compounds. In Swart, P.K., Lohmann, K.C., McKenzie, J., and Savin, S. (eds.), *Climate Change in Continental Isotopic Records.* Amer. Geophys. Union Geophys. Monogr. 78:367–374.

Schindler, D.E., Carpenter, S.R., Cole, J.J., Kitchell, J.F., and Pace, M.L., 1997, Influence of food web structure on carbon exchange between lakes and the atmosphere. *Science* 277:248–251.

Schindler, D.W., 1974, Eutrophication and recovery in experimental lakes: implications for lake management. *Science* 184:897–899.

Schindler, D.W., 1977, Evolution of phosphorus limitation in lakes. *Science* 195:260–262.

Schindler, D.W., 1985, The coupling of elemental cycles by organisms: evidence from whole-lake chemical perturbations. In Stumm, W. (ed.), *Chemical Processes in Lakes.* J. Wiley and Sons, New York, pp. 225–250.

Schindler, D.W., 1990, Experimental perturbations of whole lakes as tests of hypotheses concerning ecosystem structure and function. *Oikos* 57:25–41.

Schindler, D.W., Mills, K.H., Malley, D.F., Findlay, D.L., Shearer, J.A., Davies, I.J., Turner, M.A., Linsey, G.A., and Cruikshank, D.R., 1985, Long-term ecosystem stress: the effect of years of experimental acidification on a small lake. *Science* 228:1395–1401.

Schindler, D.W., Bayley, S.E., Parker, B.R., Beaty, K.G., Cruikshank, D.R., Fee, E.J., Schindler, E.U., and Stainton, M.P., 1996, The effect of climatic warming on the properties of boreal lakes and streams at the Experimental Lakes Area, northwestern Ontario. *Limnol. Oceanogr.* 41:1004–1017.

Schlische, R., 1991, Half-graben basin filling models: new constraints on continental extensional basin development. *Basin Res.* 3:123–141.

Schlische, R. and Olsen, P., 1990, Quantitative filling models for continental extensional basins with applications to early Mesozoic rifts of Eastern North America. *Jour. Geol.* 98:135–155.

Schlüchter, C., 1987, Talgenese im Quartär -eine Standortbestimmung. *Geograph. Helv.* 42:109–115.

Schlüter, T., Kohring, R., and Mehl, J., 1992, Hyperostotic fish bones ("Tilly bones") from presumably Pliocene phosphorites of the Lake Manyara area, northern Tanzania. *Paläont. Z.* 66:129–136.

Schmäh, A., 1993, Variation among fossil chironomid assemblages in surficial sediments of Bodensee-Untersee (SW-Germany); implications for paleolimnological interpretation. *Jour. Paleolim.* 9:99–108.

Schmid, H., 1987, Turkey's Salda Lake: a genetic model for Australia's newly discovered magnesite deposits. *Indust. Min.* August, 19–29.

Schmidt, W., 1997, Geomorphology and physiography of Florida. In Randazzo, A.F. and Jones, D.S. (eds.), *The Geology of Florida.* University Press of Florida, Gainesville, FL, pp. 1–12.

Schmude, K.L., Jennings, M.J., Otis, K.J., and Piette, R.R., 1998, Effects of habitat complexity on macroinvertebrate colonization of artificial substrates in north temperate lakes. *Jour. North Amer. Benth. Soc.* 17:73–80.

Schneider, J., Muller, J., and Sturm, M., 1986, Geology and sedimentary history of Lake Traunsee (Salzkammergut, Austria). *Hydrobiologia* 143:227–232.

Scholz, C.A., 1995, Deltas of the Lake Malawi Rift, East Africa: seismic expression and exploration implications. *Amer. Assoc. Petrol. Geol. Bull.* 79:1679–1697.

Scholz, C.A. and Rosendahl, B.R., 1988, Low lake stands in Lakes Malawi and Tanganyika, east Africa, delineated with multifold seismic data. *Science* 240:1645–1648.

Scholz, C.A. and Rosendahl, B.R., 1990, Coarse-clastic facies and stratigraphic sequence models from Lakes Malawi and Tanganyika, East Africa. In Katz, B.J. (ed.), *Lacustrine Basin Exploration: Case Studies and Modern Analogues.* Amer. Assoc. Petrol. Geol. Mem. 50:151–168.

Scholz, C.A., Johnson, T.C., and McGill, J.W., 1993, Deltaic sedimentation in a rift valley lake: new seismic reflection data from Lake Malawi (Nyasa), East Africa. *Geology* 21:395–399.

Scholz, C.A., Moore, T.C., Hutchinson, D.R., Golmshtok, A.J., Klitgord, K., and Kurotchkin, A.G., 1998, Comparative sequence stratigraphy of low-latitude versus high-latitude lacustrine rift basins: seismic data examples from the East African and Baikal rifts. *Palaeogeog., Palaeoclim., Palaeoecol.* 140:401–420.

Schoonmaker, P.K., 1998. Paleoecological perspectives on ecological scale. In Peterson, D.L. and Parker, V.T. (eds.), *Ecological Scale.* Columbia University Press, New York, pp. 79–103.

Schram, F.R., 1981, Late Paleozoic crustacean communities. *Jour. Paleo.* 55:126–137.

Schram, F.R., 1984, Fossil Syncarida. *Trans. San Diego Soc. Nat. Hist.* 20:301–312.

Schubel, K.A. and Lowenstein, T.K., 1997, Criteria for the recognition of shallow-perennial-saline lake halites based on recent sediments from the Qaidam Basin, Western China. *Jour. Sed. Res.* 67:74–87.

Schudack, M.E., 1995, Neueu mikorpaläontologische Beiträge (Ostracoda, Charophyta) zum Morrison-Ökosystem. *Berliner geowiss. Abh.* E16:389–407.

Schudack, M.E., 1996, Ostracode and charophyte biogeography in the continental Upper Jurassic of Europe and North America as influenced by plate tectonics and paleoclimate. In Morales, M. (ed.), *The Continental Jurassic.* Mus. Northern Arizona Bull. 60:333–342.

Schudack, M.E., 1998, Biostratigraphy, paleoecology and biogeography of charophytes and ostracodes from the Upper Jurassic Morrison Formation, Western Interior, USA. *Modern Geol.* 22:379–414.

Schudack, M.E., 1999, Ostracoda (marine/ nonmarine) and paleoclimate history in the Upper Jurassic of Central Europe and North America. *Marine Micropaleo.* 37:273–288.

Schulling, R.D., 1977, Source and composition of lake sediments. In Golterman, H.L. (ed.), *Interactions Between Sediments and Freshwater.* Dr. W Junk, The Hague, pp. 12–18.

Schultze, H.P., 1993, Osteichthyes: Sarcopterygii. In Benton, M.J. (ed.), *The Fossil Record 2.* Chapman and Hall, London, pp. 657–664.

Schultze, H.P. and Cloutier, R., 1996, Comparison of the Escuminac Formation ichthyofauna with other late Givetian/early Frasnian ichthyofaunas. In Schultze, H.P. and Cloutier, R. (eds.), *Devonian Fishes and Plants of Miguasha, Quebec, Canada.* Verlag Dr. Friedrich Pfeil, Munich, pp. 348–368.

Schütt, B., 1998, Reconstruction of Holocene paleoenvironments in the endorheic basin of laguna de Gallocanta, Central Spain by investigation of mineralogical and geochemical characters from lacustrine sediments. *Jour. Paleolim.* 20:217–234.

Schwalb, A. and Dean, W.E., 1998, Stable isotopes and sediments from Pickerel Lake, South Dakota, USA: a 12ky record of environmental change. *Jour. Paleolim.* 20:15–30.

Schwalb, A., Hadorn, P., Thew, N., and Straub, F., 1998, Evidence for Late Glacial and Holocene environmental changes from subfossil assemblages in sediments of Lake Neuchâtel, Switzerland. *Palaeogeog., Palaeoclim., Palaeoecol.* 140:307–323.

Schwarcz, H.P., 1989, Uranium series dating. *Quat. Intl.* 1:7–17.

Scott, A.C., 1978, Sedimentological and ecological control of Westphalian B plant assemblages from West Yorkshire. *Proc. Yorkshire Geol. Soc.* 41:461–508.

Scott, A.C., 1979, The ecology of Coal Measure floras from northern Britain. *Proc. Geol. Assoc.* 90:97–116.

Sculthorpe, C.D., 1967, *The Biology of Aquatic Vascular Plants.* St. Martins Press, New York.

Sedgwick, A. and Murchison, R.I., 1829, On the old conglomerates and other secondary deposits of the north coast of Scotland. *Proc. Geol. Soc. London* 1:1–77.

Seitzinger, S.P., 1988, Denitrification in freshwater and coastal marine ecosystems: ecological and geochemical significance. *Limnol. Oceanogr.* 33:702–724.

Self, S. and Sparks, R.S.J., 1981, *Tephra Studies.* NATO Advanced Studies Instistute, Series C. Reidel, Dordrecht.

Seltzer, G.O., Rodbell, D.T., and Abott, M., 1995. Andean glacial lakes and climate variability since the last glacial maximum. *Bull. Inst. fr. études andines.* 24:539–549.

Seltzer, G., Rodbell, D., and Burns, S., 2000, Isotopic evidence for Late Quaternary climatic change in tropical South America. *Geology* 28:35–38.

Sen, D.P. and Banerji, T., 1991, Permo-Carboniferous proglacial lake sedimentation in the Sahajuri Gondwana basin, India. *Sed. Geol.* 71:47–58.

Sennikov, A.G., 1996, Evolution of the Permian and Triassic tetrapod communities of Eastern Europe. *Palaeogeog., Palaeoclim., Palaeoecol.* 120:331–351.

Servant, M. and Fontes, J.C., 1978, Les lacs quaternaires des hautes plateaux des Andes Boliviennes. Premierè interprétations paléoclimatiques. *Cah. ORSTOM Sér. Géol.* 10:9–23.

Shafer, M.M. and Armstrong, D.E., 1988, Biogenic silica production, deposition, and dissolution in Lake Michigan. *EOS* 69:1143.

Shapiro, J. and Swain, E.B., 1983, Lessons from the silica "decline" in Lake Michigan. *Science* 221:457–459.

Sharpe, S.E., Forester, R.M., Whelan, J.F., and Moscati, R.J., 1998, How accurately does gastropod shell delta ^{18}O record hydrologic balance? *Geol. Soc. Amer. Ann. Mtg. Abstr. w/ prog.* 30:164.

Shaw, J., 1975, Sedimentary successions in Pleistocene ice-marginal lakes. In Jopling, A.V. and McDonald, B.C. (eds.), *Glaciofluvial and Glaciolacustrine Sedimentation.* Soc. Econ. Paleo. Mineral. Spec. Publ. 23:281–303.

Shaw, J., Munro-Stasiuk, M., Sawyer, B., Beaney, C., Lesemann, J.E., Musacchio, A., Rains, B., and Young, R.R., 1999, The Channeled Scablands: back to Bretz? *Geology* 27:605–608.

Shearman, D.J., McGugan, A., Stein, C., and Smith, A.J., 1989, Ikaite, $CaCO_3 \cdot H_2O$, precursor to thinolites in the Quaternary tufas and tufa mounds of the Lahontan and Mono Lake basins, western United States. *Geol. Soc. Amer. Bull.* 101:913–917.

Shemesh, A., Burkle, L.H., and Froelich, P.N., 1989, Dissolution and preservation of Antarctic diatoms and the effect of sediment thanatocoenoses. *Quat. Res.* 31:288–308.

Sheppard, R.A. and Gude, A.J., 1968, Distribution and genesis of authigenic silicate minerals in tuffs of Pleistocene Lake Tecopa, Inyo County California. *U.S. Geol. Surv. Prof. Pap.* 597.

Sheriff, R.E. and Geldart, L.P., 1995, *Exploration Seismology.* Cambridge University Press, Cambridge.

Sher-Kaul, S., Oertli, B., Castella, E., and Lachavanne, J.B., 1995, Relationship between biomass and surface area of six submerged aquatic plant species. *Aquat. Bot.* 51:147–154.

Sherwood-Pike, M.A., 1988, Freshwater fungi: fossil record and paleoecological potential. In Gray, J. (ed.), *Paleolimnology: Aspects of Freshwater Paleoecology and Biogeography.* Elsevier, Amsterdam, pp. 271–285.

Sholkovitz, E.R., 1985, Redox-related geochemistry in lakes: alkali metals, alkaline earth elements and ^{137}Cs. In Stumm, W. (ed.), *Chemical Processes in Lakes.* J. Wiley and Sons, New York, pp. 119–142.

Shukla, J., Kinter, J.L., Schneider, E.K., and Straus, D.M., 1999, Modeling of the climate system. In Martens, P. and Rotmans, J. (eds.), *Climate Change: An Integrated Perspective.* Kluwer, Dordrecht, pp. 51–104.

Sideleva, V.G., 2000, The Ichthyofauna of Lake Baikal. In Rossiter, A. and Kawanabe, H. (eds.), *Ancient Lakes: Biodiversity, Ecology and Evolution.* Advances in Ecological Research 31. Academic Press, New York, pp. 8–96.

Siegenthaler, U. and Sarmiento, J.L., 1993, Atmospheric carbon dioxide and the ocean. *Nature* 365:119–125.

Silantiev, V.V., 1998, New data on the Upper Permian non-marine bivalve *Paleomutela* in European Russia. In Johnston, P.A. and Haggart, J.W. (eds.), *Bivalves; An Eon of Evolution; Paleobiological Studies Honoring Norman D. Newell.* University of Calgary Press, Calgary, pp. 437–442.

Simcik, M.F., Eisenreich, S.J., Golden, K.A., Liu, S., Lipiatou, E., Swackhamer, D., and Long, D.T., 1996, Atmospheric loading of polycyclic aromatic hydrocarbons to Lake Michigan as recorded in the sediments. *Envir. Sci. Tech.* 30:3039–3046.

Sims, J.D. and White, D.E., 1981, Mercury in the sediments of Clear Lake. *U.S. Geol. Surv. Prof. Pap.* 1141:237–241.

Sippel, S.J., Hamilton, S.K., and Melack, J.M., 1992, Innundation area and morphometry of lakes on the Amazon River floodplain, Brazil. *Arch. Hydrobiol.* 123:385–400.

Siver, P.A., 1991, Implications for improving paleolimnological inference models utilizing scale-bearing siliceous algae: transforming scale counts to cell counts. *Jour. Paleolim.* 5:219–225.

Skelton, P.W. and Benton, M.J., 1993, Mollusca: Rostroconchia, Scaphopoda and Bivalvia. In Benton, M.J. (ed.), *The Fossil Record 2.* Chapman and Hall, London, pp. 237–264.

Sladen, C.P., 1994, Key elements during the search for hydrocarbons in lake systems. In Gierlowski-Kordesch, E. and Kelts, K. (eds.), *Global Geological Record of Lake Basins.* Cambridge University Press, New York, pp. 3–17.

Sloan, L.C., 1994, Equable climates during the early Eocene: significance of regional paleogeography for North American climate. *Geology* 22:881–884.

Sloan, L.C. and Morrill, C., 1998, Orbital forcing and Eocene continental temperatures. *Palaeogeog., Palaeoclim., Palaeoecol.* 144:21–35.

Sly, P.G., 1978, Sedimentary processes in lakes. In Lerman, A. (ed.), *Lakes: Chemistry, Geology, Physics.* Springer-Verlag, New York, pp. 65–89.

Smith, A.J., 1993a, Lacustrine ostracodes as hydrochemical indicators in lakes of the north-central United States. *Jour. Paleolim.* 8:121–134.

Smith, A.J., 1993b, Lacustrine ostracod diversity and hydrochemistry in lakes of the northern Midwest of the United States. In McKenzie, K.G. and Jones, P.J. (eds.), *Ostracoda in the Earth and Life Sciences.* Balkema, Rotterdam, pp. 493–500.

Smith, A.J., Delorme, L.D., and Forester, R.M., 1992, A lake's solute history from ostracodes: comparison of methods. In Kharaka, Y.K. and Maest, A.S. (eds.), *Water Rock Interactions.* Balkema, Rotterdam, pp. 677–680.

Smith, D.G., 1983, Anastomosed fluvial deposits: modern examples from Western Canada. In Collinson, J.D. and Lewin, J. (eds.), *Modern and Ancient Fluvial Systems.* Int. Assoc. Sediment. Spec. Publ. 6:155–168.

Smith, D.G., 1994, Glacial Lake McConnell; paleogeography, age, duration, and associated river deltas, Mackenzie River Basin, Western Canada. *Quat. Sci. Rev.* 13:829–843.

Smith, D.G. and Jol, H.M., 1992, Ground-penetrating radar investigation of a Lake Bonneville delta, Provo level, Brigham City, Utah. *Geology* 20:1083–1086.

Smith, D.G. and Jol, H.M., 1997, Radar structure of a Gilbert-type delta, Peyto Lake, Banff National Park, Canada. *Sediment. Geol.* 113:195–209.

Smith, D.M., 2000, Insect taphonomy in a recent ephemeral lake, southeastern Arizona. *Palaios* 15:152–160.

Smith, D.R. and Flegal, A.R., 1995, Lead in the biosphere: recent trends. *Ambio* 24:21–23.

Smith, G.A., 1988, Sedimentology of proximal to distal volcanoclastics dispersed across an active foldbelt: Ellensburg Formation (late Miocene), central Washington. *Sedimentology* 35:953–977.

Smith, G.A., 1993, Missoula flood dynamics and magnitudes inferred from sedimentology of slack-water deposits on the Columbia Plateau, Washington. *Geol. Soc. Amer. Bull.* 105:77–100.

Smith, G.I., Barczak, V.J., Moulton, G.F., and Liddicoat, J.C., 1980, Three million year record of climate in Searles Lake sediments. *U.S. Geol. Surv. Prof. Pap.*

Smith, G.L., 1997, Late Quaternary climates and limnology of the Lake Winnebago basin, Wisconsin, based on ostracodes. *Jour. Paleolim.* 18:249–260.

Smith, G.R., 1975, Fishes of the Plio-Pleistocene Glenns Ferry Formation, southwest Idaho. *Pap. Paleont. Mus. Paleo. Univ. Mich.* 14:1–68.

Smith, G.R., 1978, Biogeography of intermountain fishes. In Harper, K.T. and Reveal, J.L. (eds.), *Intermountain Biogeography: A Symposium.* Great Basin Mem. 2:17–42.

Smith, G.R., 1981, Late Cenozoic freshwater fishes of North America. *Ann. Rev. Ecol. Syst.* 12:163–193.

Smith, G.R., 1987, Fish speciation in a western North American Pliocene rift lake. *Palaios* 2:436–445.

Smith, G.R. and Patterson, W.P., 1994, Mio-Pliocene seasonality on the Snake River plain: comparison of faunal and oxygen isotope evidence. *Palaeogeog., Palaeoclim., Palaeoecol.* 107:291–302.

Smith, M.J., Pellatt, M.G., Walker, I.R., and Mathewes, R.W., 1998, Postglacial changes in chironomid communities and inferred climate near treeline at Mount Stoyoma, Cascade Mountains, southwestern British Columbia. *Jour. Paleolim.* 20:277–293.

Smith, N.D., 1990, The effects of glacial surging on sedimentation in a modern ice-contact lake,

Alaska. *Geol. Soc. Amer. Bull.* 102:1393–1403.

Smith, N.D. and Ashley, G., 1985, Proglacial lacustrine environment. In Ashley, G., Shaw, J., and Smith, N.D. (eds.), *Glacial Sedimentary Environments.* Soc. Econ. Paleo. Mineral. Short Course No. 16:135–216.

Smith, N.D. and Syvitski, J.P.M., 1982, Sedimentation in a glacier-fed lake: the role of pelletization on deposition of fine-grained suspensates. *Jour. Sed. Petrol.* 52:503–513.

Smith, N.D., Vendl, M.A., and Kenedy, S.K., 1982, Comparison of sedimentation regimes in four glacier-fed lakes of western Alberta. In Davidson-Arnott, R., Nickling, W., and Fahey, B.D. (eds.), *Proc. 6th Guelph Symp. on Geomorphology*, pp. 203–238.

Smith, R.C.M., 1991, Post-eruption sedimentation on the margin of a caldera lake, Taupo Volcanic Centre, New Zealand. *Sediment. Geol.* 74:89–138.

Smol, J.P., 1988, Chrysophycean microfossils in paleolimnological studies. In Gray, J. (ed.), *Paleolimnology: Aspects of Freshwater Paleoecology and Biogeography.* Elsevier, Amsterdam, pp. 287–297.

Smol, J.P., 1990, Paleolimnology—recent advances and future challenges. *Mem. Ist. Ital. Idriobiol.* 47:253–276.

Smol, J.P., 1995a, Application of Chrysophytes to problems in paleoecology. In Sandgren, C.D., Smol, J.P., and Kristiansen, J. (eds.), *Chrysophyte Algae: Ecology, Phylogeny and Development.* Cambridge University Press, Cambridge, pp. 303–329.

Smol, J.P., 1995b, Paleolimnological approaches to the evaluation and monitoring of ecosystem health: providing a history for environmental damage and recovery. In Rapport, D.J., Gaudet, C.L., and Calow, P. (eds.), *Evaluating and Monitoring the Health of Large-Scale Ecosystems.* NATO ASI Ser. I 28. Springer-Verlag, Berlin, pp. 301–318.

Smol, J.P. and Dixit, S.S., 1990, Patterns of pH change inferred from chrysophyte microfossils in Adirondack and northern New England lakes. *Jour. Paleolim.* 4:31–41.

Smol, J.P. and Glew, J.R., 1992, Paleolimnology. *Encyclopedia of Earth Science* 3:551–564.

Smol, J.P., Battarbee, R.W., Davis, R.B., and Meriläinen, J. (eds.), 1986, *Diatoms and Lake Acidity.* Dr. W Junk, Dordrecht.

Smoot, J.P., 1978, Origin of the carbonate sediments in the Wilkins Peak Member of the Green River Formation (Eocene) Wyoming, USA. In Matter, A. and Tucker, M.E. (eds.), *Modern and Ancient Lake Sediments.* Intl. Assoc. Sediment. Spec. Publ. 2:109–127.

Smoot, J.P. and Castens-Seidell, B., 1994, Sedimentary features produced by efflorescent salt crusts, Saline Valley and Death Valley, California. In Renaut, R.W. and Last, W.M. (eds.), *Sedimentology and Geochemistry of Modern and Ancient Saline Lakes.* SEPM Spec. Publ. 50:73–90.

Snowball, I.F., 1993, Geochemical control of magnetite dissolution in subarcti lake sediments and the implications for environmental magnetism. *Jour. Quat. Sci.* 8:339–346.

Sohn, I.G. and Rocha-Campos, A.C., 1990, Late Paleozoic (Gondwanan) ostracodes in the Corumbatai Fm., Parana Basin, Brazil. *Jour. Paleo.* 64:116–128.

Solem, J.O. and Birks, H.H., 2000, Late-glacial and early Holocene Trichoptera (Insecta) from Kråkenes Lake, western Norway. *Jour. Paleolim.* 23:49–56.

Solem, J.O., Solem, T., Aagaard, K., and Hanssen, O., 1997, Colonization and evolution of lakes on the central Norwegian coast following deglaciation and land uplift 9500 to 7800 years B.P. *Jour. Paleolim.* 18:269–281.

Solhøy, I.W. and Solhøy, T., 2000, The fossil oribatid mite fauna (Acari: Oribatida) in late glacial and early Holocene sediments in Kråkenes Lake, western Norway. *Jour. Paleolim.* 23:35–47.

Solhøy, T., 2001, Oribatid mites. In Smol, J.P., Birks, H.J.B., and Last, W.M. (eds.), *Tracking Environmental Change Using Lake Sediments. Vol. 4. Zoological Indicators.* Kluwer Academic Publishers, Dordrecht, pp. 81–104.

Sommaruga-Wögrath, S., Kolnig, K.A., Schmidt, R., Sommaruga, R., Tessadri, R., and Psenner, R., 1997, Temperature effects on the acidity of remote alpine lakes. *Nature* 387:64–67.

Sommer, U., 1990, The role of competition for resources in phytoplankton succession. In Sommer, U. (ed.), *Plankton Ecology: Succession in Plankton Communities.* Springer-Verlag, New York, pp. 57–106.

Søndegaard, M., Hansen, B., and Markager, S., 1995, Dynamics of dissolved organic carbon lability in a eutrophic lake. *Limnol. Oceanogr.* 40:46–54.

Soreghan, M.J., and Cohen, A.S., 1993, The effects of basin asymmetry on sand composition: Examples from Lake Tanganyika, Africa. In Johnsson, M.J. and Basu, A. (eds.) *Processes Controlling the Composition of Clastic Sediments. Geol. Soc. Amer.* Spec. Paper 284:285-301.

Soreghan, M.J. and Cohen, A.S., 1996, Textural and compositional variability across littoral segments of Lake Tanganyika: the effect of asymmetric basin structure on sedimentation in large rift lakes. *Amer. Assoc. Petrol. Geol. Bull.* 80:382–409.

Soreghan, M.J., Scholz, C.A., and Wells, J.T., 1999, Coarse-grained, deep water sedimentation along a border fault margin of Lake Malawi, Africa: seismic stratigraphic analysis. *Jour. Sed. Res.* 69:832–846.

Sorrano, P.A., Carpenter, S.R., and Elser, M.M., 1993, Zooplankton community dynamics. In Carpenter, S. and Kitchell, J.F. (eds.), *The Trophic Cascade in Lakes.* Cambridge University Press, Cambridge, pp. 116–152.

Sorvari, S., Jorhola, A., and Thompson, R., 2002, Lake diatom response to recent Arctic warming in Finnish Lapland. *Global Change Biol.* in press.

Soulié-Märsche, I., 1979, Origine et évolution des genres actuels des Characeae. *Bull. Centr. Rech. Explor.-Prod. Elf-Acquitaine* 3:821–831.

Soulié-Märsche, I., 1991, Charophytes as lacustrine biomarkers during the Quaternary in North Africa. *Jour. Afr. Earth Sci.* 12:341–351.

Soulié-Märsche, I., 1994, The paleoecological implications of the charophyte flora of the Trinity Division, Junction, Texas. *Jour. Paleo.* 68:1145–1157.

Spence, D.H.N., 1982, The zonation of plants in freshwater lakes. *Adv. Ecol. Res.* 12:37–125.

Spencer, J.E. and Patchett, P.J., 1997, Sr isotope evidence for a lacustrine origin for the upper Miocene to Pliocene Bouse Formation, lower Colorado River trough, and implications for timing of Colorado Plateau uplift. *Geol. Soc. Amer. Bull.* 109:767–778.

Spencer, J.W., 1890, Origin of the basins of the Great Lakes of America. *Am. Geol.* 15:273–295.

Spencer, R.J., Baedecker, M.J., Eugster, H.P., Forester, R.M., Goldhaber, M., Jones, B.F., Kelts, K., McKenzie, J., Madsen, D.B., Rettig, S.L., Rubin, M., and Bowser, C.J., 1984, Great Salt Lake and precursors, Utah, during the last 30,000 years. *Contr. Mineral. Petrol.* 86:321–334.

Spicer, R.A., 1981, The sorting and deposition of allochthonous plant material in a modern environment at Silwood Lake, Silwood Park, Berkshire, England. *U.S. Geol. Surv. Prof. Pap.* 1143:1-77.

Spigel, R.H. and Coulter, G.W., 1996, Comparison of hydrology and physical limnology of the East African Great Lakes: Tanganyika, Malawi, Victoria, Kivu and Turkana (with reference to some North American Great Lakes). In Johnson, T.C. and Odada, E.O. (eds.), *The Limnology, Climatology and Paleoclimatology of the East African Lakes.* Gordon and Breach, Amsterdam, pp. 103–140.

Spiker, E.C. and Hatcher, P.G., 1984, Carbon isotope fractionation of sapropelic organic matter during early diagenesis. *Org. Geochem.* 5:283–290.

Spyridakis, D.E. and Barnes, R.M., 1978, *Contemporary and historical trace metal loadings to the sediments of four lakes of the Lake Washington drainage.* Final Report. Office Water Res. Tech., Dept. Civil Eng. Univ. Washington OWRT Proj. A-083-WASH.

Srivastava, S.K. and Binda, P.L., 1984, Siliceous and silicified microfossils from the Maastrichtian Battle Formation of Southern Alberta, Canada. *Paleobiol. Contr.* 14:1–24.

Stager, J.C., 1998, Ancient analogues for recent environmental changes at Lake Victoria, East Africa. In Lehman, J.T. (ed.), *Environmental Change and Response in East African Lakes.* Kluwer, Dordrecht, pp. 37–46.

Stankiewicz, B.A., Briggs, D.E.G., Evershed, R.P., Flannery, M.B., and Wutke, M., 1997, Preservation of chitin in 25-million year old fossils. *Science* 276:1541–1543.

Starkel, L., 1991, The Vistula River Valley: a case study for Central Europe. In Starkel, L., Gregory, K.J., and Thornes, J.B. (eds.), *Temperate Paleohydrology.* J. Wiley and Sons, New York, pp. 171–188.

Steenbrink, J., van Vugt, N., Hilgen, F.J., Wijbrans, J.R., and Meulenkamp, J.E., 1999, Sedimentary cycles and volcanic ash beds in the Lower Pliocene lacustrine succession of Ptolemais (NW Greece): discrepancy between ^{40}Ar/^{39}Ar and astronomical ages. *Palaeogeog., Palaeoclim., Palaeoecol.* 152:283–303.

Steidtmann, J.R., McGee, L.C., and Middleton, L.T., 1983, Laramide sedimentation, folding and faulting in the southern Wind River Range, Wyoming. *Proc. Rocky Mtn. Assoc. Geol. Symp., Foreland Basins and Uplifts, Denver, CO,* pp. 161–167.

Stein, M., 2001, The sedimentary and geochemical record of Neogene–Quaternary water bodies in the Dead Sea Basin—inferences for the regional paleoclimatic history. *Jour. Paleolim.* 26:271–282.

Steinberg, C., Hartmann, H., Arzet, K., and Krause-Dellin, D., 1988, Paleoindication of acidification in Kleiner, Arbersee (Federal Republic of Germany, Bavarian Forest) by chydorids, chrysophytes, and diatoms. *Jour. Paleolim.* 1:149–157.

Steinhorn, I., 1985, The disappearance of the long-term meromictic stratification of the Dead Sea. *Limnol. Oceanogr.* 30:451–462.

Sten, E., Thybo, H., and Noe-Nygaard, N., 1996, Resistivity and georadar mapping of lacustrine and glaciofluvial sediments in the late-glacial to post-glacial Store Åmose basin, Denmark. *Bull. Geol. Soc. Denmark* 43:87–98.

Sternberg, L., DeNiro, M.J., and Keeley, J.E., 1984, Hydrogen, oxygen and carbon isotope ratios

of cellulose from submerged aquatic Crassulacean acid metabolism and non-Crassulacean acid metabolism plants. *Plant Physiol.* 76:68–70.

Stevenson, A.C. and Battarbee, R.W., 1991, Palaeoecological and documentary records of recent environmental change in Garaet El Ichkeul: a seasonally saline lake in NW Tunisia. *Biol. Conserv.* 58:275–295.

Stevenson, A.C. and Flower, R.J., 1991, A palaeoecological evaluation of environmental degradation in Lake Mikri Prepsa, NW Greece. *Biol. Conserv.* 57:89–109.

Stiller, M., Gat, J., and Kaushansky, P., 1997, Halite precipitation and sediment deposition as measured in sediment traps deployed in the Dead Sea: 1981–1983. In Niemi, T.M., Ben-Avraham, Z., and Gat, J.R. (eds.), *The Dead Sea: The Lake and Its Setting.* Oxford University Press, Oxford, pp. 171–183.

Stine, S., 1990, Late Holocene fluctuations of Mono Lake, eastern California. *Palaeogeog., Palaeoclim., Palaeoecol.* 78:333–381.

Stokes, S. and Swinehart, J.B., 1997, Middle and late-Holocene dune reactivation in the Nebraska Sand Hills, USA. *The Holocene* 7:263–272.

Stollhofen, H., Frommerz, B., and Stanistreet, I.G., 1999, Volcanic rocks as discriminants in evaluating tectonic versus climatic control on depositional sequences, Permo-Carboniferous continental Saar-Nahe Basin. *Jour. Geol. Soc. London* 156:801–808.

Stollhofen, H., Stainstreet, I.G., Rohn, R., Holzforster, F., and Wanke, A., 2000, The Gai-As Lake system, Northern Namibia and Brazil. In Gierlowski, E.H. and Kelts, K.R. (eds.), *Lake Basins Through Space and Time.* AAPG Stud. Geol. 46:87–108.

Storzer, D. and Poupeau, G., 1973, Ages-plateaux de mineraux et verres par la méthode des traces de fission. *Acad. Sci. C. R. Paris* 276:137–139.

Street, F.A. and Grove, A.T., 1979, Global maps of lake level fluctuations since 30,000 BP. *Quat. Res.* 12:83-118.

Street-Perrott, F.A., 1994, Palaeoperspectives: changes in terrestrial ecosystems. *Ambio* 23:37–49.

Street-Perrott, F.A. and Harrison, S.P., 1985, Lake levels and climate reconstruction. In Hecht, A.D. (ed.), *Paleoclimate Data and Modeling.* J. Wiley and Sons, New York, pp. 291–340.

Street-Perrott, F.A. and Perrott, R.A., 1990, Abrupt climate fluctuations in the tropics: the influence of Atlantic Ocean circulation. *Nature* 343:607–612.

Street-Perrott, F.A. and Perrott, R.A., 1993, Holocene vegetation, lake levels and climate of Africa. In Wright, H.E. Jr., Kutzbach, J.E., Webb III, T., Ruddiman, W.F., Street-Perrott,

F.A., and Bartlein, P.J. (eds.), *Global Climates Since the Last Glacial Maximum*. University of Minnesota Press, Minneapolis, MN, pp. 318–356.

Street-Perrott, F.A. and Roberts, N., 1983, Fluctuations in closed basin lakes as an indicator of past atmospheric circulation patterns. In Street-Perrott, F.A., Beran, M., and Ratcliffe, R. (eds.), *Variations in the Global Water Budget*. Reidel, Dordrecht, pp. 331–345.

Street-Perrott, F.A., Mitchell, J.F.B., Marchand, D.S., and Brunner, J.S., 1990, Milankovitch and albedo forcing of the tropical monsoons: a comparison of geological evidence and numerical simulations for 9000 yBP. *Trans. R. Soc. Edinburgh, Earth Sci.* 81:407–427.

Street-Perrott, F.A., Huang, Y., Perrott, R.A., Eglinton, G., Barker, P., Ben Khalifa, L., Harkness, D.D., and Olago, D.O., 1997, The impact of lower atmospheric CO_2 on tropical mountain ecosystems. *Science* 278:1422–1426.

Stuiver, M., 1968, Oxygen-18 content of atmospheric precipitation during the last 11,000 years in the Great Lakes region. *Science* 162:994–997.

Stuiver, M., 1970, Oxygen and carbon isotope ratios of freshwater carbonates as climatic indicators. *Jour. Geophys. Res.* 75:5247–5257.

Stuiver, M. and Kra, R., 1986, Calibration issue. *Radiocarbon* 28:805–1030.

Stuiver, M., Long, A., and Kra, R., 1993, Calibration 1993. *Radiocarbon* 35.

Stuiver, M. and Pollach, H.A., 1977, Discussion: reporting of ^{14}C data. *Radiocarbon* 19:355–363.

Sturm, M. and Matter, A., 1978, Turbidites and varves in Lake Brienz (Switzerland): deposition of clastic detritus by density. In Matter, A. and Tucker, M.E. (eds.), *Modern and Ancient Lake Sediments*. Intl. Assoc. Sedimentol. Spec. Publ. 2:147–168.

Suess, H.E., 1986, Secular variations of cosmogenic ^{14}C on earth. In Stuiver, M. and Kra, R. (eds.), Proc. 12th Intl. Radiocarbon Conf. *Radiocarbon* 28:259–265.

Suits, N.S. and Wilkin, R.T., 1998, Pyrite formation in the water column and sediments of a meromictic lake. *Geology* 26:1099–1102.

Supan, A.G., 1896, *Grundzuge der physische Erdkunde*, 2nd ed. Veit, Leipzig.

Surdam, R.C. and Stanley, K.O., 1979, Lacustrine sedimentation during the culminating phase of Eocene Lake Gosiute, Wyoming (Green River Formation). *Geol. Soc. Amer. Bull.* 90:93–110.

Swain, E.B., 1984, *The paucity of blue–green algae in meromictic Brownie Lake: iron limitation or heavy metal toxicity*. Ph.D. dissertation, University of Minnesota.

Swain, E.B., 1985, Measurement and interpretation of sedimentary pigments. *Freshwater Biol.* 15:53–75.

Sweets, P.R., 1992, Diatom paleolimnological evidence for lake acidification in the Trail Ridge region of Florida. *Water Air Soil Pollut.* 65:43–57.

Sweets, P.R., Bienert, R.W., Crisman, T.L., and Binford, M.W., 1990, Paleoecological investigations of recent lake acidification in Northern Florida. *Jour. Paleolim.* 4:103–137.

Swirydczuk, K., Wilkinson, B.H., and Smith, G.R., 1979, The Pliocene Glenns Ferry Oolite: lake margin carbonate deposition in the southwestern Snake River Plain. *Jour. Sed. Pet.* 49:995–1004.

Swirydczuk, K., Wilkinson, B.H., and Smith, G.R., 1980, The Pliocene Glenns Ferry Oolite-II: sedimentology of oolitic lacustrine terrace deposits. *Jour. Sed. Pet.* 50:1237–1248.

Sylvestre, F., Servant-Vildary, S., and Roux, M., 2001, Diatom-based ionic concentration and salinity models from the south Bolivian Altiplano. *Jour. Paleolim.* 25:279–295.

Syverson, K.M., 1998, Sediment record of short-lived ice-contact lakes, Burroughs Glacier, Alaska. *Boreas* 27:44–54.

Szeroczynska, K., 1998, Palaeolimnological investigations in Poland based on Cladocera (Crustacea). *Palaeogeog., Palaeoclim., Palaeoecol.* 140:335–345.

Talbot, M.R., 1990, A review of the palaeohydrological interpretation of carbon and oxygen isotopic ratios in primary lacustrine carbonates. *Chem. Geol.* 80:261–279.

Talbot, M.R., 1993, *Stable Isotope Studies of Lacustrine Carbonates*. Short Course Notes—Advances in Lacustrine Facies Analysis. CIRIT University of Barcelona/IGCP Inst. of Earth Sciences Jaume Almera, 10–11 May, 1993.

Talbot, M.R., 2001, Nitrogen isotopes in palaeolimnology. In Last, W.M. and Smol, J.P. (eds.), *Tracking Environmental Change Using Lake Sediments. Volume 2: Tracking Environmental Change Using Lake Sediments: Physical and Geochemical Methods*. Kluwer Academic, Dordrecht, pp. 401–439.

Talbot, M.R. and Allen, P.A., 1996, Lakes. In Reading, H.G. (ed.), *Sedimentary Environments: Processes, Facies and Stratigraphy*, 3rd ed. Blackwell Science, London, pp. 83–124.

Talbot, M.R. and Johannessen, T., 1992, A high-resolution palaeoclimatic record for the last 27,500 years in tropical West Africa from carbon and nitrogen isotopic composition of lacustrine organic matter. *Earth Planet. Sci. Lett.* 110:23–37.

Talbot, M.R. and Kelts, K., 1986, Primary and diagenetic carbonates in the anoxic sediments of Lake Bosumtwi, Ghana. *Geology* 14:912–916.

Talbot, M.R. and Kelts, K., 1989, Introduction. In Talbot, M.R. and Kelts, K. (eds.), *Phanerozoic Record of Lacustrine Basins and their Environmental Signals.* Palaeogeogr., Palaeoclim., Palaeoecol. 70:1–5.

Talbot, M.R. and Kelts, K., 1990, Paleolimnological signatures from carbon and oxygen isotopic ratios in carbonates from organic carbon-rich lacustrine sediments. In Katz, B.J. (ed.), *Lacustrine Basin Exploration: Case Studies and Modern Analogues.* Amer. Assoc. Petrol. Geol. Mem. 50:99–112.

Talbot, M.R. and Lærdal, T., 2000, The Late Pleistocene–Holocene palaeolimnology of Lake Victoria, East Africa, based upon elemental and isotopic analyses of sedimentary organic matter. *Jour. Paleolim.* 23:141–164.

Talbot, M.R. and Livingstone, D.A., 1989, Hydrogen index and carbon isotopes of lacustrine organic matter as lake level indicators. *Palaeogeog., Palaeoclim., Palaeoecol.* 70:121–137.

Talbot, M.R., Livingstone, D.A., Palmer, P.G., Maley, J., Melack, J.M., Delibrias, G., Gulliksen, S., 1984, Preliminary results from sediment cores from Lake Bosumtwi, Ghana. *Palaeoecol. Africa* 16:173–192.

Talma, A.S., Vogel, J., and Stiller, M., 1997, The radiocarbon content of the Dead Sea. In Niemi, T.M., Avraham, Z.B., and Gat, J.R. (eds.), *The Dead Sea: The Lake and Its Setting.* Oxford University Press, New York, pp. 193–199.

Tambareau, Y., 1982, Les ostracodes et l'histoire géologique de l'atlantique Sud au Crétacé. *Bull. Cent. Rech. Explor. Elf-Aquitaine* 6:1–37.

Tanaka, N., Monaghan, M.C., and Rye, D.M., 1986, Contribution of metabolic carbon to mollusc and barnacle shell carbonate. *Nature* 320:520–523.

Tarasov, P.E. and Harrison, S.P., 1998, Lake status records from the former Soviet Union and Mongolia: a continental-scale synthesis. In Harrison, S.P., Frenzel, B., Huckriede, U., and Weiß, M.M. (eds.), *Palaeohydrology as Reflected in Lake Level Changes as Climatic Evidence for Holocene Times.* Gustav Fischer-Verlag, Stuttgart, pp. 115–130.

Tarutani, T., Clayton, R., and Mayeda, T., 1969, The effect of polymorphism and magnesium substitution on oxygen isotope fractionation between calcium carbonate and water. *Geochim. Cosmochim. Acta* 33:987–996.

Tasch, P., 1969, Branchiopoda. In Moore, R.C. (ed.), *Treatise on Invertebrate Paleontology.* Geological Society of America, pp. R128–191.

Tasch, P., 1979, Crustacean branchiopod distribution and speciation in Mesozoic lakes of the southern continents. Terr. Biol. III. *Antarct. Res. Ser.* 30:65–74.

Tasch, P. and Gafford, E.L., 1968, Paleosalinity of Permian nonmarine deposits in Antarctica. *Science* 160:1221–1222.

Tauber, H., 1977, Investigations of aerial pollen transport in a forested area. *Dansk. Bot. Ark.* 32:1–121.

Taylor, D.W., 1988, Aspects of freshwater mollusc ecological biogeography. In Gray, J. (ed.), *Paleolimnology: Aspects of Freshwater Paleoecology and Biogeography.* Elsevier, Amsterdam, pp. 511–576.

Taylor, I.D., 1990, Evidence for high glacial-lake levels in the northeastern Lake Michigan basin and their relation to the Glenwood and Calumet phases of glacial Lake Chicago. In Schneider, A.F. and Fraser, G.S. (eds.), *Late Quaternary History of the Lake Michigan Basin.* Geol. Soc. Amer. Spec. Pap. 251:91–109.

Taylor, L.C. and Howard, K.W.F., 1993, The distribution of *Cypridopsis okeechobei* in the Duffins Creek–Rouge River drainage basins (Ontario, Canada) and its potential as an indicator of human disturbance. In McKenzie, K.G. and Jones, P.J. (eds.), *Ostracoda in the Earth and Life Sciences.* Balkema, Rotterdam, pp. 481–492.

Taylor, R.E., 1987, *Radiocarbon Dating, An Archaeological Perspective.* Academic Press, New York.

Taylor, W.D. and Sanders, R.W., 1991, Protozoa. In Thorp, J.H. and Covich, A.P. (eds.), *Ecology and Classification of North American Freshwater Invertebrates.* Academic Press, New York, pp. 37–93.

Teller, J.T., 1987, Proglacial lakes and the southern margin of the Laurentide Ice Sheet. In Ruddiman, W.F. and Wright, H.E. Jr. (eds.), *North America and Adjacent Oceans During the Last Deglaciation. The Geology of North America.* Geol. Soc. Amer. K-3:39–69.

Teller, J.T. and Mahnic, P., 1988, History of sedimentation of the northwestern Lake Superior Basin and its relation to Lake Agassiz overflow. *Can. Jour. Earth Sci.* 25:1660–1673.

Teller, J.T. and Thorliefson, L.H., 1983, The Lake Agassiz–Lake Superior connection. In Teller, J. and Clayton, L.E. (eds.), *Glacial Lake Agassiz.* Geol. Assoc. Canada Spec. Pap. 26:261–290.

Teller, J.T., Leverington, D., and Mann, J., 2001, Outbursts from Glacial Lake Agassiz and their possible impact on thermohaline circulation at the start of the Younger Dryas, Preboreal Oscillation, and 8.2ka Cold Event. *Eos Trans. AGU* 82(47):F750.

ter Braak, C.J.F., 1987, *Unimodal models to relate species to environment*. Ph.D. dissertation, University of Wageningen, Netherlands.

ter Braak, C.J.F. and Juggins, S., 1993, Weighted averaging partial least squares regression (WA-PLS): an improved method for reconstructing environmental variables from species assemblages. *Hydrobiologia* 269/270:485–502.

ter Braak, C.J.F. and Prentice, I.C., 1988, A theory of gradient analysis. *Adv. Ecol. Res.* 18:271–317.

ter Braak, C.J.F. and Van Dam, H., 1989, Inferring pH from diatoms: a comparison of old and new calibration methods. *Hydrobiologia* 178:209–223.

Tett, P., Heaney, S.I., and Droop, M.R., 1985, The Redfield Ratio and phytoplankton growth rate. *Jour. Mar. Biol. Assoc. U.K.* 65:487–504.

Thienemann, A., 1913, Der Zusammenhang zwichen dem Sauerstoffgehalt des Tiefenwassers und der Zussamensetzung der Tierfauna unsere Seen. *Int. Rev. ges. Hydrobiol.* 6:243–249.

Thienemann, A., 1920, Untersuchungen über die Beziehungen zwischen dem Sauerstoffgehalt des Wassers und der Zusammensetzung der Fauna in norddeutchen Seen. *Arch. Hydrobiol.* 12:1–65.

Thienemann, A., 1922, Die beiden Chironomusarten der Tiefenfauna der noddeutchen Seen. *Arch. Hydrobiol.* 13:609–646.

Thompson, R. and Clark, R.M., 1989, Sequence slotting for stratigraphic correlation between cores: theory and practice. *Jour. Paleolim.* 2:173–184.

Thompson, R.O.R. and Imberger, J., 1980, Response of a numerical model of a stratified lake to wind stress. *2nd Intl. Symp. Assoc. Hydrol. Res., Trondheim, Norway*, pp. 562–570.

Thompson, R. and Oldfield, F., 1986, *Environmental Magnetism*. Allen and Unwin, Winchester, MA.

Thompson, T.A., 1992, Beach ridge development and lake-level variation in southern Lake Michigan. *Sed. Geol.* 80:305–318.

Thorarinsson, S., 1939, The ice-dammed lakes of Iceland, with particular reference to their value as indicators of glacier oscillations. *Geog. Annal.* 21:216–242.

Thorp, J.H. and Chesser, R.K., 1983, Seasonal responses of lentic midge assemblages to environmental gradients. *Holarctic Ecol.* 6:123–132.

Thouveny, N., Creer, K.M., and Blunk, I., 1990, Extension of the Lac du Bouchet palaeomagnetic record over the last 120,000 years. *Earth Planet. Sci. Lett.* 97:140–161.

Tiercelin, J.J. and Mondeguer, A., 1991, The geology of the Tanganyika Trough. In Coulter, G. (ed.), *Lake Tanganyika and Its Life*. Oxford University Press, Oxford, pp. 7–48.

Tiercelin, J.J. and Vincens, A. (eds.), 1987, *Le Demi-Graben de Baringo-Bogoria, Rift Gregory, Kenya: 30 000 Ans D'Histoire Hydrologique et Sédimentaire*. Bull. Centr. Rech. Explor.-Prod. Elf-Aquitaine 11:249–540.

Tiercelin, J.J., Soreghan, M., Cohen, A., Lezzar, K.E., and Bouroullec, J.L., 1992, Sedimentation in large rift lakes: example from the Middle Pleistocene–Modern deposits of the Tanganyika Trough, East African Rift System. *Bull. Centr. Rech. Explor.-Prod. Elf-Acquitaine* 16:83–111.

Tiercelin, J.J., Cohen, A.S., Soreghan, M.J., and Lezzar, K.E., 1994, Pleistocene–modern deposits of the Lake Tanganyika rift basin East Africa: a modern analog for lacustrine source rocks and reservoirs. In Lomando, A.J., Schreiber, B.C., and Harris, P.M. (eds.), *Lacustrine Reservoirs and Depositional Systems*. SEPM Core Workshop 19:37–60.

Timms, B.V., 1992, *Lake Geomorphology*. Gleneagle Publishing, Adelaide.

Timms, B.V., 1998, Further studies on the saline lakes of the eastern Paroo, inland New South Wales, Australia. *Hydrobiologia* 381:31–42.

Tinkler, K.J. and Pengelly, J.W., 1995, Great Lakes response to catastrophic inflows from Lake Agassiz: some simulations of a hydraulic geometry for chained lake systems. *Jour. Paleolim.* 13:251–266.

Tinner, W. and Lotter, A., 2001, Central European vegetation response to abrupt climate change at 8.2 ka. *Geology* 29:551–554.

Tissot, B.P. and Welte, D.H., 1984, *Petroleum Formation and Occurrence*. Springer-Verlag, Amsterdam.

Tolonen, K., 1986, Rhizopod analysis. In Berglund, B.E. (ed.), *Handbook of Holocene Palaeoecology and Palaeohydrology*. J. Wiley and Sons, New York, pp. 645–666.

Tonn, W.M. and Magnuson, J.J., 1982, Patterns in the species composition and richness of fish assemblages in northern Wisconsin lakes. *Ecology* 63: 1149–1166.

Tonn, W.M., Magnuson, J.J., Rask, M., and Toivonen, J., 1990, Intercontinental comparison of small-lake fish assemblages: the balance between local and regional processes. *Amer. Nat.* 136:345–375.

Tracey, S., Todd, J.A., and Erwin, D.H., 1993, Mollusca: Gastropoda. In Benton, M.J. (ed.), *The Fossil Record 2*. Chapman and Hall, London, pp. 131–168.

Tracey, B., Lee, N., and Card, V., 1996, Sediment indicators of meromixis: comparison of laminations, diatoms, and sediment chemistry in Brownie Lake, Minneapolis, USA. *Jour. Paleolim.* 15:129–132.

Trauth, M.H., Deino, A., and Strecker, M., 2001, Response of the East African climate to orbital forcing during the last interglacial (130–117 ka) and the early last glacial (117–60 ka). *Geology* 29:499–502.

Trewin, N.H. and Davidson, R.G., 1999, Lake-level changes, sedimentation and faunas in a Middle Devonian basin-margin fish bed. *Jour. Geol. Sci.* 156:535–548.

Truc, G., 1978, Lacustrine sedimentation in an evaporitic environment: the Ludian (Paleogene) of the Mormoiron basin, southeastern France. In Matter, A. and Tucker, M.E. (eds.), *Modern and Ancient Lake Sediments*. Intl. Assoc. Sediment. Spec. Publ. 2:189–203.

Truze, E. and Kelts, K., 1993, Sedimentology and paleoenvironment from the maar Lac du Bouchet for the last climatic cycle, 0–120,000 years. In Negendank, J.F.W. and Zolitschka, B. (eds.), *Paleolimnology of European Maar Lakes*. Springer-Verlag, New York, pp. 237–275.

Turner, G.F., 2000, The nature of species in ancient lakes: perspectives from the fishes of Lake Malawi. In Rossiter, A. and Kawanabe, H. (eds.), *Ancient Lakes: Biodiversity, Ecology and Evolution*. Advances in Ecological Research 31. Academic Press, New York, pp. 39–60.

Turner, J., 1985, Sponge gemmules from lake sediments in the Puget Lowland, Washington. *Quat. Res.* 24:240–243.

Turner, M.A., Robinson, G.G.C., Townsend, B.E., Hann, B.J., and Amaral, J.A., 1995, Ecological effects of blooms of filamentous green algae in the littoral zone of an acid lake. *Can. Jour. Fish. Aquatic Sci.* 52:2264–2275.

Turney, C.S.M., 1998, Extraction of rhyolitic component of Vedde microtephra from minerogenic lake sediment. *Jour. Paleolim.* 19:199–206.

Turpen, J. and Angell, R., 1971, Aspects of molting and calcification in the ostracode *Heterocypris*. *Biol. Bull.* 140:331–338.

Tyler, S.W., Cook, P.G., Butt, A.Z., Thomas, J.M., Doran, P.T., and Lyons, W.B., 1998, Evidence of deep circulation in two perennially ice-covered Antarctic lakes. *Limnol. Oceanogr.* 43:625–635.

Tyrrell, J.B., 1892, *Report on northwestern Manitoba with portions of the adjacent district of Assiniboia and Saskatchewan*. Geological Survey of Canada Annual Report 5, Ottawa.

Upchurch, S.B. and Randazzo, A.F., 1997, Environmental geology of Florida. In Randazzo, A.F. and Jones, D.S. (eds.), *The*

Geology of Florida. University Press of Florida, Gainesville, FL, pp. 217–249.

Urey, H., 1947, The thermodynamic properties of ionic substances. *Jour. Chem. Soc.* 1947:562–581.

Uutala, A.J., 1990, *Chaoborus* (Diptera:Chaoboridae) mandibles—paleolimnological indicators of the historical status of fish populations in acid-sensitive lakes. *Jour. Paleolim.* 4:139–151.

Uutala, A.J., Yan, N., Dixit, A.S., Dixit, S.S., and Smol, J.P., 1994, Paleolimnological assessment of declines in fish communities in three acidic Canadian shield lakes. *Fish. Res.* 19:157–177.

Valero-Garcés, B. and Kelts, K., 1995, A sedimentary facies model for perennial and meromictic saline lakes: Holocene Medicine Lake Basin, South Dakota, USA. *Jour. Paleolim.* 14:123–149.

Valero-Garcés, B.L., Grosjean, M., Schwalb, A., Geyh, M., Messerli, B., and Kelts, K., 1996, Limnogeology of Laguna Miscanti: evidence for mid to late Holocene moisture changes in the Atacama Altiplano (Northern Chile). *Jour. Paleolim.* 16:1–21.

Vallentyne, J., 1957, The molecular nature of organic matter in lakes and oceans, with lesser reference to sewage and terrestrial materials. *Jour. Fish. Res. Bd. Canada* 14:33–82.

Vallentyne, J., 1960, Fossil pigments. In Allen, M.B. (ed.), *Comparative Biochemistry of Photoreactive Systems*. Academic Press, New York, pp. 83–105.

Van Bruggen, A.C., 1992, Biodiversity of the Mollusca: time for a new approach. In Van Bruggen, A.C., Wells, S.M., and Kemperman, T.C.M. (eds.), *Biodiversity and Conservation of the Mollusca*. Backhuys, Oestgeest-Leiden, Holland, pp. 1–19.

Van Campo, E. and Gasse, F., 1993, Pollen and diatom-inferred climatic and hydrological changes in Sumxi Co Basin (Western Tibet) since 13,000 yr B.P. *Quat. Res.* 39:300–313.

Van Dam, H., Van Geel, B., Van der Wijk, A., Geelen, J.F.M., Van der Heijden, R., and Dickman, M.D., 1988, Palaeolimnological and documented evidence for alkalization and acidification of two moorland pools (The Netherlands). *Rev. Palaeobot. Palynol.* 55:273–316.

Van Damme, D., 1984, *The Freshwater Mollusca of Northern Africa, Distribution, Biogeography and Paleoecology*. Dr. W. Junk, Dordrecht.

Van den Bold, W., 1976, Distribution of species of the tribe Cypreidini (Ostracodea, Cytherideidae) in the Neogene of the Caribbean. *Micropaleo.* 22:1–43.

Van den Bold, W., 1983, Shallow marine biostratigraphic zonation in the Caribbean Post-Eocene. In Maddocks, R. (ed.),

Applications of Ostracoda. Proc. 8th Intl. Symp. Ostracoda. University of Houston, Houston, TX, pp. 400–416.

Van den Bold, W., 1990, Stratigraphical distribution of fresh and brackish water Ostracoda in the late Neogene of Hispaniola. In Whatley, R. and Maybury, C. (eds.), *Ostracoda and Global Events*. Chapman and Hall, New York, pp. 221–232.

Van den Hoek, C., Mann, D.G., and Jahns, H.M., 1995, *Algae: An Introduction To Phycology*. Cambridge University Press, Cambridge.

Van Der Meer, J.J.M. and Warren, W.P., 1997, Sedimentology of Late Glacial clays in lacustrine basins, central Ireland. *Quat. Sci. Rev.* 16:779–791.

Van der Post, K.D., Oldfield, F., Haworth, E.Y., Crooks, P.R.J., and Appleby, P.G., 1997, A record of accelerated erosion in the recent sediments of Blenham Tarn in the English Lakes district. *Jour. Paleolim.* 18:103–120.

Van Dijk, D.E., Hobday, D.K., and Tankard, A.J., 1978, Permo-Triassic lacustrine deposits in the Eastern Kraoo Basin, Natal, South Africa. In Matter, A. and Tucker, M.E. (eds.), *Modern and Ancient Lake Sediments*. Int. Assoc. Sediment. Spec. Publ. 2:223–238.

Van Geel, B., 1986, Application of fungal and algal remains and other microfossils in palynological analysis. In Berglund, B.E. (ed.), *Handbook of Holocene Paleoecology and Paleohydrology*. J. Wiley and Sons, New York, pp. 497–505.

Van Geel, B., Mur, L.R., Ralska-Jasiewiczowa, M., and Goslar, T., 1994, Fossil akinetes of *Aphanizomenon* and *Anabaena* as indicators for medieval phosphate-eutrophication of Lake Gosciaz (Central Poland). *Rev. Palaeobot. Palynol.* 83:97–105.

Van Harten, D., 1990, The Neogene evolutionary radiation in *Cyprideis* Jones (Ostracoda: Cytheracea) in the Mediteranean area and the Paratethys. *Cour. Forsch. Inst. Senckenberg* 123:191–198.

Van Harten, D., 1996, *Cyprideis torosa* revisited. Of salinity, nodes and shell size. In Keen, M (ed.), *Proc. 2nd Euro. Ostracod. Mtg., Glasgow*, 1993, pp. 226–230.

Vanhoorne, R. and Ferguson, D.K., 1997, A palaeoecological interpretation of an Eemian floral assemblage in the Scheldt Valley at Liefkenshoek near Antwerp (Belgium). *Rev. Palaeobot. Palynol.* 97:97–107.

Van Houten, F.B., 1962, Cyclic sedimentation and the origin of analcime-rich Upper Triassic Lockatong Formation, west-central New Jersey and adjacent Pennsylvania. *Amer. Jour. Sci.* 260:561–576.

Van Huissteden, K.J., Schwan, J., and Bateman, M.D., 2001, Environmental conditions and paleowind directions at the end of the Weischelian late pleniglacial recorded in aeolian sediments and geomorphology. *Geol. Mijnbouw.* 80:1–18.

Van Kreveld, S., Sarnthein, M., Erlenkeuser, H., Grootes, P., Jung, S., Nadeau, M.J., Pflaumann, U., and Voelker, A., 2000, Potential links between surging ice sheets, circulation changes and the Dansgaard–Oeschger cycles in the Irminger Sea, 60–18kyr. *Paleoceanography* 15:425–442.

Van Landingham, S.L., 1964a, Miocene non-marine diatoms from the Yakima region in south central Washington. *Beihefte zur Nova Hedwiga* 14.

Van Landingham, S.L., 1964b, Chrysophyta cysts from the Yakima basalt (Miocene) in south-central Washington. *Jour. Paleo.* 38:729–739.

Van Landingham, S.L., 1985, Potential Neogene diagnostic diatoms from the western Snake River Basin, Idaho and Oregon. *Micropaleo.* 31:167–174.

Van Rensbergen, P., De Batist, M., Beck, C., and Manalt, F., 1998, High resolution seismic stratigraphy of late Quaternary fill of Lake Annecy (Northwestern Alps): evolution from glacial to interglacial sedimentary processes. *Sed. Geol.* 117:71–96.

Van Rensbergen, P., De Batist, M., Beck, C., and Chapron, E., 1999, High resolution seismic stratigraphy of glacial to interglacial fill of a deep glacigenic lake: Lake Le Bourget, Northwestern Alps, France. *Sed. Geol.* 128:99–129.

Van Vugt, N., Steenbrink, J., Langereis, C.G., Hilgren, F.J., and Meulenkamp, J.E., 1998, Magentostratigraphy-based astronomical tuning of the early Pliocene lacustrine sediments of Ptolemais (NW Greece) and bed-to-bed correlation with the marine record. *Earth Planet. Sci. Lett.* 164:535–551.

Vance, R.E. and Tekla, A.M., 1998, Accelerator mass spectrometry radiocarbon dating of 1994 Lake Winnipeg cores. *Jour. Paleolim.* 19:329–334.

Vance, R.E., Mathewes, R.W., and Clague, J.J., 1992, 7,000 year record of lake-level change on the northern Great Plains: a high-resolution proxy of past climate. *Geology* 20:879–882.

Vance, R.E., Clague, J.J., and Mathewes, R.W., 1993, Holocene paleohydrology of a hypersaline lake in southeastern Alberta. *Jour. Paleolim.* 8:103–120.

Vance, R.E., Last, W.M., and Smith, A.J., 1997, Hydrological and climatic implications of a multidisciplinary study of late Holocene sediment from Kenosee Lake, southeastern Saskatchewan, Canada. *Jour. Paleolim.* 18:365–393.

Vannier, J., Wang, S.Q., and Coen, M., 2001, Leperditicopid arthropods (Ordovician–Late

Devonian): functional morphology and ecological range. *Jour. Paleo.* 75:75–95.

Vardy, S.R., Warner, B.G., and Aravena, R., 1997, Holocene climate effects on the development of a peatland on the Tuktoyaktuk Peninsula, Northern Territories. *Quat. Res.* 47:90–104.

Vassiljev, J. and Harrison, S.P., 1998, Simulating the Holocene lake level record of Lake Bysjön, Southern Sweden. *Quat. Res.* 49:62–71.

Verheyen, E., Rüber, L., Snoeks, J., and Meyer, A., 1996, Mitochondrial phylogeography of rock dwelling cichlid fishes reveals evolutionary influence of historic lake level fluctuations in Lake Tanganyika, Africa. *Phil. Trans. R. Soc. London, Ser. B* 351:797–805.

Verhoeven, J.T.A., Koerselman, W., and Beltman, B., 1988, The vegetation of fens in relation to their hydrology and nutrient dynamics: a case study. In Symoens, J.J. (ed.), *Vegetation of Inland Waters*. Kluwer Academic, Dordrecht, pp. 249–282.

Vermeij, G., 1987, *Evolution and Escalation*. Princeton University Press, Princeton, NJ.

Verosub, K.L., 1979a, Paleomagnetism of varved sediments from western New England: secular variation. *Geophys. Res. Lett.* 6:245–248.

Verosub, K.L., 1979b, Paleomagnetism of varved sediments from western New England: variability of the paleomagnetic recorder. *Geophys. Res. Lett.* 6:241–244.

Verosub, K.L., 1988, Geomagnetic secular variation and the dating of Quaternary sediments. In Easterbrook, D.J. (ed.), *Dating Quaternary Sediments*. Geol. Soc. Amer. Spec. Pap. 227:123–138.

Verosub, K.L. and Roberts, A.P., 1995, Environmental magnetism: past, present and future. *Jour. Geophys. Res.* 100:2175–2192.

Verrecchia, E.P., Freytet, P., Julien, J., and Baltzer, F., 1997, The unusual hydrodynamical behaviour of freshwater oncolites. *Sed. Geol.* 113:225–243.

Verschuren, D., 1994, Sensitivity of tropical-African aquatic invertebrates to short-term trends in lake level and salinity: a paleolimnological test at Lake Oloidien, Kenya. *Jour. Paleolim.* 10:253–263.

Verschuren, D., Edgington, D.N., Kling, H.J., and Johnson, T.C., 1998, Silica depletion in Lake Victoria: sedimentary signals at offshore stations. *Jour. Great Lakes Res.* 24:118–130.

Verschuren, D., Tibby, J., Leavitt, P., and Roberts, C.N., 1999, The environmental history of a climate-sensitive lake in the former "White Highlands" of central Kenya. *Ambio* 28:494–501.

Verschuren, D., Laird, K.R., and Cumming, B.F., 2000, Rainfall and drought in equatorial east Africa during the past 1,100 years. *Nature* 403:410–414.

Veselý, J., Almquist-Jacobson, H., Miller, L.M., Norton, S.A., Appleby, P., Dixit, A., and Smol, J.P., 1993, The history and impact of air pollution at Certovo Lake, southwestern Czech Republic. *Jour. Paleolim.* 8:211–231.

Vile, M.A., Wieder, R.K., and Novák, M., 2000, 200 years of Pb deposition throughout the Czech Republic: patterns and sources. *Envir. Sci. Tech.* 34:12–21.

Vinson, D.K. and Rushforth, S.R., 1989, Diatom species composition along a thermal gradient in the Portneuf River, Idaho, USA. *Hydrobiologia* 185:41–54.

Volkman, J.K., 1988, Biological marker compounds as indicators of the depositional environments of petroleum source rocks. In Fleet, A.J., Kelts, K., and Talbot, M.R. (eds.), *Lacustrine Petroleum Source Rocks*. Geol. Soc. Lond. Spec. Publ. 40:103–122.

Vollenweider, R.A., 1976, Advances in defining critical loading levels for phosphorus in lake eutrophication. *Mem. Ist. Ital. Idrobiol.* 33:53–83.

Von Grafenstein, U., Erlenkeuser, H., Kleinmann, A., Müller, J., and Trimborn, P., 1994, High-frequency climatic oscillations during the last deglaciation as revealed by oxygen-isotope records of benthic organisms (Ammersee, southern Germany). *Jour. Paleolim.* 11:349–357.

Von Grafenstein, U., Erlenkeuser, H., Müller, J., Jouzel, J., and Johnsen, S., 1998, The cold event 8200 years ago documented in oxygen isotope records of precipitation in Europe and Greeland. *Clim. Dynam.* 14:73–81.

Von Grafenstein, U., Erlenkeuser, H., and Trimborn, P., 1999, Oxygen and carbon isotopes in modern freshwater ostracod valves: assessing vital offsets and autecological effects of interest from paleoclimate studies. *Palaeogeog., Palaeoclim., Palaeoecol.* 148:133–152.

von Gunten, H.R. and Moser, R.N., 1993, How reliable is the ^{210}Pb method? Old and new results from Switzerland. *Jour. Paleolim.* 9:161–178.

Vyverman, W. and Sabbe, K., 1995, Diatom-temperature transfer functions based on the altitudinal zonation of diatom assemblages in Papua New Guinea: a possible tool in the reconstruction of regional palaeoclimatic changes. *Jour. Paleolim.* 13:65–77.

Waitt, R.B., 1984, Periodic Jökulhlaups from Pleistocene Glacial Lake Missoula—New evidence from varved sediment in northern Idaho and Washington. *Quat. Res.* 22:46–58.

Waldman, M., 1971, Fish from the freshwater Lower Cretaceous of Victoria, Australia with comments on the palaeoenvironment. *Pal. Soc. Lond. Spec. Pap. Palaeontol.* 9.

Walker, I.R., 1995, Chironomids as indicators of past environmental change. In Armitage, P.D., Cranston, P.S., and Pinder, L.C.V. (eds.), *The Chironomidae: Biology and Ecology of Non-biting Midges*. Chapman and Hall, New York, pp. 405–422.

Walker, I.R. and Mathewes, R.W., 1989a, Early post-glacial chironomid succession in southwestern British Columbia, Canada, and its paleoenvironmental significance. *Jour. Paleolim.* 2:1–14.

Walker, I.R. and Mathewes, R.W., 1989b, Much ado about dead diptera. *Jour. Paleolim.* 2:19–22.

Walker, I.R. and Mathewes, R.W., 1989c, Chironomidae remains in surficial sediments from the Canadian Cordillera: analysis of the fauna across an altitudinal gradient. *Jour. Paleolim.* 2:61–80.

Walker, I.R., Smol, J.P., Engstrom, D.R., and Birks, H.J.B., 1991a, An assessment of Chironomidae as quantitative indicators of past climatic change. *Can. Jour. Fish. Aquat. Sci.* 48:975–987.

Walker, I.R., Mott, R.J., and Smol, J.P., 1991b, Allerød-Younger Dryas lake temperatures from midge fossils in Atlantic Canada. *Science* 253:1010–1012.

Walker, I.R., Smol, J.P., Engstrom, D.R., and Birks, H.J.B., 1992, Aquatic invertebrates, climate scale, and statistical hypothesis testing: a response to Hann, Warner, and Warwick. *Can. Jour. Fish. Aquat. Sci.* 49:1276–1280.

Walker, I.R., Levesque, A.J., Cwynar, L.C., and Lotter, A.F., 1997, An expanded surface water palaeotemperature inference model for use with fossil midges from eastern Canada. *Jour. Paleolim.* 18:165–178.

Walker, M.J.C., Griffiths, H.I., Ringwood, V., and Evans, J.G., 1993, An early-Holocene pollen, mollusc and ostracod sequence from lake marl at Llangorse Lake, South Wales, UK. *The Holocene* 3:138–149.

Walling, D.E. and Kleo, A.H.A., 1979, Sediment yields of rivers in areas of low precipitation: a global view. In *The Hydrology of Areas of Low Precipitation*. Intl. Assoc. Hydrol. Sci. Proc. Canberra Symp. IAHS-AISH Publ. 128:479–493.

Walter, R.C., 1989, Application and limitation of fission-track geochronology to Quaternary tephras. *Quat. Intl.* 1:35–46.

Wan, G.J., Santschi, P.H., Sturm, M., Farrenkothen, K., Lueck, A., Werth, E., and Schuller, C., 1987, Natural and fallout radionuclides as geochemical tracers of sedimentation in Greifensee, Switzerland. *Chem. Geol.* 63:181–196.

Wansard, G., DeDeckker, P., and Julià, R., 1998, Variability in ostracod partition coefficients *D*(Sr) and *D*(Mg). Implications for lacustrine palaeoenvironmental reconstructions. *Chem. Geol.* 146:39–54.

Ward, J.V., 1992, *Aquatic Insect Ecology*. J. Wiley and Sons, New York.

Warner, B.G. (ed.), 1990a, *Methods in Quaternary Ecology*. Geoscience Canada Reprint Ser. 5. Geological Association of Canada.

Warner, B.G., 1990b, Testate Amoebae (Protozoa). In Warner, B.G. (ed.), *Methods in Quaternary Ecology*. Geoscience Canada Reprint Ser. 5. Geological Association of Canada, pp. 65–74.

Warner, B.G., 1990c, Other fossils. In Warner, B.G. (ed.), *Methods in Quaternary Ecology*. Geoscience Canada Reprint Ser. 5. Geological Association of Canada, pp. 149–162.

Warner, B.G. and Chengalath, R., 1988, Holocene fossil *Habrotrocha angusticollis* (Bdelloidea: Rotifera) in North America. *Jour. Paleolim.* 1:141–147.

Warner, B.G., Hebda, R.J., and Hann, B.J., 1984, Postglacial palaeoecological history of a cedar swamp, Manitoulin Island, Ontario, Canada. *Palaeogeog., Palaeoclim., Palaeoecol.* 45:301–345.

Warner, B.G., Kubiw, H.J., and Karrow, P.F., 1991, Origin of a postglacial kettle-fill sequence near Georgetown, Ontario. *Can. Jour. Earth Sci.* 28:1965–1974.

Warren, J.K., 1994, Holocene Coorong Lakes, South Australia. In Gierlowski-Kordesch, E. and Kelts, K. (eds.), *Global Geological Record of Lake Basins, v. 1*. Cambridge University Press, Cambridge, pp. 387–394.

Warren, J.W., 1982, The hydrological significance of Holocene tepees, stromatolites, and boxwork limestones in coastal salinas in South Australia. *Jour. Sed. Petrol.* 52:1171–1201).

Wartes, M.A., Carroll, A.R., Greene, T.J., Keming, C., and Hu, T., 2000, Permian lacustrine deposits of northwestern China. In Gierlowski, E.H. and Kelts, K.R. (eds.), *Lake Basins Through Space and Time*. Amer. Assoc. Petrol. Geol. Stud. Geol. 46:123–132.

Warwick, W.F., 1980, Paleolimnology of the Bay of Quinte, Lake Ontario: 2800 years of cultural influence. *Can. Bull. Fish. Aquat. Sci.* 206:1–117.

Warwick, W.F., 1989, Chironomids, lake development and climate: a commentary. *Jour. Paleolim.* 2:15–17.

Watts, W.A., Allen, J.R.M., and Huntley, B., 2000, Palaeoecology of three interstadial events during oxygen-isotope Stages 3 and 4: a lacustrine record from Lago Grande di Monticchio, southern Italy. *Palaeogeog., Palaeoclim., Palaeoecol.* 155:83–93.

Waythomas, C.F., Walder, J.S., McGimsey, R.G., and Neal, C.A., 1996, A catastrophic flood caused by drainage of a caldera lake at

Aniakchak Volcano, Alaska, and implications for volcanic hazard assessment. *Geol. Soc. Amer. Bull.* 108:861–871.

Webb, J.A., 1979, A reappraisal of the paleoecology of conchostracans (Crustacea: Branchiopoda). *N. Jb. Geol. Paläont. Abh.* 158:259–275.

Webb, T. III and Bryson, R.A., 1972, Late and postglacial climatic change in northern Midwest, USA: quantitative estimates derived from fossil pollen spectra by multivariate statistical analysis. *Quat. Res.* 2:70–115.

Weedman, S.D., 1994, Upper Allegheny Group (Middle Pennsylvanian) lacustrine limestones of the Appalachian Basin, USA. In Gierlowski-Kordesch, E. and Kelts, K. (eds.), *Global Geological Record of Lake Basins.* Cambridge University Press, New York, pp. 127–134.

Wehmiller, J.F. and Miller, G.H., 2000, Aminostratigraphic dating methods in Quaternary geology. In Noller, J.S. et al. (eds.), *Quaternary Geochronology, Methods and Applications.* American Geophysical Union, Washington, DC, pp. 187–222.

Weirich, F.H., 1984, Turbidity currents: monitoring their occurrence and movement with a three dimensional sensor network. *Science* 224:384–387.

Weirich, F.H., 1986, The record of density-induced underflows in a glacial lake. *Sedimentology* 33:261–277.

Weiss, D., Shotyk, W., Appleby, P.G., Kramers, J.D., and Chburkin, A.K., 1999, Atmospheric Pb deposition since the industrial revolution recorded by five Swiss peat profiles: enrichment factors, fluxes, isotopic composition, and sources. *Envir. Sci .Tech.* 33:1340–1352.

Wells, J.T., Scholz, C.A., and Soreghan, M.J., 1999a, Processes of sedimentation on a lacustrine border fault margin: interpretation of cores from Lake Malawi, East Africa. *Jour. Sed. Res.* 69:816–831.

Wells, T.M., Cohen, A.S., Park, L.E., Dettman, D.L., and McKee, B.A., 1999, Ostracode stratigraphy and paleoecology from surficial sediments of Lake Tanganyika, Africa. *Jour. Paleolim.* 22:259–276.

Wernicke, B.P., Snow, J.K., Hodges, K.V., and Walker, J.D., 1993, Structural constraints on Neogene tectonism in the southern Great Basin. In Lahren, M.M., Trexler, J.H., and Spinosa, C. (eds.), *Crustal Evolution of the Great Basin and the Sierra Nevada.* Mackay School of Mines, pp. 453–480.

Wersin, P., Höhener, P., Giovanoli, R., and Stumm, W., 1991, Early diagenetic influences on iron transformations in a freshwater lake sediment. *Chem. Geol.* 90:233–252.

Wescott, W.A., Morley, C.K., and Karanja, F.M., 1996, Tectonic controls on the development of rift-basin lakes. In Johnson, T.C. and Odada, E. (eds.), *The Limnology, Climatology and Paleoclimatology of the East African Lakes.* Gordon and Breach, Amsterdam, pp. 3–24.

Wesselingh, F.P., Cadée, G.C., and Renema, W., 1999, Flying high: on the airborne dispersal of aquatic organisms as illustrated by the distribution histories of the gastropod genera *Tryonia* and *Planorbarius. Geol. Mijnbouw* 78:165–174.

West, K. and Cohen, A.S., 1996, Shell microstructure of gastropods from Lake Tanganyika, Africa: adaptation, convergent evolution, and escalation. *Evolution* 50:672–681.

West, K. and Michel, E., 2000, The dynamics of endemic diversification: molecular phylogeny suggests an explosive origin of the thiarid gastropods of Lake Tanganyika. In Rossiter, A. and Kawanabe, H. (eds.), *Ancient Lakes: Biodiversity, Ecology and Evolution.* Advances in Ecology Research 31. Academic Press, New York, pp. 331–354.

Westgate, J.A., 1989, Isothermal plateau fission-track ages of hydrated glass shards from silicic tephra beds. *Earth Planet Sci. Lett.* 95:226–234.

Westgate, J.A. and Gorton, M.P., 1981, Correlation techniques in tephra studies. In Self, S. and Sparks, R.S.J. (eds.), *Tephra Studies.* NATO Adv. Stud. Inst. Ser. C. Reidel, Dordrecht, pp. 73–94.

Westgate, J.A. and Naeser, N.D., 1995, Tephrochronology and fission-track dating. In Rutter, N.W. and Catto, N.R. (eds.), *Dating Methods for Quaternary Deposits.* GEOTEXT 2, Geological Society of Canada, pp. 15–28.

Wetzel, R.G., 1983, *Limnology,* 2nd ed. Saunders College Publishing, Philadelphia.

Wetzel, R.G., 1988, Water as an environment for plant life. In Symoens, J.J. (ed.), *Vegetation of Inland Waters.* Kluwer Academic, Dordrecht, pp. 1–30.

Wetzel, R.G., 2000, *Limnology: Lake and River Ecosystems,* 3rd ed. Academic Press, San Diego.

Whatley, R.C., 1988, Population structure of ostracods: some general principles for the recognition of palaeoenvironments. In De Deckker, P., Colin, J.P., and Peypouquet, J.P. (eds.), *Ostracoda in the Earth Sciences.* Elsevier, Amsterdam, pp. 245–256.

Whatley, R.C., Siveter, D.J., and Boomer, I.D., 1993, Arthropoda (Crustacea: Ostracoda). In Benton, M.J. (ed.), *The Fossil Record 2.* Chapman and Hall, London, pp. 343–356.

White, J.D.L., 1990, Depositional architecture of a maar-pitted playa: sedimentation in the Hopi

Butters volcanic field, northeastern Arizona, U.S.A. *Sed. Geol.* 67:55–84.

White, J.D.L., 1992, Pliocene subaqueous fans and Gilbert-type deltas in maar crater lakes, Hopi Buttes, Navajo Nation (Arizona) USA. *Sedimentology* 39:931–946.

White, J.R. and Gubala, C.P., 1990, Sequentially extracted metals in Adirondack lake sediment cores. *Jour. Paleolim.* 3:243–252.

White, J.W.C., Clais, P., Figge, R.A., Kenny, R., and Markgraf, V., 1994, A high resolution record of atmospheric CO_2 content from carbon isotopes in peat. *Nature* 367:153–156.

Whitehead, D.R., Charles, D.F., and Goldstein, R.A., 1990, The PIRLA project (Paleoecological Investigation of Recent Lake Acidification): an introduction to the synthesis of the project. *Jour. Paleolim.* 3:187–194.

Whiteside, M.C., 1970, Danish chydorid Cladocera: modern ecology and core studies. *Ecol. Monogr.* 40:79–118.

Whiteside, M.C. and Swindoll, M.R., 1988, Guidelines and limitations to cladoceran paleoecological analysis. In Gray, J. (ed.), *Paleolimnology: Aspects of Freshwater Paleoecology and Biogeography*. Elsevier, Amsterdam, pp. 405–412.

Whittlesey, C., 1838. Report on the geology and topography of a portion of Ohio. *Ohio Geol. Surv. Ann. Rep.* 2:41–71.

Whittlesey, C., 1850. On the natural terraces and ridges of the country bordering Lake Erie. *Amer. Jour. Sci.* (2nd ser.) 10:31–39.

Wiederholm, T. and Ericksson, L., 1979, Subfossil chironomids as evidence of eutrophication in Ekoln Bay, central Sweden. *Hydrobiologia* 62:195–208.

Wigand, P. and Rhode, D., 2002, Great Basin vegetation history and aquatic systems: the last 150,000 years. In Currey, D., Madsen, D., and Herschler, R. (eds.), *Great Basin Aquatic Systems History*, in press.

Wiggins, G.B. and Wichard, W., 1989, Phylogeny and pupation in Trichoptera, with proposals on the origin and higher classification of the order. *Jour. North Amer. Benth. Soc.* 8:260–276.

Wik, M. and Renberg, I., 1996, Environmental records of carbonaceous fly-ash particles from fossil-fuel combustion. *Jour. Paleolim.* 15:193–206.

Wilkinson, A.N., Hall, R.I., and Smol, J.P., 1999, Chrysophyte cysts as paleolimnological indicators of environmental change due to cottage development and acidic deposition in the Muskoka–Haliburton region, Ontario, Canada. *Jour. Paleolim.* 22:17–39.

Willemse, N.W. and Törnqvist, T., 1999, Holocene century-scale temperature variability from

West Greenland lake records. *Geology* 27:580–584.

Willemsen, J., Van't Veer, R., and van Geel, B., 1996, Environmental change during the medieval reclamation of the raised-bog area Watersland (The Netherlands): a palaeophytosociological approach. *Rev. Palaeobot. Palynol.* 94:75–100.

Willén, E., 1992, Long-term changes in the phytoplankton of large lakes in response to changes in nutrient loading. *Nord. J. Bot.* 12:575–587.

Williams, D.F., Peck, J., Karabanov, E.B., Prokopenko, A.A., Kravchinsky, V., King, J., and Kuzmin, M.I., 1999. Lake Baikal record of continental climate response to orbital insolation during the past 5 million years. *Science* 278:1114–1117.

Williams, N.E., 1988a, The use of caddisflies (Trichoptera) in paleoecology. In Gray, J. (ed.), *Paleolimnology: Aspects of Freshwater Paleoecology and Biogeography*. Elsevier, Amsterdam, pp. 493–500.

Williams, N.E., 1988b, Factors affecting the interpretation of caddisfly assemblages from Quaternary sediments. *Jour. Paleolim.* 1:241–248.

Williams, N.E. and Eyles, N., 1995, Sedimentary and paleoclimatic controls on caddisfly (Insecta: Trichoptera) assemblages during the last Interglacial-to-Glacial transition in southern Ontario. *Quat. Res.* 43:90–105.

Williams, W.D., 1983, *Life in Inland Waters.* Blackwell Science, Melbourne.

Williams, W.D., 1998, Salinity as a determinant of the structure of biological communities in salt lakes. *Hydrobiologia* 381:191–201.

Williamson, C.E., 1991, Copepoda. In Thorp, J.H. and Covitch, A.P. (eds.), *Ecology and Classification of North American Freshwater Invertebrates*. Academic Press, New York, pp. 787–822.

Williamson, P.G., 1981, Palaeontological documentation of speciation in Cenozoic molluscs from Turkana Basin. *Nature* 293:437–443.

Williamson, P.G., 1982, Molluscan biostratigraphy of the Koobi For a hominid-bearing deposits. *Nature* 295:140–142.

Wilson, C.C. and Hebert, P., 1996, Phylogeographic origins of lake trout (*Salvelinus namaycush*) in eastern North America. *Can. Jour. Fish. Aquat. Sci.* 53:2764–2775.

Wilson, M.V.H., 1988, Reconstruction of ancient lake environments using both autochthonous and allochthonous fossils. *Palaeogeog., Palaeoclim., Palaeoecol.* 62:609–623.

Wilson, M.V.H., Brinkman, D.B., and Neuman, A.G., 1992, Cretaceous Esocoidei (Teleostei):

early radiation of the pikes in North American freshwaters. *Jour Paleo.* 66:839–846.

Wilson, S.E., Smol, J.P., and Sauchyn, D.J., 1997, A Holocene paleosalinity diatom record from southwestern Saskatchewan, Canada: Harris Lake revisited. *Jour. Paleolim.* 17:23–31.

Winsborough, B.M. and Golubic, S., 1987, The role of diatoms in stromatolite growth: two examples from modern freshwater settings. *Jour. Phycol.* 23:195–201.

Winsborough, B.M. and Seeler, J.S., 1984, The relationship of diatom epiflora to the growth of limnic stromatolites and microbial mats. *Proc. 8th Diatom. Symp.*, pp. 395–407.

Winsborough, B.M., Seeler, J.S., Golubic, S., Folk, R.L., and Maguire, B., 1994, Recent freshwater lacustrine stromatolites, stromatolitic mats and oncoids from northeastern Mexico. In Bertrand-Sarfati, J. and Monty, C. (eds.), *Phanerozoic Stromatolites II.* Kluwer, Dordrecht, pp. 71–100.

Wintle, A.G. and Huntley, D.J., 1980, Thermoluminescence dating of ocean sediments. *Can. Jour. Earth Sci.* 17:348–360.

Witt, D.S. and Hebert, P.D.N., 2000, Cryptic species diversity and evolution in the amphipod genus *Hyalella* within central glaciated North America: a molecular phylogenetic approach. *Can. Jour. Fish. Aquat. Sci.* 57:687–698.

Wohlfarth, B., Björck, S., Possnert, G., Lemdahl, G., Brunnberg, L., Ising, J., Olsson, S., and Svensson, N.O., 1993, AMS dating Swedish varved clays of the last glacial/interglacial transition and the potential difficulties of calibrating Late Weichselian "absolute" chronologies. *Boreas* 22:113–128.

Wold, R.J., Hutchinson, D.R., and Johnson, T.C., 1982, Topography and surficial structure of Lake Superior bedrock as based on seismic reflection profiles. *Geol. Soc. Amer. Mem.* 156:257–272.

Wolfe, A.P., 1994, Late Wisconsinan and Holocene diatom stratigraphy from Amarok Lake, Baffin Island, N.W.T., Canada. *Jour. Paleolim.* 10:129–139.

Wolfe, A.P., Baron, J.S., and Cornett, R.J., 2001, Anthropogenic nitrogen deposition induces rapid ecological changes in alpine lakes of the Colorado Front Range (USA). *Jour. Paleolim.* 25:1–7.

Wolfe, B.B., Edwards, T.W.D., Aravena, R., and MacDonald, G.M., 1996, Rapid Holocene hydrologic change along boreal tree-line revealed by $\delta^{18}O$ and $\delta^{18}O$ in organic lake sediments, Northwest Territories, Canada. *Jour. Paleolim.* 15:171–181.

Wolfe, B.B., Edwards, T.W.D., and Aravena, R., 1999, Changes in carbon and nitrogen cycling during tree-line retreat recorded in the isotopic content of lacustrine organic matter, western Taimyr Peninsula, Russia. *The Holocene* 9:215–222.

Wolin, J.A. and Duthie, H.C., 1999, Diatoms as indicators of water level change in freshwater lakes. In Stoermer, E.F. and Smol, J.P. (eds.), *The Diatoms: Applications for the Environmental and Earth Sciences.* Cambridge University Press, Cambridge, pp. 183–202.

Wood, C.A., 1980, Morphometric evolution of cinder cones. *Jour. Volcanol. Geotherm. Res.* 7:387–413.

Wood, T.S., 1991, Bryozoans. In Thorp, J.H. and Covitch, A.P. (eds.), *Ecology and Classification of North American Freshwater Invertebrates.* Academic Press, New York, pp. 481–499.

Woodland, W.A., Charman, D.J., and Sims, P.C., 1998, Quantitative estimate of water tables and soil moisture in Holocene peatlands from testate amoebae. *The Holocene* 8:261–273.

Wooton, R.J., 1988, The historical ecology of aquatic insects: an overview. In Gray, J. (ed.), *Paleolimnology: Aspects of Freshwater Paleoecology and Biogeography.* Elsevier, Amsterdam, pp. 477–492.

Wooton, R.J., 1990, Major insect radiations. In Taylor, P.D. and Larwood, G.P. (eds.), *Major Evolutionary Radiations.* Syst. Assoc. Spec. 42:187–208.

Wright, H.E., 1996, Global climatic changes since the last glacial maximum: evidence from paleolimnology and paleoclimate modeling. *Jour. Paleolim.* 15:119–127.

Wright, H.E. Jr., Almendinger, J.C., and Gruger, J., 1985, Pollen diagram from the Nebraska Sandhills and the age of the dunes. *Quat. Res.* 24:115–120.

Wright, L.D., 1977, Sediment transport and deposition at river mouths: a synthesis. *Bull. Geol. Soc. Amer.* 88:857–868.

Wright, V.P., Platt, N.H., and Wimbledon, W.A., 1988, Biogenic laminar calcretes: evidence for calcified root mat horizons in paleosols. *Sedimentology* 35:603–620.

Wurster, C.M. and Patterson, W.P., 2001, Seasonal variation in stable isotope and oxygen isotope values recovered from modern lacustrine freshwater molluscs: paleoclimatological implications for sub-weekly temperature records. *Jour. Paleolim.* 26:205–218.

Yang, C., Telmer, K., and Veizer, J., 1996, Chemical dynamics of the "St. Lawrence" riverine system: δD_{H2O}, $\delta^{18}O_{H2O}$, $\delta^{13}C_{DIC}$, $\delta^{34}S_{sulfate}$, and dissolved $^{87}Sr/^{86}Sr$. *Geochim. Cosmochim. Acta* 60:851–866.

Yang, H., 2000, The Shangwang Basin (Miocene) in Shandong Province, Eastern China. In Gierlowski-Kordesch, E.H. and Kelts, K.R.

(eds.), *Lake Basins Through Space and Time.*
AAPG Stud. Geol. 46:473–479.

Yang, J.R. and Duthie, H.C., 1995, Regression and
weighted averaging models relating surfacial
diatom assemblages to water depth in Lake
Ontario. *Jour. Great Lakes Res.* 21:84–94.

Yang, J.R., Duthie, H.C., and Delorme, L.D.,
1993, Reconstruction of the recent
environmental history of Hamilton Harbour
(Lake Ontario, Canada) from analysis of
siliceous microfossils. *Jour. Great Lakes Res.*
19:55–71.

Yansa, C.H., 1998, Holocene paleovegetation and
paleohydrology of a prairie pothole in
southern Saskatchewan, Canada. *Jour.
Paleolim.* 19:429–441.

Yemane, K. and Kelts, K., 1996, Isotope
geochemistry of Upper Permian early
diagenetic calcite concretions: implications for
Late Permian waters and surface temperatures
in continental Gondwana. *Palaeogeog.,
Palaeoclim., Palaeoecol.* 125:51–73.

Yemane, K., Kahr, G., and Kelts, K., 1996, Imprints
of post-glacial climates and palaeogeography
in the detrital clay mineral assemblages of an
Upper Permian fluviolacustrine Gondwana
deposit from northern Malawi. *Palaeogeog.,
Palaeoclim., Palaeoecol.* 125:27–49.

Yo, K., 2000, The origins of modern nonmarine
ostracod faunas: evidence from the Late
Cretaceous and Early Paleogene of Mongolia.
Hydrobiologia 419:119–124.

Yonge, C.J., Goldenberg, L., and Krouse, H.R.,
1989, An isotope study of water bodies along a
traverse of SW Canada. *Jour. Hydrol.*
106:245–255.

Yoshioka, T., Wada, E., and Saijo, Y., 1988,
Isotopic characterization of Lake Kizaki and
Lake Suwa. *Japan. Jour. Limnol.* 49:119–128.

Young, R.B. and King, R.H., 1989, Sediment
chemistry and diatom stratigraphy of two high
arctic isolation lakes, Truelove Lowland,
Devon Island, N.W.T., Canada. *Jour.
Paleolim.* 2:207–225.

Yu, G., 1998, European lake status data base:
continental-scale synthesis for the Holocene. In
Harrison, S.P., Frenzel, B., Huckriede, U., and
Weiß, M.M. (eds.), *Palaeohydrology as
Reflected in Lake Level Changes as Climatic
Evidence for Holocene Times.* Gustav Fischer-
Verlag, Stuttgart, pp. 99–114.

Yu, J., Mizuno, A., and Wang, L., 1993, The
Jurassic System in the Qinshui Basin, Shanxi
Province, with notes on the bivalve province of
North China. *Palaeogeog., Palaeoclim.,
Palaeoecol.* 105:157–170.

Yu, Z. and McAndrews, J.H., 1994, Holocene
water levels at Rice Lake, Ontario, Canada:
sediment, pollen and plant macrofossil
evidence. *The Holocene* 4:141–152.

Yuretich, R., 1989, Paleocene lakes of the Central
Rocky Mountains, Western United States.
Palaeogeog., Palaeoclim., Palaeoecol. 70:53–
63.

Yuretich, R., Melles, M., Sarata, B., and Grobe, H.,
1999, Clay minerals in the sediments of Lake
Baikal: a useful climate proxy. *Jour. Sed. Res.*
69:588–596.

Yurtsever, Y., 1975, Worldwide survey of stable
isotopes in precipitation. In *Annual Report of
the Section for Isotope Hydrology.*
International Atomic Energy Agency, Vienna,
pp. 567–585.

Zak, I. and Freund, R., 1981, Asymmetry and basin
migration in the Dead Sea rift. *Tectonophysics*
80:27–38.

Zeeb, B.A. and Smol, J.P., 2001, Chrysophyte
scales and cysts. In Smol, J.P., Birks, H.J.B.,
and Last, W.M. (eds.), *Tracking
Environmental Change Using Lake Sediments.
Vol. 3. Terrestrial, Algal and Siliceous
Indicators.* Kluwer Academic Publishers,
Dordrecht, pp. 203–223.

Zernitskaya, V.P., 1997, The evolution of lakes in
the Poles'ye in the Late Glacial and Holocene.
Quat. Intl. 41/42:153–160.

Zhao, X. and Tang, Z., 2000, Lacustrine deposits
of the Upper Permian Pingdiquan Formation in
the Kelameili Area of the Juggar Basin,
Xinjiang, China. In Gierlowski, E.H. and
Kelts, K.R. (eds.), *Lake Basins Through Space
and Time.* AAPG Stud. Geol. 46:111–122.

Zhou, W., Head, M.J., Lu, X., An, Z., Jull, A.J.T.,
and Donahue, D., 1999, Teleconnection of
climate events between East Asia and polar
high latitudes during the last deglaciation.
Palaeogeog., Palaeoclim., Palaeoecol.
152:163–172.

Zhou, W., Head, M.J., An, Z., De Deckker, P., Liu,
Z., Liu, X., Donahue, D., Jull, A.J.T., and
Beck, J.W., 2001, Terrestrial evidence for a
spatial structure of tropical–polar
interconnections during the Younger Dryas
episode. *Earth Planet. Sci. Lett.* 191:231–239.

Zidek, J., 1993, Acanthodii. In Benton, M.J. (ed.),
The Fossil Record 2. Chapman and Hall,
London, pp. 589–592.

Zohary, T., Pollingher, U., Hadas, O., and
Hambright, K.D., 1998, Bloom dynamics and
sedimentation of *Peridinium gatunense* in Lake
Kinneret. *Limnol. Oceanogr.* 43:175–186.

Zolitschka, B., 1991, Absolute dating of late
Quaternary lacustrine sediments by high
resolution varve chronology. *Hydrobiologia*
214:59–61.

Zolitschka, B. and Negendank, J.F.W., 1993, Lago
Grande Di Monticchio (Southern Italy). A high
resolution sedimentary record of the last
70,000 years. In Negendank, J.F.W. and
Zolitschka, B. (eds.), *Paleolimnology of*

European Maar Lakes. Springer-Verlag, New York, pp. 277–288.

Zolitschka, B. and Negendanck, J.F.W., 1996, Sedimentology, dating and palaeoclimatic interpretation of a 76.3 ka record from Lago Grande di Monticchio, southern Italy. *Quat. Sci. Rev.* 15:101–112.

Zonenshain, L.P. and Le Pichon, X., 1986, The Black Sea and Caspian Sea deep basins as remnants of Mesozoic back arc basins. *Tectonophysics* 123:181–211.

Züllig, H., 1981, On the use of caroteinoid stratigraphy in lake sediment for detecting past development of phytoplankton. *Limnol. Oceanogr.* 26:970–976.

Index